Wave Motion in Elastic Solids

by Karl F. Graff

DOVER PUBLICATIONS, INC., *New York*

Published in Canada by General Publishing Company, Ltd., 30 Lesmill Road, Don Mills, Toronto, Ontario.
Published in the United Kingdom by Constable and Company, Ltd., 3 The Lanchesters, 162–164 Fulham Palace Road, London W6 9ER.

This Dover edition, first published in 1991, is an unabridged, corrected republication of the work originally published by Oxford University Press, London (Clarendon Press, Oxford), 1975, and by Ohio State University Press (n.p.), 1975.

Manufactured in the United States of America
Dover Publications, Inc., 31 East 2nd Street, Mineola, N.Y. 11501

Library of Congress Cataloging-in-Publication Data

Graff, Karl F.
Wave motion in elastic solids / Karl F. Graff.
p. cm.
Reprint. Originally published: Oxford : Clarendon Press, 1975. Includes bibliographical references (p.) and index.
ISBN 0-486-66745-6
1. Solids. 2. Elastic waves. I. Title.
QC176.8.W3G7 1991
530.4'12—dc20 91-9350
CIP

TO MARTHA

Preface

THE purpose of this book is to present, in one place and in a fairly comprehensive manner, an intermediate-level coverage of nearly all of the major topics of elastic wave propagation in solids. Thus subjects range from the elementary theory of waves and vibrations in strings to the three-dimensional theory of waves in thick plates. It is hoped that the result will find application not only as a textbook for a wide audience of engineering students, but also as a general reference for workers in vibrations and acoustics.

The book is organized into eight chapters and three appendices. The first four chapters cover wave motion in the simple structural shapes, namely strings (Chapter 1), longitudinal rod motion (Chapter 2), beams (Chapter 3), and membranes, plates, and (cylindrical) shells (Chapter 4). In these chapters, the so-called 'strength-of-materials' theories provide the governing equations. Chapter 1, dealing with waves in strings, is used to introduce nearly all of the basic wave propagation concepts used throughout the remainder of the book. There is also some material included in each of the first four chapters on steady vibrations of structural shapes.

Chapters 5–8 deal with wave propagation as governed by the three-dimensional equations of elasticity and cover waves in infinite media, waves in a half-space, scattering and diffraction, and waves in thick rods, plates, and shells. The appendices of the book cover the topics of the elasticity equations, integral transforms, and experimental methods.

An effort has been made to make the book as self-contained as possible. This is in part reflected by the contents of the appendices, where introductory material is included, as indicated above, on elasticity, transforms, and experimental techniques. It is further reflected by fairly complete development of a number of topics in the mechanics and mathematics of the subject, such as simple transform solutions, orthogonality conditions, approximate theories of plates, and asymptotic methods.

Throughout the book, emphasis has been placed on showing results, in addition to mere theoretical development of the subject. This has taken the form of results from both theoretical and experimental studies. The intent in doing so has been twofold. On the one hand, presentation of specific results from a range of studies should assist the student in reaching a better physical understanding of the response of systems to transients. On the other hand, the availability of a basic catalogue of results should be of some assistance to workers in the field charged with analysing waves in more complex systems. Finally, in particular reference to the inclusion of experimental results, it is hoped that the role played by better physical appreciation for wave propagation, a partial substitute for actual laboratory work and verification (or lack thereof) of theoretical results, will be brought out.

It is hoped that the book will find various uses as text or reference. The inclusion of over one hundred problems should assist its use as a textbook. As a text, it has three main uses. For use as an upper-level undergraduate book, the material would be drawn primarily from large parts of the first four chapters. For lower-level graduate use, where prior exposure to the elasticity equations could be assumed, material could be drawn from all of the first five chapters and parts of Chapters 6 and 8. For upper-level graduate use, material could be drawn from the entire book. Again, it is hoped that the somewhat self-contained nature of the book and the emphasis on results will make it of use as a reference book.

Finally, it should be noted that, as is the case for many books, this was an outgrowth of notes prepared for courses of instruction. Specifically, notes prepared over several years of teaching two graduate-level courses in the Department of Engineering Mechanics at the Ohio State University served as the basis. The courses, 'Vibrations of continuous systems' and 'Elastic wave propagation', covered both steady-state vibration phenomena and propagation of transients. Needless to say, acknowledgement to the many students taking these courses is in order. Their comments, sometimes diplomatic, sometimes otherwise, but always constructive, were of considerable help in the notes-to-book evolutionary process.

K. F. G.

Columbus, Ohio
October, 1973

Acknowledgements

Figs 2.33 and 6.35 are reproduced with the permission of Microform International Marketing Corporation, exclusive copyright licensee of Pergamon Press journal back files.

Figs 8.6, 8.8, 8.9, 8.10, and 8.11 are reproduced with permission from *Structural Mechanics* (ed. Goodier and Hoff) Pergamon Press Inc. (1960).

Fig. 6.33 is reproduced with permission from H. E. Tatel, *Journal of Geophysical Research*, vol. 75 (1970) copyright by American Geophysical Union.

Fig. 6.36 is reproduced with permission from D. Lewis and J. W. Dally, *Journal of Geophysical Research*, vol. 75 (1970) copyright by American Geophysical Union.

Figs 3.3, 4.3, 4.4, 4.8, 4.9, 4.17, and 7.2 are from P. M. Morse, *Vibration and Sound*, copyright 1948 McGraw-Hill. Fig. 6.27 is from W. M. Ewing, W. S. Jardetzky, and F. Press, *Elastic Waves in layered media* copyright 1957 McGraw-Hill. Fig. 4.13 is from I. N. Sneddon, *Fourier Transforms*, copyright 1951 McGraw-Hill. All used with permission of McGraw-Hill Book Company.

Fig. 8.52 is reproduced with permission from D. Y. Hseih and H. Kolsky, *Proceedings of the Physical Society*, vol. 71 (1958), copyright by the Institute of Physics.

Contents

Introduction

I.1 General aspects of wave propagation

The effect of a sharply applied, localized disturbance in a medium soon transmits or 'spreads' to other parts of the medium. This simple fact forms a basis for study of the fascinating subject known as *wave propagation*. The manifestation of this phenomenon are familiar to everyone in forms such as the transmission of sound in air, the spreading of ripples on a pond of water, the transmission of seismic tremors in the earth, or the transmission of radio waves. These and many other examples could be cited to illustrate the propagation of waves through gaseous, liquid, and solid media and free space.

The propagation of disturbances in the various media mentioned share many common features, so that a person versed in the science of one understands much about the others. There are sufficient differences, however, as to make a completely general development of the subject impractical and to thus require concentration on a single topic. Our attention in this development will be focused on the propagation of waves in solids. We will thus be considering solely mechanical disturbances in contrast, say, to electromagnetic or acoustic disturbances.

The physical basis for the propagation of a disturbance ultimately lies in the interaction of the discrete atoms of the solid. Investigations along such lines are more atuned to physics than mechanics, however. In solid and fluid mechanics, the medium is regarded as *continuous*, so that properties such as density or elastic constants are considered to be continuous functions representing averages of microscopic quantities. Nevertheless, in envisaging the basis for propagation of a disturbance it is helpful to first consider a model composed of discrete elements consisting of a series of interconnected masses and springs. If a disturbance is imparted to a mass particle, it is transmitted to the next mass by the intervening spring. In this manner the disturbance is soon transmitted to a remote point, although any given particle of the system will have moved only a small amount. The role of the mass and stiffness parameters in affecting the speed of propagation is quite clear in such a model. If the stiffness of the connecting springs is increased or the particle masses decreased, or both, the speed of propagation would be expected to increase. Weaker springs and/or larger masses would slow the propagation velocity. Extreme values of the parameters could lead to instantaneous or zero propagation of disturbances. So in the case of a continuous media. The mass and elastic parameters are now distributed in terms of mass density and the elastic moduli. The interaction from one part of the system to the next is the interaction of one differential element on the next. Instead of the simple push-pull motion along a series of springs and masses, the disturbance spreads outward in a three-dimensional sense. A wavefront will be associated with the outward spreading disturbance. Particles ahead of the front will have experienced no motion, while particles behind the front will have experienced motion and may in fact continue to oscillate for some time.

In the case of a solid, two distinct types of action will be possible in a wave. In one case, the solid will transmit tensile and compressive stress and the motion of particles will be in the direction of the wave motion. This behaviour is analogous to that of fluids. In addition, the solid may transmit shear stress, and the motion of particles is transverse to the direction of propagation. There is no analogue to this behaviour in fluids, although the transverse nature of the wave bears close resemblance to electromagnetic waves.

The outward propagation of waves from a disturbance is one aspect of wave motion. Inevitably, the waves encounter and interact with boundaries. In this area, the behaviour of waves in solids differs considerably from that of fluids. In a solid a single wave, be it compression or shear, will generally produce both compression and shear waves on striking a boundary, whereas acoustic and electromagnetic waves will only generate waves of their own type.

It is the continual propagation and reflection of waves in a bounded solid that brings about the state of static equilibrium. Speaking in these terms, every process of loading a solid is a dynamic process involving the propagation and reflection of waves. However, if the rate of onset of the load is slow compared with many transit times of waves within the solid, static equilibrium effectively prevails and wave effects are of no consequence. It is only when loading rates are comparable with transit times of waves that the mechanics of wave propagation must be considered.

In many problems of waves in solids, the preceding description of wave motion is too detailed, somewhat in analogy to attempting to describe waves in a solid through the motion of the atoms. The motion of simple structural shapes such as rods, beams, and plates may be described adequately in many instances without resort to considering the propagation and reflection of individual waves within the specimen. Instead, so called 'strength-of-material' theories may be devised, based on various assumptions on how these solids deform, that approximate the detailed behaviour of the solid. The first four chapters of this book will be devoted to such considerations. Such theories, while most useful, have inherent limitations as loading transients become more severe, or frequencies of excitation become higher or as the necessary assumptions on deformations become less obvious. At this stage, analysis based on the exact equations of elasticity becomes necessary, and it is this area to which the last four chapters of the book will be devoted.

The propagation of waves in solids may be divided roughly into three categories. The first is elastic waves, where the stresses in the material obey Hooke's law. The considerations of this book, whether concerned with simple structures or with analysis based on the exact equations, will assume such elastic behaviour. The two other main categories, visco-elastic waves, where viscous as well as elastic stresses act, and plastic waves in which the yield stress of the material is exceeded, will not be covered here. These subjects are sufficiently broad in themselves to warrant entire books.

The study of wave propagation is far from an abstract mathematical subject. The experimental measurements of wave phenomena furnish us with much of our information on the properties of solids, be these solids the earth, pure crystals of metal, or other substances. The deliberate introduction of stress waves in materials finds many applications in structures, electronic technology, and testing. For this reason, it is considered useful to review a range of experimental results in stress waves at the conclusion of several of the chapters.

I.2. Applications of wave phenomena

The practical applications of wave phenomena surely go back to the early history of man. The shaping of stone implements, for example, consists of striking sharp, carefully placed blows along the edges of a flint. The resulting stress waves in the 'cone of percussion' break out fragments of rock in very specific patterns. Starting at this early time, it may be safely said that interest in wave phenomena has been increasing ever since.

The motivations for the current high level of interest in the subject are the many practical applications in science and industry. In the area of *structures*, for example, the interest is mainly in the response to impact or blast loads. Under transient loads of moderate strength, completely elastic conditions may prevail throughout the structure and elastic wave theory may suffice to predict all aspects of the response. Under more severe loads, local permanent deformation, fracture, or perforation of the structure may occur. Elastic wave theory often still finds application under such conditions in predicting the response away from the area of impact.

The behaviour of structural *materials* under loads severe enough to cause permanent damage is an area of very great interest. Studies in this area generally fall in the category of *anelastic wave propagation*. Some of the techniques used in the study of these properties use elastic waves, such as waves in a high-strength steel rod, to dynamically load-test specimens of weaker materials. Most of the applications in this area are in various aspects of military and space technology. However, a number of metal forming processes, such as explosive forming, high energy rate forming, or sonic riveting and forging, find use for similar information.

Another area in the study of structures involving wave phenomena is that of crack propagation or the interaction of dynamic stress fields with existing cracks, voids, or inclusions in a material. The concept of a dynamic stress concentration factor, for example, finds application in this area. Problems in this area are the analogue of scattering and diffraction problems arising in acoustics and electromagnetics.

The field of *ultrasonics* represents another major area of application of wave phenomena. The general aspects of this area involve introducing a very low energy-level, high-frequency stress pulse or 'wave packet' into a material and observing the subsequent propagation and reflection of this energy. The means for introducing and detecting the stress waves are based on the *piezoelectric effect* in certain crystals and ceramics, whereby an electrical field applied to the material causes a mechanical strain or the inverse effect where a strain produces an electric field. Thus an electrical pulse is capable of lauching a mechanical pulse. Detection is accomplished when a mechanical pulse strikes a piezoelectric crystal and generates an electrical signal.

A host of applications are based on this reciprocal effect. For example, by studying the propagation, reflection, and attenuation of ultrasonic pulses, it is possible to determine many fundamental properties of materials such as elastic constants and damping characteristics. The field of non-destructive testing makes wide use of ultrasonics to detect defects in materials. By launching a pulse into a solid, it is possible to locate defects by the reflection of pulse energy from the defect much in the manner of underwater sonar detection. Various detection applications use longitudinal waves, shear waves, or surface waves.

Ultrasonic *delay lines* find wide application in the field of electronics. The objective of such devices is to provide a means to delay an electrical signal for a short time interval (such as a few microseconds), as dictated by signal-processing considerations. With the extremely high propagation velocity of electrical signals, it becomes impractical to accomplish this delay using purely electronic components. The relatively slow velocity of mechanical disturbances makes such time delays easily obtained. The procedure is to convert the electrical signal to a mechanical pulse using a piezoelectric transducer, to propagate the signal in some type of solid media, and to recover it a specified time later with another transducer. A wide variety of types of delay lines exist. Some are based on the propagation of longitudinal or torsional waves in thin wires, some on the propagation of shear waves in thin strips, others on the propagation of longitudinal waves in bulk solids, and still others on the propagation of waves along a surface.

The general subject of *waves in the earth* covers many interesting propagation phenomena. Earthquakes generate waves that may travel thousands of miles. Study of the propagation of such waves or tremors artificially produced have provided the most knowledge on the interior construction of the earth. Waves in the earth generated by blast are of concern from the standpoint either of blast detection or the protection of underground structures. The matter of distinguishing blast and earthquake 'signatures' is also of concern. Other aspects of waves in the earth involve less catastrophic considerations and pertain to oil and gas exploration. By studying the reflection of waves from underground discontinuities, it is possible to locate possible oil-bearing deposits.

Consider the areas of mining and quarrying where numerous applications of *waves in rock* are found. The blasting that is used in these operations has the purpose of producing intense stress waves. It is the interaction of these waves with each other and with boundaries that is responsible for the fracture and removal of the large quantities of rock. For example, in quarrying, a typical operation is to drill a series of blast holes parallel to the exposed rock face. The ensuing stress waves from the blast propagate to the exposed face and reflect, creating tensile stresses that fracture the rock. The drilling of the blast holes referred to is generally by *percussive drilling*. This process operates by transmitting longitudinal waves created by an air hammer down drill rod into the rock. Many interesting problems on reflection and transmission of waves at discontinuities exist in this application.

The phenomenon of *acoustic emission* is a producer of stress waves and hence of potential applications. It is observed that the motion of dislocations during plastic deformation produce very high-frequency, low-energy stress waves that may be detected with sensitive transducers. Waves of greater energy are created by cracking in materials in a microscopic analogue to energy release occurring in earthquakes. The study of acoustic emission waves enables some deductions to be made on the fundamental processes occurring within the material. Of at least equal interest is that monitoring of such waves enables structural integrity to be assessed by passive means.

The cataloguing of applications of elastic and anelastic wave phenomena could easily stretch on, but it is hoped that the point has been made that applications are many and widespread and that interest in the subject stems from practical considerations.

I.3. Historical background

The history of the study of wave and vibration phenomena goes back hundreds of years. Most early studies were naturally more observational than quantitative and frequently were concerned with musical tones or water waves, two of the most common associations with wave motion. From the time of Galileo onward, the science of vibrations and waves progressed rapidly in association with developments in the statics of solids. Some of the major developments in the area over the years are chronologically ordered in the following.

Sixth Century B.C.: Pythagoras studied the origin of musical sounds and the vibrations of strings.

1636: Mersenne presented the first correct published account on the vibrations of strings.

1638: Galileo described the vibrations of pendulums, the phenomenon of resonance, and the factors influencing the vibrations of strings.

1678: Robert Hooke formulated the law of proportionality between stress and strain for elastic bodies. This law forms the basis for the static and dynamic theory of elasticity.

1686: Newton investigated the speed of water waves and the speed of sound in air.

1700: Sauveur calculated vibrational frequency of a stretched string.

1713: Taylor worked out a completely dynamical solution for the vibrations of a string.

1744: Leonard Euler (1744) and Daniel Bernoulli (1751) developed the equation for the vibrations of beams and obtained the normal modes for various boundary conditions.

1747: D'Alembert derived the equation of motion of the string and solved the initial-value problem.

1755: D. Bernoulli developed the principle of superposition and applied it to the vibrations of strings.

1759: Lagrange analysed the string as a system of discrete mass particles.

1766: Euler attempted to analyse the vibrations of a bell on the basis of the behaviour of curved bars. James Bernoulli (1789) also attempted analysis of this problem.

1802: E. F. F. Chladni reported experimental investigations on the vibrations of beams and on the longitudinal and torsional vibrations of rods.

1815: Madame Sophie Germain developed the equation for the vibrations of a plate.

1821: Navier investigated the general equations of equilibrium and vibration of elastic solids. Although not all of the developments of the work met with complete acceptance, it represented one of the most important developments in mechanics.

1822: Cauchy developed most of the aspects of the pure theory of elasticity including the dynamical equations of motion for a solid. Poisson (1829) also investigated the general equations.

1828: Poisson investigated the propagation of waves through an elastic solid. He found that two wave types, longitudinal and transverse, could exist. Cauchy (1830) obtained a similar result. Poisson also solved the problem of the radial vibrations of a sphere.

1828: Poisson developed approximate theories for the vibrations of rods.

1862: Clebsch founded the general theory for the free vibrations of solid bodies using normal modes.

1872: J. Hopkinson performed the first experiments on plastic wave propagation in wires.

1876: Pochhammer obtained the frequency equation for the propagation of waves in rods according to the exact equations of elasticity. Chree (1889) carried out similar studies.

1880: Jaerisch analysed the general problem of the vibrations of a sphere. The result was obtained independently by Lamb (1882).

1882: Hertz developed the first successful theory for impact.

1883: St. Venant summarized the work on impact of earlier investigators and presented his results on transverse impact.

1887: Rayleigh investigated the propagation of surface waves on a solid.

1888: Rayleigh and Lamb (1889) developed the frequency equation for waves in a plate according to exact elasticity theory.

1904: Lamb made the first investigation of pulse propagation in a semi-infinite solid.

1911: Love developed the theory of waves in a thin layer overlying a solid and showed that such waves accounted for certain anomalies in seismogram records.

1914: B. Hopkinson performed experiments on the propagation of elastic pulses in bars.

1921: Timoshenko developed a theory for beams that accounted for shearing deformation.

1930: Donnell studied the effect of a non-linear stress–strain law on the propagation of stress waves in a bar.

1942: von Karman, Taylor (1942), and Rakmatulin developed a one-dimensional finite-amplitude plastic wave theory.

1949: Davies published an extensive theoretical and experimental study on waves in bars.

1951: Mindlin presented an approximate theory for waves in a plate that provided a general basis for development of higher-order plate and rod theories.

1951: Malvern developed a rate-dependent theory for plastic wave propagation.

1955: Perkeris presented the solution to Lamb's problem of pulse propagation in a semi-infinite solid.

Developments in wave propagation did not, of course, cease in 1955; the date only represents the author's desire not to offend more recent significant contributors to the field through inadvertent omission from a mere listing. Recent activities in the field of wave propagation have dealt with formulating various approximate theories for plates and rods and with the analysis of transient loading situations. In the latter regard, the analysis of pulse propagation in the half-space and in plates and rods has received considerable attention. The development of approximate techniques for diffraction analysis have been successfully carried out. The application of the digital computer has enabled a number of otherwise intractable problems to be solved.

1 Waves and vibrations in strings

IN beginning the analysis of wave propagation in solids, we strive for mathematical simplicity. However, many of the applications of wave phenomena involve quite complicated mathematical analyses that arise from the geometric complexities of the physical system. The taut string, on the other hand, represents a physical system whose governing equation is rather simple, yet basic to many wave propagation problems. Nearly all of the basic concepts of propagation, such as dispersion and group velocity as well as certain techniques of analysis can and will be introduced, unobscured by the complexities involved in more complicated elastic systems.

While it is sufficient to rest the case for analysing the taut string on mathematical grounds alone, it should be appreciated that practical motivations also exist. The characteristics of many musical instruments are based on the vibrations of strings. The dynamics of electrical transmission lines may be modelled on the basis of strings. Problems in the dynamics of strings arise in the manufacture of thread. Nevertheless, it is primarily for mathematical reasons that we are presently interested in the elastic string.

1.1. Waves in long strings

The basic governing equation for the taut string must first be developed. Since boundaries inevitably introduce complications in wave propagation due to reflection phenomena, the first considerations will involve 'long' strings, that is, infinite or semi-infinite strings where the problem of boundary reflection will not arise. The basic propagation characteristics of free waves and waves resulting from forced motion will be studied under these conditions.

1.1.1. *The governing equation*

Consider a differential element of taut string under tension T as shown in Fig. 1.1. It is assumed that any variation in the tension due to the displacement of the string is negligible. The mass density per unit length is ρ and the body force or external loading is $q(x, t)$. The resulting equation of motion in the

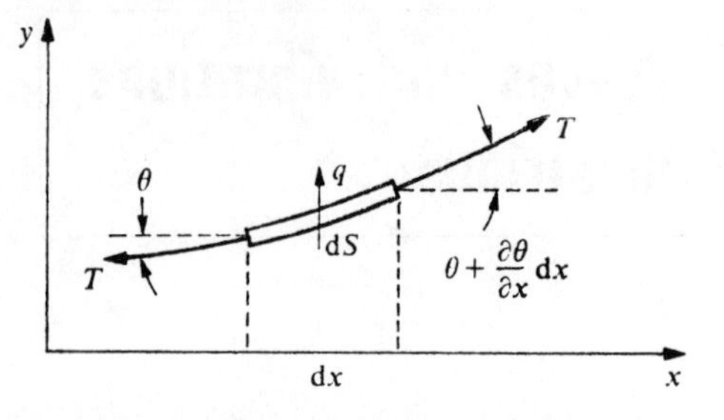

FIG 1.1. Differential element of taut string.

vertical direction is then

$$-T\sin\theta + T\sin\left(\theta + \frac{\partial\theta}{\partial x}\,\mathrm{d}x\right) + q\,\mathrm{d}s = \rho\,\mathrm{d}s\frac{\partial^2 y}{\partial t^2}. \tag{1.1.1}$$

The arc length $\mathrm{d}s$ is given by $\mathrm{d}s = (1+y'^2)^{\frac{1}{2}}\,\mathrm{d}x$. If we assume small deflections of the string, we may write that $\mathrm{d}s \simeq \mathrm{d}x$. Furthermore, for small deflections, we approximate $\sin\theta$ by θ and note that $\theta \simeq \partial y/\partial x$. The preceding equation then reduces to

$$T\frac{\partial^2 y}{\partial x^2} + q = \rho\frac{\partial^2 y}{\partial t^2}. \tag{1.1.2}$$

A number of solutions to this non-homogeneous second-order partial differential equation will be investigated in later sections. Of particular interest is the form of the homogeneous equation obtained by setting $q = 0$, giving

$$\frac{\partial^2 y}{\partial x^2} = \frac{1}{c_0^2}\frac{\partial^2 y}{\partial t^2}, \qquad c_0 = \left(\frac{T}{\rho}\right)^{\frac{1}{2}}. \tag{1.1.3}$$

This resulting equation governing the free transverse motion of the string is known as the *wave equation*. It possesses a number of interesting properties and will be found to govern the motion of a number of other elastic systems.

1.1.2. *Harmonic waves*

We shall first investigate the propagation of simple harmonic waves in a string. Using the separation of variables approach, we let $y = Y(x)T(t)$ and substitute in the wave equation (1.1.3), giving

$$\frac{Y''}{Y} = \frac{T''}{c_0^2 T} = -\gamma^2. \tag{1.1.4}$$

The resulting solution for $y(x, t)$ is then

$$y = (A_1\sin\gamma x + A_2\cos\gamma x)(A_3\sin\omega t + A_4\cos\omega t), \tag{1.1.5}$$

where the *radial frequency* is given as $\omega = \gamma c_0$. Regrouping, the solution may be written as

$$y = A_1A_4 \sin \gamma x \cos \omega t + A_2A_3 \cos \gamma x \sin \omega t + A_2A_4 \cos \gamma x \cos \omega t + \\ + A_1A_3 \sin \gamma x \sin \omega t. \qquad (1.1.6)$$

Consider a typical term of this solution written in the form

$$y = A \cos \gamma x \sin \omega t. \qquad (1.1.7)$$

The deflections of the string at successive instants of time as governed by (1.1.7) are shown in Fig. 1.2. We note that points of zero vibration amplitude,

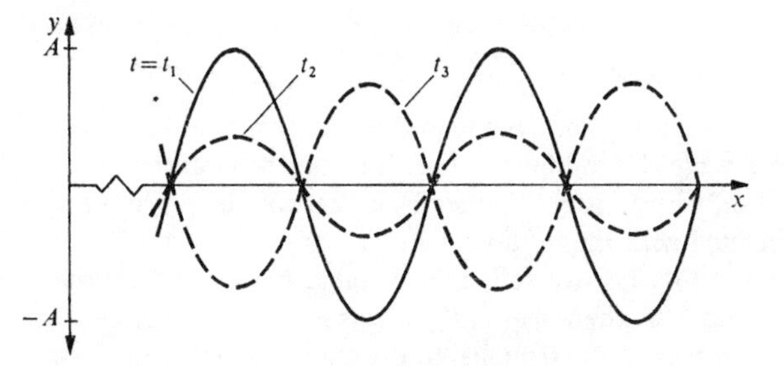

FIG. 1.2. Vibrational patterns of the string due to the standing wave $y = A \cos \gamma x \sin \omega t$.

called *nodes*, and of maximum amplitude, called the *antinodes*, occur at regular intervals along the string and remained fixed in that position with the passage of time. A vibration of this type and governed by results of the form (1.1.7) is called a stationary or *standing wave*.

Using trigonometric identities, the solution (1.1.6) may be put in the form

$$y = B_1 \sin(\gamma x + \omega t) + B_2 \sin(\gamma x - \omega t) + \\ + B_3 \cos(\gamma x + \omega t) - B_4 \cos(\gamma x - \omega t). \qquad (1.1.8)$$

Consider a typical term in the solution, given by

$$y = A \cos(\gamma x - \omega t) = A \cos \gamma(x - c_0 t). \qquad (1.1.9)$$

This may be shown to represent a wave *propagating* in the positive x direction. If we designate the argument of (1.1.9) as the *phase* ϕ, where

$$\phi = \gamma x - \omega t = \gamma(x - c_0 t), \qquad (1.1.10)$$

then we may note that for increasing time, increasing values of x are required to maintain the phase constant. The appearance of the deflection at successive instants of time would be as shown in Fig. 1.3. The propagation velocity of the constant phase is c_0, defined as the *phase velocity*. It is seen that constancy

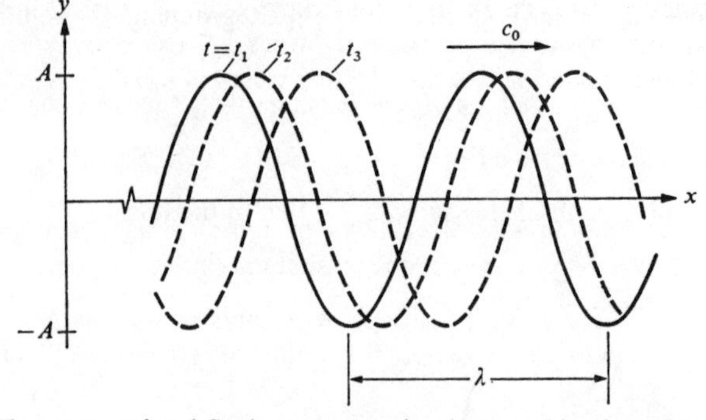

FIG. 1.3. Transverse string deflections at successive times resulting from the propagating wave $y = A\cos(\gamma x - \omega t)$.

of phase for increasing time requires $x = c_0 t$. It should be noted that (1.1.9) represents a wave of infinite length. It thus has no 'wavefront' or beginning, so that it is only by considering the phase that enables a propagation velocity to be associated with the motion.

Referring to Fig. 1.3, we define the distance between two successive points of constant phase as the *wavelength* λ. This is shown as the distance between two minimum points, or 'troughs' in the diagram, but could as well refer to successive maxima, points of zero amplitude or intermediate phase points. From (1.1.10) it is seen that the wavelength is related to γ by $\lambda = 2\pi/\gamma$. The parameter γ will be designated the *wavenumber*.

Other quantities that frequently arise in describing wave motion are the *cyclic frequency* f, where $\omega = 2\pi f$, and the *period*, T, of the wave, where $T = 1/f$. It is of value to summarize these various definitions and relations as follows:

$$\begin{aligned} A &= \text{amplitude of wave (length)} \\ \omega &= \text{radial frequency of wave (radians/time)} \\ f &= \text{cyclic frequency of wave (cycles/time)} \\ \lambda &= \text{wavelength of wave (length)} \\ \gamma &= \text{wavenumber of wave (1/length)} \\ c_0 &= \text{phase velocity of wave (length/time)} \\ T &= \text{period of wave (time)} \end{aligned} \tag{1.1.11}$$

$$\omega = \gamma c_0, \qquad \omega = 2\pi f, \qquad T = 1/f, \qquad \lambda = 2\pi/\gamma. \tag{1.1.12}$$

Having considered the characteristics of a typical harmonic wave, it is seen that the remaining terms of the solution (1.1.8) are similar in nature. The major point of difference is that terms having the argument $(\gamma x + \omega t)$ are propagating in the negative x direction. The general result (1.1.8) may be described as a propagating wave solution. It has been developed from (1.1.6), which may be described as a standing-wave solution. It may be said that the standing waves result from constructive and destructive interference of leftward and rightward propagating waves. This particular aspect is emphasized by considering two propagating waves of equal amplitude. Thus,

$$y = \frac{A}{2}\cos(\gamma x+\omega t)+\frac{A}{2}\cos(\gamma x-\omega t) = A\cos\gamma x\cos\omega t. \tag{1.1.13}$$

Finally, we consider alternate forms of representing harmonic waves. Instead of using $\sin\omega t$ or $\cos\omega t$ to represent the time dependence, we could use the exponential representation. Thus, let

$$y = Y(x)e^{i\omega t}. \tag{1.1.14}$$

Substituting in (1.1.3) and solving gives the solutions

$$y = A_1 e^{i(\gamma x+\omega t)}+B_1 e^{-i(\gamma x-\omega t)}. \tag{1.1.15}$$

These results are recognized as harmonic waves propagating in the negative and positive x direction respectively. Considering an initial time behaviour of the form $\exp(-i\omega t)$ leads to the solutions

$$y = A_2 e^{i(\gamma x-\omega t)}+B_2 e^{-i(\gamma x+\omega t)}. \tag{1.1.16}$$

1.1.3. *The D'Alembert solution*

It is possible to derive a general solution to the wave equation using Fourier superposition of harmonic waves, and this will be done in a later section. A classical solution by D'Alembert (1747) will be investigated, at this stage, that will provide considerable insight into wave-propagation phenomena. Thus, consider the wave equation (1.1.3) and introduce the change of variables

$$\xi = x-c_0 t, \qquad \eta = x+c_0 t. \tag{1.1.17}$$

By chain-rule differentiation, we have

$$\frac{\partial y}{\partial x} = \frac{\partial y}{\partial \xi}\frac{\partial \xi}{\partial x}+\frac{\partial y}{\partial \eta}\frac{\partial \eta}{\partial x} = \frac{\partial y}{\partial \xi}+\frac{\partial y}{\partial \eta},$$

$$\frac{\partial y}{\partial t} = \frac{\partial y}{\partial \xi}\frac{\partial \xi}{\partial t}+\frac{\partial y}{\partial \eta}\frac{\partial \eta}{\partial t} = -c_0\frac{\partial y}{\partial \xi}+c_0\frac{\partial y}{\partial \eta}. \tag{1.1.18}$$

The second derivatives give

$$\frac{\partial^2 y}{\partial x^2} = \frac{\partial^2 y}{\partial \xi^2}+2\frac{\partial^2 y}{\partial \xi\,\partial \eta}+\frac{\partial^2 y}{\partial \eta^2}, \qquad \frac{\partial^2 y}{\partial t^2} = c_0^2\left(\frac{\partial^2 y}{\partial \xi^2}-2\frac{\partial^2 y}{\partial \xi\,\partial \eta}+\frac{\partial^2 y}{\partial \eta^2}\right). \tag{1.1.19}$$

Substituting (1.1.19) in the wave equation gives

$$\frac{\partial^2 y(\xi,\eta)}{\partial \xi\,\partial \eta} = 0. \tag{1.1.20}$$

This may be integrated directly to give

$$\frac{\partial y}{\partial \xi} = F(\xi), \qquad y(\xi, \eta) = f(\xi) + g(\eta). \tag{1.1.21}$$

Finally, changing back to x,t variables gives the classical D'Alembert solution to the wave equation,

$$y(x, t) = f(x - c_0 t) + g(x + c_0 t). \tag{1.1.22}$$

In considering these results, we first note that f and g are arbitrary functions of integration that will be specifically determined by the initial conditions or forcing function of a given problem. More important at this stage is to note that these functions represent *propagating disturbances*. Thus, in the case of $f(x - c_0 t)$, if $x - c_0 t$ = constant, the function is obviously constant. Arguing in the same manner as for harmonic waves, increasing time requires increasing values of x to maintain the argument of the function constant. This corresponds to a wave propagating in the positive x direction. Similarly, $g(x + c_0 t)$ is a disturbance propagating in the negative x direction.

The second point to emphasize regarding the solution (1.1.22) is that whatever the shape of the disturbances $f(x - c_0 t)$, $g(x + c_0 t)$ initially, that shape is maintained during the propagation. Thus, the waves propagate *without distortion*. The resulting nature of the propagation for a disturbance moving in the positive x direction is shown in Fig. 1.4. Appreciation of the

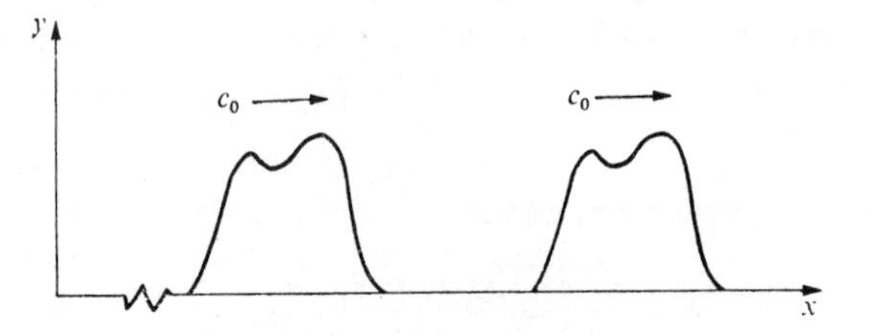

FIG. 1.4. Undistorted propagation of a pulse $f(x - c_0 t)$.

undistorted nature of the wave propagation is important for two reasons. First, it represents a fundamental characteristic of the one-dimensional wave equation. Second, it will serve as a comparison against many physical systems where the opposite is true and where pulse distortion occurs during propagation.

1.1.4. *The initial-value problem*

We now wish to determine the form of the functions arising in the general solution (1.1.22) under prescribed initial conditions in the string. Let

$$y(x, 0) = U(x), \qquad \dot{y}(x, 0) = V(x). \tag{1.1.23}$$

From the general solution we then have, at $t = 0$,

$$f(x)+g(x) = U(x), \tag{1.1.24}$$

$$-c_0 f'(x)+c_0 g'(x) = V(x), \tag{1.1.25}$$

where the prime notation indicates differentiation with respect to the argument and arises in evaluationing $\dot{y}(x, 0)$. Thus

$$\frac{\partial y}{\partial t} = \frac{\partial f(x-c_0 t)}{\partial(x-c_0 t)}\frac{\partial(x-c_0 t)}{\partial t}+\frac{\partial g(x+c_0 t)}{\partial(x+c_0 t)}\frac{\partial(x+c_0 t)}{\partial t}$$

$$= -c_0 f'(x-c_0 t)+c_0 g'(x+c_0 t). \tag{1.1.26}$$

Then

$$\dot{y}(x, 0) = -c_0 f'(x)+c_0 g'(x). \tag{1.1.27}$$

Eqn. (1.1.25) may be integrated to give

$$f(x)-g(x) = -\frac{1}{c_0}\int_b^x V(\zeta)\,\mathrm{d}\zeta. \tag{1.1.28}$$

The arbitrary lower limit of the integral merely absorbs the constant of integration. Solving (1.1.24) and (1.1.28) simultaneously gives

$$f(x) = \frac{U(x)}{2}-\frac{1}{2c_0}\int_b^x V(\zeta)\,\mathrm{d}\zeta, \tag{1.1.29}$$

$$g(x) = \frac{U(x)}{2}+\frac{1}{2c_0}\int_b^x V(\zeta)\,\mathrm{d}\zeta. \tag{1.1.30}$$

This establishes the initial values of the functions f and g for given initial conditions. For $t \neq 0$, we replace x by $x-c_0 t$ in (1.1.29) and x by $x+c_0 t$ in (1.1.30). In combining the results we note that

$$-\int_b^{x-c_0 t} V(\zeta)\,\mathrm{d}\zeta+\int_b^{x+c_0 t} V(\zeta)\,\mathrm{d}\zeta = \int_{x-c_0 t}^{x+c_0 t} V(\zeta)\,\mathrm{d}\zeta. \tag{1.1.31}$$

This may be integrated to give

$$\int_{x-c_0 t}^{x+c_0 t} V(\zeta)\,\mathrm{d}\zeta = H(x+c_0 t)-H(x-c_0 t). \tag{1.1.32}$$

Hence, the final solution to the initial value problem is

$$y(x, t) = \tfrac{1}{2}\{U(x-c_0 t)+U(x+c_0 t)\}-\frac{1}{2c_0}\{H(x-c_0 t)+H(x+c_0 t)\}. \tag{1.1.33}$$

The resulting motion thus consists of identical leftward and rightward propagating disturbances containing separate contributions from the displacement and velocity initial conditions.

Example. For simplicity of illustration, consider only displacement initial conditions. Thus, suppose

$$y(x,0) = U(x) = \begin{cases} 1, & -a < x < a \\ 0, & |x| > a \end{cases}, \tag{1.1.34}$$

$$\dot{y}(x,0) = V(x) = 0.$$

Then

$$\dot{y}(x,t) = \tfrac{1}{2}\{U(x-c_0t)+U(x+c_0t)\}. \tag{1.1.35}$$

The behaviour of the string at different instants of time is shown in Fig. 1.5.

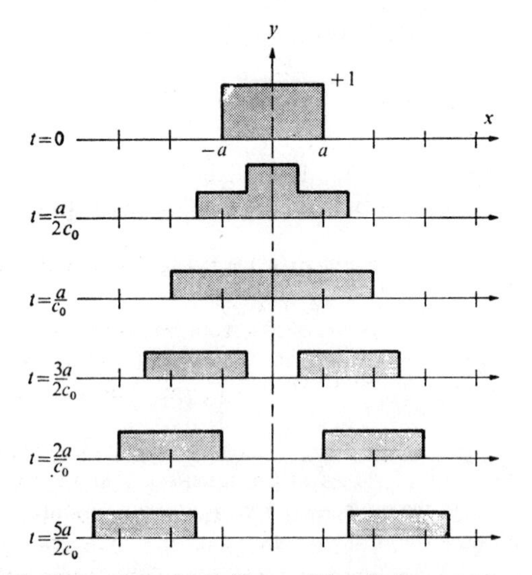

FIG. 1.5. Propagation of an initial condition displacement in a string.

We now briefly consider an alternative representation of the propagation in a string. Recalling the change of variables (1.1.17) employed in solving the wave equation, we note that $\xi = x-c_0t$ = constant represents a straight line in the x, t-plane. Along this line $f(x-c_0t) = f(\xi)$ = constant. Similarly, $\eta = x+c_0t$ = constant is a straight line, opposite in slope to ξ, along which $g(x+c_0t)$ is constant. These lines are called the *characteristics* of the solution.

The characteristics are not a special circumstance of the present problem but arise, in fact, in the general theory of hyperbolic partial differential equations, of which the wave equation is a fairly simple example. In general, the characteristics are curved lines and the quantities that are constant along these lines may be fairly involved. Methods of analysis involving characteristics find many applications in fluid mechanics in problems involving shock waves. Our considerations in this area will be quite elementary and restricted to elastic waves in one dimension.

Continuing, we plot the characteristic lines emanating from $x = a, -a$ in the x,t-plane. The results are shown in Fig. 1.6. Only four characteristic

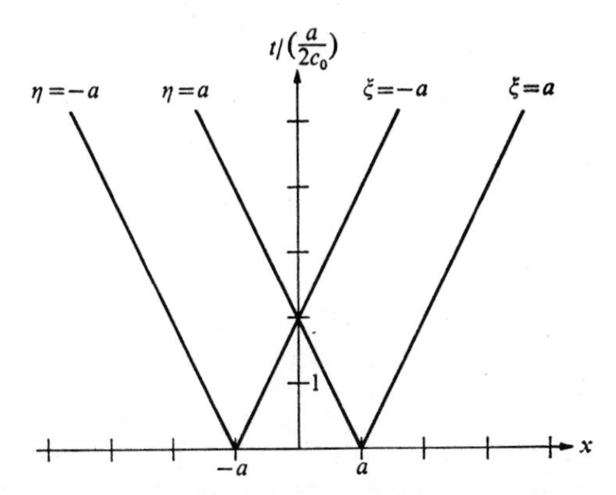

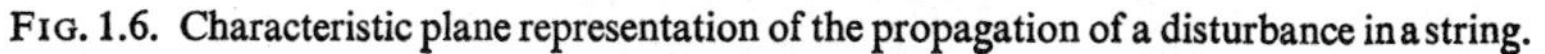
FIG. 1.6. Characteristic plane representation of the propagation of a disturbance in a string.

lines are shown, since, for lines emanating from points $|x| > a$, the values are zero along the characteristics, while for lines from $|x| < a$, the values along the lines are constant at $\frac{1}{2}$. The lines shown thus indicate the positions of the wavefronts.

Comparing the two representations of wave motion of Fig. 1.5 and Fig. 1.6, it is apparent that the characteristics representation lacks the amplitude information contained in the former figure. For the simple illustration at hand, there is thus no particular advantage to the x,t-plane representation. However, in more complicated problems involving wave reflection and transmission at boundaries, this will not be the case and the characteristics representation will prove most helpful.

1.1.5. *The initial-value problem by Fourier analysis*

The D'Alembert solution is not the only approach to the wave equation. Recalling from an earlier section that harmonic waves also satisfy the equation, it is natural to consider forming a general solution by superposition of such waves. For simplicity, assume only displacement initial conditions are prescribed, so that we have

$$y(x, 0) = U(x), \qquad \dot{y}(x, 0) = 0. \tag{1.1.36}$$

It is possible to represent $U(x)$ directly as a Fourier integral (see Appendix B.3) of the form

$$y(x, 0) = U(x) = \frac{1}{\pi}\int_{0}^{\infty} d\gamma \int_{-\infty}^{\infty} U(\zeta)\cos \gamma(x-\zeta)\, d\zeta. \quad (1.1.37)$$

The interpretation of this expression is that at $t = 0$, $U(x)$ has been represented by superimposing waves of different wavelengths, and hence of different wavenumbers γ. The various harmonic components also differ in phase angle, given by the product $\gamma\zeta$.

For $t > 0$, the various Fourier components begin moving to the right and left. We have that

$$\cos \gamma(x-\zeta) = \tfrac{1}{2}\{\cos \zeta(x-c_0t-\zeta)+\cos \gamma(x+c_0t-\zeta)\}. \quad (1.1.38)$$

The resulting wave system is then given by

$$y(x, t) = \frac{1}{2\pi}\int_{0}^{\infty} d\gamma \int_{-\infty}^{\infty} U(\zeta)\{\cos \gamma(x-c_0t-\zeta)+\cos \gamma(x+c_0t-\zeta)\}\, d\zeta. \quad (1.1.39)$$

If we replace the variables $x-c_0t$ and $x+c_0t$ by ξ and η in the preceding expression and compare the result to the Fourier integral expression (1.1.37) it is seen that we have obtained

$$y(x, t) = \tfrac{1}{2}\{U(\xi)+U(\eta)\}. \quad (1.1.40)$$

Changing the variables back to x,t gives

$$y(x, t) = \tfrac{1}{2}\{U(x-c_0t)+U(x+c_0t)\}. \quad (1.1.41)$$

This is the D'Alembert form of solution to the initial displacement problem.

Let us now examine the same problem by the Fourier transform method. We apply the exponential Fourier transform to the wave equation (1.1.3), transforming on the spatial variable. This gives

$$\frac{d^2\bar{Y}}{dt^2}+\gamma^2c_0^2\bar{Y} = 0, \quad (1.1.42)$$

where $\bar{Y}(\gamma, t)$ is the transformed displacement and γ is the transform variable. Solving (1.1.42) gives

$$\bar{Y} = A\exp(i\gamma c_0t)+B\exp(-i\gamma c_0t). \quad (1.1.43)$$

The initial conditions (1.1.36) must also be transformed, to give

$$\bar{Y}(\gamma, 0) = \bar{U}(\gamma), \qquad \frac{d\bar{Y}(\gamma, 0)}{dt} = 0. \quad (1.1.44)$$

Substituting the result (1.1.43) in these conditions gives the values of A and B as

$$A = B = \bar{U}(\gamma)/2. \tag{1.1.45}$$

The transformed solution is then

$$\bar{Y}(\gamma, t) = \frac{\bar{U}(\gamma)}{2}\{\exp(i\gamma c_0 t)+\exp(-i\gamma c_0 t)\}. \tag{1.1.46}$$

Taking the inverse Fourier transform of this result gives

$$\begin{aligned} y(x, t) &= \frac{1}{2\sqrt{(2\pi)}} \int_{-\infty}^{\infty} \bar{U}(\gamma)\{\exp(i\gamma c_0 t)+\exp(-i\gamma c_0 t)\}\exp(-i\gamma x)\,\mathrm{d}\gamma \\ &= \frac{1}{2\sqrt{(2\pi)}} \int_{-\infty}^{\infty} \bar{U}(\gamma)[\exp\{-i\gamma(x-c_0 t)\}+\exp\{-i\gamma(x+c_0 t)\}]\,\mathrm{d}\gamma. \end{aligned} \tag{1.1.47}$$

Proceeding now in a manner similar to that used in the Fourier integral approach, we temporarily replace $x-c_0 t$ and $x+c_0 t$ by ξ and η in (1.1.47). In the one case, we then have

$$\frac{1}{\sqrt{(2\pi)}} \int_{-\infty}^{\infty} \bar{U}(\gamma)\exp(-i\gamma\xi)\,\mathrm{d}\gamma = U(\xi), \tag{1.1.48}$$

simply from the definition of the inverse Fourier transform. Replacing ξ by η in (1.1.48) gives $U(\eta)$. Then the result (1.1.47) takes the form

$$y(x, t) = \tfrac{1}{2}\{U(\xi)+U(\eta)\} = \tfrac{1}{2}\{U(x-c_0 t)+U(x+c_0 t)\}, \tag{1.1.49}$$

and the D'Alembert form of the solution is again recovered.

1.1.6. *Energy in a string*

We have considered only the propagation of transverse displacements in a string up to this point. In considering the propagation of energy, we must first develop an expression for the kinetic and potential energy in a string. The kinetic energy in a differential element of string as shown in Fig. 1.1 would be given by

$$\mathrm{d}K = \rho\,\mathrm{d}s\dot{y}^2/2. \tag{1.1.50}$$

The total kinetic energy in a segment of string between x_1 and x_2 would then be

$$K(t) = \frac{1}{2}\int_{x_1}^{x_2} \rho\dot{y}^2\left\{1+\left(\frac{\partial y}{\partial x}\right)^2\right\}^{\frac{1}{2}}\,\mathrm{d}x. \tag{1.1.51}$$

It is permissible to neglect the term $(\partial y/\partial x)^2$ in comparison to unity for small deflections of the string so that

$$K(t) = \frac{1}{2}\int_{x_1}^{x_2} \rho \dot{y}^2 \, dx. \tag{1.1.52}$$

We then define the kinetic energy density $k(x, t)$ as

$$k(x, t) = \rho \dot{y}^2/2. \tag{1.1.53}$$

The computation of the potential energy in a length of string is slightly more involved. In particular, the higher-order term just neglected in the kinetic-energy calculation provides the contribution to the potential energy. Thus, the change in length of the string between x_1 and x_2 due to stretching will be given by

$$\Delta l = \int_{x_1}^{x_2} ds - (x_2 - x_1) = \int_{x_1}^{x_2} \left\{1 + \left(\frac{\partial y}{\partial x}\right)^2\right\}^{\frac{1}{2}} dx - (x_2 - x_1). \tag{1.1.54}$$

We make the approximation $\{1+(\partial y/\partial x)^2\}^{\frac{1}{2}} \cong 1+(\partial y/\partial x)^2/2$. Using this in (1.1.54) gives

$$\Delta l \cong \frac{1}{2}\int_{x_1}^{x_2} \left(\frac{\partial y}{\partial x}\right)^2 dx. \tag{1.1.55}$$

The potential energy $V(t)$ in the length x_1 to x_2 will be given by $T\Delta l$ so that

$$V(t) = \frac{1}{2}\int_{x_1}^{x_2} T\left(\frac{\partial y}{\partial x}\right)^2 dx. \tag{1.1.56}$$

We define the potential energy density $v(x, t)$ as

$$v(x, t) = \frac{T}{2}\left(\frac{\partial y}{\partial x}\right)^2. \tag{1.1.57}$$

The total energy of the system in the region x_1 to x_2 is then given by

$$E(t) = K(t) + V(t), \tag{1.1.58}$$

and the energy density is given by

$$\varepsilon(x, t) = k(x, t) + v(x, t) = \frac{\rho \dot{y}^2}{2} + \frac{T}{2}\left(\frac{\partial y}{\partial x}\right)^2. \tag{1.1.59}$$

Consider now a wave, propagating in the positive x direction, given by $y = f(x - c_0 t)$. The kinetic and potential energy densities are given by

$$k = \rho c_0^2 f'^2/2, \qquad v = Tf'^2/2. \tag{1.1.60}$$

Since $c_0^2 = T/\rho$, we see that the two expressions are equal. Thus a propagating wave in a string has its energy equally divided into kinetic and potential energies.

In addition to the energy density of the string, one is often interested in the power flow, or rate of transfer of energy past a given point or, still equivalently, the rate of doing work at a given point of the string. This quantity may be readily established by noting that the rate of doing work, or power, is given by

$$P(x_0, t) = -T\frac{\partial y}{\partial x}\dot{y}\bigg|_{x=x_0}. \tag{1.1.61}$$

Suppose that a wave having the general form $y = g(x+c_0t)$ is propagating to the left. Then we have

$$P(x_0, t) = -Tc_0g'^2 \leqslant 0. \tag{1.1.62}$$

If a wave of the form $y = f(x-c_0t)$ is travelling to the right, we have

$$P(x_0, t) = Tc_0f'^2 \geqslant 0. \tag{1.1.63}$$

We see that in the first case, power is leaving the portion of string $x > x_0$, while in the second case it is entering that region. Thus, we say that power flows in the direction of the wave.

1.1.7. *Forced motion of a semi-infinite string*

We now give our first attention to the forced motion of strings. The two basic ways of imparting energy to the string are through boundary forcing or through motion imparted in regions along the length of string. We will consider the former problem first since the governing differential equation is still homogeneous.

As a simple first example, consider the semi-infinite string $x \geqslant 0$ to be excited at the boundary by the harmonic displacement $y(0, t) = y_0 \exp(i\omega t)$. The most straightforward approach here is to assume a solution of the form previously given by (1.1.14) and substitute in the wave equation. The results are (also previously given by (1.1.15))

$$y = A_1e^{i(\gamma x+\omega t)}+B_1e^{-i(\gamma x-\omega t)}, \qquad \gamma = \omega/c_0. \tag{1.1.64}$$

The constants A_1, B_1 must be determined from the boundary conditions of the problem. From the condition at $y(0, t)$ we obtain $y_0 = A_1+B_1$, so that we have

$$y = A_1e^{i(\gamma x+\omega t)}+(y_0-A_1)e^{-i(\gamma x-\omega t)}. \tag{1.1.65}$$

The establishment of the second condition is, in fact, the whole point of this otherwise trivial example. There are no further conditions at $x = 0$. On the other hand there is, effectively, a condition at infinity. Specificially, we again note that the result (1.1.65) contains both leftward and rightward propagating waves. However, unless there is an energy source radiating waves in from infinity or a boundary that is reflecting waves back to the origin, there is no physical basis for the leftward propagating term $\exp\{i(\gamma x+\omega t)\}$ in the result. So, on the basis of a *radiation condition* we set $A_1 = 0$. Then we have

$$y = y_0 e^{-i(\gamma x-\omega t)}. \tag{1.1.66}$$

The radiation condition introduced in this problem will be used repeatedly in the analysis of future wave problems. It functions in the manner of a boundary condition, enabling certain solutions of a differential equation to be discarded. We note that the solution selected depends on the nature of the initial time dependence. Thus, if a time dependence $\exp(-i\omega t)$ had been used in the problem, the radiation condition would have resulted in $B_1 = 0$ instead of $A_1 = 0$.

We now consider the response of a string subjected to a general transient displacement

$$y(0, t) = g(t). \tag{1.1.67}$$

We again use the Fourier transform approach, applying the exponential transform to the wave equation on the time variable. This gives

$$\frac{d^2\bar{y}}{dx^2}+\frac{\omega^2}{c_0^2}\bar{y} = 0, \tag{1.1.68}$$

where $\bar{y} = \bar{y}(x, \omega)$ is the transformed displacement and ω is the transform variable. Solving (1.1.68) we obtain

$$\bar{y} = A\exp(i\omega x/c_0)+B\exp(-i\omega x/c_0). \tag{1.1.69}$$

The transformed boundary condition is given by

$$\bar{y}(0, \omega) = \bar{g}(\omega). \tag{1.1.70}$$

Substituting (1.1.69) in this condition gives $A+B = \bar{g}(\omega)$. The second boundary condition will be on the radiation, as in the preceding simple example. This is made explicit by taking the inverse transform, giving

$$y(x, t) = \frac{1}{\sqrt{(2\pi)}}\int_{-\infty}^{\infty}[\{\bar{g}(\omega)-B\}\exp(i\omega x/c_0)+B\exp(-i\omega x/c_0)]\exp(-i\omega t)\,d\omega. \tag{1.1.71}$$

We again recognize incoming radiation from infinity in the term

$$\exp\{-i(\omega x/c_0+\omega t)\}$$

and set $B = 0$. This gives

$$y(x, t) = \frac{1}{\sqrt{(2\pi)}}\int_{-\infty}^{\infty}\bar{g}(\omega)\exp\{i\omega(x/c_0-t)\}\,d\omega. \tag{1.1.72}$$

It should be immediately recognized that this result has the form

$$y(x, t) = g(x/c_0 - t). \tag{1.1.73}$$

Thus, the disturbance created at the origin propagates outward without change of shape.

We return to the solution form (1.1.72) and note for future reference that many wave-propagation problems result in solutions of this general form. Thus, given an input $y(0, t)$, find the response or output $y(x, t)$. The result is obviously the superposition of many propagating harmonic waves where the amplitudes of the various frequency components are determined by the transform $\bar{g}(\omega)$ of the input. This quantity is usually designated the *frequency spectrum* of the input. Now, in the case of the string where each frequency component propagates with the same velocity c_0 the undistorted pulse propagation (1.1.73) is predicted. However, this same solution form will provide the basis for considering more complex wave phenomena where the relationships between phase velocity, frequency, and wavenumber are not so simple as in the present case.

1.1.8. *Forced motion of an infinite string*

Our considerations thus far have been with free motion in an infinite string (the initial-value problem) and with boundary forcing of a semi-infinite string. As such we have only considered the homogeneous equation for the string. We now consider the forced motion of an infinite string such that the non-homogeneous equation (1.1.2) governs. Rewritten slightly, we thus have

$$\frac{\partial^2 y}{\partial x^2} - \frac{1}{c_0^2}\frac{\partial^2 y}{\partial t^2} = p(x, t), \tag{1.1.74}$$

where $p(x, t) = -q(x, t)/T$.

It was possible, in the analysis of the semi-infinite string, to develop a general solution for arbitrary forcing (1.1.72) and to still render some interpretation of the results. In the present problem, the forcing $p(x, t)$ may be arbitrary in both space and time. Consequently, a general solution of (1.1.74), while obtainable, is not so amenable to similar interpretation and is more a formal mathematical exercise. The approach taken in this type of problem is to replace $p(x, t)$ by special loadings, solve the resulting special problem, and obtain solutions to more general cases by superposition.

As our first special loading, replace $p(x, t)$ by the load $\delta(x-\xi)\delta(t-\tau)$. This represents a unit load occurring at time $t = \tau$ at a location $x = \xi$ on the string. The practice of replacing general loads with an impulse load and determining the system response to this load is widely used in applied mathematics. The resulting system response is usually designated as the *Green's function* of the system. For the present one-dimensional problem, this would be written as

$$G = G(x, t/\xi, \tau). \tag{1.1.75}$$

The problem is thus one of considering

$$\frac{\partial^2 G}{\partial x^2} - \frac{1}{c_0^2}\frac{\partial^2 G}{\partial t^2} = \delta(x-\xi)\,\delta(t-\tau). \tag{1.1.76}$$

A double-transform approach will be taken. We first take the Fourier transform on the space variable, giving

$$-\gamma^2\bar{G} - \frac{1}{c_0^2}\frac{\partial^2 \bar{G}}{\partial t^2} = \frac{e^{i\gamma\xi}}{\sqrt{(2\pi)}}\,\delta(t-\tau). \tag{1.1.77}$$

We assume the system is initially at rest, so that $y(x, 0) = \dot{y}(x, 0) = 0$. Taking the Laplace transform of (1.1.77) gives

$$-\gamma^2\bar{G}_L - \frac{s^2}{c_0^2}\bar{G}_L = \frac{1}{\sqrt{(2\pi)}}e^{i\gamma\xi}e^{-s\tau}, \tag{1.1.78}$$

where $\bar{G}_L = \bar{G}_L(\gamma, s/\xi, \tau)$. The transformed solution is then

$$\bar{G}_L = \frac{-c_0^2}{\sqrt{(2\pi)}}\frac{e^{i\gamma\xi}e^{-s\tau}}{s^2+c_0^2\gamma^2}. \tag{1.1.79}$$

In inverting the above, we first consider the Laplace inversion. We note from tables that

$$\mathscr{L}^{-1}\left\{\frac{1}{s^2+c_0^2\gamma^2}\right\} = \frac{1}{c_0\gamma}\sin c_0\gamma t, \tag{1.1.80}$$

and that in general

$$\mathscr{L}^{-1}\{e^{-s\tau}F_L(s)\} = F(t-\tau), \qquad F(t) = 0, \qquad t < 0. \tag{1.1.81}$$

The Laplace inverted result is then

$$\bar{G} = -\frac{c_0^2}{\sqrt{(2\pi)}}\frac{e^{i\gamma\xi}}{c_0\gamma}\sin c_0\gamma(t-\tau)H\langle t-\tau\rangle, \tag{1.1.82}$$

where $H\langle t-\tau\rangle$ is the Heaviside step function. Next taking the inverse Fourier transform we obtain

$$G = \frac{-c_0^2}{2\pi}H\langle t-\tau\rangle\int_{-\infty}^{\infty}\frac{\sin c_0\gamma(t-\tau)}{c_0\gamma}e^{-i\gamma(x-\xi)}\,d\gamma. \tag{1.1.83}$$

The inversion of this part is also fairly simple. We know† that

$$\mathscr{F}^{-1}\left\{\sqrt{\left(\frac{2}{\pi}\right)}\frac{\sin a\gamma}{\gamma}\right\} = \begin{cases}1, & |x| < a \\ 0, & |x| > a\end{cases}. \tag{1.1.84}$$

† See, for example, Appendix B.3.

To use this result, we rewrite (1.1.83) as

$$G = \frac{-c_0}{2} H\langle t'/c_0\rangle \frac{1}{\sqrt{(2\pi)}} \int_{-\infty}^{\infty} \sqrt{\left(\frac{2}{\pi}\right)} \frac{\sin t'\gamma}{\gamma} \exp(-\mathrm{i}\gamma x')\,\mathrm{d}\gamma, \qquad (1.1.85)$$

where $t' = c_0(t-\tau)$, $x' = x-\xi$. Applying (1.1.84) to this gives

$$G = -\frac{c_0}{2} H\langle t'/c_0\rangle (H\langle x'+t'\rangle - H\langle x'-t'\rangle). \qquad (1.1.86)$$

In this result, two Heaviside functions have been superimposed to provide the representation of a rectangular pulse. Now, changing back to the original x,t variables, we have

$$G(x, t/\xi, \tau) = -\frac{c_0}{2} H\langle t-\tau\rangle \{H\langle x-\xi+c_0(t-\tau)\rangle - H\langle x-\xi-c_0(t-\tau)\rangle\}. \qquad (1.1.87)$$

The ensuing motion of the string predicted by this result is shown in Fig. 1.7.

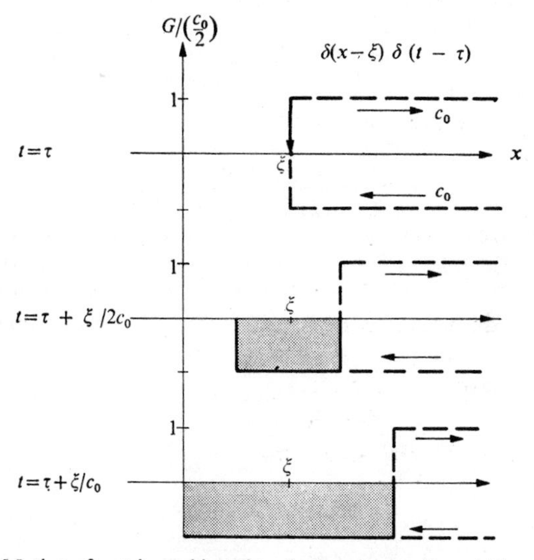

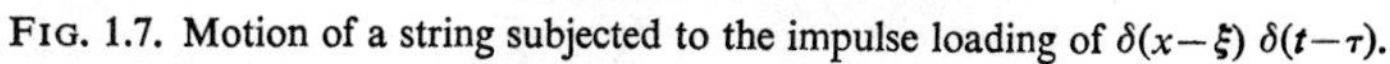

FIG. 1.7. Motion of a string subjected to the impulse loading of $\delta(x-\xi)\,\delta(t-\tau)$.

The solution for the response of the string to a general loading $p(x, t)$ is obtained from the Green's function result by the following double integral:

$$y(x, t) = \int_0^t \mathrm{d}\tau \int_{-\infty}^{\infty} G(x, t/\xi, \tau) p(\xi, \tau)\,\mathrm{d}\xi. \qquad (1.1.88)$$

For the second special case of loading on the string, we assume the time variation is simple harmonic so that $p(x,t) = f(x)\exp(-i\omega t)$. If we assume the response to have a similar form, so that $y = Y(x)\exp(-i\omega t)$, the governing equation (1.1.74) reduces to the ordinary differential equation

$$\frac{d^2Y}{dx^2}+\frac{\omega^2}{c_0^2}Y = f(x). \tag{1.1.89}$$

We now specialize the spatial variation of the load to that of a unit concentrated load acting at $x = \xi$, so that $f(x) = \delta(x-\xi)$. We thus will be seeking the Green's function of the system for harmonic loading. We denote the response as $G = G(x/\xi)$ and write (1.1.89) as

$$\frac{d^2G}{dx^2}+\frac{\omega^2}{c_0^2}G = \delta(x-\xi). \tag{1.1.90}$$

We now apply the Fourier transform to this, resulting in the transformed solution

$$\bar{G} = -\frac{1}{\sqrt{(2\pi)}}\frac{e^{i\gamma\xi}}{\gamma^2-\gamma_0^2}, \qquad \gamma_0 = \omega/c_0. \tag{1.1.91}$$

The Fourier inversion gives

$$G = \frac{-1}{2\pi}\int_{-\infty}^{\infty}\frac{e^{-i\gamma(x-\xi)}}{\gamma^2-\gamma_0^2}\,d\gamma. \tag{1.1.92}$$

We will now take the time to evaluate the above integral from first principles by carrying out the integration in the complex plane. At least two aspects of the analysis that arise in this problem occur repeatedly in more general problems of wave propagation. Before proceeding, we simplify (1.1.92) slightly by letting $\xi = 0$, corresponding to the load being applied at the origin. We thus have

$$G(x/0) = -\frac{1}{2\pi}\int_{-\infty}^{\infty}\frac{e^{-i\gamma x}}{\gamma^2-\gamma_0^2}\,d\gamma. \tag{1.1.93}$$

Considering γ complex, we have from the calculus of residues that

$$\int_C \frac{e^{-i\gamma x}}{\gamma^2-\gamma_0^2}\,d\gamma = 2\pi i\sum \text{Res}, \tag{1.1.94}$$

where C is a closed contour in the complex plane. The first step in evaluating (1.1.94) is to select the general contour in the complex plane. The two obvious possibilities are shown in Fig. 1.8 where, for the moment, the poles $\gamma = \pm\gamma_0$ located on the real axis have been ignored. Along the semicircular

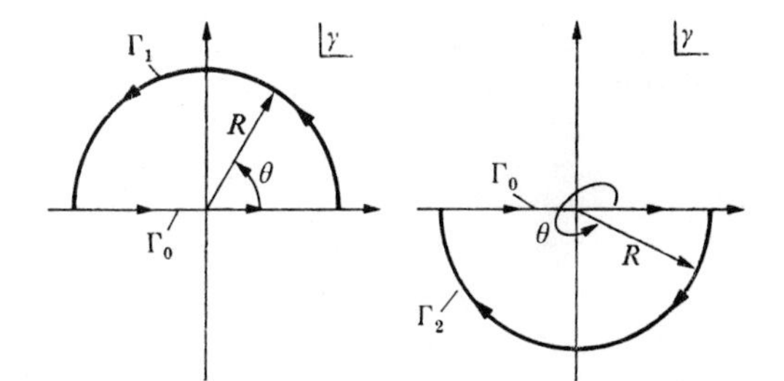

FIG. 1.8. Two possible choices of contour for evaluating (1.1.94).

path, be it Γ_1 or Γ_2, we let $\gamma = R\exp(i\theta)$. If we define I_1 to be the form of the integral (1.1.94) along Γ_1, and substitute the polar representation of γ in the integral, we have

$$I_1 = \lim_{R\to\infty} \int_0^{\pi} \frac{e^{xR\sin\theta}e^{-ixR\cos\theta}}{R^2e^{2i\theta}-\gamma_0^2} iRe^{i\theta}\,d\theta. \tag{1.1.95}$$

The important term in (1.1.95) is $\exp(xR\sin\theta)$. Along Γ_1 we have $\sin\theta > 0$ always. Then, if $x < 0$, the exponential rapidly decays for large values of R and, in fact, as $R \to \infty$ we have $I_1 \to 0$. If, on the other hand, $x > 0$ then I_1 would become infinite as $R \to \infty$. Similar arguments can be applied to the integral I_2 along Γ_2 to show that the integral vanishes for $R \to \infty$ as long as $x > 0$.

The conclusion we reach is that either contour could be selected, as long as the resulting restriction on x is specifically recognized. In our analysis of the infinite string, where $-\infty < x < \infty$, the selection of a specific contour will limit the results to $-\infty < x < 0$ or to $0 < x < \infty$ for a load applied at the origin. If the load is placed at $x = \xi$, such as for the result (1.1.92) the regions of validity would simply become $-\infty < x < \xi$ or $\xi < x < \infty$. This, then, is the first aspect of the inversion that arises in many wave propagation problems.

The second aspect relates to the poles of (1.1.94) that are located on the real axis. If this were merely a pathological circumstance of this problem, there would be possibly no need for great concern over the matter, circumventing it possibly by working a 'nicer' example. It is a fact of analysis, however, that this situation arises repeatedly in wave problems and must be accounted for.

The first step is to avoid the singularity on the path of integration by deforming the contour in a small, semicircular path either above or below the poles, as shown in Fig. 1.9 for $\gamma = \gamma_0$. Such an indentation will include or exclude a singularity from within the contour of integration depending on whether the over-all closure of the contour is above or below the real axis.

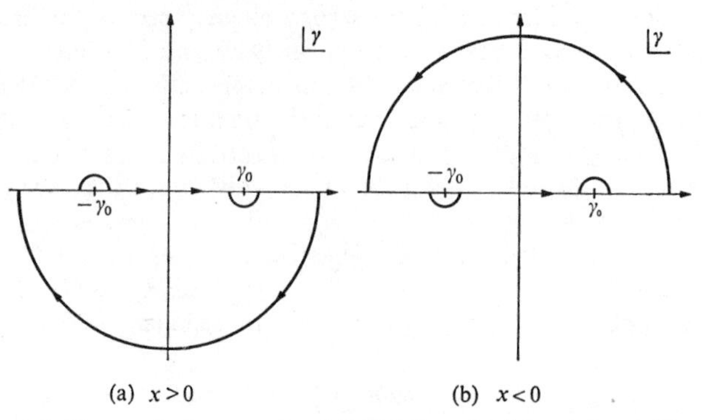

FIG. 1.9. The required contours and indentations for (a) $x > 0$ and (b) $x < 0$.

The residues of the singularities are found to be

$$\operatorname{Res}\big|_{\gamma=-\gamma_0} = -\frac{\exp(\mathrm{i}\gamma_0 x)}{2\gamma_0}, \qquad \operatorname{Res}\big|_{\gamma=\gamma_0} = \frac{\exp(-\mathrm{i}\gamma_0 x)}{2\gamma_0}. \tag{1.1.96}$$

We now recall that the time dependence in this problem is $\exp(-\mathrm{i}\omega t)$. This dependence combined with either of (1.1.96) can give leftward or rightward propagating waves, depending on whether $x \gtrless 0$ and depending on with which residue it is combined. Thus which indentations are used will be governed ultimately by the radiation condition. Indentations selected must yield outgoing waves.

To illustrate this, we suppose $x > 0$ is the region of interest. On the basis of earlier discussion, we know the general contour is closed below the real axis, as is Γ_2 of Fig. 1.8. In order to have outward propagating waves in $x > 0$, we must have behaviour of the form $\exp\{\mathrm{i}(\gamma x - \omega t)\}$. From the residues (1.1.96) it is seen that the pole at $\gamma = -\gamma_0$ must be included *within* the general contour and the pole at $\gamma = \gamma_0$ must be *excluded*. This requires indentations as shown in Fig. 1.9(a). We thus have that

$$G(x/0) = -\frac{1}{2\pi}(-2\pi\mathrm{i}\operatorname{Res}\big|_{\gamma=-\gamma_0}). \tag{1.1.97}$$

The minus sign within the parentheses of the result accounts for the fact that the path of integration of Fig. 1.9(a) is counterclockwise. As a final result, we have

$$G(x/0) = -\frac{\mathrm{i}}{2\gamma_0}\exp(\mathrm{i}\gamma_0 x), \qquad x > 0. \tag{1.1.98}$$

The proper contour for the region $x < 0$ is shown, without further discussion, in Fig. 1.9(b).

This concludes consideration of the second major aspect of the problem. We see that the proper contour selection, including indentations, is based on the range of x (thus $x > 0$ or $x < 0$ for this simple example) and the nature of the time dependence, with the radiation condition dictating the final choices. The difficulty with this problem has resulted from poles being on the axis of integration. It turns out that the whole problem vanishes nicely when there is damping in the system, as analysis in § 1.7 p. 70) will show.

The basis for extending the Green's function result to general loading situations lies in the convolution theorem (see Appendix B). If $G(x/\xi)$ is the Green's function and $f(x)$ is the loading function, we have

$$Y(x) = \frac{1}{\sqrt{(2\pi)}} \int_{-\infty}^{\infty} G(\gamma/\xi) f(x-\gamma)\, d\gamma. \tag{1.1.99}$$

If the convolution approach is not used, results for more general loadings could be considered by attempting to directly invert the transformed solution of (1.1.89), which would give an integral of the form

$$Y(x) = -\frac{1}{\sqrt{(2\pi)}} \int_{-\infty}^{\infty} \frac{\bar{f}(\gamma)}{\gamma^2-\gamma_0^2} e^{-i\gamma x}\, d\gamma. \tag{1.1.100}$$

1.2. Reflection and transmission at boundaries

Our considerations thus far have dealt with waves in infinite or semi-infinite strings. Thus, the conditions were such that the interaction of waves with boundaries did not enter. We now wish to consider waves incident on a boundary of the string. When this situation arises, wave reflection occurs with the nature of the reflected wave being determined by the nature of the boundary or 'boundary condition.'

1.2.1. *Types of boundaries*

One of the simplest types of end boundary condition would be the case of a string attached to a rigid support. Supposing this support to be at the origin, the condition would thus be $y(0, t) = 0$. Another type of constraint could be given by the end of the string sliding on a frictionless slot, shown as the free end condition in Table 1.1. The component of force in the vertical direction at $x = 0$ would be given by

$$F_v(0, t) = T\, \partial y(0, t)/\partial x. \tag{1.2.1}$$

Since the slot is frictionless and the end of the string is without mass, the net vertical force component must be zero. Thus, the case of a free end condition is specified by $\partial y(0, t)/\partial x = 0$.

It is possible to devise numerous other conditions corresponding to an attached end mass, a spring, or a dashpot. The mathematical statements of these conditions are arrived at by equating the vertical component of string tension to the forces on these elements. Table 1.1 summarizes these various

TABLE 1.1

Type	Diagram	Equation
Fixed	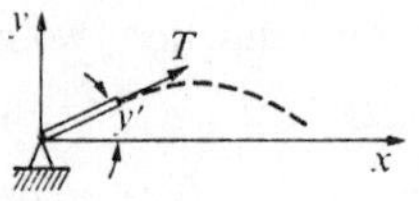	$y(0,t)=0$
Free	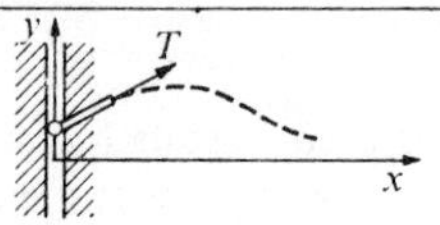	$\dfrac{\partial y(0,t)}{\partial x}=0$
Mass	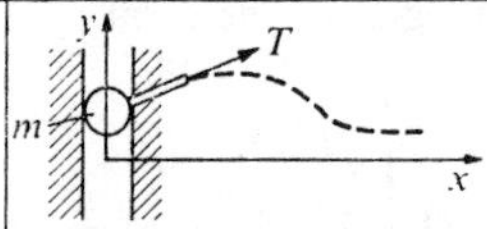	$m\ddot{y}(0,t)=T\dfrac{\partial y(0,t)}{\partial x}$
Spring	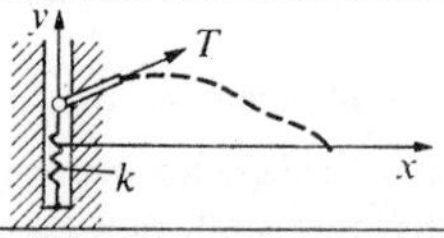	$ky(0,t)=T\dfrac{\partial y(0,t)}{\partial x}$
Dashpot	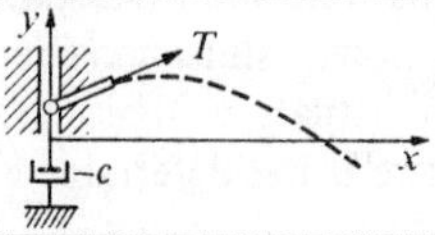	$c\dot{y}(0,t)=T\dfrac{\partial y(0,t)}{\partial x}$

conditions. More complicated conditions could be obtained by combining the spring–mass–dashpot elements.

1.2.2. *Reflection from a fixed boundary*

The reflection of an incident wave from a fixed boundary represents the simplest type of boundary interaction. All wave-reflection problems can be approached in a purely mathematical way, and the present case is no exception. However, the simple end condition permits a somewhat intuitive approach to be used, called the method of images, and saves formalism for more complicated problems.

Consider a displacement pulse $f(x+c_0t)$ propagating to the left and incident on a fixed boundary, as shown in Fig. 1.10(a). Imagine now removing

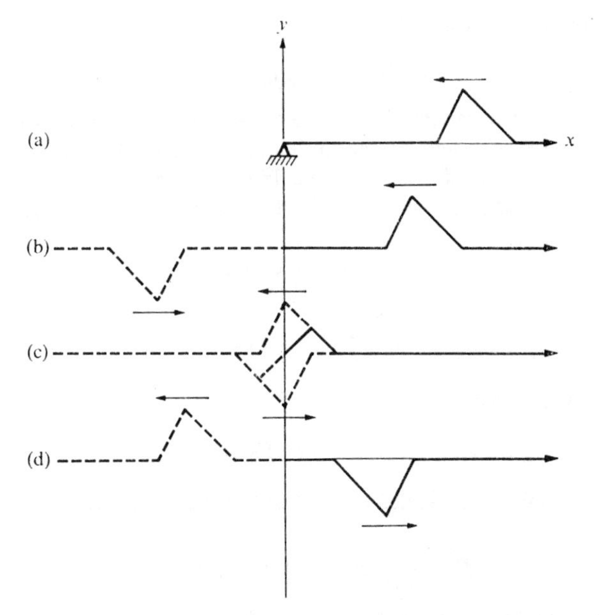

FIG. 1.10. Sequence of events during pulse reflection from a fixed boundary.

the boundary at $x = 0$ but extending the string to negative infinity. Now construct an 'image' pulse to $f(x+c_0t)$. This pulse is symmetrically placed with respect to $x = 0$, is opposite in sense to $f(x+c_0t)$, and is propagating to the right, as shown in Fig. 1.10(b). As these pulses approach the origin, they will interact, as shown in Fig. 1.10(c). It should be clear that, as they pass, their displacements will mutually cancel at $x = 0$, giving $y(0, t) = 0$ always. Thus the fixed end boundary condition for the semi-infinite string is always satisfied by the image pulse system in the infinite string. As time passes the interaction stage is completed and the image pulse propagates into the region $x > 0$, while the original 'real' pulse propagates into $x < 0$, as shown in Fig. 1.10(d). After the completion of the process, it is seen that the sign of the original pulse has been reversed. This is characteristic of the fixed boundary.

1.2.3. *Reflection from an elastic boundary*

As an example of a more complicated reflection problem, we consider the wave incident on an elastic boundary, as shown in Table 1.1 for the spring end condition. The problem is thus one of requiring the solution of the wave equation

$$y = f(c_0t-x)+g(c_0t+x) \tag{1.2.2}$$

to satisfy the boundary condition

$$ky(0, t) = T\,\partial y(0, t)/\partial x. \tag{1.2.3}$$

The ohange in the representation of the argument of the wave in (1.2.2) in contrast to the earlier representation (1.1.22) is merely for convenience in using the Laplace transform approach. It is presumed in (1.2.2) that the incident wave $g(c_0t+x)$ is prescribed. It is thus a matter of finding the reflected wave $f(c_0t-x)$.

In substituting (1.2.2) in (1.2.3) we note that

$$\left.\frac{\partial y(x, t)}{\partial x}\right|_{x=0} = -f'(c_0t-x)+g'(c_0t+x)|_{x=0} = -f'(c_0t)+g'(c_0t), \tag{1.2.4}$$

so that (1.2.3) becomes

$$k\{f(c_0t)+g(c_0t)\} = T\{-f'(c_0t)+g'(c_0t)\}. \tag{1.2.5}$$

Making the change of variables $\tau = c_0t$ and rearranging gives

$$f'(\tau)+\frac{k}{T}f(\tau) = g'(\tau)-\frac{k}{T}g(\tau). \tag{1.2.6}$$

We apply the Laplace transform to (1.2.6), obtaining

$$\left(s+\frac{k}{T}\right)\bar{f} = \left(s-\frac{k}{T}\right)\bar{g}. \tag{1.2.7}$$

We note that (1.2.6) holds only for $x = 0$. In obtaining the result (1.2.7) we have assumed xero initial conditions. If we express $s-k/T = s+k/T-2k/T$, the transformed solution may be written as

$$\bar{f} = \bar{g}-2\frac{k}{T}\frac{\bar{g}}{s+k/T}. \tag{1.2.8}$$

For the inverted solution we may formally write

$$f(\tau) = g(\tau)-\frac{2k}{T}\mathscr{L}^{-1}\left\{\frac{\bar{g}(s)}{s+k/T}\right\}. \tag{1.2.9}$$

The formal solution could be carried one step further by using the convolution integral to express the inverse transform of the product term in (1.2.9). In either case, there is little further that can be done without considering a specific form of the incident wave.

As an example, let the incident wave be a rectangular pulse defined by

$$g(c_0t+x) = H\langle c_0t+x-L\rangle-H\langle c_0t+x-(L+a)\rangle. \tag{1.2.10}$$

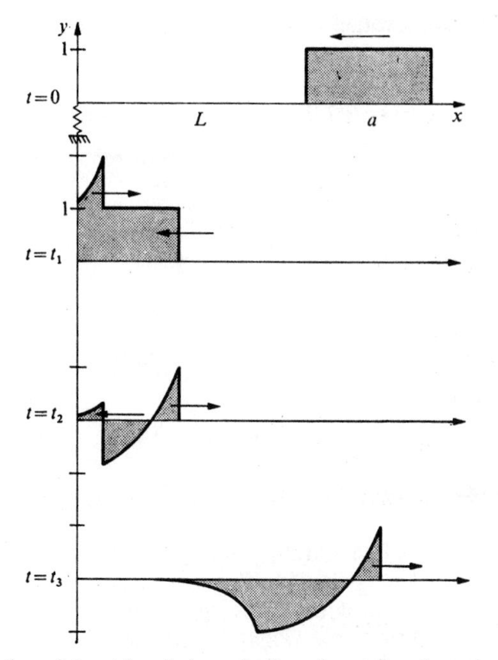

FIG. 1.11. Motion of the string during reflection of a rectangular pulse from an elastic boundary.

This combination of step functions gives a pulse of length a that, at $t = 0$, is a distance L from the origin, as shown in Fig. 1.11. It is the value of this function at $x = 0$ that is required in (1.2.9). We have that

$$g(\tau) = H\langle\tau-L\rangle - H\langle\tau-(L+a)\rangle. \tag{1.2.11}$$

Noting the Laplace transform relation

$$\mathscr{L}\{H\langle t\rangle\} = 1/s, \tag{1.2.12}$$

and recalling the previously used result (1.1.81), we obtain the transform of (1.2.10) as

$$\bar{g} = \frac{\exp(-Ls)}{s} - \frac{\exp\{-(L+a)\}}{s}. \tag{1.2.13}$$

The inverted solution (1.2.9) then becomes

$$f(\tau) = g(\tau) - \frac{2k}{T}\mathscr{L}^{-1}\left[\frac{\exp(-Ls)}{s(s+k/T)} - \frac{\exp\{-(L+a)s\}}{s(s+k/T)}\right]. \tag{1.2.14}$$

To invert the expression within the brackets, we first use the result

$$\mathscr{L}^{-1}\left\{\frac{1}{s(s+k/T)}\right\} = \frac{T}{k}\left\{1-\exp\left(-\frac{k}{T}\tau\right)\right\}. \tag{1.2.15}$$

Again using the just cited result (1.1.81), we are then able to write

$$\mathscr{L}^{-1}\left\{\frac{\exp(-Ls)}{s(s+k/T)}\right\} = \frac{T}{k}\left[1-\exp\left\{-\frac{k}{T}(\tau-L)\right\}\right]H\langle\tau-L\rangle. \tag{1.2.16}$$

The complete result, expressed in terms of t, is

$$f(c_0t) = g(c_0t)-2\left[1-\exp\left\{-\frac{k}{T}(c_0t-L)\right\}\right]H\langle c_0t-L\rangle+$$
$$+2\left[1-\exp\left\{-\frac{k}{T}(c_0t-L-a)\right\}\right]H\langle c_0t-L-a\rangle. \tag{1.2.17}$$

The response away from the origin is obtained by replacing c_0t by c_0t-x in the result (1.2.17).

The general motion of the string predicted by this result is shown in Fig. 1.11. Without assigning specific values to the parameters k, T, and a, the results must be accepted as highly qualitative. It is quite obvious, nevertheless, that the incident wave has undergone considerable distortion during the reflection process.

1.2.4. *Reflection of harmonic waves*

A procedure that is frequently used in studying the reflection characteristics of boundaries is to consider the incident wave to be a pure harmonic. Thus, instead of considering each reflection problem on an *ad hoc* basis, general, frequency-dependent relationships for the amplitude and phase of the reflected waves are obtained. We will again consider the elastic boundary to illustrate the procedure.

Consider the harmonic wave solution to the wave equation given by

$$y = Ae^{i(\gamma x+\omega t)}+Be^{-i(\gamma x-\omega t)}. \tag{1.2.18}$$

We presume the incident wave, which is the A coefficient term in the above solution, to be specified. Substituting (1.2.18) in the boundary condition (1.2.3) we obtain

$$k(A+B) = Ti\gamma(A-B). \tag{1.2.19}$$

Solving for the amplitude ratio B/A we obtain

$$\frac{B}{A} = -\frac{k-Ti\gamma}{k+Ti\gamma}, \tag{1.2.20}$$

The frequency dependence of the above may be displayed explicitly using $\omega = \gamma c_0$ to give

$$\frac{B}{A} = -\frac{kc_0 - Ti\omega}{kc_0 + Ti\omega}. \tag{1.2.21}$$

This complex quantity may be given the general representation

$$\frac{B}{A} = M(\omega)e^{i\phi(\omega)}, \tag{1.2.22}$$

where $M(\omega)$ is the real, frequency-dependent amplitude and $\phi(\omega)$ is the frequency-dependent phase angle. Thus, the reflected and incident waves will be out of phase with one another. In the present problem it is found that $M(\omega) = -1$, indicating that the amplitude of the reflected wave is the same as the incident wave. This is to be expected here since there is no loss of energy in the system.

1.2.5. *Reflection and transmission at discontinuities*

A wave propagating in a string may encounter a discontinuity such as a change in string density or an attached mass or spring. Such situations are capable of producing reflected waves. In addition, energy may also transmit across the discontinuity. Let us consider the case of a string with an abrupt change in mass per unit length, as shown in Fig. 1.12.

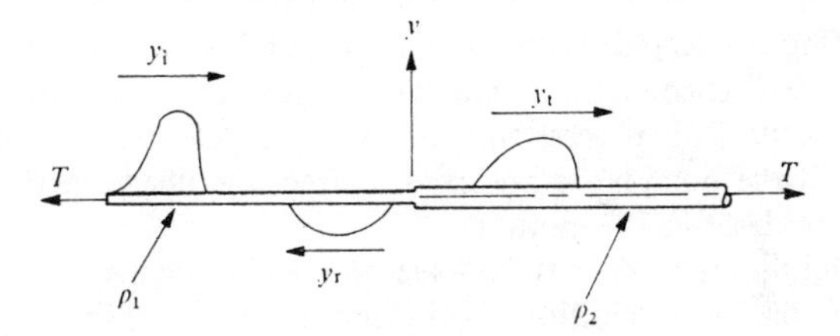

FIG. 1.12. Incident, reflected, and transmitted waves at a discontinuity in the string.

The incident, reflected, and transmitted waves may be expressed as

$$y_i = f_1(x - c_1 t), \qquad y_r = g_1(x + c_1 t), \qquad y_t = f_2(x - c_2 t). \tag{1.2.23}$$

where $c_1 = \sqrt{(T/\rho_1)}$, $c_2 = \sqrt{(T/\rho_2)}$. We presume the incident wave to be specified and wish to determine y_r and y_t. The boundary conditions at the discontinuity require that the velocity and vertical component of string tension be continuous. Thus,

$$\dot{y}_i + \dot{y}_r = \dot{y}_t, \qquad \frac{\partial y_i}{\partial x} + \frac{\partial y_r}{\partial x} = \frac{\partial y_t}{\partial x}, \qquad x = 0. \tag{1.2.24}$$

Substituting (1.2.23) in (1.2.24) gives

$$-c_1(f_1' - g_1') = -c_2 f_2', \; f_1' + g_1' = f_2'. \tag{1.2.25}$$

Solving the two equations simultaneously for f_2', g_2' gives

$$f_2' = \frac{2c_1}{c_1+c_2} f_1', \qquad g_2' = \frac{c_1-c_2}{c_1+c_2} f_1'. \tag{1.2.26}$$

This establishes the reflected and transmitted waves.

When point inhomogeneities, such as a mass, spring, or a dashpot are present, continuity of velocity and balance-of-force considerations still apply. The considerations used in expressing the force boundary conditions of the various elements in Table 1.1 can be extended for the present case. Thus, suppose a mass m is located at $x = 0$. The resulting force boundary condition would be

$$m\ddot{y}(0, t) = -T\frac{\partial y_1(0, t)}{\partial x} + T\frac{\partial y_2(0, t)}{\partial x}. \tag{1.2.27}$$

1.3. Free vibration of a finite string

We now will consider the free vibrations of a completely bounded string. The motion as described by propagating waves will be presented. However, the more useful approach to such problems will be the determination of the natural frequencies and normal modes of the system.

1.3.1. *Waves in a finite string*

Consider a finite string of length l subjected to an initial disturbance, such as an imposed displacement field. We recall from § 1.1 (see Fig. 1.5) that, when the string is released, waves will propagate both to the left and right. When each wave encounters a boundary, reflection back into the string interior will occur. This process will continually repeat itself in the case of a finite string. If the boundaries are fixed or free, the image method may be readily used to describe the motion.

Considering the string ends to be fixed, the field of image waves necessary to satisfy the boundary conditions for all time is shown in Fig. 1.13 for $t = 0$.

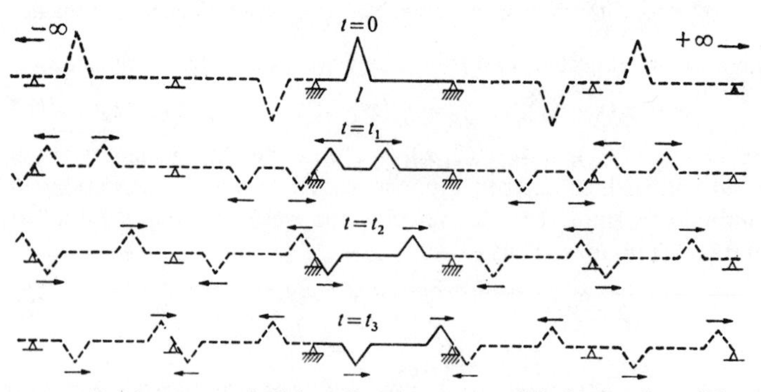

FIG. 1.13. Initial image wavefield ($t = 0$) necessary to satisfy fixed end conditions and resulting propagation at later instants of time $t = t_1, t_2, t_3$.

Thus, it is necessary to construct an infinite set of image strings in order to provide a description of the response for all time.

It is possible to give a mathematical prescription for the preceding pulse system. Thus, suppose the initial conditions are $y(x, 0) = U(x)$, $\dot{y}(x, 0) = 0$. Then we define $\bar{U}(x)$, where

$$\bar{U}(x) = \begin{cases} U(x), & 0 < x < l \\ -U(-x), & -l < x < 0, \end{cases} \tag{1.3.1}$$

and where

$$\bar{U}(x) = \bar{U}(x+2ml) \qquad (m = 0, 1, 2, \ldots). \tag{1.3.2}$$

This serves to define the pulse system shown in Fig. 1.13 for $t = 0$. For $t > 0$, we then have that

$$y(x, t) = \tfrac{1}{2}\{\bar{U}(x-c_0t)+\bar{U}(x+c_0t)\}. \tag{1.3.3}$$

As in the case of reflection from a single boundary in a semi-infinite string, a formal mathematical analysis will yield equivalent results to those obtained by 'inspection' in formulating the image approach.

The image or D'Alembert approach to waves in finite strings is only applicable to simple boundary conditions and is mainly useful to give an understanding to the basis for periodic motion in such a system. For more general analysis, other techniques are now developed.

1.3.2. *Vibrations of a fixed–fixed string*

We now apply the method of separation of variables to the free-vibrations problem. As before, consider a string of length l, fixed at both ends. Consider, as a solution to the wave equation (1.1.3),

$$y = Y(x)\psi(t). \tag{1.3.4}$$

Substituting in the wave equation gives

$$c_0^2\frac{Y''}{Y} = \frac{\ddot{\psi}}{\psi} = -\omega^2, \tag{1.3.5}$$

or

$$Y''+\frac{\omega^2}{c_0^2}Y = 0, \tag{1.3.6}$$

$$\ddot{\psi}+\omega^2\psi = 0, \tag{1.3.7}$$

where, momentarily, ω merely represents a separation of variables constant.

Considering (1.3.7) first, we have

$$\psi = A \sin \omega t+B \cos \omega t, \tag{1.3.8}$$

or, in exponential form,

$$\psi = A_1 e^{i\omega t}+B_1 e^{-i\omega t}. \tag{1.3.9}$$

This form of the solution represents simple harmonic motion at the frequency ω. If we had selected a separation constant less than zero in (1.3.5), thus

$\bar{\omega}^2 < 0$, we would then obtain the non-harmonic solution,

$$y = A_1 e^{\bar{\omega} t} + B_1 e^{-\bar{\omega} t}. \tag{1.3.10}$$

This is definitely incapable of representing the periodic motion that we know occurs in the fixed–fixed string.

Now considering (1.3.6), we have

$$Y = C \sin \beta x + D \cos \beta x,\ \beta^2 = \omega^2/c_0^2. \tag{1.3.11}$$

The boundary conditions for the problem are

$$y(0, t) = y(l, t) = 0. \tag{1.3.12}$$

Applying these to (1.3.11) gives the conditions that $D = 0$ and the *frequency equation* (also called the characteristic equation) for the system

$$\sin \beta l = 0, \qquad \beta l = n\pi \qquad (n = 1, 2, \ldots). \tag{1.3.13}$$

Thus we obtain the *natural frequencies* (also called the *eigenvalues*) of the system, given by

$$\omega_n = c_0 \beta_n = n\pi c_0/l \qquad (n = 1, 2, \ldots). \tag{1.3.14}$$

These represent the discrete frequencies at which the system is capable of undergoing harmonic motion. For a given value of n, we thus have the vibrational pattern of the string described by

$$Y_n(x) = C_n \sin \beta_n x, \qquad (n = 1, 2, \ldots), \tag{1.3.15}$$

where Y_n are usually called the *normal modes* (also called the *eigenfunctions*) of the system.

Several of the resulting vibration patterns are shown in Fig. 1.14. We note

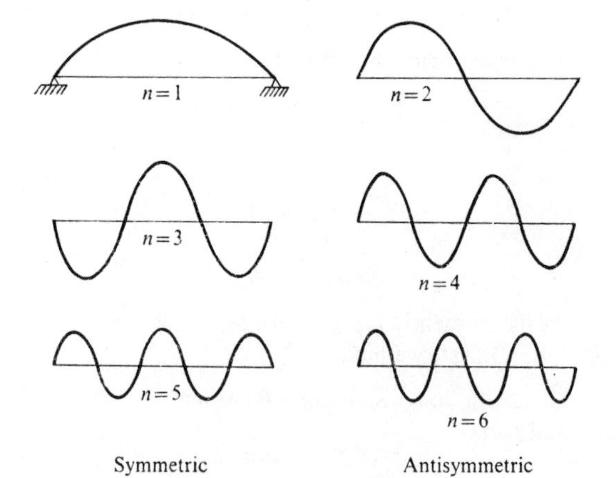

FIG. 1.14. Symmetric ($n = 1, 3, 5$) and antisymmetric ($n = 2, 4, 6$) modes of a fixed–fixed string.

that the modes for $n = 1, 3, 5, \ldots$ give *symmetric* motion while *antisymmetric* modes result from $n = 2, 4, 6 \ldots$. The points of zero displacement are called the *nodes* of vibration, while points of maximum vibration are called the *antinodes*.

Combining the time and spatial dependence for a given value of n, we obtain

$$y_n(x, t) = (A_n \sin \omega_n t + B_n \cos \omega_n t)\sin \beta_n x, \tag{1.3.16}$$

where the constant C_n of (1.3.15) has been absorbed in the constants A_n and B_n. The general solution is then obtained by the superposition of all particular solutions to give

$$y(x, t) = \sum_{n=1}^{\infty} (A_n \sin \omega_n t + B_n \cos \omega_n t)\sin \beta_n x. \tag{1.3.17}$$

The most general consideration in the free vibrations of strings is the initial value problem. Thus, we wish to determine the motion for all time given the initial conditions,

$$y(x, 0) = U(x), \qquad \dot{y}(x, 0) = V(x). \tag{1.3.18}$$

Substituting the solution (1.3.17) in the above, we obtain

$$U(x) = \sum_{n=1}^{\infty} B_n \sin \beta_n x, \tag{1.3.19}$$

$$V(x) = \sum_{n=1}^{\infty} A_n \omega_n \sin \beta_n x. \tag{1.3.20}$$

The constants A_n, B_n are found in the manner used in Fourier series analysis. Thus, multiply (1.3.19) and (1.3.20) by $\sin \beta_m x$ and integrate over the interval $0 \leqslant x \leqslant l$. Noting that

$$\int_0^l \sin \beta_n x \sin \beta_m x \, dx = \begin{cases} l/2, & m = n, \\ 0, & m \neq n, \end{cases} \tag{1.3.21}$$

we obtain

$$A_n = \frac{2}{l\omega_n} \int_0^l V(x)\sin \beta_n x \, dx, \tag{1.3.22}$$

$$B_n = \frac{2}{l} \int_0^l U(x)\sin \beta_n x \, dx. \tag{1.3.23}$$

This establishes the motion of the system.

Referring back to (1.3.17), we note that we may write

$$A_n \sin \omega_n t \sin \beta_n x = \frac{A_n}{2} \cos(\beta_n x - \omega_n t) - \frac{A_n}{2} \cos(\beta_n x + \omega_n t), \tag{1.3.24}$$

$$B_n \cos \omega_n t \sin \beta_n x = \frac{B_n}{2} \sin(\beta_n x + \omega_n t) + \frac{B_n}{2} \sin(\beta_n x - \omega_n t), \tag{1.3.25}$$

which are travelling-wave expressions. This representation merely re-emphasizes the underlying unity between wave propagation and vibrations in systems.

Finally, we note that (1.3.17) is effectively a statement that the string is a system having infinite degrees of freedom. We recall that in the vibrations of discrete element systems (that is, spring–mass systems), the problem is to describe the motion of each mass element. If there are n elements, there will be approximately n degrees of freedom to the system and the motion will be described by displacements $u_1(t), u_2(t), \ldots, u_n(t)$. In a continuous system as the number of elements increases to infinity there becomes an infinite number of degrees of freedom. In the normal mode representation (1.3.17), each mode represents a single degree of freedom. Thus, although a given mode is a continuous curve, it represents constraint for every particle on the curve to follow a certain vibration.

1.3.3. *The general normal mode solution*

The fixed–fixed string is but one of a number of combinations of boundary constraints that can be imposed on finite strings. Referring to Table 1.1, we could have free–free, fixed–free, fixed–elastic, and so on as support conditions. Rather than consider all of the possible combinations of boundary conditions, we seek to obtain a more general solution.

Again considering separation of variables, and deleting a few of the preliminary steps of the previous development, we obtain the general solution form

$$y = \sum_{n=1}^{\infty} (A_n \sin \omega_n t + B_n \cos \omega_n t) Y_n(x), \tag{1.3.26}$$

where $Y_n(x)$ are the normal modes satisfying the differential equation

$$Y_n''(x) + \beta_n^2 Y_n(x) = 0, \qquad \beta_n^2 = \omega_n^2/c_0^2, \tag{1.3.27}$$

and the particular boundary conditions of the problem. The problem is to determine the A_n and B_n for the general initial value problem. As it turns out, the principle used for the fixed–fixed string may be generalized due to the important property of *orthogonality* of the normal modes.

To develop this aspect, consider a solution $Y_m(x)$ which also, of course, satisfies (1.3.27) with n replaced by m. We may then form the following set of equations.

$$\begin{aligned} Y_m Y_n'' + \beta_n^2 Y_m Y_n &= 0, \\ Y_n Y_m'' + \beta_m^2 Y_n Y_m &= 0. \end{aligned} \tag{1.3.28}$$

The first equation has been formed by multiplying (1.3.27) by Y_m, the second in a similar fashion except for interchanging m and n. Subtracting the two equations of (1.3.27) and integrating over the interval $0 \leqslant x \leqslant l$, we obtain

$$\int_0^l (Y_m Y_n'' - Y_n Y_m'')\,dx + (\beta_n^2 - \beta_m^2)\int_0^l Y_m Y_n\,dx = 0. \tag{1.3.29}$$

Integrating the first term of the first integral by parts gives

$$\int_0^l Y_m Y_n''\,dx = [Y_m Y_n' - Y_m' Y_n]_0^l + \int_0^l Y_m'' Y_n\,dx. \tag{1.3.30}$$

Inserting this result in (1.3.29) gives

$$[Y_m(x)Y_n'(x) - Y_m'(x)Y_n(x)]_0^l + (\beta_n^2 - \beta_m^2)\int_0^l Y_m Y_n\,dx = 0. \tag{1.3.31}$$

Now, the expression in brackets will obviously vanish if the ends of the string are any combination of fixed–free conditions. In fact, the brackets vanish for any combinations of end conditions such that

$$\begin{aligned} a_1 Y(0) + b_1 Y'(0) &= 0, \\ a_2 Y(l) + b_2 Y'(l) &= 0, \end{aligned} \tag{1.3.32}$$

where fixed, free, or elastic conditions may be obtained by letting the constants a_1, a_2, b_1, b_2 be zero or non-zero, depending on the condition desired. Assuming that (1.3.32) holds, we obtain from (1.3.31)

$$(\beta_n^2 - \beta_m^2)\int_0^l Y_m Y_n\,dx = 0. \tag{1.3.33}$$

Since $\beta_n \neq \beta_m$ except for $n = m$, we obtain

$$\int_0^l Y_m Y_n\,dx = 0 \qquad (m \neq n). \tag{1.3.34}$$

This is the important orthogonality property of the normal modes.† It represents the generalization of (1.3.21) for the fixed–fixed string and enables the coefficients A_n, B_n to be established. Thus, using (1.3.34), it is readily obtained that

$$A_n = \frac{1}{\omega_n N}\int_0^l V(x)Y_n(x)\,dx, \qquad B_n = \frac{1}{N}\int_0^l U(x)Y_n(x)\,dx, \tag{1.3.35}$$

† The procedure and results (1.3.28) to (1.3.34) are, of course, standard results from the theory of Sturm–Liouville systems.

where N is the normalizing factor given by

$$N = \int_0^l Y_n^2(x)\,\mathrm{d}x. \tag{1.3.36}$$

We briefly consider the situation not accounted for by boundary conditions of the form (1.3.32). Such would be the case for mass or dashpot terminations. Thus, for a mass termination at $x=0$, we have from Table 1.1 that

$$m\ddot{y}(0,t) - T\frac{\partial y(0,t)}{\partial x} = 0. \tag{1.3.37}$$

Substituting (1.3.26) in the above gives

$$\omega_n^2 Y_n(0) + TY_n'(0) = 0. \tag{1.3.38}$$

At first glance, this appears to satisfy the general form of (1.3.32). However, ω_n is not a constant, taking on new values for different modes. Specifically, it results that the bracketed term of (1.3.31) does not vanish, and the orthogonality property of the modes does not exist. The basic difficulty arises in the *time dependent* nature of the boundary conditions. For such problems, simple orthogonality breaks down and other means of analysis must be brought to bear.†

1.4. Forced vibrations of a string

The general problem under consideration will be the forced vibrations of a finite length string, described by

$$T\frac{\partial^2 y}{\partial x^2} - \rho\frac{\partial^2 y}{\partial t^2} = -q(x,t). \tag{1.4.1}$$

The techniques developed for solving the forced-vibrations problem for the string will be applicable generally to more difficult problems arising in beam and plate vibrations. Many aspects of analysis of the finite string are similar to the techniques previously applied to forced motion of an infinite string, so that analyses will be rather briefly presented.

In (1.4.1), $q(x,t)$ may be considered a general loading in space and time. However, to arrive at a solution for such conditions, the solutions to particular loadings in space and in time will first be derived. By considering the case of harmonic loading, the solution to the case of arbitrary time variation may be found by means of the Fourier integral. Furthermore, by specializing the spatial variation of loading to the case of a concentrated load, the solution to the more general problem will also be found by a Fourier superposition.

In the case of a harmonically varying forcing function, it will be found that two distinct problem-types arise. When the forcing function is applied

† See Tong [9, pp. 260–8] and, more extensively, Berry and Naghdi [1].

at the boundaries, the differential equation becomes homogeneous but the boundary conditions are inhomogeneous. If, on the other hand, the forcing function is acting on the interior of the system, the equation is non-homogeneous, but the boundary conditions become homogeneous.

1.4.1. *Solution by Green's function*

Let us consider the case of a harmonically varying force acting at a single point $x = \xi$. Then

$$q(x, t) = P\delta(x-\xi)e^{-i\omega t}, \tag{1.4.2}$$

and the differential equation becomes

$$\frac{\partial^2 y}{\partial x^2}-\frac{1}{c_0^2}\frac{\partial^2 y}{\partial t^2} = -\frac{P}{T}\delta(x-\xi)e^{-i\omega t}. \tag{1.4.3}$$

The response may be written as

$$y(x, t) = \psi(x)e^{-i\omega t}, \tag{1.4.4}$$

giving

$$\psi''+\beta^2\psi = \delta(x-\xi), \qquad \beta^2 = \omega^2/c_0^2, \tag{1.4.5}$$

where $P/T = -1$ has been selected to provide a positive unit impulse on the right-hand side. Assuming simple, rigid supports at the ends, the boundary conditions are

$$\psi(0) = \psi(l) = 0. \tag{1.4.6}$$

Now, recalling that $\delta(x-\xi) = 0$, $x \neq \xi$, we see that two solutions to the homogeneous equation may be found: one valid for $x < \xi$, the other for $x > \xi$. Thus

$$\psi_1 = A_1 \sin \beta x+B_1 \cos \beta x \qquad (0 \leqslant x < \xi), \tag{1.4.7}$$

$$\psi_2 = A_2 \sin \beta x+B_2 \cos \beta x \qquad (\xi < x \leqslant l). \tag{1.4.8}$$

Substituting these in the boundary conditions (1.4.6) we obtain

$$\psi_1 = A_1 \sin \beta x \qquad (0 \leqslant x < \xi), \tag{1.4.9}$$

$$\psi_2 = A_2(\sin \beta x - \tan \beta l \cos \beta x) \qquad (\xi < x \leqslant l). \tag{1.4.10}$$

To match the solutions at the load discontinuity, two conditions are required, one of which is $\psi_1(\xi) = \psi_2(\xi)$, corresponding to continuity of displacement. To establish the second, we integrate (1.4.5) across the load discontinuity, giving

$$\int_{\xi-\varepsilon}^{\xi+\varepsilon} \psi''\,dx+\beta^2\int_{\xi-\varepsilon}^{\xi+\varepsilon} \psi\,dx = \int_{\xi-\varepsilon}^{\xi+\varepsilon} \delta(x-\xi)\,dx = 1. \tag{1.4.11}$$

Since $\psi(x)$ is continuous across $x = \xi$, the second integral on the left-hand side of (1.4.11) vanishes. The remaining integral gives

$$\int_{-\varepsilon}^{\xi+\varepsilon} \psi''\,dx = \psi'\Big|_{\xi-\varepsilon}^{\xi+\varepsilon} = \psi_2'(\xi)-\psi_1'(\xi) = 1. \tag{1.4.12}$$

This establishes the second boundary condition and represents the discontinuity in slope of the string. Substituting the solutions (1.4.9) and (1.4.10) in the boundary conditions at $x = \xi$ we obtain

$$\begin{aligned} A_1 \sin \beta\xi &= A_2(\sin \beta\xi - \tan \beta l \cos \beta\xi), \\ A_2(\cos \beta\xi + \tan \beta l \sin \beta\xi) - A_1 \cos \beta\xi &= 1/\beta, \end{aligned} \tag{1.4.13}$$

from which we obtain

$$\begin{aligned} \psi_1(x) &= -\frac{\sin \beta x \sin \beta(l-\xi)}{\beta \sin \beta l} \qquad (0 \leqslant x < \xi) \\ \psi_2(x) &= -\frac{\sin \beta\xi \sin \beta(l-x)}{\beta \sin \beta l} \qquad (\xi < x \leqslant l). \end{aligned} \tag{1.4.14}$$

This solution is called the Green's function for the problem and is denoted as $G(x/\xi)$. Thus

$$G(x/\xi) = \begin{cases} -\dfrac{\sin \beta x \sin \beta(l-\xi)}{\beta \sin \beta l}, & (0 \leqslant x < \xi) \\ -\dfrac{\sin \beta\xi \sin \beta(l-x)}{\beta \sin \beta l}, & (\xi < x \leqslant l). \end{cases} \tag{1.4.15}$$

The response of the system to a concentrated, harmonic load of magnitude $P = -T$ applied at $x = \xi$ is thus

$$y(x, t) = G(x/\xi)\mathrm{e}^{-\mathrm{i}\omega t}. \tag{1.4.16}$$

The above approach to the problem of forced vibration is quite powerful. The extension to general loading in space is possible using the superposition principle, but it is best to postpone this until the transform techniques are brought to bear on the problem.

1.4.2. *Solution by transform techniques*

The application of the Fourier and Laplace transform techniques to problems in forced vibrations may take several routes, applying one or the other transform or both to a given problem. Several approaches will be illustrated in the following.

1. *Application of the Laplace transform.* Consider again the case of loading $P\delta(x-\xi)\exp(-\mathrm{i}\omega t)$, so that

$$\frac{\partial^2 y}{\partial x^2} - \frac{1}{c_0^2}\frac{\partial^2 y}{\partial t^2} = \delta(x-\xi)\mathrm{e}^{-\mathrm{i}\omega t}, \tag{1.4.17}$$

where $P = -T$ has again been selected. Let $y(x, t) = \psi(x)\exp(-\mathrm{i}\omega t)$ as before, so that

$$\psi''(x) + \beta^2\psi(x) = \delta(x-\xi), \qquad \beta^2 = \omega^2/c_0^2 \tag{1.4.18}$$

must be solved. Taking the Laplace transform on the spatial variable of the preceding gives

$$s^2\bar{\Psi}(s)-s\psi(0)-\psi'(0)+\beta^2\bar{\Psi} = \mathscr{L}\{\delta(x-\xi)\}, \tag{1.4.19}$$

where

$$\mathscr{L}\{\delta(x-\xi)\} = \int_0^\infty \delta(x-\xi)e^{-sx}\,dx = e^{-s\xi}. \tag{1.4.20}$$

Solving for $\bar{\Psi}(s)$, we have

$$\bar{\Psi}(s) = \frac{s\psi(0)+\psi'(0)}{s^2+\beta^2}+\frac{e^{-s\xi}}{s^2+\beta^2}. \tag{1.4.21}$$

Inverting, we obtain

$$\psi(x) = \psi(0)\mathscr{L}^{-1}\left(\frac{s}{s^2+\beta^2}\right)+\psi'(0)\mathscr{L}^{-1}\left(\frac{1}{s^2+\beta^2}\right)+\mathscr{L}^{-1}\left(\frac{e^{-s\xi}}{s^2+\beta^2}\right). \tag{1.4.22}$$

Now

$$\mathscr{L}^{-1}\left(\frac{s}{s^2+\beta^2}\right) = \cos\beta x, \qquad \mathscr{L}^{-1}\left(\frac{1}{s^2+\beta^2}\right) = \frac{\sin\beta x}{\beta}. \tag{1.4.23}$$

To invert the last term in the solution, we regard $\exp(-s\xi)/(s^2+\beta^2)$ as the product of two functions $f_1(s)$, $f_2(s)$. Although the first impulse is to select these as $\exp(-s\xi)$ and $(s^2+\beta^2)^{-1}$, consulting tables of Laplace transforms reveals the former to have no inverse, whereas $\exp(-s\xi)/s$ does. So we let

$$f_1(s) = \frac{e^{-s\xi}}{s}, \qquad f_2(s) = \frac{s}{s^2+\beta^2}, \tag{1.4.24}$$

for which

$$\mathscr{L}^{-1}\left(\frac{e^{-s\xi}}{s}\right) = \begin{cases} 0, & 0 < x < \xi \\ 1, & \xi < x \end{cases}, \tag{1.4.25}$$

$$\mathscr{L}^{-1}\left(\frac{s}{s^2+\beta^2}\right) = \cos\beta x. \tag{1.4.26}$$

Then, by the convolution theorem,

$$\mathscr{L}^{-1}\left(\frac{e^{-s\xi}}{s^2+\beta^2}\right) = \begin{cases} 0 & (0 \leqslant x < \xi) \\ \int_0^x \cos\beta(x-u)\,du & (\xi < x \leqslant l). \end{cases} \tag{1.4.27}$$

Hence

$$\psi(x) = \psi(0)\cos\beta x+\psi'(0)\frac{\sin\beta x}{\beta}+\frac{1}{\beta}\begin{cases} 0 & (0 \leqslant x < \xi) \\ \sin\beta(x-\xi) & (\xi < x \leqslant l). \end{cases} \tag{1.4.28}$$

Assuming that the ends of the string are fixed, we have the boundary conditions

$$\psi(0) = \psi(l) = 0. \tag{1.4.29}$$

From the second condition we have

$$\psi(l) = \psi'(0)\frac{\sin \beta l}{\beta} + \frac{\sin \beta(l-\xi)}{\beta} = 0, \tag{1.4.30}$$

so that $\psi'(0)$ may be found. Then

$$\psi(x) = -\frac{\sin \beta(l-\xi)}{\beta \sin \beta l}\sin \beta x + \frac{1}{\beta}\begin{cases} 0 & (0 \leqslant x < \xi) \\ \sin \beta(x-\xi) & (\xi < x \leqslant l). \end{cases} \tag{1.4.31}$$

These results may be combined to give the alternate formulation,

$$\psi(x) = \begin{cases} -\dfrac{\sin \beta(l-\xi)}{\beta \sin \beta l}\sin \beta x & (0 \leqslant x < \xi) \\ -\dfrac{\sin \beta\xi \sin \beta(l-x)}{\beta \sin \beta l} & (\xi < x \leqslant l). \end{cases} \tag{1.4.32}$$

This result corresponds to the Green's function $G(x/\xi)$.

2. *Solution by finite Fourier transform.* Consider the loading to be of the general form

$$q(x, t) = TW(x)e^{i\omega t}, \tag{1.4.33}$$

so that the governing equation is

$$\frac{\partial^2 y}{\partial x^2} - \frac{1}{c_0^2}\frac{\partial^2 y}{\partial t^2} = -W(x)e^{i\omega t}. \tag{1.4.34}$$

Assuming $y(x, t) = Y(x)\exp(i\omega t)$, we have

$$Y''(x) + \beta^2 Y(x) + W(x) = 0. \tag{1.4.35}$$

The finite Fourier transform will be applied to the above (it must be the finite transform since x is bounded, $0 \leqslant x \leqslant l$). The nature of the boundary condition will determine whether the sine or cosine transform is appropriate. We first attempt the sine transform.

$$F_s\{Y''(x)\} = \frac{1}{l}\int_0^l Y''(x)\sin\frac{n\pi}{l}x\,dx. \tag{1.4.36}$$

Integrating by parts gives

$$F_s\{Y''(x)\} = \frac{1}{l}\left[Y'(x)\sin\frac{n\pi}{l}x - \frac{n\pi}{l}Y(x)\cos\frac{n\pi}{l}x\right]_0^l - \frac{1}{l}\left(\frac{n\pi}{l}\right)^2\int_0^l Y(x)\sin\frac{n\pi}{l}x\,dx. \tag{1.4.37}$$

For a string fixed at both ends, $Y(0) = Y(l) = 0$, while $\sin n\pi x/l = 0$, at $x = 0,l$. Thus the expression in brackets in (1.4.37) vanishes and

$$F_s\{Y''(x)\} = -\frac{1}{l}\left(\frac{n\pi}{l}\right)^2 \bar{Y}(n), \tag{1.4.38}$$

where

$$\bar{Y}(n) = \frac{1}{l}\int_0^l Y(x)\sin\frac{n\pi}{l}x\,dx. \tag{1.4.39}$$

Then the transformed equation is

$$\bar{Y}(n) = \frac{\bar{W}(n)}{\alpha_n^2-\beta^2}, \qquad \alpha_n = \frac{n\pi}{l}. \tag{1.4.40}$$

We note that, had the finite cosine transform been attempted, the expression in brackets of (1.4.37) would not vanish for the given boundary conditions. The inverse transform is then (see Appendix B)

$$Y(x) = 2\sum_{n=1}^{\infty}\frac{\bar{W}(n)\sin\alpha_n x}{\alpha_n^2-\beta^2} = \frac{2}{l}\sum_{n=1}^{\infty}\frac{\sin\alpha_n x}{\alpha_n^2-\beta^2}\int_0^l W(u)\sin\alpha_n u\,du. \tag{1.4.41}$$

Then

$$y(x, t) = Y(x)e^{i\omega t}. \tag{1.4.42}$$

The special case of $W(x) = \delta(x-\xi)$ will give the previously obtained results of (1.4.32).

3. *Laplace and finite Fourier transforms.* We again consider the forced vibration of a string fixed at $x = 0,l$, where the loading is considered to be a variable in x and t, so that

$$T\frac{\partial^2 y}{\partial x^2}-\rho\frac{\partial^2 y}{\partial t^2}+q(x, t) = 0, \tag{1.4.43}$$

and

$$y(0, t) = y(l, t) = 0. \tag{1.4.44}$$

The case of non-zero initial conditions may also be included, so that

$$y(x, 0) = h(x), \qquad \dot{y}(x, 0) = g(x). \tag{1.4.45}$$

First apply the Laplace transform,

$$c_0^2\bar{y}''(x, s)-s^2\bar{y}(x, s) = -\frac{1}{\rho}\bar{q}(x, s)-sh(x)-g(x). \tag{1.4.46}$$

Now apply the finite Fourier sine transform, giving

$$(c_0^2\alpha_n^2+s^2)\bar{Y}(n, s) = \frac{\bar{Q}(n, s)}{\rho}+sH(n)+G(n). \tag{1.4.47}$$

The transformed solution is then

$$\bar{Y}(n, s) = \frac{\bar{Q}/\rho + sH + G}{c_0^2\alpha_n^2 + s^2}, \qquad \alpha_n = \frac{n\pi}{l}. \tag{1.4.48}$$

The inverse Fourier sine transform gives

$$\bar{y}(x, s) = \frac{2}{l}\sum_{n=1}^{\infty}\frac{\sin \alpha_n x}{c_0^2\alpha_n^2+s^2}\left\{\frac{1}{\rho}\int_0^l \bar{q}(u, s)\sin \alpha_n u \, du + \right.$$

$$\left. + s\int_0^l h(u)\sin \alpha_n u \, du + \int_0^l g(u)\sin \alpha_n u \, du\right\}. \tag{1.4.49}$$

The convolution theorem is used in carrying out the Laplace inversion, since the product $\bar{Q}/(c_0^2\alpha_n^2+s^2)$ arises, as well as the function $s/(c_0^2\alpha_n^2+s^2)$. We have

$$\mathscr{L}^{-1}\left(\frac{1}{s^2+c_0^2\alpha_n^2}\right) = \frac{1}{\omega_n}\sin \omega_n t,$$

$$\mathscr{L}^{-1}\left(\frac{s}{s^2+c_0^2\alpha_n^2}\right) = \cos \omega_n t, \qquad \omega = c_0\alpha_n. \tag{1.4.50}$$

The convolution theorem is

$$\mathscr{L}^{-1}\{\bar{f}(s)\bar{g}(s)\} = \bar{f} * \bar{g} = \int_0^t f(\tau)g(t-\tau)\, d\tau. \tag{1.4.51}$$

Hence we have

$$\mathscr{L}^{-1}\left\{\frac{\bar{q}(u, s)}{s^2+\omega_n^2}\right\} = \frac{1}{\omega_n}\int_0^t q(u, \tau)\sin \omega_n(t-\tau)\, d\tau. \tag{1.4.52}$$

Applying a term-by-term inversion to $\bar{y}(x, s)$, we obtain

$$y(x, t) = \frac{2}{\rho c_0 l}\sum_{n=1}^{\infty}\frac{\sin \alpha_n x}{\alpha_n}\times$$

$$\times \int_0^l \sin \alpha_n u \, du \int_0^t q(u, \tau)\sin \omega_n(t-\tau)\, d\tau + \frac{2}{l}\sum_{n=1}^{\infty}\sin \alpha_n x \times$$

$$\times\left\{\cos \omega_n t \int_0^l h(u)\sin \alpha_n u \, du + \frac{\sin \omega_n t}{\omega_n}\int_0^l g(u)\sin \alpha_n u \, du\right\}. \tag{1.4.53}$$

The case of zero initial conditions, $g(x) = h(x) = 0$, and

$$q(x, t) = Q(t)\delta(x-\xi), \tag{1.4.54}$$

gives

$$y(x, t) = \frac{2}{\rho c_0 l}\sum_{n=1}^{\infty}\frac{\sin \alpha_n x \sin \alpha_n \xi}{\alpha_n}\int_0^t Q(\tau)\sin \omega_n(t-\tau)\, d\tau. \qquad (1.4.55)$$

If, as a further specialization, we let $Q(\tau) = \delta(\tau)$, so that the loading represents an impulse load applied at $x = \xi$, the preceding reduces to

$$y(x, t) = \frac{2}{\rho c_0 l}\sum_{n=1}^{\infty}\frac{\sin \alpha_n x \sin \alpha_n \xi}{\alpha_n}\sin \omega_n t = G(x, t/\xi). \qquad (1.4.56)$$

The above solution could be regarded as a 'double' Green's function for the string fixed at $x = 0, l$. The response to an arbitrary loading $q(x, t)$ is then

$$y(x, t) = \int_0^l d\xi \int_0^t G(x, t-\tau/\xi)q(\xi, \tau)\, d\tau. \qquad (1.4.57)$$

1.4.3. *Solution by normal modes*

Consider again the problem of forced vibration of a finite string, governed by the equation

$$T\frac{\partial^2 y}{\partial x^2}-\rho\frac{\partial^2 y}{\partial t^2}+q(x, t) = 0. \qquad (1.4.58)$$

Let us assume that the problem of free vibrations of the string has already been solved and that a set of normal modes $Y_n(x)$ $(n = 1, 2, \ldots)$ has been determined. That is, Y_n are such that

$$Y_n''(x)+\beta_n^2 Y_n = 0, \qquad \beta_n^2 = \omega_n^2/c_0^2, \qquad (1.4.59)$$

is satisfied. Furthermore, let us assume that the boundary conditions of the problem are such that $Y_n(x)$ form an orthogonal set, as described in § 1.3. Then the configuration of the string at any point x and time t may be represented by a normal-mode expansion,

$$y(x, t) = \sum_{n=1}^{\infty} q_n(t) Y_n(x), \qquad (1.4.60)$$

where the coefficients $q_n(t)$, being functions of time, represent the time-varying character of the expansion needed to represent $y(x, t)$ at successive periods of time.

Substitution of the series representation into the equation of motion gives

$$\sum_{n=1}^{\infty} (Tq_n(t)Y_n''-\rho\ddot{q}_n(t)Y_n) = -q(x, t) \qquad (1.4.61)$$

or

$$\sum_{n=1}^{\infty} \{\ddot{q}_n(t)+\omega_n^2 q_n(t)\}Y_n(x) = \frac{1}{\rho}q(x, t), \qquad (1.4.62)$$

where the relation $Y_n'' = -(\omega_n/c_0)^2 Y_n$ has been used. Then, multiplying by $Y_n(x)$ and integrating, we have

$$\sum_{n=1}^{\infty} (\ddot{q}_n + \omega_n^2 q_n) \int_0^l Y_m Y_n \, dx = \frac{1}{\rho} \int_0^l Y_m q(x, t) \, dx, \qquad (1.4.63)$$

which gives

$$\ddot{q}_n(t) + \omega_n^2 q_n(t) = \frac{1}{\rho N} \int_0^l Y_n(x) q(x, t) \, dx = Q_n(t). \qquad (1.4.64)$$

Thus, we must solve

$$\ddot{q}_n(t) + \omega_n^2 q_n(t) = Q_n(t), \qquad (n = 1, 2, 3, \ldots). \qquad (1.4.65)$$

The solution of this ordinary differential equation may be found by using the Laplace transform and the convolution theorem or by using variation of parameters. In either case, the solution is

$$q_n(t) = q_n(0) \cos \omega_n t + \frac{\dot{q}_n(0)}{\omega_n} \sin \omega_n t -$$

$$-\frac{1}{\omega_n} \int_0^t Q_n(\tau) \sin \omega_n (t - \tau) \, d\tau \qquad (n = 1, 2, \ldots), \quad (1.4.66)$$

where $q_n(0)$, $\dot{q}_n(0)$ represent the initial conditions. This establishes the total solution to the problem, since

$$y(x, t) = \sum_{n=1}^{\infty} q_n(t) Y_n(x), \qquad (1.4.67)$$

and both $q_n(t)$, $Y_n(x)$ are known.

Sometimes the solution procedure is altered slightly by assuming both $y(x, t)$, $q(x, t)$ have the normal mode expansions

$$y(x, t) = \sum_{n=1}^{\infty} q_n(t) Y_n(x), \qquad q(x, t) = \sum_{n=1}^{\infty} b_n(t) Y_n(x), \qquad (1.4.68)$$

where $b_n(t)$ are known and given by

$$b_n(t) = \int_0^l q(x, t) Y_n(x) \, dx. \qquad (1.4.69)$$

Then substitution in the differential equation gives

$$\sum_{n=1}^{\infty} \{T q_n(t) Y_n'' - \rho \ddot{q}_n Y_n\} = -\sum_{n=1}^{\infty} b_n(t) Y_n(x). \qquad (1.4.70)$$

Using (1.4.59), we obtain

$$\sum_{n=1}^{\infty}\{\ddot{q}_n(t)+\omega_n^2 q_n(t)\}Y_n(x)=\frac{1}{\rho}\sum_{n=1}^{\infty}b_n(t)Y_n(x). \tag{1.4.71}$$

Equating coefficients, we have

$$\ddot{q}_n(t)+\omega_n^2 q_n(t)=\frac{1}{\rho}b_n(t), \qquad (n=1,2,3,...), \tag{1.4.72}$$

which puts us in the same place as (1.4.65) so that the remaining steps are clear. Thus the two procedures are completely equivalent.

1.5. The string on an elastic base—dispersion

Up to this point, we have been considering vibrations and waves in a system where the governing equation, denoted as the 'wave equation', is particularly simple. One of the particular attributes of the system has been that pulses propagate without distortion, as evidenced by the D'Alembert solution. We now consider a slightly more complicated situation in which the string rests on an elastic foundation. The resulting equation will not be of wave equation form and will cause an effect known as *dispersion*. Many of the concepts developed here will find wide application to wave propagation in other elastic systems.

1.5.1. *The governing equation*

We refer back to the initial derivation of the wave equation and Fig. 1.1. The effects of an elastic foundation may be rather easily included by noting that a foundation of elastic modulus k (force length^{-2}) will result in a vertical force component of $(-ky\,dx)$ on the left-hand side of (1.1.3). Alternatively, the external load $q(x, t)$ may be interpreted as that due to the foundation, so that $q(x, t) = -ky(x, t)$. By either means, the resulting governing equation, in the absence of other body or external forces, is

$$T\frac{\partial^2 y}{\partial x^2}-ky=\rho\frac{\partial^2 y}{\partial t^2}, \tag{1.5.1}$$

or

$$\frac{\partial^2 y}{\partial x^2}-\frac{k}{T}y=\frac{1}{c_0^2}\frac{\partial^2 y}{\partial t^2}, \qquad c_0=\sqrt{(T/\rho)}. \tag{1.5.2}$$

1.5.2. *Propagation of harmonic waves*

The first and obvious remark on the result of adding foundation stiffness is that the governing equation is no longer of simple wave equation form. Thus, a solution of the form $f(x\pm c_0 t)$ will not satisfy (1.5.2). Since the major characteristic of such a solution is undistorted pulse propagation, it is now logical to expect some type of pulse distortion to occur in a system governed by (1.5.2).

A first step in assessing this effect is to determine the necessary conditions for the propagation of harmonic waves. Thus, under what conditions will a solution

$$y = Ae^{i(\gamma x-\omega t)} \tag{1.5.3}$$

satisfy (1.5.2)? The propagation direction has arbitrarily been selected in (1.5.3). Substituting in (1.5.2) gives

$$\left(-\gamma^2-\frac{k}{T}+\frac{\omega^2}{c_0^2}\right)Ae^{i(\gamma x-\omega t)} = 0. \tag{1.5.4}$$

In order for this to be satisfied non-trivially, we must have

$$\omega^2 = c_0^2(\gamma^2+k/T) \tag{1.5.5}$$

or, alternatively,

$$\gamma^2 = \frac{\omega^2}{c_0^2}-\frac{k}{T}. \tag{1.5.6}$$

Symbolically, we represent the relationships (1.5.5) and (1.5.6) respectively by

$$\omega = \omega(\gamma), \qquad \gamma = \gamma(\omega). \tag{1.5.7}$$

An alternative form of the results may be obtained by noting that

$$e^{i(\gamma x-\omega t)} = e^{i\gamma(x-ct)}, \tag{1.5.8}$$

where $c = \omega/\gamma$ is the phase velocity of the wave. In the simple string, this has the specific value of $c = c_0$, whereas in the present case it is initially unspecified. Substituting a solution

$$y = Ae^{i\gamma(x-ct)} \tag{1.5.9}$$

in (1.5.2) gives the result

$$c^2 = c_0^2(1+k/T\gamma^2), \tag{1.5.10}$$

or, alternatively,

$$\gamma^2 = \frac{k/T}{(c^2/c_0^2)-1}. \tag{1.5.11}$$

Symbolically, we represent the relations (1.5.10) and (1.5.11) respectively by

$$c = c(\gamma), \qquad \gamma = \gamma(c). \tag{1.5.12}$$

These last results, it is seen, could be obtained directly from (1.5.5) or (1.5.6) by using the basic relation that holds between frequency, wave number and phase velocity, $\omega = \gamma c$.

Finally, by using the relation $\omega = \gamma c$, a third set of relations could be obtained by eliminating γ from (1.5.5) or ω from (1.5.11) to give

$$\omega^2 = \frac{kTc^2}{(c^2/c_0^2)-1}, \qquad \gamma^2 = \frac{\omega^2 c_0^2}{\omega^2-(kc_0^2/T)}, \tag{1.5.13}$$

or in symbolic form

$$\omega = \omega(c), \qquad \gamma = \gamma(\omega). \tag{1.5.14}$$

We now inquire into the implications of the results obtained, represented as specific relations by (1.5.5), (1.5.6), (1.5.10), and (1.5.11), or in symbolic form by (1.5.7) and (1.5.12). In fact, these results give us our first insight into the mechanism of pulse distortion which we know must occur. The results show that: *a harmonic wave of frequency ω can propagate only at specific velocity c as governed by $\omega = \omega(c)$*. Suppose we consider a pulse shape at a given time $t = t_0$ to be a Fourier superposition of harmonic waves. Then, as time advances, each Fourier component of the original pulse will propagate with its own individual velocity. The various components will become increasingly out of phase relative to their original position so that the original pulse shape will become increasingly distorted.

The preceding interpretation also gives the basis for the fact that no distortion occurs in the simple string, where the foundation is absent. While the direct prediction of this lack of distortion is given by the D'Alembert solution, the indirect prediction arises in the fact that frequency, wavenumber, and wave velocity are related by $\omega = \gamma c_0$, where c_0 is a constant. Thus, each harmonic component propagates with the same velocity, so that the phase relationships of an original Fourier superposition are maintained for all time. We note, incidentally, that this result is obtained by letting $k = 0$ in the various relations between frequency, wavenumber, and wave velocity.

Consider now the roots of the preceding relations between ω, γ, and c. The most useful interpretation will come from (1.5.6). We have that

$$\gamma = \pm\left(\frac{\omega^2}{c_0^2} - \frac{k}{T}\right)^{\frac{1}{2}}. \tag{1.5.15}$$

The above roots are real if $\omega^2/c_0^2 > k/T$. Referring back to (1.5.3), we have

$$y = Ae^{-i(\pm\gamma x + \omega t)}. \tag{1.5.16}$$

Thus, the two real roots yield leftward or rightward-propagating waves, depending on which sign is selected.

We next note that if $\omega^2/c_0^2 < k/T$, then the wavenumber predicted by (1.5.15) becomes imaginary. Defining $\bar{\gamma}^2 = -\gamma^2$, we have the motion given by

$$y = Ae^{\pm\bar{\gamma}x}e^{-i\omega t}. \tag{1.5.17}$$

This corresponds to a spatially varying but non-propagating disturbance. Now, the original question posed was to establish the conditions under which a harmonic wave (1.5.3) could exist. In this context, the results for imaginary wavenumber are not acceptable since they are non-propagating. However, it will be found that imaginary wavenumbers and results of the form (1.5.17) will play important roles in problems of transient loading and boundary interaction.

Finally, the special case of $\omega^2/c_0^2 = k/T$ should be noted. This, of course, represents the transition from propagation to non-propagation. From (1.5.6) we have that $\gamma = 0$ at the frequency $\omega_c = c_0\sqrt{(k/T)}$, and the string motion to be of the form†

$$y = A\exp(-i\omega_c t). \tag{1.5.18}$$

The frequency ω_c is designated as the *cutoff* frequency of the propagating mode. There is no spatial variation in the motion, so the string is basically vibrating as a simple spring–mass system.

1.5.3. *Frequency spectrum and the dispersion curve*

The basic factors governing propagation in a string on an elastic foundation have been presented in the formulas relating frequency, wavenumber, and velocity and in the interpretations of the propagation. It is still useful, however, to display these results in graphical form for easier interpretation. Typically, two types of displays are used: one is a plot of frequency versus wavenumber and is called the *frequency spectrum* of the system; the other is a plot of phase velocity versus wavenumber and is called the *dispersion curve* of the system.

To plot the frequency spectrum, we refer to (1.5.15). We assume the frequency to be real and positive. We have established that γ is imaginary for $\omega < \omega_c$ and real for $\omega > \omega_c$. The results are shown in Fig. 1.15. The curves

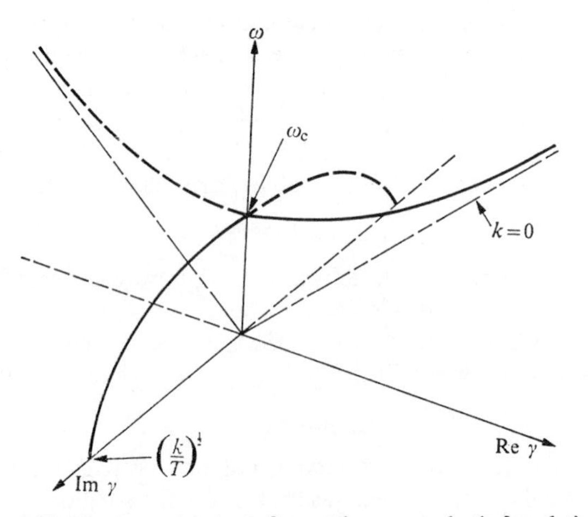

FIG. 1.15. Frequency spectrum for a string on an elastic foundation.

or branches in the real plane are hyperbolas, while the imaginary branches are ellipses. The line $k = 0$ in the figure is the non-dispersive result for the classical string. Thus, no dispersion is manifested as a straight line in the frequency spectrum.

† See problem 1.17.

It is possible to obtain phase velocity information from the frequency spectrum by the relation $\omega = \gamma c$. Thus, given a point on the real branch of the spectrum, the slope of the chord between the point and the origin is given by $\omega/\gamma = c$, the phase velocity for that particular frequency. This relation is shown in Fig. 1.16, where the frequency spectrum has been shown in slightly

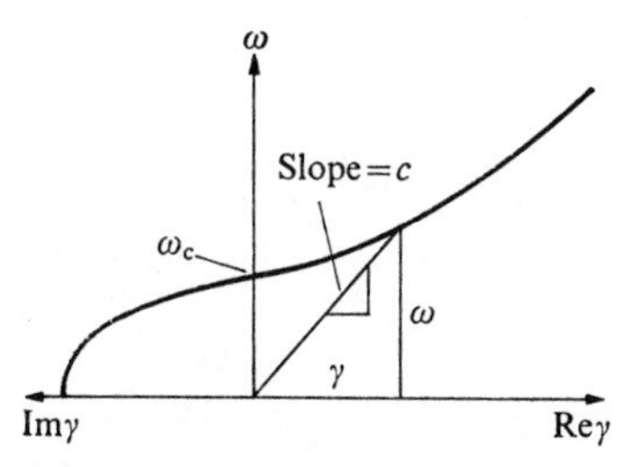

FIG. 1.16. Two-dimensional representation of the frequency spectrum showing relation between chord slope and phase velocity.

different form. Since the spectrum is usually symmetric with respect to the $\mathrm{Re}\gamma = 0$ and the $\mathrm{Im}\gamma = 0$ planes, it is sufficient to present a two-dimensional plot containing the ω, $\mathrm{Re}\gamma > 0$ and $\mathrm{Im}\gamma > 0$ axes.

Instead of deriving phase velocity information from the frequency spectrum, it is often presented independently, as previously mentioned, by dispersion curves. For this purpose, eqn (1.5.10) is required. Although it is possible to consider c as positive, negative, real, and imaginary, depending on γ, the most physically meaningful information is contained in a simple plot of $\mathrm{Re}c > 0$ versus $\mathrm{Re}\gamma > 0$ as shown in Fig. 1.17. The horizontal line of the

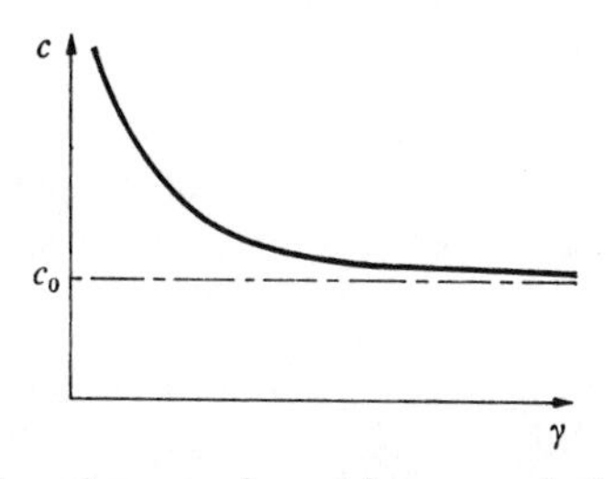

FIG. 1.17. Dispersion curve for a string on an elastic foundation.

figure is the result for the non-dispersive string where all wavelengths propagate with the same velocity c_0. At larger values of the wavenumber, corresponding to short wavelengths and high frequency, it is seen that the solid curve approaches the classical result, indicating that the foundation effect is minimal. However, at small values of wavenumber, corresponding to long wavelengths, the phase velocity increases rapidly, approaching infinity as $\gamma \to 0$. Referring back to Fig. 1.15 or Fig. 1.16, we note that as $\gamma \to 0$, $\omega \to \omega_c$, the cutoff frequency. We recall the result (1.5.18) indicating the system is in uniform vibration. The fact that $\gamma = 0$ also indicates uniform vibration since the wavelength is infinitely long. The basis for the apparently anamolous behaviour of the phase velocity is then clear, since the uniform vibration may be interpreted as a disturbance propagating with infinite speed through the medium.

1.5.4. *Harmonic and pulse excitation of a semi-infinite string*

Two examples of forced motion of a semi-infinite, dispersive string will be considered, one quite simple, the other not so simple. Each will illustrate particular aspects of propagation in dispersive media. So first consider a string $x > 0$ to be subjected to the harmonic end motion

$$y(0, t) = y_0 e^{-i\omega t}. \tag{1.5.19}$$

Considering a solution

$$y = Y(x) e^{-i\omega t}, \tag{1.5.20}$$

we rather straightforwardly can obtain

$$y = A e^{i(\gamma x - \omega t)} + B e^{-i(\gamma x + \omega t)}, \tag{1.5.21}$$

where, as previously derived,

$$\gamma^2 = \left(\frac{\omega^2}{c_0^2} - \frac{k}{T}\right) > 0. \tag{1.5.22}$$

From the radiation condition, we set $B = 0$. Applying the boundary condition (1.5.19) we obtain $A = y_0$, so that the solution is simply

$$y(x, t) = y_0 e^{i(\gamma x - \omega t)}. \tag{1.5.23}$$

The resulting motion is the propagation of simple harmonic waves, as shown in Fig. 1.18(a). Now, if the excitation drops below the cutoff frequency,

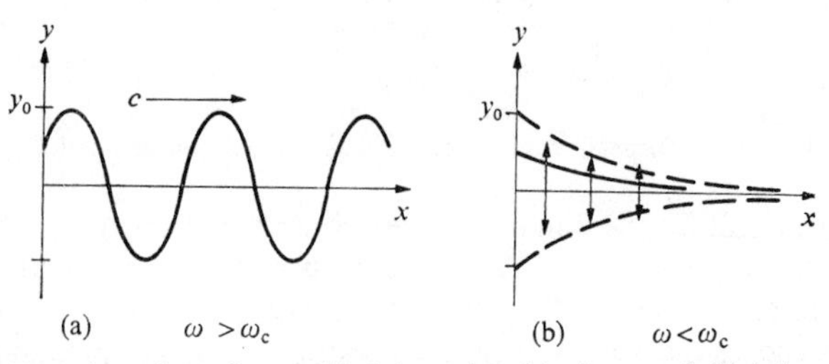

FIG. 1.18. Forced motion of a semi-infinite string (a) above and (b) below the cutoff frequency.

we have $\gamma^2 < 0$ in (1.5.22) and the resulting motion described by

$$y = y_0 e^{-\bar{\gamma}x} e^{-i\omega t}, \tag{1.5.24}$$

where $\bar{\gamma}^2 = -\gamma^2$. The motion is shown in Fig. 1.18(b). The point of this analysis is that the spatially decaying, non-propagating (or 'evanescent') mode that was not an acceptable solution for the harmonic-wave investigation is, in fact, a proper solution in the present problem for a range of forcing frequencies.

As a second example, consider that an applied force $F(t)$ is acting at the origin. The homogeneous differential equation (1.5.2) still governs the motion, now subjected to the boundary condition

$$-T\frac{\partial y(0, t)}{\partial t} = F(t), \tag{1.5.25}$$

where $F(t)$ is assumed positive in the upward direction.

The Laplace transform approach will be used. Transforming (1.5.2) gives

$$\frac{d^2\bar{y}}{dx^2} - m^2\bar{y} = 0, \tag{1.5.26}$$

where

$$m^2 = \frac{1}{c_0^2}(s^2 + k/\rho), \tag{1.5.27}$$

and $\bar{y} = \bar{y}(x, s)$. The solution of (1.5.26) is

$$\bar{y} = Ae^{-mx} + Be^{mx}. \tag{1.5.28}$$

We set $B = 0$ because of the increasing exponential behaviour. Actually, this is a bit too superficial an explanation for what actually must be ascertained by an inspection of the Laplace inversion integral for boundedness in the complex s-plane. The considerations are similar to those used in § 1.1 in closing the Fourier inversion contour. Omitting these details, we proceed to transform the force boundary condition (1.5.25), giving

$$-T\frac{d\bar{y}(0, s)}{dx} = \bar{F}(s). \tag{1.5.29}$$

Applying the solution to this, we obtain

$$A = \bar{F}(s)/mT. \tag{1.5.30}$$

This gives the transformed result

$$\bar{y}(x, s) = \frac{\bar{F}(s)}{mT}e^{-mx}. \tag{1.5.31}$$

The formal inversion of this could be expressed in terms of the convolution theorem.

To proceed further, a specific form for the forcing function must be given. Let us specify that the load be an impulse, given by

$$F(t) = P\delta(t). \tag{1.5.32}$$

Then the transform of this is $\bar{F}(s) = P$. The transformed solution (1.5.31) then takes the form

$$\bar{y}(x, s) = \frac{Pc_0}{T} \frac{\exp\{-(s^2+k/\rho)^{\frac{1}{2}}x/c_0\}}{(s^2+k/\rho)^{\frac{1}{2}}}. \tag{1.5.33}$$

The inversion of this would be no small task. Fortunately, the results

$$\mathscr{L}^{-1}\left\{\frac{1}{\sqrt{(s^2+a^2)}}\right\} = J_0(at), \tag{1.5.34}$$

have been tabulated, where $J_0(at)$ is the zero-order Bessel function. Employing the translation theorem enables (1.5.33) to be easily evaluated, giving

$$y(x, t) = \begin{cases} 0, & t < x/c_0 \\ \dfrac{Pc_0}{T} J_0\left\{\dfrac{k}{\rho}\sqrt{(t^2 - x^2/c_0^2)}\right\}, & t > x/c_0. \end{cases} \tag{1.5.35}$$

The general form of the response predicted by (1.5.35) is shown in Fig. 1.19.

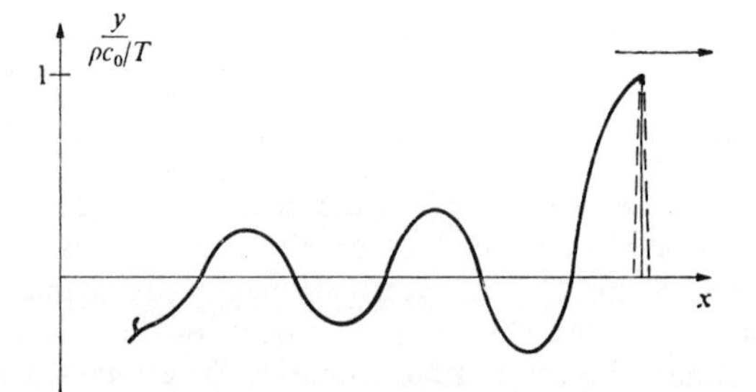

FIG. 1.19. Wave motion in a semi-infinite, elastically-supported string subjected to an impulse loading.

The main aspect to note is that the propagating pulse bears no resemblance to the input pulse, a direct consequence of dispersion. The pulse shape indicated by the dashed line in the figure is meant to illustrate a delta-function impulse, the waveform that would be propagated if the foundation were absent. We note that the head of the pulse propagates with the classical string velocity c_0.

1.6. Pulses in a dispersive media—group velocity

On the one hand, we have the wave equation and the propagation of pulses without distortion. On the other, we have the non-wave equation, the resulting frequency dependence of propagation velocity, the expected distortion of pulses, and the confirmation of this by specific example. With the string on the elastic foundation, we have been introduced to a dispersive system, the first of many to be encountered in our study of elastic waves. Because such systems are so prevalent, there is motivation for obtaining some general understanding of the consequences of dispersion and to develop techniques of analysis that will be applicable to a wide variety of situations. Work in this area represents an effort of long standing in wave-propagation theory and warrants considerable attention here. The general, unifying concept that will emerge from our study will be that of *group velocity*. The general technique of analysis will be the method of stationary phase.

1.6.1. *The concept of group velocity*

We initiate the discussion in this area by describing ripples in a pool of water, a physical situation familiar to everyone, including the earliest investigators in this area who used this example as the basis for further development. Thus it is observed that a stone dropped in a pool of still water creates an intense local disturbance (a 'splash' to the layman) which does not remain localized but spreads outward over the pool as a train of ripples as artistically depicted in Fig. 1.20. The ensuing behaviour of the ripple train

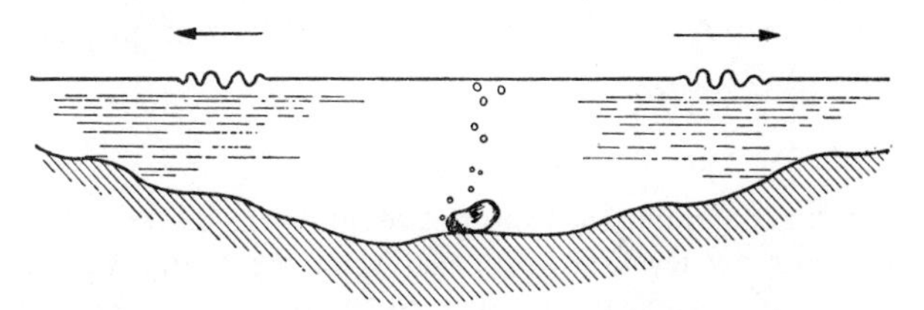

FIG. 1.20. Ripples on a pond of water resulting from a local splash.

is the aspect that drew the attention of early investigators. Quoting directly from Rayleigh.†

> It has often been remarked that, when a group of waves advances into still water, the velocity of the group is less than that of the individual waves of which it is composed; the waves appear to advance through the group, dying away as they approach its anterior limit.

Rephrasing only slightly, we say that ripples appear to originate at the rear of the group, propagate to the front, and disappear. Thus, the ripples have a higher velocity than the over-all group.

A simple analytical explanation of this behaviour was evidently first given by Stokes [8] and remains as the classical illustration of the phenomenon.

† Volume I, pp. 474 of reference [6]. A similar statement appears in Lamb's book [4, p. 380]. Evidently both statements were inspired by a paper by Scott Russell (1844), Report on waves, *Br. Assoc. Rep.* p. 369.

Thus, consider two propagating harmonic waves of equal amplitude but slightly different frequency ω_1 and ω_2, given by

$$y = A\cos(\gamma_1 x - \omega_1 t) + A\cos(\gamma_2 x - \omega_2 t), \tag{1.6.1}$$

where $\omega_1 = \gamma_1 c_1$, $\omega_2 = \gamma_2 c_2$. We may rewrite this as

$$y = 2A\cos\{\tfrac{1}{2}(\gamma_2 - \gamma_1)x - \tfrac{1}{2}(\omega_2 - \omega_1)t\}\times \times\cos\{\tfrac{1}{2}(\gamma_1 + \gamma_2)x - (\omega_1 + \omega_2)t\}. \tag{1.6.2}$$

Since the frequencies are only slightly different, the wavenumbers also will only slightly differ. We write these differences as

$$\omega_2 - \omega_1 = \Delta\omega, \qquad \gamma_2 - \gamma_1 = \Delta\gamma. \tag{1.6.3}$$

Similarly, we define the average frequency and wavenumber as

$$\omega = \tfrac{1}{2}(\omega_1 + \omega_2), \qquad \gamma = \tfrac{1}{2}(\gamma_1 + \gamma_2), \tag{1.6.4}$$

and the resulting average velocity by $c = \omega/\gamma$. Thus (1.6.2) may be written as

$$y = 2A\cos(\tfrac{1}{2}\Delta\gamma x - \tfrac{1}{2}\Delta\omega t)\cos(\gamma x - \omega t). \tag{1.6.5}$$

Now, the cosine term containing the difference terms $\Delta\gamma$,$\Delta\omega$ is a low-frequency term since $\Delta\omega$ is a small number. It will have a propagation velocity c_g where

$$c_g = \frac{\Delta\omega}{\Delta\gamma}. \tag{1.6.6}$$

The cosine term containing the average wavenumber and frequency γ and ω will be a high-frequency term, propagating at the averge velocity c. The effect of the low-frequency term will be to act as a *modulation* on the high-frequency *carrier*. The appearance of the motion is as shown in Fig. 1.21.

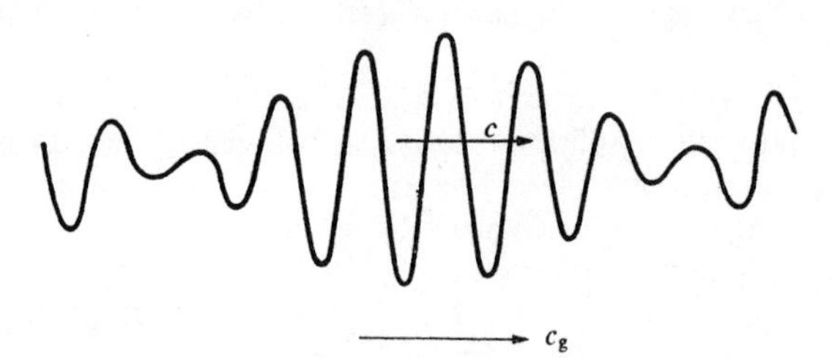

FIG. 1.21. Simple wave group formed by two waves of slightly different frequency.

It is the over-all wave group that propagates at the velocity c_g in the figure. The velocity of the high-frequency carrier may actually be greater than, equal to, or less than the velocity c_g. Whichever occurs will depend on the dispersion situation for the elastic system. For the various cases, we will have

$c > c_g$: waves will appear to originate at the rear of the group, travel to the front and disappear;

$c = c_g$: no relative motion of the group and carrier occurs and the group travels along without distortion of the wave shape.

$c < c_g$: waves will appear to originate at the front of the group, travel to the rear and disappear.

The velocity c_g that has been introduced by (1.6.6) is called the *group velocity*.

The preceding example illustrates the basic aspects of wave groups and group velocity as being formed by harmonic waves of only slightly different frequencies and wave speeds. Since only two waves were used, the results appear as a succession of groups. The existence of an isolated wave group, as illustrated in Fig. 1.20, is a consequence of many waves having slightly different frequencies and velocities interacting. Seeking to generalize along these lines, we consider a superposition of a number of waves†

$$y = \sum_{i=1}^{n} A_i \cos(\gamma_i x - \omega_i t + \phi_i), \tag{1.6.7}$$

where γ_i and ω_i differ only slightly. The phase angle of a given disturbance is ϕ_i (not to be confused with the phase $(\gamma_i x - \omega_i t + \phi_i)$).

We suppose that at some time $t = t_0$ and location $x = x_0$, the phases of the various wave trains are approximately the same so that a wave group has been formed. At a time $t = t_0 + \mathrm{d}t$ and location $x = x_0 + \mathrm{d}x$, the change in phase $\mathrm{d}P_i$ of any individual component is

$$\mathrm{d}P_i = \{\gamma_i(x_0 + \mathrm{d}x) - \omega_i(t_0 + \mathrm{d}t) + \phi_i\} - \{\gamma_i x_0 - \omega_i t_0 + \phi_i\} = \gamma_i\,\mathrm{d}x - \omega_i\,\mathrm{d}t. \tag{1.6.8}$$

In order for the wave group to be maintained, the change in phase for all of the terms must be approximately the same. This restriction is enforced by requiring that $\mathrm{d}P_j - \mathrm{d}P_i = 0$, which gives

$$(\gamma_j - \gamma_i)\,\mathrm{d}x - (\omega_j - \omega_i)\,\mathrm{d}t = 0. \tag{1.6.9}$$

Since γ_i, γ_j and ω_i, ω_j differ only slightly, we let $\mathrm{d}\gamma = \gamma_j - \gamma_i$, $\mathrm{d}\omega = \omega_j - \omega_i$. Thus (1.6.9) becomes

$$\mathrm{d}\gamma\,\mathrm{d}x - \mathrm{d}\omega\,\mathrm{d}t = 0. \tag{1.6.10}$$

The velocity of the group is then given by

$$c_g = \frac{\mathrm{d}x}{\mathrm{d}t} = \frac{\mathrm{d}\omega}{\mathrm{d}\gamma}. \tag{1.6.11}$$

This expression, which agrees in form with (1.6.6), is taken as the definition of group velocity.

† Havelock [2, p. 3].

The definition of group velocity may be expressed in alternate forms using the fundamental relation $\omega = \gamma c$. Thus

$$c_g = \frac{d(\gamma c)}{d\gamma} = c + \frac{\gamma\, dc}{d\gamma}, \tag{1.6.12}$$

where $c = c(\gamma)$. Also, since $\gamma = 2\pi/\lambda$, the results may be expressed in terms of wavelength, giving

$$c_g = c - \lambda \frac{dc}{d\lambda}. \tag{1.6.13}$$

The basic definition (1.6.11) permits a direct graphical interpretation to be made using the frequency spectrum of the system. Since the frequency spectrum is a plot of ω versus γ, the group velocity at any frequency is given by the slope of the branch of the frequency spectrum. This is illustrated in Fig. 1.22 for a hypothetical spectrum. The group velocity is indicated by the

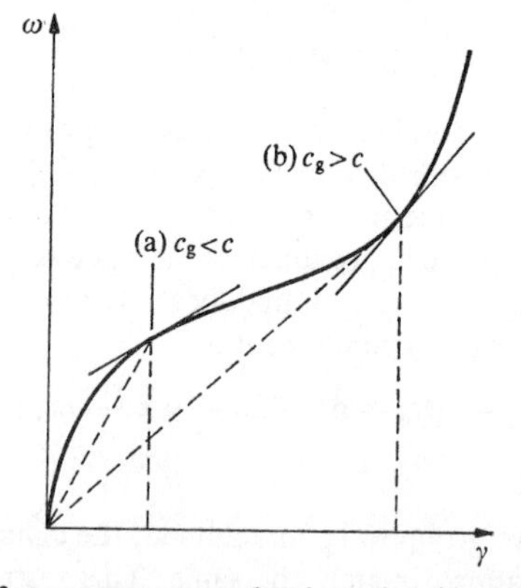

FIG. 1.22. Hypothetical frequency spectrum showing points (a) where $c_g < c$ and (b) where $c_g > c$.

local slopes to the curve of points (a) and (b). Recalling that the slope of a chord to a point is the phase velocity, it is seen that at point (a), $c_g < c$, while at point (b), $c_g > c$.

1.6.2. *Propagation of narrow-band pulses*

Having established the concepts of dispersion and group velocity, we wish to study the effects of these phenomena on pulse propagation in terms as general as possible. Now, a pulse may be characterized by its Fourier spectrum (see Appendix B) and, speaking in these terms, two possible extremes of behaviour are: (1) the pulse has a narrow frequency spectrum and is a 'narrow-band' pulse; (2) the pulse has a wide frequency band and is a 'wide-band' pulse. We will consider a technique for handling the first case in this section.

Suppose that we have applied the Fourier transform approach to the problem of forced motion of a semi-infinite string on an elastic foundation.† The inverse transform would then have the general form

$$y = \frac{1}{\sqrt{(2\pi)}} \int_{-\infty}^{\infty} f(\omega) e^{i(\gamma x - \omega t)} \, d\omega. \tag{1.6.14}$$

The above general form arises in many one-dimensional wave propagation problems. It is understood that $\gamma = \gamma(\omega)$ in the above, corresponding to a dispersive system. The nature of the forcing applied to the system is carried by $f(\omega)$. Thus, the inherent characteristics of the individual system as well as the peculiarities of the loading pulse are contained in the general solution form.

We have stated that the input is narrow-band, implying that its frequency spectrum, given by

$$f(\omega) = \frac{1}{\sqrt{(2\pi)}} \int_{-\infty}^{\infty} F(t) e^{i\omega t} \, dt, \tag{1.6.15}$$

is narrow. To interpret this more precisely, we note that a simple harmonic signal $\exp(-i\omega_0 t)$ has a spectrum given by‡

$$f(\omega_0) = \frac{1}{\sqrt{(2\pi)}} \int_{-\infty}^{\infty} \exp(-i\omega_0 t) \exp(i\omega t) \, d\omega = \sqrt{(2\pi)} \, \delta(\omega - \omega_0). \tag{1.6.16}$$

Thus, in the frequency domain, the harmonic wave shown in Fig. 1.23(a) is represented by a single line as shown in Fig. 1.23(b). Now suppose that $F(t)$ is a finite-duration oscillatory pulse having the time characteristics shown in Fig. 1.23(c). The frequency of oscillation within the 'packet' is ω_0. Then the Fourier spectrum of such a pulse is shown in Fig. 1.23(d). While not a delta function, the spectrum is sharply peaked at ω_0 with a bandwidth $\Delta\omega$ that is small relative to ω_0.

With these aspects in mind, we return to the solution form (1.6.15). We now expand $\gamma = \gamma(\omega)$ about the centre frequency ω_0. Thus

$$\gamma(\omega) = \gamma_0 + \left.\frac{d\gamma}{d\omega}\right|_{\omega_0} (\omega - \omega_0) + \ldots, \tag{1.6.17}$$

† See Problem 1.16.

‡ Most easily verified by substituting $\delta(\omega - \omega_0)$ in the inverse transform.

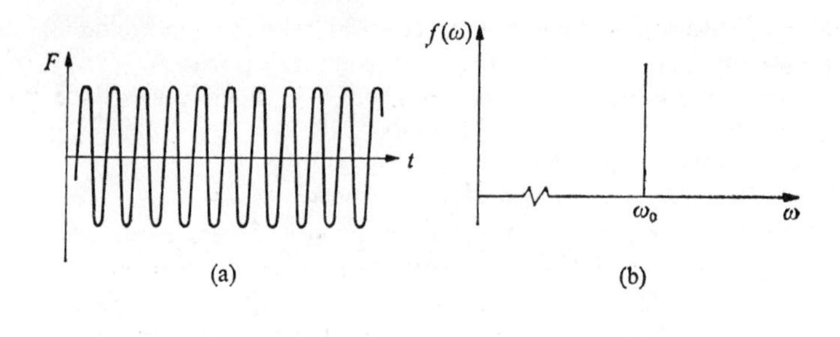

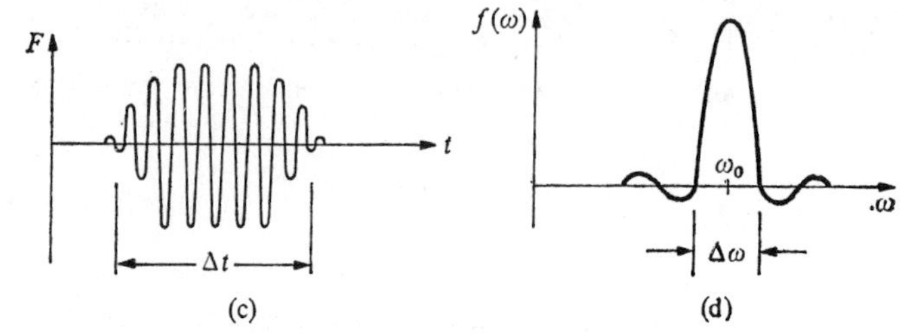

FIG. 1.23. Time behaviour and corresponding frequency spectrum for harmonic and narrow-band signals.

These first two terms will adequately represent $\gamma(\omega)$ for a narrow-band pulse. Substituting this in the solution, we have

$$y = \frac{\exp\left\{i\left(\gamma_0 - \frac{d\gamma}{d\omega}\bigg|_0 \omega_0\right)x\right\}}{\sqrt{(2\pi)}} \int_{-\infty}^{\infty} f(\omega)\exp\left\{i\left(x\frac{d\gamma}{d\omega}\bigg|_0 - t\right)\omega\right\} d\omega. \quad (1.6.18)$$

In interpreting this result, we first recall the definition of group velocity being $c_g = d\omega/d\gamma$. We thus see that $d\gamma/d\omega\,|_{\omega=\omega_0}$ is merely the reciprocal of the group velocity at the frequency ω_0. We designate this as c_g^0. We next make the change of variables,

$$t' = x/c_g^0 - t. \quad (1.6.19)$$

The integral that remains in (1.6.18) then takes the form

$$F(t') = \frac{1}{\sqrt{(2\pi)}} \int_{-\infty}^{\infty} f(\omega)\exp(i\omega t')\, d\omega. \quad (1.6.20)$$

Thus the result (1.6.18) becomes

$$y = \exp\{i(\gamma_0 - \omega_0/c_g^0)x\}F(t'). \quad (1.6.21)$$

The exponential occurring in (1.6.21) contributes nothing more than a phase angle for any given value of x. The important result is that $F(t')$, the original excitation waveform, propagates undistorted in shape and at the group velocity.

1.6.3. *Wide-band pulses—the method of stationary phase*

We have pointed out the widespread occurrence of the integral form (1.6.4) in wave-propagation problems and have presented a method for approximately evaluating it for narrow-band pulses. When the excitation is wide-band, the problem of evaluation often becomes rather formidable, even by numerical techniques, owing to the large range of the integral.

An approximate method for evaluating the integral is known as the *stationary-phase* method, or as Kelvin's method of stationary phase. The method is generally concerned with integrals of the type

$$y(x) = \int_a^b f(\omega)e^{ixh(\omega)}\,d\omega, \tag{1.6.22}$$

where $f(\omega)$ and $h(\omega)$ are real functions. The basis of the method is that for large, positive values of x, there will exist values of ω at which the major contributions to the integral occur and that, away from these critical points, the contributions of the integrand are negligible, owing to its self-cancelling oscillatory nature. This argument, and method, was first used by Stokes [7] to obtain an asymptotic representation for the function

$$y(x) = \int_0^\infty \cos x(\omega^3-\omega)\,d\omega. \tag{1.6.23}$$

Kelvin [3] first applied this principle in his consideration of a hydrodynamics problem.

In the context of wave propagation, the basis of the principle lies in the wave-group concept. Thus, the Fourier representation of the wave propagation shows a disturbance to be comprised of waves of all frequencies and wave-lengths. Initially, the wave trains superimpose to give the applied disturbance. At any subsequent time, the existing disturbance is obtained by again summing the contributions of the propagating harmonic components. Because of dispersion, it is clear that the phases of the waves will no longer agree. However, there will be positions and times at which a number of waves have the same or nearly the same phase. These elements will thus re-inforce one another and produce the predominant part of the disturbance. This is again the basic argument used in the development of this section.

The basis of the preceding argument may be better understood by considering the properties of the integrand of (1.6.22). Considering only the real part, we have

$$I = f(\omega)\cos xh(\omega). \tag{1.6.24}$$

We first plot the phase $xh(\omega)$ versus ω in Fig. 1.24(a) in the vicinity of a maximum or stationary point of $h(\omega)$. Fig. 1.24(b) shows that for ω in the vicinity of the maximum, $\cos xh(\omega)$ oscillates slowly whereas for ω away from the stationary point, there are many oscillation for small changes in ω. Fig. 1.24(c) shows the resulting behaviour of $\cos xh(\omega)$. The necessity for having x large should be apparent from Fig. 1.24(a). Large x accentuates the

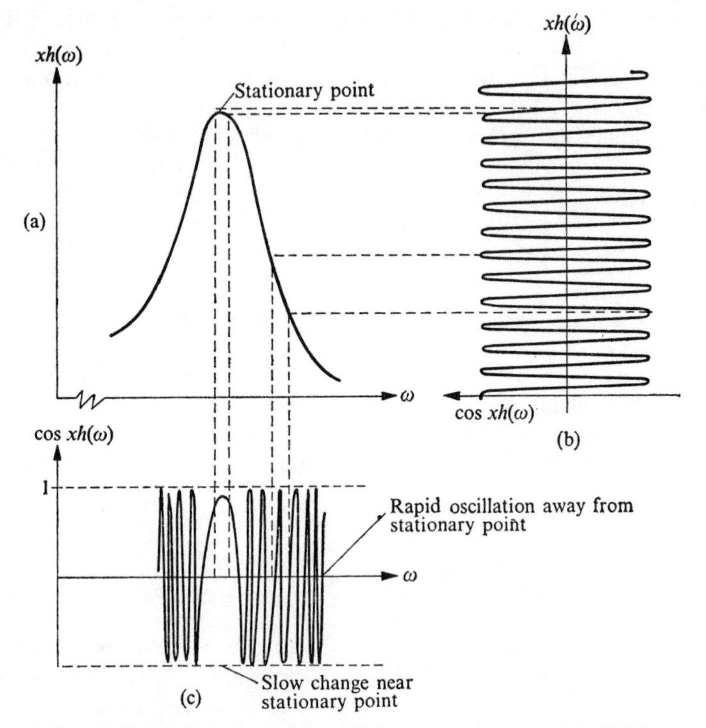

FIG. 1.24. The behaviour of $\cos xh(\omega)$ near a stationary point of $h(\omega)$.

maximum of $h(\omega)$ which increases the oscillatory nature of $\cos xh(\omega)$ away from the critical point.

To evaluate (1.6.22), making use of the ideas just described, we first expand $h(\omega)$ about ω_0. This gives

$$h(\omega) = h(\omega_0)+h'(\omega_0)(\omega-\omega_0)+(\tfrac{1}{2})h''(\omega_0)(\omega-\omega_0)^2+\dots. \quad (1.6.25)$$

If ω_0 is a stationary point, then $h'(\omega_0) = 0$ and the phase is given by

$$xh(\omega) \simeq x\{h(\omega_0)+(\tfrac{1}{2})h''(\omega_0)(\omega-\omega_0)^2\}, \quad (1.6.26)$$

where the series has been truncated at the quadratic term. Then (1.6.22) becomes

$$y(x) = \int_{-\infty}^{\infty} f(\omega_0)\exp[\mathrm{i}x\{h(\omega_0)+\tfrac{1}{2}h''(\omega_0)(\omega-\omega_0)^2\}]\,\mathrm{d}\omega, \qquad (1.6.27)$$

where the limits of integration have been extended to infinity. This is justified because of the expected negligible contributions away from the stationary point. Then

$$y(x) = f(\omega_0)\exp\{\mathrm{i}xh(\omega_0)\}\int_{-\infty}^{\infty} \exp\left\{\frac{\mathrm{i}xh''(\omega_0)}{2}(\omega-\omega_0)^2\right\}\mathrm{d}\omega. \qquad (1.6.28)$$

The integral in this result may be found in tables and is given by

$$\int_{-\infty}^{\infty} \exp\left\{\frac{\mathrm{i}xh''(\omega_0)}{2}(\omega-\omega_0)^2\right\}\mathrm{d}\omega = \sqrt{\left\{\frac{2\pi}{xh''(\omega_0)}\right\}}\exp\left(\frac{\mathrm{i}\pi}{4}\right). \qquad (1.6.29)$$

Thus we have

$$y(x) = \left\{\frac{2\pi}{xh''(\omega_0)}\right\}^{\frac{1}{2}} f(\omega_0)\exp[\mathrm{i}\{xh(\omega_0)+\pi/4\}]. \qquad (1.6.30)$$

This result represents the approximate evaluation of the integral form (1.6.22) by the method of stationary phase.

We now relate the basic result (1.6.30) with the general integral form (1.6.14) arising in wave propagation. Comparing (1.6.22) and (1.6.14), we see that

$$h(\omega) = \gamma-\omega t/x. \qquad (1.6.31)$$

To find the stationary point such that $h'(\omega_0) = 0$ we differentiate this, giving

$$\frac{\mathrm{d}h(\omega)}{\mathrm{d}\omega} = \frac{\mathrm{d}\gamma}{\mathrm{d}\omega}-\frac{t}{x} = 0 \qquad (1.6.32)$$

or

$$\frac{\mathrm{d}\omega}{\mathrm{d}\gamma} = \frac{x}{t}. \qquad (1.6.33)$$

But $\mathrm{d}\omega/\mathrm{d}\gamma$ is the definition of group velocity. Basically, the stationary-phase result states that the dominant part of the disturbance that arrives at a particular point x at time t will have travelled with a group velocity $c_g = x/t$ and will consist of the dominant frequency component ω_0 determined from (1.6.32). This aspect is illustrated in Fig. 1.25 for a given location x_0 at time t_0. Thus, in Fig. 1.25(a) a hypothetical frequency spectrum is shown and the frequency ω_0 corresponding to $c_g = x_0/t_0$ indicated. At that instant of time, a disturbance having that frequency arrives at x_0 as shown in Fig. 1.25(b). The wavelength of the disturbance is given by $\lambda_0 = 2\pi/\gamma_0$. Other details of the disturbance at that location and time will depend on the form of $h(\omega)$ and $f(\omega)$ and will be given by (1.6.30). The former quantity will characterize

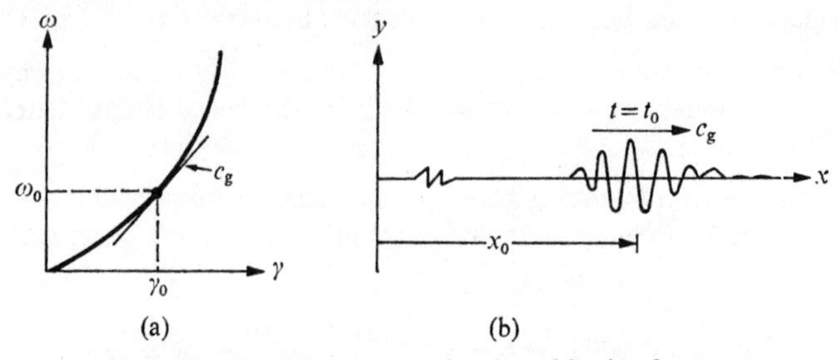

FIG. 1.25. Signal arriving at x_0 at time t_0 and having frequency ω_0.

the dispersive characteristics of the media while the latter will describe the input.

A concluding remark is in order regarding the method of stationary phase. The basis of the method is the supposed self-cancelling oscillations of the integrand. This provides the nulling mechanism for the integral away from the critical points. Other methods for obtaining asymptotic values of the integrals allow $h(\omega)$ and $f(\omega)$ to be complex and employ stronger mechanisms, involving exponential decay, than those used in the stationary-phase method. The method of steepest descent is one such technique that will be presented in a later chapter.

1.7. The string on a viscous subgrade

Our considerations thus far have been with purely elastic systems, and this will be the case for the major part of this book. Some brief attention, however, will be given here to a system exhibiting losses. Such is the case for a string on a viscous foundation. The effects of viscous damping significantly alters many basic aspects of string response.

1.7.1. *The governing equation*

The effects of a simple viscous foundation can be included in the governing equation in the same fashion used to include elastic-foundation effects. Thus, assuming a simple dashpot model for the forces, we assume a resistive force to the motion given by $q(x, t) = -\zeta\dot{y}(x, t)$. The resulting governing equation is

$$T\frac{\partial^2 y}{\partial x^2} - \zeta\frac{\partial y}{\partial t} = \rho\frac{\partial^2 y}{\partial t^2}. \tag{1.7.1}$$

There is little to remark on the equation other than it is not of the wave-equation form. It is also a matter of possible interest to note that as $\rho \to 0$, the equation passes over into the diffusion equation. Physically, this might correspond to the viscous effects of a foundation predominating over the mass effects of the string.

1.7.2. *Harmonic wave propagation*

Proceeding in somewhat the usual manner of investigating new mechanical systems, we consider the string to be infinite in length and seek to determine the conditions for propagation of free harmonic waves of the form

$$y = Ae^{i(\gamma x-\omega t)}. \tag{1.7.2}$$

Substitution in (1.7.1) gives

$$\gamma^2 = \frac{i\zeta\omega}{T}+\frac{\omega^2}{c_0^2}. \tag{1.7.3}$$

This is a complex quantity, the roots of which are given by

$$\gamma = M^{\frac{1}{2}}\exp(i\phi/2), \qquad M^{\frac{1}{2}}\exp\left\{i\left(\frac{\phi}{2}+\pi\right)\right\}, \tag{1.7.4}$$

where

$$M = \frac{1}{T}(\zeta^2\omega^2+\rho^2\omega^4)^{\frac{1}{2}}, \qquad \tan\phi = \frac{\zeta}{\rho\omega}. \tag{1.7.5}$$

We replace (1.7.4) with the more convenient notation

$$\gamma = \gamma_R+i\alpha, \quad -\gamma_R-i\alpha, \tag{1.7.6}$$

where

$$\gamma_R = M^{\frac{1}{2}}\cos\tfrac{1}{2}\phi, \qquad \alpha = M^{\frac{1}{2}}\sin\tfrac{1}{2}\phi. \tag{1.7.7}$$

Then the original waveform (1.7.2) becomes

$$y = A\exp[i\{\pm(\gamma_R+i\alpha)x-\omega t\}], \tag{1.7.8}$$

or

$$y = A\exp(\mp\alpha x)\exp\{i(\pm\gamma_R x-\omega t)\}. \tag{1.7.9}$$

If we assume $x > 0$, we are led to discard the lower (—) sign of (1.7.8), giving the final form

$$y = A\exp(-\alpha x)\exp\{i(\gamma_R x-\omega t)\}. \tag{1.7.10}$$

Interpreting this result, we conclude that the free propagation of harmonic waves of the form (1.7.2) is not possible in a damped string. Thus, in the context of the original question, the matter is closed and other aspects must be considered.

1.7.3. *Forced motion of a string*

The form of the results (1.7.10) is hardly surprising, and merely reflects the energy loss due to viscous damping. The negative answer to the matter of free wave propagation merely means that it makes no sense to consider such matters. The question of wave motion in a damped string only makes sense in the context of steady, forced vibration, where there is an energy source to maintain motion or in the context of transient response to initial conditions or applied load. With this in mind, the solution (1.7.10) is perfectly acceptable.

We note the occurrence for the first time of a complex wavenumber and see the implications on the waveform.

We can consider a very simple example of forced motion of a string to which (1.7.10) is directly applicable. Thus, consider the semi-infinite string $x > 0$ on a viscous foundation, subjected to the forced end motion

$$y(0, t) = y_0 \exp(-\mathrm{i}\omega t).$$

By inspection we can write immediately

$$y = y_0 \exp(-\alpha x)\exp\{\mathrm{i}(\gamma_R x - \omega t)\}. \tag{1.7.11}$$

A second problem to consider is that of an infinite string subjected to the harmonically varying concentrated load at the origin given by

$$q(x, t) = -\frac{1}{T}\,\delta(x)\mathrm{e}^{-\mathrm{i}\omega t}. \tag{1.7.12}$$

Consider a solution to the governing eqn (1.7.1), with this term on the right-hand side, of the form

$$y = Y(x)\mathrm{e}^{-\mathrm{i}\omega t}. \tag{1.7.13}$$

We obtain the ordinary differential equation

$$\frac{\mathrm{d}^2 Y}{\mathrm{d}x^2} + \left(\frac{\mathrm{i}\zeta\omega}{T} + \frac{\omega^2}{c_0^2}\right) Y = \delta(x). \tag{1.7.14}$$

We now state the motivation for considering this otherwise commonplace problem. Except for the addition of damping, it is identical to the problem discussed at some length in § 1.1.8. Specifically, the Fourier transform was applied and the problem was the inversion of (1.1.93). Likewise the Fourier transform can be applied to (1.7.14), giving the solution

$$Y(x) = -\frac{1}{2\pi}\int_{-\infty}^{\infty} \frac{\mathrm{e}^{-\mathrm{i}\gamma x}}{\gamma^2 - \bar{\gamma}_0^2}\,\mathrm{d}\gamma, \tag{1.7.15}$$

where $\bar{\gamma}_0^2$ in the above is given by

$$\bar{\gamma}_0^2 = \frac{\mathrm{i}\zeta\omega}{T} + \frac{\omega^2}{c_0^2}. \tag{1.7.16}$$

We recall the rather lengthy discussion concerning the poles of (1.1.93) located on the real axis, and the proper indentation about the poles, resulting in Fig. 1.9. The whole point of this exercise is to show that when damping is included in the system, the matter of pole indentation does not even arise. This is because the poles are no longer located on the real axis, but are displaced above and below it as given by (1.7.6). If the damping is quite small, the pole locations are given by†

$$\bar{\gamma} \cong \pm(\gamma_0 + i\varepsilon), \tag{1.7.17}$$

where ε is a small quantity. The resulting locations are shown in Fig. 1.26. Also shown is the contour appropriate for determining the response for $x > 0$. The dashed semicircles in the figure represent the indentations selected on the basis of the radiation condition for the undamped problem. It is seen that damping automatically shifts the poles in the proper direction.

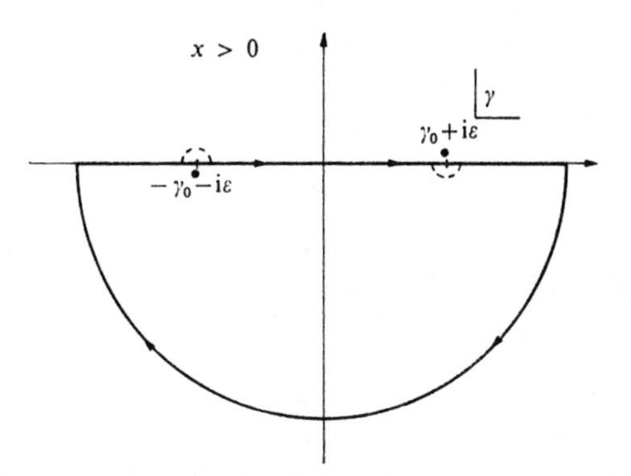

FIG. 1.26. Movement of poles off the real axis due to slight damping.

With this result pointed out, analysis will not be pursued further. It often occurs that investigators of wave problems will artificially include damping in the problem as a quick means of determining pole behaviour, then delete it from the problem.

References

1. BERRY, J. G. and NAGHDI, P. M. On the vibration of elastic bodies having time-dependent boundary conditions. *Q. J. appl. Math.* **14**, 43 (1956).
2. HAVELOCK, T. H. *The propagation of disturbances in dispersive media.* Cambridge University Press (1914).
3. KELVIN, LORD. *Phil. Mag.* **22**, 252–5 (1887).
4. LAMB, SIR H. *Hydrodynamics.* Dover Publications, New York (1945).
5. MORSE, P. M. and INGARD, K. U. *Theoretical acoustics.* McGraw-Hill, New York (1968).
6. RAYLEIGH, J. W. S. *The theory of sound,* Vols I and II. Dover Publications, New York (1945).
7. STOKES, G. G. *Mathematical and physical papers,* Vol. 2, p. 329. Cambridge University Press (1883).
8. —. *Smith's prize examination,* Cambridge, 1876. Reprinted in *Mathematics and Physics Papers,* Vol. 5, p. 362. Cambridge University Press (1905).

† Problem 1.21 pertains to specifically determining γ_0 and ε.

9. TONG, K. N. *Theory of mechanical vibration.* John Wiley and Sons, New York (1960)

Problems

1.1. Consider an infinite string which is given an initial velocity impulse over the length l as defined by

$$y(x,0)=0,\ \dot{y}(x,0)=\begin{cases}V_0 & |x|<l/2\\ 0, & |x|>l/2.\end{cases}$$

Obtain the D'Alembert solution and plot the waveform for various values of time.

1.2. The infinite string is subjected to the distributed loading,

$$q(x,t)=\begin{cases}q_0\mathrm{e}^{-\mathrm{i}\omega t}, & |x|<a\\ 0, & |x|>a.\end{cases}$$

Apply the Fourier transform method to this problem and determine the response for $x > a$. Recover the concentrated load results by letting $a \to 0$ in a 'proper' way.

1.3. Recover the D'Alembert results for the initial-value problem by applying the Laplace transform method.

1.4. Consider a semi-infinite, classical string, initially at rest and subjected to the harmonic end load $y(0, t) = y_0 \sin \omega t$ beginning at $t = 0$. Apply the Laplace transform method, obtaining the inversion by using integration in the complex plane. In particular, determine which way the Bromwich countour must be closed. Show that this depends on whether $x \gtrless c_0t$.

1.5. Consider a taut string travelling at the translational velocity V_0 to the right. Develop the equation for the transverse motion of the string by introducing a moving coordinate system. Show that a resonance effect, leading to unbounded wave amplitudes, can occur if $V_0 \to c_0$. Consider also the case of a concentrated load P moving at velocity V_0. Note that with respect to the moving coordinate system $\partial/\partial t = 0$.

1.6. Apply the method of images to the reflection of a pulse from a free-end boundary condition of a semi-infinite string. Contrast the behaviour to the case of the fixed end.

1.7. Consider an infinite string with an elastic spring located at $x = 0$ and let a step pulse be incident on this discontinuity. Determine the reflected and transmitted wave system.

1.8. Consider harmonic waves in a semi-infinite string incident, respectively, on an elastic, mass, and dashpot end condition. Obtain the reflected wave amplitude ratio B/A for each case.

1.9. Consider a rectangular pulse incident on a fixed boundary of a semi-infinite string. Apply the Laplace transform approach to the problem. Show that the image pulse arises as a natural result of the analysis.

1.10. Consider the problem of a rectangular pulse incident on an elastic boundary of a semi-infinite string. The Laplace transform was applied to this problem in § 1.2, but less powerful techniques are quite sufficient here. Solve for the reflected wave using simple first-order differential equations.

1.11. *Impedance concepts for strings.* The concept of the mechanical impedance of a system is borrowed from electrical circuit theory. The impedance of an electrical circuit at a terminal is defined as the driving voltage divided by the current flow. Making the analogue of force to voltage and velocity to current, we have the impedance given by $Z = F/v$. The driving force (and voltage) is assumed harmonic. Determine the driving-point impedance of the following systems, assuming a concentrated force of $F_0 \exp(-i\omega t)$ to be applied at the appropriate location: (a) a semi-infinite classical string driven at the end; (b) a semi-infinite string on an elastic foundation driven at the end; (c) a finite string of length l, fixed at both ends, with the load applied at $x = \xi$; (d) a finite string of length l, fixed at one end, elastically supported at the other end and driven at that point.

In certain frequency ranges, distributed systems often behave like lumped parameter systems. Write the impedance expressions for a spring, a mass, and a dashpot. With which of these elements is (a) similar? Can you give a physical explanation for this?

1.12. Determine the equations governing the natural frequencies of strings of length l constrained as follows: (a) fixed-elastically supported; (b) fixed-end mass; (c) end mass–elastic support.

1.13. A taut string of length l, fixed at each end, is subjected to the initial conditions

$$y(x, 0) = \begin{cases} hx/\xi, & (0 \leq x < \xi) \\ h(l-x)/(l-\xi), & (\xi < x \leq l); \quad \dot{y}(x, 0) = 0. \end{cases}$$

Show that the solution for this initial-value problem is given by

$$y(x, t) = \frac{2hl^2}{\pi^2 \xi(l-\xi)} \sum_{n=1}^{\infty} \frac{1}{n^2} \sin \frac{n\pi\xi}{l} \sin \frac{n\pi x}{l} \cos \frac{n\pi c_0 t}{l}.$$

1.14. A taut string of length l is subjected to the distributed steady-state forcing $q(x, t) = q_0 \sin \omega t$. Solve the problem seeking a solution of the form $y(x, t) = Y(x) \sin \omega t$.

1.15. Sketch the curve of group velocity versus wavenumber for the infinite string on an elastic foundation.

1.16. Consider the semi-infinite string on an elastic foundation subjected to the end transient $y(0, t) = y_0(t)$. Apply the Fourier transform to this problem and thus obtain the specific form of the response integral (1.6.14) for this problem. Find the stationary points for the system, and, in general, determine the form taken by the integral.

1.17. For the string on an elastic foundation, re-examine the behaviour at the cutoff frequency. Specifically, show that at $\omega = \omega_c$, a response of the form

$$y(x, t) = (A + Bx)\exp(i\omega_c t)$$

is predicted. The B term of the result is subsequently discarded on physical grounds.

1.18. When a taut wire under tension T and carrying a current I is placed between two magnetic coils, a restoring force similar to that of an elastic foundation acts on the wire when the current flows in one direction in the coils. When the current is reversed, a negative restoring force acts on the wire. Thus an effective negative spring modulus is created. Study the propagation of harmonic waves in a string on a foundation having this characteristic. Sketch the frequency spectrum. An unstable wave motion should result.

1.19. Consider two harmonic waves of slightly differing frequency propagating in a string on an elastic foundation, in the manner of the Kelvin illustration of group velocity. Is $c \gtrless c_g$ for this system? Does the particular inequality hold for the entire range of ω?

1.20. A taut, semi-infinite string rests on an elastic foundation and is subjected to the end transient

$$y(0, t) = A \exp(-4t^2/\tau^2)\cos \omega_0 t,$$

where $\omega_0 = 2\omega_c$ and ω_c is the system cutoff frequency. The modulation is the Gaussian pulse, where the time constant τ is equal to $20/\omega_c$. Other parameters are $\rho = 0{\cdot}52$ kg m^{-1}, $c_0 = 50{\cdot}8$ m s^{-1}, $k = 20{\cdot}7\pi^2$ N m^{-2}. (a) On the basis of group velocity considerations, when would the signal arrive at a station $x = 25$ cm? (b) On the basis of the phase velocity of ω_0, when would the signal arrive at $x = 25$ cm? (c) For (a), what would be the wavelength of the received signal?

1.21. In the case of a string on a viscous foundation, suppose damping is small. Show that, for these conditions,

$$\gamma_R \cong \omega/c_0, \qquad \alpha \cong \zeta\omega/2T,$$

thus indicating that the propagation velocity is unaffected by the damping and only amplitude attenuation occurs.

2 Longitudinal waves in thin rods

PHYSICALLY longitudinal wave motion in a thin rod and the transverse wave motion of a taut string are quite different. Mathematically, however, the two systems are quite similar. It turns out that the wave equation which governs the motion of the string also governs the longitudinal rod motion, at least within a range of circumstances. It is to be expected, therefore, that many results obtained for the string, such as the D'Alembert solution, will apply directly to the rod.

Despite this apparent great similarity, considerable attention is devoted to problems involving the thin rod in the following. There are several reasons for this. The first is that the inherent physical difference between the two systems results in practical applications, and hence problems of analysis, of the rod structure that have no logical counterpart in the string. Problems involving mechanical impact are a case in point. Secondly, because the rod is a practical structural system, it is important to obtain a direct appreciation of such variables as stress, particle velocity, and energy. Finally, 'other' effects which cause the governing equation to depart from the wave equation form in the string (for example, the elastic foundation) also exist for the rod (for example, lateral inertia), and these must be considered in their own right.

The thin rod represents the first of several elastic systems considered where the governing theory is based on 'strength-of-materials' considerations. Other cases are the thin beam, the thin plate, and the thin shell. Such theories do not originate from the exact equations of elasticity but from considering the motion of an element of the structure. The distinguishing characteristics of all such theories is that certain assumptions on the kinematics of deformation are made, such as 'plane sections remain plane' for the thin beam. Chapters 2–4 will be concerned with such theories.

The contents of this chapter will digress from its title in the last section, where torsional waves in thin rods are briefly covered. It will be found that the governing equation, developed on the basis of strength-of-materials considerations, is again in the form of the wave equation. Other than presenting basic data on typical wave velocities, the analysis of this system will not be pursued further.

2.1. Waves in long rods

The governing equation for the thin rod will be derived and the basic propagation characteristics considered. Since most aspects are similar to the case of waves in strings, coverage will be rather brief.

2.1.1. *The governing equation*

Consider a straight, prismatic rod as shown in Fig. 2.1(a). The coordinate

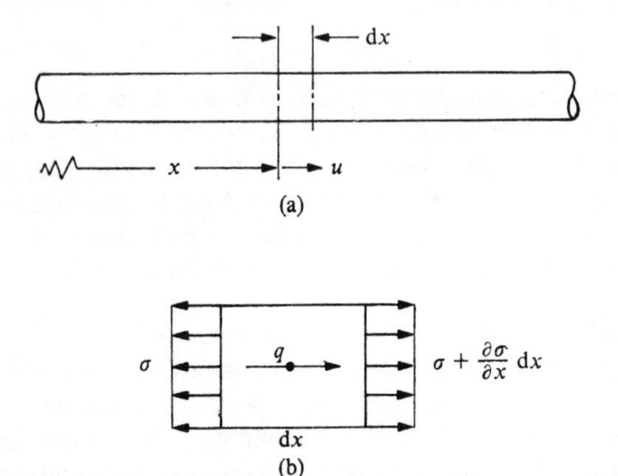

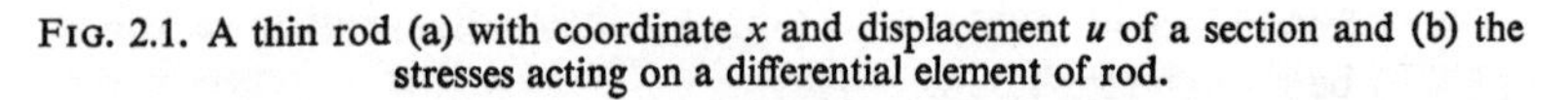

FIG. 2.1. A thin rod (a) with coordinate x and displacement u of a section and (b) the stresses acting on a differential element of rod.

x refers to a cross-section of the rod, while the longitudinal displacement of that section is given by $u(x, t)$. We presume the rod to be under a dynamically-varying stress field $\sigma(x, t)$, so that adjacent sections are subjected to varying stresses. A body force $q(x, t)$ per unit volume is also assumed present. The equation of motion in the x direction then becomes

$$-\sigma A+\left(\sigma+\frac{\partial \sigma}{\partial x}\,\mathrm{d}x\right)A+qA\,\mathrm{d}x = \rho A\,\mathrm{d}x\frac{\partial^2 u}{\partial t^2}, \tag{2.1.1}$$

where A is the cross-sectional area of the rod. We are considering a prismatic rod, so this parameter is a constant in this development. We note that tensile stress is assumed positive. Eqn (2.1.1) reduces to

$$\frac{\partial \sigma}{\partial x}+q = \rho\frac{\partial^2 u}{\partial t^2}. \tag{2.1.2}$$

Material effects have not been introduced, so at this stage the equation is applicable to non-elastic as well as elastic problems. We now presume that the material behaves elastically and follows the simple Hooke's law

$$\sigma = E\varepsilon, \tag{2.1.3}$$

where E is Young's modulus and ε is the axial strain, defined by

$$\varepsilon = \partial u/\partial x. \tag{2.1.4}$$

Using (2.1.3) and (2.1.4) in the equation of motion, we obtain

$$\frac{\partial}{\partial x}\left(E\frac{\partial u}{\partial x}\right)+q = \rho\frac{\partial^2 u}{\partial t^2}. \tag{2.1.5}$$

If the rod is homogeneous so that E (and ρ) do not vary with x, the equation reduces to

$$E\frac{\partial^2 u}{\partial x^2}+q = \rho\frac{\partial^2 u}{\partial t^2}. \tag{2.1.6}$$

We recognize this to be identical in form to (1.1.1) derived for the taut string.

There are several assumptions implicit in the development of (2.1.6), some of which have been mentioned, such as the prismatic shape and homogeneity. The former restriction will be relaxed somewhat in a later section. It is also assumed that plane, parallel cross-sections remain plane and parallel and that a uniform distribution of stress exists. Finally, we note that a very important assumption regarding lateral effects has been made. While we have assumed uniaxial stress, we have not assumed uniaxial strain. Thus, owing to the Poisson effect, there are lateral expansions and contractions arising from the axial stress. In fact, using the generalized Hooke's law we could determine these quantities. The important point is that we have neglected the *lateral inertia* effects associated with these contraction–expansions. This restriction will also be re-examined in a later section.

Continuing, we see that in the absence of body forces, eqn (2.1.6) reduces to

$$E\frac{\partial^2 u}{\partial x^2} = \rho\frac{\partial^2 u}{\partial t^2} \tag{2.1.7}$$

or

$$\frac{\partial^2 u}{\partial x^2} = \frac{1}{c_0^2}\frac{\partial^2 u}{\partial t^2}, \qquad c_0 = \sqrt{\left(\frac{E}{\rho}\right)}, \tag{2.1.8}$$

which is, of course, the familiar wave equation.

2.1.2. *Basic propagation characteristics*

It is desirable to review briefly the basic propagation aspects as governed by (2.1.8) in the context of thin rods. We know immediately that the D'Alembert solution pertains. Thus

$$u(x, t) = f(x-c_0t)+g(x+c_0t), \tag{2.1.9}$$

so that longitudinal waves propagate at the velocity c_0 in a thin rod and without distortion. Fig. 1.4 still nicely illustrates the behaviour. Typical propagation velocities in most metals are quite high, of the order of 5×10^3 m s^{-1}, compared to the velocity of sound in air of about 250 m s^{-1}.

The velocity c_0 is frequently designated the *bar velocity* or thin-rod velocity, to distinguish it from other propagation velocities for materials such as shear-wave velocity. Table 2.1 presents nominal bar velocity data for a number of materials.

TABLE 2.1
Bar velocity data for materials (nominal)

Material	Bar velocity	
	$\text{m s}^{-1}\times 10^{-3}$	$\text{in. s}^{-1}\times 10^{-4}$
Aluminum	5·23	20·6
Brass	3·43	13·5
Cadmium	2·39	9·4
Copper	3·58	14·1
Gold	2·03	8·0
Iron	5·18	20·4
Lead	1·14	4·5
Magnesium	4·90	19·3
Nickel	4·75	18·7
Silver	2·64	10·4
Steel	5·06	19·9
Tin	2·72	10·7
Tungsten	4·29	16·9
Zinc	3·81	15·0

Let us briefly consider the physical behaviour of the rod under a propagating pulse $u = f(x-c_0t)$. Recalling that $\sigma = E\,\partial u/\partial x$, we see that the stress configuration along the bar at any instant of time will be given by the slope of the displacement wave. Several pulse configurations are shown in a qualitative manner in Fig. 2.2. We see the displacement waveforms (a) and (b) of the

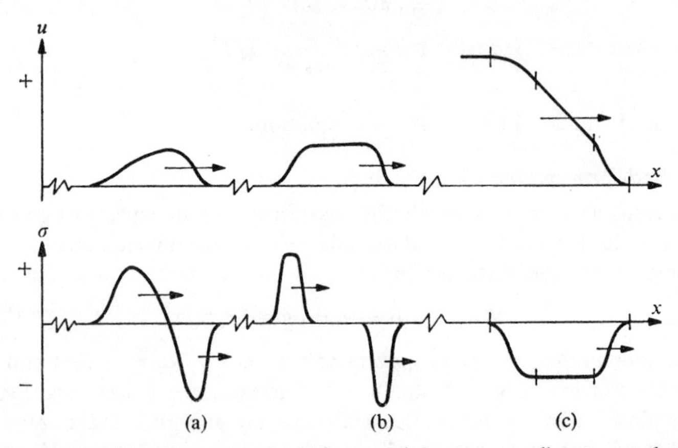

FIG. 2.2. Various displacement waveforms and the corresponding stress pulses.

figure are always positive, and that this leads to stress pulses having both tensile and compressive components. A plateau in the displacement wave, as in case (b), yields a region of zero stress. Case (c) of the figure is best considered by assuming the stress wave as prescribed, and that the displacements shown are the consequence of the stress wave. This latter waveform is commonly encountered in longitudinal impact of rods and will be studied in an idealized form in § 2.4.

As mentioned in discussing the assumptions of the theory, the Poisson effect causes lateral expansions and contractions under a propagating pulse. This effect is illustrated in greatly exaggerated form in Fig. 2.3.

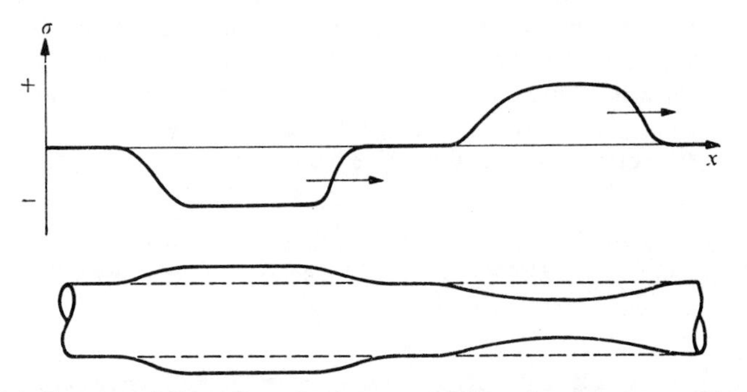

FIG. 2.3. Exaggerated illustration of Poisson expansion and contraction resulting from longitudinal stress pulses.

The particle velocity in the bar is given by

$$v(x, t) = \partial u/\partial t = -c_0 f'(x-c_0 t). \tag{2.1.10}$$

This may be expressed in terms of the stress, since

$$\sigma(x, t) = E\frac{\partial u}{\partial x} = Ef'(x-c_0 t). \tag{2.1.11}$$

Then we directly obtain

$$v(x, t) = -\frac{c_0}{E}\sigma(x, t). \tag{2.1.12}$$

Under elastic conditions, the stress is always much less than the elastic modulus, so the particle velocity will be much less than the propagation velocity. As an example, suppose a pulse of magnitude $\sigma = 10^8 \text{N m}^{-2}$ is propagating in steel. From Table 2.1 we have that $c_0 \simeq 5{\cdot}1 \times 10^3 \text{ m s}^{-1}$ and, for steel, $E \simeq 20{\cdot}7 \times 10^{10} \text{ N m}^{-2}$. This gives a particle velocity of $v = 2{\cdot}5 \text{ m s}^{-1}$. Not only is the particle velocity less than the propagation velocity, it is less by several orders of magnitude.

2.2. Reflection and transmission at boundaries

We forego repeating many of the results obtained for waves in strings, such as the initial-value problem, the characteristics plane representation, and other items covered in § 1.1, and proceed to the matter of reflection and transmission of waves. In fact, many of these results are contained in the first chapter and will be taken over rather directly. Most attention will be devoted to those areas peculiar to the rod, such as determining boundary conditions by reflected waves.

2.2.1. *Reflection from fixed and free ends*

Boundary conditions for a semi-infinite rod analogous to those for a string (see Table 1.1) can be formulated. The image method can be applied to the two simplest cases of end termination—the free and fixed ends. Again, the string results can be directly applied. However, there are one or two aspects of the phenomena in rods that deserve further emphasis. Consider first an incident pulse on a fixed boundary. The boundary condition is given by

$$u(0, t) = 0. \tag{2.2.1}$$

An image displacement pulse system is set up that will satisfy these boundary conditions and is shown in Fig. 2.4(a). Also shown is the corresponding stress

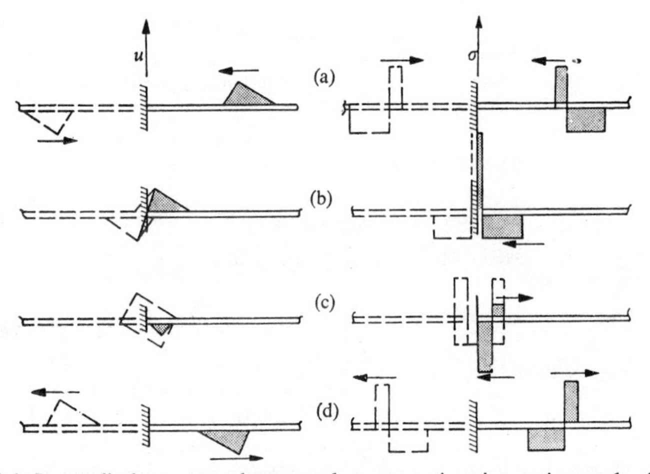

FIG. 2.4. Image displacement and stress pulse propagation, interaction, and reflection from a fixed boundary.

pulse situation. During steps (b) and (c) in the figure, interaction at the boundary occurs. The main point of this study is brought out in the stress-wave interaction in step (b). It is seen that the stresses superimpose to give double peak value. This is also occurring in step (c), where double the value of the compressive part is evident. This *stress multiplication* phenomenon is characteristic of the fixed boundary. Proceeding to part (d), after interaction has occurred, it is seen that the reflected stress pulse is identical to the incident pulse. Thus, compression has reflected as compression and tension as tension. This also is a characteristic of the fixed boundary.

Now consider a free-end boundary condition, as given by

$$\partial u(0, t)/\partial x = 0. \tag{2.2.2}$$

Again apply the image method to the reflection problem. The appropriate image system is shown in Fig. 2.5(a). As is evident in the interaction steps (b)

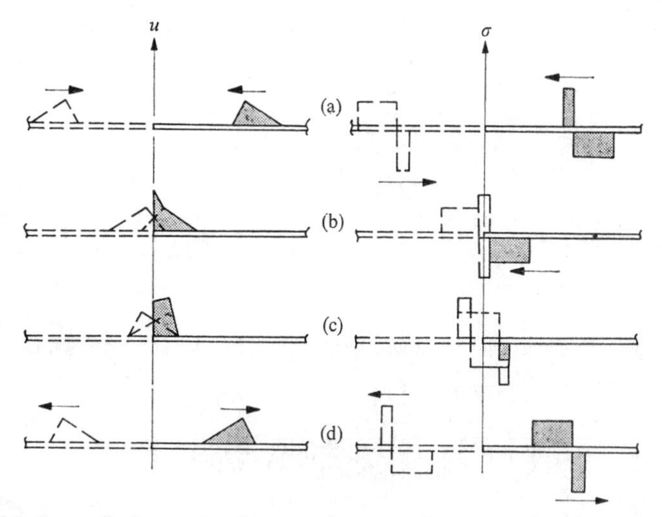

FIG. 2.5. Image displacement and stress pulse system for propagation, interaction, and reflection from a free end.

and (c), there is a doubling effect, but the doubling is associated with the *displacement* at the end and not the stress. This latter is, of course, always zero. Continuing to step (d), it is seen that the reflected stress pulse is opposite to the incident pulse; thus compression has reflected as tension and vice versa. This *stress reversal* is a characteristic of the free end.

There are several practical implications of the stress-reversal phenomenon for a free end, all basically related to the fact that if the incident wave is compressive, it will reflect as a tension wave. For example, one of the earliest experimental techniques for determining pulse characteristics employed this phenomenon.† Briefly, the method involves attaching a small pellet or 'time

† Hopkinson, B. (1914). *Phil. Trans. R. Soc.* A **213**, 437. This is mentioned in the *Introduction* (§ I.3). See Kolsky [18, pp. 87–91] for a more detailed description of the method.

piece' to the end of a rod using a thin layer of grease. Such an interface is capable of transmitting compression but not tension. When a compressive wave reaches the end of the rod, it transmits across the interface and then reverses sign. At the moment the interface stress drops from compression to zero and attempts to pass into tension, the pellet flys off due to the trapped wave momentum in the pellet. From a knowledge of the pellet length and flyoff velocity, it is possible to deduce many of the pulse characteristics.

A second application of the phenomenon is the determination of the dynamic tensile strengths of brittle materials [1, 15]. When the reflected wave causes the tensile strength of material to be reached, fracture abruptly occurs and the end of the test rod flys off in the manner of the time piece. The stress reversal phenomenon is of great importance in ballistic impact situations. Stress waves propagating through thick slabs behave in the same manner as in the rod on encountering a free boundary. It is possible, under the stress reversal, for a 'spall' to be torn from the free surface and to fly off with high velocity again due to tensile failure of the material.

2.2.2. *Reflection from other end conditions*

The elastic, mass, and dashpot end conditions for the string could be analogously formulated and analysed for the rod. Let us take a slightly more general viewpoint here that finds practical application in experimental studies involving dynamic loading of materials. Consider a semi-infinite rod in which the propagating wavefield is given by the D'Alembert solution. Thus

$$u = f(x-c_0t)+g(x+c_0t). \tag{2.2.3}$$

We suppose some force $F(t)$ to be acting on the end of the rod, such as that due to the resistance of some load against which the rod end is moving. The situation is shown in Fig. 2.6, where the dashed line indicates some hypothetical load. The velocity of the rod end is $V(t)$. The stress field in the rod will be given by

$$\begin{aligned}\sigma(x,t) &= E\{f'(x-c_0t)+g'(x+c_0t)\},\\ &= \sigma_r(x-c_0t)+\sigma_i(x+c_0t).\end{aligned} \tag{2.2.4}$$

Thus σ_i and σ_r are the incident and reflected stress pulses. Similarly, the velocity field in the rod is

$$\begin{aligned}v(x,t) &= c_0\{-f'(x-c_0t)+g'(x+c_0t)\},\\ &= \frac{c_0}{E}\{-\sigma_r(x-c_0t)+\sigma_i(x+c_0t)\}.\end{aligned} \tag{2.2.5}$$

Then, balance of force at the end of the rod requires

$$F(t) = -A\{\sigma_r(0,t)+\sigma_i(0,t)\}. \tag{2.2.6}$$

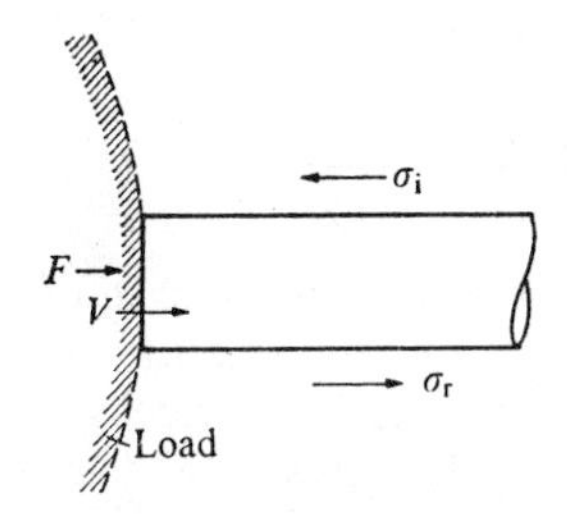

FIG. 2.6. Force F acting on the end of a rod due to the resistance of a load.

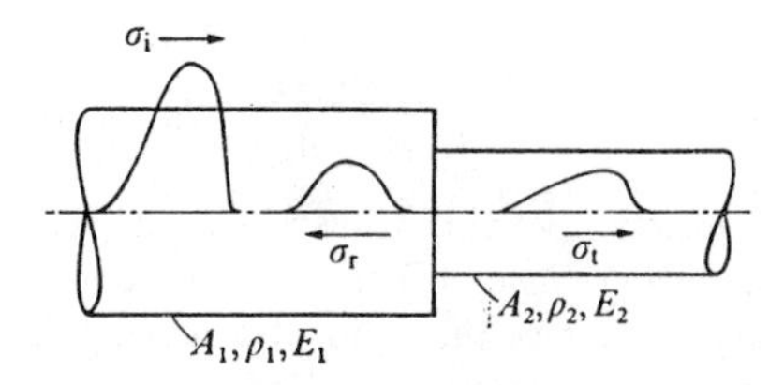

FIG. 2.7. Incident, reflected, and transmitted waves at the junction between two rods.

The velocity of the rod tip will be

$$V(t) = \frac{c_0}{E}\{-\sigma_r(0, t)+\sigma_i(0, t)\}. \tag{2.2.7}$$

These results may be used to determine the characteristics of a load by experimental means. Thus, by placing strain gauges† on the rod at some location x_0, it is a simple matter to measure the wave incident on a boundary and the wave reflected from it, so that $\sigma_i(x_0, t)$ and $\sigma_r(x_0, t)$ are known. Since the waves propagate from $x = 0$ to $x = x_0$ non-dispersively, the values of $\sigma_i(0, t)$ and $\sigma_r(0, t)$ are also known. From (2.2.6) and (2.2.7), the force and velocity at the interface are established. If the velocity results are integrated, displacement–time information is obtained. This data, combined with the force data enables the force-deformation characteristics of the load to be established.‡ While such a procedure is approximate, ignoring, for example, the inertia effects of the load, it finds application in percussive drilling studies and elsewhere.

2.2.3. *Transmission into another rod*

As in the case of the string, reflection of waves may occur at discontinuities other than a termination of the rod. In particular, we are interested here in the case of a junction between two semi-infinite rods, where there is a discontinuity in cross-section, in material properties or both. The situation is shown in Fig. 2.7. Proceeding in a manner similar to that used for the analogous case of the string (see Fig. 1.12 and eqns (1.2.23)–(1.2.26)), we have from balance of force and continuity of velocity at the junction

$$A_1(\sigma_i+\sigma_r) = A_2\sigma_t, \qquad v_i+v_r = v_t, \tag{2.2.8}$$

where the stresses and velocities are determined from the wavefields

$$u_i = f_1(x-c_1t), \qquad u_r = g_2(x+c_1t), \qquad u_t = f_2(x-c_2t). \tag{2.2.9}$$

† See Appendix C for a brief discussion of these devices.
‡ Problem 2.8 bears on this subject.

The results for the transmitted and reflected waves are then

$$\sigma_t = \frac{2A_1\rho_2c_2}{A_1\rho_1c_1+A_2\rho_2c_2}\sigma_i, \qquad \sigma_r = \frac{A_2\rho_2c_2-A_1\rho_1c_1}{A_1\rho_1c_1+A_2\rho_2c_2}\sigma_i. \tag{2.2.10}$$

Problems involving transmission across junctions are often spoken of in terms of *impedance*.† This term and concept, borrowed from electric circuit theory, expresses the ratio of a driving force to the resulting velocity at a given point of the structure. For a semi-infinite rod, it is easily shown‡ that the driving point impedance is given by

$$Z = F/V = A\sqrt{(\rho E)}. \tag{2.2.11}$$

Using this parameter, it is possible to express the results (2.2.10) as

$$\sigma_t = \frac{2(Z_2/Z_1)(A_1/A_2)}{1+(Z_2/Z_1)}\sigma_i, \qquad \sigma_r = \frac{(Z_2/Z_1)-1}{1+(Z_2/Z_1)}\sigma_i \tag{2.2.12}$$

where $Z_1 = A_1\sqrt{(\rho_1E_1)}$ and $Z_2 = A_2\sqrt{(\rho_2E_2)}$. A number of interesting reflection–transmission situations arise as Z_1, Z_2, A_1, and A_2 are varied.§

2.3. Waves and vibrations in a finite rod

Our coverage of waves and vibrations in a finite rod will be fairly brief because of much applicable work in the area of strings. Again, the distinguishing aspects of the rod versus the string will be emphasized.

2.3.1. *Waves in a finite rod—history of a stress pulse*

From the work in strings, we are aware that the image method may be extended to describe waves in a finite rod (see Fig. 1.13). Instead of an initial-value problem, we wish here to consider a rod, having free end conditions, subjected to a pressure pulse at one end. Physically, we may imagine the rod to be freely suspended by two light strings, so that it is free to swing under the action of applied loads. This configuration has no counterpart in the case of the string. Thus, while free-end boundary conditions exist, the string is incapable of translation as a whole.

Consider the applied pressure to be a step pulse described by

$$p(t) = \begin{cases} -p_0, & 0 < t < T \\ 0, & T < t, \end{cases} \tag{2.3.1}$$

† First introduced in Problem 1.11.
‡ See Problem 2.3.
§ See Problem 2.6.

where the negative sign indicates compression. The propagating stress wave will be given by

$$\sigma(x, t) = p(x - c_0 t), \tag{2.3.2}$$

before the first end reflection. Now the response of a particle at a typical point to this stress pulse can be deduced from the results presented in § 2.1. In particular, eqn (2.1.12) shows that when the step stress pulse reaches $x = x_0$, the particle will be brought from rest to the velocity

$$v(x_0, t) = -c_0 p(x_0 - c_0 t)/E$$

instantaneously and will remain at this constant velocity until the pulse passes, at which point the velocity will again be zero. The displacement will increase according to the relation $u(x_0, t) = c_0 p_0 t/E$ while the pulse passes, and will remain constant after passage.

The preceding paragraph describes the motion of a single particle under a single passage of the step pulse. The pulse will traverse the bar repeatedly, owing to end reflections, thus subjecting the particle repeatedly to similar behaviour. The response at the ends will differ slightly from this behaviour owing to the 'velocity doubling' that occurs at reflection from a free end (see Fig. 2.5). The resulting behaviour of three particles of the bar during this multiple reflection process is shown in Fig. 2.8. Also shown is the loading and suspension configuration of the bar.

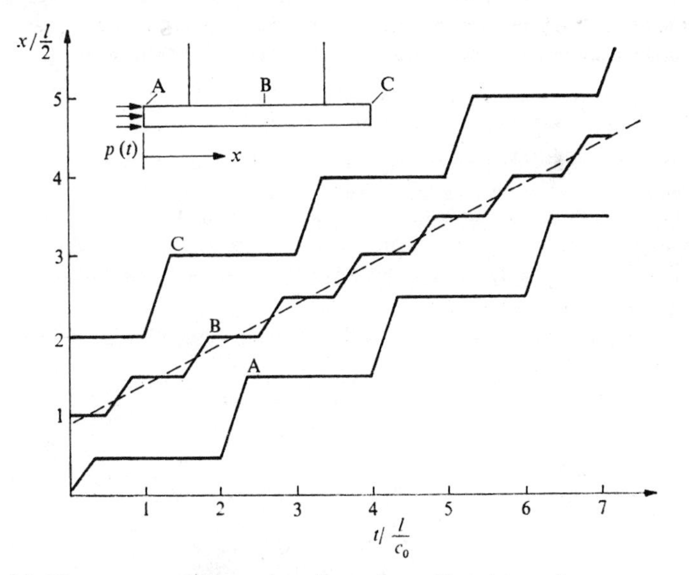

FIG. 2.8. The response of three points along a bar subjected to a step pressure pulse. (Based on Kolsky [18, p. 46].)

It is seen that point B moves to the right in a series of 'jerky' movements. The dashed line in the figure represents the average motion of B, which is located at the centre of gravity of the bar. Now, as time passes, the effects of slight damping in the material, as well as slight dispersive effects, will cause the various curves A, B, and C to be smoothed out. Specifically, curve B will approach the dashed line shown. There will then be no internal vibrations of the bar—in other words, no relative particle motions. Instead, each particle will have a uniform translational velocity given by the slope of the dashed line. This velocity will be that predicted by the simple impulse-momentum considerations of rigid-body dynamics. This example should serve to illustrate the connection between the detailed considerations of wave propagation and the gross motion considerations of rigid-body dynamics.

* 2.3.2. *Free vibrations of a finite rod*

The analysis of the free vibrations of finite rods closely follows the considerations for strings. Thus a rod of length l, constrained in some manner at $x = 0$ and $x = l$, has the motion predicted by

$$u(x, t) = \sum_{n=1}^{\infty} (A_n \sin \omega_n t + B_n \cos \omega_n t) U_n(x). \tag{2.3.3}$$

The functions $U_n(x)$ correspond to the normal modes of the rod at the natural frequencies ω_n. A variety of constraints, such as free–free, free–fixed, fixed–elastically supported, and so forth, are possible. As in the case of the string, the normal modes are orthogonal if the boundary conditions are of the form

$$aU(x_0) + bU'(x_0) = 0. \tag{2.3.4}$$

Let us consider one simple example of determining the normal modes and natural frequencies. The reasons for this are several: it is desirable to interpret the modes in terms of longitudinal rod displacements; this particular example has a slight distinguishing characteristic from anything in strings; and finally, it relates to a field of interesting applications of longitudinal resonance phenomena.

The example is the free–free rod, described by the boundary conditions

$$\frac{\partial u(0, t)}{\partial x} = \frac{\partial u(l, t)}{\partial x} = 0. \tag{2.3.5}$$

Using separation of variables in the homogeneous wave equation (for a discussion of this, see § 1.3) gives

$$U'' + \beta^2 U = 0, \qquad \beta^2 = \omega^2/c_0^2 \tag{2.3.6}$$

and

$$U = C \sin \beta x + D \cos \beta x. \tag{2.3.7}$$

Substituting this in the boundary conditions gives $C = 0$ and

$$\sin \beta l = 0, \qquad \beta l = n\pi \qquad (n = 0, 1, 2, \ldots). \tag{2.3.8}$$

The natural frequencies are thus

$$\omega_n = \frac{n\pi c_0}{l} \qquad (n = 0, 1, 2, \ldots). \tag{2.3.9}$$

The distinguishing characteristic of this result over the analogous string problem is the $n = 0$ root of the frequency equation. This is a perfectly acceptable root in the present problem, even though it predicts $U_0(x) = 0$. In fact, eqn (2.3.6) must be re-examined with $\beta = 0$, and the solution $U = U_0$, a constant, retained. This term corresponds to the rigid-body motion† of the bar, if there is any. For this particular set of boundary conditions, the general solution (2.3.7) must be supplemented by the additional U_0 term.

The normal modes for the first three natural frequencies are shown in Fig. 2.9. The longitudinal displacements along the rod are plotted on the

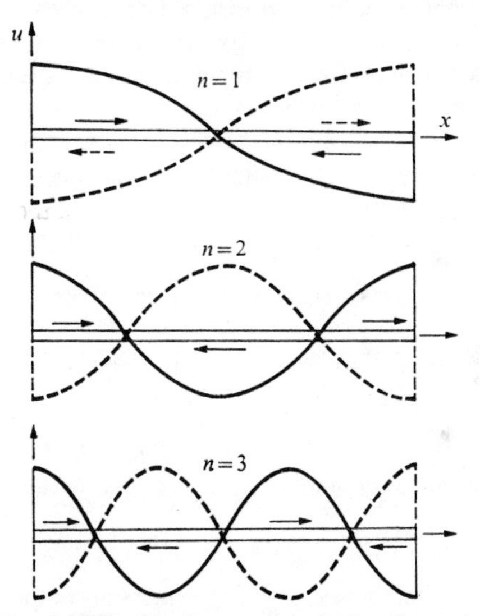

FIG. 2.9. The first three modes of a free–free rod.

vertical coordinate, while the arrows give a direct indication of particle motion. To obtain some appreciation of numbers, suppose the rod is steel and 25 cm long. For steel, $c_0 \simeq 5{\cdot}1 \times 10^3$ m s^{-1}, so (2.3.9) becomes

$$\omega_n = 20{\cdot}4 \times 10^3 \pi n \qquad (n = 1, 2, \ldots). \tag{2.3.10}$$

† See, for example, Timoshenko [38, p. 309].

The frequency in Hertz is given by $f_n = \omega_n/2\pi$, so that

$$f_n = 10{\cdot}2\times10^3 n \cong 10^4 n, \qquad (n = 1, 2, \ldots). \tag{2.3.11}$$

We thus see that the first mode ($n = 1$), called the 'half-wave' mode, has a frequency $f_1 = 10^4$ Hz. This is, of course, a rather high frequency. A bar vibrating in this range emits a very high pitched acoustical tone. The point to be made here is that longitudinal resonance phenomena generally occur at very high frequencies compared, say, to the bending vibrations (to be studied in the next chapter) of the same element.

The phenomena of longitudinal resonance of rods finds considerable practical application in the field of power ultrasonics. Electro-mechanical devices, usually called simply 'transducers' or 'resonators', that convert electrical energy to longitudinal mechanical vibrations are designed to a large extent on the basis of longitudinal rod theory. The usual procedure is to place one or several piezoelectric discs at the displacement node of a half-wavelength bar, as shown in Fig. 2.10. A piezoelectric material has the property of undergoing mechanical displacement when an electric field is applied. By applying a driving voltage at the resonant frequency of the system, the discs act as a forcing function on the system and a strong vibration is set up. The inclusion of the piezoelectric discs at the node destroys the homogeneity of the rod, of course, so that precise prediction of resonance of the system based on formulas such as (2.3.8) is not possible and more elaborate analyses are necessary.

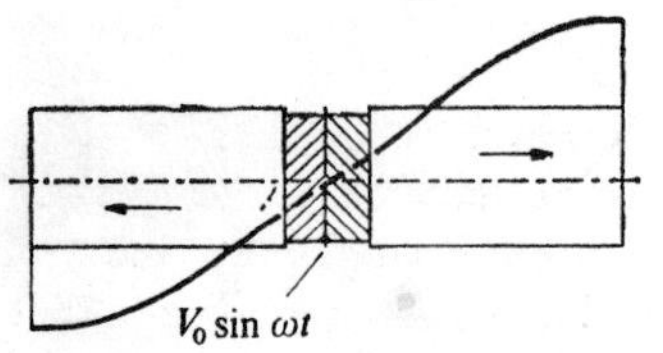
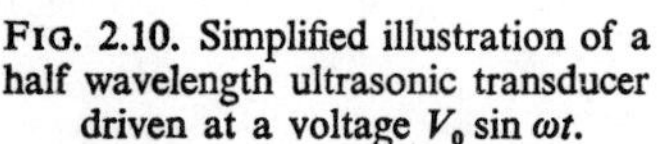

FIG. 2.10. Simplified illustration of a half wavelength ultrasonic transducer driven at a voltage $V_0 \sin \omega t$.

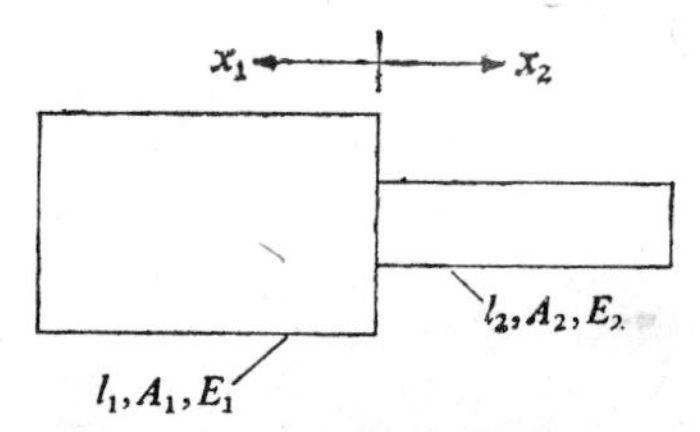

FIG. 2.11. Simple two-part composite rod.

It is possible to predict the resonant frequencies of composite rods made by connecting uniform rods together. This also has considerable application in the previously described area of power ultrasonics. To illustrate this, consider a simple two-part composite rod, as shown in Fig. 2.11. Two coordinate systems x_1 and x_2 are employed. It is presumed that the rods differ in length, cross-section, and material. Two governing equations must be used,

$$\frac{\partial^2 u_1}{\partial x_1^2} = \frac{1}{c_1^2}\frac{\partial^2 u_1}{\partial t^2}, \qquad \frac{\partial^2 u_2}{\partial x_2^2} = \frac{1}{c_2^2}\frac{\partial^2 u_2}{\partial t^2}, \tag{2.3.12}$$

where c_1, c_2 are the bar velocities of the respective materials. Seeking harmonic solutions, we write directly

$$U_1 = A \sin \beta_1 x_1 + B \cos \beta_1 x_1,$$
$$U_2 = C \sin \beta_2 x_2 + D \cos \beta_2 x_2, \tag{2.3.13}$$

where $\beta_1 = \omega/c_1$, $\beta_2 = \omega/c_2$. The boundary conditions for the problem are

$$x_1 = l_1, \qquad \mathrm{d}U_1/\mathrm{d}x_1 = 0,$$
$$x_1 = x_2 = 0, \qquad U_1 = -U_2, \qquad E_1 A_1 \frac{\mathrm{d}U_1}{\mathrm{d}x_1} = E_2 A_2 \frac{\mathrm{d}U_2}{\mathrm{d}x_2}, \tag{2.3.14}$$
$$x_2 = l_2, \qquad \mathrm{d}U_2/\mathrm{d}x_2 = 0.$$

The conditions at $x_1 = x_2 = 0$ express continuity of displacement and force at the junction. Applying these conditions to the solutions (2.3.13) gives the equations

$$A \cos \beta_1 l_1 - B \sin \beta_1 l_1 = 0,$$
$$B = -D,$$
$$E_1 A_1 A = E_2 A_2 C,$$
$$C \cos \beta_2 l_2 - D \sin \beta_2 l_2 = 0. \tag{2.3.15}$$

In order for these four homogeneous equations in A, B, C, and D to have a solution, the determinant of coefficients must vanish, giving the frequency equation

$$\frac{E_2 A_2}{E_1 A_1} \cos \beta_1 l_1 \sin \beta_2 l_2 + \cos \beta_2 l_2 \sin \beta_1 l_1 = 0. \tag{2.3.16}$$

In a general problem, this would have to be solved numerically for given values of the material and geometry parameters. A special case of interest in power ultrasonics may be easily examined, however. Suppose $E_1 = E_2$, so that the material of the two parts is the same (hence $\beta_1 = \beta_2 = \beta$) and that $l_1 = l_2$. Then (2.3.16) reduces to

$$\left(\frac{A_2}{A_1} + 1\right) \cos \beta l_1 \sin \beta l_1 = 0. \tag{2.3.17}$$

It may easily be shown that the roots corresponding to $\cos \beta l_1 = 0$ give modes of vibration antisymmetric with respect to $x_1 = x_2 = 0$, while $\sin \beta l_1 = 0$ gives symmetric modes. The first antisymmetric mode in particular is often used as a basis for transducer design. For this mode, the displacements of the two parts are given by

$$u_1 = A \sin \beta x_1, \qquad u_2 = C \sin \beta x_2, \tag{2.3.18}$$

where $C = (A_1/A_2)A$. If $A_1 > A_2$, it is seen that the displacement of u_1 is amplified relative to u_2.

2.3.3. *Forced vibrations of rods*

On the basis of the governing equation for rods, (2.1.6) where forcing effects are included, and on the basis of analogy with forced motion of strings, there would appear to be a wide class of problems associated with forced motion of rods. In a practical sense, this is not the case, however. Whereas the physical application of a transverse load at any point along the taut string is easily visualized, the means for applying longitudinal loads along a homogeneous rod are less easily imagined. While various circumstances can be contrived for such loadings, the most common means of load application is through longitudinal forcing at the ends of the rod. Thus, the concern here is with the homogeneous differential equation and inhomogeneous boundary conditions.

Consider, as a simple example of harmonic forcing, the case of a rod free at one end and subjected to the end force $F_0 \exp(i\omega t)$. Assuming a solution to the homogeneous wave equation of $U(x)\exp(i\omega t)$, we are again led to the solution form

$$U = C \sin \beta x + D \cos \beta x. \tag{2.3.19}$$

The boundary conditions are

$$EA\frac{dU}{dx}\bigg|_{x=0} = -F_0, \qquad \frac{dU}{dx}\bigg|_{x=l} = 0. \tag{2.3.20}$$

This gives

$$C = -\frac{F_0}{EA\beta}, \qquad D = -\frac{F_0}{EA\beta}\cot \beta l. \tag{2.3.21}$$

The resulting forced motion is

$$u(x, t) = -\frac{F_0}{EA\beta}(\sin \beta x + \cot \beta l \cos \beta x)e^{i\omega t}. \tag{2.3.22}$$

It is seen that the response becomes unbounded at the frequencies corresponding to $\sin \beta l = 0$, or

$$\omega_R = \frac{n\pi c_0}{l} \qquad (n = 1, 2, \ldots). \tag{2.3.23}$$

These are the resonant frequencies of the rod. We see they correspond to the natural frequencies of a free–free rod.

Velocity or displacement forcing conditions are also possible. Suppose we consider a rod, fixed at one end and subjected to the velocity forcing $V_0\exp(i\omega t)$. The boundary conditions for this case would be

$$i\omega U(0) = V_0, \qquad U(l) = 0. \tag{2.3.24}$$

It would be found that this system would be resonant at frequencies given again, coincidently, by $\sin \beta l = 0$. The natural frequency counterpart to this problem would be found to be the fixed–fixed rod. Table 2.2 shows

TABLE 2.2

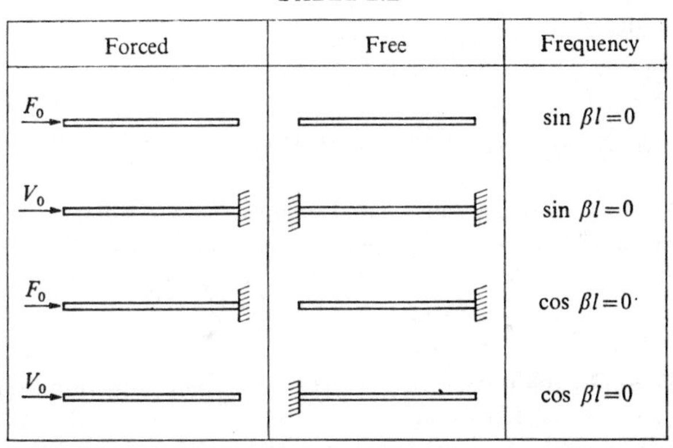

Forced	Free	Frequency
F_0		$\sin \beta l = 0$
V_0		$\sin \beta l = 0$
F_0		$\cos \beta l = 0$
V_0		$\cos \beta l = 0$

four common forced vibration situations, the resonant frequencies and the corresponding free vibration cases having the same natural frequencies. Harmonic forcing is understood.

2.3.4. *Impulse loading of a rod—two approaches*

We wish now to analyse the response of a rod to a transient load. The objective will not be to introduce new methods of analysis, but to compare presently known methods. Specifically, a wave propagation and a normal-mode method will be used and some general conclusions drawn.

The problem will be that of a rod, fixed at $x = 0$ and subjected to a transient pressure pulse at $x = l$, as shown in Fig. 2.12. The normal-mode

FIG. 2.12. Rod fixed at $x = 0$ subjected to a pressure pulse.

method of analysis will first be employed. Referring back to § 1.4, we have that

$$u(x, t) = \sum_{n=1}^{\infty} q_n(t) U_n(x), \tag{2.3.25}$$

where, for the fixed–free rod, the normal modes are given by

$$U_n(x) = \sin \beta_n x, \qquad \beta_n = \frac{n\pi}{2l} \qquad (n = 1, 3, \ldots). \tag{2.3.26}$$

The functions describing the time variation $q_n(t)$ are obtained from (see (1.4.64)),

$$\ddot{q}_n(t) + \omega_n^2 q_n(t) = \frac{2}{\rho l} \int_0^l U_n(x) q(x, t)\, dx, \tag{2.3.27}$$

where $q(x, t)$ is the longitudinal loading of the rod. It is possible to express the end loading as

$$q(x, t) = P\delta(l)f(t). \tag{2.3.28}$$

Then (2.3.27) becomes

$$\ddot{q}_n(t)+\omega_n^2 q_n(t) = \frac{2P}{\rho l}(-1)^{(n-1)/2}f(t) \qquad (n = 1, 3,\ldots). \tag{2.3.29}$$

We solve this ordinary differential equation by Laplace transforms. Noting that the initial conditions are zero, the transformed solution is

$$\bar{q}_n(s) = \frac{k\bar{f}(s)}{s^2+\omega_n^2}, \qquad k = \frac{2P(-1)^{(n-1)/2}}{\rho l}. \tag{2.3.30}$$

We now specialize the time variation $f(t)$ to the delta function $\delta(t)$. Then, since $\mathscr{L}\{\delta(t)\} = 1$, we have that

$$\bar{q}_n(s) = \frac{k}{s^2+\omega_n^2}. \tag{2.3.31}$$

From Laplace transform tables, we have that

$$\mathscr{L}^{-1}\left(\frac{1}{s^2+\omega_n^2}\right) = \frac{1}{\omega_n}\sin\omega_n t. \tag{2.3.32}$$

We thus have for our final solution

$$u(x, t) = \frac{2P}{\rho c_0 l}\sum_{1,3,\ldots}^{\infty}\frac{(-1)^{(n-1)/2}}{\beta_n}\sin\beta_n x\sin\omega_n t. \tag{2.3.33}$$

Without pausing to discuss the normal-mode results just obtained, we proceed to reconsider the same problem, applying the Laplace transform method directly. We first summarize the governing equation and conditions that set forth the problem. We have

$$\frac{\partial^2 u}{\partial x^2} = \frac{1}{c_0^2}\frac{\partial^2 u}{\partial t^2},$$

$$u(0, t) = 0, \qquad E\frac{\partial u(l, t)}{\partial x} = p(t),$$

$$u(x, 0) = \dot{u}(x, 0) = 0. \tag{2.3.34}$$

Applying the Laplace transform to the governing equation gives

$$\frac{d^2\bar{u}}{dx^2}-\frac{s^2}{c_0^2}\bar{u} = 0. \tag{2.3.35}$$

This has the solution

$$\bar{u}(x, s) = A\cosh\left(\frac{s}{c_0}x\right)+B\sinh\left(\frac{s}{c_0}x\right). \tag{2.3.36}$$

The transformed boundary conditions are

$$\bar{u}(0, s) = 0, \qquad E\,\mathrm{d}\bar{u}(l, s)/\mathrm{d}x = \bar{p}(s). \tag{2.3.37}$$

Applying these conditions to the solution (2.3.36) gives

$$A = 0, \qquad B = \frac{1}{E}\bar{p}(s)\bigg/\left(\frac{s}{c_0}\cosh\frac{s}{c_0}l\right), \tag{2.3.38}$$

resulting in the transformed solution

$$\bar{u}(x, s) = \frac{\bar{p}(s)}{\rho c_0}\left\{\sinh\left(\frac{s}{c_0}x\right)\bigg/ s\cosh\left(\frac{s}{c_0}l\right)\right\}. \tag{2.3.39}$$

To invert this, we apply the convolution integral, giving

$$u(x, t) = \frac{1}{\rho c_0}\int_0^t p(\tau)g(x, t-\tau)\,\mathrm{d}\tau, \tag{2.3.40}$$

where

$$g(x, t) = \mathscr{L}^{-1}\left\{\sinh\left(\frac{s}{c_0}x\right)\bigg/ s\cosh\left(\frac{s}{c_0}l\right)\right\}. \tag{2.3.41}$$

The problem, then, is the inversion of this function. We express $\bar{g}(x, s)$ in exponential form as

$$\begin{aligned}\bar{g}(x, s) &= \sinh\left(\frac{s}{c_0}x\right)\bigg/ s\cosh\left(\frac{s}{c_0}l\right)\\ &= \left[\exp\left\{\frac{s}{c_0}(x-l)\right\} - \exp\left\{\frac{s}{c_0}(x+l)\right\}\right]\bigg/ s\left\{1+\exp\left(-\frac{2ls}{c_0}\right)\right\}\end{aligned} \tag{2.3.42}$$

We write the expression

$$1\bigg/\left\{1+\exp\left(-\frac{2ls}{c_0}\right)\right\} = \sum_{n=0}^{\infty}(-1)^n\exp\left(-\frac{2lns}{c_0}\right). \tag{2.3.43}$$

Hence $\bar{g}(x, s)$ may be put in the form

$$\bar{g}(x, s) = \sum_{n=0}^{\infty}(-1)^n\left(\frac{1}{s}\mathrm{e}^{-\alpha s} - \frac{1}{s}\mathrm{e}^{-\beta s}\right), \tag{2.3.44}$$

where

$$\alpha = \{(2n+1)l - x\}/c_0, \qquad \beta = \{(2n+1)l + x\}/c_0. \tag{2.3.45}$$

From Laplace transform tables, we have that

$$\mathscr{L}^{-1}\left\{\frac{1}{s}\mathrm{e}^{-as}\right\} = H\langle t-a\rangle. \tag{2.3.46}$$

Then a term-by-term inversion of (2.3.44) gives

$$g(x, t) = \left\{H\left\langle t-\frac{(l-x)}{c_0}\right\rangle - H\left\langle t-\frac{(l+x)}{c_0}\right\rangle\right\} - \\ -\left\{H\left\langle t-\frac{(3l-x)}{c_0}\right\rangle - H\left\langle t-\frac{(3l+x)}{c_0}\right\rangle\right\} + \\ +\left\{H\left\langle t-\frac{(5l-x)}{c_0}\right\rangle - H\left\langle t-\frac{(5l+x)}{c_0}\right\rangle\right\} - \ldots \tag{2.3.47}$$

This corresponds to a series of rectangular pulses.

Referring back to the convolution form (2.3.40) we see that $p(t)$ must still be specified. For a general loading, the integral would have to be evaluated numerically. However, for the special case of $p(t) = P\delta(t)$, we obtain from the convolution integral directly

$$u(x, t) = \frac{P}{\rho c_0} g(x, t), \tag{2.3.48}$$

where $g(x, t)$ is given by (2.3.47).

We now have two results to the same problem, being given by (2.3.33) and (2.3.48). The first is a normal-mode solution and the second, for want of a better term, we designate a wave-propagation solution. The response at a given point $x = x_0$ of the rod as predicted by the wave solution is shown in Fig. 2.13. The response clearly is one of a wave being reflected back and

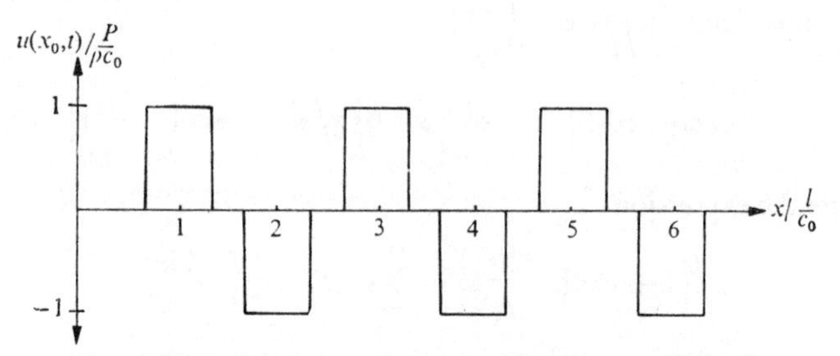

FIG. 2.13. Reflected waves in a rod subjected to a pulse load.

forth within the bar.

We note that each solution is in the form of an infinite series. Yet the two series are fundamentally different in the way in which they predict the response of the rod to this loading. Suppose we wish to predict the short-time response of the system to the load. It is seen that the normal-mode solution represents a rather slowly converging series. Thus, many terms would be required to obtain an accurate time prediction. The wave-propagation solution, on the other hand, has only to be summed to a term or so to reach the desired time at which response is sought. It provides a very accurate answer with minimal labour. On the other hand, if the long-time response is desired, many terms of the wave solution would be required. In the normal-mode case, physical intuition would lead us to discard all but the fundamental-mode term, since over long times the higher modes would damp out. Thus the normal-mode approach would prove more attractive. This example thus brings out the differing characteristics of the two approaches. For short time, the wave approach—for long time, the modal approach.

2.4. Longitudinal impact

Stress waves are produced by the mechanical impact of solids. Several situations of longitudinal impact in rods will be considered in this section and will represent first efforts at analysis in this area. Up to this stage, all of the considerations have been for prescribed loads, with no questions as to how these loads might be generated. Certainly mechanical impact represents one of the ways. In impact, however, one prescribes the material properties and geometries of the impacting solids as well as the velocity of impact. The impact force is not prescribed but is, in fact, the result to be established, as well as the ensuing wave propagation.

2.4.1. *Longitudinal collinear impact of two rods*

One of the simplest impact situations is the longitudinal impact of two flat-ended rods. It is, of course, possible to analyse the rod geometry completely, so one would suspect that such an impact situation would prove tractable. This is a technically important problem, since it provides the basis for many experimental studies of wave propagation. In the general impact cases, the bars would be of dissimilar lengths, areas, and materials, and would be travelling at two different velocities. Such a situation, at the instant of contact, is shown in Fig. 2.14(a). The analysis of such a general problem at

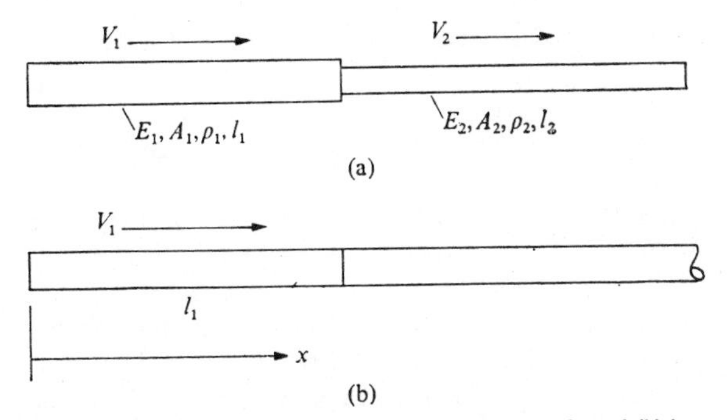

FIG. 2.14. (a) General collinear impact situation between two rods and (b) impact against a stationary semi-infinite rod.

the outset, while possible, would prove tedious. A more restricted problem will be considered first, as shown in Fig. 2.14(b).

We let the two rods be of identical materials and cross-sections, so that $E_1 = E_2$, $\rho_1 = \rho_2$, $A_1 = A_2$ in Fig. 2.14(b). Further, let the second rod be stationary ($V_2 = 0$) and of semi-infinite length ($l_2 \to \infty$). When the rods collide at $t = 0$, stress waves will propagate to the right and left from the point of contact. These wave functions may be found from the applicable initial conditions and boundary conditions for the process. From our previous work we know these functions have the form

$$u_1(x, t) = f_1(x-c_1t)+g_1(x+c_1t),$$
$$u_2(x, t) = f_2(x-c_1t)+g_2(x+c_1t). \tag{2.4.1}$$

Initially, we have that both rods are stress free, that the velocity of the long rod is zero, and that the velocity of the moving rod is V_1. Recalling that the general expressions for velocity and stress are

$$v(x, t) = -c_1f'(x-c_1t)+c_1g'(x+c_1t),$$
$$\sigma(x, t) = E\{f'(x-c_1t)+g'(x+c_1t)\}, \tag{2.4.2}$$

the initial condition statements take the form

$$-c_1f_1'(x)+c_1g_1'(x) = V_1,$$
$$E_1A_1f_1'(x)+E_1A_1g_1'(x) = 0,$$
$$-c_1f_2'(x)+c_1g_2'(x) = 0,$$
$$E_1A_1f_2'(x)+E_1A_1g_2'(x) = 0. \tag{2.4.3}$$

These simultaneous equations may be solved for f_1', g_1', f_2', g_2', giving

$$f_2' = g_2' = 0,$$
$$f_1' = -g_1' = -\frac{V_1}{2c_1} \quad (0 < x < l_1). \tag{2.4.4}$$

Thus at the instant of impact, the constant velocity–no stress situation existing in the left-hand rod is represented by two rectangular wave shapes, as shown in Fig. 2.15 for $t = 0$.

The two waves f_1', g_1' have been 'frozen' at $t = 0$, but it should be recalled that f_1' is a rightward-propagating wave and g_1' a leftward-propagating one. The wave action occurring for $t > 0$ will be governed by the interaction of these propagating waves with the various boundaries and discontinuities. In the case of f_1', this will be the junction of the two rods. However, since the rods are identical in this problem, there is in effect no discontinuity and f_1 moves across $x = l_1$. The wave g_2', on the other hand, reflects from the free end on the left in the manner established for such an end condition by the

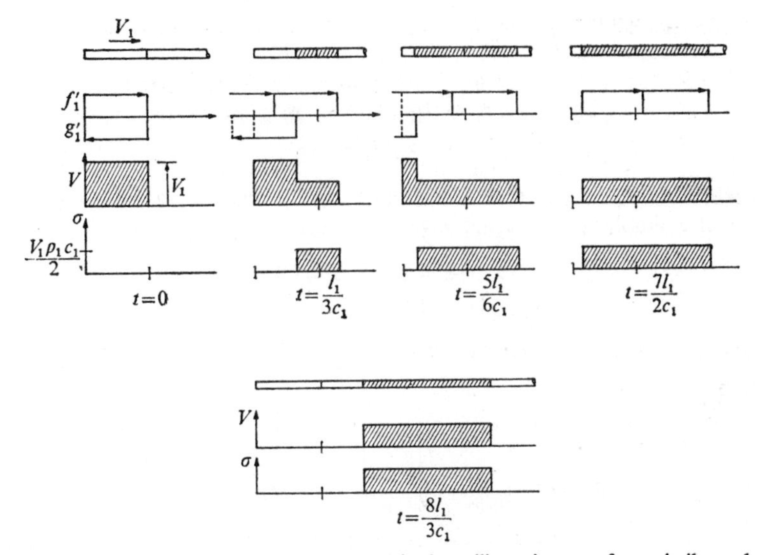

FIG. 2.15. Several stages of wave propagation in the collinear impact of two similar rods.

method of images (see Fig. 2.5). Several pertinent steps in the over-all wave process are shown in Fig. 2.15 for differing values of time. Three results are shown for each time: (1) the location of the wave functions f_1' and g_1'; (2) the resulting velocity; (3) the resulting stress. As may be seen, a rectangular pulse of magnitude $\sigma = V_1\rho_1c_1/2$ and of duration $T = 2l_1/c_1$ is propagated into the semi-infinite rod. The length of the pulse is $2l_1$, twice the length of the impact bar. The impact bar is brought to rest for this case of identical cross-sections and materials.

As was mentioned, this configuration finds application in many studies of wave propagation.† It is a very simple configuration for generating a stress pulse of predictable shape and duration. In practice, imperfections in the physical system and inadequacies of the governing theory will cause deviations of the generated pulse from the ideal shape shown. It may be observed in experiments that the corners of the pulse are rather rounded and that the rise and fall of the pulse is not nearly as sharp as shown in the figure. Or there may be observed high-frequency oscillations on the rise and fall of the pulse. The first is generally a result of rounded contact surfaces of the impact bars and finite response time of the measuring system. The second is a consequence of lateral inertia and other higher-order effects neglected in the original development of the theory.

A somewhat more elaborate example of collinear impact will now be discussed. Referring to Fig. 2.14(a), we let both rods be of the same material

† See discussion in Appendix C.

(steel). Then let

$$
\begin{aligned}
l_1 &= 0{\cdot}457 \text{ m} = 18 \text{ in}, \qquad l_2 = 2l_1, \\
d_1 &= 1{\cdot}5\, d_2, \qquad d_2 = 2{\cdot}54 \text{ cm} = 1{\cdot}0 \text{ in}, \\
c_1 = c_2 &= 5{\cdot}08 \times 10^3 \text{ m s}^{-1} = 20 \times 10^4 \text{ in s}^{-1}, \\
V_1 = V &= 6{\cdot}10 \text{ m s}^{-1} = 240 \text{ in s}^{-1}, \qquad V_2 = 0,
\end{aligned} \tag{2.4.5}
$$

in which d_1, d_2 refer to the rod diameters. Following the procedures of the previous, simpler problem, we have at $t = 0$,

$$
\begin{aligned}
f_1'(x) &= -g_1'(x) = -V/2c_1, \\
f_2'(x) &= g_2'(x) = 0.
\end{aligned} \tag{2.4.6}
$$

This initial wave structure at the instant of collision is shown in Fig. 2.16 for

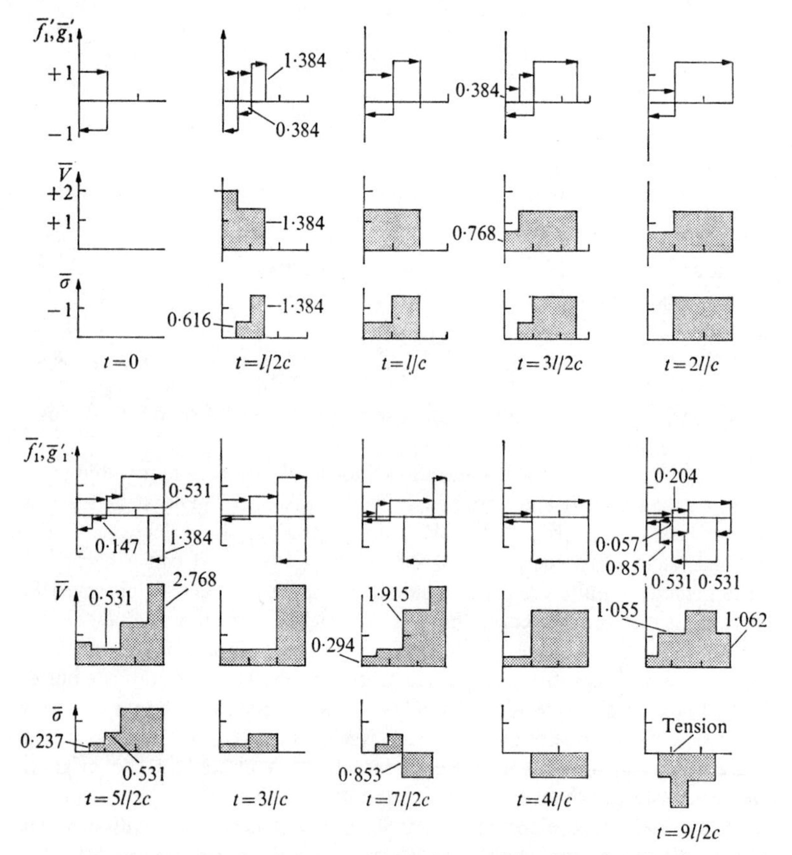

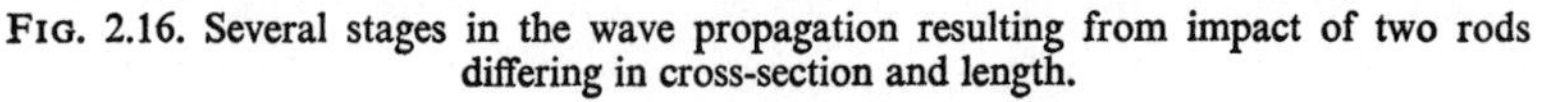

FIG. 2.16. Several stages in the wave propagation resulting from impact of two rods differing in cross-section and length.

$t = 0$. At $t > 0$, the rightward propagating component f_1' encounters the cross-section discontuinuity. The transmitted component f_2' and reflected component g_2' are given by (see § 2.2)

$$f_2' = \frac{2c_1A_1E_1}{A_1E_1c_2+A_2E_2c_1}f_1', \qquad g_2' = \frac{c_1A_2E_2-c_2A_1E_1}{A_1E_1c_2+A_2E_2c_1}f_1'. \tag{2.4.7}$$

For the given values of the physical constants, these become,

$$f_2' = 1{\cdot}384f_1', \qquad g_2' = -0{\cdot}384f_1'. \tag{2.4.8}$$

The wave structure after this first reflection and transmission is shown at $t = l_1/2c_1$. Note that the ordinates of the wavefunction, velocity, and stress diagrams have been non-dimensionalized as follows:

$$\bar{f}', \bar{g}' = \frac{2c_1}{V}(f', g'),$$

$$\bar{V} = \frac{2}{V}(f', g'), \tag{2.4.9}$$

$$\bar{\sigma} = \frac{2}{\rho_1 c_1 V}(f', g');$$

and that velocity continuity is maintained across the interface. The lack of stress continuity is a result of the area change. There is force continuity, however. A continuation of the wave process is shown in the $t = l_1/c_1$, $3l_1/2c_1$ figures. Nothing of particular significance occurs here other than the reflection of the leftward-propagating waves from the left free end.

At $t = 2l_1/c_1$, the stress waves generated by the impact have reflected from the left free end and returned to the interface. At this instant, the two bars have uniform but unequal velocities and, furthermore, the bar on the right has a greater velocity. Separation would occur except that the rightward-propagating wave in the right-hand bar (amplitude of 1·384) leaves the junction, bringing the left end of the bar momentarily to rest. The wave from the left bar immediately passes into the right bar, continuing the loading.

The situation is shown later at $t = 5l_1/2c_1$. The previously used ratios for transmitted and reflected waves still hold, only now new transmitted and reflected wave functions f_3', g_3' are computed, where

$$f_3' = 1{\cdot}384g_2', \qquad g_3' = -0{\cdot}384g_2', \tag{2.4.10}$$

and where $g_3' = 0{\cdot}384f_1'$. Thus

$$f_3' = 0{\cdot}531f_1', \qquad g_3' = 0{\cdot}148f_1'. \tag{2.4.11}$$

The continuation of the propagation process at $t = 3l_1/c_1$, $7l_1/2c_1$ is shown. Again, the only significant aspects are the 0·384 and 0·147 amplitude waves reflecting from the left free end.

At $t = 4l_1/c_1$, the critical stage of the process is reached. Again, uniform velocities momentarily prevail in the two rods, with the rod on the right having the greater velocity. However, in contrast to the previous situation at $t = 2l_1/c_1$, the rod on the right is in tension. This tension cannot transmit across the interface to continue the wave process. It is easily shown that transmission would occur into the left bar if the interface were glued by considering the wave process at $t = 9l_1/2c_1$. The wave structure now becomes quite complicated since two waves, a rightward- (amplitude = 0·147) and a leftward- (amplitude = 1.384) propagating wave encounter the boundary simultaneously. The resulting structure is shown and it is seen that the interface is in tension.

Returning to $t = 4l_1/c_1$, it is of interest to note that the subsequent separation dynamics of the two bars differ. The left bar is in a uniform velocity state—and stays in that state since the leftward- and rightward-propagating waves have the same amplitude. The right bar, on the other hand, is only instantaneously in uniform motion. After separation, it continues to vibrate owing to the internal waves of 0·531 and 1·384 amplitudes.

The point of this lengthy and involved description of this more complicated impact is to gain some appreciation of the details of the wave propagation. In even slightly more complicated impacts, this procedure would not be feasible. An approach that eases the matter of following the multiple reflections within the system is to plot the characteristic plane representation of the wavefronts (see § 1.1 and Fig. 1.6 for first presentation of this scheme). This approach does not enable wave amplitudes to be established, however. The characteristics plane diagram for the impact situation just considered is shown in Fig. 2.17. If a horizontal line is drawn at any given time t, the intersection of that line with the characteristics establishes the positions of the wavefronts. Amplitude information must be determined using the procedures given in the previous example.

In practical analysis situations, the digital computer is employed, so that problems involving several cross-section changes and an extensive number of reflections are feasible to solve. One of the areas where this finds wide application is in percussive drilling, where impact of hammers of complicated geometries occur against rods having changes in cross-section.

2.4.2. *Rigid-mass impact against a rod*

The geometry of longitudinal rod impact is rather special, and we seek to widen our considerations. In some impact situations, the striking object may be approximated as a rigid mass, thus neglecting deformation and wave propagation in that part of the system. This approximation becomes more accurate when the impact mass is much greater than the local mass of the struck rod and when the material moduli of the impact mass is greater than the struck rod. This case of mechanical impact was first considered by St. Venant [36].

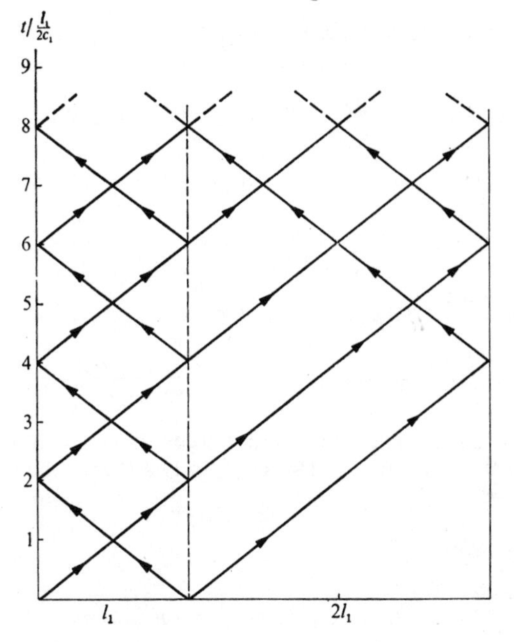

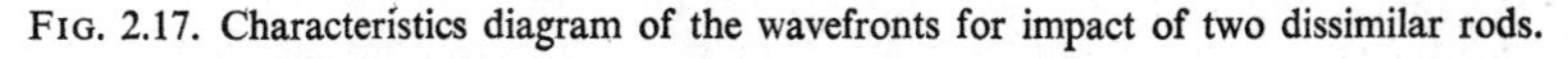
FIG. 2.17. Characteristics diagram of the wavefronts for impact of two dissimilar rods.

The case of a rigid mass impacting a semi-infinite rod is shown in Fig. 2.18. As a solution to the wave equation, we directly take

$$u(x, t) = f(c_0t-x), \tag{2.4.12}$$

corresponding to waves propagating away from the impact point. The main problem is the specification of boundary conditions. Impulse–momentum considerations applied to the mass are used for this purpose. Writing impulse-momentum, we have

$$mV_0-\int_0^t F(\tau)\,\mathrm{d}\tau = mV_f(t), \tag{2.4.13}$$

where $F(t)$ is the compressive force acting at the interface between mass and rod and t is the time after impact. In terms of the wave solution for the rod (2.4.12), we have the velocity and force at the interface given by

$$V_f(t) = c_0f'(c_0t), \qquad F(t) = -EAf'(c_0t). \tag{2.4.14}$$

In substituting (2.4.14) into (2.4.13) we note that the velocity of the mass and rod tip are of the same sign, while the forces are of opposite sign. We obtain

$$-EA\int_0^t f'(c_0\tau)\,\mathrm{d}\tau = m\{c_0f'(c_0t)-V_0\}. \tag{2.4.15}$$

The left-hand side may be integrated as follows

$$\int_0^t f'(c_0\tau)\,d\tau = \frac{1}{c_0}\int_0^{c_0 t} f'(\xi)\,d\xi = \frac{1}{c_0}\{f(c_0 t) - f(0)\}. \tag{2.4.16}$$

Then (2.4.15) gives, upon substitution of (2.4.16) and rearranging terms

$$f'(c_0 t) + \frac{EA}{mc_0^2} f(c_0 t) = \frac{EA}{mc_0^2} f(0) + \frac{V_0}{c_0}. \tag{2.4.17}$$

The first-order differential equation has the complete solution

$$f(c_0 t) = D\exp\left(-\frac{EA}{mc_0}t\right) + f(0) + \frac{mc_0}{EA}V_0. \tag{2.4.18}$$

The constant D is determined from the initial condition that the displacement is initially zero so that $f(0) = 0$. Using this in (2.4.18) we obtain

$$D = -mc_0 V_0/EA,$$

so that $u(0, t)$ is given by

$$u(c_0 t) = \frac{mc_0 V_0}{EA}\left\{1 - \exp\left(-\frac{EA}{mc_0}t\right)\right\}. \tag{2.4.19}$$

The general displacement field in the rod is given by

$$u(x, t) = \frac{mc_0 V_0}{EA} H\langle c_0 t - x\rangle [1 - \exp\{-(c_0 t - x)EA/mc_0^2\}]. \tag{2.4.20}$$

The velocity, strain, and force in the rod are given respectively by

$$v(x, t) = V_0 H\langle c_0 t - x\rangle \exp\{-(c_0 t - x)EA/mc_0^2\},$$

$$\varepsilon(x, t) = -\frac{V}{c_0} H\langle c_0 t - x\rangle \exp\{-(c_0 t - x)EA/mc_0^2\},$$

$$F(x, t) = -\frac{EAV_0}{c_0} H\langle c_0 t - x\rangle \exp\{-(c_0 t - x)EA/mc_0^2\}. \tag{2.4.21}$$

The normalized displacement, velocity, and force, evaluated at $x = 0$, are plotted in Fig. 2.19, where

$$\bar{u} = EAu(0, t)/mc_0 V_0, \qquad \bar{v} = v(0, t)/V_0,$$
$$\bar{F} = c_0 F(0, t)/EAV_0, \qquad \bar{t} = EAt/mc_0. \tag{2.4.22}$$

In interpreting the results, we note that the time constant of the response is given by $T = mc_0/EA$. We recall from § 2.2 that the impedance of a rod is

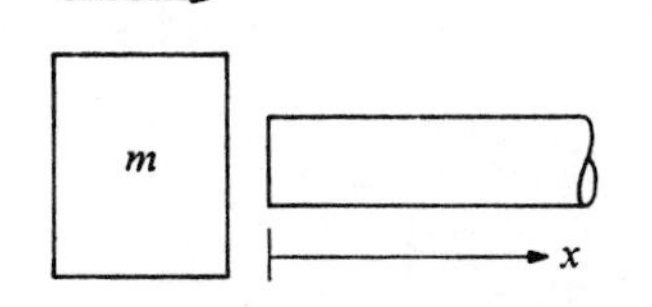

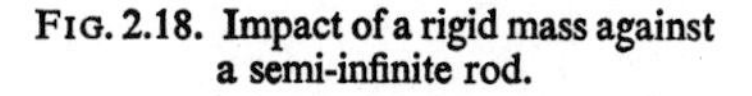

FIG. 2.18. Impact of a rigid mass against a semi-infinite rod.

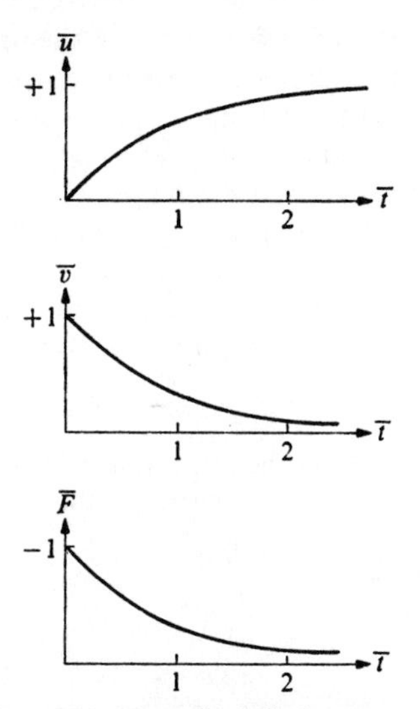

FIG. 2.19. Normalized displacement, velocity, and force at the end of a rod impacted by a rigid mass.

given by $Z = \rho A c_0$. Thus, the time constant may be expressed as $T = m/Z$. We see that a smaller and smaller time constant is produced by a smaller mass and/or a greater rod impedance. Under these conditions, the force pulse begins to resemble an impulsive load for the rod.

It is of interest to compare the case of rigid-mass impact to that of one elastic rod impacting another. On the basis of previous work on impacting elastic rods, it is possible to show that the initial force generated by such an impact is given by

$$F_E(0, 0) = -\frac{Z_1}{1+Z_1/Z_2} V_0, \tag{2.4.23}$$

whereas for the rigid-mass impact, the initial force is (see (2.4.21)),

$$F_R(0, 0) = -Z_1 V_0. \tag{2.4.24}$$

The ratio of these two is then

$$\frac{F_E(0, 0)}{F_R(0, 0)} = \frac{1}{1+Z_1/Z_2}. \tag{2.4.25}$$

Thus, the rigid-mass situation always predicts a greater impact force, a hardly surprising result.

The case of rigid-mass impact against a finite rod, fixed at one end and struck at the free end, is a situation considered by St. Venant and others.† A motivation for interest in this problem is to assess the inertia effects of the rod on the dynamic response of the system. Thus the impact of a mass on a rod could be considered as a spring–mass system, the rod being the spring of the system and having a spring constant of $K = AE/l$. By then considering the wave propagation, the results of the two analyses could be compared and inertia effects assessed.

The cited references present many detailed results; only one or two major ones will be reviewed here. Results are presented for various mass ratios M, where

$$M = m_1/m_2 = \rho_1 A_1 l_1/m_2. \tag{2.4.26}$$

Thus m_2 is the impacting mass and m_1 is the total mass of the rod. The displacements occurring at the impacted end of the rod are shown in Fig. 2.20

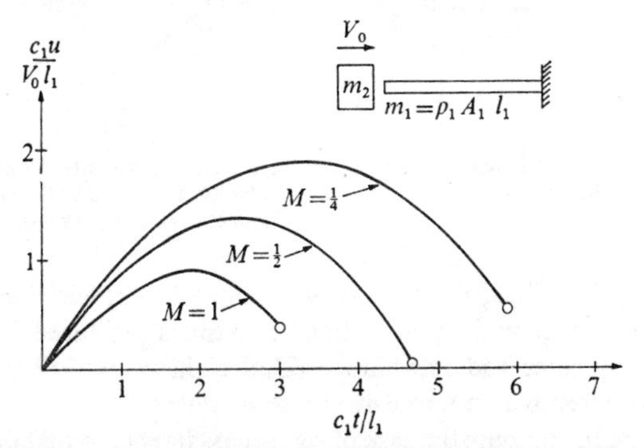

FIG. 2.20. Displacement at the end of a rod of length l struck by a rigid mass. (Based on Goldsmith [14, Fig. 26].)

for three mass ratios. The circles at the end of the curves indicate the termination of contact. To compare these results with those of a simple spring–mass model, we note that the frequency, spring constant, and period for such a system would be

$$\omega_n = \sqrt{\left(\frac{K_1}{m_2}\right)}, \qquad K_1 = \frac{A_1 E_1}{l_1}, \qquad T = 2\pi \sqrt{\left(\frac{m_2 l_1}{A_1 E_1}\right)}. \tag{2.4.27}$$

Taking one half the period, since this would be the contact time, and non-dimensionalizing it by c_1/l_1, we have for the spring–mass system

$$\bar{T}_c = \frac{c_1 T}{2 l_1} = \frac{\pi}{\sqrt{M}}. \tag{2.4.28}$$

† This problem is completely reviewed by Goldsmith [14, pp. 46–55]. Also see Timoshenko [37, pp. 498–504]. Todhunter and Pearson [39, Vol. II, Part 2, pp. 276–81], review St. Venant's work.

Calling $\tilde{T}_c$ the period of contact as established from Fig. 2.20, we have the results for the three mass ratios shown in Table 2.3.

TABLE 2.3
Contact time

M	$\tilde{T}_c$	$\bar{T}_c$
1	3·07	3·14
$\frac{1}{2}$	4·71	4·45
$\frac{1}{4}$	5·90	6·28

It is seen that periods of vibration of the two systems do not differ greatly. However, as shown in the previously cited references, more serious discrepancies occur in other parameters, such as the maximum strain.

2.4.3. *Impact of an elastic sphere against a rod*

The third impact situation that will be considered is that of a sphere in longitudinal collinear impact with a rod. In contrast to the case analysed in the preceding section, elastic deformation of the impacting solid will be accounted for using Hertz contact theory.† However, wave-propagation effects within the ball will be neglected. The approach to the problem thus combines classical longitudinal rod theory with elements of quasi-static elasticity analysis. This problem was evidently first solved by Eubanks, Muster, and Volterra [10] with experimental results later reported by Barton, Volterra, and Citron [3] and Ripperger and Abramson [35].

Consider the impact situation shown in Fig. 2.21. The displacement u_2

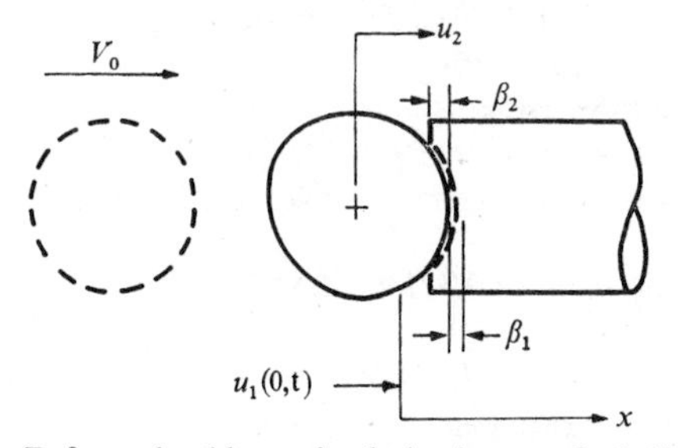

FIG. 2.21. Deformation kinematics during impact of a ball on a rod.

refers to the centre of the ball, while u_1 is the usual rod displacement. The parameters β_1 and β_2 measure the local deformation of the tip and ball respectively. As in the previous analysis of rigid-mass impact, we represent

† See, for example, Timoshenko [37, pp. 409–20].

the motion of the rod by

$$u_1(x, t) = f(c_0 t - x). \tag{2.4.29}$$

We again apply impulse-momentum considerations, giving

$$m_2 V_0 - \int_0^t F(\tau)\, d\tau = m_2 \dot{u}_2, \tag{2.4.30}$$

where the contact force on the rod is given by

$$F(t) = -EAf'(c_0 t). \tag{2.4.31}$$

The deformation–displacement condition at the contact surface now enters. The total motion of the rod contact surface consists of the gross displacement $u_1(0, t)$ plus an additional local deformation β_1. The motion of the ball contact surface is given by the gross forward motion of its centre $u_2(t)$ minus a local compression β_2. In order for the two surfaces to remain in contact, we must have

$$u_1(0, t) + \beta_1 = u_2(t) - \beta_2 \tag{2.4.32}$$

or

$$u_2(t) - u_1(0, t) = \beta_2 + \beta_1 = \alpha, \tag{2.4.33}$$

where α is defined as the 'approach,' as in Hertz contact theory.†

The governing equation for the problem is obtained by twice differentiating (2.4.33). The first differentiation gives

$$\dot{\alpha} = \dot{u}_2(t) - \dot{u}_1(0, t). \tag{2.4.34}$$

Substituting the expression for $\dot{u}_2(t)$, given by (2.4.30) and the result for $\dot{u}_1(0, t)$ obtained from (2.4.29) in (2.4.34) gives

$$\dot{\alpha} = V_0 - \frac{1}{m_2} \int_0^t F(\tau)\, d\tau - c_0 f'(c_0 t). \tag{2.4.35}$$

A second differentiation gives

$$\ddot{\alpha} = -\frac{F(t)}{m_2} - c_0^2 f''(c_0 t). \tag{2.4.36}$$

From the force expression (2.4.32) we have that

$$\dot{F}(t) = -EAc_0 f''(c_0 t). \tag{2.4.37}$$

Then (2.4.36) may be written as

$$\ddot{\alpha} - \frac{c_0}{EA} \dot{F}(t) + \frac{1}{m_2} F(t) = 0. \tag{2.4.38}$$

The relationship between the contact force and the local deformation must now be specified. This is determined from classical Hertz contact theory and

† Timoshenko [37, p. 410].

is given by

$$F = -K\alpha^{\frac{3}{2}}, \tag{2.4.39}$$

where K is a constant dependent on the elastic and geometric properties of the contact surfaces. For the case of a spherical ball in contact with a flat surface, the formula for K is

$$K = \frac{4}{3\pi}\frac{R^{\frac{1}{2}}}{k_1+k_2}, \tag{2.4.40}$$

where R is the radius of the ball and

$$k_1 = (1-\nu_1^2)/\pi E_1, \qquad k_2 = (1-\nu_2^2)/\pi E_2. \tag{2.4.41}$$

ν_1, E_1 and ν_2, E_2 are the Poisson's ratio and Young's modulus of rod and ball, respectively. Substituting in (2.4.38) gives

$$\frac{\mathrm{d}^2\alpha}{\mathrm{d}t^2}+\frac{Kc_0}{AE}\frac{\mathrm{d}\alpha^{\frac{3}{2}}}{\mathrm{d}t}+\frac{K}{m_2}\alpha^{\frac{3}{2}} = 0 \tag{2.4.42}$$

as the non-linear, ordinary differential equation governing impact against a rod. In addition, the initial conditions must be specified. In terms of the approach, we have that

$$\begin{aligned} \alpha(0) &= u_2(0)-u_1(0, 0) = 0, \\ \dot{\alpha}(0) &= \dot{u}_2(0)-\dot{u}_1(0, 0) = V_0. \end{aligned} \tag{2.4.43}$$

Given the initial conditions, (2.4.42) must be solved numerically for specific values of the material and geometry parameters. The force generated is determined from (2.4.39), with the propagated stress pulse being given by

$$\alpha_1(x, t) = \frac{1}{A}F(c_0t-x). \tag{2.4.44}$$

If desired, the rebound velocity of the ball can be computed from (2.4.30). This latter calculation enables a coefficient of restitution for the impact to be established by the ratio $\dot{u}_2/V_0$.

Some numerical results are presented for the following parameters:

$$E = 20{\cdot}7 \times 10^{10}\ \mathrm{N\,m^{-2}} = 30 \times 10^6\ \mathrm{lb\,in^{-2}};$$

$$c_0 = 5{\cdot}08 \times 10^3\ \mathrm{m\,s^{-1}} = 20 \times 10^4\ \mathrm{in\,s^{-1}}; \qquad R = 0{\cdot}64\ \mathrm{cm} = 0{\cdot}25\ \mathrm{in}$$

$$m_2 = 0{\cdot}83 \times 10^{-3}\ \mathrm{kg} = 0{\cdot}57 \times 10^{-3}\ \mathrm{slug}; \qquad A = 1{\cdot}26\ \mathrm{cm^2} = 0{\cdot}196\ \mathrm{in^2}$$

$$K = 1{\cdot}20 \times 10^{10}\ \mathrm{N\,m^{-\frac{3}{2}}} = 345\ \mathrm{lb\,ml^{-\frac{3}{2}}} \tag{2.4.45}$$

The values for E, ν, c_0, m_2, and K all correspond to values for steel. The resulting stress pulses predicted for several impact velocities are shown in Fig. 2.22. It is seen that the duration of the impact is weakly dependent on the impact velocity. The relation between rebound velocity and impact

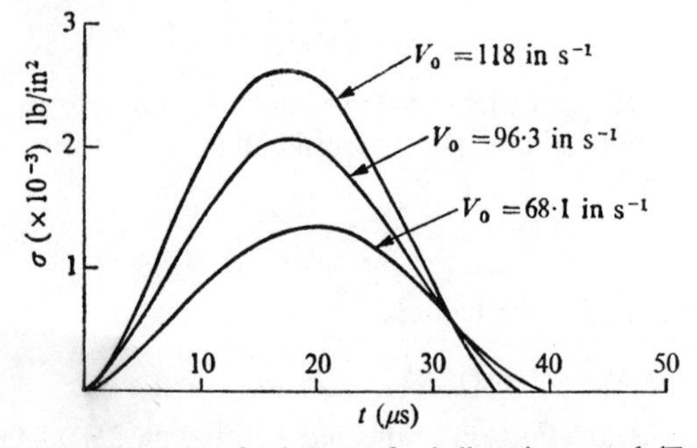

FIG. 2.22. Predicted stress pulses for impact of a ball against a rod (Feng [11], p. 18).

velocity is found to be, for all practical purposes, a linear one. For the parameters selected, it is found that $\dot{u}_2/V_0 = 0{\cdot}72$.

Because of the complicated, non-linear nature of the governing equation, it is not feasible to present general results for a wide class of impact situations. We note the relatively long period of impact compared to the ball dimensions. Thus, if impact duration were in accord with linear dimensions, as in the case of longitudinal rod impact, the duration would be an order of magnitude shorter than the resulting value of about 35 μs. This merely reflects the fundamentally different action occurring. Basically, the situation is analogous to a spring–mass system contacting the end of the rod, with the mass being that of the ball, the spring being that of the mutual deformation between the two surfaces. Considerations involving Hertz contact theory and longitudinal-rod theory have been applied to the case of collinear impact of two rods. This is reviewed extensively by Goldsmith [14, pp. 98–104].

2.5. Dispersive effects in rods

The wave equation governing the longitudinal motion of the rod predicts no distortion in a propagated wave. We shall investigate two sources of wave dispersion in this section. The first arises merely by allowing variation in cross section, but otherwise working within the assumptions of classical rod theory. The second involves including the lateral-inertia effect into the governing theory.

2.5.1. *Rods of variable cross-section—impedance*

A differential element, analogous to that shown in Fig. 2.1(b), isolated from a rod of variable cross-section is shown in Fig. 2.23. The equation of motion is given by

$$-\sigma A+\left(\sigma+\frac{\partial\sigma}{\partial x}\,\mathrm{d}x\right)\left(A+\frac{\mathrm{d}A}{\mathrm{d}x}\,\mathrm{d}x\right)=\tfrac{1}{2}\rho\left\{A+\left(A+\frac{\mathrm{d}A}{\mathrm{d}x}\,\mathrm{d}x\right)\right\}\mathrm{d}x\frac{\partial^2u}{\partial t^2}. \quad (2.5.1)$$

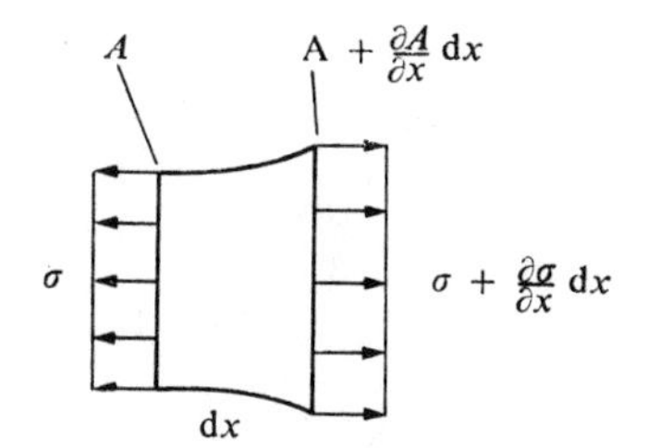

FIG. 2.23. Element from a variable cross-section rod.

The mass has been computed on the basis of the average area. Eqn (2.5.1) reduces to

$$A\frac{\partial \sigma}{\partial x}+\sigma\frac{\mathrm{d}A}{\mathrm{d}x} = \rho A\frac{\partial^2 u}{\partial t^2}. \tag{2.5.2}$$

where terms of the order of $\mathrm{d}x^2$ have been neglected. Noting that the left-hand side of (2.5.2) is given by $\partial(\sigma A)/\partial x$, we may write

$$\frac{1}{A}\frac{\partial}{\partial x}(\sigma A) = \rho\frac{\partial^2 u}{\partial t^2}. \tag{2.5.3}$$

From Hooke's law, $\sigma = E\,\partial u/\partial x$, giving

$$\frac{1}{A}\frac{\partial}{\partial x}\left(A\frac{\partial u}{\partial x}\right) = \frac{1}{c_0^2}\frac{\partial^2 u}{\partial t^2} \tag{2.5.4}$$

or

$$\frac{\partial^2 u}{\partial x^2}+\frac{1}{A}\frac{\mathrm{d}A}{\mathrm{d}x}\frac{\partial u}{\partial x} = \frac{1}{c_0^2}\frac{\partial^2 u}{\partial t^2}. \tag{2.5.5}$$

This last form clearly reveals the dispersive effects of variable cross-section. If A is constant $\mathrm{d}A/\mathrm{d}x = 0$, and the wave equation is recovered. Any variation $A = A(x)$ results in a non-wave equation. We also note that a uniform stress distribution is still assumed. In order for this assumption to remain valid, cross-sectional variation cannot be too drastic, since three-dimensional effects would enter.†

At this stage, it is seen that a cross-sectional variation must be specified before analysis can proceed further. In making some specifications, we point out that most problems in this area arise in the context of 'horn' problems. This refers to rods of tapered section that are frequently used in ultrasonics applications to achieve a high amplification of mechanical displacement (see § 2.3 and discussion pertaining to Figs. 2.10 and 2.11). Several common horn profiles, designated as linear, conical, exponential, and catenoidal are shown in Fig. 2.24. The equation governing the cross-sectional variations are

† See Morse [27, pp. 265–8] for relevant discussion of this matter in the analogous acoustic situation.

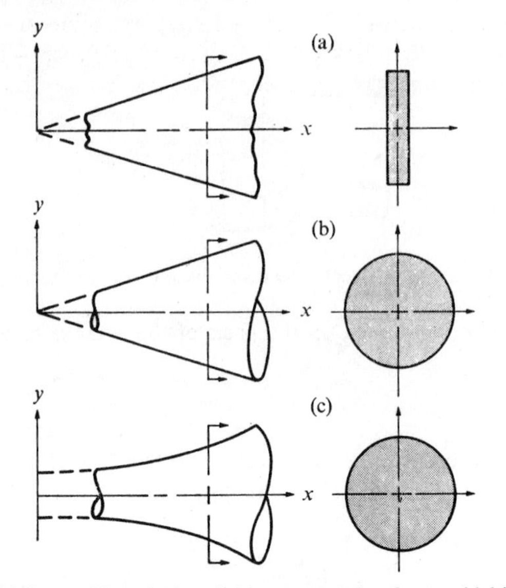

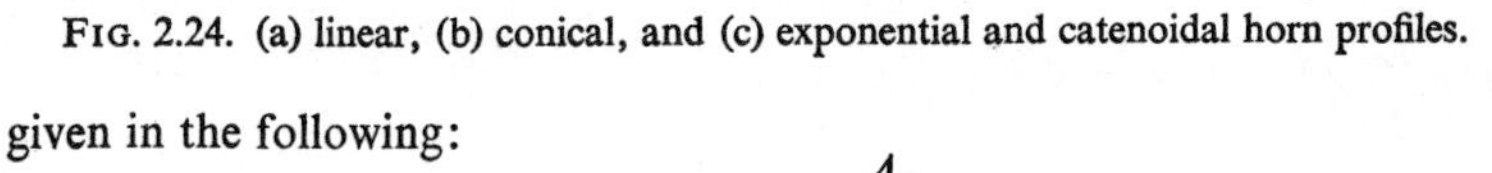

FIG. 2.24. (a) linear, (b) conical, and (c) exponential and catenoidal horn profiles.

given in the following:

$$\text{linear:} \qquad A(x) = \frac{A_0}{a}x \tag{2.5.6}$$

$$\text{conical:} \qquad A(x) = \frac{A_0}{a^2}x^2 \tag{2.5.7}$$

$$\text{exponential:} \qquad A(x) = A_0 e^{2x/h} \tag{2.5.8}$$

$$\text{catenoidal:} \qquad A(x) = A_0 \cosh^2\left(\frac{x}{h}\right). \tag{2.5.9}$$

For the linear and conical cones the one end of the horn is located at $x = a$, where the cross-sectional area is A_0 at that point. The exponential and catenoidal cases have the end of the horn located at the origin $x = 0$, where the area is A_0. The constant h governs the flair of the horn as x increases.

It is not particularly relevant to study the propagation of infinite harmonic wave trains in horn geometries, since only a few wavelengths of travel result in wavelengths comparable to the lateral dimensions, or in the area shrinking to zero, depending on the direction of propagation. It is somewhat more instructive to study the forced motion of semi-infinite horns where at least the last objection does not arise. This type of study is analogous to that carried out on acoustical horns in establishing the 'transmission coefficient'.†

† Morse [27, pp. 279–83].

As a first example, consider a horn governed by the linear area variation (2.5.6) subjected to a harmonic end forcing function $P_0 \exp(-i\omega t)$ at $x = a$. The governing equation, with the linear area variation substituted in (2.5.5), becomes

$$\frac{\partial^2 u}{\partial x^2}+\frac{1}{x}\frac{\partial u}{\partial x} = \frac{1}{c_0^2}\frac{\partial^2 u}{\partial t^2}. \tag{2.5.10}$$

Considering a solution of the form

$$u = U(x)e^{-i\omega t}, \tag{2.5.11}$$

we obtain the ordinary differential equation

$$\frac{d^2U}{dx^2}+\frac{1}{x}\frac{dU}{dx}+\beta^2 U = 0, \qquad \beta^2 = \frac{\omega^2}{c_0^2}. \tag{2.5.12}$$

This is the zeroth-order Bessel's equation, having the solution

$$U = A_1 J_0(\beta x)+B_1 Y_0(\beta x), \tag{2.5.13}$$

or, alternatively,

$$U = AH_0^{(1)}(\beta x)+BH_0^{(2)}(\beta x). \tag{2.5.14}$$

In (2.5.13), J_0 and Y_0 are Bessel's functions of the first and second kind, while (2.5.14) expresses the solution in terms of Bessel functions of the third kind, given by †

$$H_0^{(1)}(\beta x) = J_0(\beta x)+iY_0(\beta x),$$
$$H_0^{(2)}(\beta x) = J_0(\beta x)-iY_0(\beta x). \tag{2.5.15}$$

Now Bessel-function solutions will arise in a number of problems in wave propagation, with this merely being the first instance. The question that must be asked at this stage is the same arising in earlier problems of forced motion of simple systems; that is, 'which solution satisfies the radiation condition?' To establish this, the solution representation (2.5.15) is most convenient. For large values of the argument, the asymptotic representations of the functions of the third kind are‡

$$H_0^{(1)}(\beta x) \sim \sqrt{\left(\frac{2}{\pi\beta x}\right)}\exp\left\{i\left(\beta x-\frac{\pi}{4}\right)\right\}(1-\ldots),$$

$$H_0^{(2)}(\beta x) \sim \sqrt{\left(\frac{2}{\pi\beta x}\right)}\exp\left\{-i\left(\beta x-\frac{\pi}{4}\right)\right\}(1+\ldots). \tag{2.5.16}$$

Expressed in this form, it is immediately seen that the first of (2.5.16) in conjunction with the time behaviour $\exp(-i\omega t)$ gives outward-propagating waves. Consequently, this term should be retained in the solution (2.5.14) and B set equal to zero.

† See, for example, McLachlan [25, pp. 190–9].
‡ ibid., p. 198.

We now apply the boundary condition at the end of the horn, given by

$$EA_0\frac{\partial u(a,t)}{\partial x} = P_0 e^{-i\omega t}. \tag{2.5.17}$$

This gives

$$EA_0 A \frac{dH_0^{(1)}(\beta x)}{dx}\bigg|_{x=a} = P_0. \tag{2.5.18}$$

We note that

$$\frac{dH_0^{(1)}(\beta x)}{dx} = -\beta H_1^{(1)}(\beta x). \tag{2.5.19}$$

The resulting solution for the displacement is given by

$$u(x,t) = -\frac{P_0}{EA_0\beta H_1^{(1)}(\beta a)} H_0^{(1)}(\beta x) e^{-i\omega t}. \tag{2.5.20}$$

It is most useful to express the result in the form of the driving point impedance† of the horn and to refer this quantity to the impedance of a rod of uniform cross-section. Recalling that the impedance is defined as $\bar{Z} =$ force/velocity, we calculate the velocity from (2.5.20), giving

$$\dot{u}(x,t) = \frac{ic_0 P_0}{EA_0 H_1^{(1)}(\beta a)} H_0^{(1)}(\beta x) e^{-i\omega t}. \tag{2.5.21}$$

The impedance is then given by

$$Z = \frac{-iEA_0 H_1^{(1)}(\beta a)}{c_0 H_0^{(1)}(\beta a)}. \tag{2.5.22}$$

The impedance of a rod is given by $Z_R = A_0\sqrt{(\rho E)}$. Defining $\bar{Z}$ as the ratio of horn to rod impedances, we obtain

$$\bar{Z} = \frac{-iH_1^{(1)}(\beta a)}{H_0^{(1)}(\beta a)}. \tag{2.5.23}$$

We make two observations on the results. First, the impedance is complex, as opposed to the purely real impedance of a prismatic rod. Thus the horn does not act as a pure resistance, simply absorbing energy. It has reactive components in its behaviour. Secondly, the impedance will be found to exhibit a cutoff frequency, below which energy is not propagated away from the end.‡

We now consider the case of a conical horn under forced end motion. With the cross-sectional area governed by (2.5.7), the equation of motion reduces

† See Problem 1.11 and § 2.2 for earlier discussion of this concept.
‡ Recall Problem 1.11 regarding resistive, reactive impedance aspects.

to†

$$\frac{\partial^2 u}{\partial x^2}+\frac{2}{x}\frac{\partial u}{\partial x}=\frac{1}{c_0^2}\frac{\partial^2 u}{\partial t^2}. \tag{2.5.24}$$

Again taking a solution of the form (2.5.11) we obtain

$$\frac{d^2U}{dx^2}+\frac{2}{x}\frac{dU}{dx}+\beta^2 U=0. \tag{2.5.25}$$

This is again Bessel's equation having half-order Bessel functions as solutions

$$U=\frac{1}{\sqrt{x}}\{A_1 J_{\frac{1}{2}}(\beta x)+BY_{\frac{1}{2}}(\beta x)\}. \tag{2.5.26}$$

The half-order Bessel functions have rather special properties and are known as spherical Bessel functions; the governing equation (2.5.25) is known as the zeroth-order spherical Bessel's equation. The standard form of the solution of (2.5.25) is given by

$$U=Ah_0^{(1)}(\beta x)+Bh_0^{(2)}(\beta x), \tag{2.5.27}$$

where $h_0^{(1)}(\beta x)$, $h_0^{(2)}(\beta x)$ are the spherical Bessel functions of the third kind.‡ These are defined in terms of spherical functions of the first and second kinds by

$$h_n^{(1)}(\beta x)=j_n(\beta x)+in_n(\beta x),$$
$$h_n^{(2)}(\beta x)=j_n(\beta x)-in_n(\beta x), \tag{2.5.28}$$

where

$$j_n(\beta x)=\sqrt{\left(\frac{2}{\pi\beta x}\right)}J_{n+\frac{1}{2}}(\beta x),$$
$$n_n(\beta x)=\sqrt{\left(\frac{2}{\pi\beta x}\right)}Y_{n+\frac{1}{2}}(\beta x). \tag{2.5.29}$$

† We note in passing that (2.5.24) may be expressed as

$$\frac{\partial^2(xu)}{\partial x^2}=\frac{1}{c_0^2}\frac{\partial^2(xu)}{\partial t^2},$$

which is the form taken by the wave equation for spherical waves, where x is replaced by the radius r. The solution is then

$$u=\frac{1}{x}f(x-c_0t)+\frac{1}{x}g(x+c_0t).$$

This solution form will find application in Chapter 5. A particular form, given by

$$u=\frac{A}{x}e^{i(\gamma x-\omega t)}+\frac{B}{x}e^{-i(\gamma x+\omega t)}$$

could be used in the present example instead of the equivalent, but more indirect, spherical Bessel function representation.

‡ See Morse and Feshbach [28, pp. 1465–8, 1573–4] for various aspects of the spherical Bessel functions.

The special feature of the spherical Bessel functions is that they may be represented by finite series, in contrast to infinite series representations necessary for the non-half-integer order functions. Thus, for $h_n^{(1)}(\beta x)$ we have

$$h_n^{(1)}(\beta x) = \frac{e^{i\beta x}(-i)^{-n+1}}{\beta x} \sum_{r=0}^{n} \frac{(n+r)!}{r!(n-r)} \left(\frac{i}{2\beta x}\right)^n. \tag{2.5.30}$$

Again, the radiation condition will establish which Bessel function must be retained in the solution (2.5.27). The asymptotic behaviour of the function $h_n^{(1)}$ is given in general by

$$h_n^{(1)}(\beta x) \sim \frac{i^{-n}}{i\beta x} e^{i\beta x}, \tag{2.5.31}$$

which yields outgoing waves for the particular time variation of this problem. Thus our solution is reduced to

$$u(x, t) = Ah_0^{(1)}(\beta x)e^{-i\omega t}. \tag{2.5.32}$$

The condition at the end of the rod is

$$EA_0 \frac{\partial u}{\partial x}\bigg|_{x=a} = P_0 e^{-i\omega t}. \tag{2.5.33}$$

Noting that

$$dh_0^{(1)}(\beta x)/dx = -\beta h_1^{(1)}(\beta x), \tag{2.5.34}$$

we evaluate A and obtain for the solution

$$u(x, t) = -\frac{P_0}{EA_0 \beta h_1^{(1)}(\beta a)} h_0^{(1)}(\beta x)e^{-i\omega t}. \tag{2.5.35}$$

This result may be put in more specific form using (2.5.30). Thus we have that

$$h_0^{(1)}(\beta x) = -\frac{ie^{i\beta x}}{\beta x}, \qquad h_1^{(1)}(\beta x) = \frac{e^{i\beta x}}{\beta x}\left(1+\frac{i}{\beta x}\right). \tag{2.5.36}$$

Using these formulas, and calculating the impedance in the manner used for the rod of linearly varying section, it is possible to obtain

$$\bar{Z} = (1+i/\beta a), \tag{2.5.37}$$

where the impedance has been non-dimensionalized using the semi-infinite, prismatic rod.†

2.5.2. *Rods of variable section—horn resonance*

As was mentioned in the preliminary discussion of the previous section, rods of tapered section, or 'horns', find application in power ultrasonics. In this area, the interest is in the resonance characteristics of finite-length, tapered sections.

† See Problems 2.17, 2.18 for analogous studies on the exponential and catenoidal horn.

As a first and quite simple example, consider the resonance of a horn governed by the linear area variation (2.5.6), and fixed at $x = 0$ and $x = b$. The governing equation for the motion of such a system has been previously given by (2.5.10). For harmonic vibration, the ordinary differential equation (2.5.12) still pertains, as does the solution (2.5.13). We employ the Bessel functions of first and second kind instead of the third kind, since wave propagation is no longer a consideration. The boundary conditions are given by

$$U(a) = U(b) = 0. \tag{2.5.38}$$

This gives the two equations

$$\begin{aligned} A_1 J_0(\beta a) + B_1 Y_0(\beta a) &= 0, \\ A_1 J_0(\beta b) + B_1 Y_0(\beta b) &= 0, \end{aligned} \tag{2.5.39}$$

which, in turn, gives the frequency equation for the system

$$J_0(\beta a) Y_0(\beta b) - J_0(\beta b) Y_0(\beta a) = 0. \tag{2.5.40}$$

The values of a and b must be specified before the roots of this equation can be determined.

The preceding example was selected mainly for its simplicity and not for its applicability. In power ultrasonics, the interest is in determining the resonance frequencies and amplification characteristics of half-wavelength horns, under stress-free boundary conditions at either end. In practice, such a horn is driven at resonance by piezoelectric discs located at the node or, more commonly, by connecting to a half-wavelength resonator of the form shown in Fig. 2.10.

The determination of the resonance frequencies and mode shapes follows the methods used in the previous simple example. Thus, the governing equation for a particular area variation is established and a solution determined. The boundary conditions

$$\left.\frac{dU}{dx}\right|_{x=a} = \left.\frac{dU}{dx}\right|_{x=b} = 0 \tag{2.5.41}$$

are applied and a frequency equation analogous to (2.5.40) determined.

There are several ways of using the resulting formula for horn-design purposes. Fig. 2.25 presents data for conical, exponential, and catenoidal horns for a particular design procedure. This is based on specifying the ratio R of the large diameter to the small diameter of the horn and determining the resulting half-wavelength for a given resonant frequency. Thus, one specifies the desired ratio R and horn shape and obtains from the ordinate a resulting value of $2l/\lambda$. One then specifies the material of the horn and the desired operating frequency. This information is used to calculate the half-wave parameter $\lambda/2$. Thus the resulting length l of the horn is obtained.

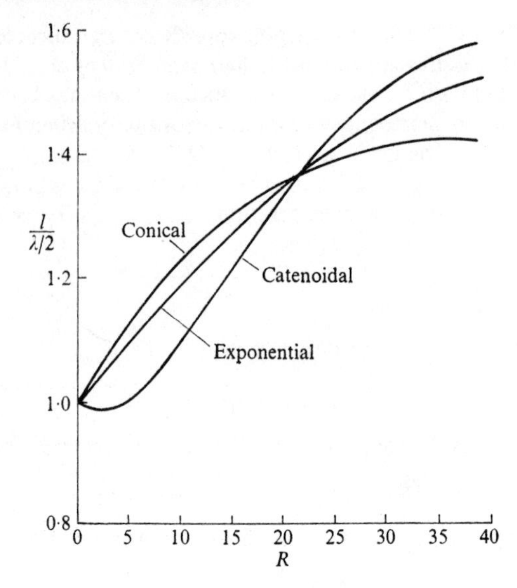

FIG. 2.25. A design chart for half-wavelength horns of conical, exponential, and catenoidal cross-section. (After Merkulov and Kharitonov [26].)

Other design parameters of interest are the amplification (ratio of small end to large end displacement), nodal location, and maximum stresses in the horn. The energy storage of the horn is also an important characteristic.

There is a large amount of literature in this area. Many aspects of horn and transducer design for ultrasonic applications are presented by Frederick [13, pp. 87–103] and by Eisner [8]. The latter author has also presented an extensive summary of the analyses of the horn problem [9].

2.5.3. *Effects of lateral inertia—dispersion*

In the derivation of the wave equation for the rod, several assumptions were mentioned, including uniform stress distribution, constant cross-section, and material homogeneity. It was also emphasized that the effects of lateral inertia were neglected. One purpose here is to re-examine this aspect and, specifically, to reformulate the theory with these effects included. Without knowing the specific form of the results, we suspect that a dispersive system will result. The fundamentally dispersive nature of the rod was first pointed out by Pochhammer† in conjunction with application of the exact theory of elasticity to wave propagation in the rod. An effort to assess the effects of lateral inertia on longitudinal vibrations was presented by Rayleigh [32,

† See Reference [31]. Note entry for 1876 in the *Introduction* (§ I.3). Also see Chapter 8, § 8.2.

Vol. 1, pp. 251–2]. Love [24, p. 428] presented the complete governing equation, and it is this development, based on energy considerations, that will be presented here.

We consider the rod and a typical cross-section as shown in Fig. 2.26, where

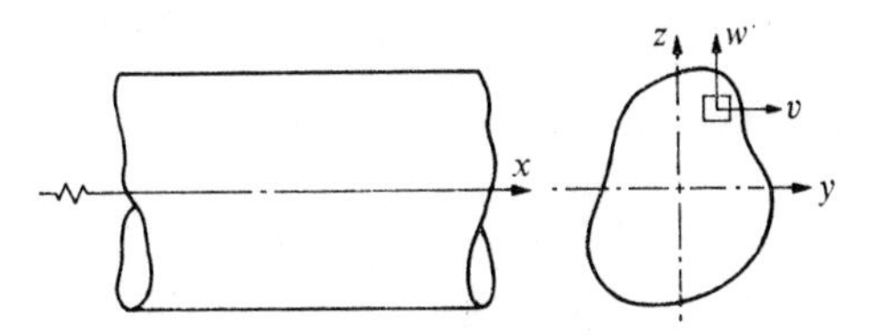

FIG. 2.26. Rod and cross-section showing lateral displacement components.

the origin is at the centroid of the rod. As a result of the longitudinal displacements u and the Poisson effect, lateral displacements v and w will occur at a typical point of the cross-section having coordinates y and z. The lateral strains existing in the rod will be given from the general Hooke's law as

$$\varepsilon_x = \frac{1}{E}\{\sigma_x - \nu(\sigma_y + \sigma_z)\},$$

$$\varepsilon_y = \frac{1}{E}\{\sigma_y - \nu(\sigma_x + \sigma_z)\},$$

$$\varepsilon_z = \frac{1}{E}\{\sigma_z - \nu(\sigma_x + \sigma_y)\}. \tag{2.5.42}$$

We assume that uniaxial stress still holds, so that $\sigma_y = \sigma_z = 0$. Hence we obtain from (2.5.42)

$$\varepsilon_y = \varepsilon_z = -\frac{\nu}{E}\sigma_x = -\nu\frac{\partial u}{\partial x}. \tag{2.5.43}$$

Now, the lateral displacements v and w are given by

$$v = \varepsilon_y y, \qquad w = \varepsilon_z z, \tag{2.5.44}$$

where y and z are the coordinates of a point in the cross-section. Substituting (2.5.43) in (2.5.44) enables the lateral displacements to be expressed in terms of the longitudinal motion as

$$v = -\nu y\frac{\partial u}{\partial x}, \qquad w = -\nu z\frac{\partial u}{\partial x}. \tag{2.5.45}$$

In applying the energy method† for the derivation of the equation of motion, expressions for the kinetic and potential energy are required. The

† Appendix A.8 briefly reviews energy considerations.

kinetic energy density is simply

$$T = \frac{\rho}{2}(\dot{u}^2+\dot{v}^2+\dot{w}^2), \tag{2.5.46}$$

while the strain energy density is given by

$$V = \sigma\varepsilon/2, \tag{2.5.47}$$

for uniaxial strain. Hamilton's equation then becomes

$$\delta\int_{t_0}^{t_1}(\tilde{T}-\tilde{V})\,\mathrm{d}t = 0, \tag{2.5.48}$$

in the absence of external loads, where

$$\begin{aligned}\tilde{T} &= \int_V \frac{\rho}{2}(\dot{u}^2+\dot{v}^2+\dot{w}^2)\,\mathrm{d}V,\\ \tilde{V} &= \int_V \frac{\sigma\varepsilon}{2}\,\mathrm{d}V.\end{aligned} \tag{2.5.49}$$

It is possible to integrate the preceding expressions over the area of the section. In first considering the kinetic energy integral, we note that

$$\dot{v} = -\nu y\frac{\partial^2 u}{\partial x\,\partial t}, \qquad \dot{w} = -\nu z\frac{\partial^2 u}{\partial x\,\partial t}. \tag{2.5.50}$$

Then

$$\tilde{T} = \int_L \mathrm{d}x \int_A \frac{\rho}{2}\left\{\dot{u}^2+\nu^2(y^2+z^2)\left(\frac{\partial^2 u}{\partial x\,\partial t}\right)^2\right\}\mathrm{d}A. \tag{2.5.51}$$

The longitudinal displacement $u(x, t)$ does not depend on the coordinates y and z, so the area integral may be evaluated exactly. The result is

$$\tilde{T} = \int_L \frac{\rho A}{2}\left\{\dot{u}^2+\nu^2k^2\left(\frac{\partial^2 u}{\partial x\,\partial t}\right)^2\right\}\mathrm{d}x, \tag{2.5.52}$$

where k^2 is the polar radius of gyration of the cross-section, given in terms of the area and polar moment of inertia as $J = Ak^2$. The potential energy density is given by

$$\tilde{V} = \int_L \mathrm{d}x \int_A \left(\frac{E\varepsilon^2}{2}\right)\mathrm{d}A = \int_L \frac{EA}{2}\left(\frac{\partial u}{\partial x}\right)^2\mathrm{d}x. \tag{2.5.53}$$

Hamilton's equation then is

$$\delta\int_{t_0}^{t_1}\mathrm{d}t\int_L A\left[\frac{\rho}{2}\left\{\left(\frac{\partial u}{\partial t}\right)^2+\nu^2k^2\left(\frac{\partial^2 u}{\partial x\,\partial t}\right)^2\right\}-\frac{E}{2}\left(\frac{\partial u}{\partial x}\right)^2\right]\mathrm{d}x = 0. \tag{2.5.54}$$

The procedures for carrying out this variation are presented in Love [24, pp. 166–7] and, in terms of the mathematics of the calculus of variations, in Langhaar [20, pp. 92–6]. Some of the steps will be presented here. The problem is one of obtaining the first variation δI of a double integral, where

$$I = \iint F\left(x, t, \frac{\partial u}{\partial x}, \frac{\partial u}{\partial t}, \frac{\partial^2 u}{\partial x\, \partial t}\right) \mathrm{d}x\, \mathrm{d}t, \tag{2.5.55}$$

and where

$$F = \frac{\rho A}{2}\left\{\left(\frac{\partial u}{\partial t}\right)^2 + \nu^2 k^2 \left(\frac{\partial^2 u}{\partial x\, \partial t}\right)^2\right\} - \frac{E}{2}\left(\frac{\partial u}{\partial x}\right). \tag{2.5.56}$$

Carrying out the indicated variation, we obtain

$$\delta I = \int_t \int_L \left(\frac{\partial F}{\partial u_x}\, \delta u_x + \frac{\partial F}{\partial \dot{u}}\, \delta \dot{u} + \frac{\partial F}{\partial \dot{u}_x}\, \delta \dot{u}_x\right) \mathrm{d}t\, \mathrm{d}x = 0, \tag{2.5.57}$$

where $u_x = \partial u/\partial x$, $\dot{u} = \partial u/\partial t$, $\dot{u}_x = \partial^2 u/\partial x\, \partial t$ in the above. We integrate by parts in the usual manner, giving for the various terms of (2.5.57)

$$\int_{t_0}^{t_1} \int_L \frac{\partial F}{\partial u_x}\, \delta u_x\, \mathrm{d}t\, \mathrm{d}x = \int_{t_0}^{t_1} \mathrm{d}t \left\{\frac{\partial F}{\partial u_x}\, \delta u \bigg|_L - \int_L \frac{\partial}{\partial x}\left(\frac{\partial F}{\partial u_x}\right) \delta u\, \mathrm{d}x\right\},$$

$$\int_{t_0}^{t_1} \int_L \frac{\partial F}{\partial \dot{u}}\, \delta \dot{u}\, \mathrm{d}t\, \mathrm{d}x = \int_L \mathrm{d}x \left\{\frac{\partial F}{\partial \dot{u}}\, \delta u \bigg|_{t_0}^{t_1} - \int_{t_0}^{t_1} \frac{\partial}{\partial t}\left(\frac{\partial F}{\partial \dot{u}}\right) \delta u\, \mathrm{d}t\right\}$$

$$\int_{t_0}^{t_1} \int_L \frac{\partial F}{\partial \dot{u}_x}\, \delta \dot{u}_x\, \mathrm{d}t\, \mathrm{d}x = \int_{t_0}^{t_1} \mathrm{d}t \left(\frac{\partial F}{\partial \dot{u}_x}\, \delta \dot{u}\right)_L -$$

$$- \int_L \mathrm{d}x \left\{\frac{\partial}{\partial x}\left(\frac{\partial F}{\partial \dot{u}_x}\right) \delta u \bigg|_{t_0}^{t_1} - \int_{t_0}^{t_1} \frac{\partial^2}{\partial x\, \partial t}\left(\frac{\partial F}{\partial \dot{u}_x}\right) \delta u\, \mathrm{d}t\right\}. \tag{2.5.58}$$

The initial conditions and boundary conditions cause the various single integral terms in (2.5.58) to vanish, leaving

$$\int_t \int_L \left\{\frac{\partial}{\partial x}\left(\frac{\partial F}{\partial u_x}\right) + \frac{\partial}{\partial t}\left(\frac{\partial F}{\partial \dot{u}}\right) - \frac{\partial^2}{\partial x\, \partial t}\left(\frac{\partial F}{\partial \dot{u}_x}\right)\right\} \delta u\, \mathrm{d}x\, \mathrm{d}t = 0. \tag{2.5.59}$$

This gives the Euler equation of the problem,

$$\frac{\partial}{\partial x}\left(\frac{\partial F}{\partial u_x}\right) + \frac{\partial}{\partial t}\left(\frac{\partial F}{\partial \dot{u}}\right) - \frac{\partial^2}{\partial x\, \partial t}\left(\frac{\partial F}{\partial \dot{u}_x}\right) = 0. \tag{2.5.60}$$

Substituting (2.5.56) in the above gives Love's equation of motion

$$\frac{\partial^2 u}{\partial x^2}+\frac{\nu^2 k^2}{c_0^2}\frac{\partial^4 u}{\partial x^2\,\partial t^2}=\frac{1}{c_0^2}\frac{\partial^2 u}{\partial t^2}. \tag{2.5.61}$$

The first step is to determine the influence of lateral-inertia effects on the dispersion characteristics of the rod by considering a solution of the form

$$u = Ae^{i\gamma(x-ct)}, \tag{2.5.62}$$

where c is the general, frequency-dependent phase velocity. Substituting (2.5.62) in (2.5.61) gives

$$-\gamma^2+\frac{\nu^2 k^2}{c_0^2}\gamma^4 c^2+\frac{\gamma^2 c^2}{c_0^2}=0. \tag{2.5.63}$$

If we define the normalized phase velocity $\bar{c}$ and wavenumber $\bar{\gamma}$, where

$$\bar{c}=\frac{c}{c_0}, \qquad \bar{\gamma}=k\nu\gamma, \tag{2.5.64}$$

then (2.5.63) becomes

$$(1+\bar{\gamma}^2)\bar{c}^2=1 \tag{2.5.65}$$

or

$$\bar{c}=1/(1+\bar{\gamma}^2)^{\frac{1}{2}}. \tag{2.5.66}$$

We have neglected the negative root of (2.5.66), since it would only give waves propagating in the negative x direction.

In assessing the result (2.5.66) we first note that if $\bar{\gamma}=0$, the classical, 'inertialess' rod result is recovered. Tracing back a few steps, we see this would occur by setting $k=0$ in (2.5.61). The dispersion curve corresponding to (2.5.66) is plotted in Fig. 2.27. Also shown in the figure as dashed lines are

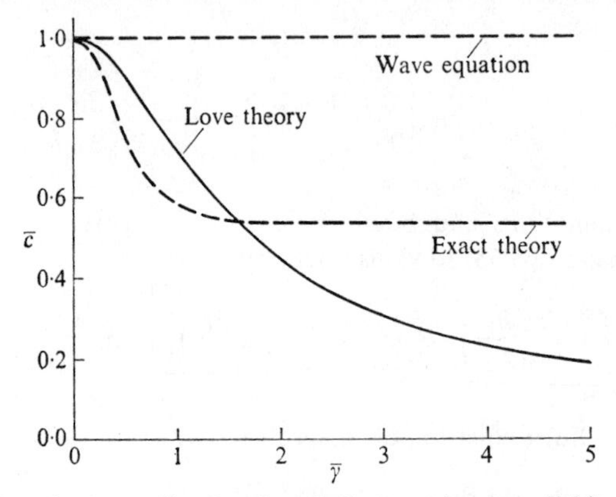

FIG. 2.27. Dispersion curve for Love's rod theory, wherein radial-inertia effects are included.

the dispersion curves for classical rod theory as governed by the wave equation, and the general shape of the curve as obtained by the exact equations of elasticity, if the cross-section of the rod were circular.

Recalling that $\bar{\gamma}$ is the reciprocal of wavelength, it is seen that, at low frequency and long wavelengths, all of the theories give the same propagation velocity c_0. For shorter wavelengths, the theories diverge, with classical rod theory breaking down seriously with respect to the Love results and exact theory for $\bar{\gamma} > 0{\cdot}3$. The Love theory at least approximates exact theory behaviour to the vicinity of $\bar{\gamma} = 2$, where departure then becomes more and more serious.

With this type of result available, it becomes possible to establish limitations of a theory in a fairly quantitative way. Thus, suppose on the basis of the dispersion curve, one imposes the restriction that $\bar{\gamma} < 0{\cdot}3$ as the wavenumber limit for the classical theory. This may be converted to frequency by noting that $\bar{\gamma}\bar{c} = \bar{\omega}$, where $\bar{\omega} = \nu k\omega/c_0$. For $\bar{\gamma} = 0{\cdot}3$, $\bar{c} \simeq 1$, so that $\bar{\omega} \simeq 0{\cdot}3$. Suppose our actual rod is of circular cross-section, 2·54 cm in diameter and made of steel. For these parameters, we would have $c_0 = 5{\cdot}08 \times 10^3$ m s^{-1}, $\nu = 0{\cdot}29$, $k = 2{\cdot}54/(2\sqrt{2})$. The resulting frequency is

$$f = 92\,800 \text{ Hz}. \tag{2.5.67}$$

In other words, the operating frequency of the system must be less than 90 000 Hz in order for classical theory to apply. In terms of a transient loading, there must not be significant components of the frequency spectrum of the loading pulse beyond 90 000 Hz. If one converts the wavenumber restriction to wavelength, it is found for the present example that $\lambda \geqslant 5$ cm must hold.

Our considerations in this section do not represent our first encounter with a dispersive system. However, they do represent our first encounter with an 'improved theory'. Thus, a basic characteristic of strength-of-materials type theories is that they contain assumptions of various forms on the kinematics of motion. In the case of the rod, they are regarding plane sections and the inertia contributions of certain components of motion. They are approximate theories. While the theories are approximate, they yield relatively simple governing equations. At the other extreme are the exact theories for various structures, such as rods and plates, that result from applying the equations of elasticity. Whilst having the virtue of exactness, they are often quite complicated. They often can serve, as was illustrated in Fig. 2.27, as a reference for an approximate theory. Attention will be given in several instances to the development of improvements on simple theories by comparing the results to those obtainable from exact analysis.

2.5.4. *Effects of lateral inertia—pulse propagation*

Davies [7, pp. 428–45] utilized Love's equation in analysing the propagation of sharp pulses in a cylindrical rod. This was only a part of a far-ranging study of the Hopkinson bar that also included extensive experiments and analyses based on the exact elasticity equations.† The problem considered was a bar of length l, free at the end $x = 0$ and subjected to a step pressure pulse P_0 at the end $x = l$. The initial conditions were taken to be zero. The basic steps of the Davies analysis will be presented in the following.

Taking the Laplace transform of (2.5.61) gives

$$\frac{d^2\bar{u}}{dx^2} - \frac{s^2}{c_0^2(1+H^2s^2)}\bar{u} = 0, \tag{2.5.68}$$

where $H^2 = \nu^2k^2/c_0^2$. This has the solution

$$\bar{u} = A\sinh\frac{s}{c_0\sqrt{(1+H^2s^2)}}x + B\cosh\frac{s}{c_0\sqrt{(1+H^2s^2)}}x. \tag{2.5.69}$$

The transformed boundary conditions of the problem are

$$\left.\frac{d\bar{u}}{dx}\right|_{x=0} = 0, \qquad E\left.\frac{d\bar{u}}{dx}\right|_{x=l} = \frac{P_0}{s}. \tag{2.5.70}$$

The first condition gives $A = 0$. Applying the second condition and solving for B yields the transformed solution

$$\bar{u} = \frac{P_0c_0\sqrt{(1+H^2s^2)}}{Es^2}\cosh\left\{\frac{sx}{c_0\sqrt{(1+H^2s^2)}}\right\}\Big/\sinh\left\{\frac{sl}{c_0\sqrt{(1+H^2s^2)}}\right\}. \tag{2.5.71}$$

The formal inversion gives

$$u(x,t) = \frac{P_0c_0}{2\pi iE}\int_{\alpha-i\infty}^{\alpha+i\infty}\frac{\sqrt{(1+H^2s^2)}}{s^2}\left[\cosh\left\{\frac{sx}{c_0\sqrt{(1+H^2s^2)}}\right\}\Big/\sinh\left\{\frac{sl}{c_0\sqrt{(1+H^2s^2)}}\right\}\right]e^{st}\,ds. \tag{2.5.72}$$

The inversion of the preceding result involves integration along the Bromwich line and closure of the contour in the left-hand plane. Three features of the integrand must be accounted for. First, at $s = 0$, a third-order pole exists owing to the s^2 and hyperbolic sine terms that approach zero. The residue of this pole is given by Davies as

$$\text{Res} = \frac{c_0t^2}{2l} + \frac{x^2}{2lc_0} - \frac{l}{6c_0} + \frac{H^2c_0}{l}. \tag{2.5.73}$$

Secondly, there are simple poles located at the zeros of the hyperbolic sine in (2.5.72), given by

$$\sinh\frac{sl}{c_0\sqrt{(1+H^2s^2)}} = 0, \qquad \frac{sl}{c_0\sqrt{(1+H^2s^2)}} = n\pi i. \tag{2.5.74}$$

† See the *Introduction* (of this book). This study will also be referred to in Chapter 8.

Solving for s gives

$$s = \frac{\pm n\pi \mathrm{i} c_0}{(l^2+n^2\pi^2H^2c_0^2)^{\frac{1}{2}}}. \tag{2.5.75}$$

Davies gives the residues at these poles as

$$\mathrm{Res} = -\frac{(-1)^n l}{\pi^2 n^2 c_0^2}\frac{l^2}{l^2+\pi^2n^2H^2c_0^2}\cos\frac{n\pi x}{l}\exp\{\pm n\pi \mathrm{i} c_0 t/(l^2+\pi^2n^2H^2c_0^2)\}^{\frac{1}{2}}. \tag{2.5.76}$$

Finally, as $n \to \infty$, it is seen from (2.5.75) that $s \to \pm\mathrm{i}/H$, corresponding to two essential singularities of the integrand. Davies states that these limit-point singularities can be excluded from within the contour by small circles and shown to make no contribution in the limit.

The resulting solution is given by $2\pi\mathrm{i}\Sigma\mathrm{Res}$, where only the real part of the solution is retained, giving

$$u(x,t) = \frac{P_0t^2}{2m}+\frac{P_0H^2}{m}+$$
$$+\frac{2P_0l^2}{\pi^2c_0^2m}\sum_{n=1}\frac{(-1)^n}{n^2}\cos\frac{n\pi x}{l}\left\{1-\frac{1}{1+n^2\psi^2}\cos\frac{n\pi c_0 t}{l\sqrt{(1+n^2\psi^2)}}\right\}. \tag{2.5.77}$$

In writing (2.5.77), the mass parameter $m = \rho l$ has been introducted and the parameter $\psi = \pi Hc_0/l$ brought in. Finally, the Fourier-series identity

$$\frac{\pi^2}{4}\left(\frac{x^2}{l^2}-\frac{1}{3}\right) = \sum_{n=1}^{\infty}\frac{(-1)^n}{n^2}\cos\frac{n\pi x}{l}, \qquad (-l < x < l), \tag{2.5.78}$$

has been used in simplifying to the form (2.5.77).

In interpreting the result, we note that the first term corresponds to a rigid-body motion. Such a term has been previously pointed out as associated with a normal-mode type of representation† of the motion of a free–free rod, so it is not surprising to see it occur here. The second term P_0H^2/m is, according to Davies, a very small, time-independent constant displacement associated with the instantaneous propagation of disturbances which is implicit in the present theory. The remaining terms represent the normal modes of the bar.

Numerous results are presented by Davies on predicted displacements and pressures and comparison of results to experiment. Because the series in (2.5.77) is rather slowly convergent, many terms are required for accuracy (80–90 terms). In Fig. 2.28, the predicted displacement of the free end of the

† Recall § 2.3 and Timoshenko [38, pp. 309].

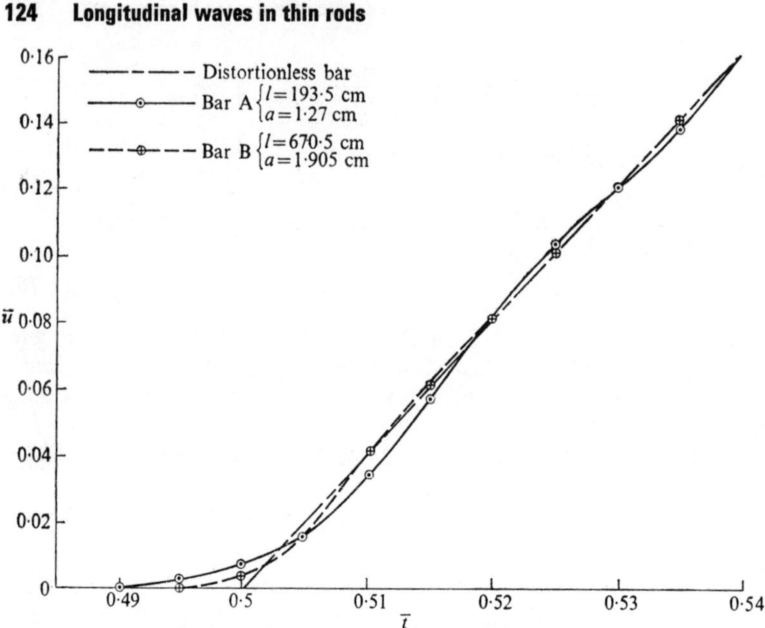

FIG. 2.28. Predicted displacement–time curves (solid lines) of the free end of a bar subjected to a step pressure pulse. [After Davies (7, Fig. 26).]

rod is shown. The displacement as predicted by classical theory is shown as the dashed line. The results have been non-dimensionalized as follows:

$$\bar{u} = \kappa \frac{u(0, t)}{u_s}, \qquad \bar{t} = \frac{t}{T_0}, \tag{2.5.79}$$

where u_s is the displacement under a static pressure of P_0 and T_0 is the time required for a pulse to travel twice the length of the bar. The parameter κ is an adjustment factor enabling data from two different bars (A and B) to be plotted on the same scale. For A, $\kappa = 1$ and for bar B, $\kappa = 3{\cdot}465$.

We have previously considered the case of a rectangular pressure pulse incident on the end of the bar and shown the displacement to be a ramp function like the dashed line of Fig. 2.28 (recall Fig. 2.8). It is seen that the effect of lateral inertia on the predicted response is to smooth the discontinuity at the base of the ramp. For later times, very slight oscillations about the classical result are predicted.

A second result presented by Davies was the predicted stress within the bar for a step pulse. This is shown in Fig. 2.29. The non-dimensional quantities are

$$\sigma = \frac{\sigma}{P_0}, \qquad \bar{t}' = \frac{2t}{T_0}. \tag{2.5.80}$$

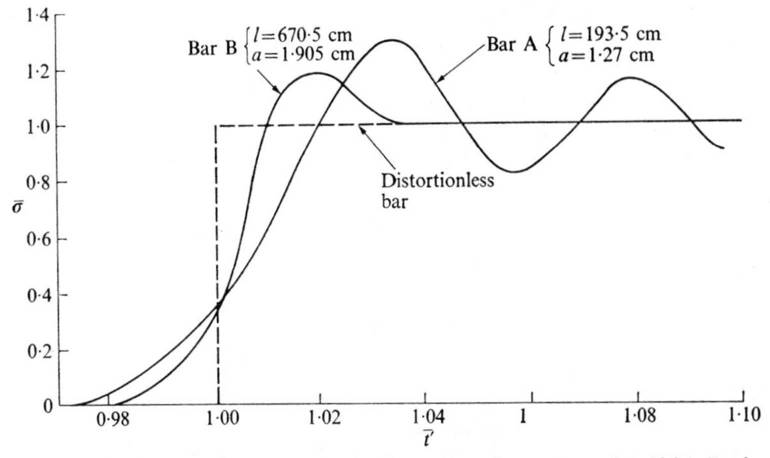

FIG. 2.29. Predicted stress–time curve of a bar subjected to a step pulse. [After Davies (7, Fig. 32).]

The predicted stress curve represents the stress that would exist at $x=0$ (the end of the bar) if the bar were continuous and not terminated at that point. Also shown as a dashed line is the undistorted step pulse that would be predicted by classical rod theory. The major item to note is the oscillatory nature of the response, with no sharp discontinuity of pulse arrival.

As mentioned earlier, the work Davies performed involved analysis based on exact theory, analysis based on Love theory, some of which has been reviewed here, and experimental studies. The experimental studies strongly confirmed the theoretical predictions shown in Fig. 2.28 and Fig. 2.29. A major consequence of the agreement of theory and experiment is that it may be concluded that lateral inertia effects alone account for the initial oscillatory behaviour of the pulse and that yet higher-order effects, such as radial shear stress, and other modes of deformation do not significantly affect this early-time behaviour. As will be shown in a later chapter, other oscillations, not accounted for by lateral inertia alone, are present in a transient and exact theory is required for their prediction.

2.6. Torsional vibrations

We now consider a digression from the study of longitudinal rod motion by reviewing the subject of torsional motion of rods. The governing equation for torsional motion in a rod, as will shortly be established, is identical in form to that of longitudinal motion, in the context of strength of materials development. In other words, the wave equation results. Without implying disfavour for this form of mechanical motion, it does not seem necessary, after the coverage of strings and longitudinal rod theory, to continue the analysis of this equation. The coverage therefore consists of development of the governing equation, plus brief remarks on application to various cross-sections.

2.6.1. *The governing equation*

Consider an element of a straight, prismatic rod subjected to variable end torques, as shown in Fig. 2.30. The equation of motion of the element, in

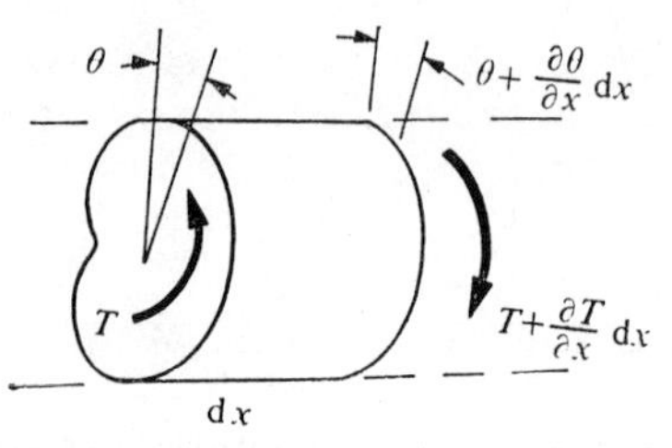

FIG. 2.30. Differential element of rod subjected to end torques.

the absence of body torques, is

$$-T+\left(T+\frac{\partial T}{\partial x}\,\mathrm{d}x\right) = \rho J\,\mathrm{d}x\frac{\partial^2\theta}{\partial t^2}, \tag{2.6.1}$$

where J is the polar moment of inertia. This reduces to

$$\frac{\partial T}{\partial x} = \rho J\frac{\partial^2\theta}{\partial t^2}. \tag{2.6.2}$$

The torque is related to the angle of twist by

$$T = C\frac{\partial\theta}{\partial x}, \tag{2.6.3}$$

where C is the torsional rigidity† of the bar. This quantity is dependent on the shear modulus G of the material and on the cross-sectional properties. Inserting (2.6.3) in (2.6.2) gives the resulting governing equation

$$\frac{\partial^2\theta}{\partial x^2} = \frac{\rho J}{C}\frac{\partial^2\theta}{\partial t^2}. \tag{2.6.4}$$

This result, except for dimensions, is the form of the wave equation derived for strings and longitudinal rod motion. It could be put into dimensional conformity merely by inserting a characteristic length of the rod, such as a diameter.

† Timoshenko [37, p. 298].

The torsional rigidity parameter appearing in the wave equation must be established from strength-of-material or elasticity considerations, depending on the complexity of the cross-section. For a rod of circular cross-section, we have $T = JG\,\partial\theta/\partial x$, where $J = \pi a^4/2$, and a is the radius of the rod. The governing equation becomes

$$\frac{\partial^2\theta}{\partial x^2} = \frac{1}{c_s^2}\frac{\partial^2\theta}{\partial t^2}, \qquad c_s = \sqrt{\left(\frac{G}{\rho}\right)}. \tag{2.6.5}$$

The propagation velocity c_s is that of shear waves in a material. The torsional rigidity of three other cross-sections as established from elasticity analysis are:

(1) ellipse: semi-major and minor axis of a and b,

$$C = \frac{\pi G a^3 b^3}{a^2+b^2}; \tag{2.6.6}$$

(2) triangle: equilateral, of height a,

$$C = \frac{Ga^4}{15\sqrt{3}}. \tag{2.6.7}$$

(3) narrow rectangular: width $2a$, depth $2b$, $b \gg a$,

$$C = \frac{a^3 bG}{3}. \tag{2.6.8}$$

For open, thin-walled cross-sections or thin-walled tubes, the torsional rigidity is fairly easily established by means of membrane-analogy analysis. For cross-sections having breadth and depth of comparable dimensions, elasticity analysis must be employed.

It should be noted that, for cross-sections lacking double axes of symmetry, it may be impossible for simple torsional vibrations to exist alone. That is torsional motion may be unavoidably coupled to bending motion through the deformation kinematics. Analysis may show the coupling to be weak and therefore negligible, but it is a consideration in torsional motion.

The case of the rod of circular cross-section is of particular interest. It may be recalled this is a case where elementary strength-of-materials considerations and those of exact elasticity yield the same results. Thus, in contrast to the theory for longitudinal motion in a circular cylindrical rod, which is approximate, the case of torsional motion in such a system should yield results in accord with exact analysis. However, exact analysis (Chapter 8) will reveal other modes of deformation not accounted for in the strength-of-materials theory.

2.7. Experimental studies in longitudinal waves

In this section, some of the experimental studies and applications of longitudinal waves in rods will be reviewed. The attempt will not be made to be comprehensive in this respect, but merely to bring out results and applications in areas related to the theoretical results of the chapter.

2.7.1. *Longitudinal impact of spheres on rods*

It will be recalled from § 2.4 that the longitudinal impact of an elastic sphere on a rod was considered on the basis of simple longitudinal rod theory and Hertzian contact theory. Theoretical results were presented for the predicted stress pulse in Fig. 2.22. Barton, Volterra, and Citron [3] obtained experimental results for this case and compared the results to theoretical predictions. A diagram of the experimental arrangement used is shown in Fig. 2.31 (a). The impacted rod was steel and 1 in. in diameter. The impact ball diameter ranged from 1 in. to 2 in. and impact velocities ranged from 121·30 cm s^{-1} to 242·61 cm s^{-1} (48–95 in. s^{-1}). The resulting stress pulses for the one-inch diameter ball impact are shown in Fig. 2.31 (b) for two different impact velocities. Waveform results are given also in Reference [3] for the case of a 2 in diameter ball. Tabular data on stress amplitudes and pulse duration is also given for other impact conditions. The quite apparent conclusion from the figure is that rather good agreement is obtained between experiment and theory. A slightly shorter pulse is predicted than observed. The peak amplitudes are only slightly different between theory and experiment.

Ripperger [34] conducted an extensive investigation of pulses in rods resulting from spherical ball impact. Various rod diameters ($\frac{1}{8}$ in., $\frac{1}{4}$ in., $\frac{1}{2}$ in. diameter) and impact ball diameters were considered as well as different impact velocities. Piezoelectric strain gauges were used to detect the strain signals. For strain-gauge stations several diameters from the impact end, the pulse shapes had the general form noted in Fig. 2.31 (b), and agreement between experiment and theory was quite good. For closer stations and for ball sizes considerably smaller than the rod diameter, a significant amount of high-frequency content was present. By using eccentric impact, bending waves were induced in some of the experimental studies.

2.7.2. *Longitudinal waves across discontinuities*

Ripperger and Abramson [35] presented experimential results for both longitudinal and flexural waves encountering a discontinuity in a rod in the form of a step change in cross-sectional area. The governing theoretical expressions, for transmitted and reflected longitudinal stress amplitudes, were derived in § 2.2, with the results being given by eqn (2.2.10). The geometry of the test bar studied by Ripperger and Abramson is shown in

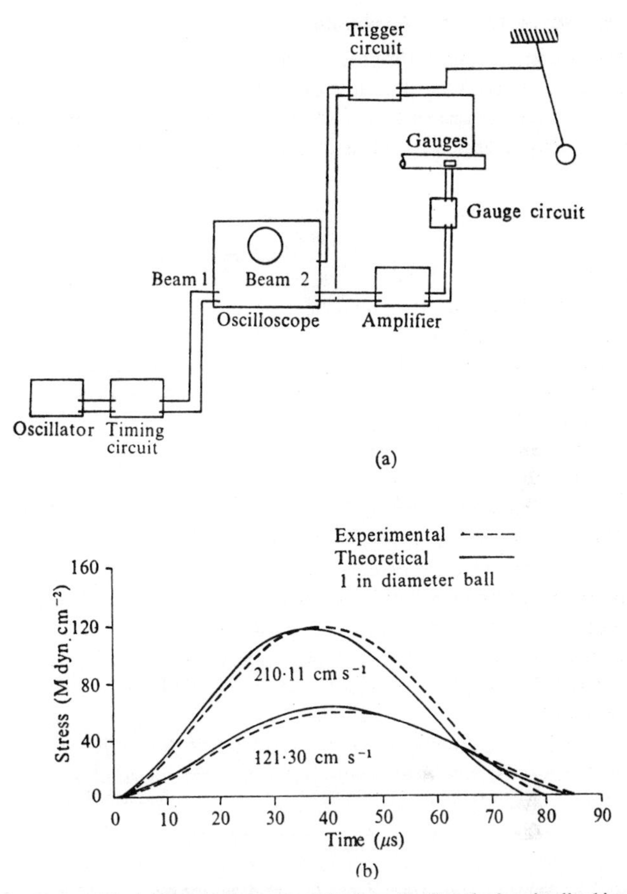

FIG. 2.31. (a) Experimental arrangement used for investigating the longitudinal impact of a ball on a rod and (b) results for the impact of a 1 in. diameter ball. (After Barton *et al.* [3], Figs. 2 and 4).

Fig. 2.32(a). In Fig. 2.32(b), the incident, reflected, and transmitted stress pulses are shown for propagation from the small end toward the large end. In Fig. 2.32(c), the case of propagation from the large end toward the small end is shown. It is seen that pulse shape and duration is maintained with only the amplitude being affected by the discontinuity.

Other studies related to this topic have also been made. Fischer [12] has obtained results for the case of a cylindrical bar having a short necked-down or swelled-out region. Some of the results obtained by Becker [4] are for a rectangular pulse reflecting off a step discontinuity.

2.7.3. *The split Hopkinson pressure bar*

An experimental apparatus for obtaining dynamic anelastic material properties that utilizes longitudinal elastic wave propagation in rods is called the split Hopkinson pressure bar. Kolsky [19] first proposed this modification of the basic Hopkinson bar. Basically, the method consists of sandwiching

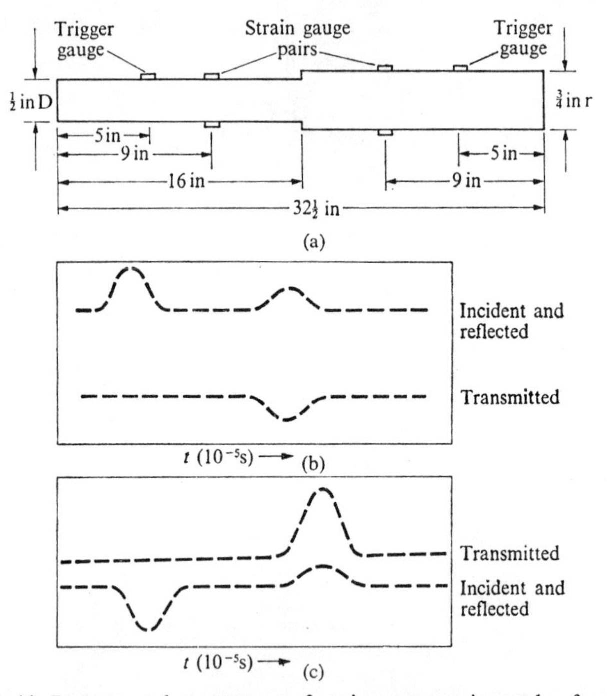

FIG. 2.32. (a) Geometry and arrangement of strain gauges on impact bar for study of longitudinal waves across a discontinuity; (b) experimental results for transmission from the small end toward the large end; (c) results for transmission from the large to the small end. (After Ripperger and Abramson [35], Figs. 2 and 4)

a short, cylindrical specimen of test material between two long rods of high-strength steel. A rectangular stress pulse is initiated in one of the elastic rods by the longitudinal impact of a striker bar. The stress wave dynamically loads the test specimen, with some of the incident wave energy being reflected from the specimen and some transmitting through the specimen to the second elastic rod. By using strain gauges to measure the incident, reflected, and transmitted waves, it is possible to determine the dynamic stress–strain relation for the material. A general schematic of the system is shown in Fig. 2.33.

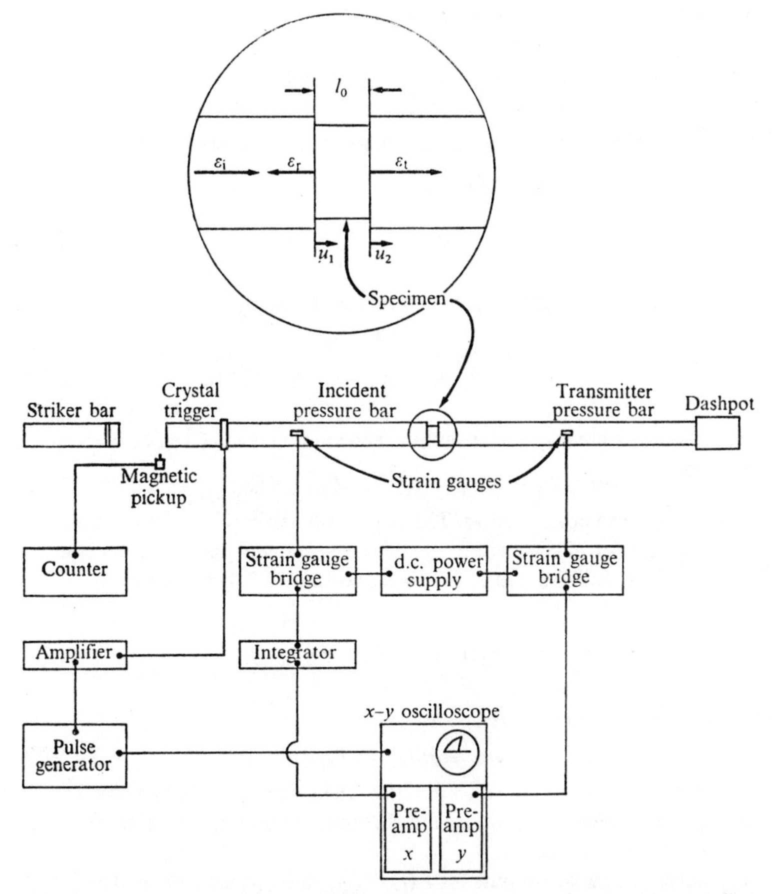

FIG. 2.33. Diagram of the split Hopkinson pressure bar apparatus. (After Lindholm [22], Fig. 1.)

The theory of the split Hopkinson bar is fairly simple and may be easily presented. From Fig. 2.33, it is supposed that the incident, reflected, and transmitted strain waves, ε_i, ε_r, ε_t are known from strain-gauge measurements. The displacements of the left and right interfaces of the specimen are denoted by u_1, u_2. From our previous work in one-dimensional rod theory, we have for the wave system in the incident bar

$$\begin{aligned} u &= f(x-c_0t)+g(x+c_0t) = u_i+u_r, \\ \epsilon &= f'+g' = \varepsilon_i+\varepsilon_r, \\ \dot{u} &= c_0(-f'+g') = c_0(-\varepsilon_i+\varepsilon_r). \end{aligned} \tag{2.7.1}$$

The displacement u_1 is then given by

$$u_1 = c_0 \int_0^t (-\varepsilon_i + \varepsilon_r)\, d\tau. \tag{2.7.2}$$

The transmitted wave system and the displacement u_2 are given by

$$u = h(x - c_0 t), \qquad \varepsilon = h' = \varepsilon_t,$$

$$\dot{u} = -c_0 h' = -c_0 \varepsilon_t, \qquad u_2 = -c_0 \int_0^t \varepsilon_t\, d\tau. \tag{2.7.3}$$

The average strain ε_s in the specimen is then given by

$$\varepsilon_s = \frac{u_2 - u_1}{l_0} = \frac{c_0}{l_0} \int_0^t (-\varepsilon_t + \varepsilon_i - \varepsilon_r)\, d\tau, \tag{2.7.4}$$

where l_0 is the original specimen length. The loads on each face of the specimen are given by

$$P_1 = AE(\varepsilon_i + \varepsilon_r), \qquad P_2 = AE\varepsilon_t, \tag{2.7.5}$$

where A is the rod cross-section. The important assumption is now made that wave-propagation effects within the short specimen may be neglected, so that $P_1 = P_2$. It follows that $\varepsilon_i + \varepsilon_r = \varepsilon_t$, so that (2.7.4) simplifies to

$$\varepsilon_s(t) = -\frac{2c_0}{l_0} \int_0^t \varepsilon_r\, d\tau. \tag{2.7.6}$$

The stress in the specimen is given by

$$\sigma_s = P_1/A_s = P_2/A_s, \tag{2.7.7}$$

where A_s is the specimen area. In practice, the strain ε_s is obtained by directly integrating the reflected strain ε_r and the stress is obtained directly from the transmitted strain signal ε_t.

Typical strain-gauge outputs for $\varepsilon_i(t)$, $\varepsilon_r(t)$, and $\varepsilon_t(t)$ are shown in Fig. 2.34.

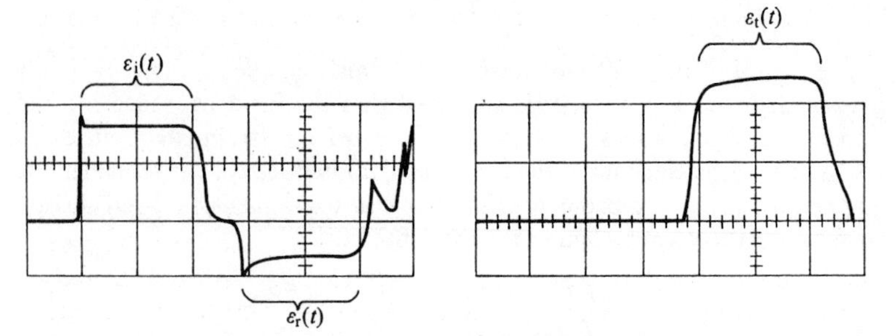

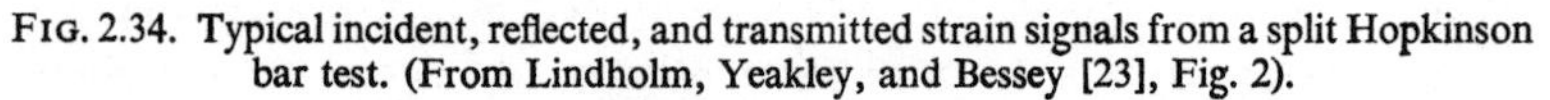

FIG. 2.34. Typical incident, reflected, and transmitted strain signals from a split Hopkinson bar test. (From Lindholm, Yeakley, and Bessey [23], Fig. 2).

Considering the incident wave in particular, it is seen that the wave is of nearly perfect rectangular shape, as predicted by simple longitudinal rod theory for the collision of two bars of identical cross-section and materials. There is, however, evidence of some high-frequency content at the start of the pulse. Results presented in Chapter 8 will bring this aspect out more clearly.

The split Hopkinson bar technique has been used by numerous investigators in studies of dynamic material properties. A fair amount of controversy has been associated with the test, mainly in conjunction with the quasi-static assumptions invoked in setting $P_1 = P_2$. The neglect of lateral-inertia effects in the specimen and the possible effects of interface friction at the specimen ends have also been the subject of discussion. Many contributions have been made here, but reference is made to Nicholas [29] for a survey of parts of this area and to Lindholm, Yeakley, and Bessey [23] and Nicholas [30] for discussion on the validity of the test technique.

2.7.4. *Lateral inertia effects*

The effect of lateral inertia on longitudinal waves was studied in § 2.5, where Love's development, also called Rayleigh's correction, was reviewed. Zemanek and Rudnick [40] have obtained results in this area confirming the validity of Love's theory for a limited range of wavelengths. Using the Love theory, it is not difficult to show that the longitudinal natural frequencies of a free–free rod are given by

$$\omega_n = n\pi c_0(l^2+n^2\pi^2\nu^2k^2)^{-\frac{1}{2}}, \tag{2.7.8}$$

where l is the rod length and k is the radius of gyration of the cross-section. Using an experimental setup as shown in Fig. 2.35(a), the rod natural frequencies (up to 140 modes) were measured. The phase velocity was then calculated from the formula

$$c = 2lf_n/n, \tag{2.7.9}$$

where $f_n = \omega_n/2\pi$. A plot of the resulting dispersion curve is shown in Fig. 2.26(b). Only a very limited range of wavenumber is shown. The case of a non-dispersive rod would be given by $\bar{c} = 1$. It is seen that the experimental results closely agree with the predictions of longitudinal theory with lateral-inertia effects included.

2.7.5. *Some other results of longitudinal wave experiments*

Many investigations of longitudinal waves in rods are aimed at determining certain material properties of the rod. For example, by measuring the decay of pulse amplitude with repeated reflection and traverse of the rod, material loss characteristics and/or visco-elastic parameters can be determined. Goldsmith, Polivka, and Yang [15] have studied pulse propagation in concrete cylinders in order to determine a material model and for assessing

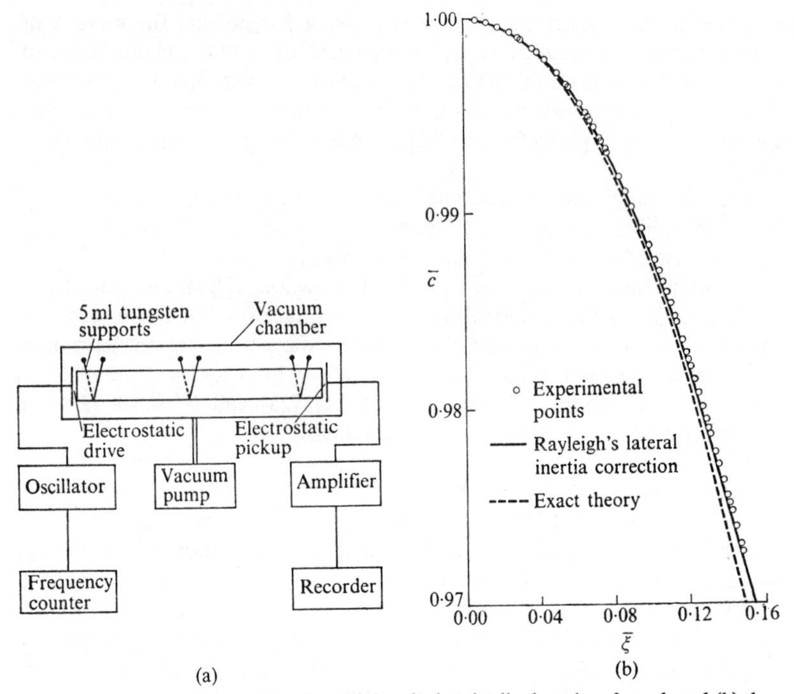

FIG. 2.35. (a) Apparatus used for determining the longitudinal modes of a rod, and (b) the dispersion curve for a limited wavenumber range, where $\bar{c} = c/c_0$, $\bar{\xi} = a\xi/2\pi$. (After Zemanek and Rudnicks [40], Figs. 1 and 5)

the effects of the highly inhomogeneous properties of concrete on pulse propagation. Similar studies were also carried out on rock cylinders by Goldsmith and Austin [16] and by Ricketts and Goldsmith [33]. The spallation phenomenon associated with the reflection of a compressive pulse from the free end of a rod has been used by Abbott and Cornick [1] to determine the tensile strength of brittle materials. Strain gauges mounted internally in the specimen bar have been used by Baker and Dove [2] to determine variations of stress distribution over the cross section. Lewis, Goldsmith, and Cunningham [21] have also used internally mounted gauges to study the wave propagation in cones. Brillhart and Dally [6] have used dynamic photo-elasticity to study waves in the conical geometry.

References

1. ABBOTT, B. W. and CORNISH, R. H. A stress-wave technique for determining the tensile strength of brittle materials. *Exp. Mech.* **22,** 148–53. (1965).
2. BAKER, W. E. and DOVE, R. C. Measurements of internal strains in a bar subjected to longitudinal impact. *Exp. Mech.* **19,** 307–11. (1962).

3. Barton, C. S., Volterra, E. G., and Citron, S. J. On elastic impacts of spheres on long rods. *Proc. 3rd U.S. natn. Cong. appl. Mech.* 89–94 (1958).
4. Becker, E. C. H. and Carl, H. Transient loading technique for mechanical impedance measurement. In *Experimental techniques in shock and vibration* (Ed. W. J. Worley). A.S.M.E., New York (1962).
5. Berlincourt, D. A., Curran, D. R., and Jaffe, H. Piezoelectric and Piezomagnetic materials and their function in transducers. In *Physical acoustics* (Ed. W. P. Mason) Chap. 3, Vol. 1, Academic Press, New York (1964).
6. Brillhart, L. V. and Dally, J. W. A dynamic photoelastic investigation of stress-wave propagation in cones. *Exp. Mech.* **25,** 1–9 (1968).
7. Davies, R. M. A critical study of the Hopkinson pressure bar. *Phil. Trans. R. Soc.* **A240,** 375–457 (1948).
8. Eisner, E. Design of sonic amplification transducers of high magnification. *J. acoust. Soc. Am.* **35,** 1367 (1963).
9. ——. Complete solutions of the 'Webster' horn equation. *J. acoust. Soc. Am.* **41,** 1126 (1967).
10. Eubanks, R. A., Muster, D., and Volterra, E. *An investigation of the dynamic properties of plastics and rubber-like materials.* Office of Nav. Res., Department of the Navy, Contract No. N7 ONR 32911, Tech. Rep. No. 1 (June 1952).
11. Feng, C. C. *Energy transfer in sonic impact coupling.* M.S. thesis, The Ohio State University (1969).
12. Fischer, H. C. Stress pulse in bar with neck or swell. *Appl. Scient. Res.* **A4,** 317–28 (1954).
13. Frederick, J. R. *Ultrasonic engineering.* John Wiley and Sons, New York (1965).
14. Goldsmith, W. *Impact: the theory and physical behaviour of colliding solids.* Edward Arnold, London (1960).
15. ——, Polivka, M., and Yang, T. Dynamic behavior of concrete. *Exp. Mech.* **23,** 65–79 (1966).
16. —— and Austin, C. F. Some dynamic characteristics of rocks. In *Stress Waves in Anelastic Solids* (Ed. H. Kolsky and W. Prager), pp. 277–303. Springer-Verlag, Berlin (1964).
17. Katz, H. W. (Ed.) *Solid state magnetic and dielectric devices.* John Wiley and Sons, New York (1959).
18. Kolsky, H. *Stress waves in solids.* Dover Publications, New York (1963).
19. ——. An investigation of the mechanical properties of materials at very high rates of loading. *Proc. phys. Soc.* **B62,** 676–700 (1949).
20. Langhaar, H. L. *Energy methods in applied mechanics.* John Wiley and Sons, New York (1962).
21. Lewis, J. L., Goldsmith, W., and Cunningham, D. M. Internal-strain measurements of longitudinal pulses in conical bars. *Exp. Mech.* **26,** 313–20 (1969).
22. Lindholm, U. S. Some experiments with the split Hopkinson pressure bar. *J. Mech. Phys. Solids* **12,** 317–35 (1964).
23. ——, Yeakley, L. M., and Bessey, R. L. *An investigation of the behavior of materials under high rates of deformation.* Tech. Rep. AFML-TR-68-194, Air Force Materials Laboratory. (1968).
24. Love, A. E. H. *A treatise on the mathematical theory of elasticity.* Dover Publications, New York (1944).

25. McLachlan, N. W. *Bessel functions for engineers.* Clarendon Press, Oxford (1961).
26. Merkulov, L. G. and Kharitonov, A. V. *Sov. Phys. Acoust.* **5**, 184 (1959).
27. Morse, P. M. *Vibration and sound.* McGraw-Hill, New York (1948).
28. —— and Feshbach, H. *Methods of theoretical physics.* Vols. I and II. McGraw-Hill, New York (1953).
29. Nicholas, T. *The mechanics of ballistic impact—A survey.* Tech. Rep. AFML-TR-67-208, Air Force Materials Laboratory. (1967).
30. ——. *An analytical study of the split Hopkinson bar technique for strain-rate dependent material behavior.* Tech. Rep. AFML-TR-71-155, Air Force Materials Laboratory (1971).
31. Pochhammer, L. Über die Fortpflanzungsgeschwindigkeiten kleiner Schwingungen in einem unbegrenzten istropen Kreiszylinder. *J. reine angew. Math.* **81**, 324–36 (1876).
32. Rayleigh, J. W. S. *The theory of sound,* Vols. I and II. Dover Publications, New York (1945).
33. Ricketts, T. E. and Goldsmith, W. Dynamic properties of rocks and composite structural materials. *Int. J. Rock Mech. Mining Sci.* **7**, 315–35 (1970).
34. Ripperger, E. A. The propagation of pulses in cylindrical bars—an experimental study. *Proc. 1st midwest. Conf. Solid Mech.* pp. 29–39 (1953).
35. —— and Abramson, H. N. Reflection and transmission of elastic pulses in a bar at a discontinuity in cross section. *Proc. 3rd midwest. Conf. Solid Mech.* p. 135 (1957).
36. St-Venant, B. D. and Flamant, M. Résistance vive ou dynamique des solides. Représentation graphique des lois du choc longitudinal. *C. r. hebd. Scéanc. Acad. Sci., Paris* **97**, 127, 214, 281, and 353 (1883).
37. Timoshenko, S. P. and Goodier, J. N. *Theory of elasticity* (3rd edn). McGraw-Hill, New York (1970).
38. ——. *Vibration problems in engineering.* Van Nostrand, New Jersey (1928).
39. Todhunter, I. and Pearson, K. *A history of the theory of elasticity,* Vol. I (1886), Vol. II, Parts 1 and 2 (1893). Dover Publications, New York (1960).
40. Zemanek, J. (Jr.) and Rudnick, I. Attenuation and dispersion of elastic waves in a cylindrical bar. *J. Acoust. Soc. Am.* **33**, 1283–8 (1961).

Problems

2.1. Derive the governing equation of motion for an inhomogeneous rod where the modulus varies as $E = E_0(1+\varepsilon x^2)$. Assume the density remains constant.

2.2. The effects of lateral inertia are neglected in the classical rod theory and uniaxial stress is assumed. Give a physical argument for the existence of radial shear stresses on the face of an element.

2.3. *Mechanical impedance.* Show the mechanical impedance of a semi-infinite rod is $Z = A\sqrt{(\rho E)}$. Determine the impedance of a mass, a spring, and a dashpot. To which of these three elements is the rod analogous?

2.4. *Effects of lateral constraint.* The lateral motion of a thin rod is unconstrained and a pressure pulse $p(t)$ applied to the end of a semi-infinite rod travels at the velocity c_0. Now instead of a semi-infinite rod, consider a semi-infinite solid or 'half-space', subjected over its entire surface to the normal pressure pulse $p(t)$. Show that the one-dimensional wave equation governs the longitudinal motion in such a media. Develop the expression for the propagation velocity. Is it faster or slower than c_0? Give a physical interpretation of the result.

2.5. *Effects of lateral constraint.* Consider a thin rod of circular cross-section imbedded in an elastic medium such that a lateral resistive pressure $f_r = -ku_r$ is developed, where u_r is the radial displacement of the surface. Derive the equation of motion for the system and sketch the general form of the frequency spectrum.

2.6. Consider the reflection–transmission expressions (2.2.12) in the form $\sigma_t = K_t\sigma_i$, $\sigma_r = K_r\sigma_i$, where K_t, K_r are appropriately defined transmission and reflection coefficients. Plot K_t and K_r as a function of Z_2/Z_1 for various ratios of A_1/A_2 ($\frac{1}{4}$, $\frac{1}{2}$, 1, 2).

2.7. Recall the discussion of the experimental procedure involving a 'time piece' of § 2.2. Consider a pulse propagating toward the end of a bar to which a time piece of length l and of identical properties and diameter has been attached with a grease layer. What must be the length of the time piece relative to the pulse length to completely 'trap' the pulse, leaving the bar at rest? Suppose a rectangular compressive pulse of magnitude σ_0 and pulse length L enters a time piece of length $L/2$. What will be the fly-off velocity of the piece? Suppose the incident pulse is triangular in shape, of amplitude σ_0 and of length L with the rising portion, or head of the pulse, being $L/3$ and the tail $2L/3$. Suppose the time piece is $L/3$ long. Determine the fly-off velocity.

2.8. The case of a rectangular pulse incident on an elastic boundary was analysed in § 1.2 and also considered in Problem 1.10. Such a boundary is 'lossless', ultimately returning all of the incident pulse energy in the reflected wave. Consider now a rectangular pulse incident on a bilinear spring boundary, as shown in Fig. P2.8(a). Loading occurs along $F = k_1\delta$ and unloading

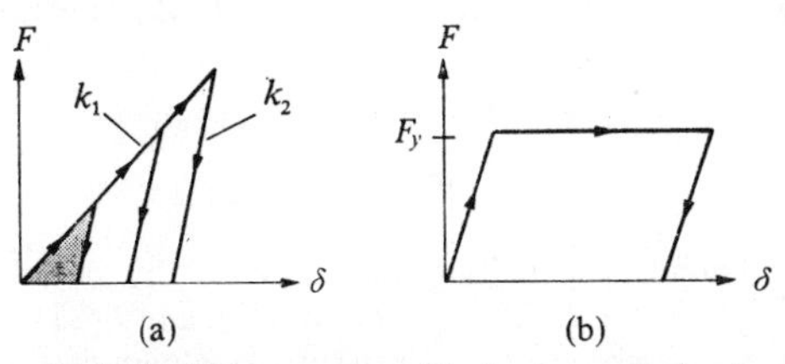

FIG. P2.8. (a) Bilinear spring; (b) elastic, perfectly plastic.

along one of the lines having slope k_2. The shaded area represents energy loss for unloading along a particular line. Such a model is a good approximation to a rock boundary. Predict the reflected wave. Consider an elastic, perfectly plastic boundary as in Fig. P2.8(b). Predict the reflected wave.

2.9. A rod of length l is fixed at one end and subjected to a compressive load P. The load is suddenly removed. Determine the resulting wave propagation in the bar. In particular, plot the displacement versus time for the end (does a sawtooth shape result?).

2.10. A rod of length l rests against a rigid base, but is not attached. A load P is applied to the rod and suddenly released (as in Problem 2.9). Determine the wave propagation in the rod. At some point, the rod will leave contact with the base. Determine the 'fly-off' velocity. Is the rod in uniform motion after fly-off?

2.11. Consider the free vibrations of a rod of length l, fixed at one end and constrained at the other by a spring of constant k. Derive the equation for the natural frequencies. Show that as $k \to 0$, ∞ that the fixed–free and fixed–fixed rod results are recovered.

2.12. The natural frequency of vibration of a simple spring–mass system is given by $\omega_n = \sqrt{(K/m)}$, where K is the spring constant and m the mass. Consider the longitudinal vibrations of an elastic rod, fixed at one end and with a mass attached at the other. Derive the frequency equation for this system. A result of the form $\alpha = \beta \tan \beta$ should be obtained, where α, β are parameters containing frequency and the properties of mass and rod. Suppose the mass of the rod is small in comparison with the attached mass. Show that the spring mass result is recovered if the approximation $\tan \beta = \beta$ is made. Show that the next order of approximation for $\tan \beta$ gives the result that adding one-third the mass of the rod to the end mass in the simple formula for ω_n predicts the natural frequency.

2.13. *Equivalent circuits.* Through analogies between mechanical and electrical quantities such as force and voltage and velocity and current, it is possible to develop electrical analogues to mechanical systems. The objective of this problem is to trace the steps for obtaining the electrical circuit for longitudinal rod vibration. Consider a rod as shown in Fig. P2.13(a), subjected to end

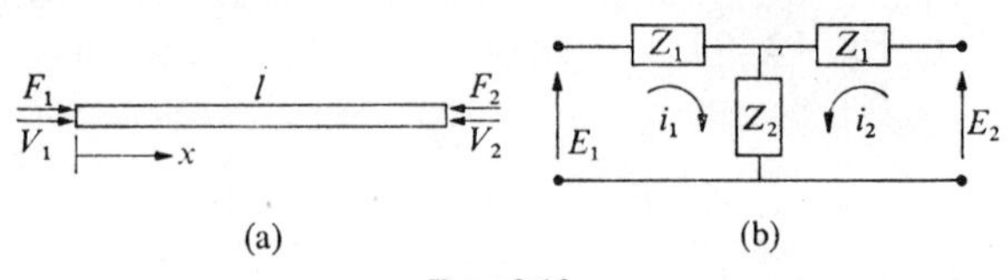

FIG. 2.13.

forces. The boundary conditions are specified by $\dot{u}(0, t) = V_1 \exp(\mathrm{i}\omega t)$, $u(l, t) = -V_2 \exp(\mathrm{i}\omega t)$, $F(0, t) = -F_1 \exp(\mathrm{i}\omega t)$, and $F(l, t) = -F_2 \exp(\mathrm{i}\omega t)$. Considering a solution $u(x, t) = U(x)\exp(\mathrm{i}\omega t)$, obtain two equations for F_1, F_2 in terms of V_1, V_2. Now write the circuit equations for the Tee circuit shown in Fig. P2.13(b) expressing E_1, E_2 in terms of i_1, i_2. Comparing the two results, you should be able to show that $Z_1 = \mathrm{i}Z_0 \tan(\gamma l/2)$, $Z_2 = Z_0/\mathrm{i} \sin \gamma l$, where Z_0 is the rod impedance and $\gamma = \omega/c_0$.

Behaviour near resonance. Suppose $x = l$ is a free end. You should be able to reduce the electrical circuit to a single impedance given by $Z_4 = \mathrm{i}Z_0 \sin \gamma l/\cos \gamma l$. The resonance of this circuit is at $\gamma l = \pi$. Express this near resonance as $\pi(\omega_0 + \delta\omega)/\omega_0$, where ω_0 is the resonance frequency. Near resonance, show that

$$Z_4 \cong \mathrm{i}Z_0\left(\pi\frac{\delta\omega}{\omega_0}-\frac{\pi^3}{6}\frac{\delta\omega^3}{\omega_0^3}+\cdots\right).$$

Finally, expand the impedance of a series LC circuit near resonance. By comparison of the two results, show $L = \rho Al/2$, $C = 2l/\pi^2 AE$. For references here, see Katz [19, p. 170], Berlincourt, Curran, and Jaffe [5, pp. 220–49].

2.14. Consider the results leading to Fig. 2.13. Sketch the displacements predicted by (2.3.48) for $x = 3l/4$, $l/2$, $l/4$, 0.

2.15. A cylindrical rod of length l collides with a semi-infinite rod of the same outside diameter. The impact rod has a longitudinal cylindrical hole drilled in the impact end. The depth of the hole is $l/4$ and the diameter is one-half the rod diameter. Construct the characteristics plane representation of the wavefront propagation for this impact problem.

2.16. Hertz contact theory gives a non-linear relation between contact force and approach. Plot this relation for the given material-geometry parameters of (2.4.44). Approximate the curve with a linear relation $F = K_1\alpha$. Incorporate this relation in the impact analysis and develop the modified form of (2.4.42). Note the greatly simplified governing equation. Obtain the predicted stress pulse and compare to the results of Fig. 2.22. Comment on the need for the non-linear relationship. Now observe the approximate half sine-wave shape of the pulses of Fig. 2.22 and make the further simplification of omitting the $\dot{\alpha}$ term of the governing equation. Determine the rebound velocity (it should equal the impact velocity). Interpret the significance of omitting the equivalent 'damping' from the system.

2.17. Determine the impedance (in non-dimensional form) of the catenoidal horn.

2.18. Determine the impedance (in non-dimensional form) of the exponential horn. Note the governing equation has constant coefficients and that somewhat simpler solution-forms apply than for the linear and conical cases.

2.19. Refer to the design chart of Fig. 2.25. It is desired that a 33·02 cm (13 in) steel exponential horn be resonant at 10^4 Hz and have a small end diameter of 1·27 cm ($\frac{1}{2}$ in). What will be the diameter of the large end?

2.20. In the Davies analysis reviewed in § 2.5, verify the calculation of the residues (2.5.73) and (2.5.76). Show that the result (2.5.77) reduces to that obtained for classical rod theory if $H = \psi = 0$.

2.21. Draw the frequency spectrum corresponding to Love's theory for longitudinal waves.

2.22. Obtain the expression for the natural frequencies of a free–free rod of length l with lateral-inertia effects included. Consider a bar of steel 25·4 cm long and of circular cross-section. Take $c_0 = 5{\cdot}08\times10^3$ m s^{-1} and Poisson's ratio as 0·3. Calculate the first and third mode frequencies for bars of 0·635 cm, 2·54 cm, and 10·16 cm ($\frac{1}{4}$ in, 1 in, and 4 in) diameters. Compare results to that of classical theory.

3 Flexural waves in thin rods

OUR considerations now turn to the transverse motion of thin rods resulting from bending action. It will be found that the simplest governing theory for such motion, based on the Bernoulli–Euler theory of beams, yields a dispersive system, in contrast to the wave equation that results for simple strings and longitudinal and torsional rod motion. Following the pattern developed in earlier chapters, harmonic waves, the initial-value problem, and forced motion for infinite beams will be considered first. Free and forced motion of finite beams will follow. A section will be devoted to the effects of an elastic and viscous foundation on beam motion, since many interesting problems enter here. Then, analogous to the development of the Love theory for longitudinal rod motion, an improved theory for beams will be developed. Finally, the effects of rod curvature on wave propagation will be studied. The theory that will be developed will include longitudinal extensional effects, as well as flexural effects.

3.1. Propagation and reflection characteristics

The governing equation for transverse rod motion will be developed and the frequency spectrum and dispersion curve associated with harmonic wave propagation presented. The initial-value problem for two types of input disturbance will be considered. The forced motion of a beam will be considered, and the concept of local energy storage described. Finally, the reflection of harmonic waves from boundaries will be studied.

3.1.1. *The governing equation*

Consider a thin rod undergoing transverse motion, as shown in Fig. 3.1(a), and consider a differential element of the rod as isolated in Fig. 3.1(b). Bending moment M, shear force V, and variations of these quantities act on the beam element, as well as a distributed force $q(x, t)$. We invoke the basic hypothesis of the Bernoulli–Euler theory of beams: namely, that plane cross-sections initially perpendicular to the axis of the beam remain plane and perpendicular to the neutral axis during bending. This assumption implies that the longitudinal strains vary linearly across the depth of the beam and that, for elastic behaviour, the neutral axis of the beam passes through the centroid of the cross-section. Further, it results that the relationship between

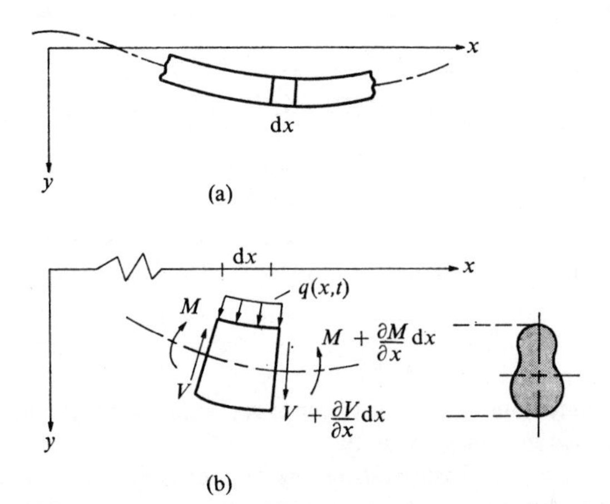

FIG. 3.1. (a) Thin rod undergoing transverse motion and (b) an element of rod subjected to various loads.

the bending moment and curvature is given by

$$\partial^2 y/\partial x^2 = -M/EI, \tag{3.1.1}$$

where y is the coordinate to the neutral surface of the beam. The result (3.1.1) carries the assumption that slopes and deflections of the beam are small.

Writing the equation of motion in the vertical direction for the element of Fig. 3.1(b), we have

$$-V+\left(V+\frac{\partial V}{\partial x}\,\mathrm{d}x\right)+q\,\mathrm{d}x = \rho A\,\mathrm{d}x\frac{\partial^2 y}{\partial t^2}, \tag{3.1.2}$$

where A is the cross-sectional area of the beam and ρ is the mass density per unit volume. This reduces to

$$\frac{\partial V}{\partial x}+q = \rho A\frac{\partial^2 y}{\partial t^2}. \tag{3.1.3}$$

Summation of moments is the second equation to be written. If we neglect the rotational-inertia effects of the element, the moment equation is, effectively, that of statics and gives

$$V = \frac{\partial M}{\partial x} \tag{3.1.4}$$

where the higher-order contributions of the loading q to the moment are neglected. Substituting (3.1.4) in (3.1.3) gives

$$\frac{\partial^2 M}{\partial x^2}+q = \rho A\frac{\partial^2 y}{\partial t^2}. \tag{3.1.5}$$

Finally, substituting (3.1.1) in (3.1.5) gives

$$\frac{\partial^2}{\partial x^2}\left(EI\frac{\partial^2 y}{\partial x^2}\right)+\rho A\frac{\partial^2 y}{\partial t^2} = q(x, t), \tag{3.1.6}$$

as the governing equation for the transverse motion of a thin rod or beam. If the loading is absent ($q = 0$), and material of the beam is homogeneous, so that E is constant and, furthermore, the cross-section is constant so that I is constant, then (3.1.6) reduces to the form

$$\frac{\partial^4 y}{\partial x^4}+\frac{1}{a^2}\frac{\partial^2 y}{\partial t^2} = 0, \qquad a^2 = \frac{EI}{\rho A}. \tag{3.1.7}$$

We immediately note this is not of the wave equation form and that a does not have the dimension of velocity.

Let us examine the basic restrictions of the theory. The restrictions on beam homogeneity and constant cross-section imposed in reducing (3.1.6) to (3.1.7) are not fundamental. In other words, beams not having these properties can be analysed within the framework of the theory, starting from (3.1.6), with only the cost of increased mathematical complexity entering. The assumption of small deflections has been specifically pointed out. Should this not be obeyed, the exact expression for curvature would have to be used in (3.1.1), instead of the second-derivative approximation. It is also assumed that the motion of the beam is in a single plane (the x,y plane). This implies that the beam has a symmetrical cross-section, with the plane of symmetry being in the x,y-plane. Any external loads must also act in this plane in order for such motion to occur. In order to relax this assumption, one must be prepared to consider the problem of coupled torsional–flexural motions. This was briefly alluded to in § 2.6.

We now consider the assumption wherein the effects of rotational inertia were neglected. To examine this assumption, we write general expressions for the translational and rotational kinetic energies of the element as

$$\mathrm{d}T_t = \tfrac{1}{2}\,\mathrm{d}m\left(\frac{\partial y^2}{\partial t^2}\right), \qquad \mathrm{d}T_r = \tfrac{1}{2}(k^2\,\mathrm{d}m)\dot{\theta}^2, \tag{3.1.8}$$

where $\mathrm{d}m$ is the mass of the element $\mathrm{d}x, k^2$ is its radius of gyration, and $\dot{\theta}$ refers to its rotational velocity. Neglecting rotary-inertia effects omits the contribution of $\mathrm{d}T_r$ to the system energy. At low frequencies, where $\dot{\theta}$ may not be large, this assumption is valid. As it turns out, however, for higher frequency motion this contribution must be included in the development. We note the similarity of this assumption to that occurring in rods, where radial-inertia effects were ignored.

The most important assumption, and hence limitation, to the theory pertains to the basic Bernoulli–Euler hypothesis that plane sections remain plane. As may be recalled from strength-of-materials theory, this assumption is valid only for pure bending. When shearing forces are present, the retention of the plane-section hypothesis in effect states that infinite shear rigidity is assumed or, equivalently, that shearing deformations are neglected. This assumption, as well as that regarding rotational inertia, will be re-examined in § 3.4 in the development of an improved theory.

3.1.2. *Propagation of harmonic waves*

We begin our study of waves and vibrations in beams in the usual fashion. Consider a beam, infinitely long and governed by the equation of motion (3.1.7). We establish conditions for the propagation of harmonic waves by assuming

$$y = Ae^{i(\gamma x-\omega t)} \tag{3.1.9}$$

and substituting in (3.1.7). The resulting relation between frequency and wavenumber is

$$\gamma^4-\omega^2/a^2 = 0 \tag{3.1.10}$$

or

$$\gamma = \pm\sqrt{\left(\frac{\omega}{a}\right)}, \qquad \pm i\sqrt{\left(\frac{\omega}{a}\right)}. \tag{3.1.11}$$

The relationship between phase velocity and wavenumber is obtained by substituting $\omega = \gamma c$ in (3.1.11) or using a propagating wave of the form $\exp\{i\gamma(x-ct)\}$ in the original substitution. By either means, one obtains

$$c = \pm a\gamma. \tag{3.1.12}$$

The frequency spectrum, showing all branches of (3.1.11) and the dispersion curve, showing only the positive branch of (3.1.12) is given in Fig. 3.2.

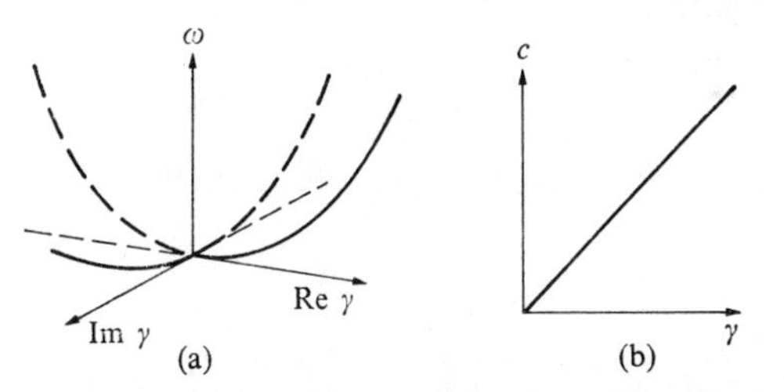

FIG. 3.2. (a) Frequency spectrum and (b) dispersion curve governing transverse harmonic waves in a rod.

Interpretation of these results is straightforward on the basis of our past work. Thus, referring to (3.1.11), we realize that the positive and negative real roots correspond to rightward- and leftward-propagating harmonic waves, while the imaginary wavenumbers give non-propagating, spatially varying vibrations. While not acceptable solutions for the case of propagating waves, we expect these solutions to play a role in forced and transient motion.

There is a striking anomaly in the results, particularly evident in the dispersion curve representation. It is seen that the phase velocity increases without limit for increasing wavenumber or shorter wavelength. This characteristic will yield some peculiar predictions in the transient response. Ultimately, it will be shown to be a consequence of neglecting rotary inertia and shear effects.

3.1.3. *The initial-value problem*

We now consider the problem of an infinite beam with an initial disturbance imposed on it of the form

$$y(x,0) = f(x), \qquad \frac{\partial y(x,0)}{\partial t} = g(x) = a\frac{\mathrm{d}^2 h(x)}{\mathrm{d}x^2}. \tag{3.1.13}$$

The special form selected for the initial velocity is purely for later convenience. Performing the Laplace transform on (3.1.7) gives

$$a^2\frac{\mathrm{d}^4\bar{y}}{\mathrm{d}x^4}(x,s)+s^2\bar{y}(x,s) = sf(x)+g(x). \tag{3.1.14}$$

Taking the Fourier integral transform of the above gives

$$a^2\xi^4\,\bar{Y}(\xi,s)+s^2\,\bar{Y}(\xi,s) = sF(\xi)-\xi^2aH(\xi). \tag{3.1.15}$$

The transformed solution is thus

$$\bar{Y}(\xi,s) = \frac{sF(\xi)-\xi^2aH(\xi)}{s^2+a^2\xi^4}. \tag{3.1.16}$$

We first perform the Laplace inversion on the result. This is easily done if it is noted that

$$\mathscr{L}^{-1}\left(\frac{s}{s^2+a^2\xi^4}\right) = \cos\xi^2at, \qquad \mathscr{L}^{-1}\left(\frac{1}{s^2+a^2\xi^4}\right) = \frac{1}{\xi^2a}\sin\xi^2at. \tag{3.1.17}$$

Hence, we obtain

$$Y(\xi,t) = F(\xi)\cos\xi^2at-H(\xi)\sin\xi^2at. \tag{3.1.18}$$

The convolution theorem may be used to invert this last result. To do so, the Fourier inversions of $\cos\xi^2at$, $\sin\xi^2at$ are required. Using the basic result† that

$$\left(\frac{1}{2\pi}\right)^{\frac{1}{2}}\int_{-\infty}^{\infty}\exp(-\xi^2a-i\xi x)\,\mathrm{d}\xi = \left(\frac{1}{2a}\right)^{\frac{1}{2}}\exp\left(\frac{-x^2}{4a}\right), \tag{3.1.19}$$

it is possible to obtain

$$F^{-1}\{\cos\xi^2at\} = \frac{1}{2\sqrt{(at)}}\left(\cos\frac{x^2}{4at}+\sin\frac{x^2}{4at}\right),$$

$$F^{-1}\{\sin\xi^2at\} = \frac{1}{2\sqrt{(at)}}\left(\cos\frac{x^2}{4at}-\sin\frac{x^2}{4at}\right). \tag{3.1.20}$$

† See Sneddon [22, p. 112].

From the convolution integral form

$$f * g = \frac{1}{\sqrt{(2\pi)}} \int_{-\infty}^{\infty} f(\zeta)g(x-\zeta)\,\mathrm{d}\zeta, \tag{3.1.21}$$

and using the results (3.1.20), we obtain for the formal solution of the problem

$$y(x,t) = \frac{1}{2\sqrt{(2\pi at)}} \int_{-\infty}^{\infty} f(x-\zeta)\left(\cos\frac{\zeta^2}{4at} + \sin\frac{\zeta^2}{4at}\right)\mathrm{d}\zeta -$$

$$-\frac{1}{2\sqrt{(2\pi at)}} \int_{-\infty}^{\infty} h(x-\zeta)\left(\cos\frac{\zeta^2}{4at} - \sin\frac{\zeta^2}{4at}\right)\mathrm{d}\zeta. \tag{3.1.22}$$

We consider a specific form for the input, given by

$$y(x,0) = f(x) = f_0\exp(-x^2/4b^2),\ \dot{y}(x,0) = 0. \tag{3.1.23}$$

This corresponds to a Gaussian distribution of displacement and, while giving a rather difficult expression to evaluate, corresponds to a physically acceptable displacement field. In developing the response to this input, Sneddon [22, p. 113] points out that it is frequently more advisable to work with the direct inverse transform of (3.1.18), given by

$$y(x,t) = \frac{1}{\sqrt{(2\pi)}} \int_{-\infty}^{\infty} \{F(\xi)\cos\xi^2 at - H(\xi)\sin\xi^2 at\}\exp(-\mathrm{i}\xi x)\,\mathrm{d}\xi, \tag{3.1.24}$$

than to work with the general result (3.1.22). Noting from tables that

$$F\{\exp(-x^2/4b^2)\} = \sqrt{2}b\exp(-\xi^2 b^2). \tag{3.1.25}$$

Then, with $H(\xi) = 0$, (3.1.24) reduces to

$$y(x,t) = \frac{f_0 b}{\sqrt{\pi}} \int_{-\infty}^{\infty} \exp(-\xi^2 b^2)\cos\xi^2 at\,\exp(-\mathrm{i}\xi x)\,\mathrm{d}\xi. \tag{3.1.26}$$

By writing $\cos\xi^2 at$ in exponential form and making use of the general result (3.1.25), one obtains

$$y(x,t) = \frac{f_0}{(1+a^2t^2/b^4)^{\frac{1}{4}}} \exp\left\{-\frac{x^2b^2}{4(b^4+a^2t^2)}\right\} \times$$

$$\times \cos\left\{\frac{atx^2}{4(b^4+a^2t^2)} - \tfrac{1}{2}\tan^{-1}\left(\frac{at}{b^2}\right)\right\}. \tag{3.1.27}$$

Fortunately, this rather complicated result has been presented by Morse [16, pp. 155–6] in a most interesting form, as shown in Fig. 3.3. The solid line represents the motion of the beam at various instants of time. The

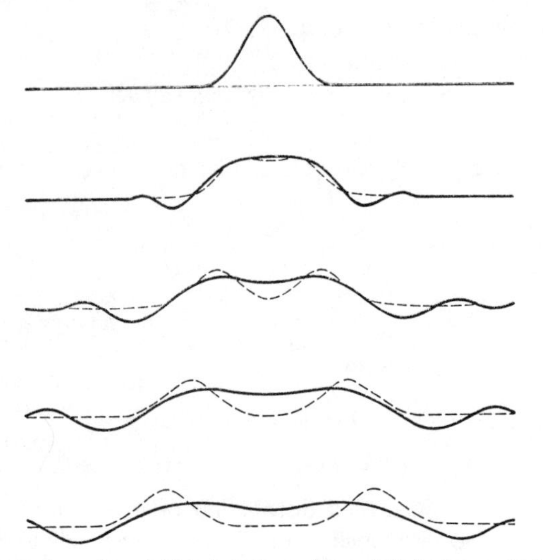

FIG. 3.3. Propagation of an initial disturbance in an infinite beam. (After Morse [16, Fig. 31].)

important item to note is the dispersion of the pulse shape at later times. Also presented in dashed lines is the non-distorted propagation of the Gaussian pulse in a string.

We now consider a more severe input to the system corresponding to an impulse in displacement. This will be specified by initial conditions given by

$$y(x, 0) = f(x) = \delta(x), \qquad \dot{y}(x, 0) = 0. \tag{3.1.28}$$

Such a displacement might be envisaged as a limiting case of the Gaussian input for smaller and smaller values of b. Using the general result (3.1.22), the solution may be written immediately as

$$y(x, t) = \frac{1}{2\sqrt{(2\pi at)}}\left(\cos\frac{x^2}{4at} + \sin\frac{x^2}{4at}\right) = \frac{1}{2\sqrt{(\pi at)}}\sin\left(\frac{x^2}{4at} + \frac{\pi}{4}\right). \tag{3.1.29}$$

The general form of this response is shown in Fig. 3.4 for a fixed location and variable time and for a fixed time at variable location. The first thing to be noted in either representation is that the high-frequency components appear at the front of the wave. For representation at $x = x_0$, this is indicated by the increased oscillation rate for smaller and smaller time, while for the representation at $t = t_0$, this is indicated by the more rapid oscillations at greater distance. This aspect is confirmed by the dispersion curve for the beam (Fig. 3.2(b)), where higher velocities are associated with higher frequencies. The second item to note is the singular behaviour of the oscillation

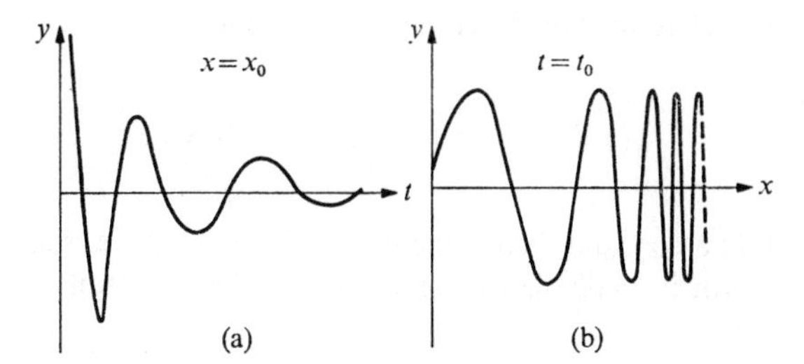

FIG. 3.4. (a) Waves at a given location x_0 in a beam for varying time and (b) at a given time t_0 for various x resulting from an initial impulse displacement field.

for small values of time. This is illustrated in the figure and is evident in the result (3.1.29). The peculiar aspect of the result is that at $t = 0+\varepsilon$, a disturbance of very high frequency is predicted instantaneously at remote locations. This is the manifestation of the infinite phase velocity anomaly pointed out in conjunction with the dispersion curve for the beam.†

A final remark on this problem is warranted. It might be postulated that the anomalous response is a consequence of the physically-unacceptable impulse initial displacement. Such a function can only be constructed by superimposing contributions from the entire wavelength (or wavenumber) spectrum, so that wavelengths approaching zero are indeed present in the initial waveform. However, the point is that it is the Bernoulli–Euler model of the beam that predicts that these components will travel at infinite velocity.

A slightly more physically realizable set of initial conditions are given by

$$y(x, 0) = 0, \qquad \dot{y}(x, 0) = \delta(x). \tag{3.1.30}$$

These conditions would approximate a sharp blow being struck at the origin of a beam initially undeformed. Applying the Laplace and then the Fourier transforms to (3.1.7) gives the transformed solution

$$\bar{Y}(\xi, s) = \frac{1}{\sqrt{(2\pi)}} \frac{1}{s^2+a^2\xi^4}. \tag{3.1.31}$$

The inverse Laplace transform of this function has already been given by (3.1.17), so that we have

$$Y(\xi, t) = \frac{1}{\sqrt{(2\pi a)}} \frac{\sin \xi^2 at}{\xi^2}. \tag{3.1.32}$$

Taking the inverse Fourier transform gives

$$y(x, t) = \frac{1}{2\pi a} \int_{-\infty}^{\infty} \frac{\sin a\xi^2 t}{\xi^2} e^{-i\xi x} \, d\xi. \tag{3.1.33}$$

† See Problem 3.7 regarding use of the stationary-phase method on this problem.

Since $(\sin a\xi^2 t)/\xi^2$ is an even function, (3.1.33) reduces to

$$y(x, t) = \frac{1}{2\pi a}\int_0^\infty \frac{\sin a\xi^2 t}{\xi^2}\cos \xi x \, d\xi, \tag{3.1.34}$$

which is, within a constant factor, the inverse Fourier cosine transform. This inverse transform is available from more comprehensive transform tables,† and is given by

$$y(x, t) = \frac{x}{4a}[S_2\{(2\pi a t)^{-\frac{1}{2}}x\} - C_2\{(2\pi a t)^{-\frac{1}{2}}x\}] + \\ + \frac{1}{2}\left(\frac{t}{\pi a}\right)^{\frac{1}{2}} \sin\left(\frac{x^2}{4at} + \frac{\pi}{4}\right), \tag{3.1.35}$$

where S_2 and C_2 are special function forms called the *Fresnel* integrals, defined by‡

$$C_2(v) = \frac{1}{\sqrt{(2\pi)}}\int_0^v \frac{\cos u}{u^{\frac{1}{2}}}\, du, \qquad S_2(v) = \frac{1}{\sqrt{(2\pi)}}\int_0^v \frac{\sin u}{u^{\frac{1}{2}}}\, du. \tag{3.1.36}$$

In order to determine the actual waveform numerical evaluation of (3.1.36) must be carried out using tabulated values of the Fresnel integrals.§ We note that for large x but small t that the high-frequency behaviour of the previous problem is still present. However, from the coefficient of the sine term it is seen that the amplitude of this 'instantaneous' high-frequency disturbance will be small.

3.1.4. *Forced motion of a beam*

We shall concern ourselves primarily with forced harmonic motion of a beam. Consider, as a first simple example, the case of a semi-infinite beam subjected to a harmonically varying end displacement. The governing equation is given by (3.1.7) for this problem. The boundary conditions are

$$y(0, t) = y_0 e^{-i\omega t}, \qquad \frac{\partial^2 y(0, t)}{\partial x^2} = 0. \tag{3.1.37}$$

The second condition results from the fact that the end of the beam is free to rotate during the motion, so that the end moment is zero. Then (3.1.1) leads to the condition of the second derivative. Considering a solution of the form

$$y(x, t) = Y(x)e^{-i\omega t}, \tag{3.1.38}$$

† Vol. I, p. 23 of Reference [6].
‡ p. 300 of Reference [1].
§ op. cit., pp. 321–2.

the governing equation reduces to

$$\frac{\mathrm{d}^4Y}{\mathrm{d}x^4}-\gamma^4Y = 0, \qquad \gamma^2 = \frac{\omega}{a}. \tag{3.1.39}$$

This has the solution

$$Y = A\mathrm{e}^{\mathrm{i}\gamma x}+B\mathrm{e}^{-\mathrm{i}\gamma x}+C\mathrm{e}^{\gamma x}+D\mathrm{e}^{-\gamma x}. \tag{3.1.40}$$

In applying boundary conditions to determine the constants, some of the considerations first presented in § 1.2 apply. Specifically, the radiation condition requires that $B = 0$ since that term, in conjunction with the time variation for the problem, gives incoming waves. The second consideration bears on the third term of (3.1.40). For $x > 0$, this results in unbounded response for increasing x, so we set $C = 0$ on physical grounds. The remaining two constants are determined from the boundary conditions (3.1.37). These give

$$A+D = y_0, \qquad -A+D = 0, \tag{3.1.41}$$

resulting in $A = D = y_0/2$. The resulting motion is given by

$$y = \frac{y_0}{2}\mathrm{e}^{\mathrm{i}(\gamma x-\omega t)}+\frac{y_0}{2}\mathrm{e}^{-\gamma x}\mathrm{e}^{-\mathrm{i}\omega t}. \tag{3.1.42}$$

The response is thus made up of a propagating and a non-propagating term. If we recall the problem of forced motion of a semi-infinite string on an elastic foundation (see Fig. 1.23), it was found for frequencies above a cutoff frequency that motion corresponding to the first term of (3.1.42) occurred, while below the cutoff frequency motion corresponding to the second term resulted. In the present problem, both occur simultaneously. In terms of energy, we may interpret the one term as continuously radiated energy and the other as *local stored energy*. This phenomenon will occur also in more complex problems. We conclude by again noting that the 'unacceptable' solutions for harmonic waves (the imaginary wavenumbers) play an important role in forced motion problems.

Let us consider a more general case of forced motion of a semi-infinite rod, as specified by

$$y(0, t) = V(t), \qquad \partial^2y(0, t)/\partial x^2 = 0, \tag{3.1.43}$$

with the beam being initially at rest. A double-transform approach will be used and a formal solution obtained. The main objective will be to demonstrate certain aspects of the Fourier cosine and sine transforms.

Apply the Laplace transform to the governing equation (3.1.7), giving

$$\frac{\mathrm{d}^4\bar{y}(x, s)}{\mathrm{d}x^4}+\frac{s^2}{a^2}\bar{y}(x, s) = 0. \tag{3.1.44}$$

We now consider application of the Fourier transform on the spatial variable. Because $0 < x < \infty$, either the Fourier cosine or sine transform must be used, and the question is—which one? This is, in fact, a common problem in transform applications and the answer is usually not obvious *a priori*. Consider using the cosine transform on (3.1.44). We must first evaluate the cosine transform of the fourth derivative term. Thus

$$F_c(\bar{y}^{\text{iv}}) = \sqrt{\left(\frac{2}{\pi}\right)} \int_0^\infty \frac{d^4\bar{y}}{dx^4} \cos \xi x \, dx. \tag{3.1.45}$$

Carrying out a repeated integration by parts gives

$$F_c(\bar{y}^{\text{iv}}) = \sqrt{\left(\frac{2}{\pi}\right)} [\bar{y}''' \cos \xi x + \bar{y}''\xi \sin \xi x - \bar{y}'\xi^2 \cos \xi x + \bar{y}\xi^3 \sin \xi x]_0^\infty - $$
$$ - \xi^4 \sqrt{\left(\frac{2}{\pi}\right)} \int_0^\infty \bar{y} \cos \xi x \, dx. \tag{3.1.46}$$

We note the integral appearing in (3.1.46) is merely the cosine transform of $\bar{y}$. The resulting transformed equation is then

$$\left(\xi^4 - \frac{s^2}{a^2}\right) \bar{Y}_c = \sqrt{\left(\frac{2}{\pi}\right)} [\bar{y}''' \cos \xi x + \bar{y}''\xi \sin \xi x - \bar{y}'\xi^2 \cos \xi x + \bar{y}\xi^3 \sin \xi x]_0^\infty. \tag{3.1.47}$$

We consider the right-hand side of this result. Since the problem is one of transient motion, it is legitimate to assume the beam at rest so that

$$\bar{y} = \bar{y}' = \bar{y}'' = \bar{y}''' = 0$$

at infinity. For the lower limit, we have that $\sin \xi x = 0$ at $x = 0$. There remain the terms $\bar{y}'''$ and $\bar{y}'$. For this problem, these are unknowns to be determined and not specified boundary conditions. Evidently the choice of the cosine transform is improper for this problem since it does not 'ask' for the right boundary conditions.

If the Fourier sine transform is applied, we obtain for the fourth derivative

$$F_s(\bar{y}^{\text{iv}}) = \sqrt{\left(\frac{2}{\pi}\right)} \int_0^\infty \frac{d^4\bar{y}}{dx^4} \sin \xi x \, dx,$$
$$ = \sqrt{\left(\frac{2}{\pi}\right)} [\bar{y}''' \sin \xi x - \bar{y}''\xi \cos \xi x - \bar{y}'\xi^2 \sin \xi x + \bar{y}\xi^3 \cos \xi x]_0^\infty + $$
$$ + \xi^4 \sqrt{\left(\frac{2}{\pi}\right)} \int_0^\infty \bar{y} \sin \xi x \, dx. \tag{3.1.48}$$

Proceeding directly to inspect the bracketed term of the result, it is seen that the difficulty that arose with the cosine transform is now removed. The upper limits are zero, based on the same arguments as used in the cosine-transform attempt. Two of the lower limit terms are removed since $\sin \xi x = 0$. The two remaining terms involve $\bar{y}''(0, s)$ and $y(0, s)$. The first of these is zero and the second is specified by the Laplace transform of the boundary condition (2.1.43). Thus, the sine transform has 'asked' for the proper conditions and is the proper transform for this problem.

The resulting transformed solution is

$$\bar{Y}_s(\xi, s) = \sqrt{\left(\frac{2}{\pi}\right)} \frac{a^2\xi^3 \bar{V}(s)}{s^2+a^2\xi^4}. \tag{3.1.49}$$

The major point of this example has now been made. The inversion of the result, first on the sine transform and then on the Laplace, using the convolution theorem in the latter step, yields the formal solution. It will be left as an exercise to obtain the result

$$y(x, t) = \frac{x}{4\pi}\sqrt{\left(\frac{2\pi}{a}\right)} \int_0^t \frac{V(\tau)}{(t-\tau)^{\frac{3}{2}}} \left\{\sin\frac{x^2}{4a(t-\tau)} + \cos\frac{x^2}{4a(t-\tau)}\right\} d\tau. \tag{3.1.50}$$

As our last example of forced motion, we will revert to a simpler loading situation. Consider an infinite beam subjected to a harmonically varying, concentrated load placed at $x = \zeta$. The inhomogeneous equation

$$\frac{\partial^4 y}{\partial x^4} + \frac{1}{a^2}\frac{\partial^2 y}{\partial t^2} = \frac{1}{EI}\delta(x-\zeta)e^{-i\omega t} \tag{3.1.51}$$

now governs the motion. The specified problem is, of course, one of determining the Green's function for the system. Considering a solution of the type $y = Y(x)\exp(-i\omega t)$, we obtain

$$\frac{d^4 Y}{dx^4} - \frac{\omega^2}{a^2} Y = \frac{1}{EI}\delta(x-\zeta). \tag{3.1.52}$$

Applying the exponential Fourier transform to (3.1.52) gives the transformed solution as

$$\bar{Y}(\xi) = \frac{1}{\sqrt{(2\pi)EI}} \frac{e^{i\xi\zeta}}{\xi^4-\xi_0^4}, \qquad \xi_0^2 = \frac{\omega}{a}. \tag{3.1.53}$$

The inverse Fourier transform gives

$$Y(x) = \frac{1}{2\pi EI} \int_{-\infty}^{\infty} \frac{e^{-i\xi(x-\zeta)}}{\xi^4-\xi_0^4} d\xi. \tag{3.1.54}$$

This result is quite simple to evaluate by residue theory. Recalling the general arguments developed in the first chapter pertaining to Fig. 1.9, we see that for $x > \zeta$, we must close the semicircular contour in the lower half of the complex ξ-plane. There are four simple poles of the integrand given by

$$\xi = \pm\xi_0, \pm i\xi_0. \tag{3.1.55}$$

Thus, two poles are on the real axis and two on the imaginary axis. The imaginary root in the upper half-plane is excluded by the contour closure. In so far as the poles on the real axis are concerned, we must recall the arguments developed in § 1.1 leading to Fig. 1.9. Thus, poles on the real axis must be included or excluded from the general contour by slight indentations as dictated by the radiation condition. Inspecting the exponential of (3.1.54) that will appear in the residues, it is seen that the pole at $\xi = -\xi_0$ in conjunction with the time behaviour $\exp(-i\omega t)$ will give waves propagating to the right for $x > \zeta$. The pole at $\xi = +\xi_0$ will give improper radiation and must be excluded from the contour by an indentation below it. The resulting contour is shown in Fig. 3.5. The residues for the simple poles at $\xi = -\xi_0$,

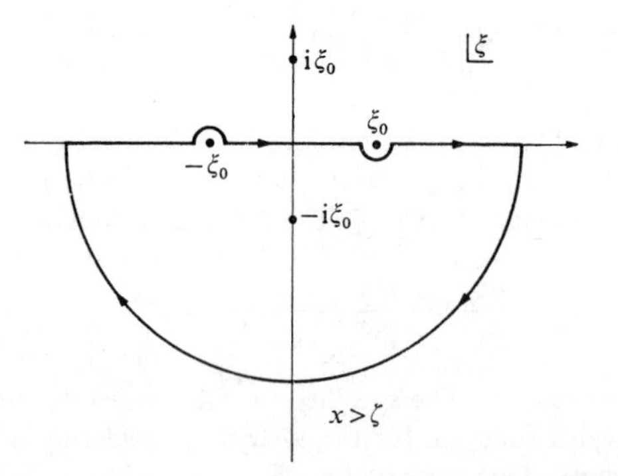

FIG. 3.5. Contour for the infinite beam subjected to a harmonically varying, concentrated load.

$-i\xi_0$ are given, respectively, by

$$-\frac{\exp\{i\xi_0(x-\zeta)\}}{4\xi_0^3}, \qquad -\frac{\exp\{-\xi_0(x-\zeta)\}}{4\xi_0^3}. \tag{3.1.56}$$

From the residue theorem, we thus obtain for the inversion of (3.1.54), the result

$$y(0, t) = \frac{i}{4\xi_0^3 EI} \exp[i\{\xi_0(x-\zeta)-\omega t\}]-$$

$$-\frac{1}{4\xi_0^3 EI} \exp\{-\xi_0(x-\zeta)\}\exp(-i\omega t), \tag{3.1.57}$$

where the time variation has been included. Superposition methods would be used to obtain the response to more general loadings. We again note the presence of propagating waves and local vibrations in the beam response.

3.1.5. *Reflection of harmonic waves*

As in our previous studies in the wave motion of strings and rods, we are interested in the interaction of propagating waves with boundaries. In approaching this problem, we see that it does not make sense to consider an incident pulse of specific shape because of the dispersive characteristics of the beam. Thus, pulse distortion due to dispersion and pulse distortion due to boundary interaction would both occur, and no general conclusions could be reached on the reflection characteristics of the boundary. Instead, we consider a steady train of harmonic waves to be incident, in the manner considered in § 1.2 for strings, and determine the frequency-dependent amplitude ratios of the reflected waves.

The various simple end conditions for beams are pinned, fixed, and free ends. Elastic boundary constraints may be incorporated in terms of deflection or torsion springs or both. These conditions must be given mathematical expression in terms of the deflection y, slope $\partial y/\partial x$, moment M and shear V where, for the last two quantities, we have

$$M = -EI\frac{\partial^2 y}{\partial x^2}, \qquad V = -EI\frac{\partial^3 y}{\partial x^3}. \tag{3.1.58}$$

End mass and dashpot constraints are also possible. Table 3.1 summarizes the various simple boundary conditions and those consisting of single elements such as a spring, dashpot, or mass.

In considering the reflective characteristics of various boundaries, we start with a solution of the governing equation (3.1.7) in the form

$$y = Ae^{-i(\gamma x+\omega t)}+Be^{i(\gamma x-\omega t)}+Ce^{-\gamma x}e^{-i\omega t}+De^{\gamma x}e^{-i\omega t}. \tag{3.1.59}$$

We presume that the incident harmonic wave, appearing as the first term of the solution, is specified in terms of amplitude A and frequency ω. It will have a corresponding wavenumber as given by (3.1.11). Next, we delete the last term of the solution, because of the unbounded response. This leaves the two amplitudes B and C to be determined by the boundary condition. As a simple first example, suppose the end of a semi-infinite beam $x > 0$ is pinned, so that the conditions, from Table 3.1, are

$$y(0, t) = \partial^2 y(0, t)/\partial x^2 = 0. \tag{3.1.60}$$

Substituting (3.1.59), with $D = 0$, in (3.1.60) gives

$$\begin{aligned} A+B+C &= 0, \\ -A-B+C &= 0. \end{aligned} \tag{3.1.61}$$

TABLE 3.1

Type	Diagram	Equation
Pinned		$y(0,t)=0$, $\dfrac{\partial^2 y(0,t)}{\partial x^2}=0$
Fixed		$y(0,t)=0$, $\dfrac{\partial y(0,t)}{\partial x}=0$
Free		$\dfrac{\partial^2 y(0,t)}{\partial x^2}=0$, $\dfrac{\partial^3 y(0,t)}{\partial x^3}=0$
Deflected spring	K_D	$\dfrac{\partial^2 y(0,t)}{\partial x^2}=0$, $-EI\dfrac{\partial^3 y(0,t)}{\partial x^3}=K_D\, y(0,t)$
Torsion spring	K_T	$y(0,t)=0$, $EI\dfrac{\partial^2 y(0,t)}{\partial x^2}=K_T\dfrac{\partial y(0,t)}{\partial x}$
Mass	m	$\dfrac{\partial^2 y(0,t)}{\partial x^2}=0$, $-EI\dfrac{\partial^3 y(0,t)}{\partial x^3}=m\dfrac{\partial^2 y(0,t)}{\partial t^2}$
Dashpot	c	$\dfrac{\partial^2 y(0,t)}{\partial x^2}=0$, $-EI\dfrac{\partial^3 y(0,t)}{\partial x^3}=c\dfrac{\partial y(0,t)}{\partial t}$

Solving these in terms of A gives $C = 0$, $B = -A$, so that the resulting disturbance is given by

$$y = Ae^{-i(\gamma x+\omega t)} - Ae^{i(\gamma x-\omega t)},$$
$$= -2iA \sin \gamma x e^{-i\omega t}. \tag{3.1.62}$$

This corresponds to a standing-wave condition.

Consider the slightly more complicated case of a free end. The boundary conditions are

$$\partial^2 y(0, t)/\partial x^2 = \partial^3 y(0, t)/\partial x^3 = 0. \tag{3.1.63}$$

Substituting (3.1.59) in the above, again with $D \doteq 0$, gives

$$\begin{aligned} -A-B+C &= 0, \\ iA-iB-C &= 0. \end{aligned} \tag{3.1.64}$$

Solving this system gives

$$\frac{B}{A} = \frac{-1+i}{1+i} = i, \qquad \frac{C}{A} = \frac{2i}{1+i} = 1+i. \tag{3.1.65}$$

The fact that the ratios are complex numbers merely indicates phase shifting relative to the incident wave. Of particular interest here is the presence of the C term. Referring back to the solution (3.1.59), it is seen that this is the local effect previously noted in the forced vibration of beams. Such an effect is frequently referred to as *end resonance* and is also noted in more complex structures.

3.2. Free and forced vibrations of finite beams

In studying the free vibrations of beams, we are interested in the natural frequencies and normal modes for various constraint conditions, whether orthogonality of these modes exists, and in applying the modal solutions to the initial-value problem. Most of the methods used in problems of forced vibrations of beams are similar to those used in strings, such as the Green's function or transform approaches. These will be reviewed in the context of beam analysis and applied to several problems.

3.2.1. *Natural frequencies of finite beams*

Numerous combinations of boundary constraints of the type shown in Table 3.1 are possible for beams of finite length. If we confine our specific attention to combinations of the simple conditions, these consisting of the pinned (P), clamped (C), and free (F) ends, we see there are six possible combinations given by P–P, P–C, P–F, C–C, C–F, and F–F. To determine the natural frequencies and normal modes for these cases, we employ separation of variables. Thus we let

$$y = Y(x)T(t), \tag{3.2.1}$$

and substituting in (3.1.7) gives

$$a^2\frac{Y^{\mathrm{iv}}}{Y} = -\frac{T''}{T} = \omega^2. \tag{3.2.2}$$

Thus, for T we have

$$T = A\cos\omega t + B\sin\omega t \tag{3.2.3}$$

The solution for Y may be expressed in exponential form. A more useful form for our present purposes is given by

$$Y = C_1\sin\beta x + C_2\cos\beta x + C_3\sinh\beta x + C_4\cosh\beta x, \qquad \beta^4 = \omega^2/a^2. \tag{3.2.4}$$

An alternative form is given by

$$Y = D_1(\cos \beta x + \cosh \beta x) + D_2(\cos \beta x - \cosh \beta x) + \\ + D_3(\sin \beta x + \sinh \beta x) + D_4(\sin \beta x - \sinh \beta x). \qquad (3.2.5)$$

The various combinations of simple boundary conditions will now be considered.

1. *Pinned–pinned:* The boundary conditions, in terms of $Y(x)$, are given by

$$Y(0) = \mathrm{d}^2Y(0)/\mathrm{d}x^2 = 0, \\ Y(l) = \mathrm{d}^2Y(l)/\mathrm{d}x^2 = 0. \qquad (3.2.6)$$

Using the solution form (3.2.4) in these boundary conditions gives

$$C_2 = C_3 = C_4 = 0$$

and the frequency equation

$$\sin \beta l = 0, \qquad \beta l = n\pi \qquad (n = 1, 2, 3...,). \qquad (3.2.7)$$

The radial and cyclic frequencies are, respectively,

$$\omega_n = a\left(\frac{n\pi}{l}\right)^2, \qquad f_n = \frac{a\pi}{2l^2}n^2 \qquad (n = 1, 2, 3,...). \qquad (3.2.8)$$

In terms of the beam parameters we have

$$f_n = \frac{n^2\pi}{2l^2}\sqrt{\left(\frac{EI}{\rho A}\right)}. \qquad (3.2.9)$$

The normal modes are then

$$Y_n = \sin \beta_n x, \qquad (3.2.10)$$

and a general solution is given by

$$y = \sum_{n=1}^{\infty} (A_n \cos \omega_n t + B_n \sin \omega_n t) \sin \beta_n x. \qquad (3.2.11)$$

2. *Free–free:* The boundary conditions are

$$\frac{\mathrm{d}^2Y(0)}{\mathrm{d}x^2} = \frac{\mathrm{d}^3Y(0)}{\mathrm{d}x^3} = 0, \\ \frac{\mathrm{d}^2Y(l)}{\mathrm{d}x^2} = \frac{\mathrm{d}^3Y(l)}{\mathrm{d}x^3} = 0. \qquad (3.2.12)$$

Using the solution form (3.2.5) in the above quickly gives $D_2 = D_4 = 0$, with the remaining two conditions at $x = l$ being given by

$$\begin{pmatrix} (\cos \beta l - \cosh \beta l) & (\sin \beta l - \sinh \beta l) \\ (\sin \beta l + \sinh \beta l) & -(\cos \beta l - \cosh \beta l) \end{pmatrix} \begin{pmatrix} D_1 \\ D_3 \end{pmatrix} = 0. \qquad (3.2.13)$$

Setting the determinant of coefficients equal to zero gives the frequency equation

$$(\cos \beta l - \cosh \beta l)^2 + (\sin^2 \beta l - \sinh^2 \beta l) = 0, \tag{3.2.14}$$

which may be reduced to

$$\cos \beta l \cosh \beta l = 1. \tag{3.2.15}$$

The first six roots of this equation are given by

$$\begin{gathered} \beta_0 l = 0, \quad \beta_1 l = 4{\cdot}730, \quad \beta_2 l = 7{\cdot}853, \\ \beta_3 l = 10{\cdot}996, \quad \beta_4 l = 14{\cdot}137, \quad \beta_5 l = 17{\cdot}279. \end{gathered} \tag{3.2.16}$$

The cyclic frequencies are expressed by

$$f_n = \frac{(\beta_n l)^2}{2\pi l^2} \sqrt{\left(\frac{EI}{\rho A}\right)}. \tag{3.2.17}$$

The root $\beta_0 l = 0$ corresponds to the rigid-body motion term for this beam.

3. *Clamped–clamped:* The boundary conditions are

$$\begin{gathered} Y(0) = \mathrm{d}Y(0)/\mathrm{d}x = 0, \\ Y(l) = \mathrm{d}Y(l)/\mathrm{d}x = 0. \end{gathered} \tag{3.2.18}$$

There results $D_1 = D_3 = 0$ and

$$\begin{pmatrix} (\cos \beta l - \cosh \beta l) & (\sin \beta l - \sinh \beta l) \\ (\sin \beta l + \sinh \beta l) & -(\cos \beta l - \cosh \beta l) \end{pmatrix} \begin{pmatrix} D_2 \\ D_4 \end{pmatrix} = 0. \tag{3.2.19}$$

This is seen to be the same matrix of coefficients that resulted in the previous problem. The roots will be the same except that the zero root is excluded. Thus $\beta_1 l = 4{\cdot}730$, $\beta_2 l = 7{\cdot}853$, etc. The mode shapes are, of course, different.

4. *Clamped–free:* The boundary conditions are

$$\begin{gathered} Y(0) = \mathrm{d}Y(0)/\mathrm{d}x = 0, \\ \mathrm{d}^2 Y(l)/\mathrm{d}x^2 = \mathrm{d}^3 Y(l)/\mathrm{d}x^3 = 0. \end{gathered} \tag{3.2.20}$$

It is found that $D_1 = D_3 = 0$ with the frequency equation being given by

$$\cos \beta l \cosh \beta l = -1. \tag{3.2.21}$$

The first few roots are

$$\begin{gathered} \beta_1 l = 1{\cdot}875, \quad \beta_2 l = 4{\cdot}694, \quad \beta_3 l = 7{\cdot}855, \\ \beta_4 l = 10{\cdot}996, \quad \beta_5 l = 14{\cdot}137, \quad \beta_6 l = 17{\cdot}279. \end{gathered} \tag{3.2.22}$$

5. *Clamped–pinned:* The boundary conditions are

$$\begin{gathered} Y(0) = \mathrm{d}Y(0)/\mathrm{d}x = 0, \\ Y(l) = \mathrm{d}^2 Y(l)/\mathrm{d}x^2 = 0. \end{gathered} \tag{3.2.23}$$

The resulting frequency equation is

$$\tan \beta l = \tanh \beta l. \tag{3.2.24}$$

The first few roots are given by

$$\beta_1 l = 3{\cdot}927, \quad \beta_2 l = 7{\cdot}069, \quad \beta_3 l = 10{\cdot}210,$$
$$\beta_4 l = 13{\cdot}352, \quad \beta_5 l = 16{\cdot}493. \tag{3.2.25}$$

6. *Pinned–free:* The boundary conditions are

$$Y(0) = \mathrm{d}^2 Y(0)/\mathrm{d}x^2 = 0, \tag{3.2.26}$$
$$\mathrm{d}^2 Y(l)/\mathrm{d}x^2 = \mathrm{d}^3 Y(l)/\mathrm{d}x^3 = 0.$$

The frequency equation turns out to be the same as for case 5 except that the zero root is an acceptable result for this case. It is ultimately interpretable in terms of rigid-body rotation.

The first few roots of the frequency equations are summarized in terms of the frequency parameter $\beta_n l$ in Table 3.2. The cyclic frequencies are given in

TABLE 3.2

Case	$\beta_0 l$	$\beta_1 l$	$\beta_2 l$	$\beta_3 l$	$\beta_4 l$	$\beta_5 l$
		1·875	4·694	7·855	10·996	14·137
		3·142	6·283	9·425	12·566	15·708
	0	3·927	7·069	10·210	13·352	16·493
		3·927	7·069	10·210	13·352	16·493
		4·730	7·853	10·996	14·137	17·279
	0	4·730	7·853	10·996	14·137	17·279

terms of the frequency parameter by (3.2.17). It is to be noted that the frequencies are irregularly spaced for most cases for the first two or three modes. However, as the mode number increases, the difference between the frequency parameter for all cases approaches π. Thus, for the clamped–free beam, $\beta_5 l - \beta_4 l = 3{\cdot}141$, for the clamped–clamped case, $\beta_5 l - \beta_4 l = 3{\cdot}142$, and so on. The pinned–pinned case has all of the modes equally spaced. We may interpret this general result as indicating an insensitivity to end constraint at higher frequencies.

3.2.2. *Orthogonality*

We investigate possible orthogonality of the normal modes in a manner similar to that employed in the study of strings (§ 1.3). Thus consider two normal modes Y_n and Y_m, where each satisfies

$$Y_n^{\text{iv}}-\frac{\omega_n^2}{a^2}Y_n=0,$$
$$Y_m^{\text{iv}}-\frac{\omega_m^2}{a^2}Y_m=0. \tag{3.2.27}$$

Multiplying the first eqn (3.2.27) by Y_m and the second eqn by Y_n, forming the difference between the two systems, and integrating over the interval of the beam gives

$$\int_0^l (Y_m Y_n^{\text{iv}}-Y_n Y_m^{\text{iv}})\,dx=-\left(\frac{\omega_m^2-\omega_n^2}{a^2}\right)\int_0^l Y_m Y_n\,dx. \tag{3.2.28}$$

Integrating the term $Y_m Y_n^{\text{iv}}$ by parts, we obtain

$$\int_0^l Y_m Y_n^{\text{iv}}\,dx=[Y_m Y_n'''-Y_m' Y_n''+Y_m'' Y_n'-Y_m''' Y_n]_0^l+\int_0^l Y_m^{\text{iv}} Y_n\,dx. \tag{3.2.29}$$

Inserting this result in (3.2.28) gives

$$[Y_m Y_n'''-Y_m' Y_n''+Y_m'' Y_n'-Y_m''' Y_n]_0^l=\left(\frac{\omega_n^2-\omega_m^2}{a^2}\right)\int_0^l Y_m Y_n\,dx. \tag{3.2.30}$$

We immediately note that combinations of the various simple boundary conditions such as clamped–free, pinned–pinned, and so on cause the left-hand side to vanish. Further, any set of homogeneous boundary conditions of the form

$$aY+bY'+cY''+dY'''=0, \tag{3.2.31}$$

where a, b, c, d are constants, will cause the left-hand side to vanish. A condition such as (3.2.31) could arise from a combination of deflection and torsion springs, for example. However, time-dependent boundary conditions, such as those resulting from mass or dashpot contributions, would not cause vanishing. So, for conditions of the form (3.2.31) we have that

$$\int_0^l Y_m Y_n\,dx=0 \qquad (m\neq n), \tag{3.2.32}$$

which establishes orthogonality.†

† op. cit., p. 41.

3.2.3. *The initial-value problem*

The initial value problem for the finite beam is specified by the homogeneous governing equation (3.1.7), the particular boundary conditions for the problem and, the initial conditions

$$y(x, 0) = f(x), \qquad \partial y(x, 0)/\partial t = g(x). \tag{3.2.33}$$

The solution is obtained by superimposing particular solutions to give

$$y(x, t) = \sum_{n=1}^{\infty}(A_n \cos \omega_n t + B_n \sin \omega_n t)Y_n(x), \tag{3.2.34}$$

where $Y_n(x)$ are the normal modes for the beam. If the modes are orthogonal, the constants A_n, B_n are determined after the manner of Fourier series investigations and give

$$A_n = \frac{1}{N}\int_0^l f(x)Y_n(x)\,dx, \qquad B_n = \frac{1}{N\omega_n}\int_0^l g(x)Y_n(x)\,dx, \tag{3.2.35}$$

where N is the normalizing factor for the interval. The general response is given by

$$y(x, t) = \sum_{n=1}^{\infty}\left\{\cos \omega_n t \int_0^l f(x)Y_n(x)\,dx + \frac{\sin \omega_n t}{\omega_n}\int_0^l g(x)Y_n(x)\,dx\right\}\frac{Y_n(x)}{N}. \tag{3.2.36}$$

3.2.4. *Forced vibrations of beams—methods of analysis*

The methods of analysis for the forced motions of finite beams are quite similar to those presented for strings (§ 1.3) and include the Green's function approach and various transform methods. These will be reviewed in the context of the beam problem.

In the Green's function approach, a harmonically varying concentrated load will be analysed, so that the governing equation is

$$\frac{\partial^4 y}{\partial x^4} + \frac{1}{a^2}\frac{\partial^2 y}{\partial t^2} = \frac{1}{EI}\delta(x-\xi)e^{-i\omega t}. \tag{3.2.37}$$

We let $y = Y(x)\exp(-i\omega t)$ and substitute in (3.3.73), giving

$$\frac{d^4 Y}{dx^4} - \beta^4 Y = \frac{1}{EI}\delta(x-\xi), \qquad \beta^4 = \frac{\omega^2}{a^2}. \tag{3.2.38}$$

Because of the properties of the Dirac delta function, we have

$$Y^{\text{iv}} - \beta^4 Y = 0, \qquad x \neq \xi. \tag{3.2.39}$$

There will be two solutions to this equation, one for the region $x < \xi$, one for $x > \xi$. Thus

$$x < \xi; \quad Y_1(x) = A_1 \sin \beta x + B_1 \cos \beta x + C_1 \sinh \beta x + D_1 \cosh \beta x, \quad (3.2.40)$$

$$x > \xi; \quad Y_2(x) = A_2 \sin \beta x + B_2 \cos \beta x + C_2 \sinh \beta x + D_2 \cosh \beta x. \quad (3.2.41)$$

The application of two boundary conditions at $x = 0$ will enable two of the constants of the set A_1, B_1, C_1, D_1 to be eliminated, while application of two conditions at $x = l$ will allow the set A_2, B_2, C_2, D_2 to be reduced to two. Four conditions are required at $x = \xi$ to determine the problem. Three of the conditions are

$$\begin{aligned} Y_1(\xi-\varepsilon) &= Y_2(\xi+\varepsilon), \\ Y_1'(\xi-\varepsilon) &= Y_2'(\xi+\varepsilon), \\ Y_1''(\xi-\varepsilon) &= Y_2''(\xi+\varepsilon), \end{aligned} \quad (3.2.42)$$

where ε is an arbitrarily small quantity. These are continuity conditions on the deflection, slope, and curvature. The fourth relation is a jump condition on the shear and may be established by integrating (3.2.38) across $x = \xi$. Thus

$$\int_{\xi-\varepsilon}^{\xi+\varepsilon} Y^{\text{iv}}\,dx - \beta^4 \int_{\xi-\varepsilon}^{\xi+\varepsilon} Y\,dx = \frac{1}{EI}\int_{\xi-\varepsilon}^{\xi+\varepsilon} \delta(x-\xi)\,dx = \frac{1}{EI}. \quad (3.2.43)$$

Since Y is continuous across $x = \xi$, the second integral vanishes and we have

$$Y'''(x)\Big|_{\xi-\varepsilon}^{\xi+\varepsilon} = 1/EI, \quad (3.2.44)$$

giving

$$Y_2'''(\xi) - Y_1'''(\xi) = 1/EI. \quad (3.2.45)$$

This establishes the necessary fourth condition at $x = \xi$.

The final details of the solution depend on actually specifying the boundary conditions and solving for the constants $A_1, B_1, \ldots, C_2, D_2$. Without going into the details of this process, the results will be of the form

$$Y(x) = \begin{cases} G(x, \xi, \omega), & (0 < x < \xi) \\ G(\xi, x, \omega), & (\xi < x < l), \end{cases} \quad (3.2.46)$$

so that

$$y(x, t) = G(x, \xi, \omega)e^{-i\omega t}. \quad (3.2.47)$$

will give the system response. The principle of integral superposition must be applied to solve for more general loadings in both time and space.

In considering methods involving integral transforms, we have at our disposal the Laplace transform and the finite Fourier sine and cosine transforms. These methods may be applied to special loading types such as concentrated loads in space and/or harmonically varying loads in time, with extension to more general loads being done by integral superposition.

Alternatively, the case of arbitrary loading may be carried through in the solution with the final results, again in an integral superposition form, obtained by applying the convolution theorem. Obviously there are numerous approaches to the same end, with the question as to the best or most convenient method being governed by the conditions of the specific problem.

In the following, several approaches using the integral transform method will be given, attempting to retain, as in the last section, some generality as to the statement of boundary conditions.

1. *Finite Fourier transform.* Consider the inhomogeneous equation

$$\frac{\partial^4 y}{\partial x^4}+\frac{1}{a^2}\frac{\partial^2 y}{\partial t^2}=\frac{1}{EI}q(x,t), \tag{3.2.48}$$

where the initial conditions are taken as zero. Let $q(x, t) = Q(x)\exp(-i\omega t)$. Then, taking $y(x, t) = Y(x)\exp(-i\omega t)$, we have

$$Y^{\text{iv}}-\beta^4 Y = Q(x)/EI, \qquad \beta^4 = \omega^2/a^2. \tag{3.2.49}$$

The finite Fourier transform will now be applied to the above. The question as to whether the sine or cosine transform is appropriate will be determined by the particular boundary conditions.† Recalling that these transforms are given by

$$\bar{F}_c(n) = \frac{1}{l}\int_0^l F(x)\cos\frac{n\pi}{l}x\,dx, \qquad \bar{F}_s(n) = \frac{1}{l}\int_0^l F(x)\sin\frac{n\pi}{l}x\,dx, \tag{3.2.50}$$

with the respective inverses

$$F(x) = \frac{\bar{F}_c(0)}{l}+2\sum_{n=1}^{\infty}\bar{F}_c(n)\cos\frac{n\pi}{l}x \tag{3.2.51}$$

and

$$F(x) = 2\sum_{n=1}^{\infty}\bar{F}_s(n)\sin\frac{n\pi}{l}x. \tag{3.2.52}$$

The Fourier sine and cosine transformed equations are, respectively,

$$(\alpha_n^4-\beta^4)\bar{Y}_s(n) = -\frac{1}{l}[Y'''(x)\sin\alpha_n x-\alpha_n Y''(x)\cos\alpha_n x-$$
$$-\alpha_n^2 Y'(x)\sin\alpha_n x+\alpha_n^3 Y(x)\cos\alpha_n x]_0^l+\frac{\bar{Q}(n)}{EI}, \tag{3.2.53}$$

and

$$(\alpha_n^4-\beta^4)\bar{Y}_c(n) = -\frac{1}{l}[Y'''(x)\cos\alpha_n x+\alpha_n Y''(x)\sin\alpha_n x-$$
$$-\alpha_n^2 Y'(x)\cos\alpha_n x-\alpha_n^3 Y(x)\sin\alpha_n x]_0^l+\frac{\bar{Q}(n)}{EI}, \tag{3.2.54}$$

† Recall that a similar question arose in applying the sine and cosine transforms to the semi-infinite beam in § 3.1.

where $\alpha_n = n\pi/l$. Now it should be evident in the sine transformed solution $\bar{Y}_s(n)$ that the following quantities must be specified:

$$\bar{Y}_s(n): \quad x = 0,\ Y(0),\ Y''(0),$$
$$x = l,\ Y(l),\ Y''(l). \tag{3.2.55}$$

In the case of a beam simply supported at both ends, we see that

$$Y(0) = Y''(0) = Y(l) = Y''(l) = 0,$$

so that the sine transform would be appropriate for such a problem. Actually, a study of the bracketed expression in $\bar{Y}_s(n)$ shows that it will vanish or be specified if the linear combinations

$$Y''(l) - \alpha_n^2 Y(l), \qquad Y''(0) - \alpha_n^2 Y(0) \tag{3.2.56}$$

are zero or known, which suggests that certain other non-ideal boundary conditions might be possible.

A similar study of $\bar{Y}_c(n)$ shows that

$$\begin{aligned} x &= 0,\ Y'(0),\ Y'''(0) \\ x &= l,\ Y'(l),\ Y'''(l) \end{aligned} \tag{3.2.57}$$

must be given, or, more generally,

$$Y'''(l) - \alpha_n^2 Y'(l), \qquad Y'''(0) - \alpha_n^2 Y'(0) \tag{3.2.58}$$

must be known.

Assuming that the conditions are such that the bracket vanishes, and following through on the sine transform case, we have

$$\bar{Y}_s(n) = \frac{\bar{Q}_s(n)}{EI(\alpha_n^4 - \beta^4)}, \tag{3.2.59}$$

where

$$\bar{Q}_s(n) = \frac{1}{l}\int_0^l Q(x)\sin\frac{n\pi}{l}x\,dx. \tag{3.2.60}$$

Taking the inverse transform we have

$$Y(x) = \frac{2}{EI}\sum_{n=1}^{\infty}\frac{\bar{Q}_s(n)}{\alpha_n^4 - \beta^4}\sin\frac{n\pi}{l}x \tag{3.2.61}$$

or

$$Y(x) = \frac{2}{EIl}\sum_{n=1}^{\infty}\left[\sin\frac{n\pi}{l}x \Big/ (\alpha_n^4 - \beta^4)\right]\int_0^l Q(u)\sin\frac{n\pi}{l}u\,du. \tag{3.2.62}$$

2. *Expansion in normal modes.* A slightly different viewpoint of this problem is to consider eqn (3.2.49) and expand $Y(x)$, $Q(x)$ in the normal modes of

the given system. Thus assume

$$Y(x) = \sum_{n=1}^{\infty} a_n Y_n(x), \qquad Q(x) = \sum_{n=1}^{\infty} b_n Y_n(x), \tag{3.2.63}$$

where the $Y_n(x)$ satisfy

$$Y_n^{\text{iv}}(x) - \beta_n^4 Y_n(x) = 0. \tag{3.2.64}$$

The coefficient b_n is given by

$$b_n = \frac{2}{l}\int_0^l Y_n(x)Q(x)\,dx. \tag{3.2.65}$$

Substituting (3.2.63) in (3.2.49) gives

$$\sum_{n=1}^{\infty}(a_n Y_n^{\text{iv}} - \beta^4 a_n Y_n) = \frac{1}{EI}\sum_{n=1}^{\infty} b_n Y_n. \tag{3.2.66}$$

Using (3.2.64) this reduces to

$$\sum_{n=1}^{\infty}(\beta_n^4 - \beta^4)a_n Y_n = \sum_{n=1}^{\infty} b_n Y_n. \tag{3.2.67}$$

Comparing coefficients gives

$$a_n = \frac{b_n}{EI(\beta_n^4 - \beta^4)}, \tag{3.2.68}$$

so that the solution becomes

$$Y(x) = \frac{2}{EIl}\sum_{n=1}^{\infty}\frac{Y_n(x)}{\beta_n^4 - \beta^4}\int_0^l Y_n(u)Q(u)\,du. \tag{3.2.69}$$

3. *The Laplace transform.* As in the last section, we again assume a loading $q(x, t) = Q(x)\exp(-i\omega t)$ and thus still consider the non-homogeneous ordinary differential equation (3.2.49). Applying the Laplace transform gives

$$\bar{Y}(s) = \frac{1}{s^4 - \beta^4}\{s^3 Y(0) + s^2 Y'(0) + sY''(0) + Y'''(0)\} + \frac{1}{EI}\frac{\bar{Q}(s)}{s^4 - \beta^4}. \tag{3.2.70}$$

To invert (3.2.70), the following Laplace inversions are required:

$$\begin{aligned}
\mathscr{L}^{-1}\left(\frac{1}{s^4 - \beta^4}\right) &= \frac{1}{\beta^3}\left(\frac{\sinh\beta x - \sin\beta x}{2}\right) = \frac{1}{\beta^3}L(\beta x),\\
\mathscr{L}^{-1}\left(\frac{s}{s^4 - \beta^4}\right) &= \frac{1}{\beta^2}\left(\frac{\cosh\beta x - \cos\beta x}{2}\right) = \frac{1}{\beta^2}M(\beta x),\\
\mathscr{L}^{-1}\left(\frac{s^2}{s^4 - \beta^4}\right) &= \frac{1}{\beta}\left(\frac{\sinh\beta x + \sin\beta x}{2}\right) = \frac{1}{\beta}N(\beta x),\\
\mathscr{L}^{-1}\left(\frac{s^3}{s^4 - \beta^4}\right) &= \left(\frac{\cosh\beta x + \cos\beta x}{2}\right) = P(\beta x).
\end{aligned} \tag{3.2.71}$$

Thus, (3.2.70) becomes

$$Y(x) = Y(0)P(\beta x)+\frac{Y'(0)N(\beta x)}{\beta}+\frac{Y''(0)M(\beta x)}{\beta^2}+\frac{Y'''(0)L(\beta x)}{\beta^3}+$$
$$+\frac{1}{EI\beta^3}\int_0^x Q(u)L\{\beta(x-u)\}\,du. \quad (3.2.72)$$

In the above, two of the constants of $Y(0)$, $Y'(0)$, $Y''(0)$, $Y'''(0)$ are zero from the conditions at $x = 0$. The other two must be found by applying the conditions at $x = l$.

We might elect to apply the Laplace transform with respect to time directly to the partial differential equation

$$\frac{\partial^4 y}{\partial x^4}+\frac{1}{a^2}\frac{\partial^2 y}{\partial t^2} = \frac{q(x, t)}{EI}, \quad (3.2.73)$$

giving

$$\bar{y}^{\text{iv}}(x, s)+\frac{s^2}{a^2}\bar{y}(x, s) = \frac{\bar{q}(x, s)}{EI}, \quad (3.2.74)$$

for the case of zero initial conditions. This still leaves us with a fourth-order equation to solve, the solution of which will contain the transform variable s in the arguments of sine, cosine, sinh, cosh functions in the form $\cos\sqrt{(sx)}$, $\sin\sqrt{(sx)}$, $\cosh\sqrt{(sx)}$, $\sinh\sqrt{(sx)}$. The determination of the inverse in such a case is somewhat complicated. This does not preclude the use of the Laplace transform with respect to time from the forced-vibration problem, but suggests it might be useful to combine it with a finite Fourier transform or an expansion in normal modes. We shall consider the latter case.

4. *Laplace transform—normal-mode expansion.* Apply the Laplace transform to the non-homogeneous equation to give

$$a^2\bar{y}^{\text{iv}}(x, s)+s^2\bar{y}(x, s) = \bar{q}(x, s)/\rho A, \quad (3.2.75)$$

where zero initial conditions have again been assumed. Now expand $\bar{y}(x, s)$, $\bar{q}(x, s)$ in a series of the normal modes $Y_n(x)$, given as

$$\bar{y}(x, s) = \sum_{n=1}^{\infty} a_n(s)Y_n(x), \qquad \bar{q}(x, s) = \sum_{n=1}^{\infty} b_n(s)Y_n(x), \quad (3.2.76)$$

where the $Y_n(x)$ satisfy (3.2.64) and the coefficients $b_n(s)$ are given by

$$b_n(s) = \frac{2}{l}\int_0^l \bar{q}(x, s)Y_n(x)\,dx. \quad (3.2.77)$$

Proceeding in the manner used to obtain (3.2.68), we obtain the result

$$a_n(s) = \frac{b_n(s)}{A\rho(a^2\beta_n^4+s^2)}. \tag{3.2.78}$$

Thus

$$\bar{y}(x, s) = \frac{2}{A\rho l}\sum_{n=1}^{\infty}\frac{Y_n(x)}{a^2\beta_n^4+s^2}\int_0^l \bar{q}(u, s)Y_n(u)\,du. \tag{3.2.79}$$

Recalling that

$$\mathscr{L}^{-1}\left(\frac{1}{s^2+a^2\beta_n^4}\right) = \frac{1}{a\beta_n^2}\sin a\beta_n^2 t, \tag{3.2.80}$$

we have, applying the convolution theorem,

$$y(x, t) = \frac{2}{A\rho l}\sum_{n=1}^{\infty}\frac{Y_n(x)}{a\beta_n^2}\int_0^l Y_n(u)\,du\int_0^t q(u, \tau)\sin a\beta_n^2(t-\tau)\,d\tau. \tag{3.2.81}$$

It might be noted that $a\beta_n^2 = \omega_n$, the natural frequency in the above.

5. *Solution by normal modes.* We have the basic equation

$$EI\frac{\partial^4 y}{\partial x^4}+\rho A\frac{\partial^2 y}{\partial t^2} = q(x, t). \tag{3.2.82}$$

Assume that $y(x, t)$ may be expanded in terms of the normal modes $Y_n(x)$ and general functions of time $q_n(t)$, so that

$$y(x, t) = \sum_{n=1}^{\infty} q_n(t)Y_n(x). \tag{3.2.83}$$

Substitution in the differential equation gives

$$\sum_{n=1}^{\infty}\left\{a^2 q_n(t)Y_n^{\text{iv}}(x)+\ddot{q}_n(t)Y_n(x)\right\} = \frac{q(x, t)}{\rho A}. \tag{3.2.84}$$

But $Y^{\text{iv}}(x) = \beta_n^4 Y_n(x)$, so that we have

$$\sum_{n=1}^{\infty}\{\ddot{q}_n(t)+a^2\beta_n^4 q_n(t)\}Y_n(x) = \frac{q(x, t)}{\rho A}. \tag{3.2.85}$$

Multiplying the above by $Y_m(x)$, integrating, and using the orthogonality property of the functions, we have

$$\ddot{q}_n(t)+a^2\beta_n^4 q_n(t) = \frac{2}{\rho A l}\int_0^l q(x, t)Y_n(x)\,dx = \frac{Q_n(t)}{\rho A} \quad (n = 1, 2, \ldots). \tag{3.2.86}$$

In the case of zero initial conditions, we know the solution of the above to be of the form

$$q_n(t) = -\frac{1}{\rho A \omega_n}\int_0^t Q_n(\tau)\sin\omega_n(t-\tau)\,d\tau, \tag{3.2.87}$$

where $\omega_n^2 = a^2\beta_n^4$. This determines the total solution.

This section has been devoted to a survey of methods of solution, avoiding consideration of special boundary conditions and considering special loadings only to the extent that these loadings enabled more general solutions to be found. In the next section, a number of special problems will be solved illustrating some of the techniques developed here.

3.2.5. *Some problems in forced vibrations of beams*

In this section, the solutions to some interesting problems in the forced vibrations of beams will be given. Several methods of analysis will be illustrated.

1. *Response of a beam to impact.* Consider the case of a simply supported beam of length l subjected to a short-duration impulse loading applied at $x = \xi$. We represent the load by

$$q(x, t) = P\delta(x-\xi)\delta(t). \tag{3.2.88}$$

From (3.2.81) the results of a Laplace-transform–normal-mode expansion solution to the vibrations of a beam with homogeneous initial conditions and under the general forcing $q(x, t)$ may be written as

$$y(x, t) = \frac{2}{\rho A l}\sum_{n=1}^{\infty}\frac{Y_n(x)}{\omega_n}\int_0^l Y_n(u)\,du\int_0^t q(u,\tau)\sin\omega_n(t-\tau)\,d\tau, \tag{3.2.89}$$

where $\omega_n = a\beta_n^2$. For the case at hand, the normal modes $Y_n(x)$ are

$$Y_n(x) = \sin\beta_n x, \tag{3.2.90}$$

where $\beta_n = n\pi/l$. Inserting the value for $q(x, t)$ we have

$$y(x, t) = \frac{2P}{\rho A l}\sum_{n=1}^{\infty}\frac{\sin\beta_n x}{\omega_n}\int_0^l \delta(u-\xi)\sin\beta_n u\,du\int_0^t \delta(\tau)\sin\omega_n(t-\tau)\,d\tau. \tag{3.2.91}$$

This gives directly

$$y(x, t) = \frac{2P}{\rho A l}\sum_{n=1}^{\infty}\frac{\sin\beta_n\xi\sin\beta_n x\sin\omega_n t}{\omega_n}. \tag{3.2.92}$$

For the case of $\xi = l/2$, $\beta_n\xi = n\pi/2$, so that

$$\sin\beta_n\xi = \sin\frac{n\pi}{2} = 1, 0, -1, 0,\ldots \quad (n = 1, 2, 3, 4,\ldots,). \tag{3.2.93}$$

Thus we have

$$y(x, t) = \frac{2P}{\rho Al} \sum_{n=1,3,5}^{\infty} \frac{(-1)^{(n-1)/2} \sin \beta_n x \sin \omega_n t}{\omega_n} \tag{3.2.94}$$

as the solution to the given problem.

The solution is in a form simple enough to enable interpretation without an elaborate evaluation of the series. Noting that $\omega_n \propto n^2$, we may write

$$y(x, t) \propto \sum_{n=1,3,5}^{\infty} (-1)^{(n-1)/2} \frac{\sin \beta_n x \sin \omega_n t}{n^2}, \tag{3.2.95}$$

or

$$y(x, t) \propto \sin \beta_1 x \sin \omega_1 t - \frac{\sin \beta_3 x \sin \omega_3 t}{9} + \frac{\sin \beta_5 x \sin \omega_5 t}{25} - \dots. \tag{3.2.96}$$

The dependence of the amplitudes on $1/n^2$ shows the dominance of the fundamental mode in the vibration. It is clear that the effect of the higher modes will be to superimpose 'ripples' on the fundamental mode. Although the theoretical solution predicts the presence of these higher frequencies in the motion for all time, observation of the motion would reveal that these components rapidly disappear after two or three oscillations, owing to damping, such as that arising from friction at the supports, internal friction in the beam, and interaction with the surrounding air.

The solution just obtained has been based on a loading impulsive in space and time. Physically, this loading is representative of the disturbance created by detonating a small explosive. A falling mass impacting on the beam, on the other hand, would actually represent a complicated interaction problem in which size of the falling mass in relation to the beam mass, vibrations of the impacting mass, and the degree to which the two bodies would 'stick' together due to plastic deformation would all be factors. The approximation of such an impact by an idealized impulse, nevertheless, provides an approximation of the response when the striking mass is somewhat less than the beam mass.

An approach to this forced-vibrations problem that brings in some of the mass–beam interaction aspects is to replace the non-homogeneous equation of forced vibrations with simple boundary conditions at $x = 0, l$ by a problem in the free vibrations of a beam where the impacting mass effects are brought in via added boundary conditions. Thus, consider†

$$\begin{aligned} EI\frac{\partial^4 y_1}{\partial x^4} + \rho A \frac{\partial^2 y_1}{\partial t^2} &= 0, \qquad y_1(0, t) = \frac{\partial^2 y_1(0, t)}{\partial x^2} = 0, \\ EI\frac{\partial^4 y_2}{\partial x^4} + \rho A \frac{\partial^2 y_2}{\partial t^2} &= 0, \qquad y_2(l, t) = \frac{\partial^2 y_2(l, t)}{\partial x^2} = 0. \end{aligned} \tag{3.2.97}$$

† See Goldsmith [7, pp. 59–60].

At $x = \xi$, continuity in deflection, slope, and curvature will give

$$y_1(\xi, t) = y_2(\xi, t), \quad \frac{\partial y_1(\xi, t)}{\partial x} = \frac{\partial y_2(\xi, t)}{\partial x}, \quad \frac{\partial^2 y_1(\xi, t)}{\partial x^2} = \frac{\partial^2 y_2(\xi, t)}{\partial x^2}. \tag{3.2.98}$$

There will be a discontinuity in the shear equal to the inertial effects of the striker, so that

$$EI\frac{\partial^3 y_2(\xi, t)}{\partial x^3} - EI\frac{\partial^3 y_1(\xi, t)}{\partial x^3} = m\frac{\partial^2 y(\xi, t)}{\partial t^2}, \tag{3.2.99}$$

where m is the striker mass. This last equation assumes that the striker remains in contact with the beam after impact. Finally, the initial conditions must be prescribed to this free-vibrations problem. If it is assumed that, at impact, a velocity equal to that of the falling mass is imparted to the beam at $x = \xi$, we have

$$y(x, 0) = 0, \qquad \partial y(x, 0)/\partial t = v_0 \delta(x-\xi). \tag{3.2.100}$$

2. *Response to a moving load.* Consider a constant force moving across a beam of length l at constant velocity V, so that

$$q(x, t) = \begin{cases} P\delta(x-Vt), & 0 \leqslant Vt < l \\ 0, & l < Vt, \end{cases} \tag{3.2.101}$$

where simple support conditions are assumed at $x = 0, l$. Thus we have

$$EI\frac{\partial^4 y}{\partial x^4} + \rho A\frac{\partial^2 y}{\partial t^2} = P\delta(x-Vt) \tag{3.2.102}$$

and

$$y(x, 0) = \dot{y}(x, 0) = 0. \tag{3.2.103}$$

Now Timoshenko† considers this problem starting from the normal-mode expansion $\sum q_n(t) Y_n(x)$. However, the general solution used in the previous impact problem is easily applied to this case. Thus we have

$$y(x, t) = \frac{2P}{\rho A l}\sum_{n=1}^{\infty}\frac{\sin \beta_n x}{\omega_n}\int_0^t \sin \omega_n(t-\tau)\, d\tau \int_0^l \delta(u-V\tau)\sin \beta_n u\, du,$$

$$= \frac{2P}{\rho A l}\sum_{n=1}^{\infty}\frac{\sin \beta_n x}{\omega_n}\int_0^t \sin \beta_n V\tau \sin \omega_n(t-\tau)\, d\tau. \tag{3.2.104}$$

Performing the integration, we obtain

$$y(x, t) = \frac{2P}{\rho A l}\sum_{n=1}^{\infty}\frac{\sin \beta_n x}{\omega_n(\beta_n^2 V^2 - \omega_n^2)}(\beta_n V \sin \omega_n t - \omega_n \sin \beta_n V t). \tag{3.2.105}$$

† p. 351 of Reference [25].

This solution has several interesting features. We see there are a series of critical velocities, given by

$$V_c = a\beta_n, \tag{3.2.106}$$

at which resonance may occur. The time of passage T_f is

$$T_f = l/V_c = l/a\beta_n = l^2/an\pi, \tag{3.2.107}$$

which can be thought of as a forcing period. The period of beam vibration, on the other hand, is $T_v = 2l^2/an^2\pi$. Considering only the fundamental period $n = 1$, it is seen that this period is twice the time required for passage of the load, $T_v = 2T_f$. Other critical conditions occur when the forcing period is

$$\begin{aligned} T_f &= T_v, \qquad n = 2, \\ T_f &= \tfrac{3}{2}T_v, \qquad n = 3, \\ T_f &= 2T_v, \qquad n = 4, \text{ etc.} \end{aligned} \tag{3.2.108}$$

Inspection of the solution for any of the critical speeds reveals that an apparently indeterminate situation of 0/0 arises. Thus, as $\beta_n V \to \omega_n$, we have

$$y(x, t) \to \frac{2P}{\rho Al}\sum_{n=1}^{\infty} \frac{\sin \beta_n x}{\omega_n^2 - \omega_n^2}(\sin \omega_n t - \sin \omega_n t) \to \frac{0}{0}. \tag{3.2.109}$$

However, by the application of L'Hospital's rule, we obtain

$$\lim_{\beta_n V \to \omega_n} y(x, t) = \frac{P}{\rho Al}\sum_{n=1}^{\infty} \frac{\sin \beta_n x}{\omega_n^2}(\sin \omega_n t - \omega_n t \cos \omega_n t), \tag{3.2.110}$$

which, in view of the finite time required for passage of the load, is definitely bounded.

The second observation has to do with the amplitude of vibration at the critical velocities and can be tied in with the limit condition just obtained. Considering just the first term we have

$$\lim_{\beta_1 V \to \omega_1} y(x, t) = \frac{P}{\rho Al\omega_1^2} \sin \beta_1 x(\sin \omega_1 t - \omega_1 t \cos \omega_1 t). \tag{3.2.111}$$

It may be easily shown that this is a maximum for

$$t = \pi/\omega_1 = l/a\beta_1 = l/V_c. \tag{3.2.112}$$

In other words, the maximum deflection occurs just as the force is leaving the beam.

3.3. Foundation and prestress effects

As in the case of transverse string motion, the motion of beams may be influenced by elastic or viscous foundations. The beam may also be subjected to tensile forces although, in contrast to the string, these are not the fundamental elastic restoring mechanisms for the beam. Finally, in complete contrast to the string, axial compressive forces may be present in the beam. We will study wave propagation and vibrations in beams as influenced by these various factors.

3.3.1. *The governing equation*

We will derive a single governing equation including the effects of a visco-elastic foundation and prestress and then reduce the general equation to several special cases. So consider the beam shown in Fig. 3.6(a) and an isolated element as shown in Fig. 3.6(b). The visco-elastic foundation yields a resistive force $f(x, t)$, as shown in Fig. 3.6(b), where

$$f(x, t) = ky + c\frac{\partial y}{\partial t}. \tag{3.3.1}$$

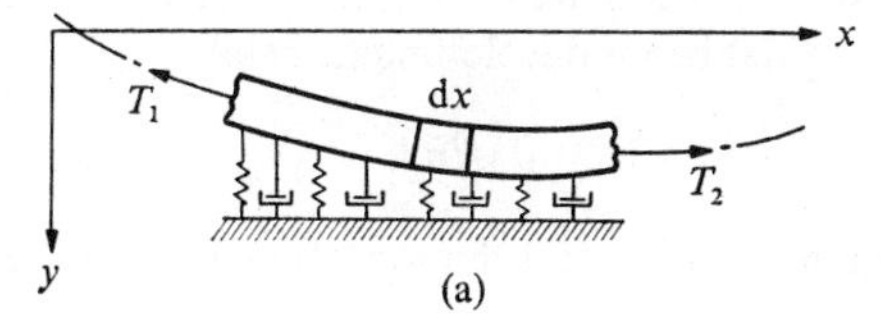

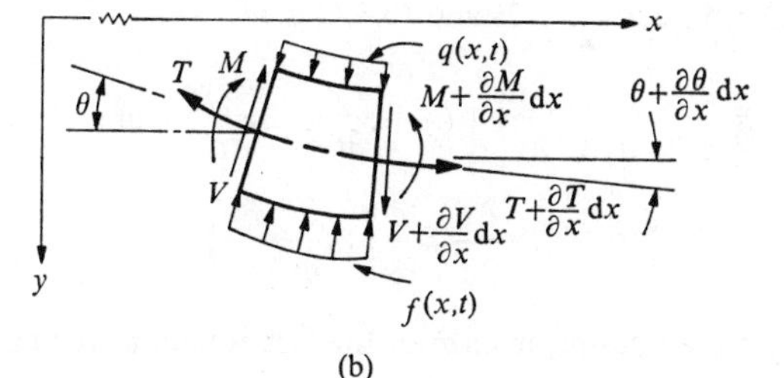

FIG. 3.6. (a) A beam under tension on a visco-elastic foundation and (b) an element of that beam.

The equation of motion in the vertical direction for the element is given by

$$-V+\left(V+\frac{\partial V}{\partial x}\,\text{d}x\right)-T\theta+\left(T+\frac{\partial T}{\partial x}\,\text{d}x\right)\left(\theta+\frac{\partial \theta}{\partial x}\,\text{d}x\right)-$$

$$-\left(ky+c\frac{\partial y}{\partial t}\right)\text{d}x+q\,\text{d}x = \rho A\,\text{d}x\frac{\partial^2 y}{\partial t^2}. \tag{3.3.2}$$

If the higher-order term in dx^2 appearing in conjunction with the tension term is neglected, this reduces to

$$\frac{\partial V}{\partial x}+\frac{\partial}{\partial x}(T\theta)-ky-c\frac{\partial y}{\partial t}+q = \rho A\frac{\partial^2 y}{\partial t^2}. \tag{3.3.3}$$

The equation for the axial motion gives

$$-T+\left(T+\frac{\partial T}{\partial x}\,dx\right)+V\theta-\left(V+\frac{\partial V}{\partial x}\,dx\right)\left(\theta+\frac{\partial \theta}{\partial x}\,dx\right) = \rho A\,dx\frac{\partial^2 u}{\partial t^2}. \tag{3.3.4}$$

In this last equation, we presume for the time being that axial-inertia effects are non-negligible. Again neglecting terms in dx^2, this last reduces to

$$\frac{\partial T}{\partial x}-\frac{\partial}{\partial x}(V\theta) = \rho A\frac{\partial^2 u}{\partial t^2}. \tag{3.3.5}$$

For the moment equation, we again neglect rotary-inertia effects. The moment effects due to the tension are of higher order, so that the results are the same as for the original beam development, being

$$V = \partial M/\partial x. \tag{3.3.6}$$

Using the relationship $\theta = \partial y/\partial x$ and the moment–curvature result

$$EI\,\partial^2 y/\partial x^2 = -M$$

as well as (3.3.6) in eqns (3.3.3) and (3.3.4) gives

$$EI\frac{\partial^4 y}{\partial x^4}-\frac{\partial}{\partial x}\left(T\frac{\partial y}{\partial x}\right)+ky+c\frac{\partial y}{\partial t}+\rho A\frac{\partial^2 y}{\partial t^2} = q(x,t) \tag{3.3.7}$$

and

$$\frac{\partial T}{\partial x}+\frac{\partial}{\partial x}\left(EI\frac{\partial^3 y}{\partial x^3}\frac{\partial y}{\partial x}\right) = \rho A\frac{\partial^2 u}{\partial t^2}. \tag{3.3.8}$$

We now neglect the non-linear term of the last equation, reducing it to

$$\frac{\partial T}{\partial x} = \rho A\frac{\partial^2 u}{\partial t^2}. \tag{3.3.9}$$

From (3.3.7) and (3.3.9) we see that we have a potentially coupled system of equations. Writing the axial force as $T = EA\,\partial u/\partial x$, we would have the longitudinal displacements appearing in the transverse equation of motion. It is seldom that axial inertia effects are of importance in transverse-motion problems, however, so we presume that (3.3.9) may be reduced to

$$\partial T/\partial x = 0. \tag{3.3.10}$$

Thus, the axial tension must be constant. The final form for the transverse equation of motion is thus

$$EI\frac{\partial^4 y}{\partial x^4}-T\frac{\partial^2 y}{\partial x^2}+ky+c\frac{\partial y}{\partial t}+\rho A\frac{\partial^2 y}{\partial t^2} = q(x, t). \tag{3.3.11}$$

The case of axial compression obtains merely by changing the sign of T in this result. We note in passing that a beam under tension could effectively become a string by reducing the stiffness EI to zero.

3.3.2. *The beam on an elastic foundation*

By requiring that $T = c = q = 0$, the case of a beam on a purely elastic foundation and free of external loading results. The equation is

$$\frac{\partial^4 y}{\partial x^4}+\frac{k}{EI}y+\frac{1}{a^2}\frac{\partial^2 y}{\partial t^2} = 0. \tag{3.3.12}$$

Let us consider the propagation of harmonic waves in an infinite beam by considering the solution

$$y = Ae^{\mathrm{i}(\gamma x-\omega t)}. \tag{3.3.13}$$

Substituting in (3.3.12) gives

$$\gamma^4 = \frac{\omega^2}{a^2}-\frac{k}{EI}. \tag{3.3.14}$$

This may be put in the dimensionless form

$$\bar{\gamma}^4 = \bar{\omega}^2-\bar{\omega}_0^2, \tag{3.3.15}$$

where

$$\bar{\gamma}^4 = I\gamma^4, \qquad \bar{\omega}^2 = \frac{\rho A}{E}\omega^2, \qquad \bar{\omega}_0^2 = \frac{\rho A}{E}\omega_0^2, \qquad \omega_0^2 = \frac{k}{\rho A}. \tag{3.3.16}$$

The frequency spectrum for this problem has a most interesting behaviour. We have that

$$\bar{\gamma} = \pm(\bar{\omega}^2-\bar{\omega}_0^2)^{\frac{1}{4}}, \pm\mathrm{i}(\bar{\omega}^2-\bar{\omega}_0^2)^{\frac{1}{4}}, \bar{\omega} > \bar{\omega}_0. \tag{3.3.17}$$

For $\bar{\omega} \to \bar{\omega}_0$ we observe a cutoff frequency, a phenomenon previously observed in strings. For $\bar{\omega} < \bar{\omega}_0$ we have that

$$\bar{\gamma} = \pm\frac{(1\pm\mathrm{i})}{\sqrt{2}}(\bar{\omega}_0^2-\bar{\omega}^2)^{\frac{1}{4}}. \tag{3.3.18}$$

We thus have complex branches to the frequency spectrum, as shown in Fig. 3.7. If we have a complex wavenumber of the general form

$$\gamma = \pm\gamma_{\mathrm{Re}}\pm\mathrm{i}\gamma_{\mathrm{Im}}, \tag{3.3.19}$$

the harmonic wave expression (3.3.13) will become

$$y = A\exp(\mp\gamma_{\mathrm{Im}}x)\exp\{\mathrm{i}(\pm\gamma_{\mathrm{Re}}x-\omega t)\}. \tag{3.3.20}$$

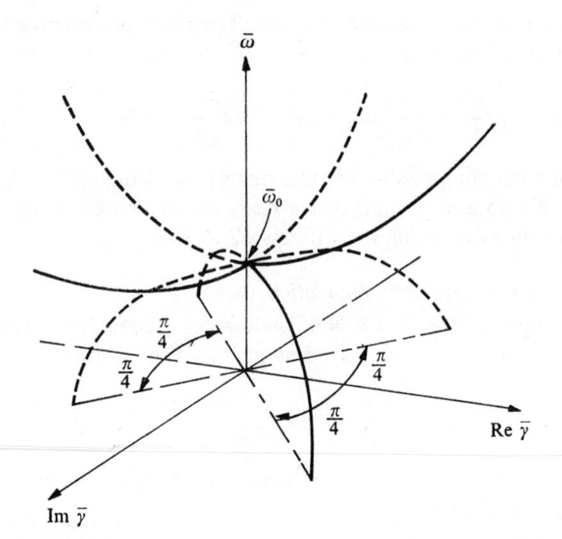

FIG. 3.7. Frequency spectrum for a beam on an elastic foundation.

The disturbance will thus be leftward or rightward propagating but also spatially varying. Again, in the context of the question on conditions for purely harmonic wave propagation, these are not acceptable solutions. Again, however, we expect these types of wavenumbers to play a role in transient and forced-motion problems.

This is not our first encounter with complex wavenumbers. They also arose in the study of harmonic waves in a damped string and were associated with the energy losses of the system. We now see that they can occur in a perfectly elastic system.

3.3.3. *The moving load on an elastically supported beam*

Problems of forced vibrations of infinite or semi-infinite beams on elastic foundations are left as exercises.† We now turn to a discussion of a most interesting effect associated with moving loads on elastically supported beams. In the areas of high-speed transportation or rocket-sledge technology, the response of rails and rail beds to high-speed moving loads is of interest. By considering the rail–rail-bed system as a beam on an elastic foundation subjected to a concentrated load moving at a constant velocity V_0, the following governing equation results:

$$EI\frac{\partial^4 y}{\partial x^4}+ky+\rho A\frac{\partial^2 y}{\partial t^2} = P_0\delta(x-V_0t). \tag{3.3.21}$$

† See Problem 3.13.

A separation-of-variables approach cannot be used on this problem owing to the inseparability of the space and time variables in the loading. The key to solution is to consider a coordinate system to be moving with the load and to realize that relative to the moving coordinate system, the beam is responding in an unchanging, steady-state manner. Thus, by introducing a new coordinate system x_1, y_1, where

$$x_1 = x - V_0 t, \qquad y_1 = y, \tag{3.3.22}$$

the governing equation (3.2.21) reduces to

$$EI\frac{d^4 y_1}{dx_1^4} + \rho A V_0^2 \frac{d^2 y_1}{dx_1^2} + k y_1 = P_0 \delta(x_1). \tag{3.3.23}$$

Thus, under steady-state conditions and a moving coordinate system, the dynamic response problem becomes a static deformation problem.

The details of the analysis of this problem are too lengthy to present here.† However, a major result that appears is the presence of a critical velocity of motion at which an infinite response is predicted. The situation corresponds to when the moving-load velocity corresponds to the propagation velocity of disturbances in the beam. The energy thus does not radiate away from the moving load but builds up in the vicinity. The value of the critical velocity turns out to be

$$V_{cr} = \left(\frac{4kEI}{\rho^2 A^2}\right)^{\frac{1}{4}}. \tag{3.3.24}$$

Fig. 3.8(a) shows the system response for the case of zero velocity, one-half the critical velocity, and twice the critical velocity. The coordinates are in non-dimensional form. The response above the critical velocity is also of interest. Note that, at the loading point, the deflection is zero. Physically, this says that the inertia of the beam instantaneously is preventing motion. Yet there is disturbance ahead of the load owing to the transmission of moment from the disturbed region behind the load.

By including damping in the foundation, infinite response is found no longer to occur. For light damping, large amplitudes of response are found near the undamped critical velocity. Depending on the magnitude of damping, responses interpretable in terms of 'underdamped' and 'overdamped' response may be established. Fig. 3.8(b) shows three situations in this regard.

3.3.4. *The effects of prestress*

Now consider the beam under the effects of prestress alone, so that the general equation (3.3.11) reduces

$$\frac{\partial^4 y}{\partial x^4} - \frac{T}{EI}\frac{\partial^2 y}{\partial x^2} + \frac{1}{a^2}\frac{\partial^2 y}{\partial t^2} = 0. \tag{3.3.25}$$

† See Kenney [10] for complete analysis, including damping effects.

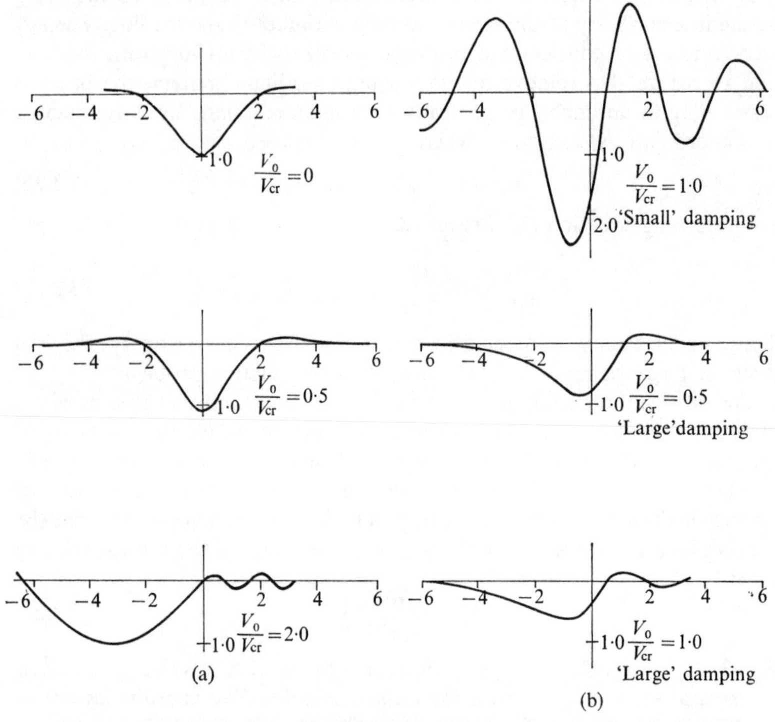

FIG. 3.8. (a) Undamped and (b) damped response of a beam to a load moving at various velocities. (Based on Kenney [10].)

Considering propagation of harmonic waves of the form $\exp\{i(\gamma x-\omega t)\}$ leads to the wavenumber–frequency relation,

$$\gamma^4+\frac{T}{EI}\gamma^2-\frac{\omega^2}{a^2}=0. \tag{3.3.26}$$

This has the roots

$$\gamma=\pm\{-\zeta\pm(\zeta^2+\omega^2/a^2)^{\frac{1}{2}}\}^{\frac{1}{2}}, \tag{3.3.27}$$

where $2\zeta = T/EI$. Now, for the case of tensile prestress, T and hence $\zeta > 0$ in (3.3.27). It should be evident under these conditions that the wavenumbers will be of the form

$$\gamma=\pm\alpha,\ \pm i\beta, \tag{3.3.28}$$

where

$$\alpha=\{-\zeta+(\zeta^2+\omega^2/a^2)^{\frac{1}{2}}\}^{\frac{1}{2}},\qquad \beta=\{\zeta+(\zeta^2+\omega^2/a^2)^{\frac{1}{2}}\}^{\frac{1}{2}}. \tag{3.3.29}$$

If the direction of the prestress is reversed, $\zeta < 0$. The resulting wavenumbers are given by

$$\gamma=\pm\beta,\ \pm i\alpha. \tag{3.3.30}$$

We note in general that $\beta > \alpha$, so that for the wavelengths associated with these wavenumbers we have that $\lambda_\beta < \lambda_\alpha$. Thus, for a given frequency, pre-compression shortens the wavelength in comparison to pre-tension. In line with this, the propagation velocity is decreased.

Some of the more interesting effects of prestress are in conjunction with the vibrations of finite beams. Consider, for example, a pinned–pinned beam subjected to axial tension. Investigating this in the manner of a normal-mode study, we let

$$y = Y(x)e^{-i\omega t} \tag{3.3.31}$$

and substitute this in (3.3.25), to obtain

$$\frac{d^4Y}{dx^4} - \frac{T}{EI}\frac{d^2Y}{dx^2} - \frac{\omega^2}{a^2}Y = 0. \tag{3.3.32}$$

Using the notation of (3.3.29), we have for the solution

$$Y = A\cos\alpha x + B\sin\alpha x + C\cosh\beta x + D\sinh\beta x. \tag{3.3.33}$$

The boundary conditions for the problem are

$$Y(0) = Y''(0) = Y(l) = Y''(l) = 0. \tag{3.3.34}$$

Applying these to the solution gives $A = C = D = 0$ and

$$\sin\alpha l = 0, \qquad \alpha l = n\pi \qquad (n = 1, 2, \ldots). \tag{3.3.35}$$

The resulting frequencies are given by

$$\omega_n = \frac{n^2\pi^2}{l^2}\sqrt{\left(\frac{EI}{\rho A}\right)}\left(1+\frac{Tl^2}{n^2\pi^2 EI}\right)^{\frac{1}{2}}. \tag{3.3.36}$$

If $T = 0$ in the above, the results for the pinned–pinned beam without prestress, previously given by (3.2.8), are recovered. We see that the effect of the tensile prestress is to increase the natural frequencies due to the stiffening effect.

If the force is reversed in direction, becoming compressive, the effect on the frequencies is directly obtained by changing the sign of T in (3.3.36). We see that a most interesting effect occurs. There results a critical load, for any value of n, such that $\omega_n \to 0$. These critical values are given by

$$T_{cr} = n^2\pi^2 EI/l^2 \tag{3.3.37}$$

and should be recognized as the Euler critical loads for a pin-ended column. Thus the instability is manifested dynamically by the vibrational frequencies approaching zero.

3.3.5. *Impulse loading of a finite, prestressed, visco-elastically supported beam*

Our objective here is to consider a single problem that includes several of the previously described effects of prestress, elastic foundation, and damping.

Several useful interpretations of the results are possible. Thus, consider the pulse loading of a simply supported beam resting on a viscous, elastic foundation and subjected to tension T on the ends, as shown in Fig. 3.9.

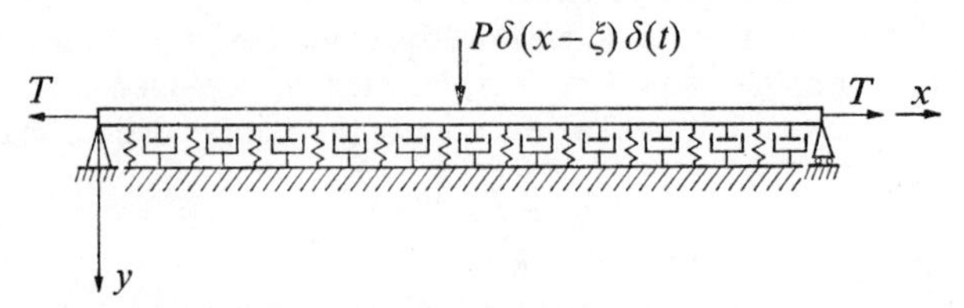

FIG. 3.9. Finite, prestressed beam on a visco-elastic foundation subjected to an impulsive load.

The governing equation for the beam is given by (3.3.11). The finite Fourier transform in conjunction with the Laplace transform will be applied. The finite Fourier sine transform is appropriate for these boundary conditions and gives

$$EI\alpha_n^4\bar{y}(n,t)+T\alpha_n^2\bar{y}(n,t)+k\bar{y}(n,t)+\rho A\ddot{\bar{y}}(n,t)+c\dot{\bar{y}}(n,t) = \bar{q}(n,t). \quad (3.3.38)$$

Applying the Laplace transform to the above yields

$$EI\alpha_n^4\bar{Y}(n,s)+T\alpha_n^2\bar{Y}(n,s)+k\bar{Y}(n,s)+s^2\rho A\bar{Y}(n,s)+sc\bar{Y}(n,s) = \bar{Q}(n,s), \quad (3.3.39)$$

where zero initial conditions have been assumed. This may be written as,

$$(s^2+2\zeta^2s+\Omega_n)\bar{Y}(n,s) = \frac{\bar{Q}(n,s)}{\rho A}, \quad (3.3.40)$$

where

$$\zeta^2 = \frac{c}{2\rho A}, \qquad \Omega_n = \omega_n^2+\eta_n^2+\gamma_n^2 \quad (3.3.41)$$

and where

$$\omega_n^2 = a^2\alpha_n^4, \qquad \eta_n^2 = \frac{T}{\rho A}\alpha_n^2, \qquad \gamma_n^2 = \frac{k}{\rho A}. \quad (3.3.42)$$

The frequencies ω_n are the natural frequencies of a simply supported beam without foundation or prestress effects. The transformed solution is then

$$\bar{Y}(n,s) = \frac{1}{\rho A}\frac{\bar{Q}(n,s)}{s^2+2\zeta^2s+\Omega_n}. \quad (3.3.43)$$

We first perform the Laplace inversion. Consider the denominator of the above. The roots are

$$s_{1,2} = -\zeta^2\pm(\zeta^4-\Omega_n)^{\frac{1}{2}} = -\zeta^2\pm i(\Omega_n-\zeta^4)^{\frac{1}{2}}. \quad (3.3.44)$$

Thus

$$s^2+2\zeta^2s+\Omega_n = (s+\alpha)(s+\beta), \quad (3.3.45)$$

where

$$\alpha = s_1, \qquad \beta = s_2. \quad (3.3.46)$$

By partial fractions we then have that

$$\frac{1}{s^2+2\zeta^2 s+\Omega_n} = \frac{1}{(s+\alpha)(s+\beta)} = \frac{1}{\beta-\alpha}\left(\frac{1}{s+\alpha}-\frac{1}{s+\beta}\right), \tag{3.3.47}$$

so that

$$\bar{Y}(n, s) = \frac{1}{\rho A(\beta-\alpha)}\left\{\frac{\bar{Q}(n, s)}{s+\alpha}-\frac{\bar{Q}(n, s)}{s+\beta}\right\}. \tag{3.3.48}$$

From tables we know

$$\mathscr{L}^{-1}\left(\frac{1}{s+\alpha}\right) = \mathrm{e}^{-\alpha t}, \tag{3.3.49}$$

so that using the convolution theorem enables us to write

$$\bar{y}(n, t) = \frac{1}{\rho A(\beta-\alpha)}\left\{\int_0^t \bar{q}(n, \tau)\mathrm{e}^{-\alpha(t-\tau)}\,\mathrm{d}\tau - \int_0^t \bar{q}(n, \tau)\mathrm{e}^{-\beta(t-\tau)}\,\mathrm{d}\tau\right\},$$

$$= \frac{1}{\rho A(\beta-\alpha)}\int_0^t \bar{q}(n, \tau)\{\mathrm{e}^{-\alpha(t-\tau)}-\mathrm{e}^{-\beta(t-\tau)}\}\,\mathrm{d}\tau. \tag{3.3.50}$$

We specify the load as $P\delta(x-\xi)\delta(t)$. Since

$$\bar{q}(n, t) = \frac{P\delta(t)}{l}\int_0^l \delta(x-\xi)\sin\alpha_n x\,\mathrm{d}x = \frac{P\delta(t)}{l}\sin\alpha_n\xi, \tag{3.3.51}$$

the result (3.3.50) becomes

$$\bar{y}(n, t) = \frac{P\sin\alpha_n\xi}{\rho Al(\beta-\alpha)}(\mathrm{e}^{-\alpha t}-\mathrm{e}^{-\beta t}). \tag{3.3.52}$$

From the definition of the Fourier inversion we have

$$y(x, t) = \frac{2P}{\rho Al}\sum_{n=1}^{\infty}\frac{\sin\alpha_n\xi}{\beta-\alpha}(\mathrm{e}^{-\alpha t}-\mathrm{e}^{-\beta t})\sin\alpha_n x. \tag{3.3.53}$$

We note the following:

$$\beta-\alpha = -2\mathrm{i}(\Omega_n-\zeta^4)^{\frac{1}{2}} \tag{3.3.54}$$

and

$$\exp(-\alpha t) - \exp(-\beta t) = -2\mathrm{i}\exp(-\zeta^2 t)\sin(\Omega_n-\zeta^4)^{\frac{1}{2}}t. \tag{3.3.55}$$

Also, if $\xi = l/2$, then

$$\sin\alpha_n\xi = (-1)^{(n-1)/2} \qquad (n = 1, 3, 5,\ldots). \tag{3.3.56}$$

Hence our solution is

$$y(x, t) = \frac{2P}{\rho Al}\exp(-\zeta^2 t)\sum_{n=1,3,\ldots}^{\infty}\frac{(-1)^{(n-1)/2}\sin\alpha_n x}{(\Omega_n-\zeta^4)^{\frac{1}{2}}}\sin(\Omega_n\zeta^4)^{\frac{1}{2}}t. \tag{3.3.57}$$

A number of observations may be made about this result.

1. In the absence of damping, elasticity, or tension, that is, if

$$c = k = T = 0,$$

the above solution reduces to that previously obtained for an impulse load (3.2.94).

2. A study of the radical $(\Omega_n - \zeta^4)^{\frac{1}{2}}$ shows that the added stiffness due to k,T tends to increase the vibrational frequency, since they are additive to the basic term ω_n^2. This has been previously observed for the free-vibrations result.

3. Consider the special case of no foundation, so that $k = c = 0$; the solution becomes

$$y(x, t) = \frac{2P}{\rho Al} \sum_{n=1,3}^{\infty} \frac{(-1)^{(n-1)/2} \sin \alpha_n x}{(\omega_n^2 + \eta_n^2)^{\frac{1}{2}}} \sin(\omega_n^2 + \eta_n^2)^{\frac{1}{2}} t \tag{3.3.58}$$

where $\omega_n^2 = a^2\alpha_n^4$, $n_n^2 = (T/\rho A)\alpha_n^2$. If we had the case $n_n^2 \gg \omega_n^2$, which would correspond to tensile forces dominating the stiffness term, the solution becomes that of the vibrating string, which is not surprising since the equation of string motion assumes no bending resistance.

4. Suppose T becomes compressive. It is only necessary to change its sign in the preceding results so that we now have $(\omega_n^2 - \eta_n^2)^{\frac{1}{2}}$. The possibility of an unstable motion now exists since if $\eta_n^2 > \omega_n^2$, we have sin → sinh and the vibration amplitudes increase without bounds as time passes.

5. Returning to the total solution (3.3.57) it is seen that critical values of ζ^2 may occur so that $\zeta^4 \gtreqless \Omega_n$. If $\zeta^4 < \Omega_n$, a damped oscillatory motion results. However, if $\zeta^4 = \Omega_n$ the case of critical damping arises, and if $\zeta^4 > \Omega_n$ the overdamped case occurs, since sin → sinh, as previously mentioned.

Of equal interest is to note that the foundation damping enters in as a simple, familiar $\exp(-\zeta^2 t)$ type term in front of the total solution. Thus, the distributed viscosity gives rise to a decay completely analogous to that existing in a lumped-parameter spring–mass–dashpot system.

3.4. Effects of shear and rotary inertia

It was pointed out in the development of the governing equation for thin beams that two major assumptions were made: first, that the effects of rotary inertia had been neglected and, secondly, that shear deformations were neglected. This latter assumption was actually stated in the form of the Bernoulli–Euler hypothesis that plane sections remain plane. The fact that this is equivalent to neglecting shear deformation is shown in the sequence of illustrations in Fig. 3.10. Thus, in Fig. 3.10(a), a typical element at a cross-section subjected to shearing force is shown. The shear stresses on the element

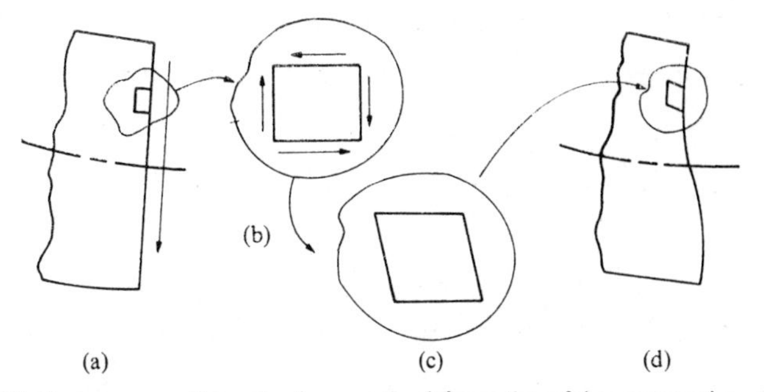

FIG. 3.10. Consequence of shearing force on the deformation of the cross-section of a beam.

are shown in (b), while in (c) the resulting shearing deformation of the element is shown. Because of variations in shear stress across the section, the various elements will be strained in differing amounts. Re-assembling these elements results in a section that is no longer perpendicular to the neutral axis and, in fact, will be 'warped' from the planar configuration.

The study of wave propagation in the Bernoulli–Euler beam showed that infinite phase velocities were predicted (Fig. 3.2(b)) and that 'instantaneous' far-field response was predicted in some problems (see Fig. 3.4), an anomaly resulting from infinite phase velocities. It was suggested at that stage that such behaviour was attributable to the rotary-inertia and shear-effect restrictions. The limitations of the theory and the probable cause were realized rather early. In 1894 Rayleigh† applied the correction for rotary inertia. Results obtained with this correction alone predicted finite propagation velocities. However, the upper bound was still greater than predicted by exact theory. In 1921 Timoshenko‡ included both effects of shear and rotary inertia and obtained results in accord with exact theory.

3.4.1. *The governing equations*

To start the development, we again consider an element of beam subjected to shear force, bending moment, and distributed load, as shown in Fig. 3.11(a). The displacement to the centroidal plane of the beam is still measured by y and the slope of the centroidal axis is still given by $\partial y/\partial x$. A new coordinate ψ is introduced to measure the slope of the cross-section due to bending. In the Bernoulli–Euler development, this is also the same as the slope of the centroidal axis $\partial y/\partial x$, so a special coordinate is not required. In Fig. 3.11(b) the essential features of the shear deformation development are shown. The slope of the centroidal axis is shown as $\partial y/\partial x$. This is now considered to be made up of two contributions. The first is ψ, due, as

† pp. 293–4 of Vol. I, Reference [18].
‡ Reference [24]; see also p. 329 of Reference [25].

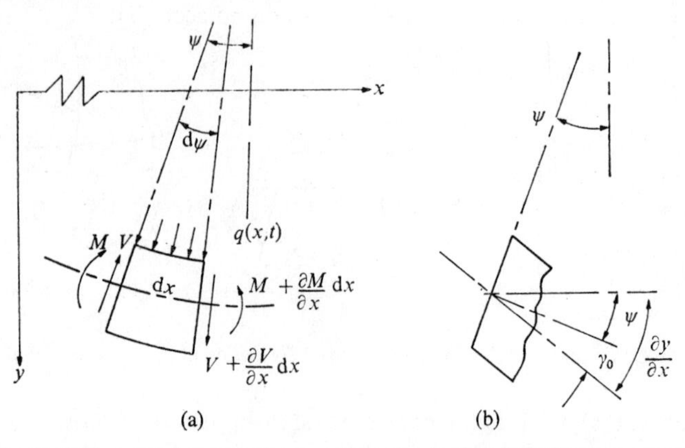

FIG. 3.11. (a) Differential element of beam subjected to load and (b) the kinematical details of additional shearing deformation.

mentioned, to the effects of bending. An additional contribution γ_0 due to shearing effects is now included. Thus we have

$$\partial y/\partial x = \psi + \gamma_0. \tag{3.4.1}$$

We should note that plane sections are still assumed to remain plane in this development, but no longer perpendicular to the centroidal plane. Thus, warping of the cross-section, as shown in Fig. 3.10(d), is still not permitted by these kinematics.

The problem now is to relate the above kinematical expressions to the loads. The assumption is made that the relationship between bending moment and curvature still holds. In terms of the parameters of the present problem this is expressed by

$$M/EI = -\partial\psi/\partial x. \tag{3.4.2}$$

This derives from the fact that $R\,\mathrm{d}\psi = \mathrm{d}x$, where R is the radius of curvature. Hence $1/R = \mathrm{d}\psi/\mathrm{d}x$. The expression (3.4.2) is the analogue of (3.4.1) for the Bernoulli–Euler development.

To determine γ_0, the essence of the Timoshenko argument is as follows. The shear force at the cross-section is given in terms of the shear stress τ or shear strain γ respectively as

$$V = \int_A \tau\,\mathrm{d}A = G\int_A \gamma\,\mathrm{d}A. \tag{3.4.3}$$

If γ_0 is the shear strain at the centroidal axis, then $G\gamma_0 A$ will give a shear force. However, it will not be equal to the value (3.4.3) obtained by integrating the variable stress distribution across the section. To bring the value into balance, an adjustment coefficient κ is introduced such that

$$V = G\int_A \gamma\,\mathrm{d}A = (G\gamma_0 A)\kappa. \tag{3.4.4}$$

The value of κ will depend on the shape of the cross-section and must be determined, usually by stress analysis means, for each cross-section. This parameter is usually designated as the (Timoshenko) shear coefficient. Substituting the expression for γ_0 obtained from (3.4.1) into (3.4.4) gives

$$V = AG\kappa\left(\frac{\partial y}{\partial x}-\psi\right). \tag{3.4.5}$$

With the preceding result in hand, development of the equations of motion is rather straightforward. Writing the equation of motion in the vertical direction for the element of Fig. 3.11(a), we obtain

$$-V+\left(V+\frac{\partial V}{\partial x}\,\mathrm{d}x\right)+q\,\mathrm{d}x = \rho A\,\mathrm{d}x\frac{\partial^2 y}{\partial t^2}, \tag{3.4.6}$$

or

$$\frac{\partial V}{\partial x}+q = \rho A\frac{\partial^2 y}{\partial t^2}. \tag{3.4.7}$$

Summing moments about an axis perpendicular to the x,y-plane and passing through the centre of the element, we obtain

$$M-\left(M+\frac{\partial M}{\partial x}\,\mathrm{d}x\right)+\frac{1}{2}V\,\mathrm{d}x+\frac{1}{2}\left(V+\frac{\partial V}{\partial x}\,\mathrm{d}x\right)\mathrm{d}x = J\frac{\partial^2 \psi}{\partial t^2}, \tag{3.4.8}$$

where J is the polar inertia of the element. For an element of mass density ρ, length $\mathrm{d}x$, and having a cross-sectional-area moment of inertia about the moment axis of I, we have that

$$J = \rho I\,\mathrm{d}x. \tag{3.4.9}$$

Then (3.4.8) reduces to

$$V-\frac{\partial M}{\partial x} = \rho I\frac{\partial^2 \psi}{\partial t^2}. \tag{3.4.10}$$

Substituting the expression for bending moment (3.4.2) and shear force (3.4.5) into the two governing equations (3.4.7) and (3.4.10) gives

$$GA\kappa\left(\frac{\partial \psi}{\partial x}-\frac{\partial^2 y}{\partial x^2}\right)+\rho A\frac{\partial^2 y}{\partial t^2} = q(x,t), \tag{3.4.11}$$

$$GA\kappa\left(\frac{\partial y}{\partial x}-\psi\right)+EI\frac{\partial^2 \psi}{\partial x^2} = \rho I\frac{\partial^2 \psi}{\partial t^2}. \tag{3.4.12}$$

The above results are the governing equations for the Timoshenko beam theory. There are, we see in this theory, two modes of deformation and these coupled governing equations represent the physical coupling that occurs between them. One mode of deformation is simply the transverse deflection of the beam as measured by $y(x, t)$. The other mode is the transverse shearing deformation γ_0, as measured by the difference $\partial y/\partial x - \psi$ (see (3.4.1)). This is our first, but far from last, encounter with coupled multi-mode elastic systems.

The effects of rotary inertia are carried solely by the inertia term of (3.4.12). Setting $I = 0$ thus removes this effect. The recovery of the Bernoulli–Euler theory by removing the shear effect can be accomplished by eliminating V between (3.4.7) and (3.4.8) and using the relationship (3.1.1).

3.4.2. *Harmonic waves*

Let us consider the propagation of harmonic waves in the infinite Timoshenko beam. Two approaches are possible. We may consider eqns (3.4.11) and (3.4.12) directly, with $q = 0$, or we may first reduce the two to a single equation. In most future investigations of this nature, one or the other approach will be used. Here, both will be presented. Considering the two equations, we assume solutions of the form

$$y = B_1 e^{i(\gamma x - \omega t)}, \qquad \psi = B_2 e^{i(\gamma x - \omega t)}. \tag{3.4.13}$$

Substituting in (3.4.11) and (3.4.12), the following two homogeneous equations in B_1, B_2 are obtained

$$\begin{aligned} (GA\kappa\gamma^2 - \rho A\omega^2)B_1 + iGA\kappa\gamma B_2 &= 0. \\ iGA\kappa\gamma B_1 - (GA\kappa + EI\gamma^2 - \rho I\omega^2)B_2 &= 0. \end{aligned} \tag{3.4.14}$$

These give the amplitude ratios

$$\frac{B_2}{B_1} = i\frac{(GA\kappa\gamma^2 - \rho A\omega^2)}{GA\kappa\gamma} = i\frac{GA\kappa\gamma}{(GA\kappa + EI\gamma^2 - \rho I\omega^2)}. \tag{3.4.15}$$

Equating the determinant of coefficients to zero gives the frequency equation

$$(GA\kappa\gamma^2 - \rho A\omega^2)(GA\kappa + EI\gamma^2 - \rho I\omega^2) - G^2A^2\kappa^2\gamma^2 = 0, \tag{3.4.16}$$

or simplifying,

$$\frac{EI}{\rho A}\gamma^4 - \frac{I}{A}\left(1 + \frac{E}{G\kappa}\right)\gamma^2\omega^2 - \omega^2 + \frac{\rho I}{GA\kappa}\omega^4 = 0. \tag{3.4.17}$$

This may be converted to the dispersion equation by using the identity $\omega = \gamma c$ in the above, giving

$$\frac{EI}{\rho A}\gamma^4 - \frac{I}{A}\left(1 + \frac{E}{G\kappa}\right)\gamma^4 c^2 - \gamma^2 c^2 + \frac{\rho I}{GA\kappa}\gamma^4 c^4 = 0. \tag{3.4.18}$$

To reduce the two governing equations to a single equation, differentiate (3.4.12) with respect to x. Then solve (3.4.11) for $\partial\psi/\partial x$, giving

$$\frac{\partial\psi}{\partial x}=\frac{\partial^2 y}{\partial x^2}-\frac{\rho}{G\kappa}\frac{\partial^2 y}{\partial t^2}. \tag{3.4.19}$$

By differentiating the above to give expressions for $\partial^3\psi/\partial x^3$, $\partial^3\psi/\partial x\,\partial t^2$ and substituting these in (3.4.12), one can obtain

$$\frac{EI}{\rho A}\frac{\partial^4 y}{\partial x^4}-\frac{I}{A}\left(1+\frac{E}{G\kappa}\right)\frac{\partial^4 y}{\partial x^2\,\partial t^2}+\frac{\partial^2 y}{\partial t^2}+\frac{\rho I}{GA\kappa}\frac{\partial^4 y}{\partial t^4}=0. \tag{3.4.20}$$

Considering a harmonic wave solution

$$y=Ce^{i(\gamma x-\omega t)} \tag{3.4.21}$$

leads directly to the result (3.4.18).

Before presenting specific numerical data on the dispersion curve we investigate the asymptotic behaviour of (3.4.18). In particular, consider the behaviour as the wavenumber becomes large. This can be easily done by factoring γ^4 from the equation and then letting $\gamma\to\infty$. The equation reduces to

$$\lim_{\gamma\to\infty}(3.4.18)=\frac{EI}{\rho A}-\frac{I}{A}\left(1+\frac{E}{G\kappa}\right)c^2+\frac{\rho I}{GA\kappa}c^4\equiv 0, \tag{3.4.22}$$

or

$$c^4-\frac{G\kappa}{\rho}\left(1+\frac{E}{G\kappa}\right)c^2+\frac{GE\kappa}{\rho^2}=0. \tag{3.4.23}$$

This may be expressed as

$$\bar{c}^4-\left(1+\frac{G\kappa}{E}\right)\bar{c}^2+\frac{G\kappa}{E}=0, \tag{3.4.24}$$

where $\bar{c}=c/c_0$ and c_0 is the bar velocity. This equation has the two roots

$$\bar{c}^2=1,\quad \frac{G\kappa}{E}, \tag{3.4.25}$$

or

$$c=c_0,\quad \sqrt{\left(\frac{G\kappa}{\rho}\right)}, \tag{3.4.26}$$

where the negative roots are ignored (giving only waves in the opposite direction) and the non-dimensionalization has been removed in the last expression. We thus have the important result that the wave velocities are bounded at large wavenumbers, in contrast to the Bernoulli–Euler behaviour. The fact that there are two limits is associated with the fact that there are two basic modes that are operative.

The long-wavelength limit ($\gamma\to 0$) should also be investigated for the possibility of cutoff frequencies. Considering (3.4.18) for this limit, we have

$$\lim_{\gamma\to 0}(3.4.18)=-\omega^2+\frac{\rho I}{GA\kappa}\omega^4=0, \tag{3.4.27}$$

which has the roots

$$\omega_c = 0, \qquad \sqrt{\left(\frac{GA\kappa}{\rho I}\right)}. \tag{3.4.28}$$

Thus, one of the modes has a finite cutoff frequency. If we consider (3.4.14), we see that as $\gamma = \varepsilon \to 0$, we have the general behaviour

$$\begin{aligned} -\rho A\omega_c^2 B_1 + \mathrm{i}O(\varepsilon)B_2 &= 0, \\ \mathrm{i}O(\varepsilon)B_1 - (GA\kappa - \rho I\omega_c^2)B_2 &= 0, \end{aligned} \tag{3.4.29}$$

where $O(\varepsilon)$ indicates a very small quantity. If $\omega_c \to 0$, we have $B_1 \neq 0$, $B_2 = 0$ as possibilities. If $\omega_c \to (GA\kappa/\rho I)^{\frac{1}{2}}$, we have $B_2 \neq 0$, $B_1 = 0$ as possibilities. For this latter behaviour, we have the motion given by

$$y = 0, \qquad \psi = B_2 \exp(-\mathrm{i}\omega_c t). \tag{3.4.30}$$

This corresponds to a shearing motion as shown in Fig. 3.12.

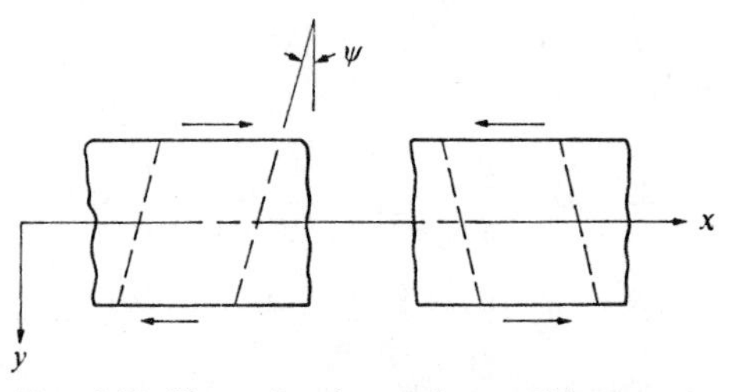

FIG. 3.12. Shear vibration at the cutoff frequency.

Before it is possible to present specific data on the dispersion curve or the frequency spectrum, it is necessary to specify the cross-section so that the shear coefficient can be determined. Values for rectangular and circular cross-sections have been given as 0·833 and $\frac{10}{9}$ respectively.

By considering a bar of circular cross-section, it is possible to compare the results for the Timoshenko beam with those from the exact theory of elasticity. These are shown in the dispersion curve of Fig. 3.13. Only the first branch of the dispersion curve for formula (3.4.17) is shown, but this is the one of greatest interest, since it is the primarily flexural mode. In addition the results for rotary-inertia effects alone (called the 'Rayleigh' theory) are shown as well as those for Bernoulli–Euler theory. The velocity and wave-number have been non-dimensionalized as

$$\bar{c} = \frac{c}{c_0}, \qquad \bar{\gamma} = \frac{a\gamma}{2\pi}, \tag{3.4.31}$$

where a is the radius of a circular bar. In addition to specifying the shear coefficient, Poisson's ratio must be given.† It is seen that the agreement

† See Problems 3.16, 3.17.

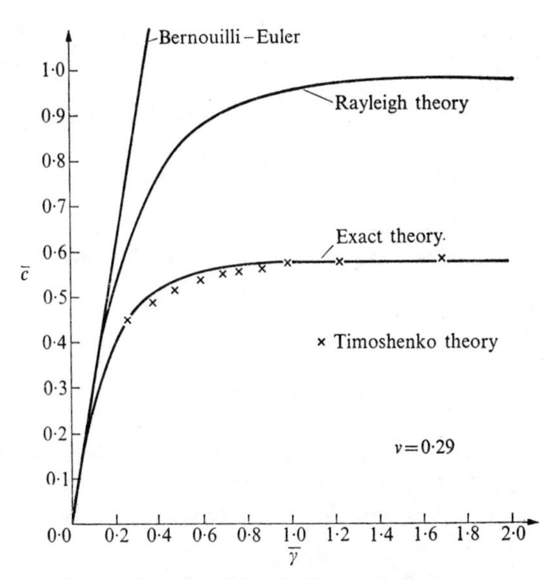

FIG. 3.13. Dispersion relations from Timoshenko, exact, Rayleigh, and Bernoulli–Euler beam theories. (After Kolsky [11, Fig. 16].)

between Timoshenko theory and exact theory is quite remarkable. As was remarked previously, the Rayleigh theory is seen to give a bounded but still too high phase velocity. The Bernoulli–Euler theory is seen to agree with Timoshenko and exact theory in only a very limited range of wavenumber (say $\bar{\gamma} \leq 0{\cdot}1$).

3.4.3. *Pulse propagation in a Timoshenko beam*

In determining the response of beams to sharp transients, it is almost necessary at the outset to use the Timoshenko beam theory in contrast to the Bernoulli–Euler theory, particularly if short-time response is desired. This is somewhat in distinction to longitudinal transients in rods, which may be adequately described by the rod wave equation even when the loading pulse is fairly broad-band.† Analysis in this area becomes rather complicated, however, and our attention must therefore be comparatively restricted because of the rather extensive analysis required.

In the following, the problem of transient loading of a semi-infinite Timoshenko beam will be reviewed in part. The review is based on the work of Boley and Chao [2]. The main objectives will be to outline the steps necessary to obtain the transformed solution to more complicated problems, to develop basic information on propagation of wavefronts by rather

† See Problem 3.18.

elementary means, to explore some of the subtleties in establishing the complete contour when branch cuts are necessary, and to present some results for a specific loading.

The governing equations for the problem are (3.4.11) and (3.4.12) with $q(x, t) = 0$. These equations may be put in non-dimensional form by introducing the dimensionless variables x_1, t_1, y_1, where

$$x_1 = x/r, \quad t_1 = c_0 t/r, \quad y_1 = y/r, \quad \gamma = E/G\kappa, \tag{3.4.32}$$

where r is the radius of gyration of the cross-section. The governing equations then reduce to

$$\psi'' + \frac{1}{\gamma}(y_1' - \psi) - \ddot{\psi} = 0, \tag{3.4.33}$$

$$\ddot{y}_1 - \frac{1}{\gamma}(y_1'' - \psi') = 0, \tag{3.4.34}$$

where the prime and dot notation refer to partial derivatives with respect to the dimensionless distance and time variables. Most of the results will be presented in terms of the shear force and bending moment. From (3.4.5) and (3.4.2) these are

$$\frac{V}{GA\kappa} = \frac{\partial y}{\partial x} - \psi = y_1' - \psi,$$

$$\frac{Mr}{EI} = -r\frac{\partial \psi}{\partial x} = -\psi'. \tag{3.4.35}$$

We apply the Laplace transform on the time variable to (3.4.33) and (3.4.34), giving

$$\bar{\psi}'' + \frac{1}{\gamma}(\bar{y}_1' - \bar{\psi}) - s^2\bar{\psi} = 0,$$

$$s^2\bar{y}_1 - \frac{1}{\gamma}(\bar{y}_1'' - \bar{\psi}') = 0, \tag{3.4.36}$$

where zero initial conditions have been assumed. These two equations are solved by eliminating one or the other of the dependent variables, giving a fourth-order equation. In terms of $\bar{y}_1$, the equation is

$$\bar{y}_1^{\text{iv}} - (1+\gamma)s^2\bar{y}_1'' + (1+\gamma s^2)s^2\bar{y}_1 = 0. \tag{3.4.37}$$

This is the Laplace transform of (3.4.20) with the non-dimensionalizations of the present problem incorporated. Considering a solution of the form $\bar{y}_1 = C\exp(\lambda x_1)$ gives the characteristic equation

$$\lambda^4 - (1+\gamma)s^2\lambda^2 + (1+\gamma s^2)s^2 = 0. \tag{3.4.38}$$

Solving this gives

$$\lambda^2 = \left(\frac{1+\gamma}{2}\right)s\left[s \pm \left\{s^2 - \frac{4(1+\gamma s^2)}{(1+\gamma)^2}\right\}^{\frac{1}{2}}\right]. \tag{3.4.39}$$

After some manipulation and again taking the square root, one obtains

$$\lambda = \pm Bs^{\frac{1}{2}}\{s \pm N(s^2 - a^2)^{\frac{1}{2}}\}^{\frac{1}{2}}, \tag{3.4.40}$$

where

$$B = \left(\frac{\gamma+1}{2}\right)^{\frac{1}{2}}, \qquad N = \frac{\gamma-1}{\gamma+1}, \qquad a = \frac{2}{\gamma-1}. \tag{3.4.41}$$

We represent (3.4.40) in simplified form as

$$\lambda = \pm\lambda_1, \pm\lambda_2, \tag{3.4.42}$$

where λ_1 corresponds to the positive sign within the bracket of (3.4.40). The resulting solution for $\bar{y}_1$ is

$$\bar{y}_1 = C_1 \exp(-\lambda_1 x_1) + C_2 \exp(-\lambda_2 x_1) + C_3 \exp(\lambda_1 x_1) + C_4 \exp(\lambda_2 x_1). \tag{3.4.43}$$

A similar solution would hold for $\bar{\psi}$ with arbitrary coefficients $B_1,\ldots, B_4$ in place of $C_1,\ldots, C_4$. These would not be independent, but related through (3.4.36). Thus, substituting (3.4.43) and the solution for $\bar{\psi}$ in the second equation of (3.4.36) leads to the following relations:

$$\frac{B_1}{C_1} = -\frac{B_3}{C_3} = \frac{(\gamma s^2 - \lambda_1^2)}{\lambda_1}, \tag{3.4.44}$$

$$\frac{B_2}{C_2} = -\frac{B_4}{C_4} = \frac{(\gamma s^2 - \lambda_2^2)}{\lambda_2}. \tag{3.4.45}$$

The solution of $\bar{\psi}$ is then given by

$$\bar{\psi} = \frac{\lambda_1^2 - \gamma s^2}{\lambda_1}\{C_3 \exp(\lambda_1 x_1) - C_1 \exp(-\lambda_1 x_1)\} + \\ + \frac{\lambda_2^2 - \gamma s^2}{\lambda_2}\{C_4 \exp(\lambda_2 x_1) - C_2 \exp(-\lambda_2 x_1)\}. \tag{3.4.46}$$

Because only semi-infinite beams $x > 0$ will be considered here, it is required that $C_3 = C_4 = 0$.

Several types of boundary conditions may be considered within the present formulation. Several of them are:

(1) step end velocity with zero moment;
(2) step end moment with zero displacement;
(3) step angular velocity with zero force;
(4) step end force with zero rotation.

Attention will be directed toward case (1), although the methods are quite similar in all cases.† The boundary conditions are given by

$$\dot{y}_1(0, t) = v_0 H\langle t\rangle, \qquad \bar{\psi}'(0, t) = 0, \tag{3.4.47}$$

† Reference [2] presents in tabular form the basic results for all of the above cases.

and in transformed form as

$$\bar{y}_1(0, s) = \frac{v_0}{s^2}, \qquad \bar{\psi}'(0, s) = 0. \tag{3.4.48}$$

Substituting (3.4.43) and (3.4.46) with $C_3 = C_4 = 0$ in these boundary conditions gives

$$C_1 = \frac{v_0(\lambda_2^2 - \gamma s^2)}{s^2(\lambda_2^2 - \lambda_1^2)}, \qquad C_2 = \frac{-v_0(\lambda_1^2 - \gamma s^2)}{s^2(\lambda_2^2 - \lambda_1^2)}. \tag{3.4.49}$$

Using (3.4.40), the expression $(\lambda_2^2 - \lambda_1^2)$ in (3.4.49) may be simplified. The resulting transformed solutions are

$$\bar{y}_1(x_1, s) = \frac{v_0 a}{2s^3(s^2 - a^2)^{\frac{1}{2}}}\{(\gamma s^2 - \lambda_2^2)\exp(-\lambda_1 x_1) - (\gamma s^2 - \lambda_1^2)\exp(-\lambda_2 x_1)\}, \tag{3.4.50}$$

$$\bar{\psi}(x_1, s) = \frac{v_0 a}{2s(s^2 - a^2)^{\frac{1}{2}}}\left\{\frac{1}{\lambda_1}\exp(-\lambda_1 x_1) - \frac{1}{\lambda_2}\exp(-\lambda_2 x_1)\right\}. \tag{3.4.51}$$

The transformed expression for the shear force is

$$\frac{\bar{V}}{GA\kappa} = \frac{v_0 \gamma a}{2s(s^2 - a^2)^{\frac{1}{2}}}\left\{-\frac{\gamma s^2 - \lambda_2^2}{\lambda_1}\exp(-\lambda_1 x_1) + \frac{\gamma s^2 - \lambda_1^2}{\lambda_2}\exp(-\lambda_2 x_1)\right\}. \tag{3.4.52}$$

In an effort to have some generality of discussion, we note that the results for $\bar{y}_1$, $\bar{\psi}$, and $\bar{V}$ all can have the general form

$$F(x_1, s) = F_1(s)\exp(-\lambda_1 x_1) + F_2(s)\exp(-\lambda_2 x_1), \tag{3.4.53}$$

where $F_1(s)$, $F_2(s)$ are obtained by inspection of the solutions. The Laplace inversion is then given by

$$f(x_1, t_1) = \frac{1}{2\pi i}\int_{c-i\infty}^{c+i\infty} F(x_1, s)\exp(st_1)\,ds = I_1 + I_2, \tag{3.4.54}$$

where I_1 is the inversion of $F_1(s)\exp(-\lambda_1 x_1)$ and I_2 the inversion of $F_2(s)\exp(-\lambda_2 x_1)$.

We first deal with the matter of closure of the Bromwich contour. This yields most interesting information on the propagation velocity of wavefronts. Thus, for I_1 we have

$$I_1 = \frac{1}{2\pi i}\int_{c-i\infty}^{c+i\infty} F_1(s)\exp(st_1 - \lambda_1 x_1)\,ds. \tag{3.4.55}$$

In closing the Bromwich contour, either to the right or left, we see that s will be of the form $R\exp(i\theta)$ along a semicircular path. For s large we have that

$$\lambda_1 = Bs\left\{1 + N\left(1 - \frac{a^2}{s^2}\right)^{\frac{1}{2}}\right\}^{\frac{1}{2}}\Bigg|_{s\to\infty} \to Bs(1+N)^{\frac{1}{2}}. \tag{3.4.56}$$

From the definitions of B and N, given by (3.4.41), this becomes

$$\lambda_1 \big|_{s\to\infty} \to \gamma^{\frac{1}{2}} s. \tag{3.4.57}$$

Then

$$\exp(st_1 - \lambda_1 x_1) \to \exp\{s(t_1 - \gamma^{\frac{1}{2}} x_1)\}. \tag{3.4.58}$$

Thus, if $t_1 < \gamma^{\frac{1}{2}} x_1$, the contour must be closed to the right, while for $t_1 > \gamma^{\frac{1}{2}} x_1$, the contour must be closed to the left. A similar type of argument for I_2 yields the result that

$$\exp(st_1 - \lambda_2 x_1) \to \exp\{s(t_1 - x_1)\}. \tag{3.4.59}$$

Thus, for $t_1 < x_1$, closure is rightward, while for $t_1 > x_1$, closure is leftward. The three types of closure are indicated in Fig. 3.14. The closure to the left

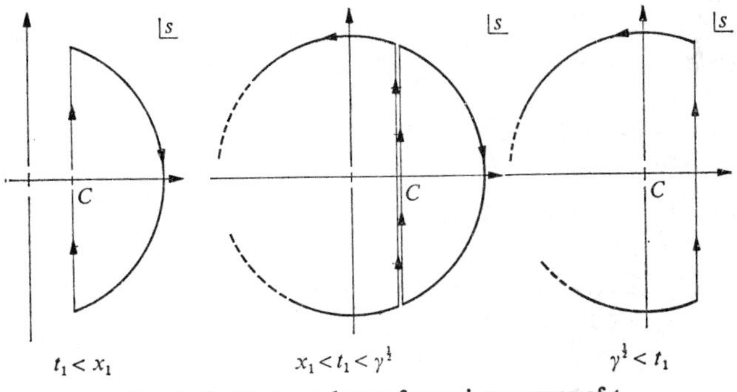

FIG. 3.14. Contour closure for various ranges of t_1.

is left purposely vague since considerable detail must be added in that region.

It does not greatly pre-empt later discussion to state here that no poles or branch points of the integrand exist in Re $s > c$. Hence, for closure to the right, we have $I_1 = I_2 = 0$. The significance of this is that it indicates signal arrival time. Thus, at a station x_1 no signal arrives until $t_1 = x_1$. In dimensional form this is given by $t = x/c_0$. Then, at $t_1 = \gamma^{\frac{1}{2}} x_1$, a second clearly distinguishable signal arrives. In dimensional form, this is at $t = (\rho/G\kappa)^{\frac{1}{2}} x$. Thus, two distinct wavefronts exist, travelling at the velocities c_0 (the faster) and $(G\kappa/\rho)^{\frac{1}{2}}$ (the slower). Referring back to (3.4.26), these are seen to be the large-wavenumber, high-frequency limits for the Timoshenko beam. As to the nature of the disturbance arriving at these times, we hope to establish more about this later.

We now consider the integral I_1 in somewhat greater detail. The complexities that arise in the evaluation of the inversion integral are in connection with the several branch points of the integrands. Specifically, consider the expression

$$I_1^{\psi} = \frac{1}{2\pi i}\int_{c-i\infty}^{c+i\infty} \frac{\exp(st_1-\lambda_1 x_1)}{s(s^2-a^2)^{\frac{1}{2}}\lambda_1}\,ds. \tag{3.4.60}$$

This corresponds to the I_1 part of the result (3.4.51) for the rotation. We have that

$$(s^2-a^2)^{\frac{1}{2}}\lambda_1 = Bs^{\frac{1}{2}}(s+a)^{\frac{1}{2}}(s-a)^{\frac{1}{2}}\{s+N(s^2-a^2)^{\frac{1}{2}}\}. \tag{3.4.61}$$

These quantities must be made single-valued by proper branch cuts in the s-plane. Using the representations

$$s-a = \rho_1\exp(i\theta_1), \qquad s = \rho\exp(i\theta), \qquad s+a = \rho_2\exp(i\theta_2), \tag{3.4.62}$$

we see that these functions may be made single-valued by restricting the arguments to the range

$$-\pi < \theta_1,\ \theta,\ \theta_2 < \pi. \tag{3.4.63}$$

The geometry of this situation in the complex plane is shown in Fig. 3.15,

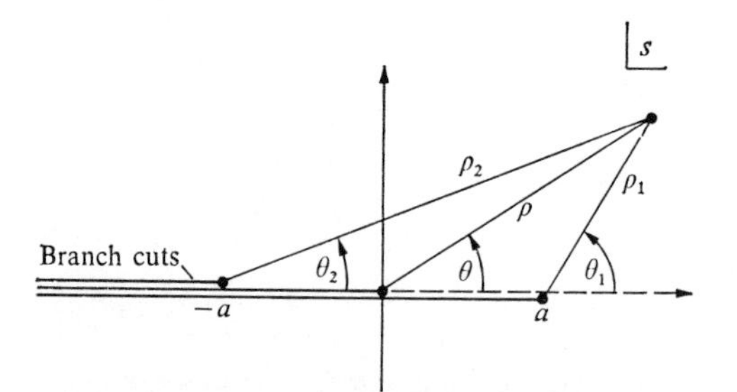

FIG. 3.15. Branch cut situation for the functions $s^{\frac{1}{2}}$, $(s+a)^{\frac{1}{2}}$, $(s-a)^{\frac{1}{2}}$.

as well as the branch cuts to the left along the real axis to establish this range. The remaining function of (3.4.61), $\{s+N(s^2-a^2)^{\frac{1}{2}}\}^{\frac{1}{2}}$, will also be single-valued if the argument of the square of this function is restricted to the range (3.4.63). However, the possibility that a branch cut might also be required for this function must be considered. This will be true if $s+N(s^2-a^2)^{\frac{1}{2}} = 0$. This will occur only if $s = \pm i/\gamma^{\frac{1}{2}}$. Considering the positive root, we write

$$s+N(s^2-a^2)^{\frac{1}{2}} = \frac{i}{\gamma^{\frac{1}{2}}}+N\left(\frac{i}{\gamma^{\frac{1}{2}}}+a\right)^{\frac{1}{2}}\left(\frac{i}{\gamma^{\frac{1}{2}}}-a\right)^{\frac{1}{2}}. \tag{3.4.64}$$

Using Fig. 3.15 as an aid to interpreting, it is seen that the real parts of the function will cancel but that the imaginary parts, being all positive, cannot be zero. Hence, $s = i/\gamma^{\frac{1}{2}}$ will not cause the function in question to be zero. A similar result holds for $s = -i/\gamma^{\frac{1}{2}}$, except that the imaginary parts are all negative. Thus, all of the branch points are accounted for.

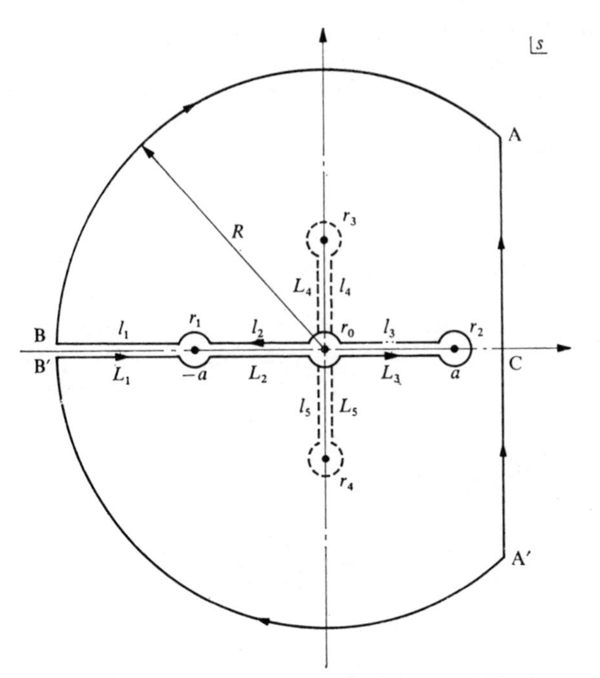

FIG. 3.16. Integration path for evaluating I_1-type integrals.

The resulting form when closed to the left, is shown in Fig. 3.16 as the solid-line contour. For this contour, the following facts hold:†

(1) the contributions from AB and A′B′ vanish as $R \to \infty$;
(2) integration along l_1+L_1 gives a zero value;
(3) integration about the circular contours r_1, r_2 gives a zero result as the radii approach zero;
(4) integration about the circular contour r_0 gives zero for the case at hand.

Since there are no poles within the contour, the entire path integral vanishes. Then the expression for I_1^{φ} may be written as

$$I_1^{\varphi} = \frac{1}{2i} \int_{L_2+L_3+l_3+l_2} \frac{\exp(st_1-\lambda_1 x_1)}{s(s^2-a^2)^{\frac{1}{2}}\lambda_1}\,ds. \tag{3.4.65}$$

The task of evaluating the result (3.4.60) thus resolves down to evaluation of four definite integrals having finite limits.

It is noted by Boley and Chao [2, p. 583] that statement (4) does not hold if the exponent of s in the denominator is greater than unity. This occurs in the problems at hand, for example, for $\bar{y}_1$. The complexity of the problem then increases because branch points of higher order arise. The authors present a

† Problem 3.20 requests the verification of these facts.

method of circumventing this difficulty by using various reciprocal relations existing among the dependent variables.

The main additional complexity that is present in the problem concerns the evaluation of the I_2 type integrals (see (3.4.54)). Specifically, the expression $\{s-N(s^2-a^2)^{\frac{1}{2}}\}^{\frac{1}{2}}$ is found to contribute additional branch points (recall the discussion of $\{s+N(s^2-a^2)^{\frac{1}{2}}\}^{\frac{1}{2}}$) within the contour. The location of these branch points and the modified contour necessary for the I_2 integrals are shown in Fig. 3.16 as the solid contour with the dashed line additions. Additional definite integrals are introduced by the paths along the additional branch cuts.

We terminate the general review of this problem with presentation of results. Fig. 3.17(a) shows the variation of shear force with time at two different

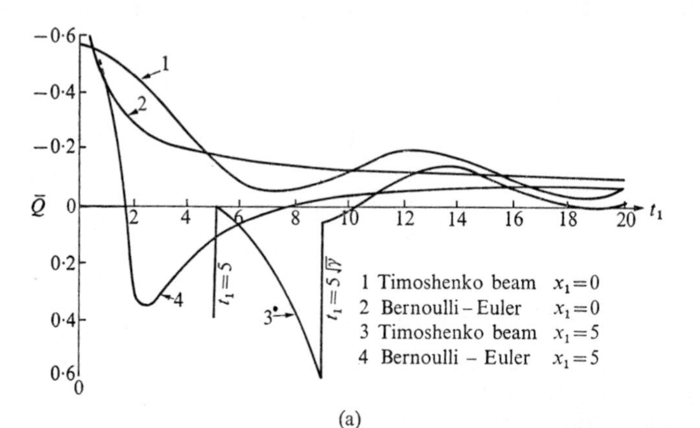

(a)

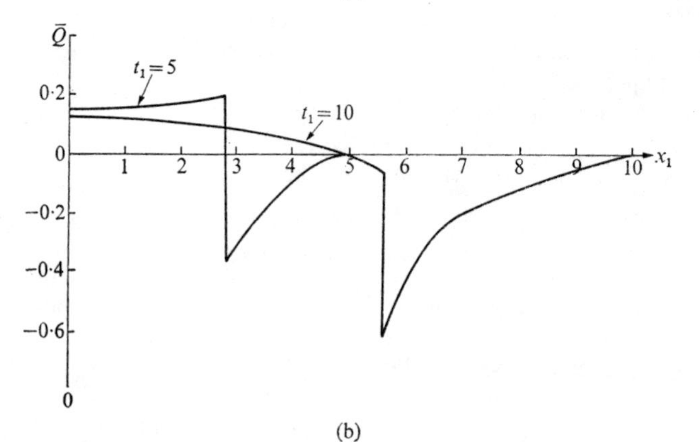

(b)

FIG. 3.17. (a) Variation in the shear force $\bar{Q}$, with time at various locations, and (b) the variation of $\bar{Q}$ with location at two times, where $\bar{Q}=Q/EAV_0$. (After Boley and Chao [2].)

locations. Also shown is the shear force as predicted by Bernoulli–Euler theory. Considering in particular the behaviour at $x_1 = 5$, it is seen that, as expected, the Bernoulli–Euler theory predicts instantaneous response. The beam is at rest, according to Timoshenko theory, until $t_1 = 5$, the arrival of the wavefront propagated at c_0. The shear increases until the arrival of the $c_0\sqrt{\gamma}$ wavefront. A discontinuity in shear occurs at this stage. Thus the step disturbance at the origin has created a shear discontinuity, and this propagates with the aforementioned limit velocity. Other results and discussion presented by the authors indicate that the velocity discontinuities propagate with this same limit velocity, while discontinuities in bending moment and angular velocity propagate at the limit velocity c_0. Fig. 3.17(b) shows the variation of shear force along the beam at two successive times. It is to be noted that the maximum value of shear force does not occur at the origin but at a later time and further location.

3.5. Wave propagation in rings

The addition of curvature to rods results in an infinite variety of shapes. The development of the governing equations for rods 'naturally curved' goes back to the 1800's (for example, to the work of Kelvin in 1859). The standard reference for the equations for rods of arbitrary curvature remains to be Love [13, Ch. 18, 21]. The dynamical equations of motion for a rod of circular curvatures are also presented by Love [13, pp. 451–4]. These basic equation forms are widely referred to. It is in the governing relations between the forces, moments, displacements, and rotations that a variety of views are set forth. A similar, but even more complicated, situation will be found to exist for shells. Our development will be restricted to rods of circular curvature.

3.5.1. *The governing equations*

Consider an element isolated from a curved rod, where the rod possesses a plane of symmetry, and the plane of curvature, symmetry, and the plane of all loads coincide, as shown in Fig. 3.18.

The tensile load is N, the shear force is V and the bending moment is M. Variations of these quantities are shown for a positive increment of angle $d\theta$ and arc length ds. The displacements of the centroidal axis in the radial and tangential directions are given by w and v respectively. The equations of motion in the w and v directions are

$$-V+\left(V+\frac{\partial V}{\partial \theta}\,d\theta\right)-\left(N+\frac{\partial N}{\partial \theta}\,d\theta\right)d\theta = \rho AR\,d\theta\frac{\partial^2 w}{\partial t^2}, \tag{3.5.1}$$

$$-N+\left(N+\frac{\partial N}{\partial \theta}\,d\theta\right)+\left(V+\frac{\partial V}{\partial \theta}\,d\theta\right)d\theta = \rho AR\,d\theta\frac{\partial^2 v}{\partial t^2}. \tag{3.5.2}$$

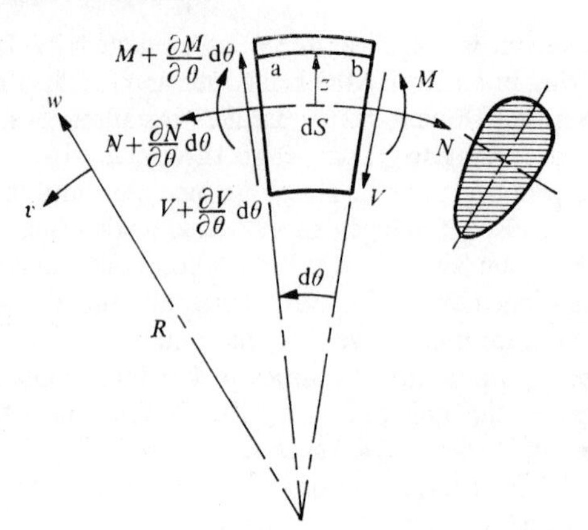

FIG. 3.18. Differential element from a curved rod.

The moment equation, if rotary-inertia effects are neglected, is given by

$$M-\left(M+\frac{\partial M}{\partial \theta}\,d\theta\right)-\left(V+\frac{\partial V}{\partial \theta}\,d\theta\right)R\,d\theta = 0. \tag{3.5.3}$$

These reduce to

$$\frac{\partial V}{\partial \theta}-N = \rho RA\frac{\partial^2 w}{\partial t^2}, \tag{3.5.4}$$

$$\frac{\partial N}{\partial \theta}+V = \rho RA\frac{\partial^2 v}{\partial t^2}, \tag{3.5.5}$$

$$\frac{\partial M}{\partial \theta}+RV = 0. \tag{3.5.6}$$

If the shearing force is eliminated from the first two equations, the result is

$$-\frac{1}{R}\frac{\partial^2 M}{\partial \theta^2}-N = \rho RA\frac{\partial^2 w}{\partial t^2}, \tag{3.5.7}$$

$$\frac{\partial N}{\partial \theta}-\frac{1}{R}\frac{\partial M}{\partial \theta} = \rho RA\frac{\partial^2 v}{\partial t^2}. \tag{3.5.8}$$

Finally, these results could be expressed in terms of the arc length s, where $ds = R\,d\theta$. Using this, we have for the preceding equations

$$\begin{aligned} -\frac{\partial^2 M}{\partial s^2}-\frac{N}{R} &= \rho A\frac{\partial^2 w}{\partial t^2}, \\ \frac{\partial N}{\partial s}-\frac{1}{R}\frac{\partial M}{\partial s} &= \rho A\frac{\partial^2 v}{\partial t^2}. \end{aligned} \tag{3.5.9}$$

Although our interest will be in rings of circular curvature, the form (3.5.9) applies to arbitrarily (plane) curved rods where R is now interpreted as a variable radius of curvature.

We now must relate the ring forces and moments to the deformations and displacements. We have shown, in Fig. 3.19, a typical lamina (ab) of the ring, located a distance z from the centroidal axis. If the axial stress in that lamina is σ, then we have for the ring that

$$N = \int_A \sigma \, \mathrm{d}A, \qquad M = -\int_A \sigma z \, \mathrm{d}A, \tag{3.5.10}$$

where $\sigma = E\varepsilon$ and ε is the axial strain in the lamina. The strain will be given by

$$\varepsilon = (l_2 - l_1)/l_1, \tag{3.5.11}$$

where l_1, l_2 are the initial and final lengths of ab. We now assume that plane sections, initially perpendicular to the central axis, remain plane and perpendicular to the central axis after deformation. This is, of course, the Bernoulli–Euler hypothesis applied to the case of rings. Then we have that

$$l_1 = (R+z)\,\mathrm{d}\theta, \qquad l_2 = (R'+z)\,\mathrm{d}\theta', \tag{3.5.12}$$

where R', $\mathrm{d}\theta'$ are the radius of curvature and subtended angle of the deformed element. We now establish these quantities in terms of the displacements.

Consider, in Fig. 3.19, the central axis of the element in the undeformed

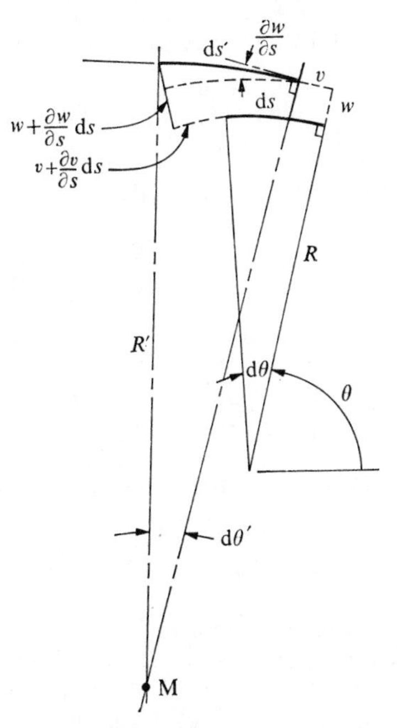

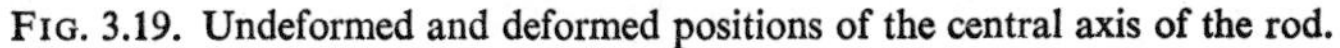

FIG. 3.19. Undeformed and deformed positions of the central axis of the rod.

and deformed positions. Point 0 undergoes the displacements v and w to the position 0′ while P undergoes the displacements $v+(\partial v/\partial s)\,\mathrm{d}s$,

$$w+(\partial w/\partial s)\,\mathrm{d}s$$

to P′. The length of the deformed element may be expressed as

$$\mathrm{d}s' = (w+R)\,\mathrm{d}\theta+\frac{\partial v}{\partial s}\,\mathrm{d}s. \tag{3.5.13}$$

The contribution of $(\partial w/\partial s)\,\mathrm{d}s$ to the change in arc length has been neglected. We now calculate $\mathrm{d}\theta'$. This angle is given by the difference in angle between the rays MO′ and MP′. The angle to MO′ is given by $\theta+v/R-\partial w/\partial s$. To MP′ the angle will be $\theta+\mathrm{d}\theta+\{v+(\partial v/\partial s)\,\mathrm{d}s\}/R-\{\partial w/\partial s+(\partial^2 w/\partial s^2)\,\mathrm{d}s\}$. Then for $\mathrm{d}\theta'$ we have

$$\begin{aligned}\mathrm{d}\theta' &= \left\{\theta+\mathrm{d}\theta+\frac{1}{R}\left(v+\frac{\partial v}{\partial s}\,\mathrm{d}s\right)-\left(\frac{\partial w}{\partial s}+\frac{\partial^2 w}{\partial s^2}\,\mathrm{d}s\right)\right\}-\left(\theta+\frac{v}{R}-\frac{\partial w}{\partial s}\right)\\ &= \mathrm{d}\theta+\frac{1}{R}\frac{\partial v}{\partial s}\,\mathrm{d}s-\frac{\partial^2 w}{\partial s^2}\,\mathrm{d}s.\end{aligned} \tag{3.5.14}$$

The results (3.5.13) and (3.5.14) may be expressed in terms of the angle $\mathrm{d}\theta$. Thus

$$\mathrm{d}s' = (w+R)\,\mathrm{d}\theta+\frac{\partial v}{\partial \theta}\,\mathrm{d}\theta, \qquad \mathrm{d}\theta' = \mathrm{d}\theta+\frac{1}{R}\frac{\partial v}{\partial \theta}\,\mathrm{d}\theta-\frac{1}{R}\frac{\partial^2 w}{\partial \theta^2}\,\mathrm{d}\theta. \tag{3.5.15}$$

We may now establish ε. Noting that $R'\,\mathrm{d}\theta' = \mathrm{d}s'$, we have that

$$l_2 = \left\{\left(w+R+\frac{\partial v}{\partial \theta}\right)+z\left(1+\frac{1}{R}\frac{\partial v}{\partial \theta}-\frac{1}{R}\frac{\partial^2 w}{\partial \theta^2}\right)\right\}\,\mathrm{d}\theta. \tag{3.5.16}$$

Then we have from (3.5.11)

$$\varepsilon = \left\{w+\frac{\partial v}{\partial \theta}+\frac{z}{R}\left(\frac{\partial w}{\partial \theta}-\frac{\partial^2 w}{\partial \theta^2}\right)\right\}\Big/R\left(1+\frac{z}{R}\right). \tag{3.5.17}$$

If z/R is neglected with respect to unity in the denominator we have

$$\varepsilon = \frac{1}{R}\left\{w+\frac{\partial v}{\partial \theta}+\frac{z}{R}\frac{\partial}{\partial \theta}\left(v-\frac{\partial w}{\partial \theta}\right)\right\}. \tag{3.5.18}$$

Upon substituting this result in the expression (3.5.10) for N and performing the integration we obtain

$$N = \frac{EA}{R}\left(w+\frac{\partial v}{\partial \theta}\right). \tag{3.5.19}$$

Thus the z contribution of (3.5.18) does not enter since the axis is the centroidal axis. Substituting in the expression for M, we obtain

$$M = -\frac{EAk^2}{R^2}\frac{\partial}{\partial\theta}\left(v-\frac{\partial w}{\partial\theta}\right), \tag{3.5.20}$$

where k^2 is the radius of gyration of the cross-section.

With the development of the ring stress–displacement equations (3.5.19) and (3.5.20), the equations of motion (3.5.7) and (3.5.8) are

$$\frac{EAk^2}{R^3}\frac{\partial^3}{\partial\theta^3}\left(v-\frac{\partial w}{\partial\theta}\right)-\frac{EA}{R}\left(w+\frac{\partial v}{\partial\theta}\right) = \rho RA\frac{\partial^2 w}{\partial t^2}, \tag{3.5.21}$$

$$\frac{EAk^2}{R^3}\frac{\partial^2}{\partial\theta^2}\left(v-\frac{\partial w}{\partial\theta}\right)+\frac{EA}{R}\frac{\partial}{\partial\theta}\left(w+\frac{\partial v}{\partial\theta}\right) = \rho RA\frac{\partial^2 v}{\partial t^2}. \tag{3.5.22}$$

3.5.2. *Wave propagation*

The propagation characteristics of thin rings may be established by considering harmonic waves of the form

$$w = A_1 e^{i(\gamma R\theta-\omega t)}, \qquad v = A_2 e^{i(\gamma R\theta-\omega t)}. \tag{3.5.23}$$

Substitution of these into the governing equations (3.5.21), (3.5.22) gives

$$\begin{bmatrix} \left(\frac{\omega^2R^2}{c_0^2}-1-\gamma^4R^2k^2\right) & -i(\gamma^3Rk^2+\gamma R) \\ i(\gamma^3Rk^2+\gamma R) & \left(\frac{\omega^2}{c_0^2}R^2-\gamma^2R^2-\gamma^2k^2\right) \end{bmatrix} \begin{bmatrix} A_1 \\ A_2 \end{bmatrix} = 0. \tag{3.5.24}$$

We introduce the non-dimensionalizations

$$\bar{k} = k/R, \qquad \bar{\gamma} = k\gamma, \qquad \bar{\omega} = \omega k/c_0, \qquad \bar{c} = c/c_0. \tag{3.5.25}$$

Then the frequency equation resulting from (3.5.24) is

$$(\bar{\omega}^2-\bar{k}^2-\bar{\gamma}^4)(\bar{\omega}^2-\bar{\gamma}^2-\bar{k}^2\bar{\gamma}^2)-\bar{k}^2\bar{\gamma}^2(\bar{\gamma}^2+1)^2 = 0, \tag{3.5.26}$$

and in expanded form is given by

$$\bar{\omega}^4-\{\bar{\gamma}^4+(1+\bar{k}^2)\bar{\gamma}^2+\bar{k}^2\}\bar{\omega}^2+\bar{\gamma}^2(\bar{k}^2-\bar{\gamma}^2)^2 = 0. \tag{3.5.27}$$

The dispersion relation is obtained by noting that $\bar{\omega} = \bar{\gamma}\bar{c}$ so that we obtain, from (3.5.27),

$$\bar{c}^4-\{\bar{\gamma}^2+(1+\bar{k}^2)+\bar{k}^2/\bar{\gamma}^2\}\bar{c}^2+\bar{\gamma}^2(1-\bar{k}^2/\bar{\gamma}^2)^2 = 0. \tag{3.5.28}$$

Before presenting the general behaviour as predicted by (3.5.27) and (3.5.28), we note the behaviour at long wavelengths ($\bar{\gamma}\to 0$) and short wavelengths ($\bar{\gamma}\to\infty$). From (3.5.27) we have that, as $\bar{\gamma}\to 0$,

$$\lim_{\bar{\gamma}\to 0}(3.5.27) = \bar{\omega}^2(\bar{\omega}^2-\bar{k}^2) = 0. \tag{3.5.29}$$

Thus, $\bar{\omega} = 0, \bar{k}$ are the limits. The latter value indicates a cutoff frequency of one of the modes. As $\bar{\gamma} \to \infty$, we have from (3.5.28) that

$$\lim_{\gamma\to\infty}(3.5.28) = \bar{\gamma}^2(\bar{c}^4/\bar{\gamma}^2 - \bar{c}^2 + 1) = 0. \tag{3.5.30}$$

If $\bar{c}$ remains finite in (3.5.30), we must have that $\bar{c} = 1$. This corresponds to the velocity of longitudinal waves in a thin straight rod. A solution to the general equation may be found for $\bar{c} = 0$. Thus (3.5.28) reduces to

$$\bar{\gamma}^2(1 - \bar{k}^2/\bar{\gamma}^2)^2 = 0,$$

from which we have $\bar{\gamma} = \bar{k}$. Another limit of interest is that obtained by allowing the radius of curvature to become large. When this occurs, we have that $\bar{k} \to 0$ as $R \to \infty$. We see that (3.5.28) reduces to

$$\lim_{\bar{k}\to 0}(3.5.28) = \bar{c}^4 - (\bar{\gamma}^2 + 1)\bar{c}^2 + \bar{\gamma}^2 = (\bar{c}^2 - \bar{\gamma}^2)(\bar{c}^2 - 1) = 0. \tag{3.5.31}$$

Thus $\bar{c} \to \bar{\gamma}, 1$. These are the classical limits for flexural waves in straight beams according to Bernoulli–Euler theory and for longitudinal waves in thin rods according to the wave equation. Thus, in dimensional form, we have

$$c \to \sqrt{\left(\frac{EI}{\rho A}\right)}\gamma,\ c_0. \tag{3.5.32}$$

The data for the frequency spectrum and dispersion curves of a curved rod must be presented in terms of the curvature parameter $\bar{k}$. If $\bar{k} = 0$, we have the case of a straight rod. The maximum value that $\bar{k}$ can attain for a rod of circular cross-section is $\frac{1}{2}$ (this gives a radius of curvature equal to the radius of the cross-section). However, restrictions similar to those for straight beams and rods hold regarding wavelength in comparison to thickness and, in this case, in comparison to radius of curvature. Consequently $\bar{k} \ll \frac{1}{2}$ must hold. The general behaviour of the real, positive branches of the frequency spectrum and dispersion curves for $\bar{k}^2 = 0{\cdot}05$ are shown in Fig. 3.20. The cutoff frequency occurring in the frequency spectrum at $\bar{\omega} = \bar{k}$ is shown. The dashed lines correspond to the case of $\bar{k} = 0$, which are the results for flexural and longitudinal waves in straight rods. The zero propagation velocity occurring at the special root of $\bar{\gamma} = \bar{k}$ corresponds to a wavelength equal to the circumference.

This concludes a rather brief treatment of waves in curved rods. There are, of course, many other aspects that could be considered. In the context of curved rings, theories including shear and rotary-inertia effects could be developed. Out-of-plane forces and motion could also be considered. The present development was based on both bending and extensional effects. In many developments, the latter effect is neglected. Love [13, p. 451] has presented the equations for out-of-plane motion of a circular ring with

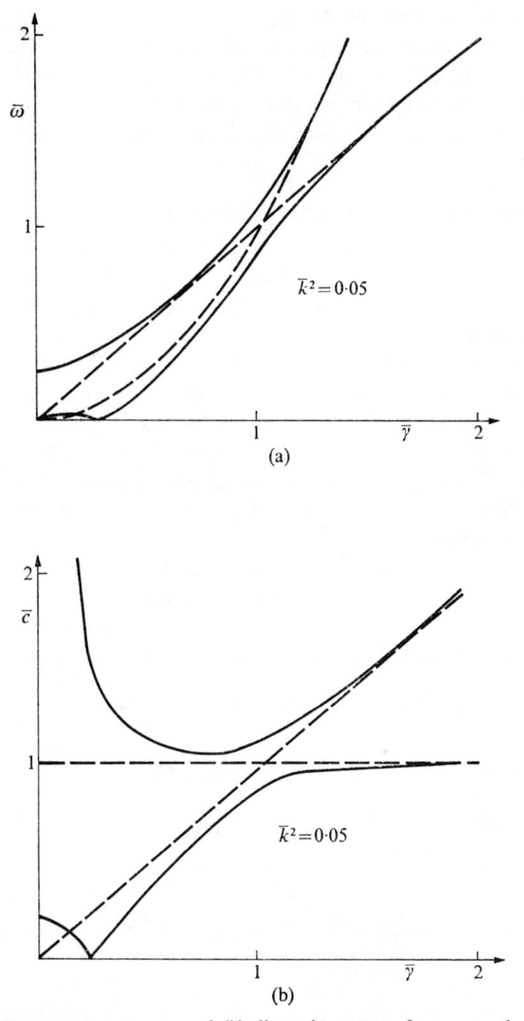

FIG. 3.20. (a) Frequency spectrum and (b) dispersion curves for a curved rod for which $\bar{k} = 0{\cdot}05$. The dashed lines correspond to the case of $\bar{k} = 0$.

extensional effects neglected. Philipson [17] has paralleled Love's development, but with extensional effects included. Morley [15] and Graff [8] have presented theories including shear effects and rotary inertia, with dispersion curves for various curvatures being given in the latter. The major portion of the technical literature on rings is devoted to the vibrations of complete rings or ring sectors, with extensional effects usually neglected.

3.6. Experimental studies on beams

A fairly large number of experimental studies have been conducted on waves and vibrations in beams, with the measurement of transient flexural waves probably receiving the greatest attention. Some of these studies, plus some results on Timoshenko beam theory and waves in curved rods, will be presented here.

3.6.1. *Propagation of transients in straight beams*

A number of studies have been made of the propagation of transient flexural disturbances in beams. Dohrenwend, Drucker, and Moore [5] presented results on waves in strings, beams, and plates. The experimental results were compared to theoretical solutions based on assumed initial velocity distributions of the form $v_0 \exp(-x^2/4b^2)$. Vigness [26] presented additional experimental data on the wave motion in a cantilever beam subjected to a step change in velocity at the built in end. Hoppmann [9] presented results for transient waves in multi-span beams. The work by Ripperger and Abramson [20], cited in § 2.7, also contained results on flexural waves in rods and followed an earlier work [19] giving results on the propagation of flexural waves in beams.

Cunningham and Goldsmith [4] reported on the oblique impact of a steel ball on a beam. Strain gauges were used to record the outer fibre strains in beams of rectangular cross-section subjected to the transverse impact. Fig. 3.21 shows the recorded waveform at various locations along the beam

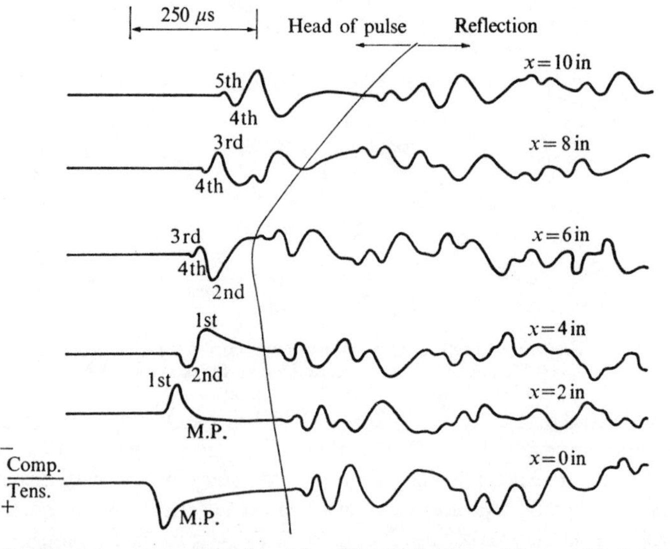

FIG. 3.21. Stress wave propagation in a $\frac{3}{8}$ in $\times \frac{1}{2}$ in $\times 22$ in clamped beam subjected to the transverse impact of a $\frac{1}{2}$ in diameter ball travelling at 30 ft s^{-1} and 7° to the vertical. (After Cunningham and Goldsmith [4, Fig. 2].)

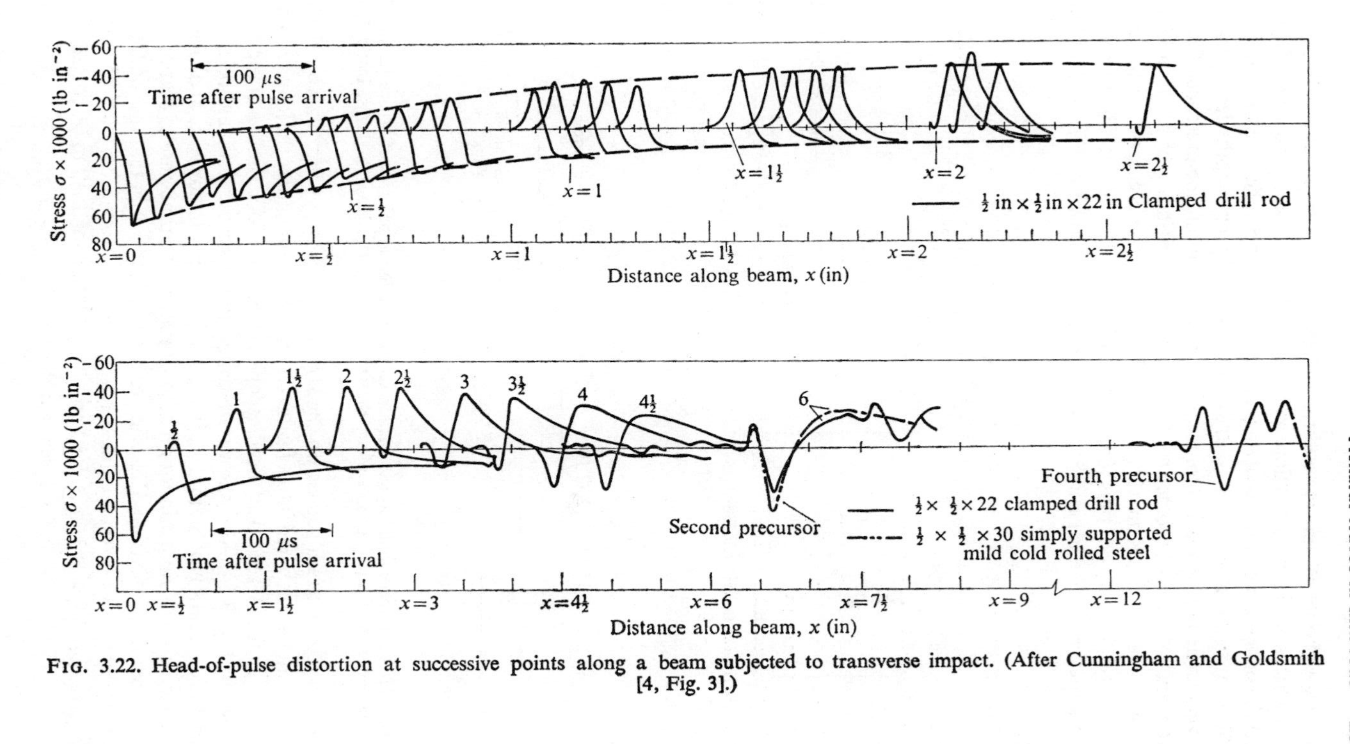

FIG. 3.22. Head-of-pulse distortion at successive points along a beam subjected to transverse impact. (After Cunningham and Goldsmith [4, Fig. 3].)

for a specific set of impact conditions. Reflections from the end of the beam comprise a substantial portion of the records and should be ignored. The details of the distortion occurring at the head of the pulse due to dispersion effects are shown in Fig. 3.22. It is seen that the peak amplitude undergoes inversion as it propagates along the beam. In addition, the presence of high-frequency wavelets becomes more noticeable with distance of propagation.

In a later study, Kuo [12] subjected a rod to the eccentric longitudinal impact of another rod, thus inducing flexural as well as longitudinal waves in the struck rod. The experimental arrangement is shown in Fig. 3.23(a) and

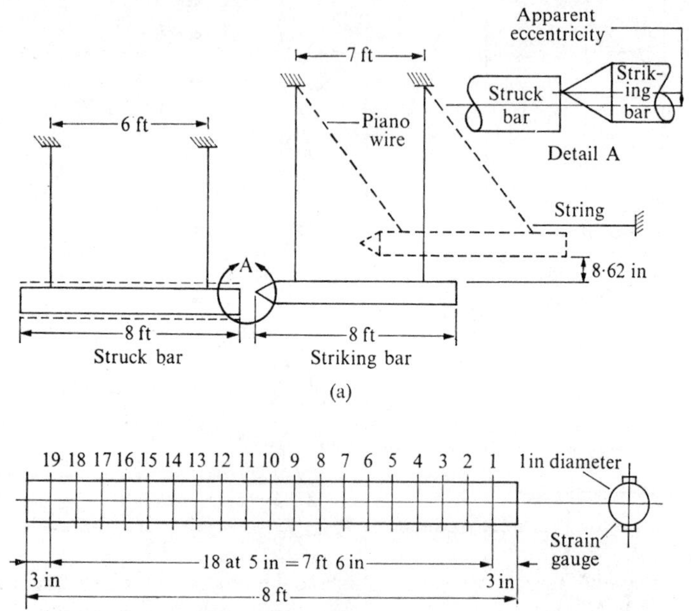

FIG. 3.23. (a) Experimental apparatus for eccentric longitudinal impact of two bars and (b) the location of strain gauges along the struck bar. (After Kuo [12, Figs. 1 and 2].)

the location of strain gauges on the struck bar shown in Fig. 3.23(b). Each pair of gauges was wired into a bridge circuit so as to eliminate the longitudinal strain signals. A record of oscilloscope traces from the various stations is shown in Fig. 3.24. Also shown are calibration records. It should be noted that the calibration of Station 16 is 400 μin per in per major vertical division, whereas it is 200 μin per in for all other stations. The time base of Station 3 is 500 μs per major division. Referring to Station 1, it is seen that the loading is practically that of a rectangular bending pulse showing a duration of about 960 μs. This time is consistent with the duration of loading expected for

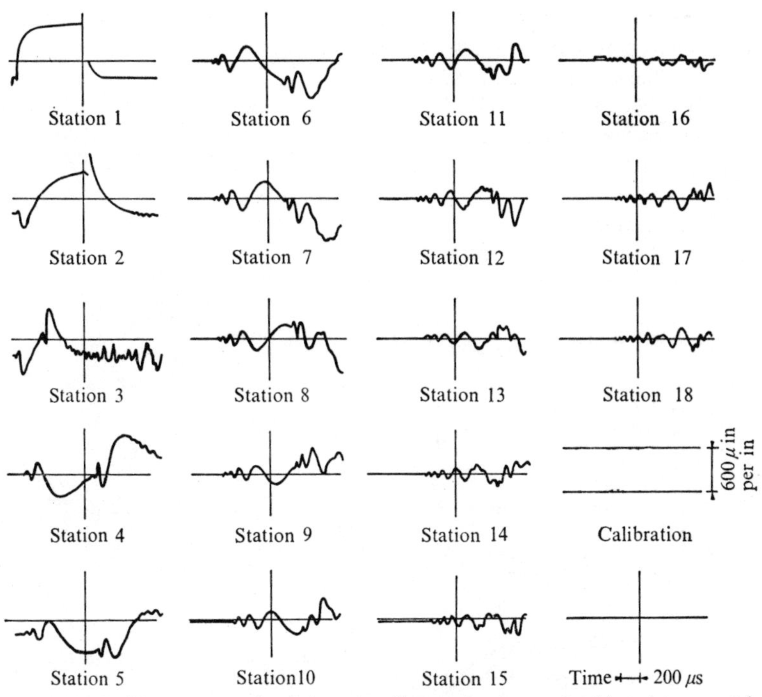

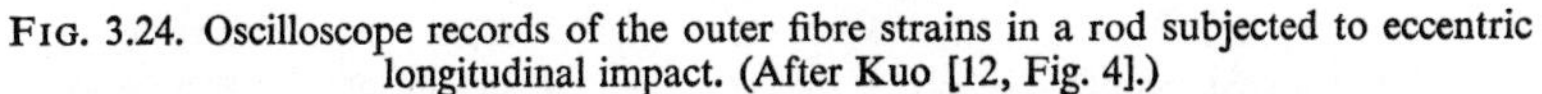

FIG. 3.24. Oscilloscope records of the outer fibre strains in a rod subjected to eccentric longitudinal impact. (After Kuo [12, Fig. 4].)

perfect longitudinal impact of two 8 ft steel rods. The degeneration of the step pulse into a variable frequency harmonic wave train is quite evident in these results.

Kuo also obtained theoretical solutions to the problem of a free–free beam subjected to an end moment. In one case, the Bernoulli–Euler model of the beam was used and a modal superposition was applied (the first 27 modes were used) to obtain the results. In the other theoretical case, the Timoshenko beam model was used, and the method of characteristics applied to obtain the solution. In Fig. 3.25 the comparison between the experimental results and the predictions of the Timoshenko beam theory are shown. Kuo also presents the comparison between the purely theoretical results of the two beam theories. The agreement of the results in the figure are generally quite good, with the main discrepancy occurring in the phase shifting.

Stephenson and Wilhoit [23] have also studied the propagation of bending transients in a rod. An ingenious scheme for producing a sudden end moment involving the rapid unloading due to fracture of a tensile loading piece was employed. A rod 30 ft in length was used with eight strain-gauge

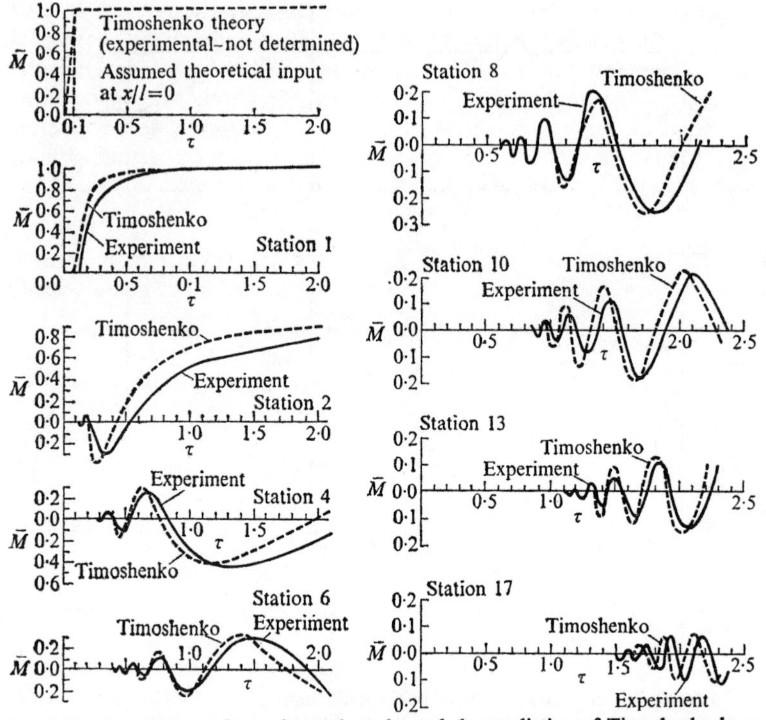

FIG. 3.25. Comparison of experimental results and the prediction of Timoshenko beam theory, where $\tau = c_0t/L$ and $\bar{M} = ML/EI$, where $\bar{M}$ is the dimensionless bending moment. (After Kuo [12, Fig. 8].)

stations in the first 21 ft recording the wave propagation. The results again showed the evolution of the pulse into a variable frequency harmonic wave train. The great length of the test specimen enabled the wavefront velocity to be established with great accuracy. Schweiger [21] has also studied the transverse impact of beams, obtaining data on the impact force and central bending strain. Photo-elastic-layer and strain-gauge data was obtained. Mori [14] obtained experimental data on the propagation of bending waves in longitudinally prestressed beams subjected to transverse impact. Finally, we note that in the previously cited work of Ripperger and Abramson [20] (§ 2.7) that experimental data on the reflection of a flexural pulse from a step discontinuity in rod cross-section was obtained.

3.6.2. *Beam vibration experiments*

Reference has already been made in the previous section to the work of Kuo [12] in comparing Timoshenko beam theory with experimental results for transient waves. Zemanek and Rudnick [28], previously cited in § 2.7 for their experimental results on longitudinal rod vibrations, have also obtained data on the transverse vibrations of beams. The experimental arrangement shown in Fig. 2.24(a) was used for the flexural wave study, except that the electrostatic drive was replaced by an electromagnetic drive unit. The first 306 resonant frequencies were excited and the resulting phase velocity computed from the formula

$$c = 2lf_n/(n+\tfrac{1}{2}). \tag{3.6.1}$$

The experimental data and the predictions of three different beam theories are shown in Fig. 3.26. The plot is one of non-dimensional phase velocity

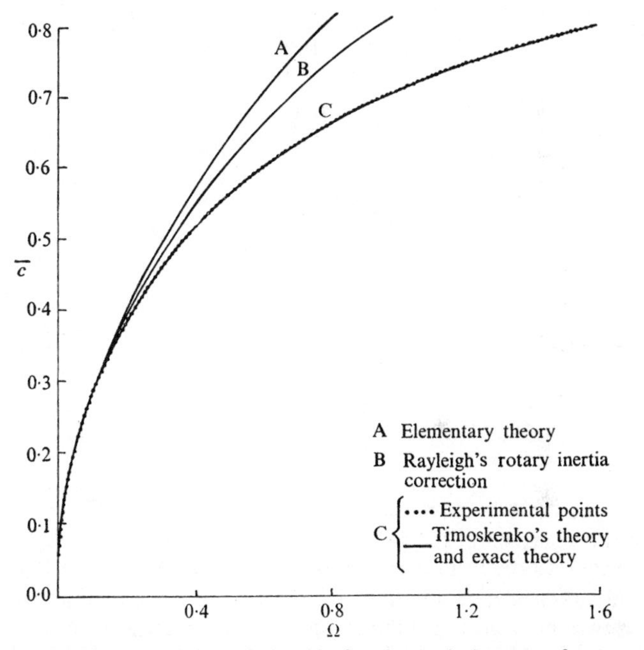

FIG. 3.26. Experimental and theoretical results for phase velocity versus frequency in the transverse motion of a beam. (After Zemanek and Rudnick [28, Fig. 6].)

versus frequency, where

$$\bar{c} = c/c_2, \qquad \Omega = \omega a/c_2, \qquad c_2 = \surd(G/\rho), \tag{3.6.2}$$

and a is the radius of the rod. Curve A is the prediction of Bernoulli–Euler beam theory. Curve B is the prediction of a beam theory indicating rotary-inertia effects alone, and curve C is the Timoshenko and exact theory prediction. The excellent agreement between the experimental results and curve C need hardly be commented on.

3.6.3. *Waves in curved rings*

Only a few theoretical or experimental studies on wave propagation in rings have been made, with reference being made to some of the theoretical work in § 3.5. Britton and Langley [3] have conducted some noteworthy experimental studies on wave propagation in helical springs. The geometry of the test specimen is shown in Fig. 3.27(a), and the technique for inducing

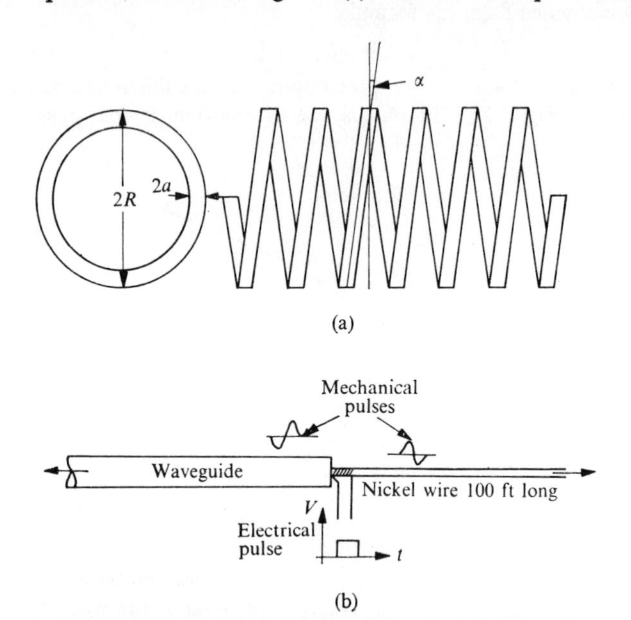

FIG. 3.27. (a) Helical ring test specimen and (b) the means for inducing stress pulses into the helix. (After Britton and Langley [3, Figs. 1 and 8].)

stress waves into the helix is shown in Fig. 3.27(b). The pulse is induced by using the strong magneto-strictive effect of nickel. The extreme length of nickel wire is merely to absorb the image pulse produced by the electrical pulse excitation. Flexural pulses may be excited by placing the nickel wire transverse to the waveguide.

The objective of the work was to compare the experimental results with those predicted by the theories put forth by Morley [15] and Wittrick [27]. In the former case, dispersion curves were obtained for a Timoshenko type of theory for the in-plane motions of a circular rod. In the latter case, out-of-plane motions were included in the development. The theoretical arrival-time curves for Morley's theory are shown in Fig. 3.28 as the solid lines for a particular value of curvature ratio ($a/R = 0{\cdot}106$). The parameters t and T_0 are time and x/c_0 respectively. T_p is the period of the arriving wave. Curves 1

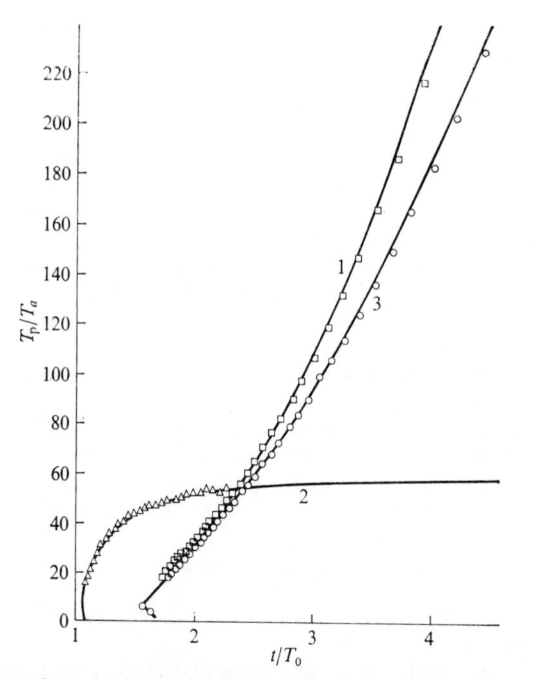

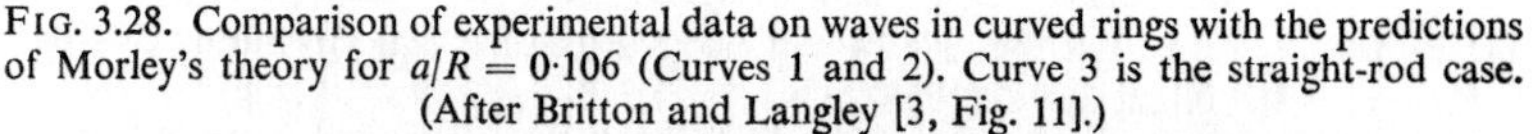
FIG. 3.28. Comparison of experimental data on waves in curved rings with the predictions of Morley's theory for $a/R = 0{\cdot}106$ (Curves 1 and 2). Curve 3 is the straight-rod case. (After Britton and Langley [3, Fig. 11].)

and 2 correspond to flexural and longitudinal modes in a curved rod, while curve 3 is the flexural mode for a straight Timoshenko beam.

The experimental technique was to induce a wide-band stress pulse by the method previously shown. The arrival time of the predominant period was measured. Data was obtained on modes 1 and 2 and for the straight rod case. The results are as shown in Fig. 3.28, with curve 3 corresponding to the straight-rod case. Agreement is seen to be excellent in all cases. Additional comparisons were made by the authors with Wittrick's theory.

References

1. ABRAMOWITZ, M. and STEGUN, I. A. *Handbook of mathematical functions: with formulas, graphs, and mathematical tables*. Dover Publications, New York (1965).
2. BOLEY, B. A. and CHAO, C. C. Some solutions of the Timoshenko beam equations. *J. appl. Mech.* **22,** 579–86 (1955).
3. BRITTON, W. G. B. and LANGLEY, G. O. Stress pulse dispersion in curved mechanical waveguides. *J. Sound Vib.* **7** (3), 417–30 (1968).

4. CUNNINGHAM, D. M. and GOLDSMITH, W. An experimental investigation of beam stresses produced by oblique impact of a steel sphere. *J. appl. Mech.* **23**, 606–11 (1956).
5. DOHRENWEND, C. O., DRUCKER, D. C., and MOORE, P. Tranverse impact Transients. *Exp. Stress Analysis* **1**, 1–10 (1944).
6. ERDÉLYI, A. (Ed.). *Tables of integral transforms.* (Bateman Manuscript Project). McGraw-Hill, New York (1954).
7. GOLDSMITH, W. *Impact: the theory and physical behaviour of colliding solids.* Edward Arnold, London (1960).
8. GRAFF, K. F. Elastic wave propagation in a curved sonic transmission line. *IEEE Trans., Sonics and Ultrasonics* **SU-17**, 1–6 (1970).
9. HOPPMANN, W. H. (II) Impulsive loads on beams. *Proc. Soc. exp. Stress Analysis* **10**, 157–64 (1952).
10. KENNEY, J. T. (Jr.). Steady-state vibrations of beam on elastic foundation for moving load. *J. appl. Mech.* **21**, 359–64 (1954).
11. KOLSKY, H. *Stress waves in solids.* Dover Publications, New York (1963).
12. KUO, S. S. Beam subjected to eccentric longitudinal impact. *Exp. Mech.* **18**, 102–8 (1961).
13. LOVE, A. E. H. *A treatise on the mathematical theory of elasticity.* Dover Publications, New York (1944).
14. MORI, D. Laterial impact of bars and plates. *Proc. Soc. exp. Stress Analysis* **15**, 171–8 (1957).
15. MORLEY, L. S. D. Elastic waves in a naturally curved rod. *Quart. Jl Mech. appl. Math.* **14**, 155–72 (1961).
16. MORSE, P. M. *Vibration and sound.* McGraw-Hill, New York (1948).
17. PHILIPSON, L. L. On the role of extension in the flexural vibrations of rings. *J. appl. Mech.* **23**, 364–6 (1956).
18. RAYLEIGH, J. W. S. *The theory of sound*, Vols. I and II. Dover Publications, New York (1945).
19. RIPPERGER, E. A. The propagation of pulses in cylindrical bars—an experimental study. *Proc. 1st midwest. Conf. Solid Mech.* pp. 29–39 (1953).
20. —— and ABRAMSON, H. N. Reflection and transmission of elastic pulses in a bar at a discontinuity in cross section. *Proc. 3rd midwest. Conf. Solid Mech.* p. 135 (1957).
21. SCHWIEGER, H. A simple calculation of the transverse impact on beams and its experimental verification. *Exp. Mech.* **22**, 378–84 (1965).
22. SNEDDON, I. N. *Fourier transforms.* McGraw-Hill, New York (1951).
23. STEPHENSON, J. G. and WILHOIT, J. C. (Jr.). An experimental study of bending impact waves in beams. *Exp. Mech.* **22**, 16–21 (1965).
24. TIMOSHENKO, S. P. On the correction for shear of the differential equation for transverse vibrations of prismatic bars. *Phil. Mag.*, Ser. 6, **41**, 744 (1921).
25. ——. *Vibration problems in engineering.* Van Nostrand, New Jersey (1928).
26. VIGNESS, I. Transverse waves in beams. *Exp. Stress Analysis* **8**, 69–82 (1951).
27. WITTRICK, W. H. On elastic wave propagation in helical springs. *Int. J. Mech. Sci.* **8**, 25–47 (1966).
28. ZEMANEK, J. (Jr.) and RUDNICK, I. Attenuation and dispersion of elastic waves in a cylindrical bar. *J. Acoust. Soc. Am.* **33**, 1283–8 (1961).

Problems

3.1. Draw the group velocity versus wavenumber curve for the Bernoulli-Euler beam.

3.2. Solve the problem of a concentrated, harmonic load applied at the origin of an infinite beam by solving the homogeneous governing equation for $x \neq 0$ and then applying proper boundary conditions at the origin.

3.3. Obtain the expression for the impedance of a semi-infinite beam subjected to the end force $F_0 \exp(-i\omega t)$ and free to rotate at the end.

3.4. Consider an infinite beam and two concentrated loads located at $x = \pm a$ and each given by $P \exp(-i\omega t)$. Solve for the steady wave propagation. Look for special cases of excitation frequency where constructive or destructive interference can occur in the region between the two loads.

3.5. Determine the amplitude ratio of harmonic waves in a semi-infinite beam incident on a fixed boundary, an elastic deflection spring boundary and a torsion spring boundary.

3.6. Consider two semi-infinite beams of differing stiffness EI (but same density) joined together at the origin. Consider harmonic waves travelling from left to right to be incident on the junction. Obtain the amplitude ratios of the reflected and transmitted waves.

3.7. Apply the method of stationary phase to analysis of the initial value problem starting with the initial conditions (3.1.28). You should obtain the same result as (3.1.29). To explain this peculiarity, go back to the stationary-phase development in § 1.6 and note the truncation of the expansion of $h(\omega)$ at the second derivative. Evaluate $h'''(\omega_0)$ for the present problem. Comment on the conditions under which the stationary-phase method can yield 'exact' results.

3.8. Obtain the normal modes $Y_n(x)$ for the cases of free–free, clamped–clamped, and clamped–free beams.

3.9. Obtain the frequency equation for a beam of length l, pinned at each end and with a torsion spring K_T at each end. Let $K_T \to 0, \infty$ and see if the pinned–pinned and clamped–clamped frequency equations are recovered. (Your general result should be in terms of the frequency parameter βl and the ratio $EI/K_T l$). Plot βl versus a properly-non-dimensionalized K_T for $3{\cdot}142 < \beta l < 4{\cdot}730$, with K_T increasing from zero. The range cited falls between the first natural frequency of a pinned–pinned and clamped–clamped beam. Is βl (frequency) sensitive to slight initial changes of K_T? Relate these results to the experimental problem of simulating pinned versus clamped boundary conditions.

3.10. It is possible to apply the Laplace transform to the problem of the natural frequencies of finite beams. Start by considering harmonic motion $y = Y \exp(-i\omega t)$ and then apply the Laplace transform on the spatial variable of the ordinary differential equation. The results given by (3.2.71) might be of help in this analysis.

3.11. Consider a beam of length l, pinned at both ends and having an intermediate deflection spring of constant K located at $x = l/2$. Determine the natural frequencies for various stiffness K. Some rather interesting results should arise in this problem. Let $K \to 0, \infty$: are the natural frequencies for a pinned–pinned beam recovered for $K \to 0$? Now let $K \to \infty$. At a certain value of K, there should be a point at which the first symmetrical mode of the beam becomes unstable and the beam switches to an antisymmetrical mode. This also occurs for the higher modes.

3.12. Draw the dispersion curve for the beam on an elastic foundation. Interpret the meaning of the double values for c and of the minimum that occurs in the curve. Interpret the behaviour as $\gamma \to 0$.

3.13. Solve the problem of forced harmonic motion of an infinite beam on an elastic foundation, where the load is described by $\delta(x-\zeta)\exp(-i\omega t)$. Use transform methods and residue theory or treat the homogeneous differential equation under proper boundary conditions.

3.14. Include slight damping in the case of an infinite beam subjected to a harmonic, concentrated load at the origin. Verify that the poles are displaced from the real axis in a manner consistent with the radiation condition (refer to Fig. 1.26 for the analogous string problem).

3.15. Determine the lowest Euler buckling load of a fixed–pinned column by the vibration method.

3.16. Calculate the second branch of the dispersion curve for the Timoshenko beam for the value of Poisson's ratio and range of wavenumber shown in Fig. 3.13.

3.17. Draw the frequency spectrum for the Timoshenko beam for a Poisson's ratio of $\nu = 0{\cdot}29$ Show both branches, and the imaginary portions, if any, of the curves. Non-dimensionalize the wavenumber, as in Fig. 3.13, and make the spectrum for $\bar{\gamma} \leq 2{\cdot}0$.

3.18. A dispersion curve for longitudinal waves according to Love's theory was shown in Fig. 2.27 and a region of validity was established in terms of frequencies. Do the same for the Timoshenko beam, using Fig. 3.13. Obtain a frequency bandwidth for a 1 in. diameter steel rod. Compare this with the bandwidth of longitudinal waves in the same rod.

3.19. Exclude the effects of rotary inertia from the Timoshenko beam results. Calculate a few points on the new dispersion curve. Results should suggest that shear effects predominate in bringing Bernoulli–Euler theory into accord with exact theory.

3.20. Establish the validity of the four 'facts' regarding the integral I_1 about the parts of the contour shown in Fig. 3.16.

3.21. Delete centre-line extensibility effects in the development of the governing equations for a curved rod. The results should be a purely flexural wave theory for the ring. Obtain the dispersion curve for this case.

3.22. Derive the governing equation for a ring having no bending resistance (that is, $M = 0$). What conditions are placed on the shear force and $\partial/\partial\theta$ variations? Using the resulting governing equation, derive the frequency equation for a complete ring. Sketch the nature of the vibrational mode.

Waves in membranes, thin plates, and shells

In our study of wave propagation and vibrations of elastic bodies, we have previously restricted our attention to one-dimensional problems. That is, only one independent spatial variable was admitted, the length along the member. Within these confines, the transverse motion of strings, longitudinal waves in rods, and transverse waves in beams have been investigated. Such structural shapes could be classified as members, one of whose dimensions (length) was large in comparison to the other two (depth, breadth).

We now will study the propagation of elastic waves under more complicated conditions, involving two independent spatial variables. Two of the structural types, membranes and thin plates, may be classified as flat elements having two dimensions (length, breadth) large in comparison to the remaining one (depth). The membrane element may be thought of as the two-dimensional analogue of the string, while the thin plate is the two-dimensional analogue of the beam. The third structural shape will be the thin shell. Again, two of the characteristic dimensions (length, radius) will be large compared to the third (thickness). There will also be the considerable added complexity of curvature.

The coverage in these areas will be somewhat more restricted than in the earlier chapters. Part of the reason for this is that nearly all of the basic methods of analysis have been illustrated at this stage and do not require further repetition. A second reason is the fact that analysis of these geometrically more complicated systems becomes, at a minimum, more tedious and frequently more complicated. Attention will be focused on the main aspects of harmonic wave propagation, initial-value problems, reflection characteristics, and vibrations of finite elements.

4.1. Transverse motion in membranes

In speaking of the membrane as the two-dimensional analogue of the string, we are speaking of an element whose restoring forces arise from the in-plane tensile or stretching forces. There is no resistance to shear and bending in such an element. It will be the two-dimensional characteristics of waves in membranes (and plates) that most distinguish this work from the previous chapters. Of particular interest with reference to later, more complicated, elastic systems will be the concept of plane waves and the reflection of oblique waves from boundaries.

4.1.1. *The governing equation*

Consider a taut membrane under in-plane tensile force T, as shown in Fig. 4.1(a), and further consider a rectangular element of the membrane in a

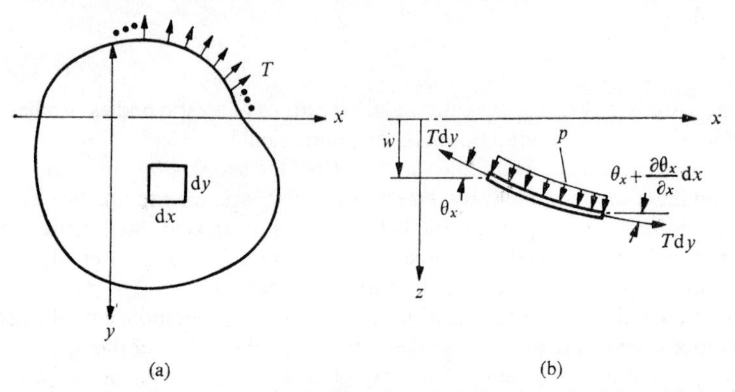

FIG. 4.1. (a) Taut membrane under tensile load, and (b) side view of a differential element of that membrane.

deflected configuration, as shown in Fig. 4.1(b). The coordinate w is used to measure the deflection of the membrane in the z direction. The equation of motion in the z-direction is given by

$$-T\,dy\theta_x + T\,dy\left(\theta_x + \frac{\partial \theta_x}{\partial x}\,dx\right) - T\,dx\theta_y + T\,dx\left(\theta_y + \frac{\partial \theta_y}{\partial y}\,dy\right) + p\,dx\,dy = \rho\,dx\,dy\frac{\partial^2 w}{\partial t^2}, \quad (4.1.1)$$

where ρ is the mass density per unit area of the membrane and the angles θ_x, θ_y are given by $\partial w/\partial x$, $\partial w/\partial y$. We note that small deflections have been assumed in making the approximations $\sin\theta_x \simeq \theta_x$, etc. in (4.1.1) and in approximating the area of the deflected element by $dx\,dy$. Equation (4.1.1) then rather directly reduces to

$$T\left(\frac{\partial^2 w}{\partial x^2} + \frac{\partial^2 w}{\partial y^2}\right) + p(x, y, t) = \rho\frac{\partial^2 w}{\partial t^2}. \quad (4.1.2)$$

Since the Laplacian operator in rectangular coordinates is defined by

$$\nabla^2 = \partial^2/\partial x^2 + \partial^2/\partial y^2,$$

we may write

$$T\,\nabla^2 w(x, y, t) + p(x, y, t) = \rho\frac{\partial^2 w(x, y, t)}{\partial t^2}, \quad (4.1.3)$$

as the governing equation for the membrane. This is the two-dimensional form of the wave equation.

Many wave-motion problems in membranes are best expressed in terms of polar coordinates due to axisymmetric loading, circular boundaries, or both. It is possible to re-derive the governing equation from equilibrium principles applied to a differential element in the polar coordinates r, θ,† or to employ a coordinate transformation on (4.1.3). Foregoing both, we merely present the results as

$$T\,\nabla^2 w(r,\theta,t)+p(r,\theta,t) = \rho\frac{\partial^2 w(r,\theta,t)}{\partial t^2}, \tag{4.1.4}$$

where ∇^2 is the Laplacian in polar coordinates, defined as

$$\nabla = \frac{1}{r}\frac{\partial}{\partial r}\left(r\frac{\partial}{\partial r}\right)+\frac{1}{r^2}\frac{\partial^2}{\partial\theta^2} = \frac{\partial^2}{\partial r^2}+\frac{1}{r}\frac{\partial}{\partial r}+\frac{1}{r^2}\frac{\partial^2}{\partial\theta^2}. \tag{4.1.5}$$

4.1.2. *Plane waves*

The study of the basic propagation characteristics of one-dimensional systems involved investigating the propagation of harmonic waves in the x direction. The analogue of this in two dimensions is to investigate the propagation of *plane waves*. Such waves are propagating disturbances in two or three dimensions, where the motion of every particle in the planes perpendicular to the direction of propagation is the same. A propagating two-dimensional plane disturbance is illustrated in Fig. 4.2. For such a disturbance,

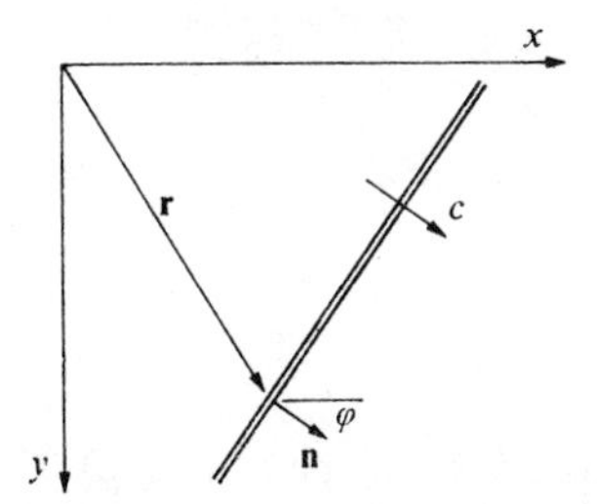

FIG. 4.2. A propagating plane disturbance in two dimensions.

the motion of every particle along the line ('plane') defined by

$$\mathbf{n}\cdot\mathbf{r}-ct = \text{constant} \tag{4.1.6}$$

will be the same. The propagation velocity of the plane is c and $\mathbf{n}$ is the normal to the plane. The position of an arbitrary point on the plane is defined by $\mathbf{r}$, where $\mathbf{n}$ and $\mathbf{r}$ are given by

$$\mathbf{n} = l\mathbf{i}+m\mathbf{j} = \cos\phi\mathbf{i}+\sin\phi\mathbf{j}, \quad \mathbf{r} = x\mathbf{i}+y\mathbf{j}. \tag{4.1.7}$$

† See Problem 4.1.

Consider, then, the propagation of plane harmonic waves in a membrane. We ask 'under what conditions can waves of the type

$$w = Ae^{i\gamma(lx+my-ct)} \tag{4.1.8}$$

exist in the membrane?' We substitute (4.1.8) in the governing equation (4.1.3) with $p(x, y, t) = 0$ and obtain

$$T(-\gamma^2 l^2 - \gamma^2 m^2)Ae^{i\gamma(lx+my-ct)} = -\rho\gamma^2 c^2 Ae^{i\gamma(lx+my-ct)}. \tag{4.1.9}$$

This reduces to

$$T\gamma^2(l^2+m^2) = \rho\gamma^2 c^2. \tag{4.1.10}$$

Since $l^2+m^2 = 1$, we have that

$$c^2 = T/\rho, \quad c = \sqrt{(T/\rho)}. \tag{4.1.11}$$

Thus we have that the propagation velocity of plane harmonic waves of any frequency is constant.

The preceding result may be generalized to a planar disturbance of arbitrary form, such as

$$w = f(\mathbf{n} \cdot \mathbf{r} - ct) = f(lx+my-ct). \tag{4.1.12}$$

Defining the argument of the function as ψ, we have

$$\begin{gathered}
\frac{\partial w}{\partial x} = \frac{\partial w}{\partial \psi}\frac{\partial \psi}{\partial x} = lw', \\
\frac{\partial^2 w}{\partial x^2} = \frac{\partial}{\partial \psi}(lw')\frac{\partial \psi}{\partial x} = l^2 w'', \\
\frac{\partial^2 w}{\partial y^2} = m^2 w'', \qquad \frac{\partial^2 w}{\partial t^2} = c^2 w''.
\end{gathered} \tag{4.1.13}$$

Thus from (4.1.3) with $p = 0$ we have

$$T(l^2+m^2)w'' = \rho c^2 w'', \tag{4.1.14}$$

which leads to the previous result that $c = \sqrt{(T/\rho)}$. Thus we conclude that it is possible for a plane disturbance of arbitrary shape to propagate without distortion in the membrane. These results are possibly not surprising in view of the effective one-dimensionalization that occurs in considering propagation in a single direction.

We now briefly consider the membrane equation in polar form (4.1.4), with $p = 0$. The natural question to raise is whether waves having a circular wave-front can exist in a membrane. Allowing polar symmetry, so that $\partial/\partial\theta = 0$, the equation to be considered is

$$\frac{\partial^2 w}{\partial r^2} + \frac{1}{r}\frac{\partial w}{\partial r} = \frac{1}{c_0^2}\frac{\partial^2 w}{\partial t^2}. \tag{4.1.15}$$

It may be quickly established that a circular wave of the form $\exp\{i(\gamma r-\omega t)\}$ will not satisfy this equation. This should not be surprising since a circular wave, moving either inward or outward, would be expected to have an amplitude that would depend on distance merely from conservation of energy considerations. However, it may be established that a solution of the form $f(r)\exp\{i(\gamma r-\omega t)\}$ will not work either. Evidently, a circular waveform will not maintain its shape in a membrane as a planar waveform does. This aspect will be remarked upon further in the next section.

4.1.3. *The initial-value problem*

We shall first consider the case of axisymmetric motion of a membrane, so that the governing equation is given by (4.4.15) and the initial conditions are

$$w(r,0) = f(r), \qquad \partial w(r,0)/\partial t = g(r). \tag{4.1.16}$$

Now, in problems involving polar or cylindrical coordinates, it is the Hankel transform that finds application. For the case at hand, we employ the zeroth-order transform given by†

$$\bar{w}(\xi,t) = \int_0^\infty rw(r,t)J_0(\xi r)\,dr. \tag{4.1.17}$$

The Hankel transform of the polar Laplacian yields a particularly simple form, giving

$$H\left(\frac{\partial^2 w}{\partial r^2}+\frac{1}{r}\frac{\partial w}{\partial r}\right) = -\xi^2\bar{w}(\xi,t). \tag{4.1.18}$$

Thus, the Hankel transform of (4.1.15) gives

$$\frac{d^2\bar{w}}{dt^2}+c_0^2\xi^2\bar{w} = 0, \tag{4.1.19}$$

which has the solution

$$\bar{w}(\xi,t) = A\cos c_0\xi t+B\sin c_0\xi t. \tag{4.1.20}$$

The transformed initial conditions (4.1.16) are

$$\bar{w}(\xi,0) = \bar{f}(\xi), \qquad \frac{\partial\bar{w}(\xi,0)}{\partial t} = \bar{g}(\xi). \tag{4.1.21}$$

Applying the solution (4.1.20) to these conditions gives the result

$$\bar{w}(\xi,t) = \bar{f}(\xi)\cos c_0\xi t+\frac{\bar{g}(\xi)}{c_0\xi}\sin c_0\xi t. \tag{4.1.22}$$

† See Appendix B.4 for brief remarks. See Chapter 2 and pp. 125–8 of Reference [16] for treatment of this problem.

Taking the inverse Hankel transform, we have

$$w(r,t) = \int_0^\infty \xi \bar{f}(\xi)\cos c_0\xi t J_0(\xi r)\,\mathrm{d}\xi + \frac{1}{c_0}\int_0^\infty \bar{g}(\xi)\sin c_0\xi t J_0(\xi r)\,\mathrm{d}\xi. \quad (4.1.23)$$

It is possible to carry the solution one step further and use a convolution-type integral to express the result. However, the generality of such an expression is not always of help in solving a particular problem, so we shall leave the solution in the preceding form.

As an example, consider a problem where the initial velocity is zero and the displacement is defined by

$$w(r,0) = \begin{cases} 1/\pi a^2, & r < a \\ 0, & r > a. \end{cases} \quad (4.1.24)$$

This results in a cylindrical membrane shape having a volume of unity. The Hankel transform of this expression is given by (see Appendix B.4)

$$\bar{w}(\xi,0) = \frac{1}{\pi a}\frac{J_1(a\xi)}{\xi}. \quad (4.1.25)$$

Now let $a \to 0$ in this result. We have for small values of argument that $J_1(z) \to z/2$, so that

$$\bar{w}(\xi,0)\big|_{a\to 0} = \frac{1}{2\pi}. \quad (4.1.26)$$

The resulting solution is given by

$$w(r,t) = \frac{1}{2\pi}\int_0^\infty \xi\cos c_0\xi t J_0(\xi r)\,\mathrm{d}\xi. \quad (4.1.27)$$

By expressing this in the form

$$w(r,t) = \frac{1}{2\pi c_0}\frac{\partial}{\partial t}\int_0^\infty \sin c_0\xi t J_0(\xi r)\,\mathrm{d}\xi \quad (4.1.28)$$

results available in tables of Hankel transforms are readily applied.† The results are

$$w(r,t) = \begin{cases} \dfrac{1}{2\pi c_0 r^{\frac{1}{2}}}\dfrac{\partial}{\partial t}\dfrac{1}{(c_0^2t^2-r^2)^{\frac{1}{2}}}, & 0 < r < c_0 t \\ 0, & c_0 t < r < \infty. \end{cases} \quad (4.1.29)$$

Upon carrying out the differentiation, we obtain

$$w(r,t) = \begin{cases} -\dfrac{1}{2\pi}\dfrac{c_0 t}{r^{\frac{1}{2}}(c_0^2t^2-r^2)^{\frac{3}{2}}}, & 0 < r < c_0 t \\ 0, & c_0 t < r < \infty. \end{cases} \quad (4.1.30)$$

† p. 528 of Reference [16].

In interpreting this result, we see that a clearly defined wavefront exists, propagating at the velocity c_0. In this respect, the response is similar to that which would exist in a string. However, it is seen that after the wavefront has passed, the disturbance persists. This differs from the string response to the analogous impulse displacement. In that case, a sharp 'spike' would propagate outward at c_0, but the medium behind the wave would be at rest once the front passed. The membrane response exhibits what is frequently referred to as a *tail* or *wake* in the response. The response of the membrane at various instants of time is shown in Fig. 4.3. Also shown is the string response to the analogous

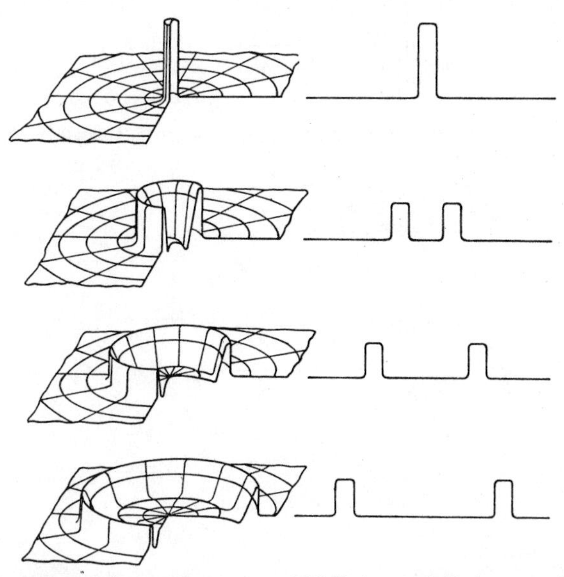

FIG. 4.3. Response of the membrane to an initially imposed displacement field and the string response to the analogous initial condition. (After Morse [12, Fig. 40].)

initial displacement.

Consider, as a next example, the case where the initial displacement is zero and the velocity is prescribed by

$$\frac{\partial w(r, 0)}{\partial t} = \begin{cases} 1/\pi a^2, & r < a \\ 0, & r > a. \end{cases} \tag{4.1.31}$$

This corresponds to a constant velocity imparted over a circular region. The Hankel transform of this expression has been given by (4.1.25) and is

$$\frac{\partial \bar{w}(r, 0)}{\partial t} = \frac{1}{\pi a} \frac{J_1(a\xi)}{\xi}. \tag{4.1.32}$$

As $a \to 0$, we again have that

$$\left.\frac{\partial \bar{w}(r, 0)}{\partial t}\right|_{a \to 0} = \frac{1}{2\pi}. \tag{4.1.33}$$

The solution (4.1.23) reduces to

$$w(r, t) = \frac{1}{2\pi c_0} \int_0^\infty \sin c_0 \xi t J_0(\xi r)\, d\xi. \tag{4.1.34}$$

The evaluation of this integral is contained in the result (4.1.29) and is

$$w(r, t) = \begin{cases} \dfrac{1}{2\pi c_0} \dfrac{1}{(c_0^2 t^2 - r^2)^{\frac{1}{2}}}, & 0 < r < c_0 t \\ 0, & c_0 t < r < \infty. \end{cases} \tag{4.1.35}$$

The response of the membrane to the velocity input is shown in Fig. 4.4. The

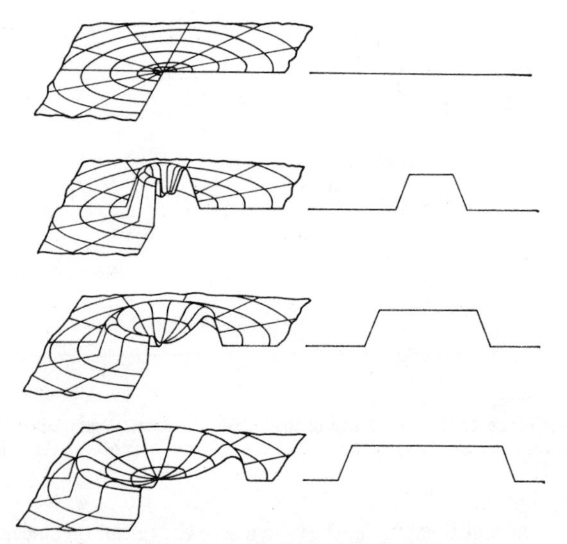

FIG. 4.4. Response of a membrane to a velocity impulse and of a string to the analogous input. (After Morse [12, Fig. 41].)

string response to the analogous input is again shown. The presence of a clearly defined wavefront and a wake to the disturbance are again to be noted in the membrane response.

4.1.4. *Forced vibration of a membrane*

Let us now consider the wave motion in a membrane resulting from a harmonically varying load acting within a circle of radius a at the origin. We may view this problem in terms of the inhomogeneous equation (4.1.4) with a prescribed loading or, alternatively, consider the homogeneous governing equation with a prescribed forcing function at $r = a$. Since our usual viewpoint in such problems has been the former, we shall adopt the latter procedure here for a change.

To establish the condition at $r = a$, we presume the load to be given by

$$p(r, t) = \begin{cases} p_0 e^{-i\omega t}, & r < a \\ 0, & r > a. \end{cases} \tag{4.1.36}$$

This will lead to a total force $\pi a^2 p_0 \exp(-i\omega t)$ acting on the circle $r < a$. This must be balanced by the vertical component of membrane force acting on the periphery of the circle $r = a$. The resulting force balance equation is given by

$$\pi a^2 p_0 e^{-i\omega t} = -2\pi a T \frac{\partial w(a, t)}{\partial r}. \tag{4.1.37}$$

Assuming a solution of the form $w(r, t) = W(r)\exp(-i\omega t)$, the boundary condition takes the form

$$dW(a)/dr = -p_0 a/2T, \tag{4.1.38}$$

and the governing equation is given by

$$\frac{d^2W}{dr^2} + \frac{1}{r}\frac{dW}{dr} + \frac{\omega^2}{c_0^2}W = 0. \tag{4.1.39}$$

We recognize (4.1.39) as Bessel's equation having the solution

$$W(r) = AH_0^{(1)}(\beta r) + BH_0^{(2)}(\beta r), \qquad \beta = \omega/c_0. \tag{4.1.40}$$

We have encountered this equation and solution form previously in conjunction with longitudinal waves in tapered horns (§ 2.5). Using the arguments developed in that section on proper selection of the Hankel function to meet the radiation condition, we set $B = 0$. Applying the condition (4.1.38), and again using results presented in the section on horns in regard to differentiation of the Hankel function, we are led to the solution

$$w(r, t) = \frac{p_0 a}{2\beta T} \frac{H_0^{(1)}(\beta r)}{H_1^{(1)}(\beta a)} e^{-i\omega t}. \tag{4.1.41}$$

Far from the region of load application, we have the response given by

$$w(r, t) \sim \frac{p_0 a}{\sqrt{(2\pi)}\beta^{\frac{3}{2}} T H_1^{(1)}(\beta a)} \frac{1}{\sqrt{r}} e^{i(\beta r - \omega t - \pi/4)}. \tag{4.1.42}$$

We note that the $r^{-\frac{1}{2}}$ amplitude dependence is the same as observed in the analysis of the initial-value problem.

4.1.5. *Reflection of waves from membrane boundaries*

The reflection of waves from the boundaries of a membrane has a new element over that of wave reflection in one-dimensional systems. In the latter systems, the direction of the incident wave on a boundary was not a variable in the process. In a two-dimensional system, the direction is a variable, since the angle of the normal of the wavefront may take on any orientation relative to the boundary.

To illustrate the basic aspects of reflection, consider a semi-infinite membrane $y \geq 0$, that is fixed along the boundary $y = 0$. The boundary condition is thus

$$w(x, 0, t) = 0. \tag{4.1.43}$$

We now wish to consider harmonic plane waves propagating in the membrane. Thus, we seek harmonic plane-wave solutions to

$$\nabla^2 w(x, y, t) = \frac{1}{c_0^2}\frac{\partial^2 w}{\partial t^2}(x, y, t). \tag{4.1.44}$$

Although we could build our analysis from the expression (4.1.8) this matter of plane-wave solutions to the system governing equation arises frequently enough in future work to trace the steps in some detail here.

Considering a separation of variables solution to the homogeneous form of (4.1.2),

$$w = X(x)Y(y)e^{-i\omega t}, \tag{4.1.45}$$

we obtain

$$\left(\frac{X''}{X}+\frac{Y''}{Y}\right) = -\frac{\omega^2}{c_0^2} = -\beta^2. \tag{4.1.46}$$

Writing this as

$$\frac{X''}{X} = -\left(\frac{Y''}{Y}+\beta^2\right) = -\xi^2, \tag{4.1.47}$$

we obtain the solutions

$$X = A_1 e^{i\xi x}+A_2 e^{-i\xi x},$$
$$Y = B_1 e^{i\zeta y}+B_2 e^{-i\zeta y}, \tag{4.1.48}$$

where

$$\zeta^2 = \beta^2-\xi^2. \tag{4.1.49}$$

Upon multiplying the results in (4.1.48) together, substituting in (4.1.45), and defining new constants A, B, C, D we obtain four plane-wave solutions

$$w = Ae^{i(\xi x-\zeta y-\omega t)}+Be^{i(\xi x+\zeta y-\omega t)}+Ce^{-i(\xi x-\zeta y+\omega t)}+De^{-i(\xi x+\zeta y+\omega t)}. \tag{4.1.50}$$

Each of these expressions corresponds to a plane wave of particular orientation and direction of propagation as illustrated in Fig. 4.5. We then proceed somewhat in the manner of wave-reflection studies in one-dimension. An incident wave is specified. In the present case, we let this be the first plane wave of (4.1.50). We then set $C = D = 0$ because they correspond to waves having a leftward direction of propagation. Thus we are reduced to considering

$$w = Ae^{i(\xi x-\zeta y-\omega t)}+Be^{i(\xi x+\zeta y-\omega t)}. \tag{4.1.51}$$

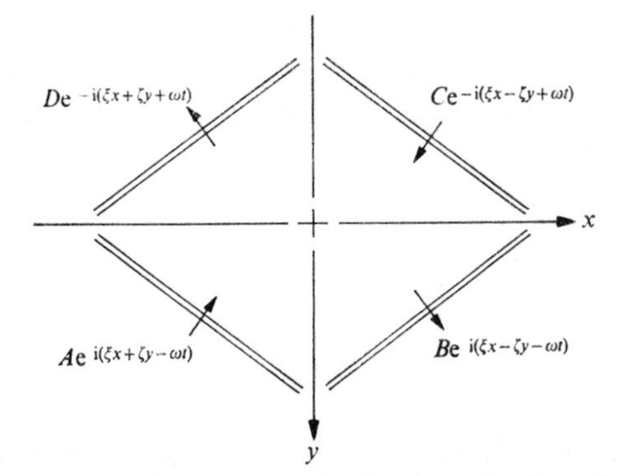

FIG. 4.5. Orientations and propagation directions of various plane waves.

Referring back to (4.1.8), we see that the plane wave was expressed in terms of the direction cosines and the propagation velocity. This can be done in the present case by defining

$$\xi = \gamma l = \gamma \cos \phi, \quad \zeta = \gamma m = \gamma \sin \phi, \tag{4.1.52}$$

where ϕ is the angle between the wave normal and the positive x-axis (see Fig. 4.2). Then (4.1.51) becomes

$$w = Ae^{i\gamma(lx-my-ct)} + Be^{i\gamma(lx+my-ct)}. \tag{4.1.53}$$

Alternatively, ξ and ζ are sometimes expressed in the form

$$\xi = \gamma l' = \gamma \sin \phi', \quad \zeta = \gamma m' = \gamma \cos \phi', \tag{4.1.54}$$

where ϕ' is the angle between the normal and the positive y-axis. The angle of incidence of a wave on an edge is then measured relative to the perpendicular, which is the usual convention in reflection–refraction studies.

To proceed to the reflection problem at hand, we now have solutions to the governing equation in the form (4.1.51) and the boundary conditions for the semi-infinite membrane given by (4.1.43). Substituting in the latter, we obtain

$$(A+B)e^{i(\xi x-\omega t)} = 0 \tag{4.1.55}$$

or

$$B/A = -1. \tag{4.1.56}$$

Thus the amplitude of the reflected wave is equal to that of the incident wave and out of phase by 180° relative to the incident wave. This aspect is shown in Fig. 4.6. The angle of incidence and reflection are, of course, equal. A second aspect of the reflection phenomenon for this problem is brought out by substituting (4.1.56) back in the solution (4.1.51), giving

$$w = A(e^{-i\zeta y} - e^{i\zeta y})e^{i(\xi x-\omega t)} = -2iA \sin \zeta y e^{i(\xi x-\omega t)}. \tag{4.1.57}$$

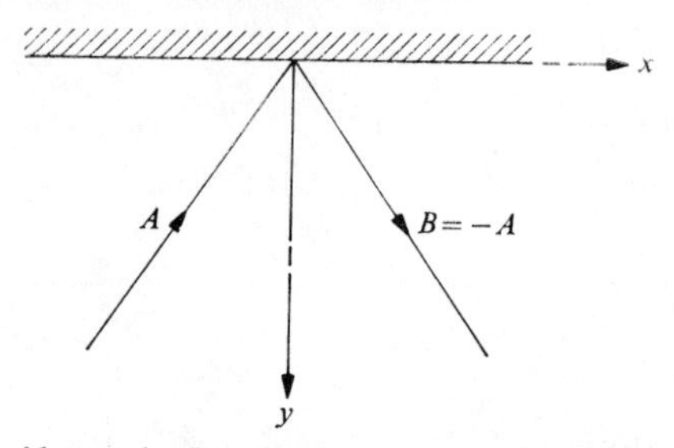

FIG. 4.6. Incident and reflected waves from a fixed membrane boundary.

The y variation in this result predicts a series of equally spaced, horizontal 'interference' bands. In between the bands, waves are propagating in the x direction.

It would be possible to speak at some length about wave reflection in membranes by considering a variety of boundary conditions. However, the fixed-edge membrane represents the most realistic situation. Furthermore, many of the considerations would tend to overlap those that will arise in Chapter 6 on plane waves in an elastic half-space.

4.1.6. *Waves in a membrane strip*

Consider now a membrane strip, bounded by parallel planes at $y = 0$ and $y = b$ and extending to infinity in the positive and negative x directions. We wish to consider the conditions for the propagation of harmonic waves in the x direction. The governing equation for the problem is again the homogeneous form of (4.1.3) with the boundary conditions

$$w(x, 0, t) = w(x, b, t) = 0. \tag{4.1.58}$$

Without repeating the separation of variables steps, we immediately consider a solution of the form

$$w(x, y, t) = Y(y)e^{i(\xi x-\omega t)}. \tag{4.1.59}$$

Substitution in the governing equation yields

$$\frac{d^2Y}{dy^2}+\zeta^2 Y = 0, \qquad \zeta^2 = \beta^2-\xi^2, \tag{4.1.60}$$

which is the same form previously encountered in considering plane waves. We write the solution of (4.1.60) as

$$Y = A\sin\zeta y + B\cos\zeta y. \tag{4.1.61}$$

The boundary conditions (4.1.58) in terms of $Y(y)$ are simply that

$$Y(0) = Y(b) = 0.$$

This gives $B = 0$ and

$$\sin \zeta b = 0, \quad \zeta b = n\pi \qquad (n = 1, 2, \ldots). \tag{4.1.62}$$

In terms of the frequency and wavenumber parameters ω and ξ, this can be expressed as

$$\omega^2 = c_0^2(\xi^2 + n^2\pi^2/b^2), \qquad (n = 1, 2, \ldots). \tag{4.1.63}$$

The interpretation of this result is as follows: for a given frequency ω and integer n, (4.1.63) may be solved for the wavenumber ξ. The resulting wavelength λ and phase velocity c of the wave are given by $\lambda = 2\pi/\xi$, $c = \omega/\xi$. These establish the parameters of the propagating harmonic disturbance in the strip. The value of n determines the y variation of the motion. Figure 4.7

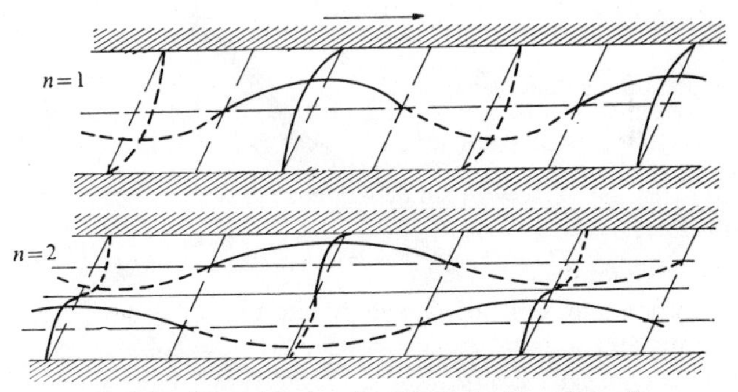

FIG. 4.7. Waves in a membrane strip for two values of n.

illustrates the situation for $n = 1, 2$. We note from (4.1.63) that, for a given frequency, the wavelength will be longer for larger values of n. This may best be seen by writing the result in the form

$$\xi^2 = \frac{\omega^2}{c_0^2} - \frac{n^2\pi^2}{b^2}. \tag{4.1.64}$$

This form also points out the fact that for a given value of n, there will be a frequency such that $\xi = 0$, and below which $\xi \to \pm i\xi$. The former will correspond to cutoff frequencies for a given mode of propagation and the latter will correspond to non-propagating behaviour.

Numerous other problems of waves in semi-infinite strips, transient motion of strips, and so on could be considered here. Again, these problems will have analogues in the area of waves in plates, and will be treated there.

4.1.7. *Vibrations of finite membranes*

Parallelling our considerations of wave motion in one-dimensional systems, we now consider the vibrations of membranes of finite size. While there is generally no difficulty in considering the vibrations of one-dimensional systems of arbitrary length, analysis of the vibrations of two-dimensional systems of arbitrary shape is generally not possible except by approximate means. Considerations must generally be restricted to simple membrane shapes conveniently described by coordinate surfaces. Hence we consider only rectangular and circular membrane shapes.

In the case of a rectangular membrane, we seek solutions to the homogeneous form of (4.1.3), subject to the boundary conditions

$$w(x, 0, t) = w(x, b, t) = w(0, y, t) = w(a, y, t) = 0. \tag{4.1.65}$$

This represents a membrane of length a, breadth b, and fixed along all edges. We consider a solution of the form

$$w = X(x)Y(y)(A\cos\omega t+B\sin\omega t). \tag{4.1.66}$$

We now express the solutions as

$$X = A_1\sin\xi x+A_2\cos\xi x, \tag{4.1.67}$$

$$Y = B_1\sin\zeta y+B_2\cos\zeta y,$$

where, as before,

$$\zeta^2 = \beta^2-\xi^2, \qquad \beta^2 = \omega^2/c_0^2. \tag{4.1.68}$$

The boundary conditions (4.1.65) are equivalent to

$$Y(0) = Y(b) = X(0) = X(a) = 0.$$

These lead to $A_2 = B_2 = 0$ and

$$\sin\xi a = 0, \quad \sin\zeta b = 0, \tag{4.1.69}$$

or

$$\xi a = n\pi, \quad \xi_n = n\pi/a \quad (n = 1, 2, ...), \tag{4.1.70}$$

$$\zeta b = m\pi, \quad \zeta_m = m\pi/b \quad (m = 1, 2, ...). \tag{4.1.71}$$

For a given value of n, m, the vibrational frequency is given from (4.1.68) as

$$\omega_{mn}^2 = c_0^2(\xi_n^2+\zeta_m^2) = \pi^2c_0^2\left(\frac{n^2}{a^2}+\frac{m^2}{b^2}\right). \tag{4.1.72}$$

The general solution is then constructed from the particular solutions by superposition. Thus

$$w = \sum_{n=1}^{\infty}\sum_{m=1}^{\infty}(A_{mn}\cos\omega_{mn}t+B_{mn}\sin\omega_{mn}t)W_{mn}, \tag{4.1.73}$$

where W_{mn} are the normal modes, given here by

$$W_{mn} = \sin\xi_n x\sin\zeta_m y. \tag{4.1.74}$$

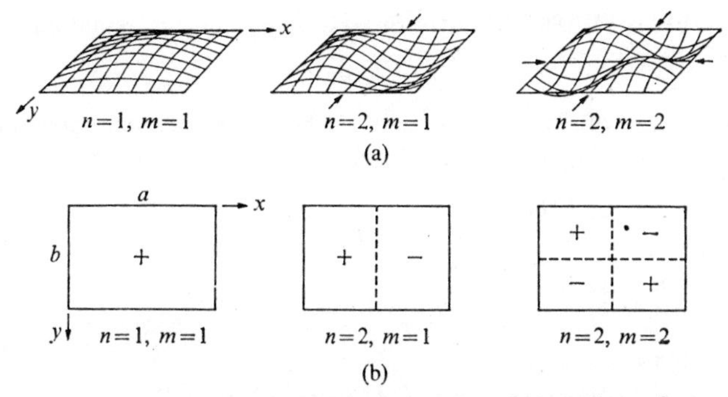

FIG. 4.8. (a) Graphic illustration of the normal modes of a membrane for $n = m = 1$, $n = 2, m = 1$, and $n = m = 2$ (based on Morse [12], Fig. 36); (b) schematic illustration of the same modes.

The mode shapes for $n = m = 1, n = 2, m = 1$, and $n = m = 2$ are shown in Fig. 4.8(a). This graphic illustration gives a clear picture of the deformation. A diagrammatic means of illustrating the same modes is shown in Fig. 4.8(b).

Let us determine the natural frequencies and normal modes of a circular membrane. The governing equation is (4.1.4), with $p(r, \theta, t) = 0$, and the boundary conditions are

$$w(a, \theta, t) = 0. \tag{4.1.75}$$

We consider a solution of the form

$$w(r, \theta, t) = R(r)\Theta(\theta)(A \cos \omega t + B \sin \omega t), \tag{4.1.76}$$

which leads to

$$\frac{R''}{R} + \frac{1}{r}\frac{R'}{R} + \beta^2 = -\frac{\Theta''}{\Theta} = \gamma^2, \qquad \beta^2 = \omega^2/c_0^2. \tag{4.1.77}$$

For Θ we have

$$\Theta = A_1 \cos \gamma\theta + A_2 \sin \gamma\theta. \tag{4.1.78}$$

Now, the condition of continuity of membrane displacement requires that $w(r, \theta, t) = w(r, \theta+2\pi, t)$. This imposes a periodicity condition on (4.1.78) that can only be met by γ being an integer n.

The governing equation for R is now given by

$$R'' + \frac{1}{r}R' + \left(\beta^2 - \frac{n^2}{r^2}\right)R = 0. \tag{4.1.79}$$

This is Bessel's equation of order n having the solution

$$R = B_1 J_n(\beta r) + B_2 Y_n(\beta r), \tag{4.1.80}$$

where J_n, Y_n are the Bessel functions of first and second kind. Now the functions $Y_n(\beta r)$ approach infinity at the origin, so for a complete membrane, we must set $B_2 = 0$. The boundary condition (4.1.75) is given in terms of R as

$R(a) = 0$. From (4.1.80) this gives the resulting frequency equation

$$J_n(\beta a) = 0. \tag{4.1.81}$$

For a given value of the integer n, there will be a series of roots corresponding to the zeros of $J_0(\beta a)$, $J_1(\beta a)$, $J_2(\beta a)$, etc. Some of the results are†

$$\begin{aligned}
&n = 0,\ J_0(\beta a) = 0: \\
&\qquad \beta a = 2{\cdot}405,\ 5{\cdot}520,\ 8{\cdot}654,\ 11{\cdot}792,\ 14{\cdot}931,\ \dots, \\
&n = 1,\ J_1(\beta a) = 0: \\
&\qquad \beta a = 3{\cdot}832,\ 7{\cdot}016,\ 10{\cdot}173,\ 13{\cdot}324,\ 16{\cdot}471,\ \dots, \\
&n = 2,\ J_2(\beta a) = 0: \\
&\qquad \beta a = 5{\cdot}136,\ 8{\cdot}417,\ 11{\cdot}620,\ 14{\cdot}796,\ 17{\cdot}960,\ \dots, \\
&n = 3,\ J_3(\beta a) = 0: \\
&\qquad \beta a = 6{\cdot}380,\ 9{\cdot}761,\ 13{\cdot}015,\ 16{\cdot}223,\ 19{\cdot}409,\ \dots .
\end{aligned} \tag{4.1.82}$$

It should be noted that $\beta a = 0$ is a root for all Bessel functions of order $n \geq 1$. However, this leads to trivial solutions for w and is excluded. The resulting natural frequencies are expressed as

$$\omega_{nm} = \beta_{nm} c_0, \qquad (n = 0, 1, \dots, m = 1, 2, \dots), \tag{4.1.83}$$

where the β_{nm} are assigned as follows: For a given value of n, $m = 1$ corresponds to the first root of J_n, $m = 2$ is the second root, and so forth. As examples

$$\omega_{01} = 2{\cdot}405(c_0/a), \quad \omega_{12} = 7{\cdot}016(c_0/a), \quad \omega_{34} = 16{\cdot}233(c_0/a), \dots . \tag{4.1.84}$$

The resulting general form of the solution becomes rather complicated, since it combines various combinations of J_n, $\sin n\theta$, $\cos n\theta$, $\cos \omega_{nm} t$, $\sin \omega_{nm} t$. It is given by

$$w(r, \theta, t) = \sum_{m=1}^{\infty} \left\{ \sum_{n=0}^{\infty} W_{nm}(A_{nm} \cos \omega_{nm} t + B_{nm} \sin \omega_{nm} t) + \right.$$
$$\left. + \sum_{n=1}^{\infty} \tilde{W}_{nm}(\tilde{A}_{nm} \cos \omega_{nm} t + \tilde{B}_{nm} \sin \omega_{nm} t) \right\}, \tag{4.1.85}$$

where the normal modes are

$$W_{nm} = J_n(\beta_{nm} r) \cos n\theta, \quad \tilde{W}_{nm} = J_n(\beta_{nm} r) \sin n\theta. \tag{4.1.86}$$

For given values of n and m ($n \neq 0$) the normal modes W_{nm}, $\tilde{W}_{nm}$ have the same shape, differing from one another only by an angular rotation of 90°. Several of the modes W_{nm} are shown in Fig. 4.9 both in graphic and diagrammatic form for several values of n and m.

We shall not investigate orthogonality or free and forced motion problems of the finite membrane. These matters will be studied in the more general case

† p. 409 of Reference [1].

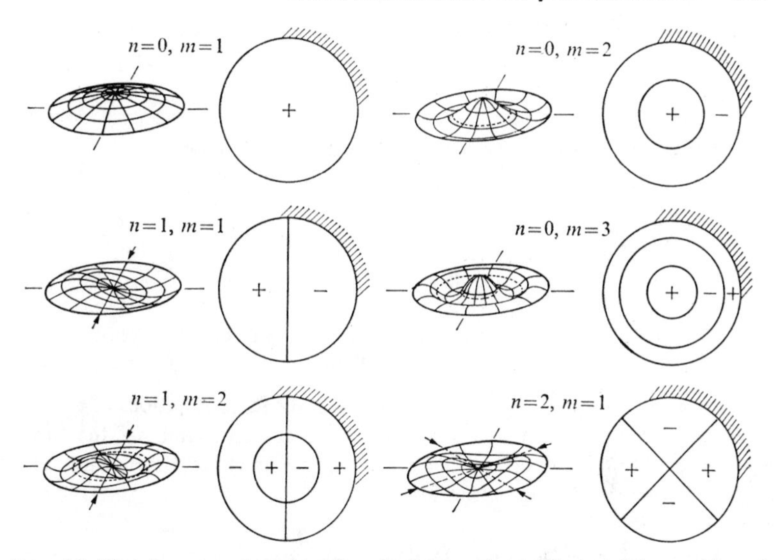

FIG. 4.9. First few normal modes of a circular membrane illustrated in graphic and schematic form. (Based, in part, on Morse [12, Fig. 42].)

of plate vibrations in the next section. Procedures developed for the plate case then easily may be modified to the simpler membrane case.

4.2. Flexural waves in thin plates

The plate is the two-dimensional analogue of the beam. We thus are considering the motion of a two-dimensional elastic system where bending moments and transverse shear forces are active, as they are in a beam. The basic kinematics of the classical theory of thin plates is the same as that of Bernoulli–Euler beams. Once the governing equations for the plate are established, the propagation of plane waves, the initial-value problem, a transient-loading problem, and wave reflection will be studied. Free and forced vibrations of finite plates will also be considered.

4.2.1. *The governing equations*

Consider a plate of thickness h and of infinite extent having its undeflected surface in the x, y-plane, as shown in Fig. 4.10(a). Consider also a differential element $h\,dx\,dy$ of the plate, also shown in Fig. 4.10(a) and in Fig. 4.10(b) where, in the latter figure, the various shear forces, bending and twisting moments, and external loads are shown acting. The bending moments per unit length M_x, M_y arise from distributions of normal stresses σ_x, σ_y, while the twisting moments per unit length M_{xy}, M_{yx} (shown as double-arrow vectors) arise from shearing stresses τ_{xy}, τ_{yx}. The shear forces per unit length Q_x, Q_y

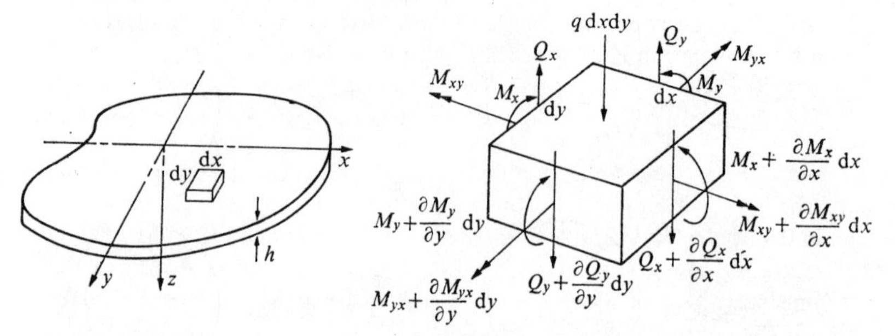

FIG. 4.10. (a) Region of a thin plate, and (b) an element of that plate subjected to forces and moments.

arise from shearing stresses τ_{xz}, τ_{yz}. There are three force equations of motion to be applied, but only that in the z direction is non-trivial. Of the three moment equations of motion, that about the z-axis is identically zero. The remaining three equations of motion give

$$-Q_x\,dy+\left(Q_x+\frac{\partial Q_x}{\partial x}\,dx\right)dy-Q_y\,dx+\left(Q_y+\frac{\partial Q_y}{\partial y}\,dy\right)dx+ \\ +q\,dx\,dy=\rho h\,dx\,dy\frac{\partial^2 w}{\partial t^2}, \quad (4.2.1)$$

$$\left(M_y+\frac{\partial M_y}{\partial y}\,dy\right)dx-M_y\,dx+M_{xy}\,dy-\left(M_{xy}+\frac{\partial M_{xy}}{\partial x}\,dx\right)dy- \\ -Q\,dx\,dy=0, \quad (4.2.2)$$

$$\left(M_x+\frac{\partial M_x}{\partial x}\,dx\right)dy-M_x\,dy+\left(M_{yx}+\frac{\partial M_{yx}}{\partial y}\,dy\right)dx-M_{yx}\,dx- \\ -Q_x\,dy\,dx=0 \quad (4.2.3)$$

The displacement $w(x, y, t)$ appearing in (4.2.1) measures the deflection of the middle plane of the plate. This will be brought out more clearly in describing the kinematics of the plate. It should be noted that rotary-inertia effects have been neglected in the moment equations (4.2.2) and (4.2.3). Higher-order contributions to the moments from the loading q have also been neglected in these equations. Cancelling terms, the equations of motion reduce to

$$\frac{\partial Q_x}{\partial x}+\frac{\partial Q_y}{\partial y}+q=\rho h\frac{\partial^2 w}{\partial t^2}, \quad (4.2.4)$$

$$\frac{\partial M_y}{\partial y}-\frac{\partial M_{xy}}{\partial x}-Q_y=0, \quad (4.2.5)$$

$$\frac{\partial M_x}{\partial x}+\frac{\partial M_{yx}}{\partial y}-Q_x=0. \quad (4.2.6)$$

Solving the last two equations for Q_x, Q_y and substituting in the first equation gives a single equation in terms of the various moments,

$$\frac{\partial^2 M_x}{\partial x^2}+\frac{\partial^2 M_{yx}}{\partial x\,\partial y}-\frac{\partial^2 M_{xy}}{\partial y\,\partial x}+\frac{\partial^2 M_y}{\partial x^2}+q=\rho h\frac{\partial^2 w}{\partial t^2}. \tag{4.2.7}$$

The relationships between the moments and the deflection must now be established.

As in the case of the beam, the kinematics of the deformation must now be introduced. Of course the situation is now more complicated owing to the two-dimensional aspects. In Fig. 4.11, the major kinematical aspects of plate

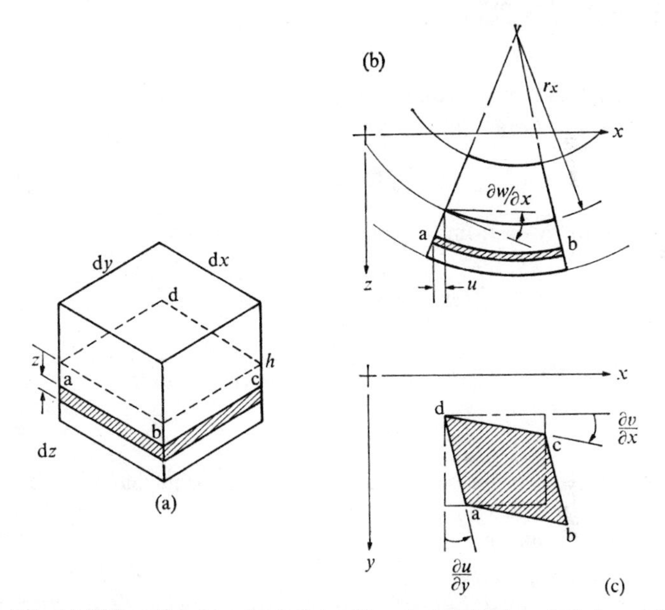

FIG. 4.11. (a) Differential element of plate with a typical lamina abcd shown shaded; (b) a side view of the element during bending; (c) a top view of the lamina showing shear deformation.

deformation that are required for our development are presented. A typical lamina (abcd) of a differential element of plate is shown in Fig. 4.11(a). The lamina is located at a distance z below the mid plane of the plate. When the element is subjected to pure bending, it is assumed to deform as shown in Fig. 4.11(b). It is assumed that plane sections remain plane and perpendicular to the mid plane, just as in the Bernoulli–Euler beam development. A behaviour similar to that shown in the above also holds in the y, z-plane. It follows that the normal strains in the lamina are given by

$$\varepsilon_x = z/r_x, \qquad \varepsilon_y = z/r_y, \tag{4.2.8}$$

where r_x, r_y are the radii of curvature in the x, z- and y, z-planes. If small deflections and slopes are assumed, the curvatures may be approximated by $-\partial^2 w/\partial x^2$, $-\partial^2 w/\partial y^2$, so that (4.2.8) becomes

$$\varepsilon_x = -z\frac{\partial^2 w}{\partial x^2}, \qquad \varepsilon_y = -z\frac{\partial^2 w}{\partial y^2}. \tag{4.2.9}$$

Now, in Fig. 4.11(c), the lamina is shown in a sheared configuration (extensional strains are not shown in this representation). The rotations of the sides are given by $\partial u/\partial y$ and $\partial v/\partial x$, where u and v are the displacement components of a particle in the x and y directions. The shear strain in the lamina is then given by†

$$\gamma_{xy} = \frac{\partial u}{\partial y} + \frac{\partial v}{\partial x}. \tag{4.2.10}$$

Referring back to Fig. 4.11(b), it is seen that the displacement component u is given by $u = -z\,\partial w/\partial x$. Similarly, $v = -z\,\partial w/\partial y$, so that γ_{xy} becomes

$$\gamma_{xy} = -2z\frac{\partial^2 w}{\partial x\,\partial y}. \tag{4.2.11}$$

The stresses are given from the general Hooke's law as

$$\sigma_x = \frac{E}{1-\nu^2}(\varepsilon_x + \nu\varepsilon_y) = -\frac{Ez}{1-\nu^2}\left(\frac{\partial^2 w}{\partial x^2} + \nu\frac{\partial^2 w}{\partial y^2}\right), \tag{4.2.12}$$

$$\sigma_y = \frac{E}{1-\nu^2}(\varepsilon_y + \nu\varepsilon_x) = -\frac{Ez}{1-\nu^2}\left(\frac{\partial^2 w}{\partial y^2} + \nu\frac{\partial^2 w}{\partial x^2}\right), \tag{4.2.13}$$

$$\tau_{xy} = G\gamma_{xy} = -2Gz\frac{\partial^2 w}{\partial x\,\partial y}. \tag{4.2.14}$$

The expressions for the bending and twisting moments may now be found. Thus, consider the face of the element in Fig. 4.10(a) defined by $h\,dy$. The bending moment on that face due to σ_x will be

$$M_x\,dy = \int_{-h/2}^{h/2} z\sigma_x\,dy\,dz, \tag{4.2.15}$$

or

$$M_x = \int_{-h/2}^{h/2} z\sigma_x\,dz. \tag{4.2.16}$$

Substituting (4.2.12) in (4.2.16) and carrying out the integration gives

$$M_x = -D\left(\frac{\partial^2 w}{\partial x^2} + \nu\frac{\partial^2 w}{\partial y^2}\right), \tag{4.2.17}$$

† The engineering definition of shear strain is used here. The mathematical elasticity definition of

$$\gamma_{xy} = \frac{1}{2}\left(\frac{\partial u}{\partial x} + \frac{\partial v}{\partial y}\right)$$

is used from Chapter 5 onward.

where

$$D = Eh^3/12(1-\nu^2). \tag{4.2.18}$$

Similarly, for M_y we obtain

$$M_y = -D\left(\frac{\partial^2 w}{\partial y^2}+\nu\frac{\partial^2 w}{\partial x^2}\right). \tag{4.2.19}$$

For M_{xy}, the integral is

$$M_{xy} = -\int_{-h/2}^{h/2} z\tau_{xy}\,dz. \tag{4.2.20}$$

The negative sign in (4.2.20) is to bring the resulting twisting moment caused by positive shear stress into accord with the convention displayed in Fig. 4.10(b). We obtain, upon substituting (4.2.14) into (4.2.20),

$$M_{xy} = D(1-\nu)\frac{\partial^2 w}{\partial x\,\partial y}. \tag{4.2.21}$$

Furthermore, we have that $M_{xy} = -M_{yx}$. Then, substituting (4.2.17), (4.2.19), and (4.2.21) in (4.2.7), we obtain

$$D\left(\frac{\partial^4 w}{\partial x^4}+2\frac{\partial^4 w}{\partial x^2\,\partial y^2}+\frac{\partial^4 w}{\partial y^4}\right)-q = -\rho h\frac{\partial^2 w}{\partial t^2}. \tag{4.2.22}$$

The expression in parenthesis may be written as

$$\frac{\partial^4 w}{\partial x^4}+2\frac{\partial^4 w}{\partial x^2\,\partial y^2}+\frac{\partial^4 w}{\partial y^4} = \left(\frac{\partial^2}{\partial x^2}+\frac{\partial^2}{\partial y^2}\right)\left(\frac{\partial^2 w}{\partial x^2}+\frac{\partial^2 w}{\partial y^2}\right) = \nabla^2\nabla^2 w. \tag{4.2.23}$$

The Laplacian of the Laplacian is designated as the biharmonic operator ∇^4. The governing equation for the motion of a thin plate is then

$$D\nabla^4 w(x, y, t)+\rho h\frac{\partial^2 w(x, y, t)}{\partial t^2} = q(x, y, t). \tag{4.2.24}$$

In terms of polar coordinates, this is given by

$$D\nabla^4 w(r, \theta, t)+\rho h\frac{\partial^2 w(r, \theta, t)}{\partial t^2} = q(r, \theta, t), \tag{4.2.25}$$

where $\nabla^4 w$ is again given by $\nabla^2\nabla^2 w$ and ∇^2 is the Laplacian in polar coordinates. The various bending and twisting moments are given, in polar coordinates, as

$$M_r = -D\left\{\frac{\partial^2 w}{\partial r^2}+\nu\left(\frac{1}{r}\frac{\partial w}{\partial r}+\frac{1}{r^2}\frac{\partial^2 w}{\partial \theta^2}\right)\right\}, \tag{4.2.26}$$

$$M_\theta = -D\left\{\frac{1}{r}\frac{\partial w}{\partial r}+\frac{1}{r^2}\frac{\partial^2 w}{\partial \theta^2}+\nu\frac{\partial^2 w}{\partial r^2}\right\}, \tag{4.2.27}$$

$$M_{r\theta} = (1-\nu)D\left(\frac{1}{r}\frac{\partial^2 w}{\partial r\,\partial \theta}-\frac{1}{r^2}\frac{\partial w}{\partial \theta}\right). \tag{4.2.28}$$

4.2.2. *Boundary conditions for a plate*

Although our immediate analyses will concern waves in infinite plates, it is relevant to consider the matter of boundary conditions at this stage. By analogy with beams, there are three simple types of plate boundaries; simply supported, fixed, and free. The mathematical statements for the first two conditions, in terms of rectangular coordinates and for an edge located at $x = x_0$, are as follows.

1. Edge simply supported at $x = x_0$. Then

$$w = M_x = 0, \tag{4.2.29}$$

or, in terms of deflections,

$$w = \frac{\partial^2 w}{\partial x^2}+\nu\frac{\partial^2 w}{\partial y^2} = 0. \tag{4.2.30}$$

2. Clamped edge at $x = x_0$. Then

$$w = \frac{\partial w}{\partial x} = 0. \tag{4.2.31}$$

The case of the free edge is not so straightforward. Since we have three plate stresses M_x, M_{xy}, Q_x to consider, it would seem that these three quantities must vanish, implying three conditions at the boundary. However, in 1850 it was shown by Kirchhoff that M_{xy}, Q_x combine to give a single condition. The development of the free edge condition can proceed by the following argument.† The equilibrium of the plate is not changed if the net twisting couple $M_{xy}\,dy$, shown in Fig. 4.12(a) is replaced by two forces, as shown in Fig. 4.12(b). There is a local perturbation in the stress field but this is assumed to rapidly decay out. If adjacent sections of the edge are considered, the situation is as shown in Fig. 4.12(c). Again replacing these couples by forces, the situation is as shown in Fig. 4.12(d). Summing forces on the shaded element, we have a net force of $(\partial M_{xy}/\partial y)\,dy$ in the upward direction. The free edge condition requires that the net vertical force be zero. Thus, introducing the quantity V_x, we have

$$V_x = Q_x-(\partial M_{xy}/\partial y) = 0. \tag{4.2.32}$$

Thus, at a free edge $x = x_0$, the boundary conditions are $V_x = M_x = 0$ or, in terms of displacements,

$$\frac{\partial^2 w}{\partial x^2}+\nu\frac{\partial^2 w}{\partial y^2} = \frac{\partial^3 w}{\partial x^3}+(2-\nu)\frac{\partial^3 w}{\partial x\,\partial y^2} = 0. \tag{4.2.33}$$

It is of interest to consider the source of this somewhat peculiar, and certainly not self-evident, condition of a free edge of a plate. The general state of stress that may exist at a point consists of the six stresses σ_x, σ_y, σ_z, τ_{xy}

† Timoshenko and Woinowsky-Krieger [17, pp. 83–4].

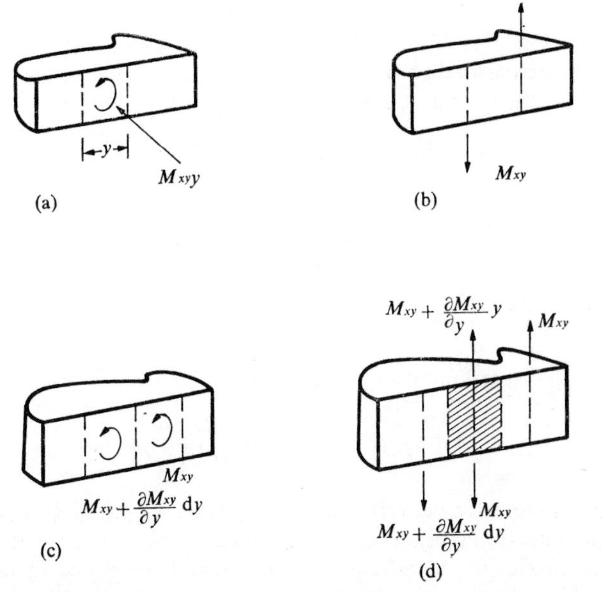

FIG. 4.12. The steps in developing the free edge boundary condition for a plate.

τ_{yz}, τ_{zx}. However, in thin plate theory it is assumed that $\sigma_z = 0$, while σ_x, σ_y, τ_{xy} are non-zero. However, τ_{zx}, τ_{zy} have not been brought into the analysis as yet, although it is obvious they must be non-zero so as to support vertical plate loads. It must be recalled, however, that plane sections have been assumed to remain plane. This implies that γ_{zy}, $\gamma_{zx} = 0$ on a typical element, since non-zero shearing strains here would yield warping of the section. Of course, this aspect has been discussed previously in conjunction with beam theory. But zero strains indicate an infinite shear rigidity. Thus, although shearing stresses are present, we are unable to determine the details of the stress distribution. The anomalous boundary condition thus has its source in a mathematical model accounting for only part of the total deformation a plate may undergo.

4.2.3. *Plane waves in an infinite plate*

It should be evident, upon drawing the parallel with simple beam theory, that waves will propagate dispersively. Let us investigate the conditions under which harmonic plane waves may propagate. Thus consider a plane wave of the form

$$w = A e^{i\gamma(\mathbf{n}.\mathbf{r}-ct)}, \tag{4.2.34}$$

where r is the position vector to a point on the plane of the wave and n is the normal to that plane and, as in the membrane investigation of the same problem,

$$\mathbf{r} = x\mathbf{i}+y\mathbf{j}, \qquad \mathbf{n} = l\mathbf{i}+m\mathbf{j}. \tag{4.2.35}$$

Letting the external load q be zero and substituting in the differential equation, we obtain for the various terms

$$\begin{aligned}\nabla^2 w &= -A(l^2+m^2)\gamma^2 e^{i\gamma(\mathbf{n}.\mathbf{r}-ct)},\\ &= -A\gamma^2 e^{i\gamma(\mathbf{n}.\mathbf{r}-ct)},\end{aligned} \tag{4.2.36}$$

$$\nabla^4 w = A\gamma^4 e^{i\gamma(\mathbf{n}.\mathbf{r}-ct)}, \tag{4.2.37}$$

$$\frac{\partial^2 w}{\partial t^2} = -A\gamma^2 c^2 e^{i\gamma(\mathbf{n}.\mathbf{r}-ct)}. \tag{4.2.38}$$

The resulting dispersion relation is given by

$$D\gamma^2-\rho hc^2 = 0, \qquad c = \sqrt{\left(\frac{D}{\rho h}\right)}\gamma. \tag{4.2.39}$$

This linear relationship between phase velocity and wavenumber is similar to that of Bernoulli–Euler beam theory. It predicts unbounded wave velocity for very short-wavelength, high-frequency conditions. This physically unacceptable situation is, as in beams, a consequence of an imperfect mathematical model in which effects of shear and rotary inertia have been neglected; just as in beam theory there are available higher-order theories of plate vibration which incorporate these effects. Finally, we note the frequency–wavenumber expression follows directly from (4.2.39) through the relation $\omega = \gamma c$, and is given by

$$\omega = \sqrt{\left(\frac{D}{\rho h}\right)}\gamma^2. \tag{4.2.40}$$

4.2.4. *An initial-value problem*

An initial-value problem in plate vibrations will be considered first, in which the case of axisymmetric vibrations will be discussed. The biharmonic equation in polar coordinates will be the governing equation, with $\partial^2/\partial\theta^2 = 0$ in the Laplacian. Then, in the absence of external loads, we have

$$D\left(\frac{\partial^2}{\partial r^2}+\frac{1}{r}\frac{\partial}{\partial r}\right)^2 w+\rho h\frac{\partial^2 w}{\partial t^2} = 0. \tag{4.2.41}$$

Assume the initial conditions

$$w(r,0) = f(r), \qquad \dot{w}(r,0) = 0. \tag{4.2.42}$$

To apply integral transforms to the present problem, the Hankel transform, which was used in the preceding section on axisymmetric membrane problems will again be used here. The zero-order Hankel transform

$$\bar{w}(\xi) = \int_0^\infty r w(r) J_0(\xi r)\, \mathrm{d}r, \tag{4.2.43}$$

with inverse

$$w(r) = \int_0^\infty \xi \bar{w}(\xi) J_0(\xi r)\, \mathrm{d}\xi, \tag{4.2.44}$$

will be relevant. In applying this transform to the biharmonic operator ∇^4 in polar coordinates, we recall the particularly simple result arising from the transform of the Laplacian in the previous membrane studies. In the case of the biharmonic, the result is†

$$H(\nabla^4 w) = \xi^4 \bar{w}(\xi, t). \tag{4.2.45}$$

Then, upon taking the Hankel transform of (4.2.41), we obtain

$$D\xi^4 \bar{w} + \rho h \frac{\partial^2 \bar{w}}{\partial t^2} = 0. \tag{4.2.46}$$

Defining $b^2 = D/\rho h$, we thus have the transformed ordinary differential equation

$$\frac{\partial^2 \bar{w}}{\partial t^2} + b^2 \xi^4 \bar{w} = 0. \tag{4.2.47}$$

This has the solution

$$\bar{w}(\xi, t) = A \cos b\xi^2 t + B \sin b\xi^2 t. \tag{4.2.48}$$

The Hankel transforms of the initial conditions (4.2.42) are

$$\bar{w}(\xi, 0) = \bar{f}(\xi), \qquad \partial \bar{w}(\xi, 0)/\partial t = 0. \tag{4.2.49}$$

Applying the initial conditions to the solution gives

$$A = \bar{f}(\xi), \qquad B = 0. \tag{4.2.50}$$

The transformed solution may thus be expressed in the form

$$\bar{w}(\xi, t) = \cos b\xi^2 t \int_0^\infty u f(u) J_0(\xi u)\, \mathrm{d}u, \tag{4.2.51}$$

where the integral representation of $\bar{f}(\xi)$ has been used and the dummy variable u employed.

By the Hankel inversion theorem, we have

$$\begin{aligned} w(r, t) &= \int_0^\infty \xi \bar{w}(\xi, t) J_0(\xi r)\, \mathrm{d}\xi, \\ &= \int_0^\infty \xi J_0(\xi r) \cos b\xi^2 t\, \mathrm{d}\xi \int_0^\infty u f(u) J_0(u\xi)\, \mathrm{d}u, \end{aligned} \tag{4.2.52}$$

† See Appendix B.4.

or, interchanging the order of integration,

$$w(r, t) = \int_0^\infty uf(u)\,du \int_0^\infty \xi J_0(\xi r)J_0(\xi u)\cos b\xi^2 t\,d\xi. \tag{4.2.53}$$

In order to complete the evaluation of this problem, a particular integral, known as 'Weber's second exponential integral' must be used. This is the known result that†

$$\int_0^\infty \xi J_0(\xi u)J_0(\xi r)\exp(-p\xi^2)\,d\xi = \frac{1}{2p}\exp\left\{-\frac{u^2+r^2}{4p}\right\}I_0\left(\frac{ur}{2p}\right), \tag{4.2.54}$$

where I_0 is the modified Bessel function of the first kind defined as

$$I_0(z) = J_0(iz). \tag{4.2.55}$$

Letting $p = -ibt$ in the above and equating the real parts gives

$$\int_0^\infty \xi J_0(\xi u)J_0(\xi r)\cos b\xi^2 t\,d\xi = \frac{1}{2bt}J_0\left(\frac{ur}{2bt}\right)\sin\left(\frac{u^2+r^2}{4bt}\right). \tag{4.2.56}$$

Hence our solution is

$$w(r, t) = \frac{1}{2bt}\int_0^\infty uf(u)J_0\left(\frac{ur}{2bt}\right)\sin\left(\frac{u^2+r^2}{4bt}\right)du. \tag{4.2.57}$$

Consider the specific initial waveform

$$f(r) = f_0\exp(-r^2/a^2), \tag{4.2.58}$$

where f_0 is a constant. Instead of using the above result directly, we reconsider the transformed solution (4.2.51) and write

$$\bar{w}(\xi, t) = f_0\cos b\xi^2 t\int_0^\infty u\exp(-u^2/a^2)J_0(\xi u)\,du. \tag{4.2.59}$$

The integral in the above has the value‡

$$\frac{a^2}{2}\exp\left(-\frac{\xi^2 a^2}{4}\right), \tag{4.2.60}$$

so that

$$\bar{w}(\xi, t) = \frac{f_0 a^2}{2}\cos b\xi^2 t\exp\left(-\frac{\xi^2 a^2}{4}\right). \tag{4.2.61}$$

† p. 393 of Reference [19].
‡ See Appendix B.5.

Hence, applying the Hankel inversion to the above gives

$$w(r, t) = \frac{f_0 a^2}{2} \int_0^\infty \xi \exp\left(-\frac{\xi^2 a^2}{4}\right) \cos b\xi^2 t J_0(\xi r)\, d\xi. \qquad (4.2.62)$$

Fortunately, this integral form has also been evaluated. It is known as the confluent hypergeometric function,† and yields the result

$$w(r, t) = \frac{f_0}{1+\tau^2} \exp\left(-\frac{\rho^2}{1+\tau^2}\right)\left\{\cos\left(\frac{\rho^2\tau}{1+\tau^2}\right) + \tau \sin\left(\frac{\rho^2\tau}{1+\tau^2}\right)\right\}, \qquad (4.2.63)$$

where

$$\tau = 4bt/a^2, \qquad \rho = r/a. \qquad (4.2.64)$$

A plot of w/f_0 for various values of non-dimensionalized time is shown in Fig. 4.13.

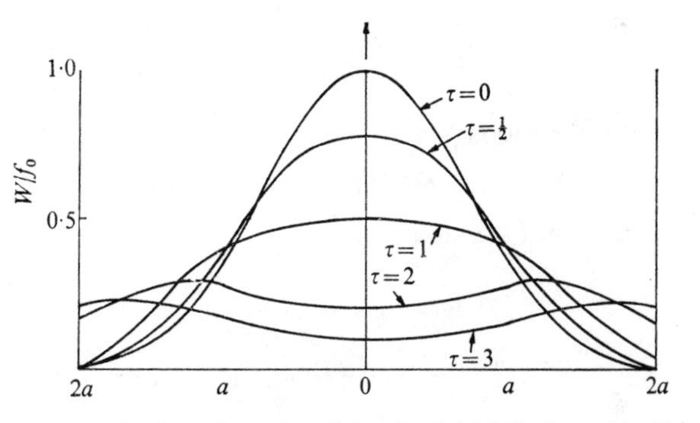

FIG. 4.13. Wave motion in a plate released from an initial displacement. (After Sneddon [16, Fig. 14].)

We will not investigate the plate motion under more severe initial conditions, such as impulse displacements or velocities. Drawing on our experience in beam theory, we would expect certain anomalies in the results, such as instantaneous response at remote locations.

4.2.5. *Forced motion of an infinite plate*

We now wish to consider an infinite plate subjected to a transient loading. Axisymmetric conditions will again be assumed. The analysis is based on the work of Medick [10].‡ The governing equation of motion for the problem is

† pp. 513–14 of Reference [16].

‡ Reference [10] derives a portion of its analytical basis from the treatment by Sneddon [16, pp. 139–47]. It contains equally important experimental studies which will be cited later in this chapter.

given by

$$D\nabla^4 w(r,t)+\rho h\frac{\partial^2 w(r,t)}{\partial t^2} = q(r,t), \tag{4.2.65}$$

with the homogeneous initial conditions

$$w(r,0) = \dot{w}(r,0) = 0 \tag{4.2.66}$$

being assumed. If we define the loading in the rather general form

$$q(r,t) = 8b\rho h f(r)\psi'(t), \tag{4.2.67}$$

where $b^2 = D/\rho h$ and $f(r)$ and $\psi'(t)$ describe the spatial and time variation of the loading respectively, then the equation of motion may be written as

$$b^2\nabla^4 w+\frac{\partial^2 w}{\partial t^2} = 8bf(r)\psi'(t). \tag{4.2.68}$$

The Hankel and Laplace transforms will be used. Applying the former, we obtain

$$\frac{\partial^2 \bar{w}(\xi,t)}{\partial t^2}+b^2\xi^4\bar{w}(\xi,t) = 8b\bar{f}(\xi)\psi'(t). \tag{4.2.69}$$

The Laplace transformation gives

$$s^2\overline{W}(\xi,s)+b^2\xi^4\overline{W}(\xi,s) = 8bs\bar{f}(\xi)\bar{\psi}(s), \tag{4.2.70}$$

so that the transformed solution is given by

$$\overline{W}(\xi,s) = \frac{8bs\bar{f}(\xi)\bar{\psi}(s)}{s^2+b^2\xi^4}. \tag{4.2.71}$$

We first apply the Laplace inversion to (4.2.71). Noting that

$$\mathscr{L}^{-1}\left(\frac{s}{s^2+b^2\xi^4}\right) = \cos b\xi^2 t, \tag{4.2.72}$$

then the Laplace inversion is

$$\bar{w}(\xi,t) = 8b\bar{f}(\xi)\int_0^t \psi(u)\cos b\xi^2(t-u)\,du. \tag{4.2.73}$$

Taking the Hankel inversion, we have

$$w(r,t) = 8b\int_0^\infty \xi J_0(\xi r)\bar{f}(\xi)\int_0^t \psi(u)\cos b\xi^2(t-u)\,du\,d\xi. \tag{4.2.74}$$

Rearranging terms slightly gives us the formal solution for $w(r,t)$ in terms of arbitrary $f(r)$, $\psi(t)$. Thus

$$w(r,t) = 8b\int_0^t \psi(u)\,du\int_0^\infty \xi\bar{f}(\xi)J_0(\xi r)\cos b\xi^2(t-u)\,d\xi. \tag{4.2.75}$$

The above solution may be further refined by putting it in the form

$$w(r,t) = 8b\int_0^t \psi(u)\,\mathrm{d}u \int_0^\infty vf(v)\,\mathrm{d}v \int_0^\infty \xi J_0(\xi r)J_0(\xi v)\cos b\xi^2(t-u)\,\mathrm{d}\xi, \quad (4.2.76)$$

where $\bar{f}(\xi)$ has been expressed as

$$\bar{f}(\xi) = \int_0^\infty vf(v)J_0(\xi v)\,\mathrm{d}v, \quad (4.2.77)$$

and the terms rearranged slightly. Using Weber's second exponential integral (see (4.2.54)), we have that

$$\int_0^\infty \xi J_0(\xi r)J_0(\xi v)\cos b\xi^2(t-u)\,\mathrm{d}\xi = \frac{1}{2b(t-u)}J_0\left(\frac{vr}{2b(t-u)}\right)\sin\frac{v^2+r^2}{4b(t-u)}. \quad (4.2.78)$$

Then the solution becomes

$$w(r,t) = 4\int_0^t \frac{\psi(u)}{t-u}\,\mathrm{d}u \int_0^\infty vf(v)J_0\left(\frac{vr}{2b(t-u)}\right)\sin\left\{\frac{v^2+r^2}{4b(t-u)}\right\}\mathrm{d}v. \quad (4.2.79)$$

The total magnitude of the force acting at the origin will be considered as $8b\rho h$. The form of $f(r)$ so that the load is at the origin is determined by noting that $f(r) = 0$, $r \neq 0$, and that

$$\int_0^\infty 2\pi r f(r)\,\mathrm{d}r = 1. \quad (4.2.80)$$

Thus $f(r) = \delta(r)/2\pi r$ will give the case of a concentrated force at the origin. The solution (4.2.79) then reduces to

$$w(r,t) = \frac{2}{\pi}\int_0^t \frac{\psi(u)}{t-u}\sin\left\{\frac{r^2}{4b(t-u)}\right\}\mathrm{d}u. \quad (4.2.81)$$

Although we are ultimately interested in the plate response to a pulse load acting at the origin, by considering a slightly different case the desired results can be more efficiently obtained. Let

$$\psi'(t) = H\langle t\rangle, \quad (4.2.82)$$

which is the Heaviside function. Then $\psi(t) = tH\langle t\rangle$, and the solution is then expressible as

$$w(r,t) = \frac{2}{\pi}\int_0^t \frac{u}{t-u}\sin\left\{\frac{r^2}{4b(t-u)}\right\}\mathrm{d}u. \quad (4.2.83)$$

To evaluate this integral, we introduce the change of variable

$$z = r^2/4b(t-u). \tag{4.2.84}$$

Noting that

$$u = 0, \qquad z = r^2/4bt; \qquad u = t,\ z = \infty, \tag{4.2.85}$$

the solution may be written as

$$w(r, t) = \frac{2}{\pi} \int_{r^2/4bt}^{\infty} \left(t - \frac{r^2}{4bz}\right) \frac{\sin z}{z}\, dz. \tag{4.2.86}$$

We now introduce a second change of variables, given by $x = r^2/4bt$. Then (4.2.86) may be written as

$$w(r, t) = \frac{2}{\pi} \int_{x}^{\infty} \left(t - \frac{xt}{z}\right) \frac{\sin z}{z}\, dz = \frac{2}{\pi} I, \tag{4.2.87}$$

where I is defined as

$$I = t \int_{x}^{\infty} \frac{\sin z}{z}\, dz - xt \int_{x}^{\infty} \frac{\sin z}{z^2}\, dz. \tag{4.2.88}$$

We now introduce the sine and cosine integrals defined as†

$$\mathrm{Si}(x) = \int_{0}^{x} \frac{\sin z}{z}\, dz, \qquad \mathrm{Ci}(x) = \int_{\infty}^{x} \frac{\cos z}{z}\, dz. \tag{4.2.89}$$

Then I may be written as

$$I = t\left(\int_{0}^{\infty} \frac{\sin z}{z}\, dz - \int_{0}^{x} \frac{\sin z}{z}\, dz\right) - xt\left(-\frac{\sin z}{z}\bigg|_{x}^{\infty} + \int_{x}^{\infty} \frac{\cos z}{z}\, dz\right). \tag{4.2.90}$$

Now‡

$$\int_{0}^{\infty} \frac{\sin z}{z}\, dz = \frac{\pi}{2} \tag{4.2.91}$$

so that I becomes

$$I = t\{\pi/2 - \mathrm{Si}(x) - \sin x + x\mathrm{Ci}(x)\} = tH(x), \tag{4.2.92}$$

where the definition of $H(x)$ is obvious. Then recalling that $x = r^2/4bt$, we have for $w(r, t)$

$$w(r, t) = \frac{2}{\pi} tH\left(\frac{r^2}{4bt}\right). \tag{4.2.93}$$

† pp. 231–3 of Reference [1].
‡ op. cit., p. 232.

Finally, we recall that $8b\rho h$ represents the total load, so that if P represents this value we have $P/4b\rho h = 2$, and $w(r, t)$ may be written as

$$w(r, t) = \frac{P}{4\pi b\rho h} tH\left(\frac{r^2}{4bt}\right), \tag{4.2.94}$$

where, from (4.2.92),

$$H\left(\frac{r^2}{4bt}\right) = \frac{\pi}{2} - \mathrm{Si}\left(\frac{r^2}{4bt}\right) - \sin\frac{r^2}{4bt} + \frac{r^2}{4bt}\mathrm{Ci}\left(\frac{r^2}{4bt}\right). \tag{4.2.95}$$

This, then, represents the response of the plate to a step load at the origin.

Other loadings may be found rather easily. Thus the response to a rectangular pulse of duration β is directly found by superposition. Thus

$$w(r, t) = \frac{P}{4\pi b\rho h}\left\{tH\left(\frac{r^2}{4bt}\right) - (t-\beta)H\left(\frac{r^2}{4b(t-\beta)}\right)\right\}. \tag{4.2.96}$$

The case of an impulse load may also be obtained. Although in theory the response to an impulse load may be found by differentiating the response to a step loading, the rather complicated form of the step solution makes it advantageous to reconsider the solution as given by (4.2.81). Thus, for an impulse loading,

$$\psi'(t) = \delta(t), \qquad \psi(t) = H\langle t\rangle, \tag{4.2.97}$$

so that

$$w(r, t) = \frac{2}{\pi}\int_0^t \frac{1}{t-u}\sin\left\{\frac{r^2}{4b(t-u)}\right\} du. \tag{4.2.98}$$

The previously used change of variables puts this in the form

$$w(r, t) = \frac{2}{\pi}\int_{r^2/4bt}^{\infty} \frac{1}{r^2/4bz}\frac{\sin z}{4bz^2/r^2} dz = \frac{2}{\pi}\int_{r^2/4bt}^{\infty} \frac{\sin z}{z} dz \tag{4.2.99}$$

or

$$w(r, t) = \frac{2}{\pi}\left(\int_0^{\infty} \frac{\sin z}{z} dz - \int_0^{r^2/4bt} \frac{\sin z}{z} dz\right). \tag{4.2.100}$$

Then

$$w(r, t) = \frac{P}{4\pi b\rho h}\left\{\frac{\pi}{2} - \mathrm{Si}\left(\frac{r^2}{4bt}\right)\right\} \tag{4.2.101}$$

represents the response to an impulse load.

Some of the theoretical results obtained by Medick for the motion of the plate are shown in Fig. 4.14. The displacement $w(r, t)$ has been non-dimensionalized as

$$\bar{w}(r, t) = \frac{w(r, t)}{(I/4\pi)\sqrt{(\rho h D)}}, \qquad I = P\beta, \tag{4.2.102}$$

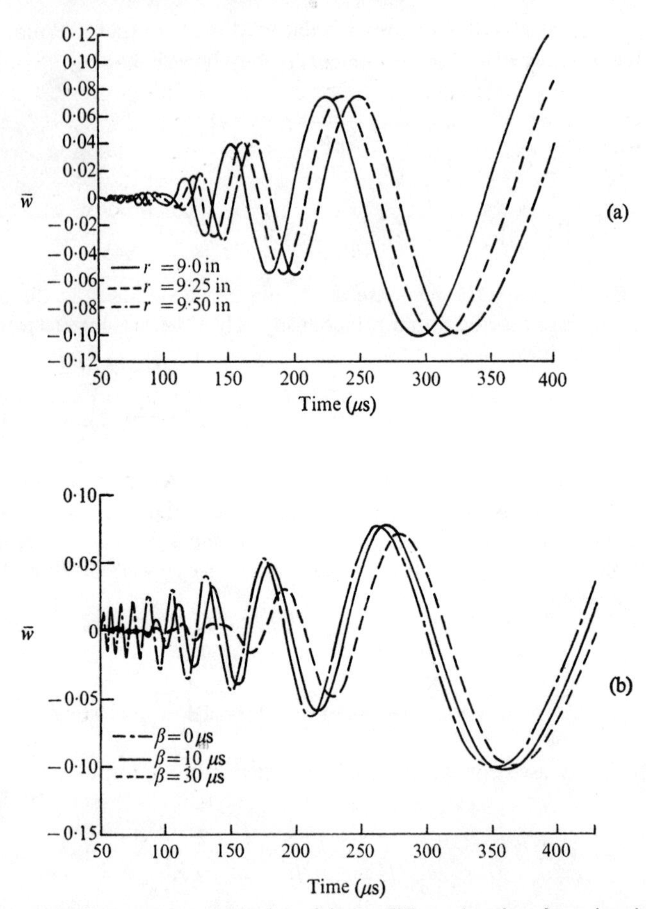

FIG. 4.14. (a) Displacement as a function of time at different locations for a given impulse, and (b) the response as a function of time at a given location for various impulse durations. (After Medick [10].)

where I is the total impulse transmitted to the plate. Precursors at every location are predicted for $t > 0$, but the scale of the drawings would not show significant amplitudes for most cases. However, in Fig. 4.14(b), the impulse response ($\beta = 0$) has been plotted, and this clearly shows high-frequency precursors of significant amplitude even for small time. It should be noted, again referring to Fig. 4.14(b), that for later times the response to impulses of varying duration do not significantly differ. Thus, as in the case of Bernoulli–Euler beam theory, classical plate theory gives poorest results for short time.

4.2.6. *Reflection of plane waves from boundaries*

We wish to study reflection phenomena in plates by considering plane waves to be incident on the straight boundary of a semi-infinite plate $y > 0$. Only plates as governed by the conditions of simply supported, clamped, and free edge boundaries will be considered. Our starting point is the homogeneous form of eqn (4.2.24). Without repeating the laborious separation of variables procedure presented for the analogous situation in membranes, we straightway consider a solution of the form

$$w = f(y)e^{i(\xi x-\omega t)}. \tag{4.2.103}$$

This represents a wave system moving in the positive x direction. Substituting in (4.2.24) gives

$$\xi^4 f(y) - 2\xi^2\frac{d^2f(y)}{dy^2} + \frac{d^4f(y)}{dy^4} - \frac{\rho h}{D}\omega^2 f(y) = 0 \tag{4.2.104}$$

or

$$\frac{d^4f}{dy^4} - 2\xi^2\frac{d^2f}{dy^2} - \left(\frac{\rho h}{D}\omega^2 - \xi^4\right)f = 0. \tag{4.2.105}$$

Introducing the definitions

$$\beta^4 = \omega^2/a^2, \qquad a^2 = D/\rho h, \qquad \eta^4 = \beta^4 - \xi^4, \tag{4.2.106}$$

we write (4.2.105) as

$$\frac{d^4f}{dy^4} - 2\xi^2\frac{d^2f}{dy^2} - \eta^4 f = 0. \tag{4.2.107}$$

For a trial solution of the type $\exp(sy)$ we obtain the characteristic roots

$$s = \pm\zeta_1, \pm i\zeta_2, \tag{4.2.108}$$

where

$$\zeta_1 = (\xi^2+\beta^2)^{\frac{1}{2}}, \qquad \zeta_2 = (\beta^2-\xi^2)^{\frac{1}{2}}. \tag{4.2.109}$$

The result gives a wave system

$$w = A\exp\{i(\xi x - \zeta_2 y - \omega t)\} + B\exp\{i(\xi x + \zeta_2 y - \omega t)\} + \\ + C\exp(-\zeta_1 y)\exp\{i(\xi x - \omega t)\} + D\exp(\zeta_1 y)\exp\{i(\xi x - \omega t)\}. \tag{4.2.110}$$

Other results are possible. Thus suppose $\beta^2 > \xi^2$ in the expression (4.2.109) for ζ_2. Then $\zeta_2 \to i\zeta_2'$, where $\zeta_2'^2 = -\zeta_2^2$. The A and B terms in the solution (4.2.110) would then take on the appearance of the C and D terms. Another possibility is that $\beta^2 = \xi^2$, so that $\zeta_2 = 0$. This yields a result

$$w = \{Ay\exp(-\zeta_1 y) + C\exp(-\zeta_1 y) + By\exp(\zeta_1 y) + D\exp(\zeta_1 y)\} \times \\ \times \exp\{-i(\xi x - \omega t)\}. \tag{4.2.111}$$

This also is not in the form of plane waves. The point is that only the form (4.2.110) permits propagating plane waves. If operative frequencies are such that either of the other possibilities mentioned arises, it is no longer possible to consider wave reflection.

Considering the solution (4.2.109) further, it is seen that the A and B terms are indeed plane harmonic waves. The A term is a wave propagating inward from the positive y direction toward the boundary at $y = 0$, while the B term is an outward propagating wave. We further see that, for a plate $y > 0$, that the D term must be discarded because it yields exponentially increasing waves.

As a first example, consider the simply supported boundary, described by

$$w(x, 0, t) = \partial^2 w(x, 0, t)/\partial y^2 = 0. \tag{4.2.112}$$

Substituting (4.2.109) with $D = 0$ in these conditions gives

$$\begin{aligned} A+B+C &= 0, \\ -\zeta_2^2 A - \zeta_2^2 B + \zeta_1^2 C &= 0. \end{aligned} \tag{4.2.113}$$

The resulting amplitude ratios are

$$B/A = -1, \qquad C = 0. \tag{4.2.114}$$

Thus a plane harmonic wave incident on a pinned boundary reflects as a plane wave with a phase shift. The behaviour is as shown in Fig. 4.15(a).

As a second example, consider a clamped boundary specified by

$$w(x, 0, t) = \partial w(x, 0, t)/\partial y = 0. \tag{4.2.115}$$

Substituting the solution (4.2.109), with $D = 0$, in the above conditions gives

$$\begin{aligned} A+B+C &= 0, \\ -\mathrm{i}\zeta_2 A + \mathrm{i}\zeta_2 B - \zeta_1 C &= 0, \end{aligned} \tag{4.2.116}$$

with the amplitude ratios

$$\frac{B}{A} = -\frac{\zeta_1 - \mathrm{i}\zeta_2}{\zeta_1 + \mathrm{i}\zeta_2}, \qquad \frac{C}{A} = -\frac{2\mathrm{i}\zeta_2}{\zeta_1 + \mathrm{i}\zeta_2}. \tag{4.2.117}$$

Thus, when a plane wave is incident on a fixed boundary, a plane wave is reflected. Also, an exponentially-attenuated wave is induced that propagates along the edge of the plate. The situation is shown in Fig. 4.15(b).

4.2.7. *Free vibrations of finite plates*

There is a vast amount of literature on the free vibrations of plates of various geometries. This has arisen, in the main, from the importance of the plate structure in aerospace applications. Only the briefest coverage can be given to the subject here.

1. *Vibrations of rectangular plates.* For our first topic, consider the free vibrations of rectangular plates. We again consider a separation-of-variable solution

$$w = X(x)Y(y)\mathrm{e}^{-\mathrm{i}\omega t}, \tag{4.2.118}$$

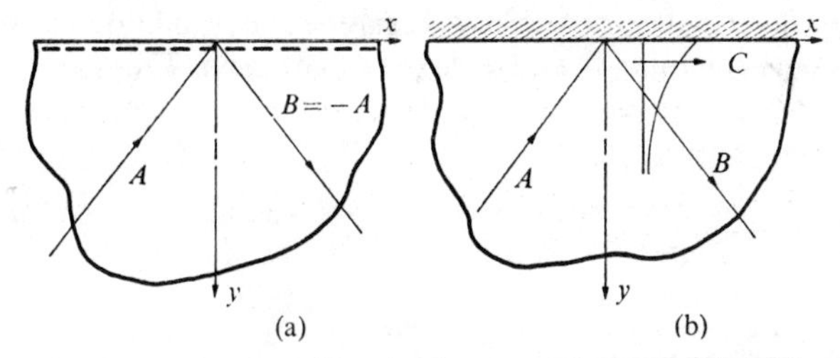

FIG. 4.15. Reflected waves from (a) a simply supported edge and (b) a clamped edge.

to the homogeneous governing equation (4.2.24). We leave boundary conditions unspecified at this stage. We obtain the resulting equation

$$X^{iv}Y+2X''Y''+XY^{iv}-\beta^4XY=0, \qquad \beta^4=\omega^2\rho h/D. \quad (4.2.119)$$

Now in order for the separation of variables to occur, we must have that

$$Y''=-\gamma^2Y, \qquad Y^{iv}=\gamma^4Y, \quad (4.2.120)$$

or

$$X''=-\alpha^2X, \qquad X^{iv}=\alpha^4X, \quad (4.2.121)$$

or both. Let us suppose the last case holds. The solution must be of the form

$$X=\sin\alpha x,\ \cos\alpha x. \quad (4.2.122)$$

If $\alpha^2<0$, hyperbolic sine and cosines could arise, but these will be excluded shortly on other grounds.

We now consider possible boundary conditions along the edges $x=0$ and $x=a$ of the plate. If we confine ourselves to the simple boundary conditions of simply supported, clamped, and free, we would have, for the various possibilities along $x=0$,

$$\begin{aligned}
&\text{SS: } X(0)Y(y)=\{-\alpha^2Y(y)+\nu Y''(y)\}X(0)=0,\\
&\text{C: } X(0)Y(y)=X'(0)Y(y)=0,\\
&\text{F: } X(0)\{-\alpha^2Y(y)+\nu Y''(y)\}=X'(0)\{-\alpha^2Y(y)+(2-\nu)Y''(y)\}=0,
\end{aligned} \quad (4.2.123)$$

with similar conditions holding on $x=a$. It should be readily apparent that the only set of boundary conditions satisfied by either one of the solutions (4.2.122) will be the case of simple support on both edges for which

$$X_n=\sin\alpha_nx, \qquad \alpha_n=n\pi/a \qquad (n=1,2,\ldots) \quad (4.2.124)$$

is a solution. Thus in order to apply separation of variables to the problem, we must restrict attention to the combinations of boundary conditions which have two opposite edges simply supported. There are six such combinations.

We now return to the y dependence. It may be that, in addition to X obeying simple support conditions, Y also will be so governed. Then

$$Y = \sin \gamma y, \cos \gamma y. \tag{4.2.125}$$

This is the simplest case and corresponds to the results obtained for X. Thus the edges $y = 0, b$ must be simply supported, and we have

$$Y_m = \sin \gamma_m y, \qquad \gamma_m = m\pi/b \qquad (m = 1, 2, ...). \tag{4.2.126}$$

This, in conjunction with (4.2.119), enables the natural frequencies of a plate simply supported on all edges to be found. Thus from (4.2.119) we have

$$(\alpha_n^4+2\alpha_n^2\gamma_m^2+\gamma_m^4-\beta^4)X_n Y_m = 0. \tag{4.2.127}$$

Solving for the frequency ω_{nm}, we obtain

$$\omega_{nm} = a(\alpha_n^2+\gamma_m^2) \tag{4.2.128}$$

or

$$\omega_{nm} = \pi^2\left(\frac{n^2}{a^2}+\frac{m^2}{b^2}\right)\sqrt{\left(\frac{D}{h}\right)} \qquad (n, m = 1, 2, ...). \tag{4.2.129}$$

The mode shapes for this simple case will be given by

$$W_{nm} = \sin \alpha_n x \sin \gamma_m y. \tag{4.2.130}$$

The case of all edges simply supported is but one of the six combinations of boundary conditions for which two opposite edges are simply supported. For the remaining five cases, we must solve

$$Y^{\text{iv}}-2\alpha_n^2 Y''-(\beta^4-\alpha_n^4)Y = 0. \tag{4.2.131}$$

The characteristic equation and resulting roots for this equation arose in the study of plane waves (see (4.2.108) and (4.2.109)). We recall that depending on whether $\beta^4 \gtreqless \alpha_n^4$, various types of solutions are possible. In the plane-wave study, only one type was admissible but, in the present problem, all are. The three cases and resulting solution forms are given in the following.

Case 1: $\beta^4 > \alpha_n^4$,

$$Y = A \sin \zeta_2 y+B \cos \zeta_2 y+C \sinh \zeta_1 y+D \cosh \zeta_1 y, \tag{4.2.132}$$

where ζ_1, ζ_2 are given by

$$\zeta_1 = (\beta^2+\alpha_n^2)^{\frac{1}{2}}, \qquad \zeta_2 = (\beta^2-\alpha_n^2)^{\frac{1}{2}} \tag{4.2.133}$$

and are of the form that previously arose in the plane-wave study.

Case 2: $\beta^4 = \alpha_n^4$.

$$Y = A \sinh \zeta_1 y+By \sinh \zeta_1 y+C \cosh \zeta_1 y+Dy \cosh \zeta_1 y. \tag{4.2.134}$$

Case 3: $\beta^4 < \alpha_n^4$,

$$Y = A \sinh \zeta_2' y + B \cosh \zeta_2' y + C \sinh \zeta_1 y + D \cosh \zeta_1 y, \quad (4.2.135)$$

where $\zeta_2'^2 = -\zeta_2^2$.

To complete the solution of any of the five remaining combinations of edge conditions on $y = 0, b$ the preceding results must be substituted into the boundary condition equations. There will result four homogeneous equations for A, B, C, D, the determinant of coefficients of which will yield a frequency equation. As an example, consider clamped edges at $y = 0, b$. Considering only the solution given by Case 1 and applying the boundary conditions

$$Y(0) = Y'(0) = Y(b) = Y'(b) = 0. \quad (4.2.136)$$

gives the equations

$$\begin{bmatrix} 0 & 1 & 0 & 1 \\ \zeta_2 & 0 & \zeta_1 & 0 \\ \sin \zeta_2 b & \cos \zeta_2 b & \sinh \zeta_1 b & \cosh \zeta_1 b \\ \zeta_2 \cos \zeta_2 b & -\zeta_2 \sin \zeta_2 b & \zeta_1 \cosh \zeta_1 b & \zeta_1 \sinh \zeta_1 b \end{bmatrix} \begin{bmatrix} A \\ B \\ C \\ D \end{bmatrix} = 0. \quad (4.2.137)$$

Equating the determinant of coefficients to zero gives the frequency equation

$$2\zeta_1\zeta_2(\cos \zeta_2 b \cosh \zeta_1 b - 1) - \alpha_n^2 \sin \zeta_2 b \sinh \zeta_2 b = 0. \quad (4.2.138)$$

For a given value of n there will be successive values of β, hence ω, which satisfy (4.2.138), and likewise for $n = 2, 3, \ldots$. Thus there will result a series of frequencies $\omega_{11}, \omega_{12}, \omega_{13}, \ldots, \omega_{21}, \omega_{22}, \ldots$. These values will, of course, depend on the aspect ratio a/b of the plate, the thickness, and the material. For completeness, the Case 2 and Case 3 solutions must be investigated. A proof that Case 3 yields no roots is rather recondite, while Case 2 gives a single root of $\omega = (D/\rho h)^{\frac{1}{2}} n^2\pi^2/a^2$. The modal curves for Case 1 are given by

$$Y_m = (\cosh \zeta_1 b - \cos \zeta_2 b)(\zeta_2 \sinh \zeta_1 y - \zeta_1 \sin \zeta_2 y) -$$
$$-(\zeta_2 \sinh \zeta_1 b - \zeta_1 \sin \zeta_2 b)(\cosh \zeta_1 y - \cos \zeta_2 y). \quad (4.2.139)$$

Thus, at a given frequency ω_{nm}, the mode shape is given by $W_{nm} = Y_m \sin \alpha_n x$.

In the following, the frequency equations and mode shapes will be summarized for all six combinations of boundary conditions. The results just discussed will be included for completeness.

SS–SS–SS–SS:

$$\sin \zeta_2 b = 0, \quad (4.2.140)$$

$$W_{nm} = \sin \alpha_n x \sin \gamma_m y. \quad (4.2.141)$$

SS–C–SS–C:

$$2\zeta_1\zeta_2(\cosh \zeta_1 b \cos \zeta_1 b - 1)\zeta_2 = \alpha_n^2 \sin \zeta_2 b \sinh \zeta_2 b, \tag{4.2.142}$$

$$W_{nm} = \{(\cosh \zeta_1 b - \cos \zeta_2 b)(\zeta_2 \sinh \zeta_1 y - \zeta_1 \sin \zeta_2 y) - $$
$$-(\zeta_2 \sinh \zeta_1 b - \zeta_1 \sin \zeta_2 b)(\cosh \zeta_1 y - \cos \zeta_2 y)\}\sin \alpha_n x. \tag{4.2.143}$$

SS–C–SS–SS:

$$\zeta_1 \cosh \zeta_1 b \sin \zeta_2 b = \zeta_2 \sinh \zeta_1 b \cos \zeta_2 b, \tag{4.2.144}$$

$$W_{nm} = (\sin \zeta_2 b \sinh \zeta_1 y - \sinh \zeta_1 b \sin \zeta_2 y)\sin \alpha_n x. \tag{4.2.145}$$

SS–F–SS–F:

$$\sinh \zeta_1 b \sin \zeta_2 b[\zeta_1^2\{\beta^2 - \alpha_n^2(1-\nu)\}^4 - \zeta_2^2\{\beta^2 + \alpha_n^2(1-\nu)\}^4]$$
$$= 2\zeta_1\zeta_2\{\beta^4 - \alpha_n^4(1-\nu)^2\}^2(\cosh \zeta_1 b \cos \zeta_2 b - 1), \tag{4.2.146}$$

$$W_{nm} = (-(\cosh \zeta_1 b - \cos \zeta_2 b)\{\beta^4 - \alpha_n^4(1-\nu)^2\}\times$$
$$\times[\zeta_2\{\beta^2 + \alpha_n^2(1-\nu)\}\sinh \zeta_1 y + \zeta_1\{\beta^2 - \alpha_n^2(1-\nu)\}\sin \zeta_2 y] +$$
$$+[\zeta_2\{\beta^2 + \alpha_n^2(1-\nu)\}^2 \sinh \zeta_1 b - \zeta_1\{\beta^2 - \alpha_n^2(1-\nu)\}^2 \sin \zeta_2 b]\times$$
$$\times[\{\beta^2 - \alpha_n^2(1-\nu)\}\cosh \zeta_1 y + \{\beta^2 + \alpha_n^2(1-\nu)\}\cos \zeta_2 y])\sin \alpha_n x. \tag{4.2.147}$$

SS–F–SS–SS:

$$\zeta_1\{\beta^2 - \alpha_n(1-\nu)\}^2 \cosh \zeta_1 b \sin \zeta_2 b$$
$$= \zeta_2\{\beta^2 + \alpha_n^2(1-\nu)\}^2 \sinh \zeta_1 b \cos \zeta_2 b, \tag{4.2.148}$$

$$W_{nm} = [\{\beta^2 - \alpha_n^2(1-\nu)\}\sin \zeta_2 b \sinh \zeta_1 y + \{\beta^2 + \alpha_n^2(1-\nu)\}\times$$
$$\times \sinh \zeta_1 b \sin \zeta_2 y]\sin \alpha_n x. \tag{4.2.149}$$

SS–F–SS–C:

$$\zeta_1\zeta_2\{\beta^4 - \alpha_n^4(1-\nu)^2\} + \zeta_1\zeta_2\{\beta^4 + \zeta_n^4(1-\nu)^2\}\cosh \zeta_1 b \cos \zeta_2 b +$$
$$+\alpha_n^2\{\beta^4(1-2\nu) - \alpha_n^4(1-\nu)^2\}\sinh \zeta_1 b \sin \zeta_2 b = 0, \tag{4.2.150}$$

$$W_{nm} = ([\{\beta^2 + \alpha_n^2(1-\nu)\}\cosh \zeta_1 b + \{\beta^2 - \alpha_n^2(1-\nu)\}\cos \zeta_1 b]\times$$
$$\times(\zeta_1 \sin \zeta_2 y - \zeta_2 \sinh \zeta_1 y) + [\zeta_2\{\beta^2 + \alpha_n^2(1-\nu)\}\sinh \zeta_1 b +$$
$$+\zeta_1\{\beta^2 - \alpha_n^2(1-\nu)\}\sin \zeta_2 b](\cosh \zeta_1 y - \cos \zeta_2 y))\sin \alpha_n x. \tag{4.2.151}$$

All of the preceding solutions are for Case 1. For Case 2, the single root $\omega = (D/\rho h)^{\frac{1}{2}} n^2\pi^2/a^2$ holds. For Case 3, it has been shown that the first three combinations of edge conditions have no roots. However, the last three combinations do have roots for this case, and these have been shown to be dependent on Poisson's ratio. A detailed discussion of these last three cases, however, is beyond the scope of the present work.

Because of the several parameters entering into the plate frequency equation, the data on the frequency parameters for plates must be based on the aspect ratio a/b. Only in the case of a plate simply supported on all edges (SS–SS–SS–SS) is frequency information and mode shape easily obtainable. Figure 4.16 shows diagrams of the modes from ω_{11} to ω_{43} for such a plate.

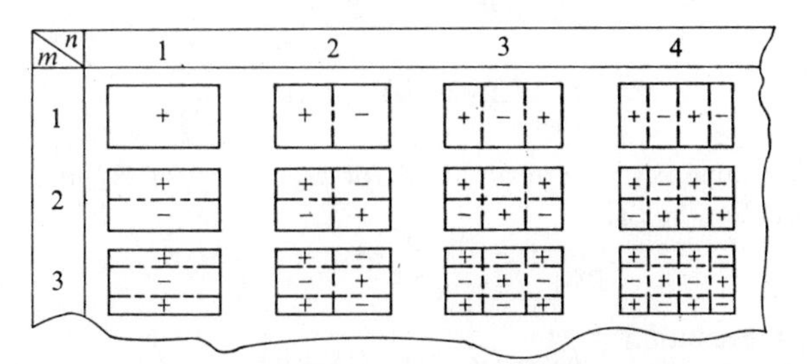

FIG. 4.16. Diagrams of the simply supported plate modes ω_{11} through ω_{43}.

2. *Vibrations of circular plates.* Let us now turn to the free vibrations of circular plates. We consider the governing eqn (4.2.25) with $q(r, \theta, t) = 0$ and assume a solution

$$w(r, \theta, t) = W(r, \theta)e^{-i\omega t}. \tag{4.2.152}$$

This leads to the result

$$\nabla^4 W(r, \theta) - \beta^4 W(r, \theta) = 0, \qquad \beta^4 = \omega^2 \rho h/D. \tag{4.2.153}$$

The solution to the above may be written as

$$W = W_1 + W_2, \tag{4.2.154}$$

where W_1, W_2 satisfy respectively

$$\nabla^2 W_1 + \beta^2 W_1 = 0; \qquad \nabla^2 W_2 - \beta^2 W_2 = 0. \tag{4.2.155}$$

Considering W_1 and assuming $W_1 = R_1\Theta_1$, we obtain

$$R_1''\Theta_1 + \frac{1}{r}R_1'\Theta_1 + \frac{1}{r^2}R_1\Theta_1'' + \beta^2 R_1\Theta_1 = 0. \tag{4.2.156}$$

Separation occurs if

$$\Theta_1'' = -\alpha^2\Theta_1. \tag{4.2.157}$$

That is,

$$\Theta_1 = \sin\alpha\theta,\ \cos\alpha\theta. \tag{4.2.158}$$

However, from continuity conditions that require

$$W(r, \theta) = W(r, \theta + 2\pi), \tag{4.2.159}$$

it is readily seen that $\alpha = n$, where n is an integer. Thus

$$\Theta_1 = \sin n\theta,\ \cos n\theta, \tag{4.2.160}$$

and

$$R_1''+\frac{1}{r}R_1'+\left(\beta^2-\frac{n^2}{r^2}\right)R_1 = 0. \tag{4.2.161}$$

This is Bessel's equation of order n, which has the solution

$$R_1 = AJ_n(\beta r)+BY_n(\beta r). \tag{4.2.162}$$

Thus

$$W_1 = \{AJ_n(\beta r)+BY_n(\beta r)\}\begin{bmatrix}\sin n\theta \\ \cos n\theta\end{bmatrix}. \tag{4.2.163}$$

For W_2 the same results are obtained for Θ_2 while, for R_2, the modified Bessel's equation holds,

$$R_2''+\frac{1}{r}R_2'-\left(\beta^2+\frac{n^2}{r^2}\right)R_2 = 0, \tag{4.2.164}$$

which has the solution

$$R_2 = CI_n(\beta r)+DK_n(\beta r). \tag{4.2.165}$$

In the above I_n, K_n are the modified Bessel functions of the first and second kind respectively, and are related to J_n, Y_n by

$$\mathrm{i}^n I_n(\beta r) = J_n(\mathrm{i}\beta r), \qquad K_n(\beta r) = Y_n(\mathrm{i}\beta r). \tag{4.2.166}$$

The total solution is thus

$$W(r,\theta) = \{AJ_n(\beta r)+BY_n(\beta r)+CI_n(\beta r)+DK_n(\beta r)\}\begin{bmatrix}\sin n\theta \\ \cos n\theta\end{bmatrix}. \tag{4.2.167}$$

If we restrict our attention to full circular plates, we must set $B = D = 0$ in the above since Y_n, K_n have singularities at $r = 0$. Thus we are reduced to

$$W(r,\theta) = \{AJ_n(\beta r)+CI_n(\beta r)\}\begin{bmatrix}\sin n\theta \\ \cos n\theta\end{bmatrix}. \tag{4.2.168}$$

There are three types of complete, circular plates having simple boundary conditions (that is, simply supported, clamped, or free on the circumference). Of the three, the clamped edge is the simplest. We have

$$W(a,\theta) = \partial W(a,\theta)/\partial r = 0. \tag{4.2.169}$$

Substituting (4.2.168) in these equations gives the following frequency equation for the clamped plate,

$$I_n(\beta a)J_n'(\beta a)-J_n(\beta a)I_n'(\beta a) = 0. \tag{4.2.170}$$

For each value of n, there will be an infinite number of roots to (4.2.172). We define the frequency parameter λ_{nm}, where

$$\lambda_{nm} = \beta_{nm}a, \tag{4.2.171}$$

n is the integer arising in (4.2.170), and m corresponds to the order of the root for a given n. Some of the first few values for λ_{nm} are

$$\begin{aligned} &\lambda_{01}^2 = 10.216, \quad &\lambda_{02}^2 = 39.771, \quad &\lambda_{03}^2 = 89.104, \\ &\lambda_{11}^2 = 21.26, \quad &\lambda_{12}^2 = 60.82, \quad &\lambda_{13}^2 = 120.08, \\ &\lambda_{21}^2 = 34.88, \quad &\lambda_{22}^2 = 84.58, \quad &\lambda_{23}^2 = 153.81. \end{aligned} \tag{4.2.172}$$

The normal modes are given by

$$W_{nm}(r, \theta) = \left\{J_n(\beta_{nm} r) - \frac{J_n(\beta_{nm} a)}{I_n(\beta_{nm} a)} I_n(\beta_{nm} r)\right\} \begin{bmatrix} \sin n\theta \\ \cos n\theta \end{bmatrix}. \tag{4.2.173}$$

A few of the mode shapes are shown in graphic form in Fig. 4.17.

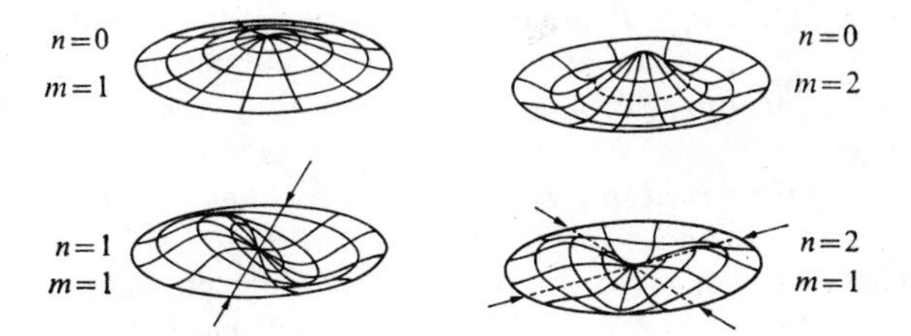

FIG. 4.17. Graphic illustration of the first few normal modes of a clamped circular plate. (After Morse [12].)

Consideration of a plate simply supported at $r = a$ yields a more complicated transcendental equation, as does the plate with a free edge. Plates with circular concentric holes may be included easily in the study. For such a case, all four terms of the solution are retained since the singular behaviour of Y_m, K_m at $r = 0$ no longer enters in. It is then necessary to specify two additional boundary conditions at the inner radius $r = b$, say.

3. *Orthogonality*. We now wish to develop the orthogonality property for the plate normal modes. We will use the rectangular plate modes for the development, although the arguments may be applied to other cases. Thus, consider the mode $W_{ij}(x, y)$ associated with the ω_{ij} natural frequency of the plate and the W_{kl} mode associated with ω_{kl}. Thus W_{ij}, W_{kl} satisfy

$$\nabla^4 W_{ij} - \beta_{ij}^4 W_{ij} = 0, \qquad \nabla^4 W_{kl} - \beta_{kl}^4 W_{kl} = 0. \tag{4.2.174}$$

Now to proceed in the usual manner to investigate whether orthogonality exists between the normal modes would require subtraction of the above two equations and integration over the region $0 \leq x \leq a$, $0 \leq y \leq b$. Considerable manipulation would be required to reach the desired results. This procedure can be circumvented if we take the following viewpoint. The equation

$$\nabla^4 W = \beta^4 W \tag{4.2.175}$$

can be considered as the equation for the deflection surface produced by the load $q = D\beta^4 W$ since, in static plate theory, the differential equation has the form

$$\nabla^4 W = q/D. \tag{4.2.176}$$

Thus, under the load $q_{ij} = D\beta_{ij}^4 W_{ij}(x, y)$, the plate assumes the deflected form $W_{ij}(x, y)$. Likewise for $q_{kl} = D\beta_{kl}^4 W_{kl}(x, y)$. We now wish to apply Betti's reciprocal theorem to this problem.

Reviewed in brief, the Betti reciprocal theorem considers an elastic body under two loads P_1 and P_2. Under P_1 alone, deflections δ_1 and δ_2 are produced, where δ_1 is in the direction of P_1 and δ_2 is in the direction of P_2, if it were acting. Under the action of P_2 alone, δ_1' and δ_2' are produced, with δ_1' being in the direction of P_1 if it were acting and δ_2' is in the direction of P_2. Then, according to the theorem, $P_1\delta_1' = P_2\delta_2$. In more formal terms, the theorem states that: *the work done by forces of the first state on the corresponding displacements of the second state is equal to the work done by the forces of the second state on the corresponding displacements of the first.*

The application of this theorem in the context of the present problem is as follows: the loading $q_{ij} = D\beta_{ij}^4 W_{ij}$ is analogous to P_1 while the deflection W_{kl} is analogous to δ_1' and $q_{kl} = D\beta_{kl}^4 W_{kl}$ is analogous to P_2 with W_{ij} corresponding to δ_2. To obtain a work expression, integration over the area of the plate is required. Thus

$$D\beta_{ij}^4 \int_0^b\int_0^a W_{ij}W_{kl}\,dx\,dy = D\beta_{kl}^4 \int_0^a\int_0^b W_{ij}W_{kl}\,dx\,dy. \tag{4.2.177}$$

This gives

$$(\beta_{ij}^4 - \beta_{kl}^4)\int_0^b\int_0^a W_{ij}W_{kl}\,dx\,dy = 0 \tag{4.2.178}$$

which, since $\beta_{ij} \neq \beta_{kl}$, establishes the orthogonality property

$$\int_0^a\int_0^b W_{ij}(x, y)W_{kl}(x, y)\,dx\,dy = 0. \tag{4.2.179}$$

It should be noted that this orthogonality condition in no way depends on the separability property $W = X(x)Y(y)$. The general motion of the plate is then conveniently expressed in the form

$$w(x, y, t) = \sum_{n=0}^{\infty}\sum_{m=1}^{\infty} q_{nm}(t)W_{nm}(x, y), \tag{4.2.180}$$

which is, of course, the logical extension of the normal-mode expansion method first encountered in string and beam problems. The initial-value problem in plate vibrations takes such a form. The orthogonality property of the normal modes enables the coefficients of the expansion to be determined, at least in theory. Of course the complicated nature of the modes and the two-dimensional nature of the expansion makes solution of even seemingly simple problems in this area quite tedious.

It was mentioned earlier that the literature on the vibrations of plates is vast. Only the briefest of results for two particular plate geometries (rectangular and circular), subject to only the simplest types of boundary constraints, have been presented. Little in the way of numerical data has been given. While this is the only practical way to proceed in a book devoted primarily to wave propagation, it should be noted that it is possible to resolve the bulk of the literature for the vibrations of homogeneous, isotropic plates into the major geometrical categories of *circular*, *elliptical*, *rectangular*, *parallelogram*, *trapezoidal*, and *triangular*, with the subject of rectangular plates receiving the greatest coverage. Reference should be made to a most useful compilation of data on plate frequencies reported by Leissa [8]. This reference presents the numerical results from several hundred papers and reports.

4.2.8. *Experimental results on waves in plates*

Several studies have been made on transient wave propagation in plates of which only a few will be mentioned here. One of the first studies was that by Dohrenwend, Drucker, and Moore [3], cited also in § 3.6 in the discussion of transient waves in beams. In that study, a large steel plate, $\frac{1}{2}$ in. thick was struck a blow with a heavy sledge hammer and the radial strain at various distances from the point of impact was measured using wire resistance strain gauges. The experimental results are shown in Fig. 4.18(a). The theoretical response of the plate was determined using an assumed initial velocity input of $v_0 \exp(-r^2/4b^2)$ and an analysis based on classical plate theory. The predicted response is shown in Fig. 4.18(b). The distance from point of impact is r and the plate thickness is h. It should be noted that this study represented one of the earlier applications of strain gauges to the measurement of sharp transients.

Press and Oliver [14] obtained results for flexural waves excited by a spark source. Their study was aimed at the broader aspects of air-coupled surface waves, with the concentrated load results being only one of several cases studied. A diagram of the experimental apparatus and the results obtained are shown in Fig. 4.19. The spark source is S and the detector, a piezoelectric device, is D. The plate was aluminium, $\frac{1}{32}$ in. (0·08 cm) thick and several feet in breadth. The baffle B was placed close to the source to prevent air-coupling over the distance d. It is seen that the response is quite similar to the far field response observed in beams.

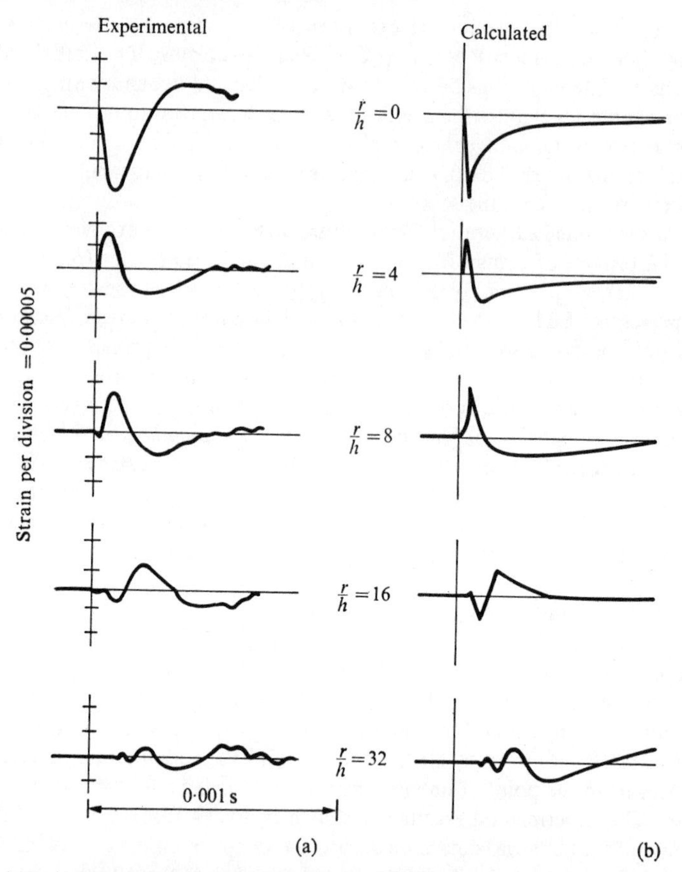

FIG. 4.18. (a) Experimental and (b) theoretical values of radial strain in a plate subjected to impact. (After Dohrenwend *et al.* [3, Fig. 10].)

The work of Medick [10] has been cited previously in this section (see, for example, Fig. 4.14) for the analysis of the transverse impact of plates. He also performed an experimental investigation of this problem using an experimental apparatus as shown in Fig. 4.20(a). Thus, a capacitance pick-up was used to measure the transient plate displacement. Typical displacement records are shown in Fig. 4.20(b). The experimental results were compared with theoretical predictions for $r = 2{\cdot}2$ in., 4·1 in., and 10·1 in. (5·59 cm, 10·91 cm, 25·65 cm) from the impact point. The results for the case of 10·1 in. is shown in Fig. 4.21. The agreement between the results is seen to be rather good. It should be recalled that the Medick analysis was based on classical plate theory. It is seen that the theoretical maxima exceed the experimental maxima

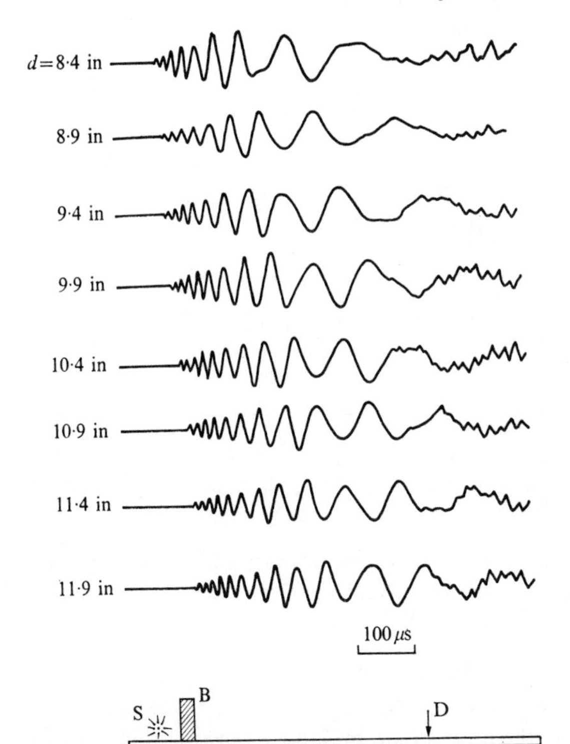

FIG. 4.19. Flexural waves excited in a thin plate by a spark source S. (After Press and Oliver [14, Fig. 1].)

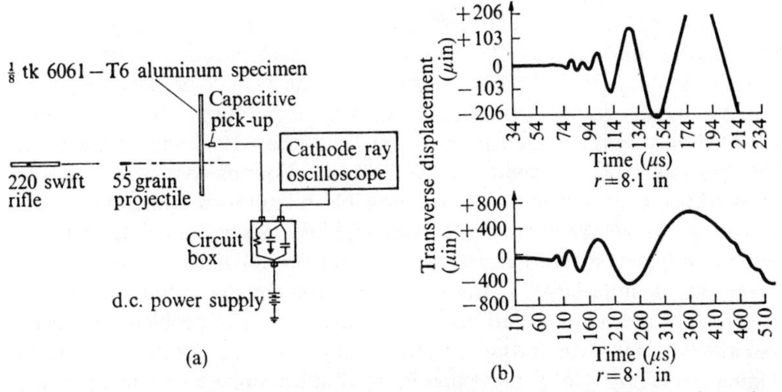

FIG. 4.20. (a) Experimental setup used to measure the transient displacement in a plate subjected to transverse impact and (b) typical displacement records. (After Medick [10, Figs. 3 and 4].)

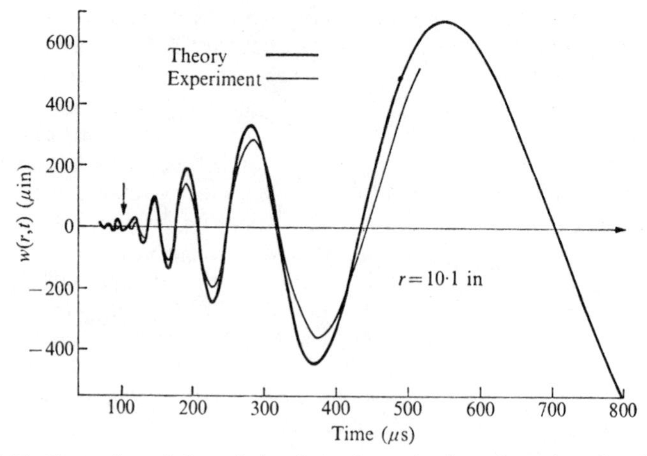

FIG. 4.21. Comparison of theoretical and experimental values of transient plate displacement for $r = 10{\cdot}1$ in. (25·65cm) (After Medick [10, Fig. 7].)

somewhat, while the phase agreement is very good. For the case of $r = 4{\cdot}1$ in. (not shown), experimental maxima slightly exceeded theoretical values and some phase shift was noted.

4.3. Waves in thin, cylindrical shells

In the investigations thus far, we have progressed from transverse waves in taut strings to the propagation of flexural waves in thin plates. Dimensionally, we have proceeded from one-dimensional to two-dimensional problems. Nearly all structural elements considered have been straight and flat, however. The situation may be further complicated by the addition of curvature, as was found in the case of rings. Thus rods and beams may become rings and plates may become shells. It is in this area that our considerations now lie.

With the additional element of curvature, the variety of elastic systems becomes unmanageably large. Thus, in the case of (uniform) rods and beams, one need only consider the length parameter of the system and its means of end-constraint, a fairly denumerable situation. For the case of plates, the shape of the plate enters as well as possible mixtures of constraint about the periphery, and the number of possibilities becomes unlimited. Even in the context of practical considerations, the number of plate geometries and constraints is quite large. It thus should be evident that adding curvature effects to the preceding situations easily creates a class of problems in elastic systems that must be brought into manageable proportions by brutal elimination. This is done in the present section by confining attention to the case of thin cylindrical shells. Fortunately, this matter of expediency covers the geometry of greatest practical importance.

The matter of the governing equations for shells should be remarked upon. Whereas the governing equations for longitudinal and flexural waves in rods and plates seem to be well agreed upon by the 'authorities', the situation is not so in the case of shells. One thus finds a large number of shell theories, usually being differentiated by slight differences in the kinematics of deformation, inertial contributions, or curvature approximations. The approach taken here to this problem will be to present the simplest shell equations, saving for a later time the more complicated theories that may be developed using procedures similar to those that will be presented for plates and rods.

The theory used in this section is known as the membrane shell theory. Under this development only forces, normal and shear, acting in the mid surface of the shell are considered. The transverse shear forces and the bending and twisting moments are assumed negligibly small. Thus the shell behaves as a curved membrane. If there is a shell analogue of the flat membrane, one expects, in analogy to the flat plate, a shell theory where moments and transverse shear forces act. The classical development in this area was put forth by Love.† Many slight modifications of this theory have appeared by Flügge [5], Vlasov [18], Donnell [4], Sanders [15], and others. One is referred to the text by Kraus [7] for extensive discussion and review of the theories, as well as numerous analyses in the statics and vibrations of shells. It is the proliferation of theories in this area that can lead to more confusion. It is characteristic of all the aforementioned developments that shear deformations are neglected, as in Bernoulli–Euler beam theory and classical plate theory. Developments of 'Timoshenko' shell theories are associated with Hildebrand *et al.* [6], Naghdi [13], and others.

With the extensive developments in shell theories, our restriction to membrane theory seems pale by comparison. There are two justifications for the present restriction, however. First, being the simplest theory, the essential features are easily presented and results obtained. More importantly, the more complicated theories associated with Love and others contain the essential weakness, from the standpoint of wave propagation and sharp transients, of neglecting shear effects. Drawing from the experience in beam analysis, one knows, practically *a priori*, that a shear theory will be required. One is justified, therefore, in considering such a theory almost immediately. These developments, which generally take the exact equations of elasticity as their starting point, are best presented in detail in Chapter 8. With this lengthy introduction, we now proceed to the study of waves in cylindrical membrane shells.

4.3.1. *Governing equations for a cylindrical membrane shell*

It is, of course, possible to develop governing equations for shells of arbitrary curvature and then to specialize the results to a particular geometry

† Chapter 23, 24, 24A of Reference [9].

such as cylindrical. If a variety of geometries are being considered, such an effort would be worthwhile. Since our attention will be confined to cylindrical membrane shells, it is much more expedient to develop the equations directly in terms of the cylindrical geometry.

Thus, in Fig. 4.22(a), a strip from a cylindrical shell, is shown indicating the

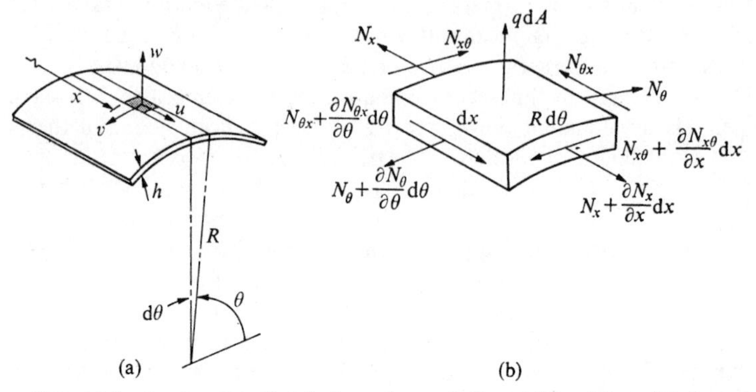

FIG. 4.22. (a) Section from a cylindrical membrane shell, and (b) a differential element of shell.

coordinates x (along the shell) and θ (the polar angle). The displacement components in the radial (w), longitudinal (u), and tangential (v) directions are shown on the shaded differential element. The forces per unit length on the element are shown in Fig. 4.22(b). Because only membrane action is being assumed, there are no bending moments or transverse shear forces. The equations of motion in the longitudinal, tangential, and radial directions are, respectively,

$$-N_x R\,d\theta+\left(N_x+\frac{\partial N_x}{\partial x}\,dx\right)R\,d\theta-N_{\theta x}\,dx+\left(N_{\theta x}+\frac{\partial N_{\theta x}}{\partial \theta}\,d\theta\right)dx$$

$$=\rho R\,d\theta\,dx h\frac{\partial^2 u}{\partial t^2},\quad (4.3.1)$$

$$-N_\theta\,dx+\left(N_\theta+\frac{\partial N_\theta}{\partial \theta}\,d\theta\right)dx-N_{x\theta}R\,d\theta+\left(N_{x\theta}+\frac{\partial N_{x\theta}}{\partial x}\,dx\right)R\,d\theta$$

$$=\rho R\,d\theta\,dx h\frac{\partial^2 v}{\partial t^2},\quad (4.3.2)$$

$$-N_\theta\frac{d\theta}{2}\,dx-\left(N_\theta+\frac{\partial N_\theta}{\partial \theta}\,d\theta\right)\frac{d\theta}{2}\,dx+qR\,d\theta\,dx=\rho R\,d\theta\,dx h\frac{\partial^2 w}{\partial t^2}.\quad (4.3.3)$$

These yield the equations of motion for a cylindrical membrane shell as

$$\frac{\partial N_x}{\partial x}+\frac{1}{R}\frac{\partial N_{\theta x}}{\partial \theta} = \rho h\frac{\partial^2 u}{\partial t^2}$$

$$\frac{1}{R}\frac{\partial N_\theta}{\partial \theta}+\frac{\partial N_{x\theta}}{\partial x} = \rho h\frac{\partial^2 v}{\partial t^2} \tag{4.3.4}$$

$$-\frac{N_\theta}{R}+q = \rho h\frac{\partial^2 w}{\partial t^2}.$$

The expressions for the membrane stresses N_x, N_θ, $N_{x\theta}$, $N_{\theta x}$ are obtained by integrating the usual stresses across the shell thickness. Thus we may write the general formula

$$N_x, N_\theta, N_{x\theta}, N_{\theta x} = \int_{-h/2}^{h/2} (\sigma_x, \sigma_\theta, \tau_{x\theta}, \tau_{\theta x})\,\mathrm{d}z, \tag{4.3.5}$$

where z is the distance measured outward from the mid surface of the shell. It is the same as the coordinate introduced in the curved-rod development (see Fig. 3.18). From Hooke's law, we have that

$$\sigma_x = \frac{E}{1-\nu^2}(\varepsilon_x+\nu\varepsilon_\theta), \qquad \sigma_\theta = \frac{E}{1-\nu^2}(\varepsilon_\theta+\nu\varepsilon_x), \qquad \tau_{x\theta} = \tau_{\theta x} = \gamma G, \tag{4.3.6}$$

where ε_x, ε_θ are the axial strains of the middle surface of the shell element and γ is the shear strain of the element with $\gamma = \gamma_{x\theta} = \gamma_{\theta x}$. Under the assumption of membrane-type stresses only, the stresses σ_x, σ_θ, $\tau_{x\theta}$, $\tau_{\theta x}$ are constant through the shell thickness, so that (4.3.5) becomes

$$N_x = \frac{Eh}{1-\nu^2}(\varepsilon_x+\nu\varepsilon_\theta), \qquad N_\theta = \frac{Eh}{1-\nu^2}(\varepsilon_\theta+\nu\varepsilon_x),$$

$$N_{x\theta} = N_{\theta x} = Gh\gamma = \frac{Eh}{2(1+\nu)}\gamma. \tag{4.3.7}$$

The kinematics of deformation must now be considered. Again, under the present conditions of membrane behaviour and cylindrical shape, these aspects are quite simple. In the axial direction, we have, exactly as in longitudinal rod theory, that

$$\varepsilon_x = \partial u/\partial x. \tag{4.3.8}$$

For ε_θ we have $\varepsilon_\theta = (\mathrm{d}s'-\mathrm{d}s)/\mathrm{d}s$, where $\mathrm{d}s = R\,\mathrm{d}\theta$ is the initial length of the arc segment. Now, an expression for $\mathrm{d}s'$, the length of an arc segment after deformation, was previously obtained in the analysis of waves in rings (see Fig. 3.19 and (3.5.13)), and this result is directly applicable here. Thus, we have that $ds' = (w+R)\,d\theta+(\partial v/\partial\theta)\,d\theta$. The result is that

$$\varepsilon_\theta = \frac{1}{R}\left(w+\frac{\partial v}{\partial\theta}\right). \tag{4.3.9}$$

The expression for the shear strain results directly from considering small changes in angle of the sides dx and $R\,d\theta$ of the element due to $\partial v/\partial x$ and $\partial u/\partial\theta$. We have that

$$\gamma = \frac{\partial v}{\partial x}+\frac{1}{R}\frac{\partial u}{\partial\theta}. \tag{4.3.10}$$

The membrane stresses are thus given by

$$\begin{gathered}
N_x = \frac{Eh}{1-\nu^2}\left\{\frac{\partial u}{\partial x}+\frac{\nu}{R}\left(w+\frac{\partial v}{\partial\theta}\right)\right\},\\
N_\theta = \frac{Eh}{1-\nu^2}\left(\frac{w}{R}+\frac{1}{r}\frac{\partial v}{\partial\theta}+\nu\frac{\partial u}{\partial x}\right),\\
N_{x\theta} = N_{\theta x} = \frac{Eh}{2(1+\nu)}\left(\frac{\partial v}{\partial x}+\frac{1}{R}\frac{\partial u}{\partial\theta}\right).
\end{gathered} \tag{4.3.11}$$

The displacement equations of motion result from substituting (4.3.11) in (4.3.4). If we note that the coefficient of $N_{x\theta}$ may be written as

$$Eh(1-\nu)/2(1-\nu^2),$$

then we have

$$\left\{\frac{\partial^2 u}{\partial x^2}+\frac{\nu}{R}\left(\frac{\partial w}{\partial x}+\frac{\partial^2 v}{\partial x\,\partial\theta}\right)\right\}+\frac{(1-\nu)}{2R}\left(\frac{\partial^2 v}{\partial\theta\,\partial x}+\frac{1}{R}\frac{\partial^2 u}{\partial\theta^2}\right) = \frac{\rho(1-\nu^2)}{E}\frac{\partial^2 u}{\partial t^2}, \tag{4.3.12}$$

$$\frac{1}{R}\left(\frac{1}{R}\frac{\partial w}{\partial\theta}+\frac{1}{R}\frac{\partial^2 v}{\partial\theta^2}+\nu\frac{\partial^2 u}{\partial\theta\,\partial x}\right)+\frac{(1-\nu)}{2}\left(\frac{\partial^2 v}{\partial x^2}+\frac{1}{R}\frac{\partial^2 u}{\partial x\,\partial\theta}\right) = \frac{\rho(1-\nu^2)}{E}\frac{\partial^2 v}{\partial t^2}, \tag{4.3.13}$$

$$-\frac{1}{R}\left(\frac{w}{R}+\frac{1}{R}\frac{\partial v}{\partial\theta}+\nu\frac{\partial u}{\partial x}\right)+\frac{1-\nu^2}{Eh}q = \frac{\rho(1-\nu^2)}{E}\frac{\partial^2 w}{\partial t^2}. \tag{4.3.14}$$

We recall the remark that cylindrical membrane shells represent about the simplest theory for the simplest shell geometry. Yet even under these circumstances, coupled equations of some complexity arise.

The governing equations for a shell, including bending effects on the deformation and bending moments and shear forces in the equations of motion, yield considerably more complicated equations than the preceding. However, the Donnell formulation,† including these effects with, additionally, some slight simplifications related to the influence of transverse shearing force on tangential motion, and related to the expressions for curvature and twist, yields equations quite like (4.3.12)–(4.3.14). There is, in fact, only the additional term $(-h^2\nabla^4 w/12)$ on the left-hand side of (4.3.14).

† See pp. 200–4 of Reference [7].

4.3.2. *Wave propagation in the shell*

We wish to consider the propagation of harmonic waves in the membrane shell. We could immediately substitute harmonic wave expressions for u, v, w in eqns (4.3.12)–(4.3.14) and obtain a frequency equation. A rather complicated expression would result, however, and interpretation of the results would be difficult. A better approach is to consider various special modes of motion and study the simplified frequency equation.

One of the most important special cases results from considering motion independent of θ.† Thus, if $\partial/\partial\theta = 0$ (also set $q = 0$) in the displacement equations of motion we obtain

$$\frac{\partial^2 u}{\partial x^2}+\frac{\nu}{R}\frac{\partial w}{\partial x} = \frac{\rho(1-\nu^2)}{E}\frac{\partial^2 u}{\partial t^2}, \tag{4.3.15}$$

$$\frac{(1-\nu)}{2}\frac{\partial^2 v}{\partial x^2} = \frac{\rho(1-\nu^2)}{E}\frac{\partial^2 v}{\partial t^2}, \tag{4.3.16}$$

$$-\frac{w}{R^2}-\frac{\nu}{R}\frac{\partial u}{\partial x} = \frac{\rho(1-\nu^2)}{E}\frac{\partial^2 w}{\partial t^2}. \tag{4.3.17}$$

The first observation to make is that the equation for the tangential motion (4.3.16) has uncoupled from the remaining two equations. This equation may be written as

$$\frac{\partial^2 v}{\partial x^2} = \frac{2\rho(1+\nu)}{E}\frac{\partial^2 v}{\partial t^2}. \tag{4.3.18}$$

Since $G = E/2(1+\nu)$, we have the result

$$\frac{\partial^2 v}{\partial x^2} = \frac{1}{c_s^2}\frac{\partial^2 v}{\partial t^2}, \qquad c_s = \sqrt{\left(\frac{G}{\rho}\right)}. \tag{4.3.19}$$

This is the familiar wave equation, and it is governing the purely torsional motion of the shell. We note the propagation velocity of the torsional disturbance is $\sqrt{(G/\rho)}$, the same as found for such waves in a solid circular rod (see § 2.6).

Now consider the motion as governed by the remaining coupled equations in u and w. We let

$$u = Ae^{i(\gamma x-\omega t)}, \qquad w = Be^{i(\gamma x-\omega t)}, \tag{4.3.20}$$

and substitute in (4.3.15) and (4.3.17). This gives

$$\begin{bmatrix} \left(\dfrac{\omega^2}{c_p^2}-\gamma^2\right) & i\gamma\dfrac{\nu}{R} \\ -i\gamma\dfrac{\nu}{R} & \left(\dfrac{\omega^2}{c_p^2}-\dfrac{1}{R^2}\right) \end{bmatrix} \begin{bmatrix} A \\ B \end{bmatrix} = 0, \tag{4.3.21}$$

† You are asked to consider x-independent motion in Problem 4.16.

where the 'thin-plate' velocity has been introduced, given by†

$$c_p = (E/(1-\nu^2)\rho)^{\frac{1}{2}}. \tag{4.3.22}$$

The determinant of coefficients of (4.3.21) gives the frequency equation

$$\omega^4 - c_p^2\left(\gamma^2 + \frac{1}{R^2}\right)\omega^2 + (1-\nu^2)\frac{c_p^4}{R^2}\gamma^2 = 0. \tag{4.3.23}$$

In terms of phase velocity and wavenumber, we have

$$c^4 - c_p^2\left(1 + \frac{1}{R^2\gamma^2}\right)c^2 + (1-\nu^2)\frac{c_p^4}{R^2\gamma^2} = 0. \tag{4.3.24}$$

Introducing the non-dimensionalized quantities

$$\bar{c} = c/c_p, \quad \bar{\gamma} = h\gamma, \quad \bar{h} = h/R, \quad \bar{\omega} = h\omega/c_p, \tag{4.3.25}$$

the above equations reduce to

$$\bar{\omega}^4 - (\bar{h}^2 + \bar{\gamma}^2)\bar{\omega}^2 + (1-\nu^2)\bar{h}^2\bar{\gamma}^2 = 0, \tag{4.3.26}$$

and, using $\bar{\omega} = \bar{\gamma}\bar{c}$,

$$\bar{c}^4 - (1 + \bar{h}^2/\bar{\gamma}^2)\bar{c}^2 + (1-\nu^2)\bar{h}^2/\bar{\gamma}^2 = 0. \tag{4.3.27}$$

The long- and short-wavelength limits are easily obtained. At long wavelengths, we have

$$\begin{aligned} &\lim_{\bar{\gamma}\to 0}(4.3.26) = \bar{\omega}^4 - \bar{h}^2\bar{\omega}^2 = 0, \qquad \bar{\omega} = 0, \bar{h}, \\ &\lim_{\bar{\gamma}\to 0}(4.3.27) = \frac{1}{\bar{\gamma}^2}\left\{\bar{c}^4\bar{\gamma}^2 - \bar{h}^2\bar{c}^2 + (1-\nu^2)\bar{h}^2\right\} = 0. \end{aligned} \tag{4.3.28}$$

In the last case, if $\bar{c}^2$ is finite, we must have the result that $\bar{c}^2 = (1-\nu^2)$. If one inspects this result in its dimensional form, it gives $c(\gamma \to 0) = (E/\rho)^{\frac{1}{2}}$, which is the longitudinal bar velocity. Further, we have

$$\lim_{\bar{\gamma}\to\infty}(4.3.27) = \bar{c}^4 - \bar{c}^2 = 0, \qquad \bar{c} = 0, 1. \tag{4.3.29}$$

We also note, as $R \to \infty$, that $\bar{h} \to 0$ and the result (4.3.29) is again recovered from (4.3.27). The complete dispersion curves are shown in Fig. 4.23 for two different curvature ratios $\bar{h}$ and for a material having a Poisson's ratio of 0·3.

To obtain more general results on the propagation characteristics of waves in the longitudinal direction, we consider solutions for u, v, w of the form $f(\theta)\exp\{i(\gamma x - \omega t)\}$. Conditions on continuity of the displacements for θ and $\theta + 2\pi$, similar to those arising in the analysis of circular membranes and plates, requires that $f(\theta) = \sin n\theta, \cos n\theta$. In order for separation of variables to occur, inspection of the governing equations (4.3.12)–(4.3.14) shows that

† See Problem 4.13.

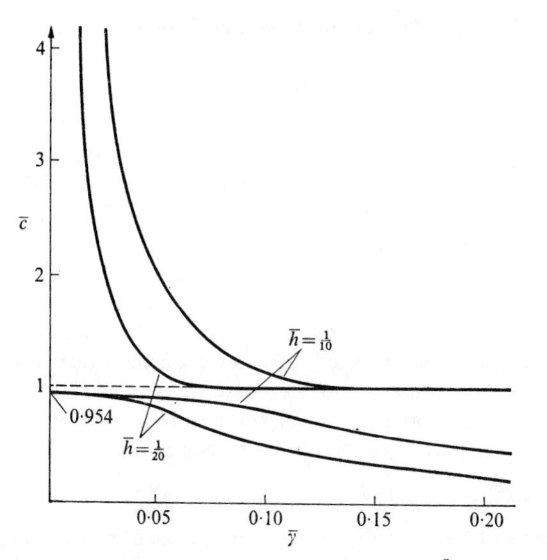

FIG. 4.23. Dispersion curves for a membrane shell for $\bar{h} = \frac{1}{10}$ and $\frac{1}{20}$

u, v, w must have the form

$$u = A\cos n\theta e^{i(\gamma x-\omega t)}, \qquad v = B\sin n\theta e^{i(\gamma x-\omega t)},$$
$$w = C\cos n\theta e^{i(\gamma x-\omega t)}. \tag{4.3.30}$$

A solution with the $\sin n\theta$, $\cos n\theta$ terms interchanged would also be valid. Proceeding to substitute (4.3.30) in the governing equations gives, after some simplification, rearrangement and the introduction of the previous non-dimensionalizations (4.3.25),

$$\begin{bmatrix} \left\{\bar{\omega}^2-\bar{\gamma}^2-\frac{(1-\nu)\bar{h}^2n^2}{2}\right\} & \frac{i(1+\nu)\bar{h}n\bar{\gamma}}{2} & i\nu\bar{h}\bar{\gamma} \\ \frac{i(1+\nu)}{2}\bar{h}n\bar{\gamma} & -\left\{\bar{\omega}^2+\bar{h}^2n^2-\frac{(1-\nu)}{2}\bar{\gamma}^2\right\} & \bar{h}^2n \\ i\nu\bar{h}\bar{\gamma} & n\bar{h}^2 & -(\bar{\omega}^2-\bar{h}^2) \end{bmatrix} \begin{bmatrix} A \\ B \\ C \end{bmatrix} = 0. \tag{4.3.31}$$

We first note that the case of $n = 0$ gives

$$\left\{\bar{\omega}^2-\frac{(1-\nu)}{2}\bar{\gamma}^2\right\}\left\{(\bar{\omega}^2-\bar{\gamma}^2)(\bar{\omega}^2-\bar{h}^2)-\nu^2\bar{h}^2\bar{\gamma}^2\right\} = 0. \tag{4.3.32}$$

The second bracketed term in the above is the previously-studied frequency eqn (4.3.25) for the axisymmetric longitudinal modes. The first bracketed term gives $\bar{\omega}^2 = (1-\nu)\bar{\gamma}^2/2$ or, in dimensional form,

$$\omega^2 = c_p^2\frac{(1-\nu)}{2}\gamma^2 = \frac{G}{\rho}\gamma^2, \tag{4.3.33}$$

or, using $\omega = \gamma c$, we have that $c = c_s$. This is the result for the torsional mode. Of course, if $n = 0$ the torsional term in (4.3.30) vanishes. However, this would not be the case if the alternate solution involving the interchanged sines and cosines were used.

Investigation of the general result (4.3.31) for $n \neq 0$ becomes rather complex. The general motion of the membrane shell is transverse, such as in the transverse motion of a string. However, the deformation of the cross-section can be fairly complicated, depending on the value of n. While this aspect of analysis will not be further pursued, we remark that the study of waves in circular rods and shells according to the exact theory (Chapter 8) will have many similarities to the membrane-shell analysis.

4.3.3. *Longitudinal impact of a membrane shell*

The analysis of transient disturbances in the membrane shell is, of course, difficult, due to the complicated nature of the governing equations. Moreover, there does not appear to have been a great deal of analysis in this area. The coverage here will be restricted to a brief review and presentation of results obtained by Berkowitz [2] on the longitudinal impact of a semi-infinite cylindrical shell against a rigid surface. The basic situation is shown in Fig. 4.24 at the instant of impact of the shell, travelling to the left with velocity

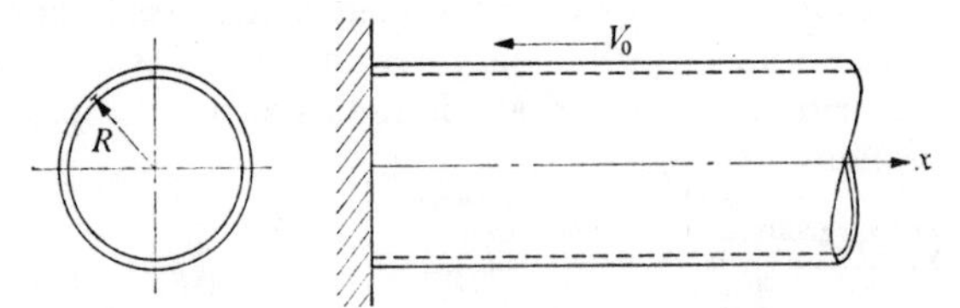

FIG. 4.24. Longitudinal impact of a cylindrical shell against a rigid barrier.

V_0, with the rigid surface.

Axisymmetric conditions pertain, so that the governing equations (4.3.15) and (4.3.17) describe the motion. The initial conditions of the problem are given by

$$u(x, 0) = w(x, 0) = \frac{\partial w(x, 0)}{\partial t} = 0, \qquad \frac{\partial u(x, 0)}{\partial t} = -V_0, \tag{4.3.34}$$

and the boundary conditions are

$$\partial u(0, t)/\partial t = 0, \qquad \partial u(\infty, t)/\partial t = -V_0. \tag{4.3.35}$$

Taking the Laplace transform of (4.3.15) and (4.3.17) gives

$$\frac{\mathrm{d}^2\bar{u}}{\mathrm{d}x^2}+\frac{\nu}{R}\frac{\mathrm{d}\bar{w}}{\mathrm{d}x} = \frac{s^2}{c_p^2}\bar{u}+\frac{V_0}{c_p^2},$$
$$-\frac{1}{R^2}\bar{w}-\frac{\nu}{R}\frac{\mathrm{d}\bar{u}}{\mathrm{d}x} = \frac{s^2}{c_p^2}\bar{w}, \tag{4.3.36}$$

where $\bar{u} = \bar{u}(x, s)$, $\bar{w} = \bar{w}(x, s)$, and the plate velocity c_p previously used has been introduced. Eliminating $\bar{w}$ in the first equation gives

$$\frac{\mathrm{d}^2\bar{u}}{\mathrm{d}x^2}-s^2\zeta^2\bar{u} = V_0\zeta^2, \tag{4.3.37}$$

where

$$\zeta^2 = \frac{s^2+c_p^2/R^2}{c_p^2(s^2+c_0^2/R^2)}, \qquad c_0^2 = \frac{E}{\rho}. \tag{4.3.38}$$

This has the solution

$$\bar{u} = A\mathrm{e}^{-s\zeta x}+B\mathrm{e}^{s\zeta x}-\frac{V_0}{s^2}. \tag{4.3.39}$$

The transformed boundary conditions (4.3.35) are

$$\bar{u}(0, s) = 0, \quad \bar{u}(\infty, s) = -V_0/s^2. \tag{4.3.40}$$

These conditions give $B = 0$ and the resulting transformed solution

$$\bar{u} = -\frac{V_0}{s^2}+\frac{V_0}{s^2}\mathrm{e}^{-s\zeta x}. \tag{4.3.41}$$

The solution for $\bar{w}$ is obtained by substituting the result for $\bar{u}$ in the second of (4.3.36), giving

$$\bar{w} = \frac{\nu V_0}{R}\frac{\zeta\mathrm{e}^{-s\zeta x}}{s(s^2+c_p^2/R^2)}. \tag{4.3.42}$$

The transformed expression for the axial shell stress is

$$\bar{N}_x = \frac{Eh}{1-\nu^2}\left(\frac{\mathrm{d}\bar{u}}{\mathrm{d}x}+\frac{\nu}{R}\bar{w}\right). \tag{4.3.43}$$

Substituting the transformed solutions for $\bar{u}$, $\bar{w}$ in $\bar{N}_x$ gives

$$\bar{N}_x = \frac{-EhV_0}{(1-\nu^2)c_p^2}\frac{\mathrm{e}^{-s\zeta x}}{s\zeta}. \tag{4.3.44}$$

We shall consider only the inversion of $\bar{N}_x$. We have

$$N_x(x, t) = \frac{-EhV_0}{(1-\nu^2)c_p^2}\frac{1}{2\pi\mathrm{i}}\int_{c-\mathrm{i}\infty}^{c+\mathrm{i}\infty}\frac{\mathrm{e}^{s(t-\zeta x)}}{s\zeta}\,\mathrm{d}s. \tag{4.3.45}$$

Thus an evaluation of the integral I is sought, where

$$I = \int_{c-i\infty}^{c+i\infty} \frac{e^{s(t-\zeta x)}}{s\zeta} ds. \tag{4.3.46}$$

Applying the same arguments as used in the Timoshenko beam analysis (§ 3.4), it is seen that the Bromwich contour must be closed to the right for $t-\zeta x < 0$ and to the left for $t-\zeta x > 0$. For large s, inspection of the expression (4.3.38) for ζ shows that $\zeta \to 1/c_p$. Since (as it turns out) there are no poles for Re $s > c$, the result is that

$$I = 0, \qquad t < x/c_p. \tag{4.3.47}$$

This establishes the arrival time of the first possible disturbance. It corresponds to the previously-established short-wavelength limit velocity for the shell.

It is seen that the integrand of I has a simple pole at $s = 0$. Again considering ζ, we may write

$$\zeta = \frac{(s+ic_p/R)^{\frac{1}{2}}(s-ic_p/R)^{\frac{1}{2}}}{c_p(s+ic_0/R)^{\frac{1}{2}}(s-ic_0/R)^{\frac{1}{2}}}. \tag{4.3.48}$$

Thus branch points are located at $\pm ic_0/R$, $\pm ic_p/R$. Appropriate branch cuts are introduced to make the integrand single-valued. The resulting form of the contour, closed to the left and deformed about the branch cuts, is shown in Fig. 4.25. Now the integral along the arc AB may be shown to vanish.

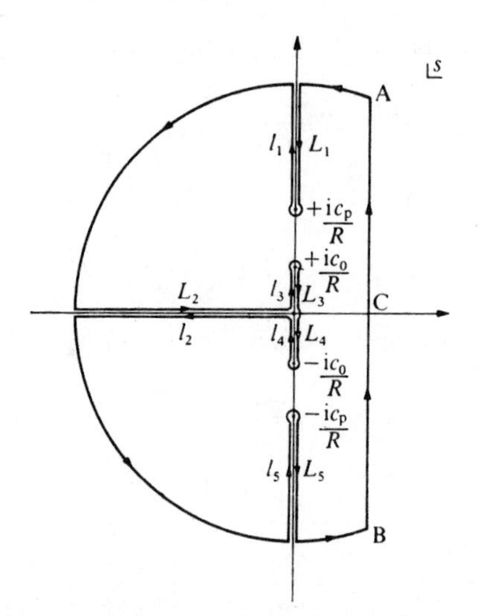

FIG. 4.25. Bromwich contour for evaluation of the integral I.

Furthermore, the integrals along L_2 and l_2 cancel. The remaining integrals contribute. Fortunately, they combine, giving the result that

$$I = I_1 + I_2, \tag{4.3.49}$$

where

$$I_1 = -\int_{c_p/R}^{\infty} \frac{F(s)}{s\zeta'}\,ds, \qquad I_2 = -\int_{0}^{c_0/R} \frac{F(s)}{s\zeta'}\,ds, \tag{4.3.50}$$

and where

$$F(s) = \sin(st+s\zeta'x) + \sin(st-s\zeta'x),$$
$$\zeta' = \frac{(c_p^2/R^2-s^2)^{\frac{1}{2}}}{c_p(c_0^2/R^2-s^2)^{\frac{1}{2}}}, \tag{4.3.51}$$

and where s is now real in the above expressions.

The result (4.3.49) represents the effective starting point of the analysis of Berkowitz [2]. The problem, of course, is to evaluate the integrals I_1 and I_2. The method of stationary phase is employed to obtain results for large time. The analysis is rather recondite and, while not necessarily beyond the scope of the present work, would prove quite lengthy to review here..The major result of the analysis is the predicted stress N_x for large values of time. This is shown in Fig. 4.26, where the dimensionless quantities

$$\hat{N}_x = N_x/DVh, \qquad \hat{x} = x/R, \qquad T = c_pt/R, \tag{4.3.52}$$

and

$$D = E/(1-\nu^2), \qquad V = V_0/c_p \tag{4.3.53}$$

have been introduced. The most interesting aspect of the result is the arrival time of the major part of the pulse. Although the first signal arrives at a given point according to the propagation velocity c_p, the main wavefront arrives according to the bar velocity c_0. The signal actually starts to rise slightly ahead

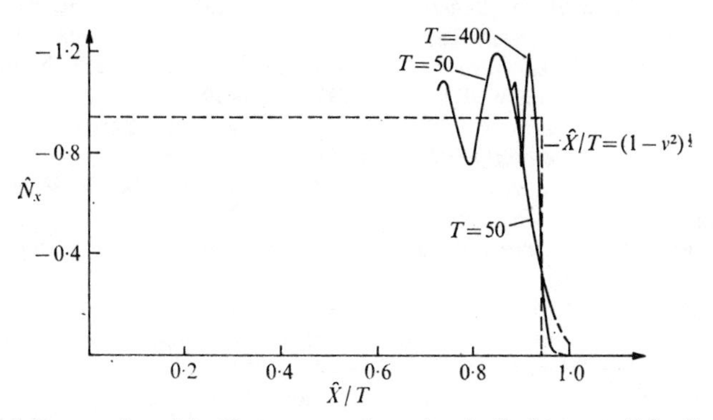

FIG. 4.26. Propagation of the N_x stress wave due to longitudinal impact. (After Berkowitz [2].)

of this time and peaks slightly after this time. The second aspect to note is the oscillatory character of the signal. The general behaviour of the oscillation is governed by the Airy integral. This type of response will be noted in several future analyses of step inputs to elastic systems. Some care must be used in interpreting the oscillatory behaviour for the values of $T = 50$ and 400. One might be tempted to conclude that the further the signal progressed, the more oscillatory it becomes. Exactly the opposite is true. To establish this, the width of the rising portion of the pulse should be calculated for the two values of time shown. The rise time at $T = 400$ is much slower, as it should be. It is the presentation of this data on the same scale that causes this illusion. No results are obtained for $\hat{x}/T < 0{\cdot}7$. However, the pulse quite likely decays to the dashed line. Finally, mention should be made as to other results not shown in the figure. The author's analysis showed a sharp spike pulse occurring at the c_p arrival time ($\hat{x}/T = 1$ in the figure). Evidently analysis by Miklowitz [11] of a somewhat similar problem in impact of rods also showed such a spike but this was discarded owing to considerations of the exact theory. The possibility exists that the spike stress arising in the analysis reviewed here should be similarly discarded but, as pointed out by Berkowitz, this awaits analysis of the shell-impact problem by exact theory.

References

1. Abramowitz, M. and Stegun, I. A. *Handbook of mathematical functions: with formulas, graphs, and mathematical tables*. Dover Publications, New York (1965).
2. Berkowitz, H. M. Longitudinal impact of a semi-infinite elastic cylindrical shell. *J. appl. Mech.* **30,** 347–54 (1963).
3. Dohrenwend, C. O., Drucker, D. C., and Moore, P. Transverse impact transients. *Exp. Stress Analysis* **1,** 1–10 (1944).
4. Donnell, L. H. *Stability of thin walled tubes under torsion*. NACA *Rep.* No. 479 (1933).
5. Flügge, W. *Statik und Dynamik der Schalen*. Springer-Verlag, Berlin (1934).
6. Hildebrand, F. B., Reissner, E., and Thomas, G. B. *Notes on the foundation of the theory of small displacements of orthotropic shells*. Tech. Notes natn. advis. Comm. Aeronaut., Wash. No. 1833, pp. 1–59 (1949).
7. Kraus, H. *Thin elastic shells*. John Wiley and Sons, New York (1967).
8. Leissa, A. W. *Vibrations of plates*. NASA S.P. 160 (1969).
9. Love, A. E. H. *A treatise on the mathematical theory of elasticity*. Dover Publications, New York (1944).
10. Medick, M. A. On classical plate theory and wave propagation. *J. appl. Mech.* **28,** 223–8 (1961)
11. Miklowitz, J. On the use of approximate theories of an elastic rod in problems of longitudinal impact. *Proc. 3rd U.S. natn. Congr. appl. Mech.* pp. 215–24. A.S.M.E. New York (1958).
12. Morse, P. M. *Vibration and sound*. McGraw-Hill, New York (1948).
13. Naghdi, P. M. On the theory of thin elastic shells. *Q. appl. Math.* **14,** 369–80 (1957).

14. PRESS, F. and OLIVER, J. Model study of air-coupled surface waves. *J. Acoust. Soc. Am.* **27**, 43–6 (1955).
15. SANDERS, J. L. An improved first approximation theory for thin shells. NASA tech. Rep. R24 (1959). Also see BUDIANSKY, B. and SANDERS, J. L. On the 'best' first order linear shell theory. *Progress in applied Mechanics*, Prager Anniversary Volume, p. 129. Macmillan, London (1963).
16. SNEDDON, I. N. *Fourier transforms*. McGraw-Hill, New York (1951).
17. TIMOSHENKO, S. and WOINOWSKY-KRIEGER, S. *Theory of plates and shells*. McGraw-Hill, New York (1959).
18. VLASOV, V. Z. *General theory of shells and its application to engineering*. Moscow-Leningrad (1949). NASA tech. Trans. 99 (1964).
19. WATSON, G. N. *A treatise on the theory of bessel functions* (2nd edn). Cambridge University Press (1966).

Problems

4.1. Derive the governing equation for the motion of a membrane in polar coordinates. Start from 'first principles', that is consider a polar element initially instead of a coordinate transformation.

4.2. Starting with a separation of variables solution to the membrane equation (4.1.3), with $p = 0$, develop the solution form (4.1.8).

4.3. Consider an infinite membrane resting on an elastic foundation. Obtain the dispersion curve governing propagation of plane waves in such a membrane.

4.4. Extending the considerations of Problem 4.3, consider a semi-infinite membrane $y > 0$, resting on an elastic foundation. Assuming fixed boundary conditions along $y = 0$, consider plane harmonic waves to be incident on the boundary. Determine the amplitude ratio of the reflected waves.

4.5. Consider the case of waves in a membrane strip defined by $x \geq 0, y = \pm b/2$. Consider the case of forced motion at the boundary $x = 0$, given by $w(0, y, t) = w_0 \cos py \exp(i\omega t)$, where $p = \pi/b$. Determine the resulting membrane response for frequencies above, at, and below the cutoff frequency.

4.6. Draw diagrams of the ω_{35} and ω_{42} modes of a rectangular membrane and the ω_{32} mode of a circular membrane.

4.7. Consider a circular membrane fixed along the outer radius b and having a rigid cylindrical mass of radius a attached in the centre. Derive the governing equation for the natural frequencies of this system.

4.8. Consider a membrane consisting of a circular sector fixed on all edges, where the sector radius is a and the angle is ϕ. Develop the equation for the natural frequencies.

4.9. Attempt to derive the governing equations and boundary conditions for a plate using energy considerations.

4.10. Attempt to derive the expression for the driving-point impedance (that is, the ratio of force to velocity) of a plate subjected to the harmonic pressure $p(r, \theta, t) = p_0 \exp(i\omega t), r \leq a$.

4.11. Consider a semi-infinite plate having a free edge and consider plane harmonic waves to be impinging on the edge. Determine the reflected wave amplitude ratios B/A, C/A, where A is the amplitude of the incident wave.

4.12. Consider wave propagation in a plate strip of width $2a$, where the two edges of the strip are simply supported. Obtain the frequency equation governing the propagation of harmonic waves.

4.13. Consider longitudinal (instead of flexural) plane waves in a plate. Show that the propagation velocity of such waves is $c_p = \{E/\rho(1-\nu^2)\}^{\frac{1}{2}}$.

4.14. Derive the frequency equation for the natural frequencies of a clamped, circular plate.

4.15. Sketch the deformation mode of a cylindrical shell cross-section for the case of $n = 1$.

4.16. Consider the propagation of waves in membrane shells under the conditions $u = \partial/\partial x = 0$. Obtain the frequency equation governing the propagation. Compare the results to those obtained for waves in a curved ring. In what way does the model for the shell differ from that of the ring for this mode of propagation?

Waves in infinite media

WE NOW embark on a new phase of study of elastic wave propagation and vibrations. Our considerations will now be ruled by the exact equations and boundary conditions of infinitesimal isotropic elasticity theory. In our prior investigations of Chapters 1–4 the considerations were on strength-of-material theories for rods, plates, and shells. Inherent in all such theories were assumptions on the kinematics of deformation. Since the kinematics were generally only approximations of the true deformations, the resulting theories were approximate. Improvements in the theories were possible (for example, the Love rod theory and the Timoshenko beam theory), but only limited additional information was obtainable.

Thus, one objective in turning to the exact elasticity equations will be to develop exact theories for previously considered structural shapes such as rods, plates, and shells. Another objective, of equal importance, will be to consider classes of problems that have no counterparts in the elementary theories. In other words, problems that are inherently three-dimensional problems. These concern waves in extended media and the interaction of elastic waves with surfaces and boundaries.

As the first step in our investigation, we shall consider the class of problems for which boundary interactions are not possible, namely those involving infinite media. Fundamental insight will be obtained on the nature of elastic waves that may exist by considering special solutions to the equations of motion. Waves emanating from point disturbances will then be covered followed by the study of waves from various cavity sources.

5.1. Wave types

The basic elasticity equations will be presented, followed by resolution into scalar and vector potential equations. It will be shown that two basic types of wave, dilatational and distortional, can propagate in an infinite medium, with each being characterized by a specific velocity. Furthermore, these wave types can exist independent—or uncoupled—from one another. The propagation of plane waves will be studied, with the nature of the resulting displacement and stress fields being determined and shown graphically. Typical propagation velocities of dilatational and distortional waves are presented.

5.1.1. *The governing equations*

The equations for a homogeneous isotropic elastic solid may be summarized in Cartesian tensor notation as†

$$
\begin{aligned}
\tau_{ij,j}+\rho f_i &= \rho \ddot{u}_i \\
\tau_{ij} &= \lambda \varepsilon_{kk}\delta_{ij}+2\mu\varepsilon_{ij}, \\
\varepsilon_{ij} &= \tfrac{1}{2}(u_{i,j}+u_{j,i}), \\
\omega_{ij} &= \tfrac{1}{2}(u_{i,j}-u_{j,i}),
\end{aligned}
\tag{5.1.1}
$$

where τ_{ij} is the stress tensor at a point and u_i is the displacement vector of a material point. The stress tensor is symmetric, so that $\tau_{ij} = \tau_{ji}$. The mass density per unit volume of the material is ρ and f_i is the body force per unit mass of material. The strain and rotation tensors are given by ε_{ij} and ω_{ij} respectively. It should he noted for the former that the factor of $\frac{1}{2}$ is now present in the shear strain, in contrast to the engineering definition of shear strain used in the earlier chapters. The elastic constants for the material are λ and μ, the Lamé constants. The latter is the usual shear modulus and both constants may be expressed in terms of the other elastic constants, such as Young's modulus, Poisson's ratio, and the bulk modulus.

The governing equations in terms of displacements are obtained by substituting the expression for strain into the stress–strain relation and that result into the stress equations of motion, giving Navier's equations for the media

$$(\lambda+\mu)u_{j,ji}+\mu u_{i,jj}+\rho f_i = \rho \ddot{u}_i. \tag{5.1.2}$$

The vector equivalent of this expression is

$$(\lambda+\mu)\nabla\nabla.\mathbf{u}+\mu\nabla^2\mathbf{u}+\rho\mathbf{f} = \rho\ddot{\mathbf{u}}. \tag{5.1.3}$$

In terms of rectangular scalar notation, this represents the three equations

$$
\begin{aligned}
(\lambda+\mu)\left(\frac{\partial^2 u}{\partial x^2}+\frac{\partial^2 v}{\partial x\,\partial y}+\frac{\partial^2 w}{\partial x\,\partial z}\right)+\mu\,\nabla^2 u+\rho f_x &= \rho\frac{\partial^2 u}{\partial t^2}, \\
(\lambda+\mu)\left(\frac{\partial^2 u}{\partial y\,\partial x}+\frac{\partial^2 v}{\partial y^2}+\frac{\partial^2 w}{\partial y\,\partial z}\right)+\mu\,\nabla^2 v+\rho f_y &= \rho\frac{\partial^2 v}{\partial t^2}, \\
(\lambda+\mu)\left(\frac{\partial^2 u}{\partial z\,\partial x}+\frac{\partial^2 v}{\partial z\,\partial y}+\frac{\partial^2 w}{\partial z^2}\right)+\mu\,\nabla^2 w+\rho f_z &= \rho\frac{\partial^2 w}{\partial t^2},
\end{aligned}
\tag{5.1.4}
$$

where u, v, w are the particle displacements in the x, y, z directions. Returning to the vector notation, we note that the dilatation of a material is defined by

$$\Delta = \nabla.\mathbf{u} = \varepsilon_x+\varepsilon_y+\varepsilon_z = \varepsilon_{kk}, \tag{5.1.5}$$

† See Appendix A for a review of the elasticity equations.

so that (5.1.3) may also be written as

$$(\lambda+\mu)\nabla\,\Delta+\mu\nabla^2\mathbf{u}+\rho\mathbf{f} = \rho\ddot{\mathbf{u}}. \tag{5.1.6}$$

The results (5.1.3) and (5.1.6) are the most commonly employed forms of the equations. An alternative form that also finds application is obtainable by using the vector identity

$$\nabla^2\mathbf{u} = \nabla\nabla.\mathbf{u}-\nabla\times\nabla\times\mathbf{u}. \tag{5.1.7}$$

Substituting the above result for $\nabla^2\mathbf{u}$ in (5.1.3) gives

$$(\lambda+2\mu)\nabla\nabla.\mathbf{u}-\mu\nabla\times\nabla\times\mathbf{u}+\rho\mathbf{f} = \rho\ddot{\mathbf{u}}. \tag{5.1.8}$$

Recalling that the rotation vector $\boldsymbol{\omega}$ is defined by

$$\boldsymbol{\omega} = \tfrac{1}{2}\nabla\times\mathbf{u}, \tag{5.1.9}$$

and again using the dilatation Δ, we may express the last result (5.1.8) as

$$(\lambda+2\mu)\nabla\,\Delta-2\mu\nabla\times\boldsymbol{\omega}+\rho\mathbf{f} = \rho\ddot{\mathbf{u}}. \tag{5.1.10}$$

One of the advantages of the last form is that it explicitly displays the dilatation and rotation. A greater advantage is that the result is valid in any curvilinear coordinate system, whereas the results (5.1.3) and (5.1.6) are valid only in rectangular coordinates.

We note the highly complex nature of the displacement equations of motion. It is possible to obtain a simpler set of equations by introducing the scalar and vector potentials Φ and $\mathbf{H}$ such that

$$\mathbf{u} = \nabla\Phi+\nabla\times\mathbf{H}, \qquad \nabla.\mathbf{H} = 0. \tag{5.1.11}$$

The resolution of a vector field into the gradient of a scalar and the curl of a zero-divergence vector is due to a theorem by Helmholtz.† The condition $\nabla.\mathbf{H} = 0$ provides the necessary additional condition to uniquely determine the three components of $\mathbf{u}$ from the four components of Φ, $\mathbf{H}$. We also express

$$\mathbf{f} = \nabla f+\nabla\times\mathbf{B}, \qquad \nabla.\mathbf{B} = 0. \tag{5.1.12}$$

Substituting (5.1.11) and (5.1.12) in (5.1.3) gives

$$(\lambda+\mu)\nabla\nabla.(\nabla\Phi+\nabla\times\mathbf{H})+\mu\nabla^2(\nabla\Phi+\nabla\times\mathbf{H})+\nabla f+\nabla\times\mathbf{B} = \rho(\nabla\ddot{\Phi}+\nabla\times\ddot{\mathbf{H}}). \tag{5.1.13}$$

These regroup to‡

$$\nabla\{(\lambda+2\mu)\nabla^2\Phi+\rho f-\rho\ddot{\Phi}\}+\nabla\times(\mu\nabla^2\mathbf{H}+\rho\mathbf{B}-\rho\ddot{\mathbf{H}}) = 0. \tag{5.1.14}$$

This equation will be satisfied if each bracketed term vanishes, thus giving

$$(\lambda+2\mu)\nabla^2\Phi+\rho f = \rho\ddot{\Phi}, \tag{5.1.15}$$

$$\mu\,\nabla^2\mathbf{H}+\rho\mathbf{B} = \rho\ddot{\mathbf{H}}. \tag{5.1.16}$$

† See Morse and Feshbach [16, pp. 52–3] for discussion and proof.

‡ One must use $\nabla.\nabla\Phi = \nabla^2\Phi$, the fact that $\nabla^2(\nabla\Phi) = \nabla(\nabla^2\Phi)$ and the fact that $\nabla.\nabla\times\mathbf{H} = 0$ to achieve the results.

An alternative form of the scalar and vector potential equations are obtained by substituting (5.1.11) into the general vector form (5.1.8). The result is

$$(\lambda+2\mu)\nabla^2\Phi+\rho f = \rho\ddot{\Phi}, \tag{5.1.17}$$

$$-\mu\, \nabla\times\nabla\times\mathbf{H}+\rho\mathbf{B} = \rho\ddot{\mathbf{H}}. \tag{5.1.18}$$

While it is evident that (5.1.14) will be satisfied if eqns (5.1.15) and (5.1.16) hold, it would also appear that values of Φ and $\mathbf{H}$ not satisfying these equations might still cause the original equation to be satisfied. This aspect, in fact, has been investigated by Sternberg [19] and Sternberg and Gintin [20], and it has been established that the complete solution is given by the solution of (5.1.15) and (5.1.16).

5.1.2. *Dilatational and distortional waves*

Consider the governing displacement equations in the absence of body forces, given by

$$(\lambda+\mu)\nabla\nabla.\mathbf{u}+\mu\, \nabla^2\mathbf{u} = \rho\ddot{\mathbf{u}}. \tag{5.1.19}$$

If the vector operation of divergence is performed on the above, we obtain

$$(\lambda+\mu)\nabla.(\nabla\nabla.\mathbf{u})+\mu\nabla.(\nabla^2\mathbf{u}) = \rho\nabla.\ddot{\mathbf{u}}. \tag{5.1.20}$$

Since $\nabla\cdot\nabla \sim \nabla^2$, $\nabla.(\nabla^2\mathbf{u}) = \nabla^2(\nabla.\mathbf{u})$, and $\nabla.\mathbf{u} = \Delta$, the dilatation, (5.1.20) reduces to

$$(\lambda+2\mu)\, \nabla^2\Delta = \rho\frac{\partial^2\Delta}{\partial t^2}. \tag{5.1.21}$$

This we recognize as the wave equation, expressible in the form

$$\nabla^2\Delta = \frac{1}{c_1^2}\frac{\partial^2\Delta}{\partial t^2}, \tag{5.1.22}$$

where the propagation velocity c_1 is given by†

$$c_1 = \left(\frac{\lambda+2\mu}{\rho}\right)^{\frac{1}{2}}. \tag{5.1.23}$$

We thus conclude that a change in volume, or dilatational disturbance, will propagate at the velocity c_1.

We now perform the operation of curl on the governing equation (5.1.20). Since the curl of the gradient of a scalar is zero, this gives

$$\mu\, \nabla^2\boldsymbol{\omega} = \rho\frac{\partial^2\boldsymbol{\omega}}{\partial t^2}, \tag{5.1.24}$$

† Alternative expressions in terms of E, k, ν are

$$c_1 = \left(\frac{E(1-\nu)}{\rho(1+\nu)(1-2\nu)}\right)^{\frac{1}{2}} = \left(\frac{3k(3k+E)}{\rho(9k-E)}\right)^{\frac{1}{2}} = \left(\frac{\mu(4\mu-E)}{\rho(3\mu-E)}\right)^{\frac{1}{2}}.$$

where $\boldsymbol{\omega} = \nabla \times \mathbf{u}/2$ is the previously-defined rotation vector. This result is in the form of the vector wave equation and may be expressed as

$$\nabla^2 \boldsymbol{\omega} = \frac{1}{c_2^2} \frac{\partial^2 \boldsymbol{\omega}}{\partial t^2}, \tag{5.1.25}$$

where the propagation velocity c_2 is given by

$$c_2 = \sqrt{(\mu/\rho)}. \tag{5.1.26}$$

Thus, rotational waves propagate with a velocity c_2 in the medium. This characteristic velocity has been previously encountered in Chapters 2 and 4 as the velocity of torsional waves in circular rods and membrane shells, and was indicated to be the propagation velocity of shear waves.

Reasoning slightly differently, suppose that the rotation $\boldsymbol{\omega}$ is zero. Then the displacement vector may be represented as the gradient of a scalar, namely, $\mathbf{u} = \nabla \psi$, so that (5.1.8) becomes, with $\mathbf{f} = 0$,

$$(\lambda + 2\mu) \nabla^2(\nabla \psi) = \rho \frac{\partial^2}{\partial t^2}(\nabla \psi). \tag{5.1.27}$$

This result also follows directly from (5.1.21). The interpretation here is that an irrotational disturbance propagates at the velocity c_1. In a similar manner, suppose that the dilatation Δ is zero. Then (5.1.6), with $\mathbf{f} = 0$, reduces directly to

$$\mu \nabla^2 \mathbf{u} = \rho \frac{\partial^2 \mathbf{u}}{\partial t^2}. \tag{5.1.28}$$

Here the velocity c_2 arises again. The interpretation of the result is that equivoluminal waves propagate at the velocity c_2.

Finally we refer to eqns (5.1.15) and (5.1.16), the equations that resulted from introducing the scalar and vector potentials Φ and $\mathbf{H}$. If the body forces are zero, we have $\mathbf{f} = \mathbf{B} = 0$, and the two equations again give the scalar and vector wave equations and contain the velocities c_1 and c_2. The significance of the Helmholtz resolution of $\mathbf{u}$ becomes even more apparent at this stage. The scalar potential is seen to be associated with the dilatational part of the disturbance, and the vector potential is associated with the rotational part.

We have thus found that waves may propagate in the interior of an elastic solid at two different speeds c_1 and c_2. Volumetric waves, involving no rotation, propagate at c_1 while rotational waves, involving no volume changes, propagate at c_2. The ratio of the two wave speeds may be expressed as

$$k = \frac{c_1}{c_2} = \left(\frac{\lambda + 2\mu}{\mu}\right)^{\frac{1}{2}} = \left(\frac{2 - 2\nu}{1 - 2\nu}\right)^{\frac{1}{2}}. \tag{5.1.29}$$

Since $0 \leq \nu \leq \frac{1}{2}$ always, we see that $c_1 > c_2$. Nominal values of the propagation velocities for several materials are given in Table 5.1.

TABLE 5.1

Propagation velocities c_1, c_2 for various materials†

	c_1		c_2	
Material	$\text{m s}^{-1}\times 10^{-3}$	$\text{in. s}^{-1}\times 10^{-4}$	$\text{m s}^{-1}\times 10^{-3}$	$\text{in. s}^{-1}\times 10^{-4}$
Aluminum	6·15	24·2	3·10	12·2
Brass	4·24	16·7	2·14	8·42
Copper	4·27	16·8	2·15	8·47
Gold	3·14	12·4	1·17	4·59
Iron	5·06	19·9	3·19	12·6
Lead	2·12	8·35	0·74	2·93
Magnesium	6·44	25·3	3·09	12·2
Nickel	5·59	22·0	2·93	11·6
Silver	3·45	13·6	1·57	6·17
Steel	5·71	22·5	3·16	12·4
Tin	2·96	11·6	1·49	5·87
Tungsten	4·78	18·8	2·64	10·4
Zinc	3·86	15·2	2·56	10·1

† Nominal values calculated from data on elastic constants and density.

A variety of terminology exists for the two wave-types. Dilatational waves are also called irrotational and primary (P) waves. The rotational waves are also called equi-voluminal, distortional, and secondary (S) waves. The P and S wave designations have arisen in seismology, where they are also occasionally picturesquely designated as the 'push' and 'shake' waves. Other respective designations frequently used are longitudinal and shear waves, although this aspect of their behaviour is yet to be investigated.

5.1.3. *Plane waves*

Let us now investigate the conditions under which plane waves may propagate in an infinite elastic solid. Our approach will be to generalize to three dimensions the approach taken in the study of plane waves in membranes (see, for example, Fig. 4.2 and eqn (4.1.8)). Thus, consider a plane wave, as shown in Fig. 5.1, and expressed by

$$\mathbf{u} = \mathbf{A}f(\mathbf{n}.\mathbf{r}-ct) \tag{5.1.30}$$

or, in index notation,

$$u_i = A_i f(n_k x_k - ct). \tag{5.1.31}$$

Thus, in (5.1.31), the vector A_i gives the particle displacement along the plane of the wave while n_i is the wave normal. We define the phase as ψ, where

$$\psi = n_k x_k - ct. \tag{5.1.32}$$

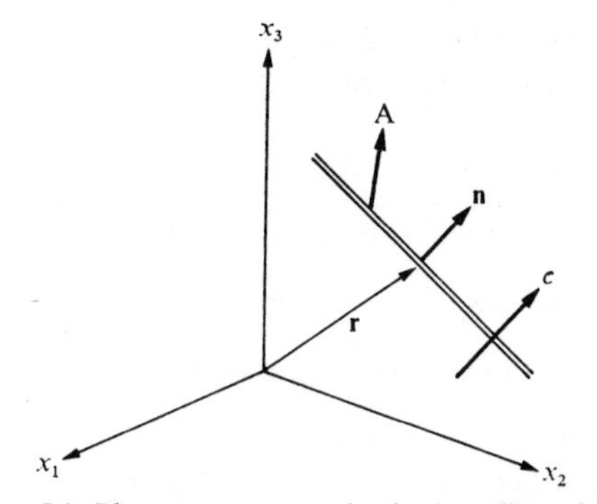

FIG. 5.1. Plane wave propagating in three dimensions.

In substituting (5.1.31) in the equations of motion (5.1.2) (with body forces zero), we must evaluate $u_{i,jj}$, $u_{j,ji}$, and $\ddot{u}_i$. We have that

$$\begin{aligned} u_{i,j} &= A_i\frac{\partial f}{\partial x_j} = A_i\frac{\partial f}{\partial \psi}\frac{\partial \psi}{\partial x_j} = A_i f' n_j, \\ u_{i,jj} &= A_i f'' n_j n_j = A_i f'', \qquad n_j n_j = 1, \\ u_{j,ji} &= A_j f'' n_j n_i, \qquad \ddot{u}_i = c^2 A_i f''. \end{aligned} \tag{5.1.33}$$

Thus the governing equations reduce to

$$(\lambda+\mu)A_j n_j n_i + \mu A_i = \rho c^2 A_i \qquad (i = 1, 2, 3). \tag{5.1.34}$$

This represents three homogeneous equations in the amplitude components A_1, A_2, A_3. Upon expanding the determinant of coefficients, there results

$$(\lambda+2\mu-\rho c^2)(\mu-\rho c^2)^2 = 0, \tag{5.1.35}$$

which gives the two roots

$$c_1 = [(\lambda+2\mu)/\rho]^{\frac{1}{2}}, \qquad c_2 = (\mu/\rho)^{\frac{1}{2}}, \tag{5.1.36}$$

which, again, are the velocities of dilatational and distortional waves. Thus, plane waves propagate at one or the other velocity in a media.

Let us now consider the nature of the displacements and tractions for plane waves. Without attempting a general development, we merely first postulate that u_i is parallel to n_i. If this is the case, we have that

$$A_j n_j = A, \qquad n_i = A_i/a, \tag{5.1.37}$$

so that (5.1.34) becomes

$$(\lambda+2\mu-\rho c^2)A_i = 0. \tag{5.1.38}$$

It may be further shown for this condition that $\nabla\times\mathbf{u} = 0$. The tractions are given by $t_i = \tau_{ij}n_j$. Using the stress–strain and the strain–displacement relations, this may be written as

$$t_i = \{\lambda u_{k,k}\,\delta_{ij} + \mu(u_{i,j} + u_{j,i})\}n_j. \tag{5.1.39}$$

Substituting expressions for $u_{i,j}$, $u_{k,k}$ derived as in (5.1.33) in the traction expression gives

$$t_i = \{(\lambda+\mu)A_k n_k n_i + \mu A_i\}f'. \tag{5.1.40}$$

Again, if u_i is parallel to n_i, using the result (5.1.37) in the above gives

$$t_i = (\lambda+2\mu)Af'n_i. \tag{5.1.41}$$

Thus, the tractions are parallel to the wave normal. Hence we have established that, if the displacements are in the direction of the wave normal, the propagation velocity of such a wave is c_1 and the stresses along the wave are normal to the front.

A second case is to consider the displacements to be perpendicular to the wave normal. Under these conditions, we have $A_j n_j = 0$, so that (5.1.34) reduces to

$$(\mu - \rho c^2)A_i = 0. \tag{5.1.42}$$

It may also be shown that $\nabla.\mathbf{u} = 0$ for this case. The traction expression (5.1.40) reduces immediately to

$$t_i = \mu A_i f'. \tag{5.1.43}$$

This indicates that the tractions are also perpendicular to the wavefront. Thus, for this case, the disturbance propagates at the velocity c_2 and only shearing stresses are acting along the wavefront.

All of the basic aspects of plane waves have now been demonstrated. At the risk of some redundancy, but for emphasis on the nature of the wave motion occurring, suppose that simple harmonic plane waves are propagating in the media. For simplicity, suppose they are propagating along a coordinate axis. Then, dropping the index notation, we write

$$u = A\mathrm{e}^{\mathrm{i}\gamma(x-ct)}, \qquad v = B\mathrm{e}^{\mathrm{i}\gamma(x-ct)}, \qquad w = C\mathrm{e}^{\mathrm{i}\gamma(x-ct)}, \tag{5.1.44}$$

where u, v, w are displacements in the x, y, z (or x_1, x_2, x_3) directions. Substitution in the governing eqn (5.1.4) gives

$$\begin{aligned} \gamma^2\{(\lambda+2\mu)-\rho c^2\}A &= 0,\\ \gamma^2(\mu-\rho c^2)B &= 0,\\ \gamma^2(\mu-\rho c^2)C &= 0. \end{aligned} \tag{5.1.45}$$

If $c = c_1$, which satisfies the first of (5.1.45), we see that $B = C = 0$ is required, whereas if $c = c_2$, satisfying the last two of (5.1.45), then $A = 0$ is required. In the first case, where $A \neq 0$, the motion is purely longitudinal, while in the last case, with $A = 0$, the motion is purely transverse. Also, in the last case, the transverse components B and C are independent of one another. For final emphasis, the basic nature of the particle motion for the two types of waves is shown in Fig. 5.2 for the present case of harmonic waves.

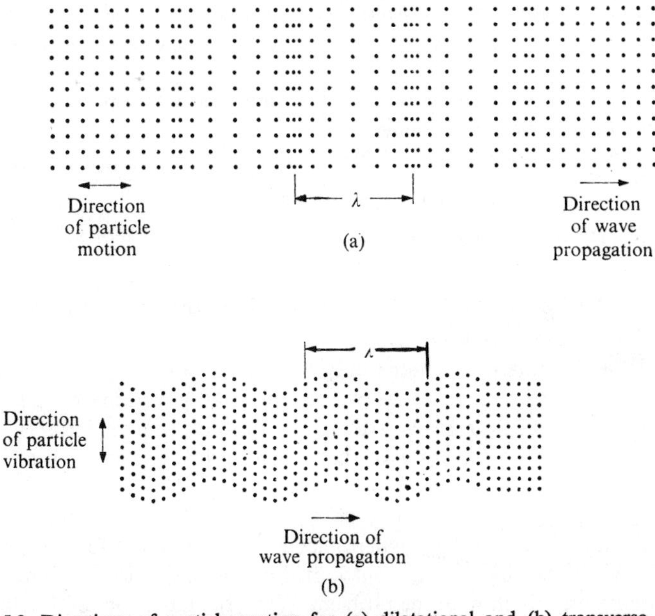

FIG. 5.2. Directions of particle motion for (a) dilatational and (b) transverse waves. (After Fredrick [14, Figs. 2.1, 2.3].)

5.2. Waves generated by body forces

Waves in a continuous and infinite elastic solid may result from imposing certain initial conditions of displacement and velocity on the media, with the disturbances propagating into the undisturbed media with increasing time. Mathematically, this requires consideration of the homogeneous equations of motion. Another source of waves results from time-varying body forces in the interior. This requires consideration of the non-homogeneous governing equations. Another wave source arises from time-varying forces on the walls of cavities embedded in the material, a type of problem that will be considered in the next section.

Solutions of body-force problems represented some of the earliest investigations of dynamical elasticity. Hence certain classical solutions contributed by Poisson, Kirchhoff, and others will be discussed briefly. Consideration will then be given to the analysis of two problems in wave propagation from body forces, one fairly simple and one not so simple.

5.2.1. *Certain classical solutions*

We have only considered various special solutions to the governing equations thus far, such as solutions demonstrating the existence of dilatational and rotational waves and the propagation of plane waves. In seeking solutions to the governing equations for arbitrary initial conditions and body forces, the equations in terms of the potentials Φ and $\mathbf{H}$ ((5.1.17) and (5.1.18)) are usually considered.

The classical initial-value problem, with body forces absent, was given the following formulation by Cauchy.

Given the values of a function $\Phi(x, y, z, t)$ and its time derivatives at $t = 0$,

$$\Phi(x, y, z, 0) = \Phi_0(x, y, z),$$
$$\left.\frac{\partial\Phi(x, y, z, t)}{\partial t}\right|_{t=0} = \Phi_1(x, y, z), \tag{5.2.1}$$

determine the function $\Phi(\bar{x}, \bar{y}, \bar{z}, t)$ which satisfies the given conditions and the wave equation $\nabla^2\Phi = \ddot{\Phi}/c^2$.

Since the formulation pertains only to a single scalar equation, it is in reality a problem in acoustics. However, since the governing equations in potential form are four scalar wave equations of similar form (in rectangular coordinates), solution of the Cauchy problem has application in elasticity.

Both Poisson (1820) and Kirchhoff (1883) contributed solutions to this problem. In addition to the original papers, refer to Love† for a review of these works. Rayleigh‡ also has contributed in this field. Kirchhoff presented the more general solution to the problem. While the intricacies of developing this classical solution will not be presented here, the results will be given.

Consider a volume V of material bounded by the surface S as shown in Fig. 5.3.

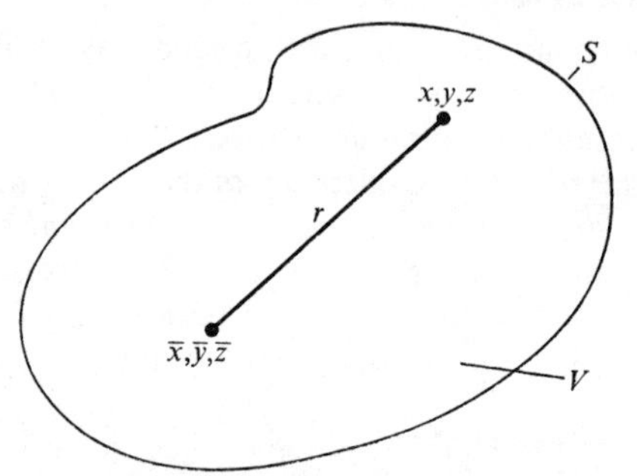

FIG. 5.3. A volume V of material bounded by the surface S.

† pp. 300–2 of Reference [12]. In addition, arbitrary initial conditions involving both dilatational and distortional disturbances are discussed on pp. 302–4 of that Reference.

‡ Vol. 2 of Reference [17].

The Kirchhoff analysis considers a single scalar equation with body force included, given by

$$c^2\nabla^2\Phi+f = \partial^2\Phi/\partial t^2. \tag{5.2.2}$$

The solution to this, often called the retarded potential solution, is given by

$$\Phi(\bar{x},\bar{y},\bar{z},t) = \frac{1}{4\pi c^2}\int_V \frac{1}{r}[f]\,\mathrm{d}V+\frac{1}{4\pi}\int_S \left(\frac{1}{r}\left[\frac{\partial\Phi}{\partial n}\right]-[\Phi]\frac{\partial r^{-1}}{\partial n}+\frac{1}{cr}\frac{\partial r}{\partial n}\left[\frac{\partial\Phi}{\partial t}\right]\right)\mathrm{d}S, \tag{5.2.3}$$

where $[f]$, $[\partial\Phi/\partial n]$, and other bracketed terms are given according to

$$[f] = f \quad \text{at} \quad t = t-r/c. \tag{5.2.4}$$

Thus, the solution at any given point and time $(\bar{x},\bar{y},\bar{z},t)$ is found from the solution at an earlier (or 'retarded') time $t-r/c$. The result may be looked upon as embodying Huygen's principle, which states that every point on a wavefront acts as a source, emitting waves travelling at velocity c.

The solution to problems involving motion due to body forces are obtainable by proper application of the Kirchhoff results. In 1849 Stokes evidently used this method to solve the problem of a concentrated force acting at the origin in a media (the solution to this problem preceded solution to the static counterpart). By superposition of the results for a single force, the case of a centre of twist or dilatation can be obtained. Love† reviews the analyses in this field.

5.2.2. *The simple SH wave source*

For our first analysis of waves generated by body forces in an infinite media, we shall consider the 'SH' wave source. The designation is somewhat in anticipation of later studies of waves in a half-space. Basically, we assume that the following holds for displacements and body forces:

$$\begin{aligned} u_x &= u_y = 0, \qquad & u_z &= u_z(x,y,t),\\ f_x &= f_y = 0, \qquad & f_z &= f_z(x,y,t). \end{aligned} \tag{5.2.5}$$

Under these circumstances, the governing equations (5.1.4) reduce to

$$\mu\,\nabla^2 u_z+\rho f_z = \rho\ddot{u}_z, \qquad \nabla^2 = \frac{\partial^2}{\partial x^2}+\frac{\partial^2}{\partial y^2}. \tag{5.2.6}$$

The particle motions are polarized in a single direction and the resulting waves will be shear waves propagating at the velocity c_2. The SH designation pertains to 'horizontally polarized shear (or secondary) waves'. The governing equation, it should be noted, is the same as for the transverse motion of membranes. We shall consider several aspects of solution in the following.

† Pp. 304–6 of Reference [12].

As a first illustration, suppose the body force has a harmonic time variation given by

$$f_z = c_2^2 F(x, y)e^{-i\omega t}. \tag{5.2.7}$$

If we assume that $u_z = U(x, y)\exp(-i\omega t)$, then the governing equation (5.2.6) reduces to

$$\nabla^2 U + \beta^2 U = -F(x, y), \qquad \beta^2 = \omega^2/c_2^2. \tag{5.2.8}$$

We will use the double Fourier transform defined by

$$\bar{h}(\xi, \eta) = \frac{1}{2\pi}\int_{-\infty}^{\infty}\int_{-\infty}^{\infty} h(x, y)e^{-i(\xi x+\eta y)}\,dx\,dy. \tag{5.2.9}$$

Applying this transform to (5.2.8) gives the transformed solution

$$\bar{U}(\xi, \eta) = \frac{\bar{F}(\xi, \eta)}{\xi^2+\eta^2-\beta^2}, \tag{5.2.10}$$

with the inverse given by

$$U(x, y, \omega) = \frac{1}{2\pi}\int_{-\infty}^{\infty}\int_{-\infty}^{\infty} \frac{\bar{F}(\xi, \eta)}{\xi^2+\eta^2-\beta^2} e^{i(\xi x+\eta y)}\,d\xi\,d\eta. \tag{5.2.11}$$

Suppose the loading is given by the concentrated line load at x_0, y_0,

$$F(x, y) = \delta(x-x_0)\delta(y-y_0). \tag{5.2.12}$$

Then the solution (5.2.11) is the Green's function

$$G(\mathbf{r}, \mathbf{r}_0, \omega) = \frac{1}{2\pi}\int_{-\infty}^{\infty}\int_{-\infty}^{\infty} \frac{\exp\{i\xi(x-x_0)+i\eta(y-y_0)\}}{\xi^2+\eta^2-\beta^2}\,d\xi\,d\eta. \tag{5.2.13}$$

The above is a formal result. The axisymmetry existing about the axis x_0, y_0 enables an explicit solution form to be obtained. This may be better recognized with the aid of Fig. 5.4. Thus, we may write

$$\mathbf{l} = \xi\mathbf{i}+\eta\mathbf{j}, \qquad \mathbf{R} = (x-x_0)\mathbf{i}+(y-y_0)\mathbf{j}. \tag{5.2.14}$$

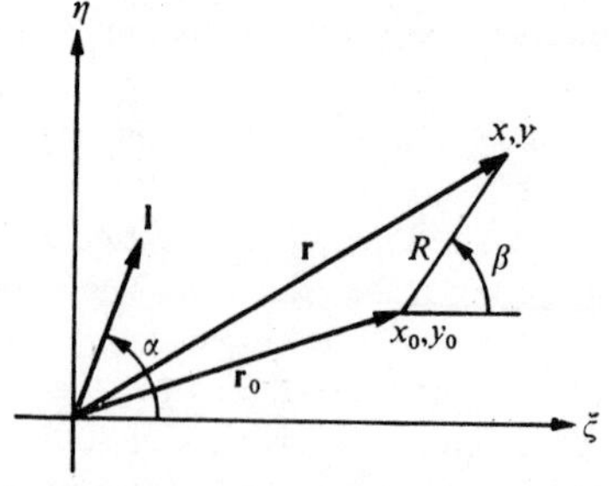

FIG. 5.4. ξ,η-plane representation of $\exp\{i\xi(x-x_0)+i\eta(y-y_0)\}$.

Then (5.2.13) may be expressed in the form

$$G(\mathbf{r}, \mathbf{r}_0, \omega) = \frac{1}{2\pi}\int_{-\infty}^{\infty}\int_{-\infty}^{\infty} \frac{\exp(\mathrm{i}\mathbf{l}\,.\,\mathbf{R})}{l^2-\beta^2}\,\mathrm{d}\xi\,\mathrm{d}\eta. \tag{5.2.15}$$

This may be converted to a polar-coordinate representation l, α if it is noted that $\mathrm{d}\xi\,\mathrm{d}\eta = l\,\mathrm{d}l\,\mathrm{d}\alpha$ and that $\mathbf{l}\,.\,\mathbf{R} = lR\cos(\alpha-\beta)$. Then (5.2.15) becomes

$$G(\mathbf{r}, \mathbf{r}_0, \omega) = \frac{1}{2\pi}\int_0^{\infty} l\,\mathrm{d}l \int_0^{2\pi} \frac{\exp\{\mathrm{i}lR\cos(\alpha-\beta)\}}{l^2-\beta^2}\,\mathrm{d}\alpha. \tag{5.2.16}$$

The integral with respect to α is in the form that defines the Bessel function† $2\pi J_0(lR)$. So we have

$$G(\mathbf{r}, \mathbf{r}_0, \omega) = \int_0^{\infty} \frac{J_0(lR)}{l^2-\beta^2}\,l\,\mathrm{d}l. \tag{5.2.17}$$

This integral also is of a known form. Thus it is known that

$$\int_0^{\infty} \frac{J_0(lR)}{l^2+\beta^2}\,l\,\mathrm{d}l = K_0(\beta R), \tag{5.2.18}$$

where $K_0(\beta R)$ is the modified Bessel function of the second kind. We also have that

$$K_0(z) = \frac{\pi\mathrm{i}}{2}H_0^{(1)}(\mathrm{i}z). \tag{5.2.19}$$

Thus we have that

$$\int_0^{\infty} \frac{J_0(lR)}{l^2-\beta^2}\,l\,\mathrm{d}l = \frac{\pi\mathrm{i}}{2}H_0^{(1)}(\beta R), \tag{5.2.20}$$

and

$$G(\mathbf{r}, \mathbf{r}_0, \omega) = \frac{\pi\mathrm{i}}{2}H_0^{(1)}(\beta R). \tag{5.2.21}$$

The objective here is to illustrate general analysis procedures. In the case of a concentrated, harmonic load acting at the origin, working in polar co-ordinates is most direct. Thus, the governing equation still has the form

$$\nabla^2 u_z - \frac{1}{c_2^2}\frac{\partial^2 u_z}{\partial t^2} = -F_z(r)\mathrm{e}^{-\mathrm{i}\omega t}, \tag{5.2.22}$$

where

$$\nabla^2 = \frac{\partial^2}{\partial r^2} + \frac{1}{r}\frac{\partial}{\partial r}. \tag{5.2.23}$$

† See pp. 1361–2 of Reference [16] for this result and other aspects of the present development.

Again letting $u_z(r, t) = U(r)\exp(-\mathrm{i}\omega t)$ we have

$$\nabla^2 U + \beta^2 U = -F(r), \qquad \beta^2 = \omega^2/c_2^2. \tag{5.2.24}$$

Applying the Hankel transform† to this equation, we obtain the transformed solution

$$\bar{U} = \bar{F}(\xi)/(\xi^2-\beta^2), \tag{5.2.25}$$

with the inverse given by

$$U(r) = \int_0^\infty \frac{\xi \bar{F}(\xi)}{\xi^2-\beta^2} J_0(\xi r)\,\mathrm{d}r. \tag{5.2.26}$$

For a concentrated line load, we let‡ $F(r) = \delta(r)/2\pi r$ so that $\bar{F}(\xi) = \frac{1}{2}\pi$. Then

$$U(r) = \frac{1}{2\pi}\int_0^\infty \frac{\xi J_0(\xi r)}{\xi^2-\beta^2}\,\mathrm{d}\xi. \tag{5.2.27}$$

The results for this integral have just been given by (5.2.20), so that we have

$$U(r) = \frac{\mathrm{i}}{4} H_0^{(1)}(\beta r). \tag{5.2.28}$$

Instead of considering a harmonic load, it is possible to use the triple Fourier transform, defined by

$$\bar{h}(\xi, \eta, \omega) = \frac{1}{(2\pi)^{\frac{3}{2}}} \int_{-\infty}^{\infty}\int_{-\infty}^{\infty}\int_{-\infty}^{\infty} h(x, y, t)\mathrm{e}^{-\mathrm{i}(\xi x+\eta y-\omega t)}\,\mathrm{d}x\,\mathrm{d}y\,\mathrm{d}t, \tag{5.2.29}$$

with the inverse given by

$$h(x, y, t) = \frac{1}{(2\pi)^{\frac{3}{2}}} \int_{-\infty}^{\infty}\int_{-\infty}^{\infty}\int_{-\infty}^{\infty} \bar{h}(\xi, \eta, \omega)\mathrm{e}^{\mathrm{i}(\xi x+\eta y-\omega t)}\,\mathrm{d}\xi\,\mathrm{d}\eta\,\mathrm{d}\omega. \tag{5.2.30}$$

Applying this transform to the governing equation (5.2.6) gives the result

$$\bar{u}(\xi, \eta, \omega) = \frac{\bar{F}(\xi, \eta, \omega)}{(\xi^2+\eta^2-\omega^2)/c_2^2}, \tag{5.2.31}$$

where the body force has been taken to be $f(x, y, t) = c_2^2 F(x, y, t)$. The inverse is given by

$$u(x, y, t) = \frac{1}{(2\pi)^{\frac{3}{2}}} \int_{-\infty}^{\infty}\int_{-\infty}^{\infty}\int_{-\infty}^{\infty} \frac{\bar{F}(\xi, \eta, \omega)}{(\xi^2+\eta^2-\omega^2)/c_2^2} \mathrm{e}^{\mathrm{i}(\xi x+\eta y-\omega t)}\,\mathrm{d}\xi\,\mathrm{d}\eta\,\mathrm{d}\omega. \tag{5.2.32}$$

† See Appendix B.4; also see §§ 4.1, 4.2 for previous applications.
‡ See § 4.2 for the basis for obtaining this result.

The result is particularly instructive to consider. The exponential term in the above should be recognized as corresponding to plane harmonic waves. The solution may thus be regarded as a superposition of plane waves. The previously obtained solution for the harmonic source may also be interpreted in this manner. The particular result (5.2.13), for example, may be regarded as the plane-wave representation of the Bessel function.

Let us now consider the case of an impulse source. Thus, suppose

$$f(x, y, t) = c_2^2\delta(x-x_0)\delta(y-y_0)\delta(t). \tag{5.2.33}$$

The solution to this problem will yield the Green's function $g(\mathbf{r}, \mathbf{r}_0, t)$. We could consider obtaining this result directly. However, since the result $G(\mathbf{r}, \mathbf{r}_0, \omega)$ has been obtained already, we shall use a basic relationship holding between the two Green's functions. Thus, it may be shown that the Laplace transform of $g(\mathbf{r}, \mathbf{r}_0, t)$ gives $G(\mathbf{r}, \mathbf{r}_0, \omega)$, namely,

$$\mathscr{L}\{g(\mathbf{r}, \mathbf{r}_0, t)\} = G(\mathbf{r}, \mathbf{r}_0, \omega). \tag{5.2.34}$$

The approach taken will be indirect and will illustrate, for the first time, an elegant means of inverting Laplace transforms that will be used in later studies. Returning to (5.2.17), we replace ω by is, thus giving

$$G(\mathbf{r}, \mathbf{r}_0, \mathrm{i}s) = \int\limits_0^\infty \frac{J_0(lR)}{l^2+s^2/c_2^2}\, l\, \mathrm{d}l = K_0\left(\frac{sR}{c_2}\right). \tag{5.2.35}$$

The integral representation for K_0 may be rearranged† to the form

$$K_0\left(\frac{sR}{c_2}\right) = \int\limits_1^\infty \frac{\exp(-sR\tau/c_2)}{(\tau^2-1)^{\frac{1}{2}}}\, \mathrm{d}\tau. \tag{5.2.36}$$

The form of the exponential is suggestive of the Laplace transform definition

$$\bar{f}(s) = \int\limits_0^\infty f(t)\mathrm{e}^{-st}\, \mathrm{d}t. \tag{5.2.37}$$

We introduce the change of variables $t = R\tau/c_2$, so that (5.2.36) becomes

$$K_0\left(\frac{sR}{c_2}\right) = \int\limits_{R/c_2}^\infty \frac{\mathrm{e}^{-st}\, \mathrm{d}t}{\{t^2-(R/c_2)^2\}^{\frac{1}{2}}}. \tag{5.2.38}$$

† See pp. 1363 of Reference [16].

This is nearly of the form of the Laplace transformation except for the lower limit. We introduce the step function $H\langle t-R/c_2\rangle$, so that (5.2.38) becomes

$$K_0\left(\frac{sR}{c_2}\right) = \int_0^\infty \frac{e^{-st}H\langle t-R/c_2\rangle}{\{t^2-(R/c_2)^2\}^{\frac{1}{2}}}\,dt = \mathscr{L}\left[\frac{H\langle t-R/c_2\rangle}{\{t-(R/c_2)^2\}^{\frac{1}{2}}}\right]. \tag{5.2.39}$$

Thus we have expressed the Green's function $G(\mathbf{r}, \mathbf{r}_0, \omega)$ in terms of a Laplace transformation. Referring to the basic relationship (5.2.34) we have, by inspection of the result (5.2.38), that

$$g(\mathbf{r}, \mathbf{r}_0, t) = H\langle t-R/c_2\rangle\{t^2-(R/c_2)^2\}^{-\frac{1}{2}}. \tag{5.2.40}$$

The result shows no disturbance prior to $t = R/c_2$. A sharp front arrives at that time and a wake exists behind the front. This is, of course, the result obtained in the analysis of the membrane initial-value problem.† This scheme for obtaining the inverse transformation of a function is known as Cagniard's method‡ and will find application in the analysis of the half-space problem and elsewhere.

5.2.3. *More general body forces*

We now will consider more general body-force loading than the SH wave source considered in the previous section. The analysis presented will basically follow a portion of the analysis by Eason, Fulton, and Sneddon [5]. Thus, consider again the displacement equations of motion (5.1.2). We introduce the four-dimensional Fourier transform, defined by

$$\bar{\phi}(\xi_1, \xi_2, \xi_3, \omega) = \frac{1}{4\pi^2}\int_{V_4} \phi(x_1, x_2, x_3, \tau)\exp\{i(\xi_p x_p+\omega\tau)\}\,dV_4 \quad (p = 1, 2, 3), \tag{5.2.41}$$

where $dV_4 = dx_1\,dx_2\,dx_3\,d\tau$ and V_4 denotes the entire $x_1x_2x_3t$ space. The inverse is simply

$$\phi(x_1, x_2, x_3, \tau) = \frac{1}{4\pi^2}\int_{\bar{V}_4} \bar{\phi}(\xi_1, \xi_2, \xi_3, \omega)\exp\{-i(\xi_p x_p+\omega\tau)\}\,d\bar{V}_4, \tag{5.2.42}$$

where $d\bar{V}_4 = d\xi_1\,d\xi_2\,d\xi_3\,d\omega$. The symbol τ corresponds to a spatial coordinate defined by $\tau = c_1 t$. Thus, the spatial coordinate is determined by the time.

† See § 4.1.
‡ See Reference [3].

We apply the transform (5.2.41) to the governing equations. Retaining the index notation, we obtain the transformed equations

$$(\lambda+\mu)\xi_p\xi_q\bar{u}_q+\mu\xi_q\xi_q\bar{u}_p-\rho c_1^2\omega^2\bar{u}_p = \rho\bar{f}_p, \tag{5.2.43}$$

where the time t has been replaced by τ/c_1 before transformation. The determinant of coefficients for this set of equations gives

$$D = \{(\lambda+2\mu)\gamma^2-\rho c_1^2\omega^2\}(\mu\gamma^2-\rho c_1^2\omega^2)^2, \tag{5.2.44}$$

where $\gamma^2 = \xi_1^2+\xi_2^2+\xi_3^2$. The solution for $\bar{u}_1$ is given by

$$\bar{u}_1 = (\mu\gamma^2-\rho c_1^2\omega^2)[\{(\lambda+2\mu)\gamma^2-\rho c_1^2\omega^2\}\rho\bar{f}_1-(\lambda+\mu)\xi_1(\rho\bar{f}_q\xi_q)]/D. \tag{5.2.45}$$

The solution for all $\bar{u}_p$ is obtained merely by replacing the unity subscripts of $\bar{f}_1$, ξ_1 by p. Introducing the definition

$$\beta^2 = (\lambda+2\mu)/\mu, \tag{5.2.46}$$

it is possible to write the transformed solution for $\bar{u}_p$ as

$$\bar{u}_p = \frac{\beta^2(\gamma^2-\omega^2)\bar{f}_p-(\beta^2-1)\xi_p\xi_q\bar{f}_q}{c_1^2(\gamma^2-\omega^2)(\gamma^2-\beta^2\omega^2)}. \tag{5.2.47}$$

Expressions for the transformed strains $\bar{\varepsilon}_{pq}$, dilatation $\bar{\Delta}$, stresses $\bar{\tau}_{pq}$, and principal stresses $\bar{\tau}_{pp}$ are also given by Eason *et al.* [5]. The inverted expressions for the displacements are

$$u_p = \frac{1}{4\pi^2}\int\limits_{V_4}\frac{\beta^2(\gamma^2-\omega^2)\bar{f}_p-(\beta^2-1)\xi_p\xi_q\bar{f}_q}{c_1^2(\gamma^2-\omega^2)(\gamma^2-\beta^2\omega^2)}\exp\{-\mathrm{i}(\xi_r x_r+\omega\tau)\}\,\mathrm{d}V_4. \tag{5.2.48}$$

Now, the analysis by Eason *et al.* is quite extensive and covers two-dimensional and three-dimensional axisymmetric problems under a variety of loading conditions (harmonic, impulse, step). A large portion of the work pertains to various moving-load problems. Even the results for a static loading are recovered as a special case. Only the two-dimensional problem of the displacements produced by harmonic and impulse loadings will be pursued here in detail.

To two-dimensionalize the results (5.2.48), we first note that $f_3 = 0$ and all quantities will be a function of x_1, x_2 only. The three-dimensional Fourier transform would be appropriate and would result in $u_\alpha(\alpha = 1, 2)$ being given by

$$u_\alpha = \frac{1}{(2\pi)^{\frac{3}{2}}}\int\limits_{\bar{V}_3}\frac{\beta^2(\gamma^2-\omega^2)\bar{f}_\alpha-(\beta^2-1)\xi_\alpha\xi_\eta\bar{f}_\eta}{c_1^2(\gamma^2-\omega^2)(\gamma^2-\beta^2\omega^2)}\exp\{-\mathrm{i}(\xi_\eta x_\eta+\omega\tau)\}\,\mathrm{d}\bar{V}_3, \tag{5.2.49}$$

where now $\gamma^2 = \xi_1^2+\xi_2^2$ and α, $\eta = 1, 2$. In the following, we shall revert to rectangular coordinate notation, where $x = x_1$, $y = x_2$, $u = u_1$, $v = u_2$. If we confine our attention to those problems where the body force $f_y = 0$, we have the two-dimensional displacement solutions given by

$$u = \frac{1}{c_1^2(2\pi)^{\frac{3}{2}}} \int\limits_{V_3} \frac{\{\beta^2(\xi^2+\eta^2-\omega^2)-(\beta^2-1)\xi^2\}}{(\xi^2+\eta^2-\omega^2)(\xi^2+\eta^2-\beta^2\omega^2)} \bar{f}_x e^{-i(\xi x+\eta y+\omega\tau)}\, dV_3, \quad (5.2.50)$$

$$v = \frac{-(\beta^2-1)}{c_1^2(2\pi)^{\frac{3}{2}}} \int\limits_{V_3} \frac{\xi\eta \bar{f}_x e^{-i(\xi x+\eta y+\omega\tau)}}{(\xi^2+\eta^2-\omega^2)(\xi^2+\eta^2-\beta^2\omega^2)}\, dV_3. \quad (5.2.51)$$

As a first type of loading, suppose that f_x is a harmonic force concentrated at the origin and given by

$$f_x = \frac{F}{p}\delta(x)\delta(y)e^{i\lambda\tau}, \qquad \lambda = p/c_1, \quad (5.2.52)$$

and p is the usual radial frequency. The three-dimensional transform of f_x is given by

$$\bar{f}_x = \frac{1}{(2\pi)^{\frac{3}{2}}} \int\limits_{V_3} f_x e^{i(\xi x+\eta y+\omega\tau)}\, dV_3. \quad (5.2.53)$$

Substituting (5.2.52) in the above gives†

$$\bar{f}_x = \frac{F}{\sqrt{(2\pi)}p}\delta(\omega+\lambda). \quad (5.2.54)$$

Before substituting $\bar{f}_x$ in the inversion integrals, we use partial fractions and rewrite (5.2.50) and (5.2.51) as

$$u = \frac{1}{c_1^2(2\pi)^{\frac{3}{2}}} \int\limits_{V_3} \frac{\bar{f}_x}{\xi^2+\eta^2}\left(\frac{\xi^2}{\xi^2+\eta^2-\omega^2}+\frac{\beta^2\eta^2}{\xi^2+\eta^2-\beta^2\omega^2}\right)e^{-i(\xi x+\eta y+\omega\tau)}\, dV_3, \quad (5.2.55)$$

$$v = \frac{1}{c_1^2(2\pi)^{\frac{3}{2}}} \int\limits_{V_3} \frac{\xi\eta\bar{f}_x}{\xi^2+\eta^2}\left(\frac{1}{\xi^2+\eta^2-\omega^2}-\frac{\beta^2}{\xi^2+\eta^2-\beta^2\omega^2}\right)e^{-i(\xi x+\eta y+\omega\tau)}\, dV_3. \quad (5.2.56)$$

When (5.2.54) is substituted into these last results for u and v, the integration with respect to ω follows immediately by virtue of the properties of the Dirac delta function. The results may be expressed in the relatively simple form

$$u = -\frac{Fe^{i\lambda\tau}}{4\pi^2\mu\beta^2}\left[\frac{\partial^2}{\partial x^2}I(x, y, \lambda)+\beta^2\frac{\partial^2}{\partial y^2}I(x, y, \beta\lambda)\right], \quad (5.2.57)$$

$$v = -\frac{Fe^{i\lambda\tau}}{4\pi^2\mu\beta^2}\left[\frac{\partial^2}{\partial x\,\partial y}\{I(x, y, \lambda)-\beta^2 I(x, y, \beta\lambda)\}\right], \quad (5.2.58)$$

† We have previously established the Fourier transform of a harmonic function by an inverse argument. Thus

$$\int_{-\infty}^{\infty} \delta(\omega+\lambda)e^{-i\omega\tau}\, d\omega = e^{i\lambda\tau}.$$

where

$$I(x, y, \lambda) = \int_{-\infty}^{\infty}\int_{-\infty}^{\infty} \frac{e^{-i(\xi x+\eta y)}\,d\xi\,d\eta}{(\xi^2+\eta^2)(\xi^2+\eta^2-\lambda^2)}. \tag{5.2.59}$$

We now introduce a change of variables quite similar in form to that used on the SH wave source problem (see Fig. 5.4) previously analysed. Thus, let $\xi = \rho\cos\phi$, $\eta = \rho\sin\phi$, $x = r\cos\theta$, $y = r\sin\theta$. The integral I then is given by

$$I(r, \theta, \lambda) = \int_0^{2\pi}\int_0^{\infty} \frac{e^{-i\rho r\cos(\phi-\theta)}}{\rho(\rho^2-\lambda^2)}\,d\rho\,d\phi. \tag{5.2.60}$$

The integration with respect to ϕ makes use of the result that†

$$J_0(z) = \frac{1}{2\pi}\int_0^{2\pi} e^{\pm iz\sin\phi}\,d\phi. \tag{5.2.61}$$

We thus obtain

$$I(r, \lambda) = 2\pi\int_0^{\infty} \frac{J_0(\rho r)}{\rho(\rho^2-\lambda^2)}\,d\rho. \tag{5.2.62}$$

We now note that

$$\frac{\partial I}{\partial r} = -2\pi\int_0^{\infty} \frac{J_1(\rho r)}{\rho^2-\lambda^2}\,d\rho = \frac{2\pi}{\lambda^2}\int_0^{\infty}\left(1-\frac{\rho^2}{\rho^2-\lambda^2}\right)J_1(\rho r)\,d\rho. \tag{5.2.63}$$

The evaluation of this last integral form uses two special results on Bessel functions. The first is that‡

$$\int_0^{\infty} J_1(\rho r)\,d\rho = 1/r. \tag{5.2.64}$$

The second result is more recondite, and is given by Watson as§

$$\int_0^{\infty} \frac{\rho^2}{\rho^2-\lambda^2}J_1(\rho r)\,d\rho = \frac{\pi i\lambda}{2}H_1^{(1)}(\lambda r). \tag{5.2.65}$$

Thus, we have that

$$\frac{\partial I(r, \lambda)}{\partial r} = \frac{2\pi}{\lambda^2}\left\{\frac{1}{r}-\frac{\pi i\lambda}{2}H_1^{(1)}(\lambda r)\right\}. \tag{5.2.66}$$

† P. 190 of Reference [13].

‡ See, for example, McLachlan [13, p. 194, formula 68].

§ Watson [21, p. 424, eqn (1)].

The resulting expressions for the displacements are given by

$$u = \frac{\mathrm{i}Fe^{\mathrm{i}pt}}{4\mu\beta^2 r}\left[\frac{1}{p}\left\{c_1 H_1^{(1)}\left(\frac{pr}{c_1}\right)+c_2 H_1^{(1)}\left(\frac{pr}{c_2}\right)\right\}-\frac{1}{r}\left\{x^2 H_2^{(1)}\left(\frac{pr}{c_1}\right)+\beta^2 y^2 H_2^{(1)}\left(\frac{pr}{c_2}\right)\right\}\right],$$

$$v = -\frac{\mathrm{i}Fe^{\mathrm{i}pt}xy}{4\mu\beta^2 r^2}\left\{H_2^{(1)}\left(\frac{pr}{c_1}\right)-\beta^2 H_2^{(1)}\left(\frac{pr}{c_2}\right)\right\}. \tag{5.2.67}$$

Consider now the case of an impulse load where f_x is given by

$$f_x = \frac{F}{\rho}\delta(x)\delta(y)\delta(t). \tag{5.2.68}$$

The transformed body force is given by

$$\bar{f}_x = Fc_1/\rho(2\pi)^{\frac{3}{2}}, \tag{5.2.69}$$

where the fact that $\delta(t) = c_1\delta(\tau)$ has been used. The solutions for u and v follow again from (5.2.55) and (5.2.56), and may be expressed as

$$u = -\frac{Fc_1}{8\pi^3\mu\beta^2}\left(\frac{\partial^2 I_1}{\partial x^2}+\beta^2\frac{\partial^2 I_2}{\partial y^2}\right), \tag{5.2.70}$$

$$v = -\frac{Fc_1}{8\pi^3\mu\beta^2}\frac{\partial^2}{\partial x\,\partial y}(I_1-\beta^2 I_2), \tag{5.2.71}$$

where

$$I_1 = \int_{V_3}\frac{e^{-\mathrm{i}(\xi x+\eta y+\omega\tau)}}{(\xi^2+\eta^2)(\xi^2+\eta^2-\omega^2)}\,\mathrm{d}V_3, \tag{5.2.72}$$

$$I_2 = \int_{V_3}\frac{e^{-\mathrm{i}(\xi x+\eta y+\omega\tau)}}{(\xi^2+\eta^2)(\xi^2+\eta^2-\beta^2\omega^2)}\,\mathrm{d}V_3. \tag{5.2.73}$$

Again, it is found that $\partial I_1/\partial r$, $\partial I_2/\partial r$ are functions only of r. Thus

$$\frac{\partial I_2}{\partial r} = -\frac{4\pi^2}{\beta}\int_0^\infty \frac{\sin(\rho\tau/\beta)}{\rho}J_1(\rho r)\,\mathrm{d}\rho. \tag{5.2.74}$$

Again using a special result on the Bessel function, it is found that†

$$\frac{\partial I_2}{\partial r} = \begin{cases} -4\pi^2\tau/\beta^2 r, & \tau < \beta r \\ -4\pi^2\{\tau-(\tau^2-\beta^2 r^2)^{\frac{1}{2}}\}/\beta^2 r, & \tau > \beta r. \end{cases} \tag{5.2.75}$$

† Formula 73, p. 195, of Reference [13].

A similar result may be obtained for $\partial I_1/\partial r$ by letting $\beta = 1$ in (5.2.75). The resulting expressions for the displacements are given by

$$\frac{2\pi\mu\beta^2 u}{c_1 F} = \begin{cases} 0, & r > \tau \\ \dfrac{x^2}{r^2}(\tau^2-r^2)^{-\frac{1}{2}} + \dfrac{x^2-y^2}{r^4}(\tau^2-r^2)^{\frac{1}{2}}, & \tau' < r < \tau \\ \dfrac{x^2}{r^2}(\tau^2-r^2)^{-\frac{1}{2}} + \dfrac{\beta y^2}{r^2}(\tau'^2-r^2)^{-\frac{1}{2}} + & \\ \quad + \dfrac{x^2-y^2}{r^4}\{(\tau^2-r^2)^{\frac{1}{2}} - \beta(\tau'^2-r^2)^{\frac{1}{2}}\}, & r < \tau', \end{cases} \tag{5.2.76}$$

$$\frac{2\pi\mu\beta^2 v}{c_1 F} = \begin{cases} 0, & r > \tau \\ \dfrac{xy}{r^2}\left\{(\tau^2-r^2)^{-\frac{1}{2}} + \dfrac{2}{r^2}(\tau^2-r^2)^{\frac{1}{2}}\right\}, & \tau' < r < \tau \\ \dfrac{xy}{r^2}\Big[(\tau^2-r^2)^{-\frac{1}{2}} - \beta(\tau'^2-r^2)^{-\frac{1}{2}} + & \\ \quad + \dfrac{2}{r^2}\{(\tau^2-r^2)^{\frac{1}{2}} - \beta(\tau'^2-r^2)^{\frac{1}{2}}\}\Big], & r < \tau', \end{cases} \tag{5.2.77}$$

where $\tau = c_1 t$, $\tau' = c_2 t$. The various intervals in (5.2.76) and (5.2.77) correspond to the arrival of the dilatational and distortional waves. Thus, for $r > \tau(c_1 t)$, no disturbance is felt. At $r = \tau$, a sharp wavefront from the dilatational part arrives, and at $r = \tau'$, the distortional wave arrives. The radial displacement is shown for various orientations relative to the source in Fig. 5.5. It is to be noted that a discontinuity occurs in the radial displacement u_r for $\theta = 0°$ when the P wave arrives, but not when the S wave arrives. At

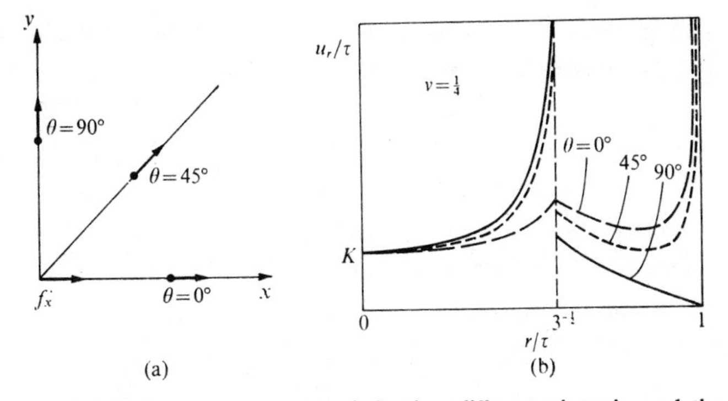

FIG. 5.5. Radial displacement parameter u_r/τ for three different orientations relative to the applied force ($K = c_1F/2\pi\beta^2\mu\tau^2$). (After Eason *et al.* [5].)

$\theta = 90°$, perpendicular to the direction of the applied force, a discontinuity exists for u_r on the arrival of the S wave, but not for the earlier P wave. At $\theta = 45°$, discontinuities exist for both wavefronts.

5.3. Cavity source problems

Let us now consider the case of cavities, spherical or cylindrical, buried in an infinite medium and subjected to internal pressures of one type or another. Because the resulting waves will emanate from the cavity and propagate outward, boundary effects are still avoided in this class of problems. Our considerations mainly will be with dilatational wave sources. That is, waves resulting from time-varying but uniformly-applied normal pressures on the walls of the cavity. Such problems are essentially those of acoustic wave propagation. Waves from more general loading will be briefly discussed.

5.3.1. *Harmonic dilatational waves from a spherical cavity*

Consider a spherical cavity of radius a subjected to internal pressure, as shown in Fig. 5.6. The loading will be presumed to have spherical symmetry in

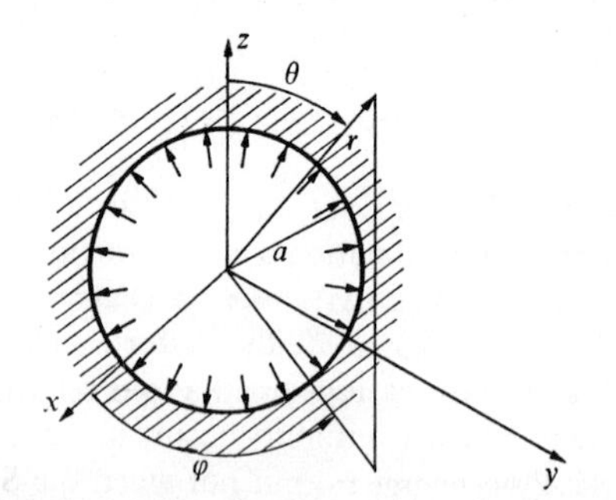

FIG. 5.6. Spherical cavity of radius a subjected to internal pressure.

the present analysis. Consequently the wave propagation will also have spherical symmetry, and will be described by

$$u_r = u(r, t), \qquad u_\theta = u_\phi = \frac{\partial}{\partial\theta} = \frac{\partial}{\partial\phi} = 0. \tag{5.3.1}$$

Consider now the Helmholtz resolution of the displacement vector in terms of spherical coordinates. This is given in general by

$$\mathbf{u} = \left(\frac{\partial\Phi}{\partial r}\mathbf{e}_r + \frac{1}{r}\frac{\partial\Phi}{\partial\theta}\mathbf{e}_\theta + \frac{1}{r\sin\theta}\frac{\partial\Phi}{\partial\phi}\mathbf{e}_\phi\right) + \left(\frac{1}{r}\frac{\partial H_\phi}{\partial\theta} - \frac{1}{r\sin\phi}\frac{\partial H_\theta}{\partial\phi}\right)\mathbf{e}_r + \\ + \left(\frac{1}{r\sin\theta}\frac{\partial H_r}{\partial\phi} - \frac{\partial H_\phi}{\partial r}\right)\mathbf{e}_\theta + \left(\frac{\partial H_\theta}{\partial r} - \frac{1}{r}\frac{\partial H_r}{\partial\theta}\right)\mathbf{e}_\phi. \tag{5.3.2}$$

As a consequence of the assumed symmetry of the displacement field, (5.3.2) simply reduces to

$$\mathbf{u} = \frac{\partial \Phi}{\partial r}\mathbf{e}_r \tag{5.3.3}$$

With this considerable simplification, the potential equations reduce to

$$(\lambda+2\mu)\frac{1}{r^2}\frac{\partial}{\partial r}\left(r^2\frac{\partial \Phi}{\partial r}\right) = \rho\frac{\partial^2 \Phi}{\partial t^2} \tag{5.3.4}$$

or

$$\frac{\partial^2 \Phi}{\partial r^2}+\frac{2}{r}\frac{\partial \Phi}{\partial r} = \frac{1}{c_1^2}\frac{\partial^2 \Phi}{\partial t^2}. \tag{5.3.5}$$

Our immediate considerations will be with harmonic loading having the time variation $\exp(-\mathrm{i}\omega t)$. Then, assuming a solution $\Phi = \Phi(r)\exp(-\mathrm{i}\omega t)$, we have (5.3.5) reducing to

$$\frac{\mathrm{d}^2\Phi(r)}{\mathrm{d}r^2}+\frac{2}{r}\frac{d\Phi(r)}{\mathrm{d}r}+\beta^2\Phi(r) = 0, \qquad \beta^2 = \omega^2/c_1^2. \tag{5.3.6}$$

This equation has been previously encountered in the analysis of longitudinal waves in conical horns and is known as the spherical Bessel's equation having the solution†

$$\Phi(r) = Ah_0^{(1)}(\beta r)+Bh_0^{(2)}(\beta r). \tag{5.3.7}$$

The condition of outgoing waves from the cavity dictates selection of $h_0^{(1)}$, since its asymptotic behaviour

$$h_n^{(1)}(\beta r) \sim \frac{\mathrm{i}^{-n}}{\mathrm{i}\beta r}\mathrm{e}^{\mathrm{i}\beta r} \tag{5.3.8}$$

in conjunction with the time dependence for this problem satisfies the radiation condition. We thus set $B = 0$ in (5.3.7).

The determination of A must come from the boundary conditions, given by

$$\tau_{rr}(a, \theta, \phi, t) = p_0\mathrm{e}^{-\mathrm{i}\omega t}, \qquad \tau_{r\theta} = \tau_{r\phi} = 0. \tag{5.3.9}$$

For the condition of spherical symmetry, we have that

$$\tau_{rr} = \lambda\left(\frac{\partial u}{\partial r}+\frac{2}{r}u\right)+2\mu\frac{\partial u}{\partial r}. \tag{5.3.10}$$

Since $u = \partial\Phi/\partial r$, this last expression becomes

$$\tau_{rr} = \lambda\left(\frac{\partial^2 \Phi}{\partial r^2}+\frac{2}{r}\frac{\partial \Phi}{\partial r}\right)+2\mu\frac{\partial^2 \Phi}{\partial r^2}. \tag{5.3.11}$$

If we compare the terms in the above to the equation of motion (5.3.5), we see that we may write

$$\tau_{rr} = \frac{\lambda}{c_1^2}\frac{\partial^2 \Phi}{\partial t^2}+2\mu\left(\frac{1}{c_1^2}\frac{\partial^2 \Phi}{\partial t^2}-\frac{2}{r}\frac{\partial \Phi}{\partial r}\right). \tag{5.3.12}$$

† See eqns (2.5.24)–(2.5.27).

Recognizing the harmonic time dependence, we write this as

$$\tau_{rr} = -(\lambda+2\mu)\beta^2\Phi(r) - \frac{4\mu}{r}\frac{\mathrm{d}\Phi(r)}{\mathrm{d}r}. \tag{5.3.13}$$

Substituting the solution (5.3.7) in the boundary condition gives

$$p_0 = -A\mu\left\{k^2\beta^2 h_0(\beta r) + \frac{4}{r}\frac{dh_0(\beta r)}{\mathrm{d}r}\right\}\Bigg|_{r=a}, \tag{5.3.14}$$

where $k^2 = (\lambda+2\mu)/\mu$. Since $\mathrm{d}h_0(\beta r)/\mathrm{d}r = -\beta h_1(\beta r)$, we have

$$p_0 = -A\mu\left\{k^2\beta^2 h_0(\beta a) - \frac{4\beta}{a}h_1(\beta a)\right\}. \tag{5.3.15}$$

The resulting solution is given by

$$\Phi(r) = -\frac{p_0 a^2}{\mu M(\beta a)} h_0^{(1)}(\beta r), \tag{5.3.16}$$

where

$$M(\beta a) = k^2(\beta a)^2 h_0^{(1)}(\beta a) - 4\beta a h_1^{(1)}(\beta a). \tag{5.3.17}$$

The displacement field is given by

$$u(r) = \frac{p_0 a^2 \beta}{\mu M(\beta a)} h_1^{(1)}(\beta r), \tag{5.3.18}$$

and the stress field by

$$\tau_{rr}(r) = \frac{p_0}{M(\beta a)}\left(\frac{a}{r}\right)^2\left\{k^2(\beta r)^2 h_0^{(1)}(\beta r) - 4\beta r h_1^{(1)}(\beta r)\right\}. \tag{5.3.19}$$

Equation (5.3.19), with the time factor $\exp(-\mathrm{i}\omega t)$ inserted, represents the formal solution to the problem. As previously discussed,† the spherical Hankel functions may be represented by finite series. For $h_0^{(1)}$, $h_1^{(1)}$ we have

$$h_0^{(1)}(\beta r) = -\frac{\mathrm{i}e^{\mathrm{i}\beta r}}{\beta r}, \qquad h_1^{(1)}(\beta r) = \frac{e^{\mathrm{i}\beta r}}{\beta r}\left(1+\frac{\mathrm{i}}{\beta r}\right). \tag{5.3.20}$$

If these expressions are inserted in (5.3.19), we obtain

$$\tau_{rr}(r) = -\frac{p_0}{M(\beta a)}\left(\frac{a}{r}\right)^2\left\{\mathrm{i}k^2\beta r + 4\left(1+\frac{\mathrm{i}}{\beta r}\right)\right\}e^{\mathrm{i}\beta r}. \tag{5.3.21}$$

It is of interest to consider the result as $a \to 0$, so that the cavity size becomes small in comparison with the wavelength at a given frequency. We first define the 'strength' of the source as

$$S_0 = 4\pi a^3 p_0/3, \tag{5.3.22}$$

† See (2.5.30).

and require that this quantity be constant. For $M(\beta a)$ we have

$$M(\beta a) = -\frac{1}{\beta a}\left\{k^2(\beta a)^2\mathrm{i}+4\beta a\left(1+\frac{\mathrm{i}}{\beta a}\right)\right\}\mathrm{e}^{\mathrm{i}\beta a}. \tag{5.3.23}$$

Then $\tau_{rr}(r)$ is given by

$$\tau_{rr}(r) = p_0 a^3 \beta \frac{\{\mathrm{i}k^2\beta/r+4(1+\mathrm{i}/\beta r)/r^2\}\mathrm{e}^{\mathrm{i}\beta(r-a)}}{\{k^2(\beta a)^2\mathrm{i}+4\beta a(1+\mathrm{i}/\beta a)\}}. \tag{5.3.24}$$

As $a \to 0$, $\beta a \to 0$ and the denominator becomes 4i. For large r compared to the wavelength, the numerator is approximated by $\mathrm{i}k^2\beta/r$. The result, with the time dependence inserted, reduces to

$$\tau_{rr}(r, t) \to \frac{3S_0}{16\pi}\frac{k^2\beta^2}{r}\mathrm{e}^{\mathrm{i}(\beta r-\omega t)}. \tag{5.3.25}$$

This represents the far-field radiation for a 'point' source of dilatational waves.

5.3.2. *Dilatational waves from a step pulse*

We now consider the waves generated by a transient load applied to the surface of the cavity. Again, spherical symmetry will be assumed. Our considerations may start with eqn (5.3.5) which can be written in the form

$$\frac{1}{r^2}\frac{\partial}{\partial r}\left(r^2\frac{\partial \Phi}{\partial r}\right) = \frac{1}{c_1^2}\frac{\partial^2 \Phi}{\partial t^2}. \tag{5.3.26}$$

This may be rewritten in the form

$$\frac{\partial^2(r\Phi)}{\partial r^2} = \frac{1}{c_1^2}\frac{\partial^2(r\Phi)}{\partial t^2}. \tag{5.3.27}$$

This is in the wave equation form in terms of $r\Phi$. The solution may be written immediately as

$$\Phi(r, t) = \frac{1}{r}f(r-c_1 t)+\frac{1}{r}g(r+c_1 t). \tag{5.3.28}$$

These represent outgoing and incoming spherical waves, propagating without distortion of shape, but changing only in amplitude. It should be emphasized that it is the potential Φ that is behaving in this manner. Since, in the present problem, we will have only outgoing waves, the solution $g(r+c_1 t)$ will be discarded. To complete the prescription of the problem, consider the initial conditions to be homogeneous, so that

$$u(r, 0) = \partial u(r, 0)/\partial t = 0. \tag{5.3.29}$$

We shall assume a specific loading at this stage, prescribed by a step pressure pulse, given by

$$\tau_{rr}(a, t) = -p_0 H\langle t\rangle, \qquad \tau_{r\theta} = \tau_{r\phi} = 0. \tag{5.3.30}$$

As a first step, express the solution (5.3.28) in the form

$$\Phi = \frac{1}{r}F(\tau), \qquad \tau = t - \frac{r-a}{c_1}, \tag{5.3.31}$$

where $\tau = 0$, for $t < (r-a)/c_1$ and where $c_1 t = r-a$ corresponds to the position of the wavefront for all time. The displacement is given by

$$u = \frac{\partial\Phi}{\partial r} = -\frac{1}{r^2}F(\tau) - \frac{1}{rc_1}F'(\tau). \tag{5.3.32}$$

For the stress, we have

$$\tau_{rr} = \lambda\nabla^2\Phi + 2\mu\frac{\partial^2\Phi}{\partial r^2} = \frac{\lambda+2\mu}{c_1^2}\frac{F''(\tau)}{r} + 4\mu\frac{F(\tau)}{r^3} + \frac{4\mu}{c_1}\frac{F'(\tau)}{r^2}, \tag{5.3.33}$$

where the prime notation indicates differentiation with respect to τ. We now substitute τ_{rr} into the boundary condition (5.3.30), giving after a slight amount of manipulation

$$F''(t) + \frac{4c_2^2}{ac_1}F'(t) + \frac{4c_2^2}{a^2}F(t) = -\frac{ac_2^2}{\mu}p_0. \tag{5.3.34}$$

Thus, upon setting $r = a$, we see that τ reduces to t.

The solution to the governing equation for F is given by

$$F(t) = e^{-\zeta t}(A\cos\omega t + B\sin\omega t) - a^3 p/4\mu, \tag{5.3.35}$$

where the last term is the particular solution and where

$$\zeta = 2c_2^2/ac_1, \qquad \omega = \zeta\{(c_1^2/c_2^2) - 1\}^{\frac{1}{2}}. \tag{5.3.36}$$

Applying the initial conditions to determine A and B, we obtain

$$F(t) = \frac{a^3 p}{4\mu}\left\{-1 + e^{-\zeta t}\left(\cos\omega t + \frac{\zeta}{\omega}\sin\omega t\right)\right\}. \tag{5.3.37}$$

For $r > a$, we merely replace t by τ to obtain the general solution. We note that $F(\tau) = 0$ for $\tau < 0$. The resulting expression for $u(r)$ is given by

$$u(r,t) = \begin{cases} 0, & \tau < 0 \\ \dfrac{a^3 p}{4\mu r^2}\left[1 - e^{-\zeta\tau}\left\{\left(\dfrac{2r}{a} - 1\right)\dfrac{\zeta}{\omega}\sin\omega\tau - \cos\omega\tau\right\}\right], & \tau > 0. \end{cases} \tag{5.3.38}$$

The displacement field at three different values of r/a is shown in Fig. 5.7. The first and obvious observation is that an oscillatory disturbance is produced from a step pulse. In view of the non-dispersive nature of the governing equation for $r\Phi$ (see (5.3.27)), this might have been unexpected. The reason, of course, is that, while $r\Phi$ propagates without distortion, the displacements

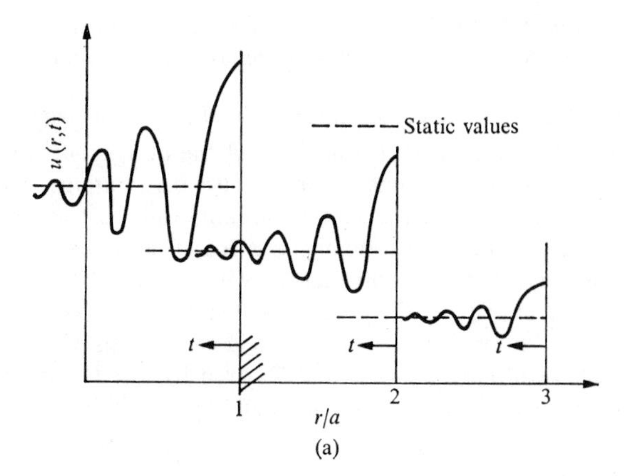

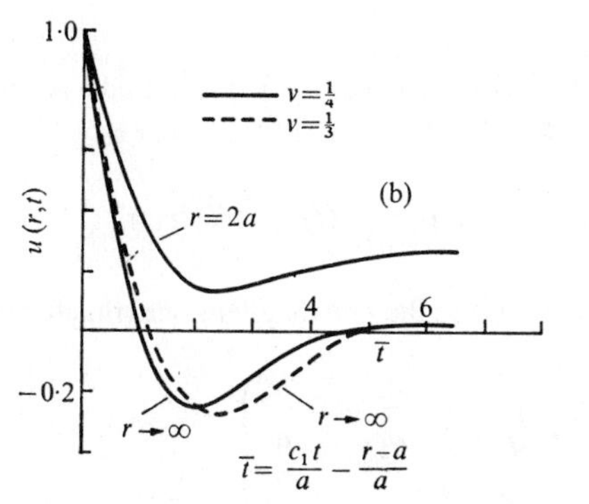

FIG. 5.7. (a) General form of the displacements resulting from a step pressure pulse applied at the surface ($r = a$) of a spherical cavity and (b) the radial stress obtained by Selberg [18] for a specific case.

and stresses (5.3.32) and (5.3.33) are derivatives or combinations of derivatives of Φ and may assume a different pulse shape. We note for $r \gg 1$, that

$$u(r, t) \rightarrow a^3p/4\mu r^2, \tag{5.3.39}$$

which corresponds to the static displacement field (shown by the dashed lines in the figure). A further interpretation of interest is to suppose ζ large so that ω is large and, for $\tau > 0$, $\zeta\tau$ is large. The result is that, at the head of the pulse, the oscillations become more rapid and decay more quickly. Of course, we must bear in mind that the exact nature of the oscillation is dependent on ζ and ω, which depends on the materials and cavity size. For realistic materials and cavity sizes on the order of several centimetres, the behaviour is much more highly damped than indicated in the figure. Figure 5.7(b) shows the results obtained by Selberg [18] for the radial stresses under an applied step loading. As mentioned, a highly damped behaviour is evident.

5.3.3. *General case of dilatational waves with spherical symmetry*

We now wish to generalize the analysis of the cavity-source problem to include arbitrary time variations in the pressure distributions in the cavity. However, by assuming the pressure to be uniformly distributed over the interior of the cavity and to be normal pressure only, as in the last section, the spherically symmetric nature of the problem will be retained. Only dilatational waves will be generated and the problem will still be effectively a one-space-dimension one. Many elements of the previous analysis will be retained. Specifically, we still have the potential Φ given by (5.3.31) and the homogeneous initial conditions (5.3.29). The boundary conditions are given by

$$\tau_{rr}(a,t) = -p(t), \quad t>0, \tag{5.3.40}$$

where $p(t)$ is an arbitrary function of time.

The previously given expression for τ_{rr} (5.3.33) still holds. Applying this to the boundary condition (5.3.40) gives

$$F''(t)+\frac{4c_2^2}{ac_1}F'(t)+\frac{4c_2^2}{a^2}F(t) = \frac{-ac_2^2}{\mu}p(t). \tag{5.3.41}$$

We apply the Fourier transform to the above, giving the transformed solution

$$\bar{F}(\omega) = \frac{ac_2^2}{\mu}\bar{p}(\omega)\bigg/\left(\omega^2-\frac{4c_2^2}{ac_1}\mathrm{i}\omega-\frac{4c_2^2}{a^2}\right). \tag{5.3.42}$$

Hence

$$F(t) = \frac{ac_2^2}{\mu\sqrt{(2\pi)}}\int_{-\infty}^{\infty}\bar{p}(\omega)\mathrm{e}^{-\mathrm{i}\omega t}\bigg/\left(\omega^2-\frac{4c_2^2}{ac_1}\mathrm{i}\omega-\frac{4c_2^2}{a^2}\right)\,\mathrm{d}\omega. \tag{5.3.43}$$

We note several items. First, that $F(\tau)$ is given from the above by replacing t by τ. Second, that $\Phi(\tau)$ is given by $F(\tau)/r$. Third, we may incorporate the integral definition of $\bar{p}(\omega)$ directly in the result. Thus we may write

$$\Phi(\tau) = \frac{ac_2^2}{2\pi\mu}\frac{1}{r}\int_{-\infty}^{\infty}\int_{-\infty}^{\infty}p(t')\exp\{\mathrm{i}\omega(t'-\tau)\}\bigg/\left(\omega^2-\frac{4c_2^2}{ac_1}\mathrm{i}\omega-\frac{4c_2^2}{a^2}\right)\,\mathrm{d}t'\,\mathrm{d}\omega, \tag{5.3.44}$$

where t' is a dummy variable. Indulging in one final modification, we let $\omega = c_1x/a$. Then we have that

$$\omega^2-\frac{4c_2^2}{ac_1}\mathrm{i}\omega-\frac{4c_2^2}{a^2} = \frac{c_1^2}{a^2}\left\{x^2-\frac{2(1-2\nu)}{(1-\nu)}(\mathrm{i}x+1)\right\}. \tag{5.3.45}$$

In deriving this, we have used the identity $(\lambda+2\mu)/\mu = (2-2\nu)/(1-2\nu)$. Then (5.3.44) may be written as

$$\Phi(\tau) = \frac{a^2}{2\pi\rho c_1}\frac{1}{r}\int_{-\infty}^{\infty}\int_{-\infty}^{\infty} p(t')\exp\left\{i\frac{c_1 x}{a}(t'-\tau)\right\}\Big/\left\{x^2-\frac{2(1-2\nu)}{1-\nu}(ix+1)\right\}\,dt'\,dx. \tag{5.3.46}$$

The integration with respect to x is achieved by contour integration. Thus define

$$I = \int_{-\infty}^{\infty}\frac{\exp\{ic_1 x(t'-\tau)/a\}}{x^2-\kappa(ix+1)}\,dx, \tag{5.3.47}$$

where $\kappa = 2(1-2\nu)/(1-\nu)$. Considering x complex, we see that the poles are located at

$$x = \tfrac{1}{2}\{\kappa i \pm (4\kappa-\kappa^2)^{\frac{1}{2}}\}. \tag{5.3.48}$$

Typical values of κ are $1(\nu = \frac{1}{3})$, $1\cdot14(\nu = 0\cdot3)$, and $2(\nu = \frac{1}{4})$, so that $4\kappa-\kappa^2 > 0$ for common materials. The poles will thus be located in the upper half-plane symmetric with respect to the imaginary axis. The appropriate contour will be along the real axis and closed in the upper or lower half-plane. For $t' < \tau$, the contour must be closed below the real axis and $I = 0$. For $t' > \tau$, closure is above and the residues must be computed. It may easily be shown that the residues are respectively

$$\mathrm{Res}[\tfrac{1}{2}\{\kappa i \pm (4\kappa-\kappa^2)^{\frac{1}{2}}\}] = \pm\exp\left[\frac{ic_1(t'-\tau)}{2a}\{\kappa i \pm (4\kappa-\kappa^2)^{\frac{1}{2}}\}\right]\Big/2(4\kappa-\kappa^2)^{\frac{1}{2}}. \tag{5.3.49}$$

Then we obtain that

$$-2\pi i\sum \mathrm{Res} = 4\pi\frac{\exp\{-c_1(t'-\tau)\kappa/2a\}}{(4\kappa-\kappa^2)^{\frac{1}{2}}}\sin\left\{\frac{c_1(t'-\tau)}{2a}(4\kappa-\kappa^2)^{\frac{1}{2}}\right\}. \tag{5.3.50}$$

Before substituting this in (5.3.46), we note that $(4\kappa-\kappa^2)^{\frac{1}{2}} = 2(1-2\nu)^{\frac{1}{2}}/(1-\nu)$. Then we obtain

$$\Phi(\tau) = \frac{a^2(1-\nu)}{\rho c_1(1-2\nu)^{\frac{1}{2}}}\frac{1}{r}\int_{\tau}^{\infty} p(t')\exp\left\{-\frac{1-2\nu}{1-\nu}\frac{c_1}{a}(t'-\tau)\right\}\times$$
$$\times\sin\left\{\frac{(1-2\nu)^{\frac{1}{2}}}{1-\nu}\frac{c_1}{a}(t'-\tau)\right\}\,dt'. \tag{5.3.51}$$

The lower limit has been changed to τ, since $\Phi(\tau) = 0$ for $t' < \tau$.†

A number of results have appeared in the literature depicting waveforms resulting from various pressure pulses. The general form of the response to a

† See Hopkins [10] for analysis of the cavity problem, including numerous references to past work. This result (5.3.51) varies from the result of Hopkin's formula (5.31) in the limits of the integral. This appears to stem from a sign difference between the iω term in the present work versus that of formula (5.25) of Hopkins. A considerable portion of Hopkins's work is devoted to anelastic waves from cavities.

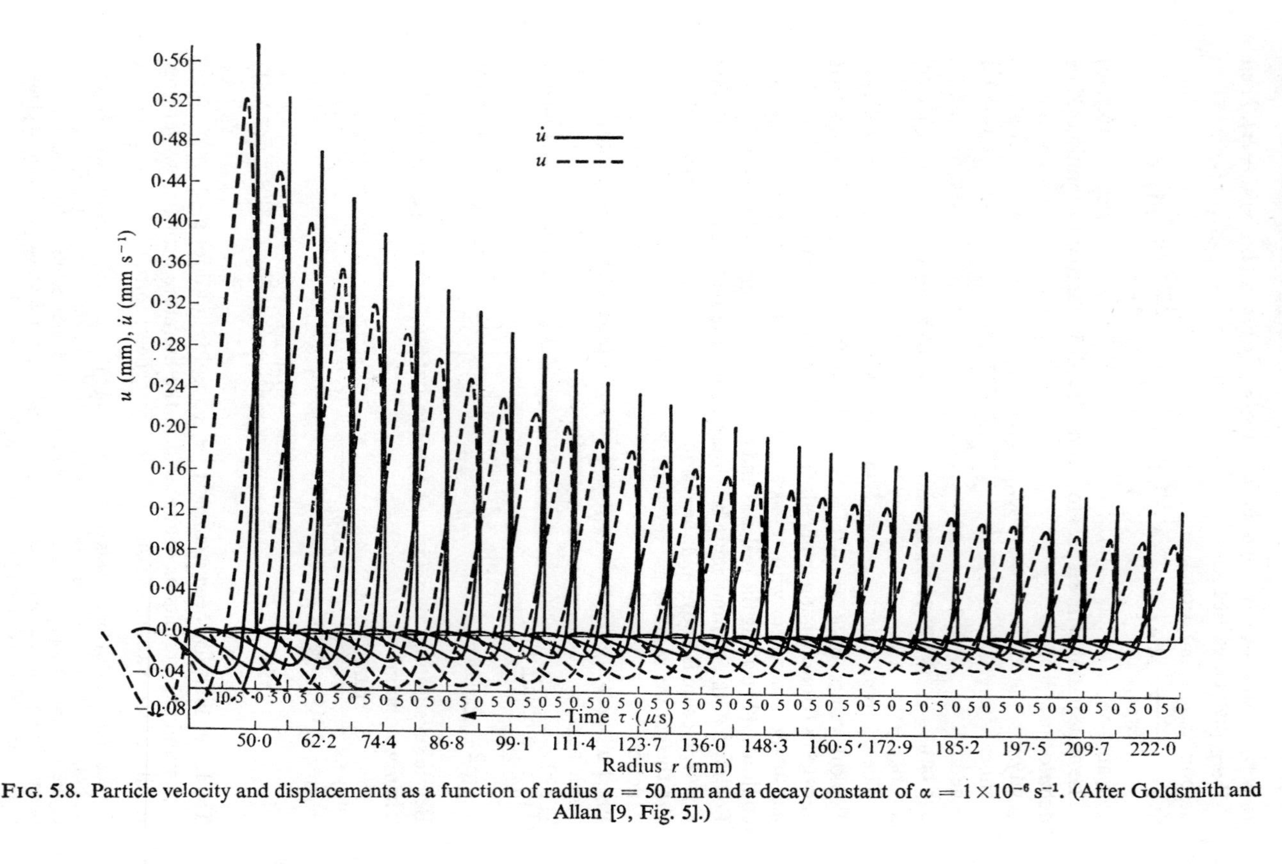

FIG. 5.8. Particle velocity and displacements as a function of radius $a = 50$ mm and a decay constant of $\alpha = 1\times10^{-6}\ \mathrm{s}^{-1}$. (After Goldsmith and Allan [9, Fig. 5].)

step pulse was given in the previous section, as well as the specific result by Selberg [18]. Goldsmith and Allen [9] have presented extensive results for the response to an input pulse given by

$$\tau_{rr}(a) = -p_0 e^{-\alpha t}. \tag{5.3.52}$$

The analysis of this problem was presented earlier by Blake [2]. This type of pulse approximates explosive loading. Figure 5.8 presents but one of a number of velocity and stress results presented by Goldsmith *et al.* [9] for a variety of cavity sizes and decay constants α. The figure is interpreted in the same manner as Fig. 5.7(a), with the time response at a given radius obtained for time increasing to the left. Other results are presented by Goldsmith *et al.* [9] for cavities of 20 mm, 1000 mm, with decay constants of $1\times10^{-6}\,\mathrm{s}^{-1}$ and $2\times10^{-6}\,\mathrm{s}^{-1}$.

The previous results have been for spherical symmetry, so that only dilatational waves are generated. If the pressure distribution is not symmetric or if shear stresses are present, shear waves also will be generated and, in general, the analysis will be more complicated. Das Gupta [4], Wright and Carpenter [22], Meyer [14], Achenbach and Sun [1], and others have contributed in this field. Eringen [6] has given the formal solution to the general problem, where the cavity wall is subjected to arbitrary normal and shear stresses. The results for the displacements and stresses are in terms of infinite series in Hankel functions and spherical harmonics. The means for reducing the general results for a number of special cases (ten in all) of loading, including concentrated loads and ring loads, are given. Specific results, however, are presented only for the displacements and stresses in the case of a spherically symmetric impulse loading.

5.3.4. *Harmonic waves from a cylindrical cavity*

Suppose now we have an infinite cylindrical cavity of radius a in an elastic media, with the axis of the cavity coinciding with the z-axis. If we presume all loads on the cavity surface to be z independent, the problem will be two dimensional in terms of x and y. Within this broad class, two simple cases of harmonic waves will be considered in this section.

The governing scalar and vector wave equations are

$$\nabla^2\Phi = \frac{1}{c_1^2}\frac{\partial^2\Phi}{\partial t^2}, \qquad \nabla^2\mathbf{H} = \frac{1}{c_2^2}\frac{\partial^2\mathbf{H}}{\partial t^2}, \tag{5.3.53}$$

where the scalar and vector Laplacians are given by

$$\nabla^2\Phi = \frac{1}{r}\frac{\partial}{\partial r}\left(r\frac{\partial\Phi}{\partial r}\right)+\frac{1}{r^2}\frac{\partial^2\Phi}{\partial\theta^2}+\frac{\partial^2\Phi}{\partial z^2}, \tag{5.3.54}$$

$$\nabla^2\mathbf{H} = \left(\nabla^2 H_r-\frac{H_r}{r^2}-\frac{2}{r^2}\frac{\partial H_\theta}{\partial\theta}\right)\mathbf{e}_r+\left(\nabla^2 H_\theta-\frac{H_\theta}{r^2}+\frac{2}{r^2}\frac{\partial H_r}{\partial\theta}\right)\mathbf{e}_\theta+\nabla^2 H_z\mathbf{e}_z, \tag{5.3.55}$$

and where

$$\mathbf{u} = \nabla\Phi + \nabla\times\mathbf{H}. \tag{5.3.56}$$

The stresses are given by

$$\tau_{ij} = \lambda\varepsilon_{kk}\delta_{ij} + 2\mu\varepsilon_{ij}, \tag{5.3.57}$$

where

$$e_{rr} = \frac{\partial u_r}{\partial r}, \qquad e_{\theta\theta} = \frac{1}{r}\frac{\partial u_\theta}{\partial \theta} + \frac{u_r}{r}, \qquad e_{zz} = \frac{\partial u_z}{\partial z},$$

$$e_{\theta z} = \frac{1}{2}\left(\frac{\partial u_\theta}{\partial z} + \frac{1}{r}\frac{\partial u_z}{\partial \theta}\right), \qquad e_{zr} = \frac{1}{2}\left(\frac{\partial u_r}{\partial z} + \frac{\partial u_z}{\partial r}\right), \tag{5.3.58}$$

$$e_{r\theta} = \frac{1}{2}\left(\frac{1}{r}\frac{\partial u_r}{\partial \theta} + \frac{\partial u_\theta}{\partial r} - \frac{u_\theta}{r}\right).$$

Consider first the case of dilatational waves, which will arise if

$$u_\theta = \frac{\partial}{\partial \theta} = u_z = \frac{\partial}{\partial z} = 0. \tag{5.3.59}$$

Then (5.3.56) reduces to

$$\mathbf{u} = \frac{\partial \Phi}{\partial r}\mathbf{e}_r, \tag{5.3.60}$$

where Φ is governed by

$$\frac{1}{r}\frac{\partial}{\partial r}\left(r\frac{\partial \Phi}{\partial r}\right) = \frac{1}{c_1^2}\frac{\partial^2 \Phi}{\partial t^2} \tag{5.3.61}$$

or

$$\frac{\partial^2 \Phi}{\partial r^2} + \frac{1}{r}\frac{\partial \Phi}{\partial r} = \frac{1}{c_1^2}\frac{\partial^2 \Phi}{\partial t^2}. \tag{5.3.62}$$

Suppose a uniform harmonically varying pressure is acting on the cavity surface $r = a$. Thus

$$\tau_{rr}(a) = -p_0 e^{-i\omega t}. \tag{5.3.63}$$

Then we may let $\Phi(r, t) = \Phi(r)\exp(-i\omega t)$, so that (5.3.62) becomes

$$\frac{d^2\Phi}{dr^2} + \frac{1}{r}\frac{d\Phi}{dr} + \beta^2\Phi = 0, \qquad \beta^2 = \omega^2/c_1^2. \tag{5.3.64}$$

This represents the third instance that the zeroth-order Bessel's equation has arisen in our study of wave motion.† We know immediately that the solution for the time behaviour in this problem must be given by

$$\Phi(r) = AH_0^{(1)}(\beta r). \tag{5.3.65}$$

The constant A is determined from the boundary condition at $r = a$. The stress is

$$\tau_{rr} = \lambda\left(\frac{\partial u_r}{\partial r} + \frac{u_r}{r}\right) + 2\mu\frac{\partial u_r}{\partial r},$$

$$= \lambda\left(\frac{\partial^2 \Phi}{\partial r^2} + \frac{1}{r}\frac{\partial \Phi}{\partial r}\right) + 2\mu\frac{\partial^2 \Phi}{\partial r^2}. \tag{5.3.66}$$

† Previously in the study of membranes (§ 4.1) and earlier yet in the case of longitudinal waves in tapered rods (§ 2.5). See these sections for discussion of $H_0^{(1)}$, $H_0^{(2)}$, and the asymptotic behaviour.

This may also be written as

$$\tau_{rr} = (\lambda+2\mu)\left(\frac{\partial^2\Phi}{\partial r^2}+\frac{1}{r}\frac{\partial\Phi}{\partial r}\right)-\frac{2\mu}{r}\frac{\partial\Phi}{\partial r}. \tag{5.3.67}$$

Using (5.3.64) in the above, re-organizing the harmonic time dependence, and factoring out μ gives

$$\tau_{rr} = -\mu\left(k^2\beta^2\Phi+\frac{2}{r}\frac{\mathrm{d}\Phi}{\mathrm{d}r}\right), \tag{5.3.68}$$

where $k^2 = (\lambda+2\mu)/\mu$. Applying this to the boundary condition (5.3.63) gives

$$\mu A\left\{k^2\beta^2H_0^{(1)}(\beta r)+\frac{2}{r}\frac{\mathrm{d}H_0^{(1)}(\beta r)}{\mathrm{d}r}\right\}\bigg|_{r=a} = p_0. \tag{5.3.69}$$

Using the relation $\mathrm{d}H_0^{(1)}(\beta r)/\mathrm{d}r = -\beta H_1^{(1)}(\beta r)$, A is easily found, with the result being given by

$$\Phi(r, t) = \frac{p_0}{\mu N(\beta a)}H_0^{(1)}(\beta r)\mathrm{e}^{-\mathrm{i}\omega t}, \tag{5.3.70}$$

where

$$N(\beta a) = k^2\beta^2H_0^{(1)}(\beta a)-2\frac{\beta}{a}H_1^{(1)}(\beta a). \tag{5.3.71}$$

The case of torsional waves is also easily obtained. We assume

$$\frac{\partial}{\partial\theta} = u_z = \frac{\partial}{\partial z} = 0. \tag{5.3.72}$$

Then (5.3.56) reduces to

$$\mathbf{u} = \frac{\partial\Phi}{\partial r}\mathbf{e}_r-\frac{\partial H_z}{\partial r}\mathbf{e}_\theta, \tag{5.3.73}$$

where Φ is given by (5.3.70), the same as in the study of dilatational waves, and H_z is given by

$$\nabla^2H_z = \frac{1}{c_2^2}\frac{\partial^2H_z}{\partial t^2}. \tag{5.3.74}$$

We note that $\mathbf{u}$ has two independent components u_r and u_θ. A purely dilatational disturbance similar to that just considered would be described by u_r if normal stresses were acting on the cavity surface. However, we will assume the boundary conditions to be purely shear, given by

$$\tau_{r\theta}(a, t) = \tau_0\mathrm{e}^{-\mathrm{i}\omega t}. \tag{5.3.75}$$

Thus only a shear wave will be produced. We have, for the symmetry conditions of the problem, that

$$\tau_{r\theta} = \mu\left(\frac{\partial u_\theta}{\partial r}-\frac{u_\theta}{r}\right), \tag{5.3.76}$$

where $u_\theta = -\partial H_z/\partial r$. Thus we may write

$$\tau_{r\theta} = -\mu\left(\frac{\partial^2H_z}{\partial r^2}-\frac{1}{r}\frac{\partial H_z}{\partial r}\right). \tag{5.3.77}$$

Following procedures similar to that used for dilatational waves, this may be written as

$$\tau_{rr} = \mu\left\{\beta^2 H(r) + \frac{2}{r}\frac{\partial H(r)}{\partial r}\right\}, \tag{5.3.78}$$

where $H_z(r, t) = H(r)\exp(-i\omega t)$ and $\beta^2 = \omega^2/c_2^2$. At this point the methods of analysis are quite similar to those for dilatational harmonic waves, and there is little to be gained by pursuing the problem further.

5.3.5. *Transient dilatational waves from a cylindrical cavity*

We now consider a transient disturbance applied to walls of a cylindrical cavity. Other than the nature of the loading, the displacements, stresses, and governing equation will be as given in the previous section on harmonic waves from a cylindrical cavity. Although this problem is the cylindrical analogue of the previously considered spherical cavity under a step pressure, the analysis is not correspondingly simple. Basically, this is because the scalar wave equation (5.3.62) in cylindrical coordinates does not admit a solution of the type $r^n f(r-c_1 t)$, such as applied in spherical coordinates ($n = -1$) and also in rectangular coordinates ($n = 0$). The analysis of this problem has been given by Selberg [18].† The following is based on his work.

Instead of using the potentials Φ and $\mathbf{H}$, we work directly with the dilatation Δ, given by

$$\Delta = \nabla . \mathbf{u} = \frac{1}{r}\frac{\partial(ru_r)}{\partial r}, \tag{5.3.79}$$

for the present symmetry conditions. As we found in our initial investigation of types of waves in an infinite media, the dilatation is also governed by the scalar wave equation

$$\nabla^2\Delta = \frac{1}{c_1^2}\frac{\partial^2\Delta}{\partial t^2}, \tag{5.3.80}$$

where, still,

$$\nabla^2 = \frac{r^{-1}\partial(r\partial/\partial r)}{\partial r}$$

for the axial symmetry and z-independence of the present problem.

We apply the Laplace transform to (5.3.80), giving

$$\frac{d^2\bar{\Delta}}{dr^2} + \frac{1}{r}\frac{d\bar{\Delta}}{dr} - \beta^2\bar{\Delta} = 0, \qquad \beta = s/c_2, \tag{5.3.81}$$

where $\bar{\Delta} = \bar{\Delta}(r, s)$ and homogeneous initial conditions have been assumed. This has the solution

$$\bar{\Delta} = AK_0(\beta r) + BI_0(\beta r), \tag{5.3.82}$$

† Selberg's paper covers both the spherical and cylindrical cavity problems. Figure 5.7(b) has presented some of his results for the former. The bulk of the paper is devoted to the cylindrical-cavity analysis.

where I_0, K_0 are modified Bessel functions of the first and second kinds of zero order.† These are related to the Bessel functions of the first kind and the Hankel functions by‡

$$K_\nu(z) = \frac{\pi i}{2} e^{\nu\pi i/2} H_\nu^{(1)}(iz), \qquad I_\nu(z) = e^{-\nu\pi i/2} J_\nu(iz). \tag{5.3.83}$$

For increasing r, it is found that $I_0(\beta r)$ is unbounded, whereas $K_0(\beta r) \to 0$. We thus set $B = 0$ in (5.3.82).

Consider now the boundary conditions. We may write for the transformed stress

$$\bar{\tau}_{rr} = \lambda\bar{\Delta} + 2\mu\frac{\partial \bar{u}_r}{\partial r} = (\lambda+2\mu)\bar{\Delta} - 2\mu\frac{\bar{u}_r}{r}. \tag{5.3.84}$$

We note from (5.3.80) that we may write u_r as

$$\bar{u}_r = -\frac{1}{r}\int_r^\infty \eta\,\bar{\Delta}(\eta, s)\,d\eta. \tag{5.3.85}$$

Now, we have that §

$$\frac{1}{\beta}\frac{d}{dr}\{rK_1(\beta r)\} = rK_0(\beta r), \tag{5.3.86}$$

so that

$$\bar{u}_r = \frac{A}{\beta} K_1(\beta r). \tag{5.3.87}$$

Then we have

$$\bar{\tau}_{rr}(r, s) = A\left\{(\lambda+2\mu)K_0(\beta r) - \frac{2\mu}{\beta r}K_1(\beta r)\right\}. \tag{5.3.88}$$

Let the boundary conditions be prescribed by

$$\tau_{rr}(a, t) = p(t), \qquad \bar{\tau}_{rr}(a, s) = \bar{p}(s). \tag{5.3.89}$$

Then A is easily found from (5.3.88) for $r = a$. The resulting solution, in inverted form, is given by

$$\tau_{rr}(r, t) = \frac{1}{2\pi i}\int_{c-i\infty}^{c+i\infty} \frac{F(\beta r)}{F(\beta a)}\bar{p}(s)e^{st}\,ds, \tag{5.3.90}$$

where

$$F(\beta r) = (\lambda+2\mu)K_0(\beta r) - 2\mu K_1(\beta r)/\beta r. \tag{5.3.91}$$

Evaluation of the inversion integral requires knowledge of the zeros of $F(\beta a)$. Selberg establishes, through fairly sophisticated considerations on the properties of Bessel functions, that only one root exists and that it is located in

† P. 190 of Reference [13].
‡ Formula 161 and formula 206 of Reference [13].
§ Formula 32 of Reference [13].

the second quadrant of the s-plane. We label that root as s_0 and the corresponding value of β is labelled β_0.

Now consider the particular case of $p(t) = H\langle t\rangle$, so that $\bar{p}(s) = 1/s$. Then Selberg expresses his result in the following form

$$\tau_{rr} = \frac{1}{\pi}\,\mathrm{Im}\int_0^{-\infty} \frac{F(\beta r)}{sF(\beta a)}\exp(st)\,\mathrm{d}s + 2\,\mathrm{Re}\,\frac{F(\beta_0 r)}{s_0 F'(\beta_0 a)}\exp(s_0 t) + \left(\frac{a_0}{r}\right)^2, \quad (5.3.92)$$

where Im, Re denote the imaginary and real parts of the respective expressions. Asymptotic expressions are obtainable for large values of r. Thus, using the approximate formula† for large z,

$$K_\nu(z) \sim \sqrt{\left(\frac{\pi}{2z}\right)}\mathrm{e}^{-z}. \quad (5.3.93)$$

Selberg obtains for $r/a \to \infty$

$$\sqrt{\left(\frac{r}{a}\right)}\tau_{rr} = \frac{\lambda+2\mu}{\sqrt{(2\pi)}}\,\mathrm{Im}\int_0^{\infty} \frac{\exp\left\{\dfrac{c_2\xi}{a}\left(t-\dfrac{r}{c_2}\right)\right\}}{\xi^{\frac{3}{2}}F(\xi)}\,\mathrm{d}\xi + $$

$$+(\lambda+2\mu)\sqrt{(2\pi)}\,\mathrm{Re}\,\frac{\exp\left\{\dfrac{c_2\xi_0}{a}\left(t-\dfrac{r}{c_2}\right)\right\}}{\xi_0^{\frac{3}{2}}F'(\xi_0)}, \quad (5.3.94)$$

where $\xi = \beta a$, $\xi_0 = \beta_0 a$.

Results were determined for the case of $\nu = \frac{1}{4}$ (corresponding to $\lambda = \mu$). The resulting value of $\beta_0 a$ was found to be $\beta_0 a = -0{\cdot}442057+0{\cdot}447357\mathrm{i}$. Figure 5.9(a) shows the result for a step loading, where $\bar{\sigma}_r$, $\bar{t}$ are given by

$$\bar{\sigma}_r = \sqrt{\left(\frac{r}{a}\right)}\sigma_r, \qquad \bar{t} = t - \frac{r-a}{a}. \quad (5.3.95)$$

Results are also given, without further theoretical discussion, for the case of a pressure pulse of the form

$$\tau_{rr}(a, t) = \mathrm{e}^{-\kappa t}. \quad (5.3.96)$$

These cases are shown in Fig. 5.9(b) and (c). In (b), all curves are for $r/a \to \infty$, for various values of κ. In (c), $\kappa = 0{\cdot}25$, and differing positions are shown.

We close this section by commenting on the more general problem of the cylindrical cavity subjected to tractions τ_{rr} and $\tau_{r\theta}$, arbitrarily prescribed with respect to θ, but invariant with respect to z. As in the case of the spherical cavity, Eringen [7] has contributed a general formal solution to the problem, with discussions of several special cases. Miklowitz [15], Jordan [11], and others have also contributed in this area.

† Formula 204 of Reference [13].

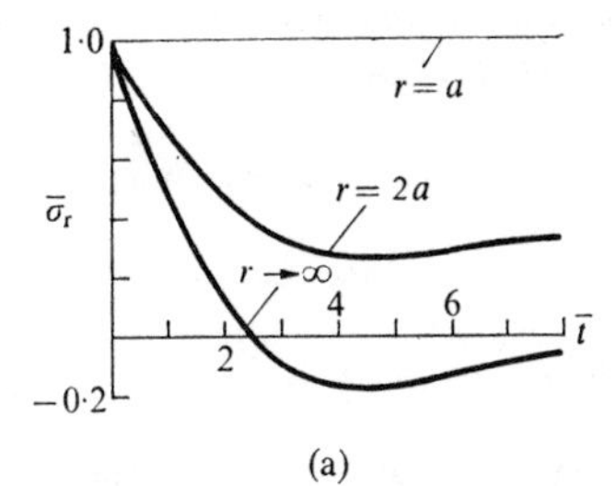

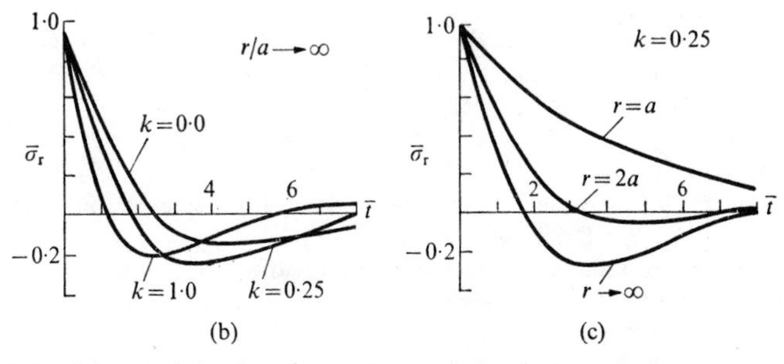

FIG. 5.9. Radial stress τ_{rr} resulting from a step pressure pulse in a cylindrical cavity (a), and from exponential decay pulses, (b) and (c). (After Selberg [18].)

References

1. ACHENBACH, J. D. and SUN, C. T. Propagation of waves from a spherical surface of time-dependent radius. *J. Acoust. Soc. Am.* **40,** 877–82 (1966).
2. BLAKE, F. G. Spherical wave propagation in solid media. *J. acoust. Soc. Am.* **24,** 211 (1952).
3. CAGNIARD, L. *Reflection et refraction des ondes seismiques progressive.* Gauthiers-Villars, Paris (1935).
4. DAS GUPTA, S. C. Waves and stresses produced in an elastic medium due to impulsive radial forces and twist on the surface of a spherical cavity. *Geofis. pura. Appl.* **27,** 3–8 (1954).
5. EASON, G., FULTON, J., and SNEDDON, I. N. The generation of waves in an infinite elastic solid by variable body forces. *Phil. Trans. R. Soc.* **A248,** 575–607 (1955–6).
6. ERINGEN, A. C. Elasto-dynamic problem concerning the spherical cavity. *Q. Jl Mech. appl. Math.* **10,** 257–70 (1957).
7. ——. Propagation of elastic waves generated by dynamical loads on a circular cavity. *J. appl. Mech.* **28,** 218–22 (1961).
8. FREDERICK, J. R. *Ultrasonic engineering.* John Wiley and Sons, New York (1965).
9. GOLDSMITH, W. and ALLEN, W. A. Graphical representation of the spherical propagation of explosive pulses in elastic media. *J. acoust. Soc. Am.* **27,** 47–55 (1955).

10. HOPKINS, H. G. Dynamic expansion of spherical cavities in metals. In *Progress in Solid Mechanics*, Vol. I, Chap. 3, (Ed. I. N. Sneddon and R. Hill) North Holland (1960).
11. JORDAN, D. W. The stress wave from a finite, cylindrical explosive source. *J. Math. Mech.* **11**, 503–51 (1962).
12. LOVE, A. E. H. *A treatise on the mathematical theory of elasticity*. Dover Publications, New York (1944).
13. MCLACHLAN, N. W. *Bessel functions for engineers*. Clarendon Press, Oxford (1961).
14. MEYER, M. L. On spherical near fields and far fields in elastic and viscoelastic solids. *J. Mech. Phys. Solids* **12**, 77–111 (1964).
15. MIKLOWITZ, J. Plane-stress unloading waves emanating from a suddenly punched hole in a stretched elastic plate. *J. appl. Mech.* **27**, 165–7 (1960).
16. MORSE, P. and FESHBACH, H. *Methods of theoretical physics* Vols. I and II. McGraw-Hill, New York (1953).
17. RAYLEIGH, J. W. S. *The theory of sound* Vols. I and II. Dover Publications, New York (1945).
18. SELBERG, H. L. Transient compression waves from spherical and cylindrical cavities. *Ark. Fys.* **5**, 97–108 (1952).
19. STERNBERG, E. On the integration of the equations of motion in the classical theory of elasticity. *Archs ration Mech. Analysis* **6**, 34 (1960).
20. —— and GINTIN, *Proc. 4th U.S. Congr. appl. Mech.* p. 793 (1962).
21. WATSON, G. N. *A treatise on the theory of Bessel functions* (2nd edn). Cambridge University Press (1966).
22. WRIGHT, J. K. and CARPENTER, E. W. The generation of horizontally polarized shear waves by underground explosions. *J. geophys. Res.* **67**, 1957–63 (1962).

Problems

5.1. Consider a plane harmonic dilatational wave to be propagating in an infinite medium. Determine the expression for the dilatation.

5.2. Consider two SH wave sources f_1, f_2, where $f_1 = c_2^2\, \delta(x-a)\delta(y)\exp(-i\omega t)$ and $f_2 = c_2^2\, \delta(x+a)\, \delta(y)\exp(-i\omega t)$. Using superposition of the simple SH wave source results, obtain an expression for $u_z(x, y, t)$. Assume the wavelength is large in comparison to $2a$. See if an approximate expression for u_z can be obtained for these circumstances.

5.3. Consider the problem of a spherical cavity subjected to a uniformly distributed pressure pulse $\tau_{rr}(a, t) = P_0\, \delta(t)$ and obtain the solution for the wave propagation. Sketch the waveform at various time intervals.

5.4. Consider a solid sphere of radius a. Determine the frequency equation governing the spherically symmetric 'expansion' modes of the sphere.

5.5. Consider the step pressure pulse solution (5.3.38). Discuss the consequences on the pulse shape for ζ becoming large. In the limit, does the wave begin to resemble a step pulse, with a sharp 'spike' disturbance at the wavefront?

5.6. Consider a cylindrical cavity, defined by $r = a$, in an infinite medium. Let the surface of the cavity be subjected to the prescribed displacements $u_r = u_z = 0$, $u_\theta(a, \theta, t) = u_0 \exp(i\omega t)$. Determine the resulting wave propagation.

6 Waves in semi-infinite media

As the next step in the study of elastic waves, we consider the propagation and reflection of waves in a semi-infinite media. It is, of course, the inclusion of a boundary that distinguishes this problem from those of the last chapter. Waves in semi-infinite solids, either homogeneous or inhomogeneous, have been of long-standing interest in seismology. Problems in ultrasonics, delay lines, soil dynamics, blast, and impact have also led to analysis of waves in a half-space.

The first area of study will be the propagation of plane harmonic waves in a half-space. This study will establish the characteristics of mode conversion that occurs when waves encounter a free boundary and the existence of surface waves. Understanding of the behaviour of simple harmonic waves will provide insight into the interaction of waves with more complicated boundary shapes, will find application in the study of waves in plates and rods, and will aid in the analysis of transient-loading problems.

The analysis of waves from surface and buried sources will then be carried out. The technical literature in this area is quite extensive, and the methods of analysis are quite sophisticated. Here, it will only be possible to give rather limited coverage to this topic. As an introduction, the simplest case of SH wave sources will be considered. This problem will be used to develop the method of steepest descent, an analytical tool representing an extension of the stationary-phase method developed in Chapter 1.

Finally, consideration will be given to wave propagation in a layered media. This topic also is extremely broad, with a large amount of technical literature and entire books devoted to the subject. The reflection and refraction of plane waves between two media in contact will be covered, a new type of surface wave considered, and a limited number of results presented for pulse propagation.

6.1. Propagation and reflection of plane waves in a half-space

When an elastic wave encounters a boundary between two media, energy is reflected and transmitted from and across the boundary. This is, of course, no different to phenomena occurring in acoustics and optics. If the boundary is a free surface, a pure reflection process will occur. It will be found that a distinguishing characteristic of the wave–boundary interaction process for elastic waves in a solid is that mode conversion occurs. This describes the behaviour by which an incident wave, either pressure or shear, is converted into two waves on reflection. This behaviour, along with the fact that two types of waves may exist in an elastic solid, accounts for the relative complexity of elastic wave problems in solids compared to equivalent problems in acoustics and electromagnetics. It will also be established that surface waves may propagate in a half-space. These waves, of long-standing interest in seismology, have counterparts in hydrodynamics and electromagnetics.

6.1.1. *Governing equations*

Let us consider plane harmonic waves propagating in the half-space $y \geqslant 0$. No generality will be lost if we assume the wave normal **n** to lie in the x,y-plane. This will be called the 'vertical' plane and the x,z-plane, which is the surface of the half-space, will be called the 'horizontal' plane. On the basis of our studies of plane waves in infinite media, we should realize that the particle motion due to dilatational effects will be in the direction of the wave normal and will thus lie completely in the vertical plane. The transverse particle motion due to shear may have components in the vertical plane and, also, parallel to the horizontal plane. The general situation is shown in Fig. 6.1. The normal displacement component is u_n and the transverse

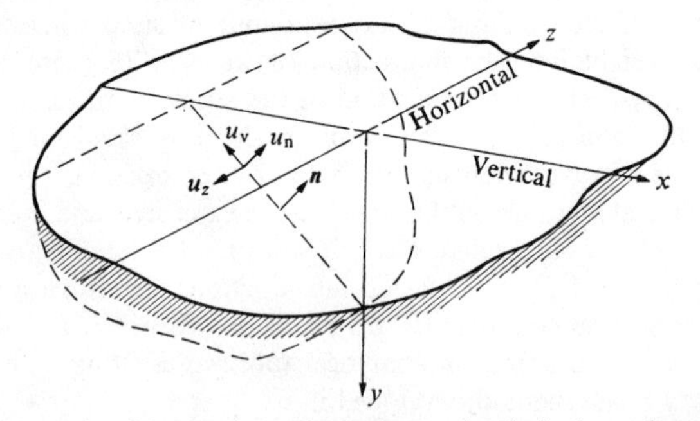

FIG. 6.1. Plane wave, with wave normal **n** in the x,y-(vertical) plane advancing toward a free surface.

components are u_v and u_z which are, respectively, in the vertical and horizontal planes. Finally, since every point along the plane of the wave is executing the same motion, we have that the motion is invariant with respect to z if the wave normal is in the vertical plane.

The governing equations for our investigation are

$$u_x = \frac{\partial \Phi}{\partial x} + \frac{\partial H_z}{\partial y}, \qquad u_y = \frac{\partial \Phi}{\partial y} - \frac{\partial H_z}{\partial x}, \tag{6.1.1}$$

$$u_z = -\frac{\partial H_x}{\partial y} + \frac{\partial H_y}{\partial x}, \qquad \frac{\partial H_x}{\partial x} + \frac{\partial H_y}{\partial y} = 0, \tag{6.1.2}$$

$$\nabla^2 \Phi = \frac{1}{c_1^2} \frac{\partial^2 \Phi}{\partial t^2}, \qquad \nabla^2 H_p = \frac{1}{c_2^2} \frac{\partial^2 H_p}{\partial t^2} \qquad (p = x, y, z). \tag{6.1.3}$$

where the z independence of all quantities has been used. The second equation of (6.1.2) results from $\nabla.\mathbf{H} = 0$. The stress–displacement expressions are given by

$$\tau_{xx} = (\lambda+2\mu)\left(\frac{\partial u_x}{\partial x}+\frac{\partial u_y}{\partial y}\right)-2\mu\frac{\partial u_y}{\partial y}, \tag{6.1.4}$$

$$\tau_{yy} = (\lambda+2\mu)\left(\frac{\partial u_x}{\partial x}+\frac{\partial u_y}{\partial y}\right)-2\mu\frac{\partial u_x}{\partial x}, \tag{6.1.5}$$

$$\tau_{zz} = \frac{\lambda}{2(\lambda+\mu)}(\tau_{xx}+\tau_{yy}), \tag{6.1.6}$$

$$\tau_{xy} = \mu\left(\frac{\partial u_x}{\partial y}+\frac{\partial u_y}{\partial x}\right), \tag{6.1.7}$$

$$\tau_{yz} = \mu\frac{\partial u_z}{\partial y}, \qquad \tau_{xz} = 0. \tag{6.1.8}$$

In terms of the potentials Φ, H_x, H_y, H_z, these are

$$\tau_{xx} = (\lambda+2\mu)\left(\frac{\partial^2\Phi}{\partial x^2}+\frac{\partial^2\Phi}{\partial y^2}\right)-2\mu\left(\frac{\partial^2\Phi}{\partial y^2}-\frac{\partial^2 H_z}{\partial y\,\partial x}\right), \tag{6.1.9}$$

$$\tau_{yy} = (\lambda+2\mu)\left(\frac{\partial^2\Phi}{\partial x^2}+\frac{\partial^2\Phi}{\partial y^2}\right)-2\mu\left(\frac{\partial^2\Phi}{\partial x^2}+\frac{\partial^2 H_z}{\partial x\,\partial y}\right), \tag{6.1.10}$$

$$\tau_{xy} = \mu\left(2\frac{\partial^2\Phi}{\partial x\,\partial y}+\frac{\partial^2 H_z}{\partial y^2}-\frac{\partial^2 H_z}{\partial x^2}\right), \tag{6.1.11}$$

$$\tau_{yz} = \mu\left(-\frac{\partial^2 H_x}{\partial y^2}+\frac{\partial^2 H_y}{\partial y\,\partial x}\right), \qquad \tau_{xz} = 0. \tag{6.1.12}$$

Finally, we have the boundary conditions given by

$$\tau_{yy} = \tau_{yx} = \tau_{yz} = 0, \qquad y = 0. \tag{6.1.13}$$

We make the observation, at this stage, that we are actually dealing with two uncoupled problems of wave motion. Thus, we see that u_x, u_y depend only on Φ and H_z, which are in turn governed by the independent scalar wave equations (6.1.3), where $p = z$ in the second of the two equations. Further, the stresses τ_{xx}, τ_{yy}, τ_{xy} depend only on u_x, u_y, and hence only on Φ and H_z. The component u_z depends on H_x, H_y which are governed by the second of (6.1.3), where $p = y,z$. Further, the stress τ_{yz} depends only on u_z, and hence only on H_x, H_y. This makes it possible to resolve the motion into two parts, where one is plane strain with $u_z = \partial/\partial z = 0$, u_x, $u_y \neq 0$ and the other is SH wave motion, where $\partial/\partial z = u_x = u_y = 0$, $u_z \neq 0$. If we did not recognize the uncoupling at this stage, it would still be brought out in the boundary condition equations (this will be shown in the analysis of waves in plates). Thus we have the following:

plane strain: $u_z = \partial/\partial z = 0$,

$$u_x = \frac{\partial \Phi}{\partial x} + \frac{\partial H_z}{\partial y}, \qquad u_y = \frac{\partial \Phi}{\partial y} - \frac{\partial H_z}{\partial x}, \tag{6.1.14}$$

$$\nabla^2 \Phi = \frac{1}{c_1^2}\frac{\partial^2 \Phi}{\partial t^2}, \qquad \nabla^2 H_z = \frac{1}{c_2^2}\frac{\partial^2 H_z}{\partial t^2}. \tag{6.1.15}$$

$\tau_{xx}, \tau_{yy}, \tau_{zz}, \tau_{xy}$: eqns (6.1.4)–(6.1.7), (6.1.9)–(6.1.11).

$$\tau_{yy} = \tau_{xy} = 0, \qquad y = 0. \tag{6.1.16}$$

SH waves: $u_x = u_y = \partial/\partial z = 0$,

$$u_z = -\frac{\partial H_x}{\partial y} + \frac{\partial H_y}{\partial x}, \qquad \frac{\partial H_x}{\partial x} + \frac{\partial H_y}{\partial y} = 0, \tag{6.1.17}$$

$$\nabla^2 H_x = \frac{1}{c_2^2}\frac{\partial^2 H_x}{\partial t^2}, \qquad \nabla^2 H_y = \frac{1}{c_2^2}\frac{\partial^2 H_y}{\partial t^2}, \tag{6.1.18}$$

τ_{yz}: eqns (6.1.8), (6.1.12),

$$\tau_{yz} = 0, \qquad y = 0. \tag{6.1.19}$$

We also note in the case of SH waves that we could directly consider the displacement equation of motion

$$\nabla^2 u_z = \frac{1}{c_2^2}\frac{\partial^2 u_z}{\partial t^2}. \tag{6.1.20}$$

Consider now the solution to the case of plane strain and let

$$\Phi = f(y)e^{i(\xi x - \omega t)}, \qquad H_z = h_z(y)e^{i(\zeta x - \omega t)}. \tag{6.1.21}$$

Substitution in the governing equations (6.1.3) gives

$$\frac{d^2 f}{dy^2} + \alpha^2 f = 0, \qquad \frac{d^2 h_z}{dy^2} + \beta^2 h = 0, \tag{6.1.22}$$

where

$$\alpha^2 = \frac{\omega^2}{c_1^2} - \xi^2, \qquad \beta^2 = \frac{\omega^2}{c_2^2} - \zeta^2. \tag{6.1.23}$$

The plane wave solutions for Φ and H_z are then given by

$$\Phi = A_1 e^{i(\xi x - \alpha y - \omega t)} + A_2 e^{i(\xi x + \alpha y - \omega t)}, \tag{6.1.24}$$

$$H_z = B_1 e^{i(\zeta x - \beta y - \omega t)} + B_2 e^{i(\zeta x + \beta y - \omega t)}. \tag{6.1.25}$$

If we define θ_1, θ_2 as the angles between the y-axis and the wave normal of the dilatational and shear waves, we may write

$$\xi = \gamma_1 \sin \theta_1, \qquad \alpha = \gamma_1 \cos \theta_1, \tag{6.1.26}$$

$$\zeta = \gamma_2 \sin \theta_2, \qquad \beta = \gamma_2 \cos \theta_2, \tag{6.1.27}$$

where γ_1, γ_2 are the wavenumbers along the respective waves. Then Φ and H_z may also be written as

$$\Phi = A_1 \exp\{i\gamma_1(\sin \theta_1 x - \cos \theta_1 y - c_1 t)\} + \\ + A_2 \exp\{i\gamma_1(\sin \theta_1 x + \cos \theta_1 y - c_1 t)\}, \tag{6.1.28}$$

$$H_z = B_1 \exp\{i\gamma_2(\sin \theta_2 x - \cos \theta_2 y - c_2 t)\} + \\ + B_2 \exp\{i\gamma_2(\sin \theta_2 x + \cos \theta_2 y - c_2 t)\}. \tag{6.1.29}$$

The resulting plane-wave situation is shown in Fig. 6.2 for the Φ wave.

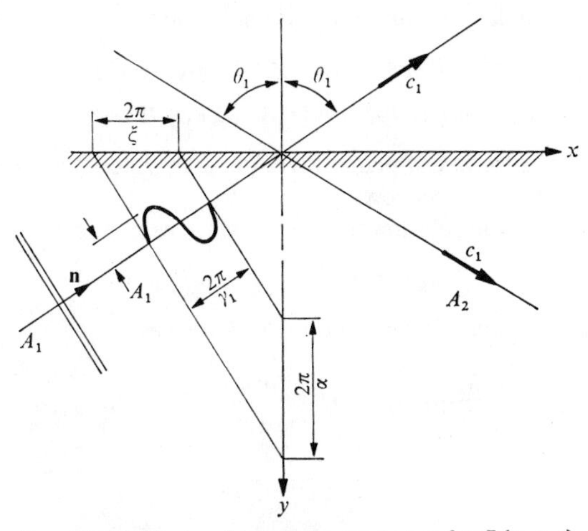

FIG. 6.2. Incident and reflected wave system for $\Phi(x, y, t)$.

Since γ_1 is the wavenumber in the wave direction, $2\pi/\gamma_1$ is the wavelength in the direction of travel. It is seen that ξ and α may be interpreted as the horizontal and vertical wavenumbers, with $2\pi/\xi$, $2\pi/\alpha$ being the horizontal and vertical wavelengths. The horizontal wavelength is of interest since this would be a quantity readily measured by surface transducers.

We now apply the solutions (6.1.28) and (6.1.29) to the plane-strain boundary conditions (6.1.16), obtaining

$$(\tau_{yy})_{y=0}: \quad \gamma_1^2(2 \sin^2\theta_1 - k^2)(A_1 + A_2)\exp\{i\gamma_1(\sin \theta_1 x - c_1 t)\} - \\ \gamma_2^2 \sin 2\theta_2(B_1 - B_2)\exp\{i\gamma_2(\sin \theta_2 x - c_2 t)\} = 0, \tag{6.1.30}$$

$$(\tau_{xy})_{y=0}: \quad \gamma_1^2 \sin 2\theta_1(A_1 - A_2)\exp\{i\gamma_1(\sin \theta_1 x - c_1 t)\} - \\ \gamma_2^2 \cos 2\theta_2(B_1 + B_2)\exp\{i\gamma_2(\sin \theta_2 x - c_2 t)\} = 0, \tag{6.1.31}$$

where we have again introduced the ratio of the wave velocities,

$$k^2 = c_1^2/c_2^2 = (\lambda+2\mu)/\mu.$$

Of course, we could immediately factor out $\exp(-i\omega t)$ from the above. We now take note of the following argument. If the preceding results are to hold for arbitrary x, then we must be able to factor $\exp(i\gamma_1 \sin\theta_1 x)$, $\exp(i\gamma_2 \sin\theta_2 x)$ from the results. This can only occur if we have

$$\gamma_1 \sin\theta_1 = \gamma_2 \sin\theta_2. \tag{6.1.32}$$

Since $\omega = \gamma_1 c_1 = \gamma_2 c_2$, so that $\gamma_2/\gamma_1 = c_1/c_2 = k$, we may write the above also as

$$\gamma_2/\gamma_1 = \sin\theta_1/\sin\theta_2 = k. \tag{6.1.33}$$

This result may be interpreted as the form of Snell's law for elastic waves. With this, the boundary condition equations reduce to

$$\gamma_1^2(2\sin^2\theta_1-k^2)(A_1+A_2)-\gamma_2^2 \sin 2\theta_2(B_1-B_2) = 0, \tag{6.1.34}$$

$$\gamma_2^2 \sin 2\theta_1(A_1-A_2)-\gamma_2^2 \cos 2\theta_2(B_1+B_2) = 0. \tag{6.1.35}$$

This governs the reflection of plane waves in a half-space. Various cases will be considered in the next section.

We now consider the solution for the SH wave case, and let

$$H_x = h_x(y)e^{i(\xi x-\omega t)}, \qquad H_y = h_y(y)e^{i(\xi x-\omega t)}. \tag{6.1.36}$$

Substitution in the governing equations (6.1.18) gives

$$\frac{d^2h_x}{dy^2}+\eta^2 h_x = 0, \qquad \frac{d^2h_y}{dy^2}+\eta^2 h_y = 0, \tag{6.1.37}$$

where

$$\eta^2 = \frac{\omega^2}{c_2^2}-\xi^2. \tag{6.1.38}$$

The solutions are

$$H_x = C_1 e^{i(\xi x-\eta y-\omega t)}+C_2 e^{i(\xi x+\eta y-\omega t)}, \tag{6.1.39}$$

$$H_y = D_1 e^{i(\xi x-\eta y-\omega t)}+D_2 e^{i(\xi x+\eta y-\omega t)}. \tag{6.1.40}$$

These results may also be expressed in terms of an incidence angle θ_3, as was previously done for Φ and H_z.

Not all of the above quantities are independent. This is a consequence of the $\nabla.\mathbf{H} = 0$ condition that attaches to the Helmholtz vector resolution. As mentioned in the previous chapter, this enables the three components of the displacement to be established from the four components of Φ and $\mathbf{H}$. In the present study of plane strain and SH waves, it was found that u_x and u_y depend only on the two quantities Φ and H_z and are thus uniquely determined. Not so for u_z, which depends on H_x and H_y. For the present case of z independence, the divergence condition is given by the second eqn (6.1.2).

Applying this condition to the results (6.1.39) and (6.1.40) gives

$$i\xi(C_1 e^{-i\eta y}+C_2 e^{i\eta y})+i\eta(-D_1 e^{-i\eta y}+D_2 e^{i\eta y}) = 0, \tag{6.1.41}$$

where $\exp\{i(\xi x-\omega t)\}$ has been factored out. This re-groups to

$$(\xi C_1-\eta D_1)e^{-i\eta y}+(\xi C_2+\eta D_2)e^{i\eta y} = 0. \tag{6.1.42}$$

In order for the above result to hold for all y, we must have C_1, D_1, C_2, D_2 related by

$$\xi C_1 = \eta D_1, \qquad \xi C_2 = -\eta D_2. \tag{6.1.43}$$

Thus, two of the constants of (6.1.39) and (6.1.40) may be eliminated. Arbitrarily choosing to eliminate D_1, D_2, we thus have

$$H_x = C_1 e^{i(\xi x-\eta y-\omega t)}+C_2 e^{i(\xi x+\eta y-\omega t)}, \tag{6.1.44}$$

$$H_y = \frac{\xi}{\eta}C_1 e^{i(\xi x-\eta y-\omega t)}-\frac{\xi}{\eta}C_2 e^{i(\xi x+\eta y-\omega t)}. \tag{6.1.45}$$

Substituting these results into the boundary condition $\tau_{yz} = 0$, $y = 0$ gives

$$(\tau_{yz}): \qquad \eta^2(C_1+C_2)+\xi^2(C_1+C_2) = 0. \tag{6.1.46}$$

This result governs the reflection of SH waves in a half-space. Specific cases will be considered in the next section.

6.1.2. *Waves at oblique incidence*

We now wish to determine the reflection characteristics of various types of incoming waves. The order of investigation will be (1) P waves at oblique incidence, (2) SV waves at oblique incidence, (3) SH waves at oblique incidence, and (4) Pairs of incident waves. By oblique incidence, we mean that θ_1 or θ_2 of the incident wave may be in the range of $0 < \theta_1, \theta_2 < \pi/2$.

1. *Incident P waves.* Let us assume that the incident wave is compressional only. This results if $B_1 = 0$ in the wave expressions (6.1.28) and (6.1.29). Further, we presume that the angle θ_1, frequency ω, and amplitude A_1 of the wave are specified. The boundary condition equations (6.1.30), (6.1.31) become

$$\gamma_1^2(2\sin^2\theta_1-k^2)(A_1+A_2)+\gamma_2^2\sin 2\theta_2 B_2 = 0, \tag{6.1.47}$$

$$\gamma_1^2\sin 2\theta_1(A_1-A_2)-\gamma_2^2\cos 2\theta_2 B_2 = 0. \tag{6.1.48}$$

Solving, we obtain the amplitude ratios

$$\frac{A_2}{A_1} = \frac{\sin 2\theta_1\sin 2\theta_2-k^2\cos^2 2\theta_2}{\sin 2\theta_1\sin 2\theta_2+k^2\cos^2 2\theta_2}, \tag{6.1.49}$$

$$\frac{B_2}{A_1} = \frac{2\sin 2\theta_1\cos 2\theta_2}{\sin 2\theta_1\sin 2\theta_2+k^2\cos^2 2\theta_2}. \tag{6.1.50}$$

The reflection angle θ_2 and the wavenumber γ_2 are given by (6.1.33).

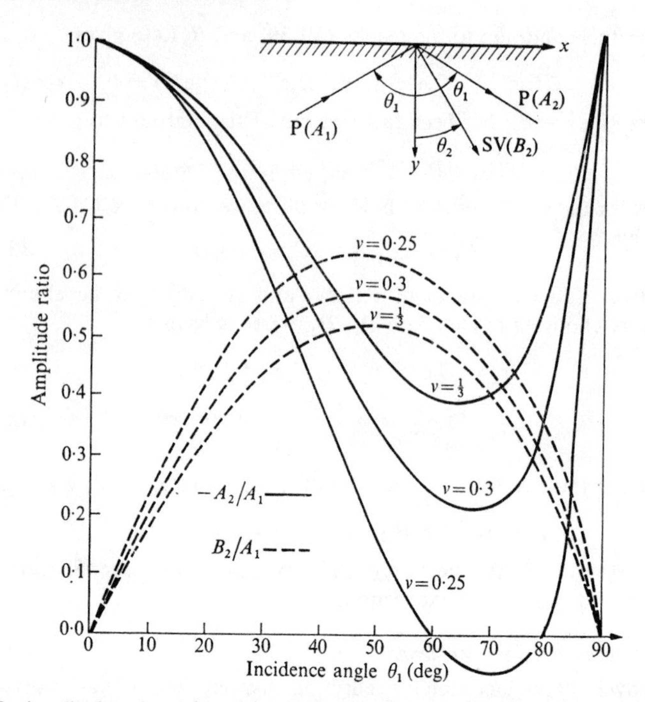

FIG. 6.3. Amplitude ratios A_2/A_1, B_2/A_1 for incident P waves, for various Poisson's ratios, with a ray representation of the reflection also shown.

One of the major features of this result is that, for a single P wave incident, two waves, P and SV, are reflected. This phenomenon is referred to as *mode conversion.* We recall that P and S waves propagate independently, or uncoupled, in a material. However, when a free surface is present, coupling of the two wave systems occurs through the boundary conditions.

The nature of the reflection is shown as part of Fig. 6.3, where a simple ray representation has been used to designate the various waves and directions of propagation. We note from (6.1.33) that $\gamma_2 = c_1\gamma_1/c_2 > \gamma_1$. Hence from $\sin\theta_2 = \gamma_1 \sin\theta_1/\gamma_2$ we conclude that $\theta_2 < \theta_1$ always. From (6.1.49) and (6.1.50), it is seen that the amplitude ratios A_2/A_1, B_2/A_1 depend only on the incidence angle and the Poisson's ratio of the material (since k^2 depends only on Poisson's ratio). These ratios are plotted in Fig. 6.3 for $\nu = \frac{1}{3}, 0{\cdot}3, \frac{1}{4}$. We note an apparent anomaly for B_2/A_1 when $\nu = \frac{1}{3}$. Thus, $B_2/A_1 > 1$ for $\theta \approx 50\,°$, possibly implying violation of energy conservation. However, energy transfer depends on propagation velocity as well as amplitude. The SV waves are slower than the P waves, so it is possible for amplitude ratios greater than unity to occur.

A special case of interest is that of normal incidence ($\theta_1 = 0°$). From (6.1.49) and (6.1.50) we obtain

$$A_2/A_1 = -1, \qquad B_2/A_1 = 0. \tag{6.1.51}$$

This corresponds to an incident compression wave reflecting as tension and tension reflecting as compression, a result first noted in the case of longitudinal waves in rods. A second special case occurs for $A_2/A_1 = 0$, implying no reflected P wave. This occurs if

$$\sin 2\theta_1 \sin 2\theta_2 - k^2 \cos^2\theta_2 = 0. \tag{6.1.52}$$

This equation can be satisfied only for selected ranges of ν. Thus, if $\nu = \frac{1}{4}$, we obtain $\theta_1 = 60°$, $77{\cdot}5°$ as critical values of incidence for which no P wave is reflected. These values also appear in Fig. 6.3.

2. *Incident SV waves.* This case is obtained by setting $A_1 = 0$ in (6.1.28) and the boundary condition equations (6.1.34) and (6.1.35). The resulting amplitude ratios are

$$\frac{B_2}{B_1} = \frac{\sin 2\theta_1 \sin 2\theta_2 - k^2 \cos^2 2\theta_2}{\sin 2\theta_1 \sin 2\theta_2 + k^2 \cos^2 2\theta_2}, \tag{6.1.53}$$

$$\frac{A_2}{B_1} = \frac{-2k^2 \sin 2\theta_2 \cos 2\theta_2}{\sin 2\theta_1 \sin 2\theta_2 + k^2 \cos^2 2\theta_2}. \tag{6.1.54}$$

As for incident P waves, mode conversion again occurs for incident SV waves. The ray representation of the wave reflection is shown in Fig. 6.4. The reflection angle θ_1 and wavenumber γ_1 are determined by (6.1.33). We see for both types of incident wave, P or SV, that the ray of the reflected P wave is 'outside' that of the SV wave. The amplitude ratios A_2/B_1, B_2/B_1 are shown in Fig. 6.4 for several values of Poisson's ratio.

Various special cases of wave reflection are easily found. The case of normal incidence is given by $\theta_2 = 0$. We have from (6.1.53) and (6.1.54) that

$$A_2/B_1 = 0, \qquad B_2/B_1 = -1. \tag{6.1.55}$$

The case of $\theta_2 = 45°$ gives $A_2/B_1 = 0$, $B_2/B_1 = 1$. This result finds application in plate theory. It is also possible to have an incident SV wave, with only a reflected P wave ($B_2 = 0$). This occurs for

$$\sin 2\theta_1 \sin 2\theta_2 = k^2 \cos^2 2\theta_2. \tag{6.1.56}$$

A special case of particular interest occurs for θ_2 beyond the *critical angle*. Thus, referring to Fig. 6.4, we note that for increasing θ_2 there will occur a value of θ_2 such that the reflected P wave will be tangental to the surface. The value of θ_2 for which this occurs will be given from (6.1.33) as

$$\sin \theta_1 = k \sin \theta_2 = 1, \tag{6.1.57}$$

where $k > 1$ always. For example, if $\nu = \frac{1}{3}$, we have that $k = 2$ and θ_2

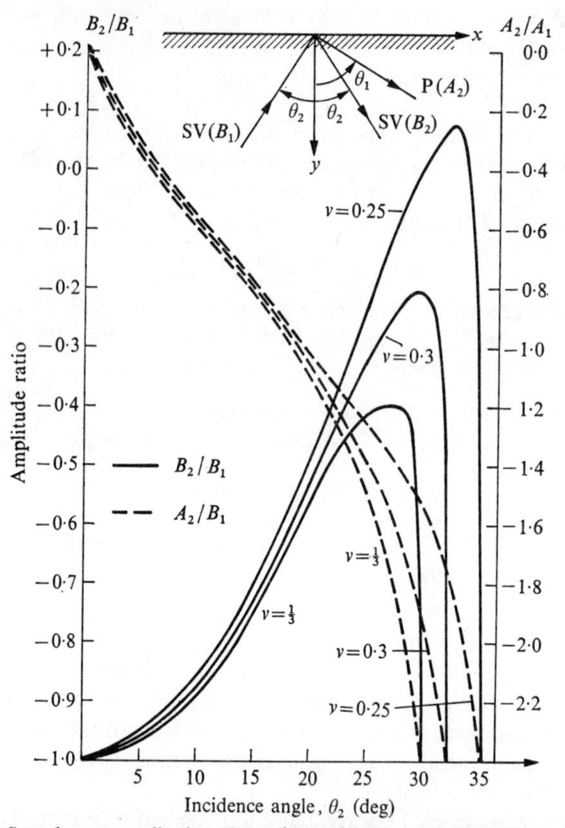

FIG. 6.4. Reflected wave amplitude ratios A_2/B_1, B_2/B_1 for incident SV waves and various Poisson's ratios, with the ray representation of the reflection also shown.

(critical) = 30 °. Now, since SV waves may be incident at any angle, values of θ_2 greater than the critical value are possible. But this would predict values of $\sin\theta_1 > 1$, an impossible situation. To understand the nature of the reflection phenomenon for this case, we must go back to the original differential equation determining Φ, specifically the first of (6.1.22), and note that $\alpha^2 > 0$ was assumed in arriving at the results (6.1.24) and (6.1.28). Now we may express α^2 as

$$\alpha^2 = \frac{\omega^2}{c_1^2} - \xi^2 = \frac{\omega^2}{c_1^2} - \gamma_1^2 \sin^2\theta_1. \tag{6.1.58}$$

Since $\gamma_1 \sin\theta_1 = \gamma_2 \sin\theta_2$ and $\omega = \gamma_1 c_1 = \gamma_2 c_2$, this may be put in the form

$$\alpha^2 = \frac{\gamma_2^2}{k^2}(1 - k^2 \sin^2\theta_2). \tag{6.1.59}$$

If (6.1.57) holds, $\alpha^2 = 0$, and $\alpha^2 < 0$ if $k^2 \sin^2\theta_2 > 1$. Although $\sin\theta_1 > 1$ is not possible, $\alpha^2 \leqslant 0$ is possible. Our approach then is to re-examine the governing equation for these cases. We have

$$\alpha^2 = 0, \qquad \mathrm{d}^2 f/\mathrm{d}y^2 = 0, \qquad f = A_1 y + A_2, \tag{6.1.60}$$

$$\Phi = (A_1 y + A_2)\mathrm{e}^{\mathrm{i}(\xi x - \omega t)}. \tag{6.1.61}$$

The A_1 term must be discarded on physical grounds in the present half-space problem, leaving a plane wave of constant amplitude travelling parallel to the free surface. The behaviour is shown in Fig. 6.5(a). If

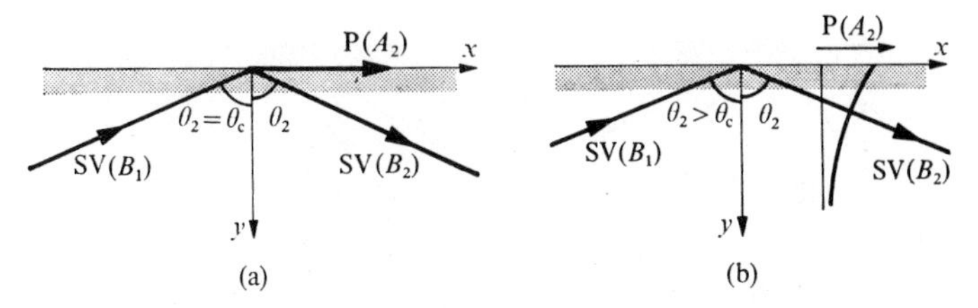

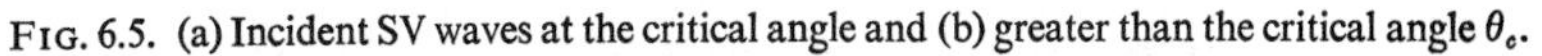

FIG. 6.5. (a) Incident SV waves at the critical angle and (b) greater than the critical angle θ_c.

$$\alpha^2 < 0, \qquad \frac{\mathrm{d}^2 f}{\mathrm{d}y^2} - \bar{\alpha}^2 f = 0, \qquad \bar{\alpha}^2 = -\alpha^2, \tag{6.1.62}$$

$$\Phi = A_1 \mathrm{e}^{\bar{\alpha} y}\mathrm{e}^{\mathrm{i}(\xi x - \omega t)} + A_2 \mathrm{e}^{-\bar{\alpha} y}\mathrm{e}^{\mathrm{i}(\xi x - \omega t)}. \tag{6.1.63}$$

Again, we discard A_1 because of the unbounded wave amplitude for increasing y. The result is an exponentially decaying wave, as shown in Fig. 6.5(b).

3. *Incident SH waves*. This is the simplest case of wave reflection. Thus, from (6.1.46), we have

$$(\eta^2 + \xi^2)(C_1 + C_2) = 0, \qquad C_2 = -C_1, \tag{6.1.64}$$

and the reflection angle is equal to the incidence angle. Thus, the SH wave reflects as itself, with no mode conversion, quite analogous to acoustic wave reflection. Now, if a shear wave of arbitrary polarization impinges on a free surface, the SV portion of the wave will lose a portion of its energy to P waves, whereas the SH portion of the amplitude and energy will reflect with only change in phase.

4. *Reflections of pairs of waves*. Under special circumstances, P and SV waves may reflect as themselves. This can occur when pairs of P and SV waves with specific amplitude ratios are incident on the free boundary. The necessary conditions may be obtained directly from (6.1.34), (6.1.35). However, by first noting that $\gamma_2^2/\gamma_1^2 = k^2$ and $\sin\theta_1 = k\sin\theta_2$, we may reduce the

cited equations to

$$\cos 2\theta_2(A_1+A_2)+\sin 2\theta_2(B_1-B_2) = 0, \tag{6.1.65}$$

$$\sin 2\theta_1(A_1-A_2)-k^2\cos 2\theta_2(B_1+B_2) = 0. \tag{6.1.66}$$

These relations may be satisfied by either

$$A_1 = -A_2, \qquad B_1 = B_2, \qquad A_1/B_1 = k^2 \cos 2\theta_2 \csc 2\theta_1 \tag{6.1.67}$$

or

$$A_1 = A_2, \qquad B_1 = -B_2, \qquad A_1/B_1 = -\tan 2\theta_2. \tag{6.1.68}$$

The two resulting cases of wave reflection are shown in Fig. 6.6(a) and (b)

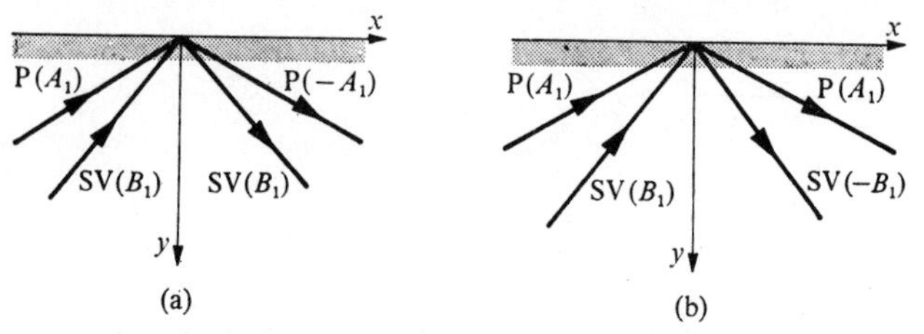

FIG. 6.6. (a) Reflections of pairs of waves, with $A_1 = -A_2$, $B_1 = B_2$, and (b) $A_1 = A_2$, $B_1 = -B_2$.

respectively. This behaviour finds application in the study of plates.

6.1.3. *Waves at grazing incidence*

A study of the previously-derived amplitude ratios for P and SV waves shows that for $\theta_1 \to 90°$ (in case of P waves) or $\theta_2 \to 90°$ (in case of SV waves) that no mode conversion occurs so that a P wave incoming at grazing incidence reflects as itself, and likewise for a grazing SV wave. This is also shown in the results of Figs. 6.3 and 6.4. However, it has been shown by a proper limiting process involving the amplitude ratios that other cases of reflection are possible.† In our discussion of this case, however, we shall follow the procedure used in investigating SV waves incident beyond the critical angle.

Thus, in obtaining (6.1.24) and (6.1.25) for Φ and H_z, it was assumed that $\alpha^2, \beta^2 > 0$ in (6.1.22) and (6.1.23). However, as we realize from the study of the special case of SV waves, as $\theta_1 \to 90°$, $\alpha \to 0$ and as $\theta_2 \to 90°$, $\beta \to 0$. We thus have for $\alpha \to 0$ (P waves grazing) that

$$\frac{d^2f}{dy^2}+\alpha^2 f = 0 \to \frac{d^2f}{dy^2} = 0, \qquad f = A_1+A_2y, \tag{6.1.69}$$

$$\Phi = (A_1+A_2y)e^{i(\xi x-\omega t)}, \tag{6.1.70}$$

† Goodier and Bishop [14].

and for $\beta \to 0$ (SV waves grazing) that

$$\frac{d^2h_z}{dy^2}+\beta^2 h_z = 0 \to \frac{d^2h_z}{dy^2} = 0, \qquad h_z = B_1+B_2y, \tag{6.1.71}$$

$$H_z = (B_1+B_2y)e^{i(\xi x-\omega t)}, \tag{6.1.72}$$

$$\Phi = A_2 e^{-\bar{\alpha}y} e^{i(\xi x-\omega t)}. \tag{6.1.73}$$

We note that for grazing SV waves, the critical angle has been excluded so that Φ is given by (6.1.63), as shown in the previous section.

For the case of grazing P waves, we set $B_1 = 0$ in (6.1.25) and, using (6.1.70), we have that

$$\Phi = (A_1+A_2y)e^{i(\xi x-\omega t)}, \tag{6.1.74}$$

$$H_z = B_2 e^{i(\xi x+\beta y-\omega t)}. \tag{6.1.75}$$

The wave reflection situation is as shown in Fig. 6.7(a). For the case of grazing

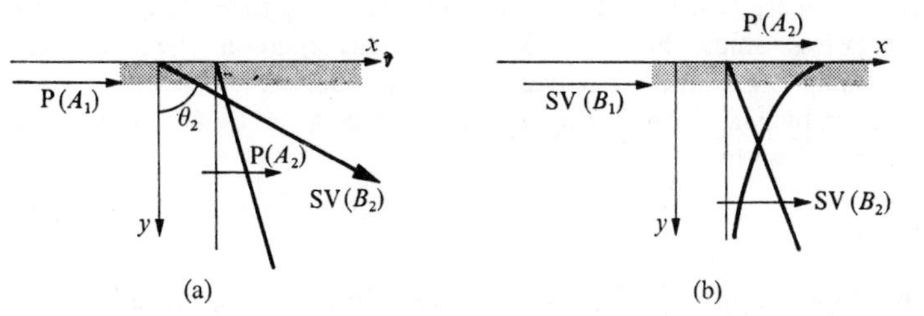

FIG. 6.7. Reflection of waves at grazing incidence with (a) P waves incident and (b) SV waves incident.

SV waves, we directly have (6.1.72) and (6.1.73) as describing the wave propagation. The wave system is as shown in Fig. 6.7(b).

The possibility of wave systems with linearly increasing amplitudes, as occur in the preceding solutions, certainly violates physical intuition, as have other mathematically correct solutions obtained in other studies. Goodier and Bishop suggested that SV wavefronts having some of the characteristics just illustrated are encountered in some cases of impact, although the continuations of these fronts into the interior was left open to question. In the context of the half-space problem, such waves are of little interest. However, it will be found that these waves have a role in the theory of vibrations of finite elastic plates as governed by the equations of elasticity.

6.1.4. *Surface waves*

We have shown in the work of Chapter 5 that in an unbounded elastic media two and only two types of waves can be propagated. In the preceding sections we have investigated the interaction of these waves with a boundary and have noted the mode conversion that occurs when P or SV waves impinge, also noting that at most only two waves of the P and SV type are produced. However, when there is a boundary, as in the half-space problem, a third type of wave may exist whose effects are confined closely to the surface. These waves were first investigated by Lord Rayleigh [37], who showed that their effect decreases rapidly with depth and that their velocity of propagation is smaller than that of body waves.

The discovery of this wave-type was closely related to seismology, where it was early observed that earthquake tremors consisted of two early, rather minor disturbances corresponding to P and SV wave arrivals, followed closely by a significant damage-causing tremor. Such a disturbance was not consistent with existing understanding of elastic wave theory. The first question was, therefore, whether another wave-type could exist. Further consideration strongly suggested that such a wave should be a surface wave. Thus, the relative insignificance of P and SV waves was considered to be a consequence of volumetric dispersion of their energy into the earth's interior. The significant energy associated with the third wave suggested that it dissipated its energy less rapidly than P and SV waves. This could only be accounted for by assuming it was essentially confined to the surface.

We start by again considering Φ and H_z to be given by (6.1.3) with, now, $f(y)$, $h_z(y)$ given by

$$\frac{\mathrm{d}^2 f}{\mathrm{d}y^2}-\bar{\alpha}^2 f=0, \qquad \frac{\mathrm{d}^2 h_z}{\mathrm{d}y^2}-\bar{\beta}^2 h_z=0, \tag{6.1.76}$$

where $\bar{\alpha}^2=-\alpha^2$, $\bar{\beta}^2=-\beta^2$, and α^2, β^2 are as previously defined by (6.1.23). We recognize the first of (6.1.76) as arising in the study of incident SV waves beyond the critical angle. The solutions give waves with exponentially increasing and decreasing parts. Discarding the increasing terms, we have

$$\Phi = A\mathrm{e}^{-\bar{\alpha}y}\mathrm{e}^{\mathrm{i}\xi(x-ct)}, \qquad H_z = B\mathrm{e}^{-\bar{\beta}y}\mathrm{e}^{\mathrm{i}\xi(x-ct)}. \tag{6.1.77}$$

The expressions for the displacements and stresses are

$$u_x = (\mathrm{i}\xi A\mathrm{e}^{-\bar{\alpha}y}-\bar{\beta}B\mathrm{e}^{-\bar{\beta}y})\mathrm{e}^{\mathrm{i}\xi(x-ct)}, \tag{6.1.78}$$

$$u_y = -(\bar{\alpha}A\mathrm{e}^{-\bar{\alpha}y}+\mathrm{i}\xi B\mathrm{e}^{-\bar{\beta}y})\mathrm{e}^{\mathrm{i}\xi(x-ct)}, \tag{6.1.79}$$

$$\tau_{xx} = \mu\{(\bar{\beta}^2-\xi^2-2\bar{\alpha}^2)A\mathrm{e}^{-\bar{\alpha}y}-2\mathrm{i}\bar{\beta}\xi B\mathrm{e}^{-\bar{\beta}y}\}\mathrm{e}^{\mathrm{i}\xi(x-ct)}, \tag{6.1.80}$$

$$\tau_{yy} = \mu\{(\bar{\beta}^2+\xi^2)A\mathrm{e}^{-\bar{\alpha}y}+2\mathrm{i}\bar{\beta}\xi B\mathrm{e}^{-\bar{\beta}y}\}\mathrm{e}^{\mathrm{i}\xi(x-ct)}, \tag{6.1.81}$$

$$\tau_{xy} = \mu\{-2\mathrm{i}\bar{\alpha}\xi A\mathrm{e}^{-\bar{\alpha}y}+(\xi^2+\bar{\beta}^2)B\mathrm{e}^{-\bar{\beta}y}\}\mathrm{e}^{\mathrm{i}\xi(x-ct)}. \tag{6.1.82}$$

At the free surface we have $\tau_{yy}=\tau_{xy}=0$. From (6.1.81) and (6.1.82) this gives

$$\begin{aligned} (\bar{\beta}^2+\xi^2)A+2\mathrm{i}\bar{\beta}\xi B &= 0,\\ -2\mathrm{i}\bar{\alpha}\xi A+(\bar{\beta}^2+\xi^2)B &= 0. \end{aligned} \tag{6.1.83}$$

These give the amplitude ratios

$$\frac{A}{B} = -\frac{2i\bar{\beta}\xi}{\bar{\beta}^2+\xi^2} = \frac{\bar{\beta}^2+\xi^2}{2i\bar{\alpha}\xi}, \tag{6.1.84}$$

and the frequency equation for surface waves

$$(\bar{\beta}^2+\xi^2)^2-4\bar{\alpha}\bar{\beta}\xi^2 = 0, \tag{6.1.85}$$

where $\bar{\alpha}^2 = \xi^2-\omega^2/c_1^2$, $\bar{\beta}^2 = \xi^2-\omega^2/c_2^2$. This result may be expressed in terms of wave velocity by noting that $\omega = \xi c$. The result is

$$(2-c^2/c_2^2)^2 = 4(1-c^2/c_1^2)^{\frac{1}{2}}(1-c^2/c_2^2)^{\frac{1}{2}}. \tag{6.1.86}$$

Finally, we may rationalize this last equation to give

$$\frac{c^2}{c_2^2}\left\{\left(\frac{c}{c_2}\right)^6-8\left(\frac{c}{c_2}\right)^4+(24-16k^{-2})\left(\frac{c}{c_2}\right)^2-16(1-k^{-2})\right\} = 0. \tag{6.1.87}$$

To investigate the roots of (6.1.87) we first note that it is a reduced cubic equation in $(c/c_2)^2$ and, secondly, that the roots are dependent on Poisson's ratio, since as we have shown earlier $k^2 = 2(1-\nu)/(1-2\nu)$. There will be three roots to the wave velocity equation. Previous studies have shown that the nature of roots (that is, real, imaginary, and complex) is dependent on the range of Poisson's ratio; thus

$$\begin{aligned} &\nu > 0{\cdot}263\ \ldots,\ 1\ \text{real, 2 complex conjugate roots,} \\ &\nu < 0{\cdot}263\ \ldots,\ 3\ \text{real roots.} \end{aligned} \tag{6.1.88}$$

However, any resulting complex roots will not be acceptable in the present situation, since this will yield behaviour of the type

$$\Phi,\ H_z \sim e^{-ay}e^{-bt}e^{i(\xi x+dy-et)}, \tag{6.1.89}$$

which is indicative of attenuation with respect to time, such as if damping were present, which is not the case in the present problem. Furthermore, we cannot have roots for which $c/c_2 > 1$, since $\bar{\beta}^2 > 0$ and this condition would be violated. It has been shown† that for all real media ($0 < \nu < 0{\cdot}5$) there is only one real root meeting this last requirement. The resulting surface wave propagating with the velocity established from (6.1.87) is usually called the *Rayleigh surface wave.*

As an example, consider the surface wave for $\lambda = \mu$. This is a special case, known as 'Poisson's relation', and corresponds to $\nu = \frac{1}{4}$, a value often used for rock. The roots are

$$(c/c_2)^2 = 4,\ 2+2/\sqrt{3},\ 2-2/\sqrt{3}. \tag{6.1.90}$$

The question is: which corresponds to the Rayleigh velocity c_R? By putting

† According to Viktorov [43, p. 3].

the two largest values back into the expressions for $\bar{\alpha}^2, \bar{\beta}^2$, it is found that the requirement of $\bar{\alpha}^2, \bar{\beta}^2 > 0$ is violated. This leaves

$$c/c_2 = (2-2/\sqrt{3})^{\frac{1}{2}} = 0{\cdot}9194 \tag{6.1.91}$$

or

$$c_R = 0{\cdot}9194c_2. \tag{6.1.92}$$

A plot of c_R for all values of Poisson's ratio is shown in Fig. 6.8. The fact

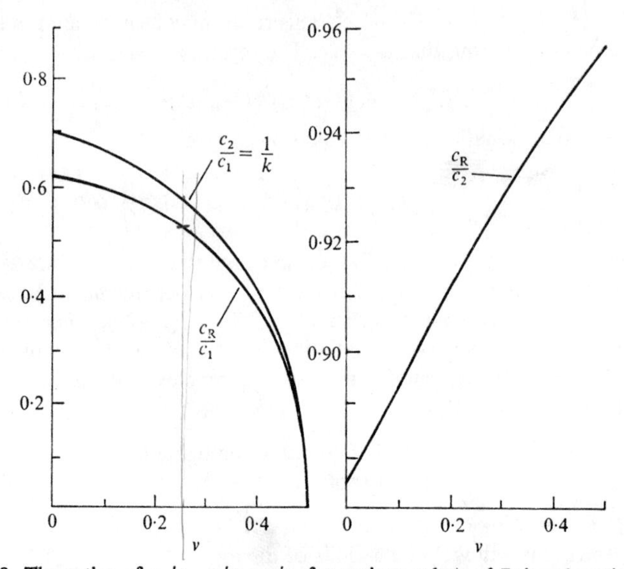

FIG. 6.8. The ratios of c_R/c_1, c_2/c_1, c_R/c_2 for various values of Poisson's ratio. (After Knopoff [21].)

that the propagation velocity is independent of frequency indicates that a surface pulse propagates non-dispersively, with a velocity somewhat less than the shear velocity. An approximate expression that has been developed for the Rayleigh velocity is†

$$c_R/c_2 = (0{\cdot}87+1{\cdot}12\nu)/(1+\nu). \tag{6.1.93}$$

The particle motion may be found from (6.1.78) and (6.1.79), using also the amplitude ratio (6.1.84). At $y = 0$, these become

$$u_x = Ai\left(\gamma - \frac{\bar{\beta}^2+\gamma^2}{2\gamma}\right)e^{i\gamma(x-ct)},$$
$$u_y = A\left(-\bar{\alpha} + \frac{\bar{\beta}^2+\gamma^2}{2\bar{\beta}}\right)e^{i\gamma(x-ct)}. \tag{6.1.94}$$

† op cit., p. 3.

Taking real parts, these give

$$u_x = a(\gamma)\sin \gamma(x-ct), \qquad u_y = b(\gamma)\cos \gamma(x-ct). \qquad (6.1.95)$$

The particle motion is elliptical in nature and retrograde with respect to the direction of propagation (that is, it is counter clockwise for a wave travelling to the right) which is in contrast to the case of water waves. The vertical component of the displacement is greater than the horizontal component at the surface (for example, vertical = 1·5 × horizontal is typical). The motion decreases exponentially in amplitude away from the surface. However, at a slight depth (given as 0·192 wavelength), the direction of particle rotation reverses. A diagram of the particle motion is shown in Fig. 6.9(a). In Fig. 6.9(b), the displacement components at various locations at and below the free surface are shown. The displacements have been normalized with respect to the free surface displacement in the vertical direction u_{y_0}. The depth has been normalized with respect to the Rayleigh wave length λ_R,

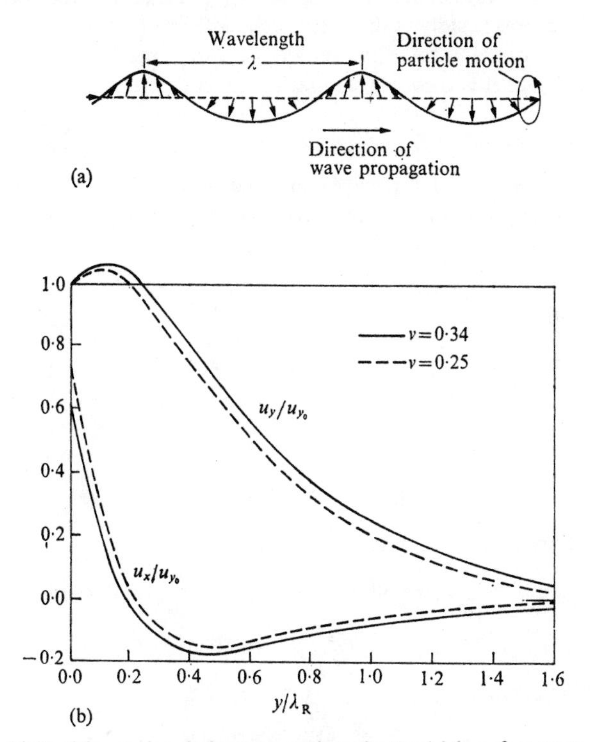

FIG. 6.9. (a) Particles motion during propagation of a Rayleigh surface wave. (Based on Frederick [12, Fig. 2.3]). (b) Normalized displacements under surface wave excitation. (After Viktorov [43, Fig. 2].)

given by $\lambda_R = c_R/f$. Note that the results have been plotted for two values of Poisson's ratio ($\nu = 0{\cdot}25$, $0{\cdot}34$). Although ν may theoretically take on values of $0 \leqslant \nu \leqslant 0{\cdot}5$, the values shown cover a range representative of many practical materials ($\nu = 0{\cdot}25 \sim$ rock; $\nu \leqslant 0{\cdot}34 \sim$ many metals).

Finally, we remark on the extraneous root of the surface wave equation. Only one of the three roots of (6.1.87) corresponds to Rayleigh surface waves and the question of the significance, if any, of the remaining roots remains. These are extraneous roots introduced by rationalizing the original wave velocity equation (6.1.86).

In fact, a meaningful interpretation attaches to the remaining roots, as pointed out by Ewing, Jardetsky, and Press [11, p. 33]. They correspond to previously discussed special cases of incident harmonic P and SV waves, corresponding to a P wave incident, an SV wave only reflected and to an incident SV wave, a P wave only reflected.

As inferred by early seismologists, surface, or Rayleigh, waves are essentially two dimensional. Hence, energy associated with these waves will not disperse as rapidly as the energy associated with the three-dimensional waves of dilatation and rotation. Such waves are of particular importance in seismology, since it is these waves that are most destructive in earthquakes.

6.1.5. *Wave reflection under mixed boundary conditions*

We close this section on waves in a half-space by briefly considering boundary conditions other than traction-free ones. Consider the case of mixed boundary conditions, given for the plane-strain case by

$$u_y = \tau_{xy} = 0, \qquad y = 0. \tag{6.1.96}$$

Such conditions could correspond to an elastic half-space constrained by a rigid, lubricated boundary. As previously given, the potential functions governing the waves are

$$\Phi = A_1 e^{i(\xi x - \alpha y - \omega t)} + A_2 e^{i(\xi x + \alpha y - \omega t)}, \tag{6.1.97}$$

$$H_z = B_1 e^{i(\xi x - \beta y - \omega t)} + B_2 e^{i(\xi x + \beta y - \omega t)}. \tag{6.1.98}$$

The displacements and stresses of interest are

$$u_y = -i\alpha(A_1 e^{-i\alpha y} - A_2 e^{i\alpha y})e^{i(\xi x - \omega t)} - i\xi(B_1 e^{-i\beta y} + B_2 e^{i\beta y})e^{i(\xi x - \omega t)}, \tag{6.1.99}$$

$$\tau_{xy} = \mu\{2\xi\alpha(A_1 e^{-i\alpha y} - A_2 e^{i\alpha y}) + (\xi^2 - \beta^2)(B_1 e^{-i\beta y} + B_2 e^{i\beta y})\}e^{i(\xi x - \omega t)}. \tag{6.1.100}$$

Applying the conditions $u_y = \tau_{xy} = 0$ at $y = 0$ gives

$$\begin{aligned} 2\xi\alpha(A_1 - A_2) + (\xi^2 - \beta^2)(B_1 + B_2) &= 0, \\ \alpha(A_1 - A_2) + \xi(B_1 + B_2) &= 0, \end{aligned} \tag{6.1.101}$$

which in turn may be written as

$$\begin{bmatrix} 2\xi\alpha & (\xi^2-\beta^2) \\ \alpha & \xi \end{bmatrix}\begin{bmatrix} (A_1-A_2) \\ (B_1+B_2) \end{bmatrix} = 0. \tag{6.1.102}$$

Thus

$$A_1 = A_2, \qquad B_1 = -B_2, \tag{6.1.103}$$

unless the determinant is zero. Investigating this possibility, we have

$$\alpha\{2\xi^2-(\xi^2-\beta^2)\} \overset{?}{=} 0. \tag{6.1.104}$$

Since $\beta^2 = \omega^2/c_2^2-\xi^2$, this reduces to $\alpha\omega^2 \overset{?}{=} 0$. But we cannot have $\omega = 0$, and if $\alpha = 0$, we will not have the plane wave case. Hence, we have that

$$A_2/A_1 = 1, \quad B_2/B_1 = -1. \tag{6.1.105}$$

It is seen that no mode conversion occurs since, if $B_1 = 0$ (P wave in), $B_2 = 0$, and if $A_1 = 0$ (SV wave in), $A_2 = 0$.

The possibility of surface waves in such a half-space is also of interest. We use the solutions (6.4.77) directly in the boundary conditions for u_y, τ_{xy}, giving

$$\begin{aligned} \bar{\alpha}A+i\xi B &= 0, \\ 2i\bar{\alpha}\xi A+(\xi^2+\bar{\beta}^2)B &= 0. \end{aligned} \tag{6.1.106}$$

This gives the frequency equation

$$\bar{\alpha}\{(\xi^2+\bar{\beta}^2)-2\xi^2\} = 0, \tag{6.1.107}$$

where $\bar{\alpha}$, $\bar{\beta}$ are as previously defined. The solution $\bar{\alpha} = 0$ is not acceptable since this would require $\bar{\beta}^2 < 0$ and give plane waves instead of surface waves. The other root is $\omega^2/c_2^2 = 0$, or $c^2 = 0$, which is not acceptable since wave propagation would not occur. Hence, surface waves cannot exist in a half-space under such boundary conditions. This case of boundary constraint, while of less physical interest than the traction-free case, will play a role in certain plate studies.

6.2. SH wave source—method of steepest descent

We now consider the first of several problems involving waves from sources located on the surface of a half-space. The simplest case will be considered in this section—that of a harmonic SH wave source. Such a source is created by an applied shearing load, where the load is acting in, say, the z direction and variations in that direction are zero. The case of prescribed displacements may also be considered. The important facts about such a source, as was found in Chapter 5, are that pure shear waves in the displacement variable u_z are propagated and that a single scalar equation governs the motion. The situation is thus quite analogous to an acoustics problem.

The first portion of the section will be devoted to obtaining several general results for SH wave propagation. The second portion will be devoted to introducing the method of steepest descent, an analytical technique for obtaining approximate evaluations of integrals. This method represents a generalization of the stationary-phase technique introduced in the first chapter. It finds wide application in many problems of elastic wave analysis.

6.2.1. *The SH wave source—formal solution*

Let us consider the problem of waves generated by a distribution of load given, in general terms, by

$$\begin{aligned} \tau_{yz}(x, 0, z, t) &= \tau(x)e^{i\omega t}, \\ \tau_{yy}(x, 0, z, t) &= \tau_{yx}(x, 0, z, t) = 0. \end{aligned} \tag{6.2.1}$$

Such a loading will generate SH waves where the only displacement component is $u_z = u_z(x, y, t)$. Under these conditions, the displacement equations of motion reduce to

$$\nabla^2 u_z = \frac{1}{c_2^2}\frac{\partial^2 u_z}{\partial t^2}, \qquad \nabla^2 = \frac{\partial^2}{\partial x^2}+\frac{\partial^2}{\partial y^2}. \tag{6.2.2}$$

If we assume a harmonic time variation in u_z such that

$$u_z(x, y, t) = U(x, y)\exp(i\omega t),$$

then (6.2.2) reduces to

$$\nabla^2 U+\beta^2 U = 0, \qquad \beta^2 = \omega^2/c_2^2. \tag{6.2.3}$$

We apply the Fourier transform† to the last, giving

$$\frac{d^2\bar{U}}{dy^2}-(\xi^2-\beta^2)\bar{U} = 0. \tag{6.2.4}$$

This has the solution

$$\bar{U}(\xi, y, \omega) = Ae^{-\gamma y}+Be^{\gamma y}, \qquad \gamma^2 = \xi^2-\beta^2. \tag{6.2.5}$$

On the basis of finite response for increasing y, we set $B = 0$ in the last result. For the case of SH waves, as we found in the last section, the stress τ_{yz} is given by

$$\tau_{yz} = \mu\frac{\partial u_z}{\partial y} = \mu\frac{\partial U}{\partial y}e^{i\omega t}. \tag{6.2.6}$$

The first boundary condition of (6.2.1) then reduces to

$$\mu\frac{\partial U}{\partial y}\bigg|_{y=0} = \tau(x). \tag{6.2.7}$$

† We take as our transform definitions

$$\bar{f}(\xi) = \frac{1}{\sqrt{(2\pi)}}\int_{-\infty}^{\infty} f(x)e^{-i\xi x}\,dx, \qquad f(x) = \frac{1}{\sqrt{(2\pi)}}\int_{-\infty}^{\infty} \bar{f}(\xi)e^{i\xi x}\,d\xi.$$

The transformed condition is

$$\mu\frac{\mathrm{d}}{\mathrm{d}y}\bigg|_{y=0} = \bar{\tau}(\xi). \tag{6.2.8}$$

Substituting (6.2.5) with $B = 0$ in the last expression, we obtain for the transformed solution

$$\bar{U}(\xi, y) = -\frac{\bar{\tau}(\xi)}{\mu\gamma}\mathrm{e}^{-\gamma y}. \tag{6.2.9}$$

Finally, the formal inverted solution is

$$U(x, y) = -\frac{1}{\mu\sqrt{(2\pi)}}\int_{-\infty}^{\infty}\frac{\bar{\tau}(\xi)}{\gamma}\mathrm{e}^{-\gamma y}\mathrm{e}^{\mathrm{i}\xi x}\,\mathrm{d}\xi. \tag{6.2.10}$$

Several results for specific load distributions are easily obtained. Suppose $\tau(x)$ is of constant magnitude and is acting on a finite width strip such that

$$\tau(x) = \begin{cases}\tau_0, & |x| < a\\ 0, & |x| > a.\end{cases} \tag{6.2.11}$$

Applying the Fourier transform to this, we obtain

$$\bar{\tau}(\xi) = \frac{2\tau_0}{\sqrt{(2\pi)}}\frac{\sin\xi a}{\xi}. \tag{6.2.12}$$

Then we have

$$U(x, y) = -\frac{\tau_0}{\mu\pi}\int_{-\infty}^{\infty}\frac{\sin\xi a\mathrm{e}^{-\gamma y}\mathrm{e}^{\mathrm{i}\xi x}}{\gamma\xi}\,\mathrm{d}\xi. \tag{6.2.13}$$

If $a \to 0$ but the total load per unit length remains constant, the results for a line load are obtained. Thus, multiply and divide (6.2.13) by $2a$, let $2a\tau_0 = P$, and let $a \to 0$. Noting that $\sin\xi a/\xi a \to 1$ as $a \to 0$, we obtain the Green's function

$$G(\mathbf{r}, 0, \omega) = -\frac{P}{2\pi\mu}\int_{-\infty}^{\infty}\frac{\mathrm{e}^{-\gamma y}\mathrm{e}^{\mathrm{i}\xi x}}{\gamma}\,\mathrm{d}\xi. \tag{6.2.14}$$

The case of a line source located at $x = x_0$ is given by

$$G(\mathbf{r}, x_0, \omega) = -\frac{P}{2\pi\mu}\int_{-\infty}^{\infty}\frac{\exp(-\gamma y)\exp\{\mathrm{i}\xi(x-x_0)\}}{\gamma}\,\mathrm{d}\xi. \tag{6.2.15}$$

These last results could be obtained also by originally considering a loading

$$\tau_{yz}(x, 0, z, t) = P\delta(x-x_0)\mathrm{e}^{\mathrm{i}\omega t}, \tag{6.2.16}$$

so that

$$\bar{\tau}(\xi, 0) = \frac{P}{\sqrt{(2\pi)}}\exp(-\mathrm{i}\xi x_0). \tag{6.2.17}$$

6.2.2. *Direct evaluation of the integral*

As will soon be found, the various integral results for the SH source problem can be evaluated exactly by certain changes of variable. Before so doing we consider some of the aspects involved in directly carrying out the evaluation for a line load at the origin. A number of the considerations involving the contour integral are quite similar to those for more complicated half-space problems.

Consider, then, the result (6.2.14). It should be evident that a semicircular contour including the real ξ-axis would be appropriate. If $x > 0$, the contour should be closed in the upper half-ξ-plane, while for $x < 0$, closure is in the lower half-plane. The main difficulty in evaluation arises from the existence of branch points, and it is this aspect that we now consider. Thus, the function $\gamma = (\xi^2-\beta^2)^{\frac{1}{2}}$ has branch points at $\xi = \pm\beta$, which are on the real axis.

It will be useful to generalize slightly here and let $\beta^2 = \beta'^2$, where β' is complex, given by

$$\beta' = \alpha - \mathrm{i}\lambda, \qquad (\alpha, \lambda > 0). \tag{6.2.18}$$

This is similar to the inclusion of damping, as previously done in the analysis of strings. With β'^2 complex, it is seen that the branch points $\xi = \pm\beta'$ are now moved off the real axis into the second and fourth quadrants.

The necessary branch cuts will be selected according to the requirement that $\mathrm{Re}\,\gamma > 0$. This arises from $\exp(-\gamma y)$, where $y > 0$ always. If we let

$$\gamma = X+\mathrm{i}Y, \qquad \gamma^2 = X^2-Y^2+2\mathrm{i}XY, \tag{6.2.19}$$

then the cuts will be defined by $\mathrm{Re}\,\gamma = 0$, which in turn requires that

$$\mathrm{Re}\,\gamma^2 \leqslant 0, \qquad \mathrm{Im}\,\gamma^2 = 0. \tag{6.2.20}$$

Letting $\xi = \zeta+\mathrm{i}\eta$ and β' as given by (6.2.18), we write

$$\gamma^2 = \zeta^2-\eta^2+2\mathrm{i}\zeta\eta-\alpha^2+\lambda^2+2\mathrm{i}\alpha\lambda. \tag{6.2.21}$$

From (6.2.20), this leads to

$$\zeta^2-\eta^2-\alpha^2+\lambda^2 \leqslant 0, \qquad \zeta\eta+\alpha\lambda = 0. \tag{6.2.22}$$

The second condition yields the two hyperbolas (one for α, $\lambda > 0$, the other for α, $\lambda < 0$) passing through the branch points. The first conditions show which portion of the hyperbolas the cut must be along. The resulting branch points and branch cuts are shown in Fig. 6.10(a). Suppose now that β' is real, so that $\lambda \to 0$ in (6.2.22). This gives

$$\zeta^2-\eta^2+\alpha^2 \leqslant 0, \qquad \zeta\eta = 0. \tag{6.2.23}$$

If $\zeta = 0$, then $-\eta^2+\alpha^2 \leqslant 0$, so that $\eta^2 \leqslant \alpha^2$ and if $\eta = 0$, $-\zeta^2 \leqslant \alpha^2$. These combine to give the cuts shown in Fig. 6.10(b). These are the limiting cases of the hyperbolas for the presence of damping.

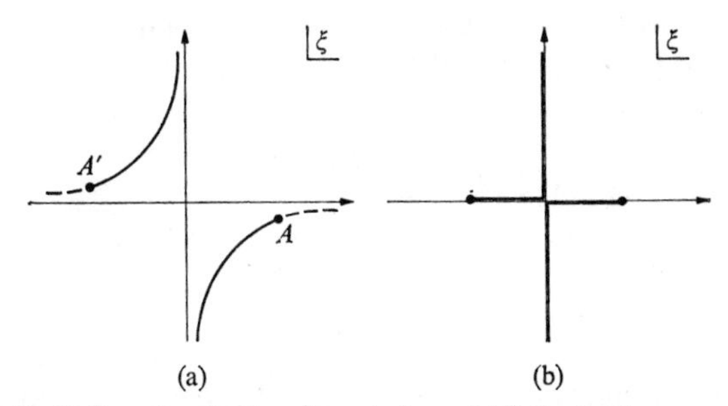

FIG. 6.10. (a) Branch cuts along hyperbolas and (b) the limiting case as $\lambda \to 0$.

On the basis of the branch cuts indicated, the contour used for evaluating (6.2.14) has the form shown in Fig. 6.11. We have

$$\int_{-R}^{R} + \int_{\Gamma_1+\Gamma_2} + \int_{\Gamma} = 0, \tag{6.2.24}$$

where no poles have been enclosed within the contour. The contributions along Γ_1, Γ_2 may be shown to go to zero as $R \to \infty$. The result is that

$$I = \int_{-\infty}^{\infty} \frac{e^{-\gamma y} e^{i\xi x}}{\gamma} d\xi = \lim_{R\to\infty} \int_{-R}^{R} \frac{e^{-\gamma y} e^{i\xi x}}{\gamma} d\xi = -\int_{\Gamma}. \tag{6.2.25}$$

Now, the integral around the branch cut may be broken into five parts along AO, OB, BO, ON, and a small circular path about the branch point at B. For the latter integral, it is found that the integrand is of $O(\varepsilon^{\frac{1}{2}})$, where ε is the radius of a small circle about the branch point. As $\varepsilon \to 0$, this contribution vanishes. If we let

$$\xi - \beta = \rho_1 \exp(i\theta_1), \qquad \xi + \beta = \rho_2 \exp(i\theta_2), \tag{6.2.26}$$

where θ_1, θ_2 are measured from the real ξ-axis, we find that the argument of

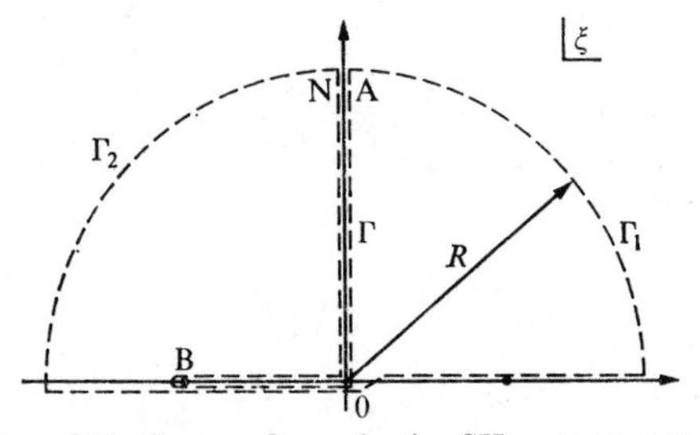

FIG. 6.11. Contour for evaluating SH wave response.

γ is $\pi/2$, $\pi/2$, $-\pi/2$, $-\pi/2$ along, respectively, AO, OB, BO, and ON. The resulting expression for $G(\mathbf{r}, 0, \omega)$ is given by

$$G(\mathbf{r}, 0, \omega) = \frac{P}{\pi\mu}\left\{\int_0^\infty \frac{e^{-\xi x}}{(\beta^2+\xi^2)^{\frac{1}{2}}} \cos(\beta^2+\xi^2)^{\frac{1}{2}} y \, d\xi - \right.$$

$$\left. -i\int_0^\infty \frac{e^{-i\xi x}}{(\beta^2-\xi^2)^{\frac{1}{2}}} \cos(\beta^2-\xi^2)^{\frac{1}{2}} y \, d\xi\right\}, \quad (6.2.27)$$

where ξ is a real quantity. The first integral results from combining the contributions of OA and ON, the second from OB, BO. We shall not pursue this evaluation further except to note that the task of evaluating these integrals would remain in a general problem. In this example, when the solution is known to be a Hankel function (this will be shown next), the integrals could be manipulated into the integral identities for the Hankel function. The main purpose here is to illustrate the contour integration along the single branch cut that arises for SH waves.

Rather than pursuing the integral results (6.2.27) further, we can obtain an explicit solution for the SH wave problem by certain changes of variables similar to those used in considering SH waves in an infinite media. To start, we express (6.2.10) as

$$U(x, y) = -\frac{1}{2\pi\mu}\int_{-\infty}^{\infty} \tau(x')\, dx' \int_{-\infty}^{\infty} \frac{\exp(-\gamma y)\exp\{i\xi(x-x')\}}{\gamma}\, d\xi. \quad (6.2.28)$$

We may express $\exp(-\gamma y)$ as an inverse Fourier transform through the following steps:

$$\mathscr{F}^{-1}\left(\frac{\gamma}{\gamma^2+\zeta^2}\right) = \frac{1}{2\pi}\int_{-\infty}^{\infty} \frac{\gamma}{\gamma^2+\zeta^2} e^{i\zeta y}\, d\zeta$$

$$= \frac{1}{\sqrt{(2\pi)}}\left\{\sqrt{\left(\frac{2}{\pi}\right)}\int_0^\infty \frac{\gamma}{\gamma^2+\zeta^2} \cos \zeta y \, dy\right\}$$

$$= \frac{1}{\sqrt{(2\pi)}}\mathscr{F}_c^{-1}\left(\frac{\gamma}{\gamma^2+\zeta^2}\right). \quad (6.2.29)$$

From tables we have

$$\mathscr{F}_c^{-1}\left(\frac{\gamma}{\gamma^2+\zeta^2}\right) = \sqrt{\left(\frac{\pi}{2}\right)} e^{-\gamma y}, \quad (6.2.30)$$

so that

$$e^{-\gamma y} = 2\mathscr{F}^{-1}\left(\frac{\gamma}{\gamma^2+\zeta^2}\right). \quad (6.2.31)$$

Then (6.2.28) becomes

$$U(x, y) = -\frac{1}{2\pi^2\mu}\int_{-\infty}^{\infty}\tau(x')\,dx'\int_{-\infty}^{\infty}\int_{-\infty}^{\infty}\frac{\exp[i\{\zeta y+\xi(x-x')\}]}{\gamma^2+\zeta^2}\,d\xi\,d\zeta. \quad (6.2.32)$$

At this stage, the double integral (call it q) is recognizable as similar to the one encountered for the SH source in an infinite medium. As before, we let

$$\mathbf{l} = \xi\mathbf{i}+\zeta\mathbf{j}, \qquad \mathbf{R} = (x-x')\mathbf{i}+y\mathbf{j}, \quad (6.2.33)$$

and convert the double integral q to polar coordinates, giving

$$q(R) = \int_0^{\infty} l\,dl\int_0^{2\pi}\frac{\exp(i\mathbf{l}.\mathbf{R})}{l^2-\beta^2}\,d\alpha = \pi^2 iH_0^{(1)}(\beta R), \quad (6.2.34)$$

where

$$R^2 = (x-x')^2+y^2. \quad (6.2.35)$$

Then

$$U(x, y) = -\frac{1}{2\pi^2\mu}\int_{-\infty}^{\infty}\tau(x')q(R)\,dx'. \quad (6.2.36)$$

For the special case of a point-loading $\tau(x') = P_0\delta(x')$, we obtain

$$G(\mathbf{r}, \omega) = -\frac{Pi}{2\mu}H_0^{(1)}(\beta R). \quad (6.2.37)$$

The first thing to note in this result is that it is identical (within constants) to the solution for an infinite medium. This is because the infinite-medium solution automatically satisfies the boundary conditions for a half-space on planes passed through the source (this may be intuitively evident, after the fact). However, a careful comparison of the coefficients would show that identical sources, S_0, placed in the infinite medium and on the surface of the half-space would yield twice the response in the latter case. This is interpreted as the result of half the source output being reflected back into the semi-infinite media, whereas the source effectively radiates into two semi-infinite media in the former case.

Because of the close analogy of SH waves to acoustic waves, the solution of source problems by using image-source techniques naturally suggests itself. This is quite applicable here. Thus, suppose we have a 'buried' SH wave source in a semi-infinite media. This problem may be solved by considering an infinite medium with a source and an image source placed such that the perpendicular plane bisecting the line between the two sources coincides with the boundary of the half-space. As the two sources approach the boundary, a source of double strength is obtained, which also explains twice the response of a surface source compared to a single source in an infinite medium.

6.2.3. *Method of steepest descent*

In the preceding section, we have obtained exact results for the SH wave source or obtained explicit integral forms to be evaluated. The exact evaluation was, of course, satisfying in terms of being explicit. However, the results were quite atypical, since most wave-propagation problems are usually too complex for this to occur and must be evaluated using various approximate techniques. Typically, the problem may be resolved into evaluation of a contour integral of the type

$$I = \int_C e^{Rf(\xi)} g(\xi)\, d\xi, \tag{6.2.38}$$

where a portion of the path is along the real axis (Fourier transforms) or along the Bromwich contour (Laplace transforms).

We have previously approximately evaluated integrals of the type

$$\int_a^b e^{ixh(\omega)} g(\omega)\, d\omega, \tag{6.2.39}$$

by the method of stationary phase. In such integrals, ω, $g(\omega)$, and $h(\omega)$ were real. It was shown for x large, that the major contribution to the integral was from a small region near a stationary point, given by $h'(\omega) = 0$. Away from such a point, the increasing rapid oscillations of the integrand were assumed to cancel by interference. In the following, we shall develop the method for approximately evaluating the contour integral of a complex variable ξ, for large values of R. This will represent a generalization of the stationary-phase method.

1. *General basis of the method.* The basic idea of the steepest-descent method is to deform the contour of integration to pass through a strategic point (or points) such that the major contribution to the integral comes from a short portion of path in the vicinity of that point. If we write

$$f(\xi) = f_1 + if_2, \tag{6.2.40}$$

then

$$\exp(Rf) = \exp(Rf_1)\exp(iRf_2). \tag{6.2.41}$$

The objective will be to maximize the contribution of $\exp(Rf_1)$ at a point along the deformed contour. Away from the point, it is desired that this contribution rapidly decrease.

However, a potential weakness of this argument is that $\exp(iRf_2)$ is oscillatory and, for large R, these dense oscillations may negate the contributions of the real part. It would be desirable, therefore, to have a path such that f_2 is (relatively) constant.

2. *Saddle points.* The selection of critical points and paths will be aided by a brief review of topological aspects of analytic functions of a complex variable. Thus consider the real and imaginary parts of $f(\xi)$, given by

$$f_1 = f_1(\zeta, \eta), \qquad f_2 = f_2(\zeta, \eta), \tag{6.2.42}$$

The gradient of f_1 is given by

$$\nabla f_1 = \frac{\partial f_1}{\partial \zeta}\mathbf{i} + \frac{\partial f_1}{\partial \eta}\mathbf{j}. \tag{6.2.43}$$

This represents the direction of maximum variation of f_1. The unit vector in this direction could be $\mathbf{n}_1$ given by

$$\mathbf{n}_1 = \nabla f_1 / |\nabla f_1|. \tag{6.2.44}$$

Now, the direction of $\mathbf{n}_1$ is always at right angles to the contour lines of f_1 (the lines of constant f_1). This latter direction is $\mathbf{s}_1$, where

$$\mathbf{n}_1 . \mathbf{s}_1 = 0, \qquad \mathbf{s}_1 = \frac{1}{|\nabla f_1|}\left(-\frac{\partial f_1}{\partial \eta}\mathbf{i} + \frac{\partial f_1}{\partial \zeta}\mathbf{j}\right). \tag{6.2.45}$$

It may be easily shown that the direction of maximum variation of f_1 (that is, $\mathbf{n}_1$) is parallel to the direction of constant f_2. Thus we have

$$\nabla f_2 = \frac{\partial f_2}{\partial \zeta}\mathbf{i} + \frac{\partial f_2}{\partial \eta}\mathbf{j}, \tag{6.2.46}$$

and

$$\mathbf{n}_2 = \frac{\nabla f_2}{|\nabla f_2|}, \qquad \mathbf{s}_2 = \frac{1}{|\nabla f_2|}\left(-\frac{\partial f_2}{\partial \eta}\mathbf{i} + \frac{\partial f_2}{\partial \zeta}\mathbf{j}\right), \tag{6.2.47}$$

where $\mathbf{s}_2$ is the direction of f_2 constant. However, from the Cauchy–Riemann conditions

$$\frac{\partial f_1}{\partial \zeta} = \frac{\partial f_2}{\partial \eta}, \qquad \frac{\partial f_1}{\partial \eta} = -\frac{\partial f_2}{\partial \zeta}. \tag{6.2.48}$$

Hence

$$\mathbf{s}_2 = \frac{-1}{|\nabla f_2|}\left(\frac{\partial f_1}{\partial \zeta}\mathbf{i} + \frac{\partial f_1}{\partial \eta}\mathbf{j}\right). \tag{6.2.49}$$

Thus, $\mathbf{s}_2$ is parallel to $\mathbf{n}_1$. This aspect is of great importance in removing the possible problem of oscillations of $\exp(\mathrm{i}Rf_2)$ along the strategic path.

Now, extreme values of f_1 occur for $\nabla f_1 = 0$ or

$$\partial f_1/\partial \zeta = \partial f_2/\partial \eta = 0. \tag{6.2.50}$$

It also holds that the Cauchy–Riemann conditions are satisfied in any direction $\mathbf{n}$ and $\mathbf{s}$, so that, at an extreme, we have

$$\partial f_1/\partial n = \partial f_1/\partial s = 0. \tag{6.2.51}$$

Also, by the Cauchy–Riemann conditions, it immediately follows that f_2 is stationary, and, in general, $\mathrm{d}f/\mathrm{d}\xi = 0$.

The nature of the extremes of f_1 defined by (6.2.50) are of a particular type. By the maximum-modulus theorem of complex variables, f_1 (and f_2) cannot attain local maximum or minimum values within a contour C, but in fact attain the extreme values on C. The only way that f_1 can satisfy (6.2.50) and yet not violate the maximum-modulus theorem is to pass through a 'saddle point', as shown in Fig. 6.12(a). We see that at P_s, the slope is

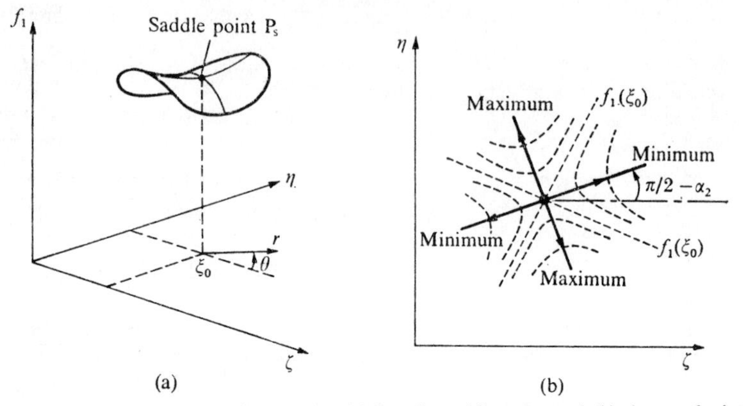

FIG. 6.12. (a) The surface $f_1(\beta, \eta)$ in the vicinity of a saddle point and (b) the topological behaviour of f_1 in the vicinity of the saddle point ξ_0.

locally zero. However, in passing through P_s in one direction, f_1 attains a local maximum while, in another direction, it reaches a local minimum.

The local topology of a saddle point may be better understood by expanding $f(\xi)$ about ξ_0 as

$$f(\xi) = f(\xi_0)+f'(\xi_0)(\xi-\xi_0)+\tfrac{1}{2}f''(\xi_0)(\xi-\xi_0)^2+\dots. \tag{6.2.52}$$

However, $f'(\xi_0) = 0$. We further assume, in this level of development of the theory, that $f''(\xi_0) \neq 0$, and let

$$\tfrac{1}{2}f''(\xi_0) = a_2 \exp(\mathrm{i}\alpha_2), \qquad \xi-\xi_0 = r\exp(\mathrm{i}\theta). \tag{6.2.53}$$

Then we have for $f(\xi)$ and $f_1(\xi)$ in the vicinity of ξ_0,

$$f(\xi) = f(\xi_0)+a_2r^2\exp\{\mathrm{i}(2\theta+\alpha_2)\}, \tag{6.2.54}$$

$$f_1(\xi_0) = \mathrm{Re}\, f(\xi_0)+a_2r^2\cos(2\theta+\alpha_2). \tag{6.2.55}$$

Noting the variation of $\cos(2\theta+\alpha_2)$, it is seen that f_1 will attain two maxima and two minima as θ varies from 0 to 2π about ξ_0. The directions are shown in Fig. 6.12(b). This behaviour is in accord with that shown qualitatively in Fig. 6.12(a).

The local contour lines of f_1 are given by $\cos(2\theta+\alpha_2) = 0$, so that $f_1 = \text{constant} = \mathrm{Re}\, f(\xi_0)$. There are two such lines, and they are at 45° with respect to the maximum and mininum directions. These are shown in Fig. 6.12, as are additional contours above and below $f_1(\xi_0)$. In the maximum direction these are contours of $f_1 > f_1(\xi_0)$, while in the minimum directions they are for $f_1 < f_1(\xi_0)$. Locally, they are families of hyperbolas.

Finally we have that

$$f_2 = \text{Im}\, f(\xi_0) + a_2 r^2 \sin(2\theta + \alpha_2). \tag{6.2.56}$$

Several analogous remarks regarding the behaviour of f_2 in the vicinity of the saddle point are possible, but the most important thing to note is that the directions of constant f_2 coincide with the maximum and minimum directions of f_1.

3. *The deformed contour*. The preceding has shown the detailed behaviour in the vicinity of a saddle point. We now assume that the original contour C may be deformed into a new one C' along the steepest descent path through the saddle point. The path C', we realize, is defined by the characteristic that $f_2 =$ constant (that is, it is a contour line of f_2). This may be used to determine the specific form of C' in a given problem. It is often important to know this form, since in deforming the original path to the new one, various poles and branch points may be passed. It then becomes necessary to account for these by adding the necessary residues or additional paths along branch cuts. We shall not immediately concern ourselves with such additions, however.

4. *Change in variables*. Along the deformed contour C', which passes through ξ, we realize that $f(\xi)$ may be written as

$$f(\xi) = f_1(\xi) + \text{i} f_2(\xi_0), \tag{6.2.57}$$

since f_2 is constant. It is possible to express f_1 in terms of its maximum value at ξ_0 as

$$f_1(\xi) = f_1(\xi_0) - m, \tag{6.2.58}$$

where m is always positive. A graphical interpretation of this is shown in Fig. 6.13(a). Then for $f(\xi)$, we may write

$$f(\xi) = f_1(\xi_0) - m + \text{i} f_2(\xi_0) = f(\xi_0) - m. \tag{6.2.59}$$

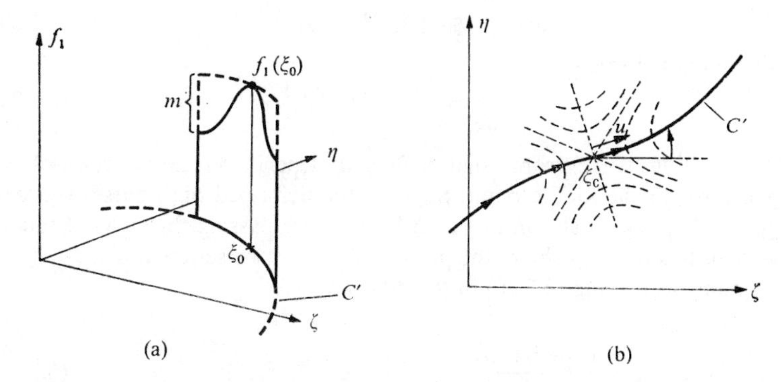

FIG. 6.13. (a) Representation of f_1 along the contour C' and (b) a particular path direction of C' through the saddle point at ξ_0.

It is convenient, from the standpoint of later developments, to let $m = u^2$, so that we have

$$f(\xi) = f(\xi_0) - u^2. \tag{6.2.60}$$

The original integral is then

$$I = \exp\{Rf(\xi_0)\}\int_{-\infty}^{\infty}\exp(-Ru^2)g\{\xi(u)\}\left(\frac{d\xi}{du}\right)du, \tag{6.2.61}$$

where the range of integration has been extended to $\pm\infty$, since negligible contributions are expected away from the saddle point.

We must now evaluate $d\xi/du$. Now in the vicinity of ξ_0 we have, from (6.2.52), that

$$f(\xi) - f(\xi_0) = \tfrac{1}{2}f''(\xi_0)(\xi-\xi_0)^2. \tag{6.2.62}$$

Comparing this to (6.2.55), we then have

$$u^2 = -f''(\xi_0)(\xi-\xi_0)^2. \tag{6.2.63}$$

Letting $\xi - \xi_0 = r\exp(i\theta)$, as before, we have

$$u^2 = -r^2 f''(\xi_0)e^{2i\theta}. \tag{6.2.64}$$

We write

$$f''(\xi_0) = |f''(\xi_0)|\, e^{i\beta}, \tag{6.2.65}$$

so that

$$u^2 = -r^2|f''|\, e^{i(\beta+2\theta)} = -r^2|f''|\, e^{i\phi}. \tag{6.2.66}$$

In order for u^2 to be real and positive, we must have $\exp(i\phi)$ real and negative. This fixes $\phi = \pm\pi$. Then

$$u = r|f''|^{\frac{1}{2}}\, e^{i\phi/2}e^{\pi i/2}, \qquad \phi = \pm\pi, \tag{6.2.67}$$

so that

$$u = \pm r|f''|^{\frac{1}{2}}. \tag{6.2.68}$$

Then, substituting for r, we obtain

$$u = \pm(\xi-\xi_0)e^{-i\theta}|f''(\xi_0)|^{\frac{1}{2}}. \tag{6.2.69}$$

Differentiating this gives

$$\frac{du}{d\xi} = \pm e^{-i\theta}|f''(\xi_0)|^{\frac{1}{2}}. \tag{6.2.70}$$

The selection of the proper sign in (6.2.70) depends on the direction of the path through the saddle point. Suppose the deformed path passes through the saddle point as shown in Fig. 6.13(b). After passing through the saddle point, if u is to be positive, the positive sign must be selected.

Upon substituting (6.2.70) in (6.2.61), we have

$$I = \frac{\exp\{Rf(\xi_0)\}\exp(i\theta)}{|f''(\xi_0)|^{\frac{1}{2}}}\int_{-\infty}^{\infty}\exp(-Ru^2)g\{\xi(u)\}\, du. \tag{6.2.71}$$

5. *Watson's lemma.* A simplified form of Watson's lemma pertains to the existence of an asymptotic expression of the integral†

$$\bar{I} = \int_{-\infty}^{\infty} \exp(-\tfrac{1}{2}a^2z^2)f(z)\,\mathrm{d}z. \tag{6.2.72}$$

Writing the series expansion for $f(z)$ as

$$f(z) = a_0 + a_1 z + a_2 z^2 + \dots \tag{6.2.73}$$

and noting that

$$\int_{-\infty}^{\infty} \exp(-\tfrac{1}{2}a^2z^2)z^{2n}\,\mathrm{d}z = \sqrt{(2\pi)}\frac{(2n)!}{2^n n!\,a^{2n+1}}, \tag{6.2.74}$$

we have

$$\bar{I} \sim \sqrt{(2\pi)}\left(\frac{a_0}{a}+\frac{a_2}{a^3}+\frac{1.3}{a^5}a_4+\dots\right), \tag{6.2.75}$$

with odd powers zero.

This result may be applied to the present integral, where we let $a = (2R)^{\frac{1}{2}}$. Letting $g\{\xi(u)\}$ be given by $G(u)$, we have

$$\int_{-\infty}^{\infty} \exp(-Ru^2)G(u)\,\mathrm{d}u = \sqrt{(2\pi)}\left\{\frac{G(0)}{(2R)^{\frac{1}{2}}}+\frac{1}{2!}\frac{G''(0)}{(2R)^{\frac{3}{2}}}+\dots\right\}. \tag{6.2.76}$$

Considering only the first term, we obtain as a final result

$$I = \frac{\sqrt{(\pi)}\exp(\mathrm{i}\theta)\exp\{Rf(\xi_0)\}G(0)}{|Rf''(\xi_0)|^{\frac{1}{2}}}. \tag{6.2.77}$$

This completes our general development of the steepest-descent method. No consideration has been given to complications that may occur if poles lie on the steepest-descent path, or at a saddle point, or due to vanishing of higher derivatives in the expansion of $f(\xi)$.‡

6.2.4. *SH wave source solution by steepest descent*

Let us apply the method of steepest descent to determination of the far-field response of the line SH wave source acting at the origin. Specifically, we consider (6.2.14) or, writing just the integral,

$$I = \int_{-\infty}^{\infty} \frac{\mathrm{e}^{-\gamma y}\mathrm{e}^{\mathrm{i}\xi x}}{\gamma}\,\mathrm{d}\xi. \tag{6.2.78}$$

† Jeffreys and Jeffreys [20, pp. 501–6].

‡ Numerous developments of this appear in the literature of mathematics, physics, and engineering. Two most useful references are by Jeffreys and Jeffreys [20, pp. 473] and Carrier, Krook, and Pearson [2, pp. 257]. Ewing, Jardetsky, and Press [11] and Brekhvoskikh [1] also contain developments of the method.

We let $x = R\cos\theta$, $y = R\sin\theta$. Then

$$I = \int_{-\infty}^{\infty} \frac{e^{R(i\xi\cos\theta - \gamma\sin\theta)}}{\gamma}\, d\xi. \tag{6.2.79}$$

Comparing this expression to (6.2.38), we let

$$f(\xi) = i\xi\cos\theta - \gamma\sin\theta, \qquad g(\xi) = (\xi^2-\beta^2)^{-\frac{1}{2}}. \tag{6.2.80}$$

The saddle points are then given by

$$df/d\xi = -(\xi^2-\beta^2)^{-\frac{1}{2}}\xi\sin\theta + i\cos\theta = 0. \tag{6.2.81}$$

so that

$$\xi = \pm\beta\cos\theta. \tag{6.2.82}$$

Also

$$\frac{d^2f}{d\xi^2} = \frac{-\sin\theta}{(\xi^2-\beta^2)^{\frac{1}{2}}} + \frac{\xi^2\sin\theta}{(\xi^2-\beta^2)^{\frac{3}{2}}}. \tag{6.2.83}$$

We then have, at the saddle point $\xi = -\beta\cos\theta$,

$$f(\xi_0) = -i\beta, \qquad f''(\xi_0) = i\,\mathrm{cosec}^2\theta/\beta. \tag{6.2.84}$$

In addition, we must determine $G(0)$ or $g\{\xi(0)\}$. From (6.2.63) we have

$$-u^2 = f''(\xi_0)(\xi-\xi_0)^2 = \frac{i\,\mathrm{cosec}^2\theta}{\beta}(\xi+\beta\cos\theta)^2, \tag{6.2.85}$$

which gives for u

$$u = \frac{\mathrm{cosec}\,\theta}{\sqrt{\beta}} e^{-\pi i/4}(\xi+\beta\cos\theta), \qquad \frac{\mathrm{cosec}\,\theta}{\sqrt{\beta}} e^{3\pi i/4}(\xi+\beta\cos\theta). \tag{6.2.86}$$

Solving for ξ,

$$\xi = \sqrt{(\beta)}\sin\theta e^{\pi i/4}u - \beta\cos\theta, \qquad \sqrt{(\beta)}\sin\theta e^{-3\pi i/4}u - \beta\cos\theta. \tag{6.2.87}$$

For $u = 0$, we have

$$\xi(0) = -\beta\cos\theta, \tag{6.2.88}$$

so that

$$g\{\xi(0)\} = \frac{1}{(\beta^2\cos^2\theta-\beta^2)^{\frac{1}{2}}} = \frac{-i}{\beta\sin\theta}. \tag{6.2.89}$$

Finally, we must evaluate the term $\exp(i\theta)$. This will pertain to the direction of the path through the saddle point. Referring back to (6.2.53), we may write

$$\frac{f''(\xi_0)}{2} = a_2\exp(i\alpha_2) = \frac{i\,\mathrm{cosec}^2\theta}{2k} = \frac{\mathrm{cosec}^2\theta}{2k}\exp\left(\frac{\pi}{2}i\right), \tag{6.2.90}$$

$$\xi - \xi_0 = \xi + k\cos\theta = re^{i\theta}. \tag{6.2.91}$$

Then, following (6.2.54) and (6.2.55), we have

$$\begin{aligned} a_2r^2\cos(2\theta+\alpha_2) &= \frac{\mathrm{cosec}^2\theta r^2}{2k}\cos\left(2\theta+\frac{\pi}{2}\right), \\ &= \frac{\mathrm{cosec}^2\theta r^2}{2k}\cos\phi. \end{aligned} \tag{6.2.92}$$

The steepest-descent direction occurs for $\phi = \pi$. Then $2\theta + \pi/2 = \pi$, or $\theta = \pi/4$ so that $\exp(\mathrm{i}\theta) = \exp(\pi\mathrm{i}/4)$. Substituting the various quantities obtained in (6.2.71) we have

$$I = \frac{-\sqrt{(\pi)}\mathrm{i}e^{-\mathrm{i}\beta R}e^{\pi\mathrm{i}/4}}{\beta\sin\theta\left|\dfrac{\mathrm{i}\,\mathrm{cosec}^2\theta R}{\beta}\right|^{\frac{1}{2}}} = \frac{-\sqrt{(\pi)}e^{-\mathrm{i}\beta R}e^{3\pi\mathrm{i}/4}}{\sqrt{(\beta R)}}. \tag{6.2.93}$$

Then

$$G(\mathbf{r}, 0, \omega) = \frac{Pe^{-\mathrm{i}\beta R}e^{3\pi\mathrm{i}/4}}{2\mu\sqrt{(\pi\beta R)}}. \tag{6.2.94}$$

Previously, we had obtained (see (6.2.37))

$$G(\mathbf{r}, 0, \omega) = \frac{P\pi\mathrm{i}}{\mu}H_0^{(1)}(\beta R) \tag{6.2.95}$$

as the exact solution. This has the asymptotic form

$$G(\mathbf{r}, 0, \omega) \sim \frac{P}{\mu}\sqrt{\left(\frac{2\pi}{\beta R}\right)}e^{\mathrm{i}(\beta R - 3\pi/4)}, \tag{6.2.96}$$

a result in agreement with the steepest-descent result.

This problem represents a fortunate circumstance—where the problem is rather simple, an exact solution is obtainable, and asymptotic results are already available. The steepest-descent method, it is found, yields the same asymptotic behaviour. In the more complicated problems to be considered, this asymptotic method is frequently the only means of evaluation.

6.3. Surface source problems

Our attention now turns to more general problems of waves from surface sources. Under general conditions of normal and shear loads applied to the surface of the half-space, both dilatational and shear waves are generated, with the resulting mathematical expressions for the wave propagation being considerably more complicated than those for the simple SH wave source of the last section.

The classical analysis in this area was done by Lamb [22] in 1904. He considered the half-space subjected to line and point loads on the surface and to buried sources. Harmonic loadings were considered, and superposition techniques were used to obtain results for pulse loadings. Since the early analysis of Lamb, a great many contributions have appeared pertaining to what is usually referred to as 'Lamb's' problem. Ewing, Jardetsky, and Press† have given a very thorough review of the analysis of this problem. Miklowitz‡ also thoroughly reviews the literature, including contributions by Mitra [33], Pekeris [34], Chao [3], Lang [23], and many others that appeared subsequent to the book by Ewing *et al.*

† Pp. 34–70 of Reference [11].
‡ Pp. 821–7 of Reference [29].

The large class of problems involving waves in a half-space from surface and buried sources makes it impossible to present analytical details of the various treatments. The approach taken will be to consider the case of a harmonic normal loading in the half-space in some detail, drawing on the work presented in Ewing *et al.* [11] and on the work of Miller and Pursey [30, 31]. Some consideration will then be given to the analysis of the transient-load problem. The main emphasis in the transient problem will be on presenting the results, in terms of displacement and stress waveforms, as obtained by a number of investigators. Included here will be the case of studies involving impact on a half-space and results obtained for buried sources.

6.3.1. *Waves from a harmonic, normal line force*

We consider a half-space subjected to a loading normal to the surface, applied parallel to the z-axis, and invariant with respect to z. The load is considered to act on the strip $|x| < a$ and to be of unity magnitude. The resulting situation is one of plane strain, where $u_z = \partial/\partial z = 0$.

1. *Basic equations.* Following Miller and Pursey [30], we use the displacement equations of motion (see (5.1.8) with $\mathbf{f} = 0$),

$$(\lambda+2\mu)\nabla\nabla\cdot\mathbf{u}-\mu\nabla\times\nabla\times\mathbf{u} = \rho\partial^2\mathbf{u}/\partial t^2. \tag{6.3.1}$$

For the conditions of plane strain, we have

$$\Delta = \nabla\cdot\mathbf{u} = \frac{\partial u_x}{\partial x}+\frac{\partial u_y}{\partial y}, \qquad W\mathbf{k} = \nabla\times\mathbf{u} = \left(\frac{\partial u_x}{\partial y}-\frac{\partial u_y}{\partial x}\right)\mathbf{k}. \tag{6.3.2}$$

We take note of the harmonic time variation of the loading $\exp(i\omega t)$ and write

$$u_x(x, y, t) = u_x(x, y)e^{i\omega t}, \qquad u_y(x, y, t) = u_y(x, y)e^{i\omega t}. \tag{6.3.3}$$

Then the vector equation (6.3.1) may be written as

$$(\lambda+2\mu)\frac{\partial\Delta}{\partial x}+\mu\frac{\partial W}{\partial y}+\rho\omega^2u_x = 0, \tag{6.3.4}$$

$$(\lambda+2\mu)\frac{\partial\Delta}{\partial y}-\mu\frac{\partial W}{\partial x}+\rho\omega^2u_y = 0. \tag{6.3.5}$$

Eliminating W and Δ successively from the preceding we obtain

$$\nabla^2\Delta+k_1^2\Delta = 0, \qquad \nabla^2W+k_2^2W = 0, \tag{6.3.6}$$

where $\nabla^2 = \partial^2/\partial x^2+\partial^2/\partial y^2$ and

$$k_1^2 = \omega^2/c_1^2, \qquad k_2^2 = \omega^2/c_2^2. \tag{6.3.7}$$

The stresses

$$\tau_{yy} = \lambda\Delta+2\mu\frac{\partial u_y}{\partial y}, \qquad \tau_{xy} = \mu\left(\frac{\partial u_y}{\partial x}+\frac{\partial u_x}{\partial y}\right), \tag{6.3.8}$$

with $\tau_{yz} = 0$, may be expressed in terms of Δ and W as

$$\tau_{yy} = \frac{\mu^2}{\rho\omega^2}\left\{2\frac{\partial^2 W}{\partial x\,\partial y} - k^2(k^2-2)\frac{\partial^2 \Delta}{\partial x^2} - k^4\frac{\partial^2 \Delta}{\partial y^2}\right\}, \tag{6.3.9}$$

$$\tau_{xy} = \frac{\mu^2}{\rho\omega^2}\left(\frac{\partial^2 W}{\partial x^2} - \frac{\partial^2 W}{\partial y^2} - 2k^2\frac{\partial^2 \Delta}{\partial x\,\partial y}\right), \tag{6.3.10}$$

where $k^2 = (\lambda+2\mu)/\mu = k_2^2/k_1^2$.

2. *The formal solution.* The Fourier transform on the spatial variable x will be applied, where

$$\bar{f}(\zeta) = \int_{-\infty}^{\infty} f(x)\mathrm{e}^{-\mathrm{i}\zeta x}\,\mathrm{d}x, \qquad f(x) = \frac{1}{2\pi}\int_{-\infty}^{\infty} \bar{f}(\zeta)\mathrm{e}^{\mathrm{i}\zeta x}\,\mathrm{d}\zeta \tag{6.3.11}$$

are taken as the transform and inverse transform definitions, respectively. Applying the transform to (6.3.9) and (6.3.5) we obtain for the transformed displacements

$$\bar{u}_x = -\frac{1}{k_2^2}\left(\frac{\mathrm{d}\overline{W}}{\mathrm{d}y} - \mathrm{i}\zeta k^2\bar{\Delta}\right), \tag{6.3.12}$$

$$\bar{u}_y = -\frac{1}{k_2^2}\left(k^2\frac{\mathrm{d}\bar{\Delta}}{\mathrm{d}y} - \mathrm{i}\zeta\overline{W}\right). \tag{6.3.13}$$

The transformed wave equations (6.3.7) are

$$\frac{\mathrm{d}^2\bar{\Delta}}{\mathrm{d}y^2} - (\zeta^2-k_1^2)\bar{\Delta} = 0, \qquad \frac{\mathrm{d}^2\overline{W}}{\mathrm{d}y^2} - (\zeta^2-k_2^2)\overline{W} = 0, \tag{6.3.14}$$

and the transformed stresses (6.3.9) and (6.3.10) are

$$\bar{\tau}_{yy} = \frac{\mu^2}{\rho\omega^2}\left\{-k^4\frac{\mathrm{d}^2\bar{\Delta}}{\mathrm{d}y^2} + 2\mathrm{i}\zeta\frac{\mathrm{d}\overline{W}}{\mathrm{d}y} + k^2(k^2-2)\zeta^2\bar{\Delta}\right\}, \tag{6.3.15}$$

$$\bar{\tau}_{xy} = -\frac{\mu^2}{\rho\omega^2}\left(\frac{\mathrm{d}^2\overline{W}}{\mathrm{d}y^2} + 2\mathrm{i}\zeta k^2\frac{\mathrm{d}\bar{\Delta}}{\mathrm{d}y} + \zeta^2\overline{W}\right). \tag{6.3.16}$$

The boundary conditions for the problem are (omitting the time variation),

$$\tau_{yy}(x, 0) = \begin{cases} 1, & |x| < a \\ 0, & |x| > a \end{cases} \qquad \tau_{xy}(x, 0) = 0 \tag{6.3.17}$$

The transformed boundary conditions are then

$$\bar{\tau}_{yy}(\zeta, 0) = 2\sin\zeta a/\zeta, \qquad \bar{\tau}_{xy}(\zeta, 0) = 0. \tag{6.3.18}$$

Now, the solutions of the governing equations (6.3.14) are given by

$$\bar{\Delta} = A\exp\{-(\zeta^2-k_1^2)^{\frac{1}{2}}y\}, \qquad \overline{W} = B\exp\{-(\zeta^2-k_2^2)^{\frac{1}{2}}y\}, \tag{6.3.19}$$

where the exponentially increasing terms have been discarded. When these solutions are substituted in the expression for $\bar{\tau}_{yy}$ and $\bar{\tau}_{xy}$ and the results into the boundary conditions (6.3.18), there results

$$\frac{\mu^2}{\rho\omega^2}\{-k^4(\zeta^2-k_1^2)A-2i\zeta(\zeta^2-k_2^2)^{\frac{1}{2}}B+k^2(k^2-2)\zeta^2A\} = \frac{2\sin\zeta a}{\zeta}, \quad (6.3.20)$$

$$\{(\zeta^2-k_2^2)B-2i\zeta k^2(\zeta^2-k_1^2)^{\frac{1}{2}}A+\zeta^2B\} = 0. \quad (6.3.21)$$

Solving these two equations for A and B and putting the results back into (6.3.19), we have

$$\bar{\Delta} = \frac{2\rho\omega^2(k_2^2-2\zeta^2)}{\mu^2k^2\zeta F(\zeta)}\sin\zeta a\exp\{-(\zeta^2-k_1^2)^{\frac{1}{2}}y\}, \quad (6.3.22)$$

$$\overline{W} = \frac{4i\rho\omega^2(\zeta^2-k_1^2)}{\mu^2F(\zeta)}\sin\zeta a\exp\{-(\zeta^2-k_2^2)^{\frac{1}{2}}y\}, \quad (6.3.23)$$

where

$$F(\zeta) = (2\zeta^2-k_2^2)^2-4\zeta^2(\zeta^2-k_1^2)^{\frac{1}{2}}(\zeta^2-k_2^2)^{\frac{1}{2}}. \quad (6.3.24)$$

The transformed displacements are given by

$$\bar{u}_x = \frac{2\sin\zeta a}{i\mu F(\zeta)}[2(\zeta^2-k_1^2)^{\frac{1}{2}}(\zeta^2-k_2^2)^{\frac{1}{2}}\exp\{-(\zeta^2-k_2^2)^{\frac{1}{2}}y\}+$$
$$+(k_2^2-2\zeta^2)\exp\{-(\zeta^2-k_1^2)^{\frac{1}{2}}y\}], \quad (6.3.25)$$

$$\bar{u}_y = \frac{2(\zeta^2-k_1^2)^{\frac{1}{2}}\sin\zeta a}{\mu\zeta F(\zeta)}[2\zeta^2\exp\{-(\zeta^2-k_2^2)^{\frac{1}{2}}y\}+(k_2^2-2\zeta^2)\exp\{-(\zeta^2-k_1^2)^{\frac{1}{2}}y\}]. \quad (6.3.26)$$

The displacement components u_x, u_y are obtained from the inversion integral, the second of (6.3.11). Before writing the results, we note that k_1, k_2 have the units of reciprocal length. We choose k_1 as a normalizing factor for all length parameters in the results. Then the displacements are

$$u_x(x, y) = \frac{1}{i\mu\pi}\int_{-\infty}^{\infty}\frac{\sin\zeta a}{F_0(\zeta)}[2(\zeta^2-1)^{\frac{1}{2}}(\zeta^2-k^2)^{\frac{1}{2}}\exp\{-(\zeta^2-k^2)^{\frac{1}{2}}y\}+$$
$$+(k^2-2\zeta^2)\exp\{-(\zeta^2-1)^{\frac{1}{2}}y\}]\exp(i\zeta x)\,d\zeta, \quad (6.3.27)$$

$$u_y(x, y) = \frac{1}{\mu\pi}\int_{-\infty}^{\infty}\frac{(\zeta^2-1)^{\frac{1}{2}}\sin\zeta a}{\zeta F_0(\zeta)}[2\zeta^2\exp\{-(\zeta^2-k^2)^{\frac{1}{2}}y\}+$$
$$+(k^2-2\zeta^2)\exp\{-(\zeta^2-1)^{\frac{1}{2}}y\}]\exp(i\zeta x)\,d\zeta, \quad (6.3.28)$$

where

$$F_0(\zeta) = (2\zeta^2-k^2)^2-4\zeta^2(\zeta^2-1)^{\frac{1}{2}}(\zeta^2-k^2)^{\frac{1}{2}}. \quad (6.3.29)$$

It should be understood that $x \sim k_1x$, $y \sim k_1y$, $\zeta \sim \zeta/k_1$ in the preceding.

The case of a line source (the first problem considered by Lamb) recovers from the results for u_x, u_y by letting $a \to 0$ but letting the total load per unit length $2a$ be a constant. Thus, in (6.3.27), $\sin \zeta a$ would be replaced by $a\zeta$ and $\sin \zeta a/\zeta$ in (6.3.28) would be replaced by a.

3. *Poles of the integrands.* We denote the integrands of $u_x(x, y)$ and $u_y(x, y)$ by $\Phi(\zeta), \Psi(\zeta)$ respectively. The poles of these functions are given by the zeros of $F_0(\zeta) = 0$. This equation is, in fact, the previously-encountered frequency equation for Rayleigh surface waves, as may be seen by comparing (6.3.29) to (6.1.86). A direct correspondence is obtained if the comparison is made between (6.3.24) and the earlier equation, where

$$\bar{\alpha}^2 = \xi^2 - \omega^2/c_1^2, \qquad \bar{\beta}^2 = \xi^2 - \omega^2/c_2^2$$

in the latter. Thus the equation governing the propagation of free surface waves has arisen in the course of analysing the forced motion of a half-space. The zeros of $F_0(\zeta)$ are given by $\zeta = \pm\zeta_R$, where ζ_R is a real number.† We have

$$\nu = \tfrac{1}{4}, \qquad \zeta_R = \pm 1{\cdot}8839,$$
$$\nu = \tfrac{1}{3}, \qquad \zeta_R = \pm 2{\cdot}1447. \tag{6.3.30}$$

Recall that ζ_R is non-dimensionalized, so that the wave velocity would be found from the preceding by replacing ζ_R by ζ_R/k_1.

4. *Branch points and branch cuts.* We see that the functions $\Phi(\zeta)$, $\Psi(\zeta)$ will have branch points located at $\zeta = \pm 1, \pm k$, due to $(\zeta^2-1)^{\frac{1}{2}}$, $(\zeta^2-k^2)^{\frac{1}{2}}$. Now the considerations as to the appropriate branch necessary to make the integrands single valued about the contour follow quite closely those presented for the single set of branch points arising in the SH wave-source problem.‡ Thus, briefly, the functions $\alpha(\zeta) = (\zeta^2-1)^{\frac{1}{2}}$, $\beta(\zeta) = (\zeta^2-k^2)^{\frac{1}{2}}$ must have their real parts always greater than zero owing to the terms $\exp(-\alpha y)$ and $\exp(-\beta y)$ appearing in the solution. With small damping included in the system, the branch cuts lie along hyperbolas which then degenerate to cuts along the real and imaginary axes as the damping becomes zero. The resulting situation is shown in Fig. 6.14. The contour is closed in the upper half-plane on the assumption that $x > 0$. The entire contour integral is given by

$$\oint = \int_{\Gamma_1} + \int_{\Gamma_\alpha} + \int_{\Gamma_\beta} + \int_{\Gamma_2} + \int_{-R}^{R} = -2\pi i \sum \text{Res}. \tag{6.3.31}$$

† Recall in the discussion of surface waves that other, extraneous roots arose from solving the rationalized equation.

‡ See (6.2.18)–(6.2.23) and Fig. 6.17(a) and (b). See also Ewing *et al.* [11, pp. 45–54].

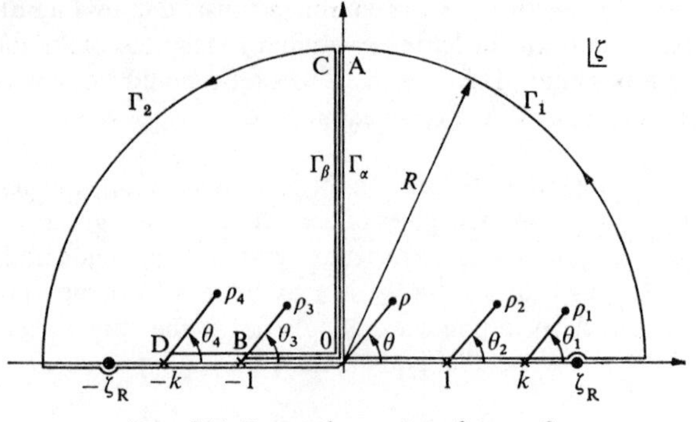

FIG. 6.14. Integration contour for $x > 0$.

The integrals along the branch cuts Γ_α, Γ_β are given by the segments

$$\Gamma_\alpha = \text{AO}+\text{OB}+\text{BO}+\text{OA},$$
$$\Gamma_\beta = \text{CO}+\text{OD}+\text{DO}+\text{OC}. \tag{6.3.32}$$

Furthermore, the poles at $\zeta = -\zeta_R$, $+\zeta_R$ are, respectively, included and excluded from the contour by slight indentations as dictated by the radiation condition.

To check that the branch cuts Γ_α, Γ_β do indeed yield the proper behaviour of $\alpha(\zeta)$, $\beta(\zeta)$, we introduce the polar coordinates ρ_1, θ_1, ρ_2, θ_2, ρ, θ, etc. as shown in Fig. 6.14. With these coordinates, we have that

$$\alpha(\zeta) = (\rho_2\rho_3)^{\frac{1}{2}} \exp\{i(\theta_2+\theta_3)/2\}, \qquad \beta(\zeta) = (\rho_1\rho_4)^{\frac{1}{2}} \exp\{i(\theta_1+\theta_4)/2\}. \tag{6.3.33}$$

The values of these functions along the various parts of Γ_α, Γ_β are given in Table 6.1. It is seen that Re α, Re $\beta \geqslant 0$ always. Thus the formal integral

TABLE 6.1
Value of α, β along Γ_α, Γ_β

Path	$\theta_2+\theta_3$	$\theta_1+\theta_4$	α	β	Re α	Im β	Re β	Im α
AO	π	π	$(1+\rho^2)^{\frac{1}{2}}i$	$(k^2+\rho^2)^{\frac{1}{2}}i$	0	>0	0	>0
OB	π	π	$(1+\rho^2)^{\frac{1}{2}}i$	$(k^2-\rho^2)^{\frac{1}{2}}i$	0	>0	0	>0
BO	$-\pi$	π	$-(1-\rho^2)^{\frac{1}{2}}i$	$(k^2-\rho^2)^{\frac{1}{2}}i$	0	<0	0	>0
OA	$-\pi$	π	$-(1+\rho^2)^{\frac{1}{2}}i$	$(k^2+\rho^2)^{\frac{1}{2}}i$	0	<0	0	>0
CO		Same as OA						
OB		Same as BO						
BD	0	π	$(\rho^2-1)^{\frac{1}{2}}$	$(k^2-\rho^2)^{\frac{1}{2}}i$	>0	0	0	>0
DB	0	$-\pi$	$(\rho^2-1)^{\frac{1}{2}}$	$-(k^2-\rho^2)^{\frac{1}{2}}i$	>0	0	0	<0
BO	$-\pi$	$-\pi$	$-(1-\rho^2)^{\frac{1}{2}}i$	$-(k^2-\rho^2)^{\frac{1}{2}}i$	0	<0	0	<0
OC	$-\pi$	$-\pi$	$-(1+\rho^2)^{\frac{1}{2}}i$	$-(k^2+\rho^2)^{\frac{1}{2}}i$	0	<0	0	<0

solutions (6.3.27) and (6.3.28) are determined from the residues at the poles and the evaluation of the integrals along the various portions of the contour. The contributions from Γ_1, Γ_2 vanish, so determination of the branch-cut integrals is the main task in obtaining an exact evaluation.

5. *Application of the steepest-descent method.* The task of evaluating the branch-cut integrals remains most formidable, so approximate results are sought. The method of steepest descent will be used to determine the field at infinity. We note that the integrals occurring in the displacement expressions (6.3.27) and (6.3.28) have the general form

$$I_1 = \int_{-\infty}^{\infty} \chi(\zeta)\exp\{i\zeta x-(\zeta^2-m^2)^{\frac{1}{2}}y\}\, d\zeta, \tag{6.3.34}$$

where $m = 1$ or k. We introduce the polar coordinates

$$x = R\sin\theta, \qquad y = R\cos\theta, \tag{6.3.35}$$

where θ is measured from the y-axis. Then I_1 may be written as

$$I_1 = \int_{-\infty}^{\infty} \chi(\zeta)e^{Rf(\zeta)}\, d\zeta, \tag{6.3.36}$$

where

$$f(\zeta) = i\zeta\sin\theta-(\zeta^2-m^2)^{\frac{1}{2}}\cos\theta. \tag{6.3.37}$$

We see that I_1 is now in the form (6.2.38), used in the development of the steepest-descent method.

We determine the saddle points of $\operatorname{Re} f(\zeta)$ from $df(\zeta)/d\zeta = 0$, and obtain

$$i\sin\theta = \frac{\zeta\cos\theta}{(\zeta^2-m^2)^{\frac{1}{2}}}, \qquad \zeta_0 = \pm m\sin\theta. \tag{6.3.38}$$

The point at $\zeta_0 = -m\sin\theta$ is the one of interest here; the point at $+m\sin\theta$ would be used for consideration on the negative real axis, in the range $-m < \zeta_0 < 0$. At the saddle point, we obtain that

$$f(\zeta_0) = -im, \qquad f''(\zeta_0) = i\sec^2\theta/m. \tag{6.3.39}$$

The contour of integration is now deformed to pass through the saddle point along the steepest-descent path. The general shape of the deformed contour will now be established. We know that near the saddle point that (see (6.2.62))

$$f(\zeta)-f(\zeta_0) = \tfrac{1}{2}(\zeta-\zeta_0)^2 f''(\zeta_0) \tag{6.3.40}$$

Also $f(\zeta)-f(\zeta_0) = -u^2$ (see (6.2.63)) along the path. Since $f''(\zeta_0)$ is given by the second of (6.3.39), we have from the preceding,

$$-u^2 = \tfrac{1}{2}(\zeta-\zeta_0)^2\frac{i\sec^2\theta}{m}, \tag{6.3.41}$$

from which we obtain

$$\zeta-\zeta_0 = \pm 2\sqrt{m}\,u\cos\theta e^{\pi i/4}. \tag{6.3.42}$$

The sign will be determined by the direction in which the contour passes through the saddle point. Thus, the positive sign represents a path at $+45°$, the negative sign a path at $-135°$, both with respect to the real ζ-axis.

To determine the path behaviour away from the saddle point, we note that the points at which the path crosses the real ζ-axis are given by

$$\text{Im}\{f(\zeta)-f(\zeta_0)\} = 0, \qquad \zeta \sim \text{real}. \tag{6.3.43}$$

The first condition holds for all points along the steepest-descent path, the second condition gives the real-axis intersections. Substituting (6.3.40) and $f(\zeta_0) = -im$ in the preceding condition gives $\zeta_1 = -\text{m cosec } \theta$ as the only other crossover point (the branch point also being a crossover point).

The asymptotic behaviour of the path may be determined. Writing the first condition of (6.3.43) as $\text{Im} f(\zeta) = \text{Im} f(\zeta_0)$ we have

$$\text{Im}\{i\zeta \sin \theta - (\zeta^2 - m^2)^{\frac{1}{2}} \cos \theta\} = -m. \tag{6.3.44}$$

For large $|\zeta|$ we may neglect m in the radical above and obtain

$$\text{Im}\{\zeta(i \sin \theta \mp \cos \theta\} = -m. \tag{6.3.45}$$

Letting $\zeta = \xi + i\eta$, we obtain

$$\xi \sin \theta \mp \eta \cos \theta = -m, \tag{6.3.46}$$

as the lines for the path asymptotes. We note that $0 < \theta < \pi/2$, so that $\sin \theta$, $\cos \theta > 0$. Also we note that both asymptotes pass through

$$\zeta = -m \text{ cosec } \theta.$$

The general form of the contour is shown in Fig. 6.15(a). On the basis of the direction of passage, we select the positive sign in (6.3.46).

We now consider the deformed contour relative to the pole and branch points along the negative real axis, located at $\zeta = -\zeta_R, -k, -1$, where $\zeta_R > k > 1$. Consider the case where $m = -k$. The saddle point will lie in the range $-k < \zeta_0 < 0$, and the branch point at $\zeta = -1$ is located within this range. For $-1 < \zeta_0 < 0$, the contour is as shown in Fig. 6.15(a). For $\theta > \text{cosec}^{-1}k$, $\zeta_0 < -1$ and the steepest-descent contour differs in its manner of circulating the branch point from that shown in the figure. Basically, a

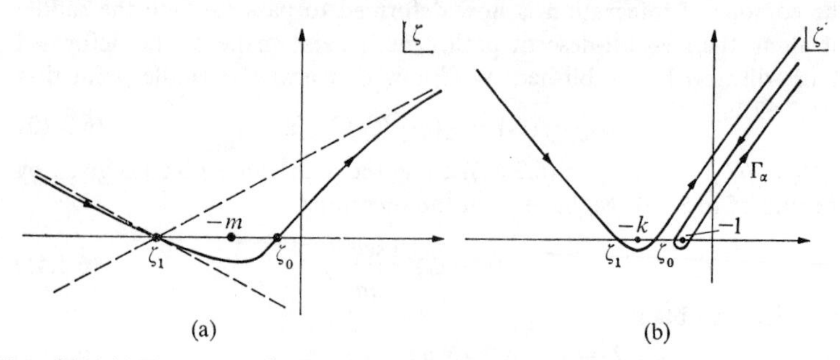

FIG. 6.15. (a) The steepest descent contour passing through ζ_0, and (b) modification of the steepest descent path to include a branch point.

'return loop' must be added that returns from infinity, passes about the branch point in the proper manner, and goes off again to infinity, as shown in Fig. 6.15(b). When $\zeta_0 = -1$, so that the steepest-descent path passes through the branch point, the path Γ_α is made to coincide with the steepest-descent path. When $m = -1$, there is no problem of the positive sloping portion of the path not properly passing the branch point at $\zeta = -k$. However, when $\theta > \operatorname{cosec}^{-1}k$, we have $\zeta_1 > -k$, so that the negative sloping portion is not properly circulating that branch point. A return loop is incorporated to account for $-k$ similar to that used in the previous case. If these additional loops yielded significant contributions, their exact evaluation would be necessary. However, Miller and Pursey state that contributions from these loops are asymptotically negligible.

There remains the matter of the pole $-\zeta_R$. If $\theta < \operatorname{cosec}^{-1}(\zeta_R/m)$, then the crossover point ζ_1 will be to the right of ζ_R and the pole will no longer be within the general contour. For $\theta > \operatorname{cosec}^{-1}(\zeta_R/m)$, the pole will be within the contour and must be accounted for by the additional residue term. Referring back to (6.3.36) and (6.3.37), we see that the resulting residue will contain the term

$$\exp\{-R(\zeta_R^2-m^2)^{\frac{1}{2}}\}\cos\theta. \tag{6.3.47}$$

For R large, this will result in a negligible contribution and may be neglected except when $\theta = \pi/2$. This circumstance will yield the surface wave contribution and will, in fact, be the dominant term. This will be given separate consideration.

With the saddle points established and the steepest-descent path determined, the resulting asymptotic value of the integral I_1 is given by

$$I_1 \sim \sqrt{\left(\frac{2\pi m}{R}\right)}\exp\left\{i\left(\frac{\pi}{4}-mR\right)\right\}\cos\theta[\chi(-m\sin\theta)], \tag{6.3.48}$$

where $0 < \theta < \pi/2$.

6. *The far-field results.* Using the result (6.3.48) in the evaluation of the integrals (6.3.27) and (6.3.28), where we let $a \to 0$ and thus have the line source case, one obtains

$$u_x \sim \frac{ae^{\pi i/4}\cos\theta}{\mu}\left(\frac{2}{\pi R}\right)^{\frac{1}{2}}\left\{-\frac{k^{\frac{5}{2}}\sin 2\theta(k^2\sin^2\theta-1)^{\frac{1}{2}}}{F_0(k\sin\theta)}e^{-ikR}+\right.$$
$$\left.+\frac{i\sin\theta(k^2-2\sin^2\theta)}{F_0(\sin\theta)}e^{-iR}\right\}, \tag{6.3.49}$$

$$u_y \sim \frac{ae^{\pi i/4}\cos\theta}{\mu}\left(\frac{2}{\pi R}\right)^{\frac{1}{2}}\left\{\frac{2k^{\frac{5}{2}}\sin^2\theta(k^2\sin^2\theta-1)^{\frac{1}{2}}}{F_0(k\sin\theta)}e^{-ikR}+\right.$$
$$\left.+\frac{i\cos\theta(k^2-2\sin^2\theta)}{F_0(\sin\theta)}e^{-iR}\right\}. \tag{6.3.50}$$

To obtain the radial and tangential components of the field, we use the relations

$$u_R = u_y \cos\theta + u_x \sin\theta,$$
$$u_\theta = u_x \cos\theta - u_y \sin\theta. \tag{6.3.51}$$

This gives

$$u_R \sim \frac{a\exp\left\{i\left(\frac{3\pi}{4}-R\right)\right\}}{\mu}\left(\frac{2}{\pi R}\right)^{\frac{1}{2}}\frac{\cos\theta(k^2-2\sin^2\theta)}{F_0(\sin\theta)}, \tag{6.3.52}$$

$$u_\theta \sim \frac{a\exp\left\{i\left(\frac{5\pi}{4}-kR\right)\right\}}{\mu}\left(\frac{2k^5}{\pi R}\right)^{\frac{1}{2}}\frac{\sin 2\theta(k^2\sin^2\theta-1)^{\frac{1}{2}}}{F_0(k\sin\theta)}. \tag{6.3.53}$$

The resulting displacement fields u_R, u_θ are shown as a function of θ $(0 \leq \theta < \pi/2)$ for Poisson's ratio of $\frac{1}{3}$ in Fig. 6.16.

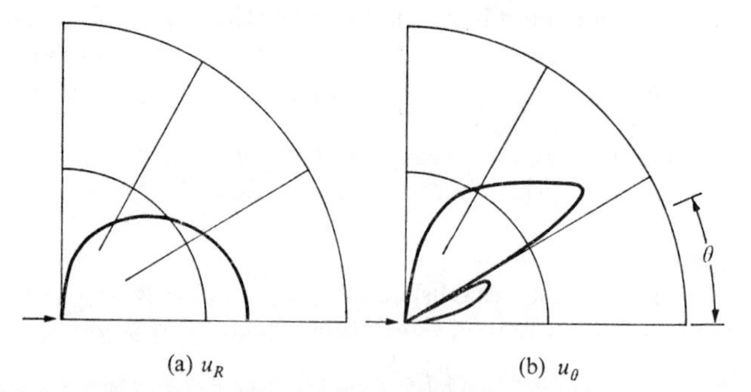

FIG. 6.16. Polar plots of (a) u_R, and (b) u_θ for Poisson's ratio of $\frac{1}{3}$. (After Miller and Pursey [30, Figs. 5 and 7].)

7. *Surface waves.* The case of surface waves from a line source is given by setting $y = 0$ in the results (6.3.27) and (6.3.28) for u_x, u_y and letting $a \to 0$ in the manner previously prescribed. We obtain

$$u_x(x,0) = \frac{a}{i\mu\pi}\int_{-\infty}^{\infty}\frac{\zeta}{F_0(\zeta)}\{2(\zeta^2-1)^{\frac{1}{2}}(\zeta^2-k^2)^{\frac{1}{2}}+(k^2-2\zeta^2)\}e^{i\zeta x}\,d\zeta, \tag{6.3.54}$$

$$u_y(x,0) = \frac{ak^2}{\mu\pi}\int_{-\infty}^{\infty}\frac{(\zeta^2-1)^{\frac{1}{2}}}{F_0(\zeta)}e^{i\zeta x}\,d\zeta. \tag{6.3.55}$$

If, in performing the contour integration of Fig. 6.14, we consider only the residue contribution and neglect the contribution of the branch cut integrals Γ_α, Γ_β, we will obtain the surface-wave effect. Writing the general forms of the integrands of $u_x(x, 0)$, $u_y(x, 0)$ as

$$\chi(\zeta) = \psi(\zeta)/F_0(\zeta), \tag{6.3.56}$$

we have the residue at $\zeta = -\zeta_R$ given by

$$\text{Res} = \frac{\psi(-\zeta_R)\exp(-i\zeta_R x)}{F_0'(-\zeta_R)}. \tag{6.3.57}$$

Thus the general waveform will be given by

$$u_x(x, 0) \sim \exp\{-i(\zeta_R x-\omega t)\}, \qquad u_y(x, 0) \sim \exp\{i(\zeta_R x-\omega t)\}. \tag{6.3.58}$$

The propagation velocity will be c_R, that of Rayleigh surface waves. We note that the amplitudes are not affected by distance of propagation, whereas $u_x(x, y)$, $u_y(x, y)$, or $u_R(R, \theta)$, $u_\theta(R, \theta)$ go as $R^{-\frac{1}{2}}$. This, again, is a consequence of the waves propagating along the surface. We emphasize that this result is for the plane-strain, line-load case. If the loading were a point load, the Rayleigh wave amplitude would be attenuated with distance, but not as severely as the waves into the interior. Miller and Pursey give specific results for two values of Poisson's ratio. Thus

$\nu = \frac{1}{4}$:

$$u_x(x, 0) \sim 0{\cdot}250\frac{a}{\mu} \exp(-1{\cdot}884ix),$$

$$u_y(x, 0) \sim 0{\cdot}367\frac{ia}{\mu} \exp(-1{\cdot}884ix). \tag{6.3.59}$$

$\nu = \frac{1}{3}$:

$$u_x(x, 0) \sim 0{\cdot}198\frac{a}{\mu} \exp(-2{\cdot}145ix),$$

$$u_y(x, 0) \sim 0{\cdot}311\frac{ia}{\mu} \exp(-2{\cdot}145ix). \tag{6.3.60}$$

6.3.2. *Other results for harmonic sources*

Many analyses have been done on periodic loading of a half-space, of which the results presented in the previous section are somewhat representative. The previous analysis was concerned with a line normal load. The case of a line tangential loading, where the action of the load is perpendicular to the load line has also been considered by Lamb [22], Miller and Pursey [30], and, no doubt, others. Of greater practical interest is the case of a point normal load, or a normal load applied over a small (circular) region. The case of torsional loads applied at a point or over a small region have also been considered. The bases for interest in harmonic wave excitation are several fold. The first is that harmonic wave solutions provide a basis, through superposition techniques, of obtaining solutions to transient problems. There are several areas, however, where the results are of direct application. In ultrasonics, transducers transmit wave trains that are often very narrow-band and may be approximated as purely harmonic. Many problems involving foundation vibrations involve steady-state excitation of an extended medium.

The major additional results given here for harmonic loads are also from Miller and Pursey [30, 31]. Thus in [30] the problems of a tangential line load, a normal load applied to a circular region $r \leqslant a$, and the case of a torque applied about the y-axis were all analysed. Far-field radiation patterns for the tangential loading were also obtained, and are shown in Fig. 6.17.

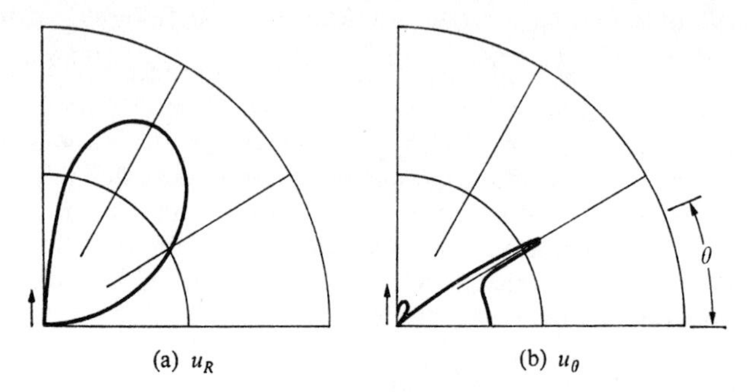

FIG. 6.17. The far-field displacements (a) u_R and (b) u_θ for a line tangential load applied at the origin for $0 \leqslant \theta < \pi/2$. Poisson's ratio $\nu = \frac{1}{3}$. (After Miller and Pursey [30, Figs. 9 and 11].)

The analytical expressions for these given by Miller and Pursey are

$$u_R \sim \frac{a \exp\left\{i\left(\frac{3\pi}{4} - R\right)\right\}}{\mu}\left(\frac{2}{\pi R}\right)^{\frac{1}{2}} \frac{\sin 2\theta(k^2 - \sin^2\theta)}{F_0(\sin\theta)}, \tag{6.3.61}$$

$$u_\theta \sim \frac{a \exp\left\{i\left(\frac{3\pi}{4} - kR\right)\right\}}{\mu}\left(\frac{2k^7}{\pi R}\right) \frac{\cos\theta\cos 2\theta}{F_0(k\sin\theta)}. \tag{6.3.62}$$

As mentioned, Miller and Pursey also presented the analysis of a normal stress, applied to the circular area $r \leq a$, and varying harmonically with time. The analysis is in terms of the coordinates r, y, but the results for the displacements are also given in terms of u_R, u_θ. For the far field, where $0 < \theta < \pi/2$, these are

$$u_R \sim -\frac{a^2}{2\mu}\frac{e^{-iR}}{R}\frac{\cos\theta(k^2 - 2\sin^2\theta)}{F_0(\sin\theta)}, \tag{6.3.63}$$

$$u_\theta \sim \frac{ia^2k^3}{2\mu}\frac{e^{-ikR}}{R}\frac{\sin 2\theta(k^2\sin^2\theta - 1)}{F_0(k\sin\theta)}. \tag{6.3.64}$$

The surface wave results are also given. Thus, for $\nu = \frac{1}{4}$,

$$u_r(r, 0) \sim 0{\cdot}215 \frac{a^2 \exp(\pi i/4)}{\mu\sqrt{r}} \exp(-1{\cdot}884 ir), \tag{6.3.65}$$

$$u_y(r, 0) \sim -0{\cdot}316 \frac{a^2 \exp(\pi i/4)}{\mu\sqrt{r}} \exp(-1{\cdot}884 ir). \tag{6.3.66}$$

For $\nu = \frac{1}{3}$, replace 0·215, 0·316, and 1·884 by 0·182, 0·286, and 2·145 respectively. It should be noted that the surface wave now undergoes amplitude attention with distance as $R^{-\frac{1}{2}}$, while the body waves attenuate as R^{-1}, a more severe attenuation. Lord [25] has computed the far-field radiation diagrams for this case of loading. The results, in terms of stresses for a Poisson's ratio of 0·20 are shown in Fig. 6.18. The scale is arbitrary for the

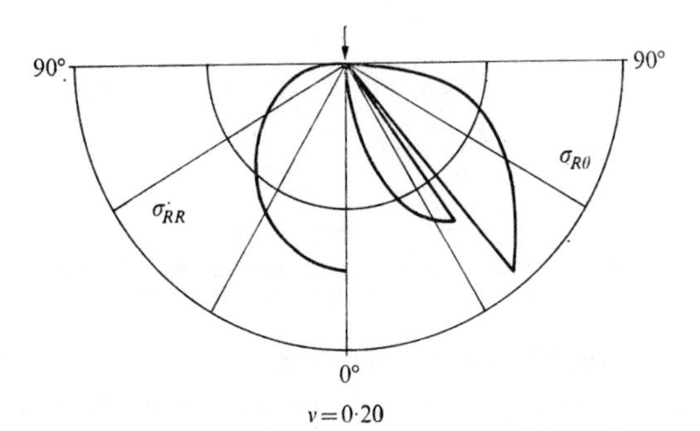

FIG. 6.18. Polar diagram for the stresses σ_{RR}, $\sigma_{R\theta}$ resulting from a point normal load at the origin for $\nu = 0{\cdot}20$. (After Lord [25, Figs. 7 and 8].)

plots, although both quadrants are to the same scale.

In another aspect of their work, Miller and Pursey [31] have computed the partition of energy among the dilatational, shear, and surface waves due to an oscillating normal point force. Woods [44] has presented this data in a most informative manner, as shown in Fig. 6.19. Thus the compressional and shear waves are shown spreading out in hemispherical wavefronts. The spacing of the wavefronts is in accord with their differing velocities. The relative amplitude of particle motion is shown. Also shown is the Rayleigh surface wave, with the vertical and horizontal displacement components shown on the leftward- and rightward-propagating parts of the wave. The various powers of r^{-n} ($n = 0{\cdot}5, 1, 2$) give the geometric attentuation of the displacement amplitudes with radial distance r. The shear window indicates the portion of the shear wave along which amplitudes are greatest. The partition of energy is shown in the table of Fig. 6.19. The predominance of the Rayleigh wave containing 67 per cent of the input energy and undergoing more gradual amplitude attentuation is clearly evident.

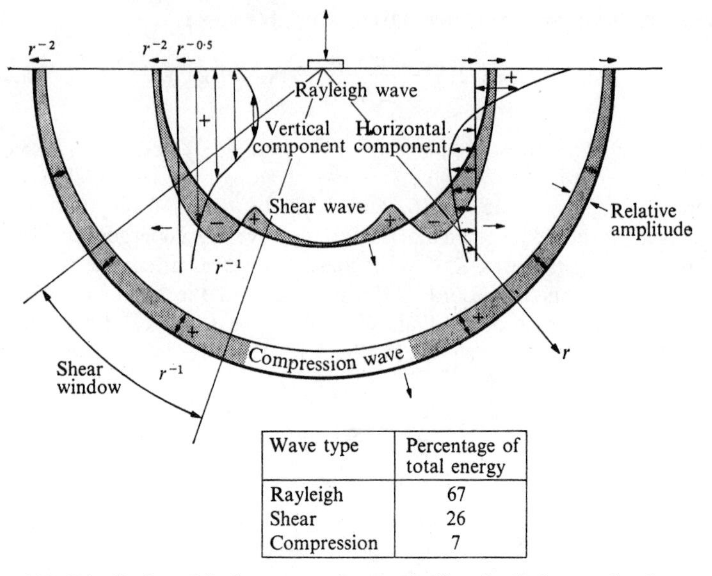

Wave type	Percentage of total energy
Rayleigh	67
Shear	26
Compression	7

FIG. 6.19. Distribution of displacement and energy in dilatational, shear, and surface waves from a harmonic normal load on a half-space for $\nu = \frac{1}{4}$. (After Woods [44, Fig. 1].)

6.3.3. *Transient normal loading on a half-space*

The original analysis by Lamb of the half-space problem employed Fourier superposition of harmonic waves to obtain the response to transient loading. Others have approached the problem directly by assuming zero initial conditions, and considering step or impulse loading. In the following, the case of a step normal loading, applied at the origin of the half-space will be formulated using integral transforms. The inversion of the results for two special cases will be partially carried out. The technique for performing the inversion, known as Cagniard's method, will be shown. The following section will be devoted to presenting several results of transient-surface or buried-source analyses.

1. *Governing equations and boundary conditions.* Consider the half-space $z > 0$ loaded by a concentrated, upward-directed force. Polar coordinates will be used, and consistency with the usual r, θ, z notation requires that the z-axis now be the vertical instead of the y-axis. Owing to the axisymmetry existing for this problem, the displacement component $u_\theta = 0$, so that

$$\mathbf{u}(r, z, t) = u_r \mathbf{e}_r + u_z \mathbf{e}_z. \tag{6.3.67}$$

In terms of the scalar and vector potentials Φ, $\mathbf{H}$, we have

$$\mathbf{u} = \nabla\Phi + \nabla\times(H_\theta \mathbf{e}_\theta). \tag{6.3.68}$$

The governing equations for Φ, H_θ are then

$$\nabla^2\Phi = \frac{1}{c_1^2}\frac{\partial^2\Phi}{\partial t^2}, \qquad \nabla^2 H_\theta - \frac{1}{r^2}H_\theta = \frac{1}{c_2^2}\frac{\partial^2 H_\theta}{\partial t^2}, \tag{6.3.69}$$

where $\nabla^2 = r^{-1}\,\partial(r\partial/\partial r)/\partial r + \partial^2/\partial z^2$. It is possible to reduce the second equation of (6.3.69) to the scalar wave equation by defining the function Ψ, where

$$H_\theta = -\partial\Psi/\partial r. \tag{6.3.70}$$

Then we obtain

$$\nabla^2\Psi = \frac{1}{c_2^2}\frac{\partial^2\Psi}{\partial t^2}. \tag{6.3.71}$$

The displacements and stresses are given by

$$u_r = \frac{\partial\Phi}{\partial r} + \frac{\partial^2\Psi}{\partial r\,\partial z}, \qquad u_\theta = 0,$$

$$u_z = \frac{\partial\Phi}{\partial z} + \frac{\partial^2\Psi}{\partial z^2} - \frac{1}{c_2^2}\ddot{\Psi}. \tag{6.3.72}$$

The stresses are

$$\tau_{rr} = \lambda\nabla^2\Phi + 2\mu\left(\frac{\partial^2\Phi}{\partial r^2} + \frac{\partial^3\Psi}{\partial r^2\,\partial z}\right),$$

$$\tau_{\theta\theta} = \lambda\nabla^2\Phi + \frac{2\mu}{r}\left(\frac{\partial\Phi}{\partial r} + \frac{\partial^2\Psi}{\partial r\,\partial z}\right),$$

$$\tau_{zz} = \lambda\nabla^2\Phi + 2\mu\frac{\partial}{\partial z}\left(\frac{\partial\Phi}{\partial z} + \frac{\partial^2\Psi}{\partial z^2} - \frac{1}{c_2^2}\ddot{\Psi}\right),$$

$$\tau_{rz} = \mu\frac{\partial}{\partial r}\left(2\frac{\partial\Phi}{\partial z} + 2\frac{\partial^2\Psi}{\partial z^2} - \frac{1}{c_2^2}\ddot{\Psi}\right),$$

$$\tau_{\theta z} = \tau_{r\theta} = 0. \tag{6.3.73}$$

The appearance of $\ddot{\Psi}$ terms in the above is a consequence of using the identity (6.3.71) when the appropriate combinations of spatial derivatives occur.

For the boundary conditions, we have

$$\tau_{zr} = 0, \qquad \tau_{zz} = F(r)H\langle t\rangle, \qquad z = 0, \tag{6.3.74}$$

where $F(r)$ is taken to be an arbitrary function for the moment. It may be represented as a Fourier–Bessel integral of the form

$$F(r) = \int_0^\infty \xi J_0(\xi r)\,d\xi \int_0^\infty F(\sigma)\sigma J_0(\sigma\xi)\,d\sigma. \tag{6.3.75}$$

This is, of course, the polar-coordinate analogue of the Fourier integral. In the present problem, we wish $F(r)$ to represent a point load. This is the case if

$$F(r) = \frac{\delta(r)}{2\pi r} Z, \tag{6.3.76}$$

where Z represents the magnitude of the applied force. Then (6.3.75) becomes

$$F(r) = \frac{Z}{2\pi} \int_0^\infty \xi J_0(\xi r)\, d\xi. \tag{6.3.77}$$

Finally, we note that homogeneous initial conditions are assumed.

2. *Transformed equations.* The Laplace and Hankel transforms will be applied to the present problem. The Laplace transform of various quantities will be indicated as

$$U_i(r, z, p) = \mathscr{L}\{u_i(r, z, t)\}, \qquad T_{ij}(r, z, p) = \mathscr{L}\{\tau_{ij}(r, z, t)\},$$
$$\bar{\Phi}, \bar{\Psi} = \mathscr{L}\{\Phi, \Psi\}. \tag{6.3.78}$$

The transformed governing equations are

$$\nabla^2\bar{\Phi} - \frac{p^2}{c_1^2}\bar{\Phi} = 0, \qquad \nabla^2\bar{\Psi} - \frac{p^2}{c_2^2}\bar{\Psi} = 0. \tag{6.3.79}$$

The transformed displacements and stresses are

$$U_r = \frac{\partial\bar{\Phi}}{\partial r} + \frac{\partial^2\bar{\Psi}}{\partial r\,\partial z},$$
$$U_z = \frac{\partial\bar{\Phi}}{\partial z} + \frac{\partial^2\bar{\Psi}}{\partial z^2} - \frac{p^2}{c_2^2}\bar{\Psi}, \tag{6.3.80}$$

$$T_{rr} = \lambda\nabla^2\bar{\Phi} + 2\mu\left(\frac{\partial^2\bar{\Phi}}{\partial r^2} + \frac{\partial^3\bar{\Psi}}{\partial r^2\,\partial z}\right),$$
$$T_{\theta\theta} = \lambda\nabla^2\bar{\Phi} + \frac{2\mu}{r}\left(\frac{\partial\bar{\Phi}}{\partial r} + \frac{\partial^2\bar{\Psi}}{\partial r\,\partial z}\right),$$
$$T_{zz} = \lambda\nabla^2\bar{\Phi} + 2\mu\frac{\partial}{\partial z}\left(\frac{\partial\bar{\Phi}}{\partial z} + \frac{\partial^2\bar{\Psi}}{\partial z^2} - \frac{p^2}{c_2^2}\bar{\Psi}\right),$$
$$T_{rz} = \mu\frac{\partial}{\partial r}\left(2\frac{\partial\bar{\Phi}}{\partial z} + 2\frac{\partial^2\bar{\Psi}}{\partial z^2} - \frac{p^2}{c_2^2}\bar{\Psi}\right). \tag{6.3.81}$$

The transformed boundary conditions are

$$T_{zr} = 0, \qquad T_{zz} = F(r)/p, \qquad z = 0. \tag{6.3.82}$$

The procedure we could follow at this point would be to solve (6.3.79) explicitly subject to (6.3.82). This would lead directly to a transformed solution U_r, U_z. The problem of the Laplace inversion would then leave us with little choice but to tackle it head-on. An alternative procedure, used by Chao [3] will be used here. Thus, the Hankel transform will now be applied to the preceding Laplace-transformed equations. Defining

$$\bar{\phi}(\xi, z, p) = \int_0^\infty \xi J_0(\xi r)\tilde{\Phi}(r, z, p)\,\mathrm{d}r, \tag{6.3.83}$$

as the zero-order Hankel transform of $\tilde{\Phi}$, with similar definitions for $\bar{\psi}$, $\bar{U}_r$, etc., we have for the Hankel transform of (6.3.79)

$$\frac{\mathrm{d}^2\bar{\phi}}{\mathrm{d}z^2} - (\xi^2+h^2)\bar{\phi} = 0, \qquad \frac{\mathrm{d}^2\bar{\psi}}{\mathrm{d}z^2} - (\xi^2+k^2)\bar{\psi} = 0, \tag{6.3.84}$$

where

$$h^2 = p^2/c_1^2, \qquad k^2 = p^2/c_2^2 = (c_1^2/c_2^2)h^2. \tag{6.3.85}$$

We now apply the Hankel transform to U_r, U_z. However, because of the way the transform operates on derivatives with respect to r, the first-order transform is appropriate for U_r, while the zero-order transform is appropriate for U_z, giving

$$\bar{U}_r = \int_0^\infty rU_rJ_1(\xi r)\,\mathrm{d}r, \qquad \bar{U}_z = \int_0^\infty rU_zJ_0(\xi r)\,\mathrm{d}r. \tag{6.3.86}$$

This gives

$$\begin{aligned} \bar{U}_r &= -\xi(\bar{\phi}+\bar{\psi}'), \\ \bar{U}_z &= \bar{\phi}'+\bar{\psi}''-k^2\bar{\psi}, \end{aligned} \tag{6.3.87}$$

where the primes indicate derivatives with respect to z. In a similar manner. we transform the stresses according to

$$\bar{T}_{zz} = \int_0^\infty rT_{zz}J_0(\xi r)\,\mathrm{d}r, \qquad \bar{T}_{rz} = \int_0^\infty rT_{rz}J_1(\xi r)\,\mathrm{d}r, \tag{6.3.88}$$

giving

$$\begin{aligned} \bar{T}_{zz} &= \lambda h^2\bar{\phi}+2\mu(\bar{\phi}''+\bar{\psi}'''-k^2\bar{\psi}'), \\ \bar{T}_{rz} &= -\mu\xi(2\bar{\phi}'+2\bar{\psi}''-k^2\bar{\psi}). \end{aligned} \tag{6.3.89}$$

The boundary conditions are thus

$$\begin{aligned} \lambda h^2\bar{\phi}+2\mu(\bar{\phi}''+\bar{\psi}''-k^2\bar{\psi}') &= Z/2\pi p, \qquad z = 0, \\ 2\bar{\phi}'+2\bar{\psi}''-k^2\bar{\psi} &= 0, \qquad z = 0. \end{aligned} \tag{6.3.90}$$

3. *Transformed solutions and Hankel inversion.* We have for the solutions of the transformed equations (6.3.84),

$$\bar{\phi} = A_1\mathrm{e}^{-\alpha z}+B_1\mathrm{e}^{\alpha z}, \qquad \bar{\psi} = A_2\mathrm{e}^{-\beta z}+B_2\mathrm{e}^{\beta z}, \tag{6.3.91}$$

where

$$\alpha^2 = \xi^2+h^2, \qquad \beta^2 = \xi^2+k^2. \tag{6.3.92}$$

We discard the B_1, B_2 terms since they will lead to unbounded results, and substitute the remaining in the boundary conditions (6.3.90), giving

$$(\lambda h^2+2\mu\alpha^2)A_1-2\mu\beta(\beta^2-k^2)A_2 = Z/2\pi p,$$
$$-2\alpha A_1+(2\beta^2-k^2)A_2 = 0. \tag{6.3.93}$$

Noting that

$$\lambda h^2+2\mu\alpha^2 = \mu(k^2+2\xi^2), \qquad \beta^2-k^2 = \xi^2, \qquad 2\beta^2-k^2 = 2\xi^2+k^2, \tag{6.3.94}$$

then (6.3.93) may be written as

$$\begin{bmatrix} (k^2+2\xi^2) & -2\beta\xi^2 \\ -2\alpha & (2\xi^2+k^2) \end{bmatrix} \begin{bmatrix} A_1 \\ A_2 \end{bmatrix} = \begin{bmatrix} \dfrac{Z}{2\pi p\mu} \\ 0 \end{bmatrix}. \tag{6.3.95}$$

Solving for A_1, A_2 we obtain

$$A_1 = \frac{Z}{2\pi\mu p}\frac{2\xi^2+k^2}{D}, \qquad A_2 = \frac{Z}{2\pi\mu p}\frac{2\alpha}{D},$$
$$D = (2\xi^2+k^2)^2-4\alpha\beta\xi^2. \tag{6.3.96}$$

With the coefficients in hand, we have the transformed solutions. Thus, from (6.3.87),

$$\bar{U}_r = \frac{-Z}{2\pi\mu p}\frac{\xi\{(2\xi^2+k^2)e^{-\alpha z}-2\alpha\beta e^{-\beta z}\}}{D},$$
$$\bar{U}_z = \frac{Z}{2\pi\mu p}\frac{\{-\alpha(2\xi^2+k^2)e^{-\alpha z}+2\alpha\xi^2 e^{-\beta z}\}}{D}. \tag{6.3.97}$$

The preceding are the Laplace–Hankel transformed solutions. Performing the Hankel inversion, we have

$$U_r(r, z, p) = \frac{Z}{2\pi\mu p}\int_0^\infty \frac{\xi^2\{(2\xi^2+k^2)e^{-\alpha z}-2\alpha\beta e^{-\beta z}\}}{D}J_1(\xi r)\,d\xi, \tag{6.3.98}$$

$$U_z(r, z, p) = \frac{Z}{2\pi\mu p}\int_0^\infty \frac{\{-\alpha\xi(2\xi^2+k^2)e^{-\alpha z}+2\alpha\xi^3 e^{-\beta z}\}}{D}J_0(\xi r)\,d\xi. \tag{6.3.99}$$

We shall now seek to carry out the evaluation of the above, including the Laplace inversion aspect. Unfortunately, integrations of the above, valid for all r and z, are most difficult to obtain. Instead, only special cases have been evaluated, such as the values for $r = 0$ (directly under the load) or for $z = 0$ (along the surface). For these cases, the integrals are somewhat simplified. Fortunately, these special cases are also cases of considerable practical interest.

4. *Evaluation for $r = 0$.* The case of $r = 0$ is the simplest. From (6.3.98) we see that $U_r = 0$, since $J_1(0) = 0$, and that

$$U_z = \frac{Z}{2\pi\mu p}\int_0^\infty \frac{\{-\alpha\xi(2\xi^2+k^2)e^{-\alpha z}+2\alpha\xi^3 e^{-\beta z}\}}{D}\,d\xi, \tag{6.3.100}$$

where, we recall, α and β contain p, the Laplace transform parameter. We will apply Cagniard's method to this problem.† The essence of this method is as follows: Suppose we wish to Laplace invert $U(p)$ given as some integral

$$U(p) = \int_a^\infty g(t)e^{-f(p)t}\,dt. \tag{6.3.101}$$

The procedure is, by proper changes of variables, to manipulate the preceding into a form

$$U(p) = \int_0^\infty h(t)e^{-pt}\,dt = \mathscr{L}\{h(t)\}. \tag{6.3.102}$$

Then, by inspection,

$$u(t) = \mathscr{L}^{-1}\{U(p)\} = h(t). \tag{6.3.103}$$

The key to this technique is making the proper changes of variables to bring the integral into the desired form.

Proceeding, we break the integral into two parts and take the portion $\exp(-\alpha z)$. Let $\alpha z = pt$, which effectively treats α as an independent variable. Then

$$pt = (p/c_2)c_2 t = kc_2 t, \tag{6.3.104}$$

so that $\alpha = kc_2t/z = k\tau$, where $\tau = c_2t/z$. The consequences of the above change, which is meant to transform $\exp(-\alpha z)$ into $\exp(-pt)$ or the form of the Laplace transform, must now be incorporated in the remaining terms. Thus, from (6.3.92),

$$\alpha^2 = \xi^2+p^2/c_1^2 = k^2\tau^2. \tag{6.3.105}$$

Defining $\epsilon = (c_2/c_1)^2$, this gives

$$\xi^2 = k^2\tau^2-p^2/c_1^2 = k^2\tau^2-k^2\varepsilon = k^2(\tau^2-\varepsilon) \tag{6.3.106}$$

and

$$d\xi = \frac{p^2\tau}{c_2\xi z}\,dt. \tag{6.3.107}$$

Similarly, defining $\delta_1 = 1-\varepsilon$, $\delta_2 = 1-2\varepsilon$, we have

$$\beta^2 = k^2(\tau^2+\delta_1), \qquad 2\xi^2+k^2 = k^2(2\tau^2+\delta_2). \tag{6.3.108}$$

† Cagniard, L., *Reflexion et refraction des seismiques progressives*, Gauthiers–Villar, Paris (1935). Fung [13, pp. 218–25] gives a rather extensive presentation of the method as applied to Lamb's problem of the suddenly applied line load.

Manipulations on the second integral, containing $\exp(-\beta z)$ follow a similar pattern; thus let

$$\beta z = pt, \qquad \beta = k\tau, \qquad \xi^2 = k^2(\tau^2-1),$$
$$d\xi = \frac{p^2\tau}{c_2\xi z}\,dt, \qquad 2\xi^2+k^2 = k^2(2\tau^2-1), \qquad \alpha^2 = k^2(\tau^2-\delta_1). \tag{6.3.109}$$

The limits must now be considered. In particular, at $\xi = 0$, we have from (6.3.105)

$$\tau = p/c_1k = c_2/c_1 = c_2t/z, \qquad t = z/c_1, \tag{6.3.110}$$

while from (6.3.109) we have for $\xi = 0$, $t = z/c_2$ for the lower limit. The two integrals associated with U_z have thus taken the form

$$U_z \propto \int\limits_{z/c_1}^{\infty} (\ldots)e^{-pt}\,dt + \int\limits_{z/c_2}^{\infty} (\ldots)e^{-pt}\,dt. \tag{6.3.111}$$

We now introduce the step functions $H\langle t-z/c_1\rangle$ and $H\langle t-z/c_2\rangle$ inside the integrals. This enables the lower limits to be written as zero. Thus (6.3.100) becomes

$$U_z(0, z, p) = \frac{-Z}{2\pi\mu}\int\limits_0^{\infty} \frac{\tau(2\tau^2+\delta_2)e^{-pt}H\langle t-z/c_1\rangle}{(2\tau^2+\delta_2)^2-4\tau(\tau^2+\delta_1)^{\frac{1}{2}}(\tau^2-\varepsilon)}\left(\frac{\tau}{z}\right)dt +$$
$$+\frac{Z}{2\pi\mu}\int\limits_0^{\infty} \frac{2(\tau^2-\delta_1)^{\frac{1}{2}}(\tau^2-1)e^{-pt}H\langle t-z/c_2\rangle}{(2\tau^2-1)^2-4\tau(\tau^2-\delta_1)^{\frac{1}{2}}(\tau^2-1)}\left(\frac{\tau}{z}\right)dt. \tag{6.3.112}$$

The result is now in a form suitable for inversion by inspection and is

$$u_z(0, z, t) = \frac{Z}{2\pi\mu}\left\{\frac{-\tau^2(2\tau^2+\delta_2)H\langle t-z/c_1\rangle}{(2\tau+\delta_2^2)^2-4\tau(\tau^2-\varepsilon)(\tau^2+\delta_1)^{\frac{1}{2}}}+\right.$$
$$\left.+\frac{2\tau(\tau^2-1)(\tau^2-\delta_1)^{\frac{1}{2}}H\langle t-z/c_2\rangle}{(2\tau^2-1)-4\tau(\tau^2-1)(\tau^2-\delta_1)^{\frac{1}{2}}}\right\}, \tag{6.3.113}$$

where, we recall, $\tau = c_2t/z$.

5. *Evaluation for $z = 0$.* By letting $z = 0$, we will obtain the surface behaviour. The solutions (6.3.98) and (6.3.99) then become

$$U_r(r, 0, p) = \frac{Z}{2\pi\mu p}\int\limits_0^{\infty} \frac{(2\xi^2+k^2-2\alpha\beta)}{D}\xi^2 J_1(\xi r)\,d\xi,$$
$$U_z(r, 0, p) = \frac{-Z}{2\pi\mu p}\int\limits_0^{\infty} \frac{\alpha\xi k^2}{D}J_0(\xi r)\,d\xi. \tag{6.3.114}$$

While considerably simplified over the original expressions, the presence of the Bessel functions makes the inversion process rather difficult. Only the inversion of $U_z(r, 0, p)$ will be partially outlined. The procedure follows that of Pekeris [34], where complete details may be found, as well as consideration of $U_r(r, 0, p)$.

We first introduce a change of variables; let $\xi = kx = px/c_2$. We now assume a specific value of Poisson's ratio $\nu = \frac{1}{4}$. Then

$$\alpha^2 = k^2(x^2+\tfrac{1}{3}), \qquad \beta^2 = k^2(x^2+1). \tag{6.3.115}$$

Then the second of (6.3.114) becomes

$$U_z(r, 0, p) = \frac{-Z}{2\pi\mu c_2}\int_0^\infty \frac{x(x^2+\frac{1}{3})^{\frac{1}{2}}J_0\left(\frac{p}{c_2}xr\right)\,dx}{(2x^2+1)^2-4x^2(x^2+\frac{1}{3})^{\frac{1}{2}}(x^2+1)^{\frac{1}{2}}}. \tag{6.3.116}$$

We define

$$N(pr) = \int_0^\infty xm(x)J_0\left(\frac{prx}{c_2}\right)\,dx, \tag{6.3.117}$$

where

$$m(x) = \frac{(x^2+\frac{1}{3})^{\frac{1}{2}}}{(2x^2+1)^2-4x^2(x^2+\frac{1}{3})^{\frac{1}{2}}(x^2+1)^{\frac{1}{2}}}. \tag{6.3.118}$$

Then we have that

$$U_z(r, 0, p) = -\left(\frac{Z}{2\pi\mu c_2}\right)N(pr). \tag{6.3.119}$$

We have in mind the application of Cagniard's method of inversion. However, as (6.3.116) is presently posed, the necessary ingredient of $\exp(-pt)$ is absent. In order to introduce this, we use the integral relation for the Bessel function, which is†

$$J_0(z) = \frac{2}{\pi}\int_0^\infty \sin(z\cosh\theta)\,d\theta,$$

$$= \frac{1}{\pi i}\int_0^\infty (e^{iz\cosh\theta}-e^{-iz\cosh\theta})\,d\theta. \tag{6.3.120}$$

Then $N(pr)$ becomes

$$N(pr) = \frac{1}{\pi i}\int_0^\infty xm(x)\int_0^\infty \exp\left(\frac{ip}{c_2}rx\cosh\theta\right)\,d\theta\,dx-$$

$$-\frac{1}{\pi i}\int_0^\infty xm(x)\int_0^\infty \exp\left(-\frac{ip}{c_2}rx\cosh\theta\right)\,d\theta\,dx \tag{6.3.121}$$

† P. 56 of Reference [28].

or, defining the expressions N_1, N_2 in the obvious manner,

$$N(pr) = \frac{1}{\pi i}(N_1 - N_2). \tag{6.3.122}$$

We now wish to carry out the integration with respect to x in (6.3.121). The results will then be a single integral that may be treated by Cagniard's method. Integration in the complex plane will be used. To carry this out, we must determine locations of poles, take note of the branch points, and select a proper contour. For the poles, we consider

$$(2x^2+1)^2 - 4x^2(x^2+\tfrac{1}{3})^{\frac{1}{2}}(x^2+1)^{\frac{1}{2}} = 0, \tag{6.3.123}$$

which is the equivalent of the Rayleigh surface wave equation for this problem. Rationalizing the above gives

$$x^6 + \tfrac{7}{4}x^4 + \tfrac{3}{4}x^2 + \tfrac{3}{32} = 0, \tag{6.3.124}$$

which has roots

$$x^2 = -\frac{3+\sqrt{3}}{4}, \quad -\frac{3-\sqrt{3}}{4}, \quad -\frac{1}{4}. \tag{6.3.125}$$

The roots $x = \pm i(3-\sqrt{3})/2$, $\pm i/2$ are extraneous, arising from the rationalization. Thus, the poles of the integrand are those located at $x = \pm i(3+\sqrt{3})/2$. Branch points exist at $x = \pm i/\sqrt{3}$, $\pm i$, so appropriate branch cuts must be made to make the integrands single-valued.

The contours selected for the evaluation of N_1, N_2 are shown in Fig. 6.20. In the figure the upper contour must be associated with evaluating N_1

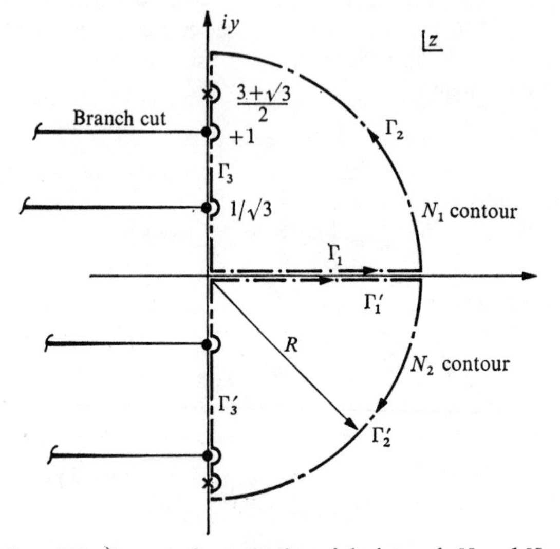

FIG. 6.20. Contours for evaluation of the integrals N_1 and N_2.

because of the positive sign of the exponential in (6.3.121) and vice versa for N_2. We thus have for N_1

$$\int_{\Gamma_1}+\int_{\Gamma_2}+\int_{\Gamma_3}=0, \tag{6.3.126}$$

since no poles exist within the contour. As $R\to\infty$, the second integral goes to zero, and we have

$$\lim_{R\to\infty}\left(\int_0^R+\int_{iR}^0\right)=0. \tag{6.3.127}$$

The first integral is that which we are seeking to evaluate. The presence of the singularities on the path of integration for the second integral shows that this must be interpreted in the Cauchy principal-value sense. Similar considerations hold for integration about Γ_1', Γ_2', Γ_3'. In particular, integration about the poles at $\pm i(3+\sqrt{3})/2$ will yield contributions equal to one-half the residue at that point.

The resulting expressions for N_1, N_2 will then be given by

$$N_1=-\int_0^\infty ym(iy)\int_0^\infty \exp\left(\frac{-p}{c_2}ry\cosh\theta\right)\,d\theta\,dy+\pi i\text{Res}, \tag{6.3.128}$$

$$N_2=-\int_0^\infty ym(-iy)\int_0^\infty \exp\left(\frac{-p}{c_2}ry\cosh\theta\right)\,d\theta\,dy+\pi i\text{Res}, \tag{6.3.129}$$

where the slash mark on the integral sign indicates interpretation in the principal value sense. We combine N_1, N_2 according to (6.3.122), giving

$$N(pr)=\frac{i}{\pi}\int_0^\infty iy\{m(iy)-m(-iy)\}\int_0^\infty \exp\left(\frac{-p}{c_2}ry\cosh\theta\right)\,d\theta\,dy, \tag{6.3.130}$$

where the residue expressions have cancelled. Finally, we note that $m(-iy)$ is the complex conjugate of $m(iy)$, so that

$$i\{m(iy)-m(-iy)\}=-2\text{Im}\,m(iy), \tag{6.3.131}$$

giving

$$N(pr)=-\frac{2}{\pi}\,\text{Im}\int_0^\infty ym(iy)\int_0^\infty \exp\left(-p\frac{ry}{c_2}\cosh\theta\right)\,d\theta\,dy. \tag{6.3.132}$$

The point of these operations now emerges. The θ-integral of (6.3.132) is coincident with the integral definition of the modified Bessel function of the second kind $K_0(pry)$, where

$$K_0(pry)=\int_0^\infty \exp\left(-\frac{pry}{c_2}\cosh\theta\right)\,d\theta. \tag{6.3.133}$$

Making the change of variable $(ry/c_2)\cosh\theta = t$, we have

$$\cosh\theta = \frac{c_2}{ry}t, \qquad \sinh\theta = \left\{\left(\frac{c_2}{ry}t\right)^2 - 1\right\}^{\frac{1}{2}}, \qquad d\theta = dt\Big/\frac{ry}{c_2}\sinh\theta. \tag{6.3.134}$$

Then (6.3.133) becomes

$$K_0(pry) = \int_{ry/c_2}^{\infty} e^{-pt}\left[c_2\,dt\Big/ry\left\{\left(\frac{c_2}{ry}t\right)^2 - 1\right\}^{\frac{1}{2}}\right]. \tag{6.3.135}$$

Cagniard's method may now be used. Thus

$$\mathscr{L}^{-1}(K_0) = \frac{c_2}{ry\left\{\left(\frac{c_2}{ry}t\right)^2 - 1\right\}^{\frac{1}{2}}} H\langle t - ry/c_2\rangle, \tag{6.3.136}$$

where

$$\tau = \frac{c_2}{r}t, \tag{6.3.137}$$

so that

$$\mathscr{L}^{-1}\{K(pry)\} = \frac{c_2}{r(\tau^2 - y^2)^{\frac{1}{2}}} H\langle \tau - y\rangle. \tag{6.3.138}$$

Hence (6.3.132) is of the form

$$N(pr) = -\frac{2}{\pi}\,\mathrm{Im}\int_0^{\infty} ym(iy)\,dy\,\mathscr{L}\{K(pry)\}, \tag{6.3.139}$$

so that

$$\mathscr{L}^{-1}\{N(pr)\} = -\frac{2c_2}{r\pi}\,\mathrm{Im}\int_0^{\infty}\frac{ym(iy)\,dy}{(\tau^2 - y^2)^{\frac{1}{2}}}, \tag{6.3.140}$$

$$m(iy) = \frac{(\frac{1}{3} - y^2)^{\frac{1}{2}}}{(1 - 2y^2)^2 + 4y^2(\frac{1}{3} - y^2)^{\frac{1}{2}}(1 - y^2)^{\frac{1}{2}}}. \tag{6.3.141}$$

The treatment of the principal-value integral in the above is all that remains. We first observe that the imaginary part of $m(iy)$ changes for various ranges of y; thus

$$y < \frac{1}{\sqrt{3}}; \quad \mathrm{Im}\, m(iy) = 0,$$

$$\frac{1}{\sqrt{3}} < y < 1; \qquad m(iy) = i(y^2 + \tfrac{1}{3})^{\frac{1}{2}}\frac{\{(1 - 2y^2)^2 - 4iy(y^2 - \frac{1}{3})^{\frac{1}{2}}(1 - y^2)^{\frac{1}{2}}\}}{1 - 8y^2 + \frac{56}{3}y^4 - \frac{32}{3}y^6},$$

$$1 < y; \qquad m(iy) = i(y^2 - \tfrac{1}{3})^{\frac{1}{2}}\frac{\{(1 - 2y^2)^2 + 4y^2(y^2 - \frac{1}{3})^{\frac{1}{2}}(y^2 - 1)^{\frac{1}{2}}\}}{1 - 8y^2 + \frac{56}{3}y^4 - \frac{32}{3}y^6}. \tag{6.3.142}$$

We then note that y varies with respect to τ in that $(\tau^2 - y^2)^{\frac{1}{2}}$ may become imaginary for $y > \tau$. Hence, the solution for τ in the various regions given in (6.3.142) is (see Pekeris [34]),

$$u_z(r,0,t) = \begin{cases} 0, & (\tau < 1/\sqrt{3}) \\ \dfrac{3Z}{\pi^2 \mu r} \displaystyle\int_{1/\sqrt{3}}^{\tau} \dfrac{y(y^2-\frac{1}{3})^{\frac{1}{2}}(1-2y^2)^2}{(\tau^2-y^2)^{\frac{1}{2}}F(y)}\,\mathrm{d}y, & (1/\sqrt{3} < \tau < 1), \\ \dfrac{3Z}{\pi^2 \mu r} \displaystyle\int_{1}^{\tau} \dfrac{y(y^2-\frac{1}{3})^{\frac{1}{2}}\{(1-2y^2)^2+4y^2(y^2-\frac{1}{3})^{\frac{1}{2}}(y^2-1)^{\frac{1}{2}}\}}{(\tau^2-y^2)^{\frac{1}{2}}F(y)}\,\mathrm{d}y, & (1 < \tau), \end{cases} \quad (6.3.143)$$

where

$$F(y) = 3-24y^2+56y^4-32y^6 = \tfrac{1}{2}(1-4y^3)(4y^2-3+\sqrt{3})(4y^2-3-\sqrt{3}), \quad (6.3.144)$$

where the integral for $\tau > 1$ is in the principal-value sense because of the pole at $(3+\sqrt{3})/2$. Also recall $\tau = c_2 t/r$.

Pekeris then carries out the evaluation of the integrals of the solution by a method involving partial fractions and obtains closed form results. Thus, for the integral appearing in (6.3.143),

$$G_1(\tau) = \int_{1/\sqrt{3}}^{\tau} \frac{y(y^2-\frac{1}{3})^{\frac{1}{2}}(1-2y^2)^2\,\mathrm{d}y}{(\tau^2-y^2)^{\frac{1}{2}}F(y)}, \quad (6.3.145)$$

the change of variables

$$y^2 = \tfrac{1}{3}+\omega^2\sin^2\theta, \qquad \omega^2 = \tau^2-\tfrac{1}{3}, \quad (6.3.146)$$

is made. The integral is thus transformed into

$$G_1(\tau) = \frac{1}{96}\int_0^{\pi/2}\left\{-12+\frac{1}{(\frac{1}{12}+\omega^2\sin^2\theta)}-\frac{B}{(-b+\omega^2\sin^2\theta)}-\frac{C}{(c+\omega^2\sin^2\theta)}\right\}\mathrm{d}\theta, \quad (6.3.147)$$

where

$$B = 3+5/\sqrt{3}, \quad b = 5/12+\sqrt{3}/4, \quad C = 3-5/\sqrt{3}, \quad c = 3/4-5/12. \quad (6.3.148)$$

Using the results that

$$\int_0^{\pi/2}\frac{\mathrm{d}\theta}{(\alpha^2+\omega^2\sin^2\theta)} = \frac{\pi}{2\alpha(\alpha^2+\omega^2)^{\frac{1}{2}}},$$

$$\int_0^{\pi/2}\frac{\mathrm{d}\theta}{(-\beta^2+\omega^2\sin^2\theta)} = \begin{cases} 0, & \beta < \omega \\ \dfrac{-\pi}{2\beta(\beta^2-\omega^2)^{\frac{1}{2}}}, & \beta > \omega \end{cases} \quad (6.3.149)$$

enables the integrals of (6.3.143) to be evaluated. The final results for the horizontal displacement field are

$$u_z(r,0,t) = \begin{cases} 0, & \tau < 1/\sqrt{3} \\ \dfrac{Z}{32\mu\pi r}\left\{6-\left(\dfrac{3}{\tau^2-\frac{1}{4}}\right)^{\frac{1}{2}}-\left(\dfrac{3\sqrt{3}+5}{\frac{3}{4}+\frac{\sqrt{3}}{4}-\tau^2}\right)^{\frac{1}{2}}+\left(\dfrac{3\sqrt{3}-5}{\tau^2+\frac{\sqrt{3}}{4}-\frac{3}{4}}\right)^{\frac{1}{2}}\right\}, & 1/\sqrt{3} < \tau < 1 \\ \dfrac{Z}{16\pi\mu r}\left\{6-\left(\dfrac{3\sqrt{3}+5}{\frac{3}{4}+\frac{\sqrt{3}}{4}-\tau^2}\right)^{\frac{1}{2}}\right\}, & 1 < \tau < \gamma \\ \dfrac{3Z}{8\pi\mu r}, & \tau > \gamma, \end{cases} \tag{6.3.150}$$

where $\gamma = (3+\sqrt{3})^{\frac{1}{2}}/2$. The various intervals are associated with the arrival of P, S, and surface waves. As will be shown in the next section, a singularity exists in the predicted response at the Rayleigh-wave arrival time.

6.3.4. *Results for transient loads on a half-space*

A number of transient-loading situations have been analysed by various investigators. The intent in this section will be to present the results, in terms of displacement and stress waveforms, of some of these studies.

Consider first the case of a transient, normal load applied at the origin. Near the conclusion of his extensive analysis of the half-space problem, Lamb [22] considered a line loading having the time variation

$$Z(t) = \frac{Z_0}{\pi}\frac{\tau}{t^2+\tau^2}, \tag{6.3.151}$$

where τ is a constant. For τ small, a sharp, impulse-line loading function is described. His results for the horizontal and vertical displacement components on the surface, far from the source, are shown in Fig. 6.21. Time and amplitude scales are not given, but the first disturbance is due to the arrival of the P wave, the second corresponds to the S wave, and the major response is from the arrival of the Rayleigh wave.

Pekeris [34], Lang [23], and Mitra [33] have also considered the half-space under a normal transient loading having a step behaviour in time. Some aspects of the Pekeris analysis have been under consideration in the previous section. His results for the horizontal and vertical displacements are shown in Fig. 6.22. Lang's analysis, although based on a different method than that of Pekeris, yielded essentially the same results. Mitra obtained the response

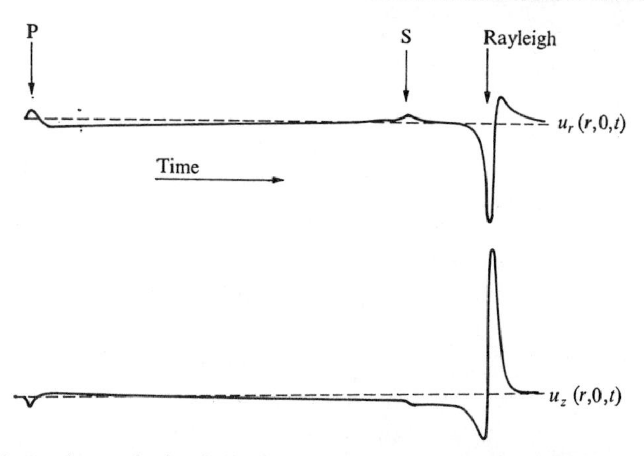

FIG. 6.21. Lamb's results for the horizontal and vertical surface displacements from an impulse-type point loading. (After Lamb [22, Fig. 10].)

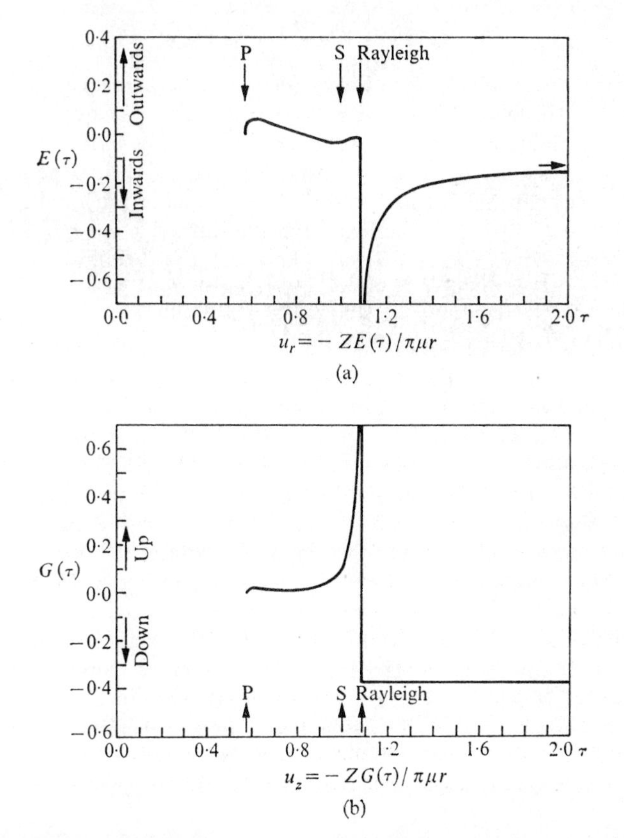

FIG. 6.22. (a) The vertical displacements $u_r(r, 0, t)$ and (b) the horizontal displacements $u_z(r, 0, t)$ resulting from a step loading. Note that $u_r = -ZE(\tau)/\pi\mu r$, $u_z = -ZG(\tau)/\pi\mu r$, where $\tau = c_2t/r$. (After Pekeris [34, Figs. 3 and 4].)

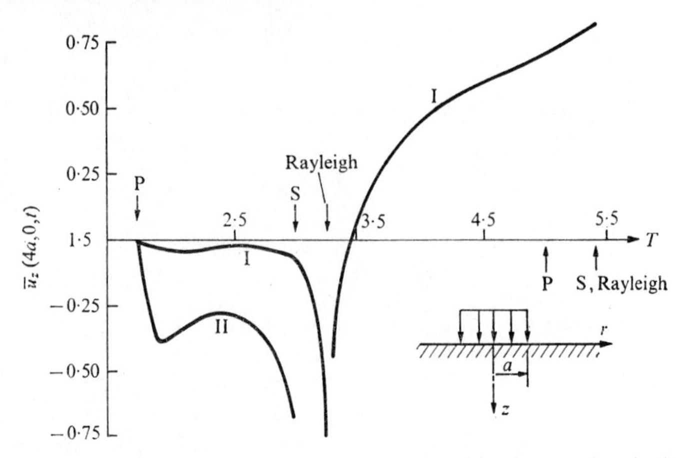

FIG. 6.23. Vertical displacement $\bar{u}_z(r, 0, t)$ at $r = 4a$ resulting from an impulse loading $P\delta(t)$ applied to a circular area $r \leqslant a$ on the surface of a half-space, where $\bar{u}_z = \pi^2\mu u_z/Pc_2$ and $T = c_2t/a$. (After Mitra [33, Fig. 3].)

to a normal load applied impulsively ($\delta(t)$ time behaviour) to a circular region $r \leqslant a$. His method of analysis was the Cagniard technique. The results for the vertical displacement $u_z(r, 0, t)$ are shown in Fig. 6.23 for $r = 4a$ as curve I. The response is more complicated owing to the finite source size. Thus, P wave contributions first start at $T = \sqrt{3}$ and continue until $T = 5$, when the first S wave contributions from the furthest point $r = 5a$ of the pressure area arrive. The first S waves arrive at $T = 3$ followed closely at $T = 6/(3+\sqrt{3})^{\frac{1}{2}}$ by the first Rayleigh wave contributions.

Another aspect of the response of the half-space may be brought out in conjunction with the earlier Fig. 6.19. Shown are the hemispherical P and S wavefronts followed by the Rayleigh surface wave. In addition, there is a wavefront known as the 'head wave' or *von Schmidt wave*, having a wavefront originating from the intersection point of the P wave with the free surface and being tangent to the SV wave surface. Fung [13, p. 225] points out that these waves can be understood on the basis of Huygen's principle, with the P wave acting as a continual wave source, generating both P and SV waves. Another possible explanation is that, near the surface, the wave system is that resulting from a P wave at grazing incidence (see Fig. 6.17).

The case of normal impact on a half-space has been analysed. Thus Hunter [19] has studied the case of a spherical ball impacting the surface of a half-space.† Using Hertz contact theory and a Fourier synthesis based on the results of Miller and Pursey [30], he determines the energy absorbed during the impact. It is found that the loss of impact energy to stress waves

† This problem is analogous to that considered in § 2.4 for the impact of a ball on a rod.

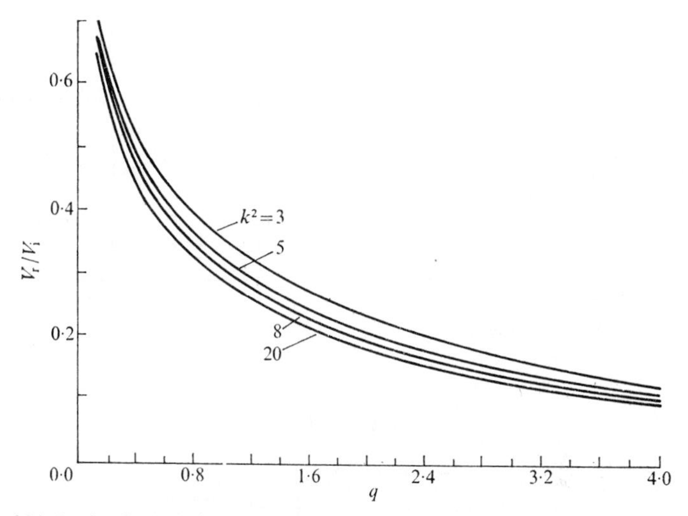

FIG. 6.24. Ratio of rebound to impact velocity, V_r/V_i, versus disc parameter $q = 2\rho_s r_0/\rho_d h$ for various k^2. (After Gutzwiller [17, Fig. 6].)

in an elastic collision process is very small (less than 1 per cent).† Gutzwiller [17] has considered the impact of a rigid, circular disc on a half-space. He introduces the concept of 'mode of vibration', whereby a transient in the coupled disc half-space can be described by a superposition of modes. His main results are for the ratio of rebound to impact velocity (V_r/V_i) for various discs impacting different elastic media and are shown in Fig. 6.24, where the parameter q is given by $q = 2\rho_s r_0/\rho_d h$. Here r_0, h, ρ_d are the radius, height, and density of the disc and ρ_s is the density of the elastic media. The parameter $k^2 = (\lambda+2\mu)/\mu$.

The analysis of a concentrated, tangential force, with step behaviour in time, has been carried out by Chao [3]. He obtained the response along $r = 0$ and on the surface $z = 0$. The problem possesses only x,z-plane symmetry, so that there are u_r, u_θ, and u_z displacements. Fig. 6.25(a)–(c) shows his results for the surface displacements $\bar{u}_r$, $\bar{u}_\theta$, $\bar{u}_z$, where

$$\bar{u}_r = \pi\mu r u_r/F\cos\theta, \quad \bar{u}_\theta = \pi\mu r u_\theta/F\sin\theta, \quad \bar{u}_z = \pi\mu r u_z/F\cos\theta, \qquad (6.3.152)$$

and F is the magnitude of the applied load. The non-dimensional time τ is $\tau = c_2 t/r$. Chao points out that, in directions perpendicular to the applied force ($\theta = \pm\pi/2$), $u_r = u_z = 0$ and u_θ decays rapidly with distance, so that no strong surface waves are noted in these directions.

Problems of torsional loads applied to the surface of the half-space have been considered by several investigators. As mentioned earlier, Miller and

† See § 6.5 on experimental results.

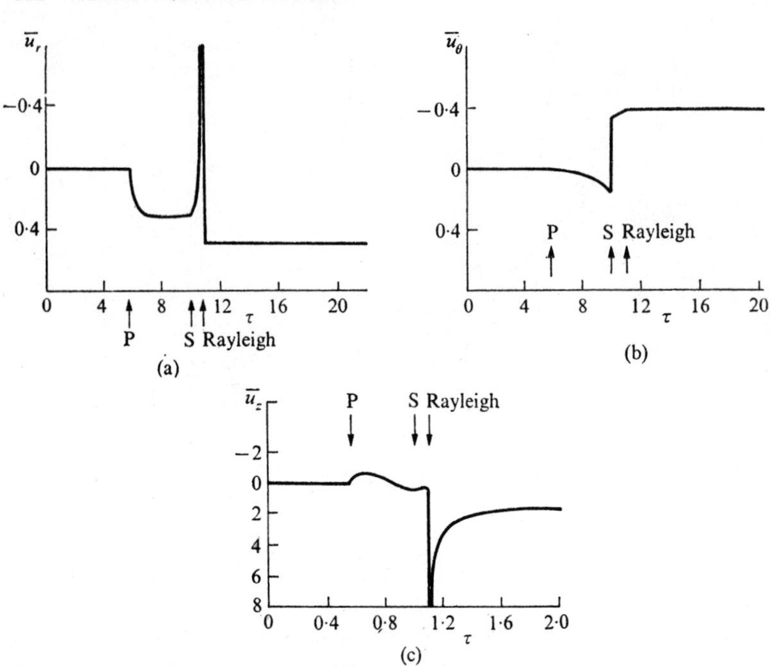

FIG. 6.25. The surface displacements (a) $\bar{u}_r$, (b) $\bar{u}_\theta$, (c) $\bar{u}_z$ versus τ, resulting from a tangential point load F suddenly applied at the origin, where $\bar{\tau} = c_2t/r$. (After Chao ([3, Figs. 3, 4, and 5].)

Pursey [30] considered a harmonic torsional loading. Mitra [32] has considered the case of an impulsive twisting moment applied to a rigid die attached to the half-space. Hill [18] solved the case of a rigid sphere imbedded in the surface of an elastic half-space, where an impulse torque is applied to the sphere. Eason [10] has considered an impulse torque applied to a half-space. Several distributions of stress yielding the resulting torque are considered. His Cases 1, 2, 3 are as follows:

$$\text{Case 1:}\ \tau_{r\theta} = \begin{cases} Qr, & r < a \\ 0, & r > a, \end{cases}$$

$$\text{Case 2:}\ \tau_{r\theta} = \begin{cases} Q/r, & r < a \\ 0, & r > a, \end{cases}$$

$$\text{Case 3:}\ \tau_{r\theta} = \begin{cases} Qr/(a^2-r^2)^{\frac{1}{2}}, & r < a \\ 0, & r > a, \end{cases} \tag{6.3.153}$$

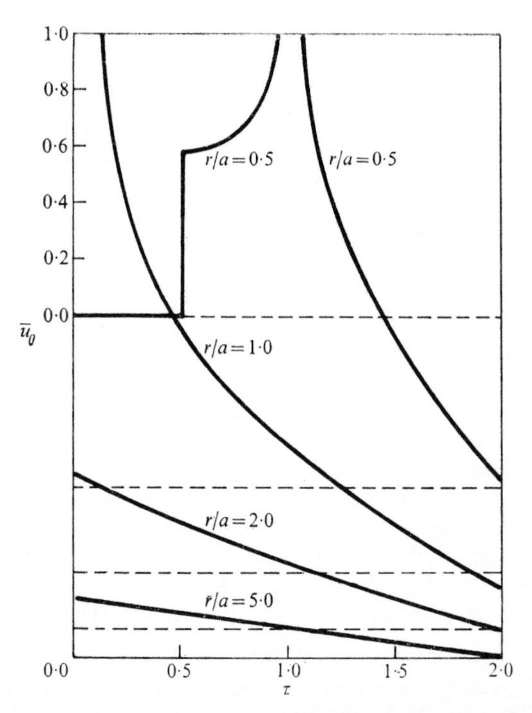

FIG. 6.26. Surface displacement $u_\theta(r, 0, t)$ resulting from a torsional impulse applied at the origin for the Case 3 stress distribution. (After Eason [10, Fig. 3].)

where Q is a constant. The time variation in all cases is that of the Dirac delta function. The first case was previously considered by Mitra. Eason's results for Case 3 are shown in Fig. 6.26. The results are given in terms of $\bar{u}_\theta$ versus τ, where

$$\bar{u}_\theta = -\mu u_\theta/Qa, \qquad \tau = (c_2 t+a-r)/a. \tag{6.3.154}$$

In the figure the response is shown at $r/a = 0·5$, $1·0$, $2·0$, and $5·0$. All curves are to the same scale, but have been displaced vertically for clarity of presentation. The dashed lines in each case correspond to $\bar{u}_r = 0$. Note that $r/a = 0·5$ is within the circle of the applied load.

All of the considerations thus far have been for the case of a surface source. Buried-source problems are of considerable interest and have been studied by many, including, initially, Lamb [22]. Results have been obtained for both line and point loadings. Ewing, Jardetsky, and Press [11] review many aspects of the analysis in this area. Although the analysis is quite complicated, it is of interest to consider certain results for the buried line source. Consider

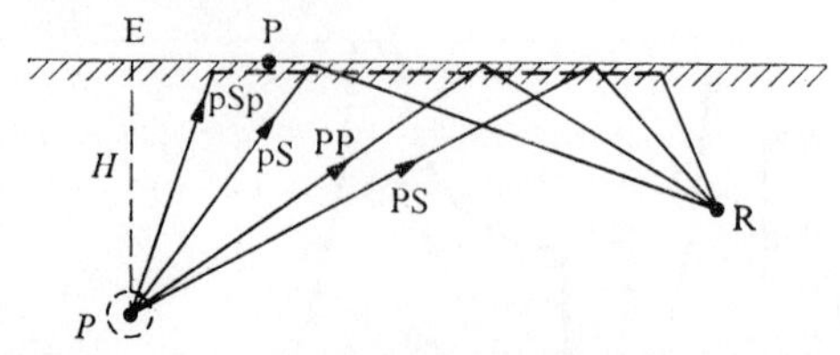

FIG. 6.27. Various propagation–reflection paths for P and S waves arriving at a receiver R as generated by a P wave source. (Based on Ewing, Jardetsky, and Press [11, Fig. 2.12].)

the case of a P wave source, as in Fig. 6.27, with a receiver R located some distance away and also below the surface. Analysis of the branch-cut integrals shows that the signal arriving at the receiver is in terms of P and S wave contributions arriving through various propagation, reflection paths. Thus, the first four arrival paths are shown, labelled as PP, PS, pS, pSp. The PP wave is formed from the initial source P wave and a P wave reflection. The PS wave is the initial P wave and an S wave reflection. The pS wave is also the initial P wave and an S wave reflection. The PS and pS cases differ in that the PS wave travels most of the path as a P wave, the pS wave travels mostly as an S wave. Finally, the pSp wave starts as a P wave, travels part way along the surface as an S wave, and then along the final portion as a P wave.

Another aspect of interest reviewed in Reference [11] concerns the first appearance of the Rayligh wave from a buried source, and results from a steepest descent analysis. Referring to Fig. 6.27, it is found that the minimum distance is EP, where $\mathrm{EP} = c_R H/(c_1^2 - c_R^2)^{\frac{1}{2}}$. This represents the location such that the time $\mathrm{EP}/c_R = (H^2 + \mathrm{EP}^2)^{\frac{1}{2}}/c_1$.

Pekeris [35] has considered the case of a buried point load applied vertically and varying in a step fashion with time. In a later paper, Pekeris and Lifson [36] presented a number of results for this case. All results were obtained for the vertical and horizontal displacements at the surface due to the source at a depth H. Before considering these, it is useful to further consider the waves that may arrive at a point from a buried source. As shown in Fig. 6.28(a), an S wave arriving at a point r on the surface at less than the

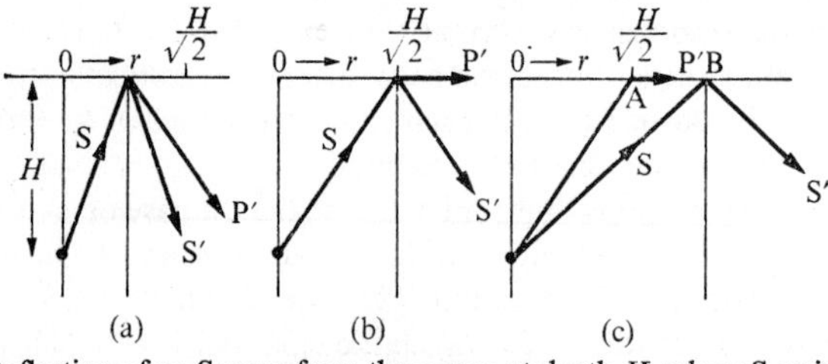

FIG. 6.28. Reflection of an S wave from the source at depth H, where S arrives at (a) less than, (b) equal to, and (c) greater than the critical angle. (After Pekeris and Lifson [36, Fig. 1].)

critical angle reflects as a P′ and S′ wave. At the critical angle given by (see (6.1.57)) $k \sin \theta_2 = 1$ the P wave is parallel to the surface, as shown in Fig. 6.28(b). For $\lambda = \mu$, the case considered by Pekeris and Lifson, $k = \sqrt{3}$ and the critical radius is $r = H/\sqrt{2}$. For values of r greater than the critical value, an S wave continues to arrive directly. However, an SP′ wave, where P′ is along the surface, can precede the S wave arrival. In addition to the S wave behaviour, a P wave arriving directly from the source represents the first signal. Much of this is similar to the behaviour presented in Fig. 6.27 for a buried source and a buried receiver.

The vertical and horizontal surface displacements are shown in Fig. 6.29(a)–(d) for various distances, r/H from the epicentre of the loading. The time base is given by τ, where $\tau = c_2 t/R$ and $R = (r^2+H^2)^{\frac{1}{2}}$. The normalized displacements $\bar{u}_z(r, 0, t)$, $\bar{u}_r(r, 0, t)$ are plotted, where

$$u_z(r, 0, t) = \frac{3Z}{\pi^2 \mu R}\bar{u}_z, \qquad u_r(r, 0, t) = -\frac{3Z}{\pi^2 \mu R}\bar{u}_r \tag{6.3.155}$$

and where Z is the applied force, acting in the downward direction at a depth H. The arrival of P, S, SP, and Rayleigh waves are marked on the displacement curves. The arrival of the P wave is unambiguous in the various cases. However, depending on location, the S or the SP wave may arrive first, as suggested by the previous discussion. It is to be noted that the Rayleigh wave does not appear when r is small, in accord with the 'minimum distance' discussion pertaining to Fig. 6.28. Also shown in Fig. 6.29(b) and (d) are the results for a surface source ($H = 0$).

6.4. Waves in layered media

The propagation and reflection of waves in a homogeneous, isotropic half-space represents a large class of problems of practical interest, particularly in the field of seismology. In many situations waves originate and propagate in media having a layered structure, where interfaces between dissimilar materials exist. Again, seismic waves in a layered earth are an example, but important applications in structures exist, such as waves in composite plates and shells. The large class of problems for a 'simple' half-space, such as reflection of plane waves and source problems, now have their more complicated and numerous counterparts in layered media problems. The very extent of the resulting class of problems precludes more than a cursory review in this presentation. One is referred to Ewing, Jardetsky, and Press [11] and Brekhovskikh [1] as source books devoted nearly in their entirety to this subject.

The simplest situation in this area consists of two semi-infinite media in contact. Such a situation, strictly speaking, is not a layered system since neither media possesses two parallel boundaries. Nevertheless, this natural extension of the single, semi-infinite media problem is a pre-requisite to the

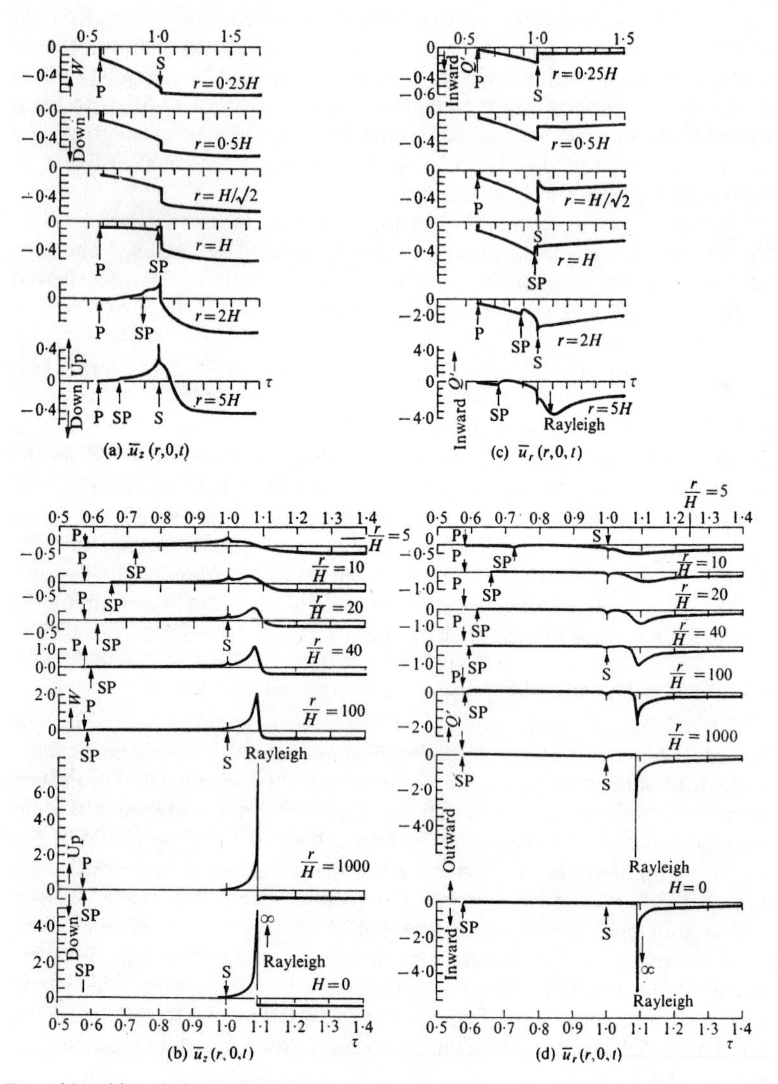

FIG. 6.29. (a) and (b) Vertical displacement components $\bar{u}_z$ and (c) and (d) horizontal displacements $\bar{u}_r$ for a concentrated force applied downward at a depth **H** with a step-function time behaviour. (After Pekeris and Lifson [36, Figs. 2–5].)

analysis of layered systems. The basic aspects of the propagation and transmission of plane waves across the interface between two semi-infinite media will be reviewed, including the various interface conditions that may exist. Certain results of waves from sources will be discussed. In the case of distinctly layered media, one situation will be analysed in some detail. This will be the case of SH waves in a layer overlaying a half-space. Such waves are also known as Love waves. Some discussion will be given to other situations.

6.4.1. *Two semi-infinite media in contact—plane waves*

When propagating waves encounter a boundary between two media, reflected waves occur, somewhat as in the case of encountering a free surface. In addition, energy is transmitted across the boundary in the form of refracted waves. The approach taken in the study of wave reflection–refraction in two semi-infinite media in contact is to consider propagating plane harmonic waves encountering the boundary.

Thus, consider two media in contact along the plane $y = 0$. The properties of the lower media are given by λ, μ, ρ and those of the upper media by λ', μ', ρ'. Proceeding in a manner quite analogous to the case of waves in a half-space, we have the governing equations for the lower media given by (6.1.1)–(6.1.12). For the upper media, we may reference the same equations, except that everywhere the quantities u_i, Φ, H_i, c_1, c_2, τ_{ij}, λ, μ, must be replaced by primed quantities. If we reduce our considerations to those of plane strain, eqns (6.1.14), (6.1.15) pertain, with the lower and upper media again being differentiated by unprimed and primed notation.

Considering now the case of plane harmonic waves, we parallel eqns (6.1.21)–(6.1.25). Thus for the lower media we still have the plane-wave expressions (6.1.24) and (6.1.25). For the upper media, we will have similar expressions except that Φ, A_1, A_2, H_z, B_1, B_2, α, β are replaced by primed quantities. Expression of all plane-wave results in terms of incidence, reflection, and refraction angles θ_1, θ_2, θ_1', θ_2', such as in (6.1.28) and (6.1.29) is also possible, but will not be used here. If we take the viewpoint that plane waves are propagating toward the boundary from the lower media, then all of the terms A_1, A_2, B_1, B_2 remain in the solutions. However, only two terms A_1', B_1' would remain in the primed solution, since the A_2', B_2' expressions would represent waves approaching the boundary from negative infinity. Thus, to summarize, the plane hamonic wave solutions are

$$\Phi = A_1 e^{i(\xi x - \alpha y - \omega t)} + A_2 e^{i(\xi x + \alpha y - \omega t)}, \tag{6.4.1}$$

$$H_z = B_1 e^{i(\xi x - \beta y - \omega t)} + B_2 e^{i(\xi x + \beta y - \omega t)}, \tag{6.4.2}$$

$$\Phi' = A_1' e^{i(\xi x - \alpha' y - \omega t)}, \qquad H_z' = B_1' e^{i(\xi x - \beta' y - \omega t)}, \tag{6.4.3}$$

where

$$\alpha^2 = \frac{\omega^2}{c_1^2} - \xi^2, \qquad \beta^2 = \frac{\omega^2}{c_2^2} - \xi^2, \tag{6.4.4}$$

$$\alpha'^2 = \frac{\omega^2}{c_1'^2} - \xi^2, \qquad \beta'^2 = \frac{\omega^2}{c_2'^2} - \xi^2. \tag{6.4.5}$$

The boundary conditions existing at the interface must now be specified. Two conditions are of particular interest in practical situations. The first is the case where the two media are bonded together. Under such conditions, continuity of displacement and stress across the interface is required. Thus, for a bonded interface, we have

$$u_x(x, 0, t) = u_x'(x, 0, t), \qquad u_y(x, 0, t) = u_y'(x, 0, t),$$
$$\tau_{yy}(x, 0, t) = \tau_{yy}'(x, 0, t), \qquad \tau_{xy}(x, 0, t) = \tau_{xy}'(x, 0, t). \tag{6.4.6}$$

The second condition is the case of a lubricated interface where transverse slip may occur. For this condition, we have

$$u_y(x, 0, t) = u_y'(x, 0, t), \qquad \tau_{yy}(x, 0, t) = \tau_{yy}'(x, 0, t),$$
$$\tau_{xy}(x, 0, t) = \tau_{xy}'(x, 0, t) = 0. \tag{6.4.7}$$

This last case is of particular interest in ultrasonics, where transducers used to launch and receive waves into and from a solid are often coupled by an oil or grease film to the media.

Considering, then, the case of a bonded boundary, so that conditions (6.4.6) apply, we substitute the solutions (6.4.1)–(6.4.3) in the expressions for the displacements and stresses as given by (6.1.10), (6.1.11), and (6.1.14). Thus

$$u_x = u_x': \quad \xi(A_1+A_2)-\beta(B_1-B_2) = \xi A_1'-\beta' B_1', \tag{6.4.8}$$

$$u_y = u_y': \quad \alpha(A_1-A_2)+\xi(B_1+B_2) = \alpha' A_1'+\xi B_2', \tag{6.4.9}$$

$$\tau_{yy} = \tau_{yy}': \quad \mu\{(\beta^2-\xi^2)(A_1+A_2)+2\xi\beta(B_1-B_2)\}$$
$$= \mu'\{(\beta'^2-\xi^2)A_1'+2\xi\beta' B_1'\}, \tag{6.4.10}$$

$$\tau_{xy} = \tau_{xy}': \quad \mu\{2\xi\alpha(A_1-A_2)-(\beta^2-\xi^2)(B_1+B_2)\}$$
$$= \mu'\{2\xi\alpha' A_1'-(\beta'^2-\xi^2)B_1'\}. \tag{6.4.11}$$

The left-hand sides of (6.4.10) and (6.4.11) are identical to the boundary-condition equations (6.1.30) and (6.1.31), except for the multiplying parameter μ. Amplitude ratios immediately follow from the preceding equations. If we presume that we have an incident P wave, so that $B_1 = 0$, there follows the ratios A_2/A_1, B_2/A_1, A_1'/A_1, B_1'/A_1. If we presume an incident SV wave, so that $A_1 = 0$, the amplitude ratios A_2/B_1, B_2/B_1, A_1'/B_1, B_1'/B_1 follow. Thus, for an incident wave, either P or SV, there are generally two reflected and two refracted waves.

The angles of reflection, refraction may be readily obtained. We may write (6.4.1) and (6.4.2) in the forms (6.1.28) and (6.1.29), and for Φ', H_z' write

$$\Phi' = A_1' \exp\{i\gamma_1'(\sin\theta_1' x - \cos\theta_1' y - c_1' t)\}, \tag{6.4.12}$$

$$H_z' = B_1' \exp\{i\gamma_2'(\sin\theta_2' x - \cos\theta_2' y - c_2' t)\}, \tag{6.4.13}$$

where θ_1', θ_2' are the angles of the refracted wave normals relative to the vertical axis. We then have, from the requirement that the boundary conditions at the interface be independent of x and t, that

$$\gamma_1 \sin\theta_1 = \gamma_2 \sin\theta_2 = \gamma_1' \sin\theta_1' = \gamma_2' \sin\theta_2' \tag{6.4.14}$$

and also $\gamma_1 c_1 = \gamma_2 c_2 = \gamma_1' c_1' = \gamma_2' c_2' = \omega$. We presume that the frequency ω and incidence angle θ_1 or θ_2 of the incoming wave is specified. Thus, suppose the values ω, θ_1 are given. Then $\gamma_2, \gamma_1', \gamma_2'$ are found from $\gamma_2 = \omega/c_2, \gamma_1' = \omega/c_1'$, $\gamma_2' = \omega/c_2'$. The reflection and refraction angles follow from (6.4.14). The amplitude ratios follow from results obtained from (6.4.8) to (6.4.11). To obtain the parameters α, β, α', β', ξ we merely use

$$\alpha = \gamma_1 \cos\theta_1, \quad \beta = \gamma_2 \cos\theta_2, \quad \alpha' = \gamma_1' \cos\theta_1', \quad \beta' = \gamma_2' \cos\theta_2', \tag{6.4.15}$$

and the fact that $\xi = \gamma_1 \sin\theta_1 = \gamma_2 \sin\theta_2 = \gamma_1' \sin\theta_1' = \gamma_2' \sin\theta_2'$. Alternatively, the amplitude ratio expressions originally may be derived in terms of $\theta_1, \theta_2, \theta_1', \theta_2'$ as was done for the half-space. The resulting general reflection–refraction situation involving an incident wave, two reflected waves and two refracted waves, with known angles of incidence, reflection and refraction, is shown in Fig. 6.30 for the case of incident P and SV waves. There are a number of special circumstances of wave reflection–refraction that can occur,

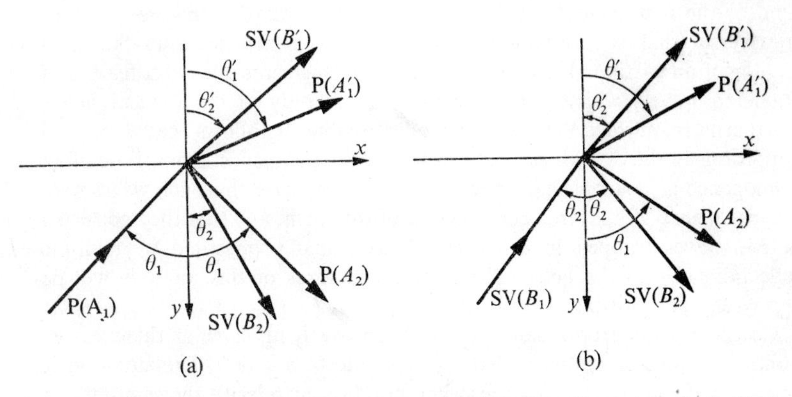

FIG. 6.30. Reflection–refraction of plane waves at the boundary of two media for incident; (a) P waves and (b) SV waves.

depending on the type of incident wave, the incidence angle, the material properties of the two media, and in which of the two media the incident wave is travelling. In particular, for incident SV waves, critical angles may exist beyond which the reflected P wave may disappear.

In addition to the wave-reflection problem, one might inquire whether the analogue of Rayleigh waves for a half-space can exist for two semi-infinite media in contact. Evidently Love [26] investigated the possibility and found that such waves could exist. Stonely [40] more thoroughly investigated this wave-type and found that the existence of such a wave required that the shear-wave velocities of the two media had to be nearly the same. This generalized Rayleigh wave is usually called the Stonely wave.

6.4.2. *Waves in layered media—Love waves*

Suppose we consider a homogeneous, isotropic semi-infinite media. Consider then a layer of some thickness h_1 and different material properties λ_1, μ_1, ρ_1 to be attached to the surface of the semi-infinite media. Consider the additional layers h_2, h_3,... of properties λ_2, μ_2, ρ_2, λ_3, μ_3, ρ_3,... to be added. Such a system constitutes the general layered media. The propagation of waves in such a system, with multiple reflections and refractions occurring at the interfaces according to the laws for a single interface are, of course, most complicated. In fact, the case of waves in a single layered system represents a more complicated situation than the waves in plates, an area to be considered in Chapter 8. References [1] and [11] again give an extensive treatment. One special case, easily analysed, of classic interest and having considerable practical application will now be considered.

In particular, we shall consider the propagation of SH waves in a half-space overlaid by a thin solid layer. As is the case for many problems in wave propagation the original impetus for study in this area came from seismology. One of the first established facts of seismology was the presence of large transverse (that is, horizontal) components of displacement in the main tremor of an earthquake. However, such displacements are not a feature of Rayleigh waves, which contain displacements only in the vertical plane. Furthermore, surface SH waves are not possible. It follows that the actual conditions in the earth must differ in some essential respect from those of an homogeneous, isotropic half-space. Love [26] suspected that such waves were a consequence of a layered construction of the earth, and that they consisted of SH waves trapped in a superficial layer and propagated by multiple reflections within the layer. The essential features of this analysis will be shown in the following.

Consider an isotropic half-space with an overlying layer of thickness T, bonded to the semi-infinite media at the interface $y = 0$. The situation is shown in Fig. 6.31. We use the prime notation to refer to the semi-infinite media having properties μ', c_2' and unprimed notation for the layer, having

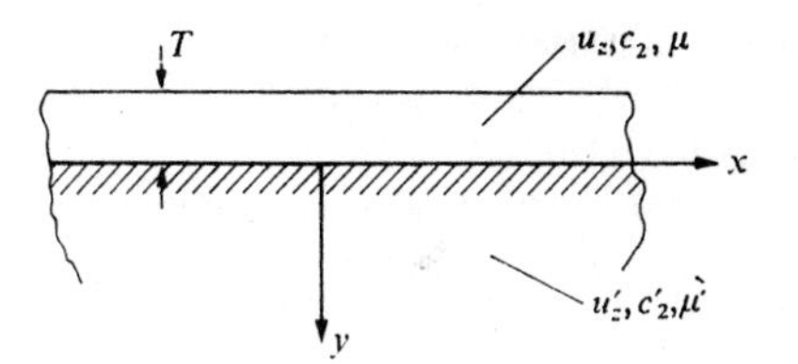

FIG. 6.31. A layer of thickness T over the semi-infinite media.

properties μ, c_2. We will consider only u_z', u_z displacements so that

$$u_z = u_z(x, y, t), \qquad u_x = u_y = 0, \tag{6.4.16}$$

$$u_z' = u_z'(x, y, t), \qquad u_x' = u_y' = 0. \tag{6.4.17}$$

The governing equations may be written in terms of potential functions. However, as was found earlier for SH waves, it is possible and more direct to work with the displacement equations of motion, which reduce simply to

$$\nabla^2 u_z = \frac{1}{c_2^2}\frac{\partial^2 u_z}{\partial t^2}, \qquad \nabla^2 u_z' = \frac{1}{c_2'^2}\frac{\partial^2 u_z'}{\partial t^2}, \tag{6.4.18}$$

were $\nabla^2 = \partial^2/\partial x^2 + \partial^2/\partial y^2$. The non-trivial boundary conditions for the problem are

$$\begin{aligned} \tau_{yz} &= 0, \qquad y = -T, \\ \tau_{yz} &= \tau_{yz}', \qquad u_z = u_z', \qquad y = 0. \end{aligned} \tag{6.4.19}$$

For solutions to (6.4.18) we let

$$u_z = U(y)e^{i\xi(x-ct)}, \qquad u_z' = U'(y)e^{i\xi(x-ct)}, \tag{6.4.20}$$

and obtain

$$\frac{d^2U}{dy^2} + \beta^2 U = 0, \qquad \frac{d^2U'}{dy^2} - \beta'^2 U' = 0, \tag{6.4.21}$$

where

$$\beta^2 = \xi^2(c^2/c_2^2 - 1), \qquad \beta'^2 = \xi^2(1 - c^2/c_2'^2). \tag{6.4.22}$$

The resulting solutions are

$$u_z = A_1 e^{i(\xi x - \beta y - \omega t)} + A_2 e^{i(\xi x + \beta y - \omega t)}, \tag{6.4.23}$$

$$u_z' = B_1 e^{-\beta' y} e^{i(\xi x - \omega t)}. \tag{6.4.24}$$

The solution (6.4.23) represents plane waves propagating back and forth within the layer. The solution (6.4.24) (an $\exp(+\beta' y)$ term has been discarded) gives a wave that retains its energy close to the interface. A plane wave solution, based on $\beta'^2 < 0$, could be selected instead for u_z'. However, such a wave system would not be capable of giving the behaviour observed in seismology, since it would represent refracted waves carrying energy away from the layer. Such a wave system would quickly lose its energy and not be of significance at any distance. Thus the question is, are the solutions (6.4.23) and (6.4.24), capable of describing the physical phenomena, able to satisfy the boundary conditions.

We substitute the solutions in the boundary conditions (6.4.19) and obtain

$$\tau_{yz} = 0: \qquad e^{i\beta T}A_1 - e^{-i\beta T}A_2 = 0, \tag{6.4.25}$$

$$u_z = u_z': \qquad A_1 + A_2 - B_1 = 0, \tag{6.4.26}$$

$$\tau_{yz} = \tau_{yz}': \quad -i\mu A_1 + i\mu\beta A_2 - \mu'\beta' B_1 = 0. \tag{6.4.27}$$

The resulting determinant of coefficients gives

$$\mu'\beta' - \mu\beta \tan \beta T = 0. \tag{6.4.28}$$

Using (6.4.22) this may be put in the form

$$\mu'(1-(c/c_2')^2)^{\frac{1}{2}} - \mu((c/c_2)^2-1)^{\frac{1}{2}} \tan \xi T((c/c_2)^2-1)^{\frac{1}{2}} = 0. \tag{6.4.29}$$

By using the relation $\omega = \xi c$, this result may be expressed in terms of frequency and wavenumber.

The first observation is simply that Love waves, as SH waves in a layer are generally called, are dispersive. That is, the roots of (6.4.29) for successive values of ξ will result in $c = c(\xi)$ or in $\omega = \omega(\xi)$. The multiple branches of the tangent function also suggest that multiple roots will exist for any given ξ. Thus the dispersion curves and frequency spectrum should have multiple branches, corresponding to various modes of propagation. We note that, as $\gamma \to 0$ in (6.4.29), $c \to c_2'$. Thus, as the wavelength becomes large compared, say to the thickness T, the Love waves take on the velocity of the lower media.

In addition to applications in seismology, Love waves also find application in delay lines, as do Rayleigh waves. Thus, by depositing thin layers over substrates, another type of surface-wave delay line results.

6.5. Experimental studies on waves in semi-infinite media

Experimental investigations on waves in semi-infinite media are very wide in scope, ranging from ultrasonic excitation of high-frequency waves in small specimens to seismological studies of the earth. Only a most limited assortment of results will be given here, with the main attention being given to various aspects of surface waves.

6.5.1. *Waves into a half-space from a surface source*

Photo-elasticity has been used in a number of studies of waves propagated into a half-space from an impulsive surface source. One of the first was reported by Dally, Durelli, and Riley [5], where a low-modulus urethane rubber plate was dynamically loaded by a small explosive charge and the dynamic fringe propagation recorded by a high-speed (6000 frames per

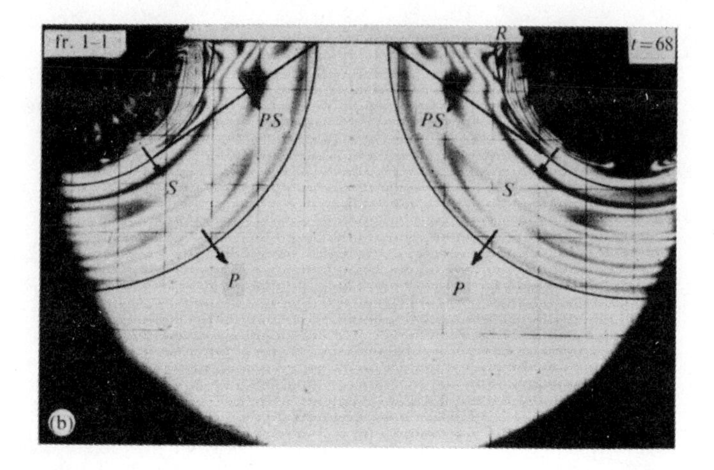

FIG. 6.32 (b)

second) camera. In a later photo-elastic study, Dally [4] considered a half-plane subjected to two loads placed some distance apart and obtained results on the development and interaction of the two waves. Before wave interaction occurs, the individual wave systems are the same as for the half-space under point loading, so it is of interest to study the results from that standpoint. It should be noted here that several of the later cited photo-elastic studies on waves in wedges or past discontinuities also present, for reference purposes, the results for a simple half-space.

The geometry of the plate and the placement of small charges of lead azide (PbN_6) are shown in Fig. 6.32(a). The charges were simultaneously

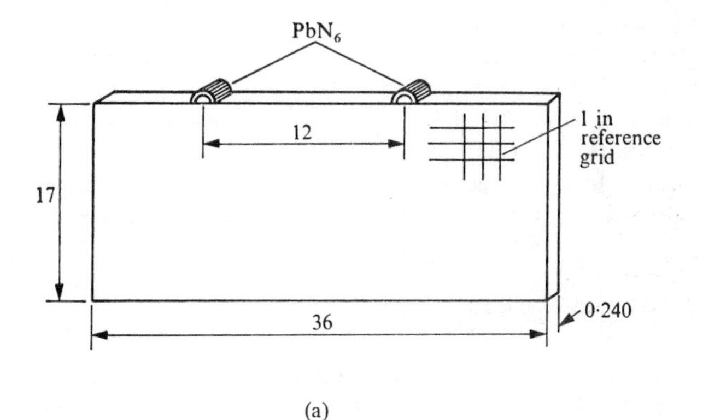

FIG. 6.32. (a) Photo-elastic plate and loading configuration employed in study of a double-loaded half-plane and (b: see separate plate) fringe pattern from a double-loaded half-plane before wave interaction has occurred. (After Dally [4, Fig. 1B, 6A].)

detonated and the dynamic fringe patterns recorded by a multiple-spark Cranz–Schardin camera operating at 127 000 frames per second. An early stage in the development of the wave system, before interaction has occurred, is shown in Fig. 6.32(b). Superimposed on the fringe system are solid lines showing positions of the various wavefronts. Thus, the cylindrical-fronted P and S wave systems are so indicated in the figure. The straight-crested PS wave is the von Schmidt head wave, arising from reflection of a grazing-incidence P wave. The Rayleigh surface wave is indicated by R. Because of the rather early stage of wave development at the instant shown, the Rayleigh wave has not yet clearly separated from the S wave. For the material constants of the CR-39 photo-elastic plate, the various wave velocities are $c_1 = 7{\cdot}63\times10^4$ in. s^{-1}, $c_2 = 4{\cdot}35\times10^4$ in. s^{-1}, and $c_R = 4{\cdot}0\times10^4$ in. s^{-1} (or 1·94, 1·10, and $1{\cdot}01\times10^3$ m s^{-1}, respectively). It is seen that the photo-elastic technique, with the capability of viewing the entire field, offers an excellent technique for qualitatively evaluating the stress field. Data on the stresses is easily obtainable along the boundary where one of the principal stresses is zero and the stress–optic law $\sigma_1 - \sigma_2 = Nf_\sigma/h$ reduces to $\sigma_1 = Nf_\sigma/h$.

The experimental conditions of edge-loading a plate used in photo-elastic studies do not exactly correspond to a line-loading of a half-space, since plane-strain conditions in the thickness direction do not prevail. In static photo-elastic studies, plane-stress conditions are assumed to hold in the plate thickness direction. Under dynamic loading conditions, wavelengths of the order of the plate thickness may occur, so that the plane-stress assumption is less valid. Dally and Riley [7] have attempted to circumvent this problem by using the embedded polariscope technique. This latter approach has enabled them to study directly the three-dimensional problem of a point load on a half-space using photo-elastic techniques.

In addition to the simple half-space, photo-elastic investigations have been done on waves in layered media. Thus Riley and Dally [39] have considered the case of a photo-elastic layer overlying an aluminium subgrade, while Daniel and Marino [9] have used a model consisting of two transparent photo-elastic layers. The geometry of the latter study was actually consistent with the case of two semi-infinite media in contact. A Moiré fringe technique was also used in the latter study to supplement the photo-elastic aspects.

6.5.2. *Surface waves on a half-space*

Measurement of surface strains and displacements in a half-space actually preceded entire-field measurements by photo-elastic means. Thus, Tatel [41] used a 'model seismogram' approach to compare the theoretical predictions of Lamb on surface motion to experimental measurements. A wave was induced on the surface of a large block of steel using a pulsed piezoelectric transducer. A detector placed several centimetres away measured the vertical component of displacement. The geometry of the apparatus is shown in Fig. 6.33(a), and the resulting vertical component displacement and velocity records are shown in (b) and (c). The basic waveform is seen to strongly

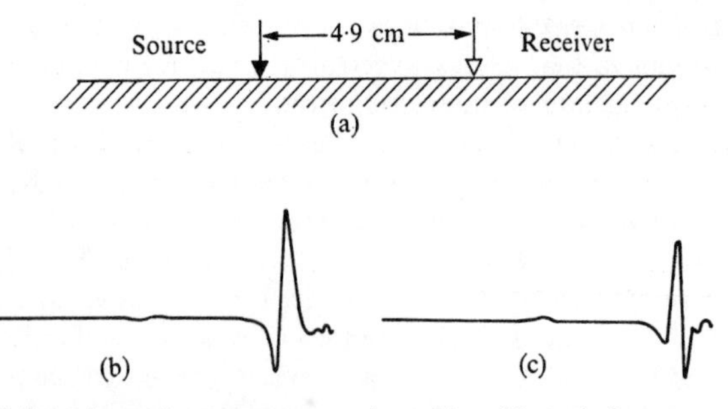

FIG. 6.33. (a) Experimental arrangement, and resulting (b) vertical component displacement and (c) velocity records obtained by Tatel [41].

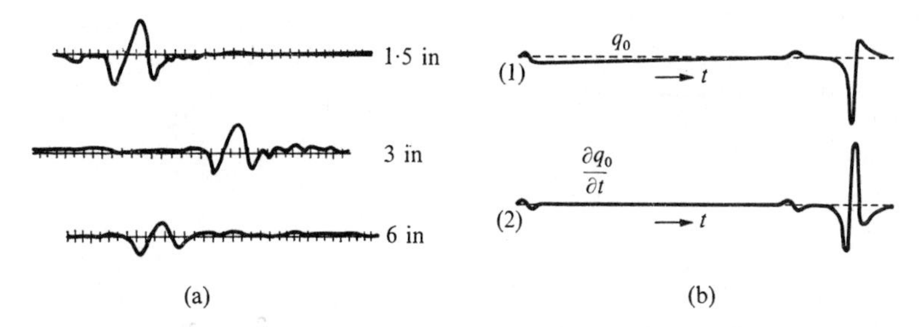

FIG. 6.34. (a) Strain–time records at three distances from the point of impact and (b) the radial displacement predicted by Lamb and the time derivative of that displacement. (After Goodier *et al.* [15, Figs. 1, 2].)

resemble that predicted by Lamb [22], shown in Fig. 6.21 and in the following Fig. 6.34. Barely perceptible in the figures is the arrival of the P wave.

Goodier, Jahsman, and Ripperger [15] studied the surface waves produced by the impact of a spherical ball on a half-space as a possible technique for recording the force–time curve oft he impact. The piezoelectric transducer used in the study recorded the sum of the principal strains ϵ_0, given by

$$\epsilon_0 = \partial q_0/\partial r + q_0/r,$$

where $q_0 = q_0(r, t)$ is the radial displacement at a given distance r. The resulting strain record at three different locations is shown in Fig. 6.34(a). The stress waves were caused by the impact of a $\frac{1}{16}$ in (0·16 cm) diameter ball. Goodier *et al.* were able to approximately compare the results to the theoretical predictions of Lamb [22]. Thus, Lamb's result for the radial displacement q_0, first shown in Fig. 6.34(a), is again shown in Fig. 6.34(b). By treating the disturbance as one-dimensional, the derivative $\partial q_0/\partial r$ has the same form as $\partial q_0/\partial t$, the latter which is also shown in Fig. 6.34(b). Since the contribution of q_0/r is small far from the source, the comparison of waveforms between Fig. 6.34(a) and Fig. 6.34(b) may be made and a strong similarity noted. Other aspects of the work by Goodier *et al.* dealt with obtaining the force–time curves of impact from the surface-wave measurements.

Dally and Thau [8] conducted a photo-elastic study of surface-wave propagation on a half-plane and compared results with theoretical predictions. General agreement was noted for effects due to the P wave but discrepancies were found between the theoretical and experimental results for the S and Rayleigh waves. The differences were ascribed to differences in the experimental loading situation and the loading considered in the mathematical analysis. In addition, three-dimensional effects due to the high-frequency content in the waves was believed contributory to the discrepancies. In a later paper, Thau and Dally [42] studied the subsurface characteristics of the

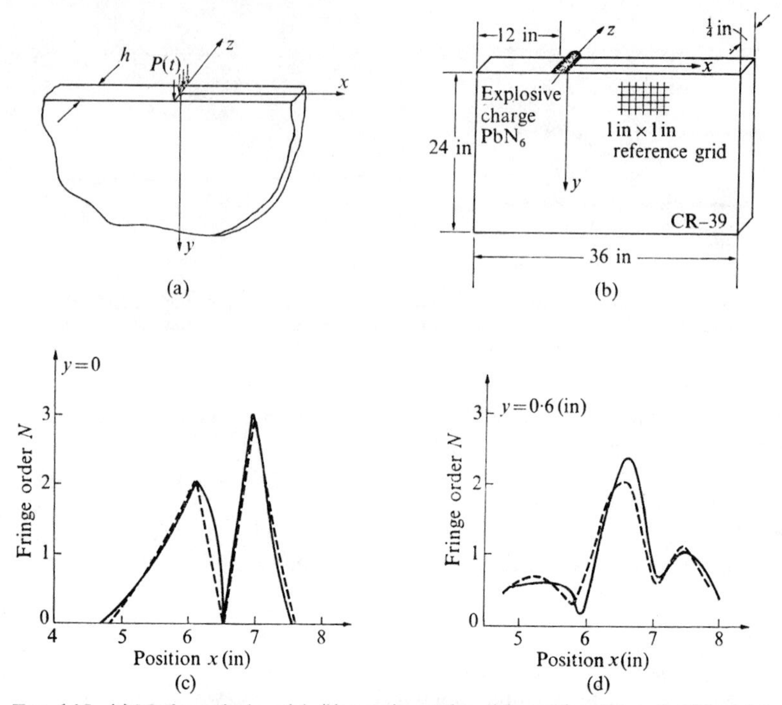

FIG. 6.35. (a) Mathematical model, (b) experimental model considered in study of Rayleigh waves, and (c), (d) comparisons of predicted and measured fringe-order versus propagation distance x for two values of y (at $t = 203$ μs after detonation). (After Thau and Dally [42, Figs. 1, 6].)

Rayleigh wave and obtained excellent agreement between theory and experiment. The theoretical model and experimental model used in the study are shown in Fig. 6.35(a), (b). The dynamic fringe pattern resulting from detonating the lead azide charge was recorded by the Cranz–Schardin camera operating at 68 000 frames per second. The general fringe field was as shown previously in Fig. 6.32(b). Theoretical predictions of the fringe order were made for various positions along the surface and at various depths. Fig. 6.35(c), (d) shows the comparison of results for a particular instant ($t = 203$ μs after detonation) at two different depths.

6.5.3. *Other studies on surface waves*

The previously-cited studies have pertained to surface waves in a half-space. Many other studies have been carried out aimed at determining the propagation of surface waves along curved boundaries or their interactions with surface irregularities. While such problems have not been covered theoretically in this book, they represent interesting extensions of such work.

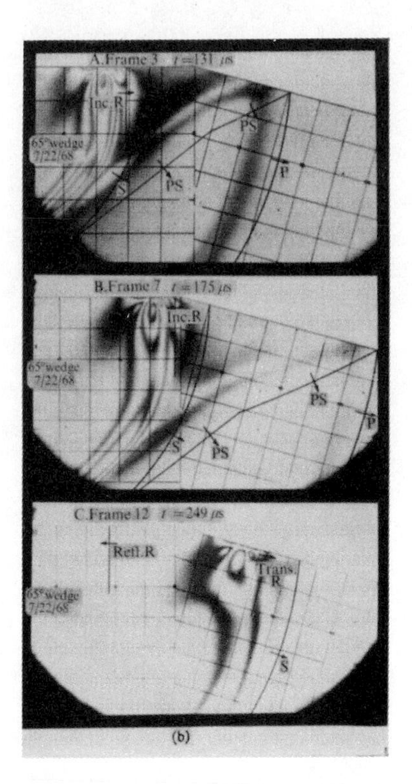

FIG. 6.36 (b)

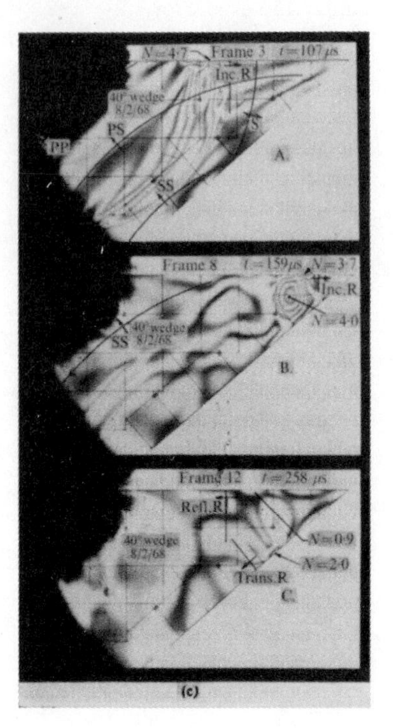

FIG. 6.36 (c)

The case of Rayleigh wave propagation along a curved surface has been studied using photo-elastic means by Marino and Dally [27]. Using techniques similar to those previously described for photo-elastic half-space studies, wave propagation along both concave and convex surfaces was measured. The main observations were (1) increased attenuation of the Rayleigh wave with a decrease in the radius of curvature for both convex and concave surfaces; (2) a decrease in surface wave amplitude with decreasing curvature for concave surfaces; (3) an increase in the compression portion of the surface wave with decreasing curvature of convex surfaces; and (4) the formulation of secondary Rayleigh pulses with both concave and convex surfaces of sharp curvatures. Viktorov [43, pp. 29–42] reports on surface-wave propagation along curved surfaces, including ultrasonic studies in this area.

The propagation of surface waves in wedges has also been investigated using both photo-elastic and ultrasonic techniques. Lewis and Dally [24] have reported on the former aspect. A typical specimen geometry is shown in Fig. 6.36(a). Fig. 6.36(b) shows three stages in the propagation of a wave

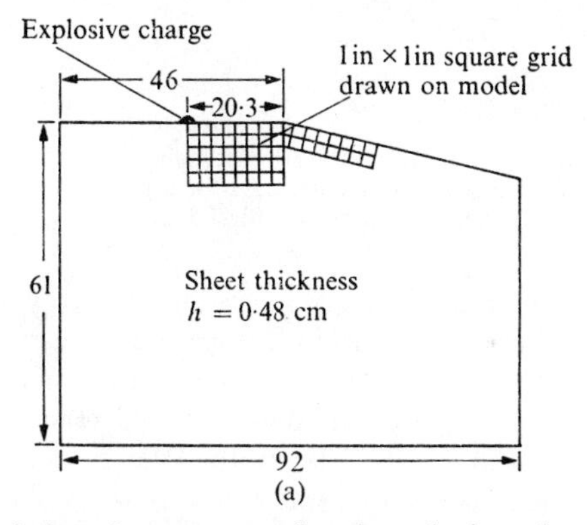

FIG. 6.36. (a) Typical specimen geometry for photo-elastic study of waves in a wedge (dimensions in centimetres), and transmission and reflection of waves in a (b: see separate plate) 165° wedge and (c: see seperate plate) 40° wedge. (After Lewis and Dally [24, Figs. 1A, 5, 7].)

from a point loading past a 165° wedge. The rather shallow wedge angle causes very little reflection of the Rayleigh wave. The propagation of waves in a very sharp (40° wedge angle) wedge is shown in Fig. 6.36(c). A more significant surface-wave reflection from the sharp corner is seen to occur. Some transmission of the Rayleigh wave is also noted. Viktorov [43, pp. 42–6] also reports on ultrasonic investigations of surface waves in wedges and rounded corners.

The propagation and reflection of surface waves from other surface discontinuities is also of interest. The effectiveness of trenches in screening

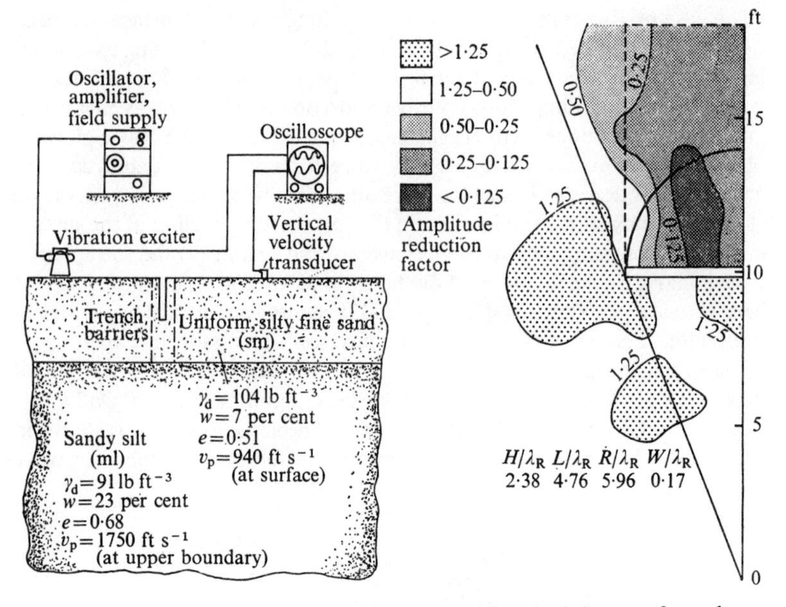

FIG. 6.37. (a) Experimental apparatus used for studying the influence of trenches on screening surface waves and (b) experimental data for a particular trench configuration. (After Woods [44, Figs. 5, 20].)

structural foundations from vibratory surface-wave energy has been studied by Woods [44]. A diagram of the field test apparatus is shown in Fig. 6.37(a). Trenches of various depths placed close to and far from the vibration source were considered. An example of the data obtained for a particular trench configuration is shown in Fig. 6.37(b). In the later figure, the trench depth was 4 ft (1·22 m), the trench length was 8 ft (2·44 m), and the trench width was 4 in. (10·16 cm). The vertical axis is the axis of symmetry of the system so that only half the test area is shown. The data is presented in terms of an 'amplitude reduction factor'. The vibration levels at various points in the test site were measured before the trench was dug. After the trench was prepared, vibration levels were again measured and the ratio of 'after' to 'before' measurements taken to obtain the reduction factor. Vibration frequencies of 200 Hz, 250 Hz, 300 Hz, and 350 Hz were used, with Fig. 6.37(b) being for 250 Hz. At this frequency it was established that the Rayleigh wave velocity was 420 ft s^{-1} (128 m s^{-1}) and that the wavelength was 1·68 ft (0·51 m). A wide variety of tests were reported by Woods with conclusions drawn as to effectiveness of various screening configurations.

A photo-elastic investigation on the reflection and transmission of surface waves from surface discontinuities have been carried out by Dally and Lewis

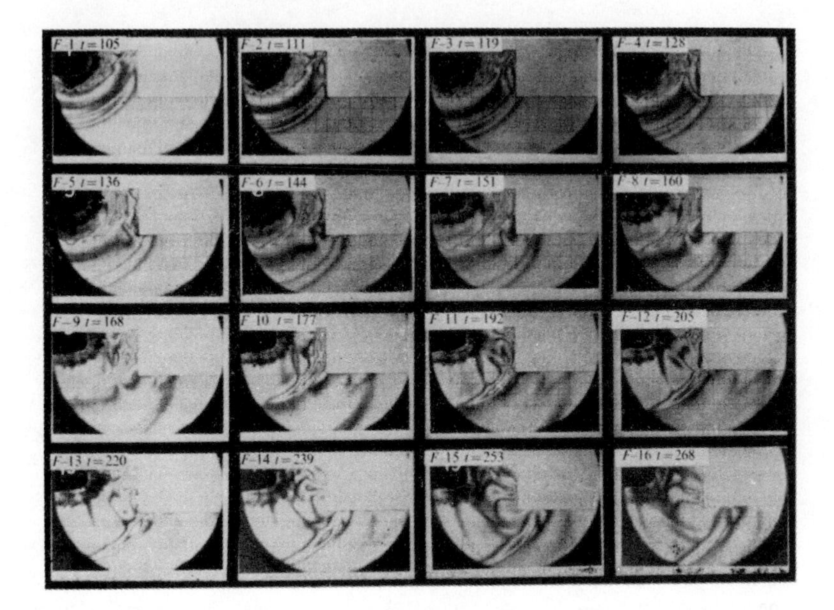

F-1 t=105
F-2 t=111
F-3 t=119
F-4 t=128
F-5 t=136
F-6 t=144
F-7 t=151
F-8 t=160
F-9 t=168
F-10 t=177
F-11 t=192
F-12 t=205
F-13 t=220
F-14 t=239
F-15 t=253
F-16 t=268

FIG. 6.38 (b)

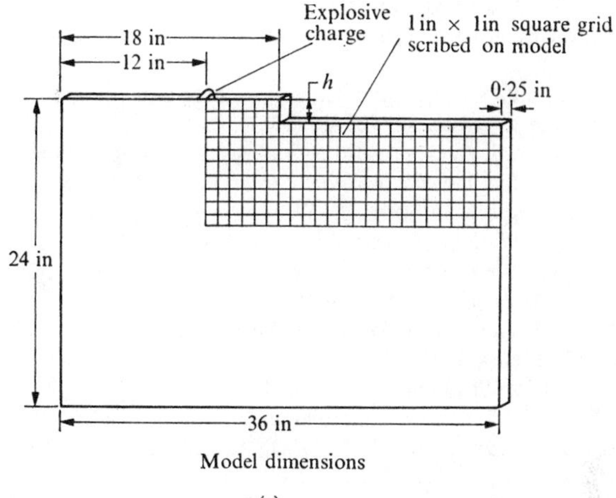

Fig. 6.38. (a) Geometry of the photo-elastic test model and (b: see separate plate) dynamic fringe pattern for wave propagation past a 4 in. step change in elevation. (After Dally and Lewis [6, Figs. 1,4].)

[6] and Reinhardt and Dally [38]. In the former study, the discontinuity was in the form of a step change in elevation in a half-space. The geometry of the photo-elastic model is shown in Fig. 6.38(a) and the dynamic fringe patterns recorded for the case of $h = 4$ in. (10·2 cm) shown in Fig. 6.38(b). Transmission and reflection coefficients for the Rayleigh wave were determined for several step changes in elevation. In the latter study by Reinhardt and Dally, the discontinuity was in the form of a vertical slot in a half-plane. Transmission and reflection coefficients for the Rayleigh wave were determined for various slot depths. A number of ultrasonic studies have been conducted on the influence of surface defects on Rayleigh waves and are reported by Viktorov [43, pp. 57–65].

References

1. Brekhovskikh, Leonid M. *Waves in layered media.* Academic Press, New York (1960).
2. Carrier, G. F., Krook, M., and Pearson, C. E. *Functions of a complex variable: theory and technique.* McGraw-Hill, New York (1966).
3. Chao, C.-C. Dynamical response of an elastic half-space to tangential surface loadings. *J. appl. Mech.* **27**, 559–67 (1960).
4. Dally, J. W. A dynamic photoelastic study of a doubly loaded half-plane. *Develop. Mech.* **4**, 649–64 (1968).
5. ——, Durrelli, A. J., and Riley, W. F. Photoelastic study of stress wave propagation in large plates. *Proc. Soc. exp. Stress Analysis* **17**, 33–50 (1960).

6. DALLY, J. W. and LEWIS, D. (III). A photoelastic analysis of propagation of Rayleigh waves past a step change in elevation. *Bull. seism. Soc. Am.* **58,** 539–63 (1968).
7. —— and RILEY, W. F. Initial studies in three-dimensional dynamic photoelasticity. *J. appl. Mech.* **34,** 405–10 (1967).
8. —— and THAU, S. A. Observations of stress wave propagation in a half-plane with boundary loading. *Int. J. Solids Struct.* **3,** 293–307 (1967).
9. DANIEL, I. M. and MARINO, R. L. Wave propagation in layered model due to point-source loading in low-impedance medium. *Exp. Mech.* **28,** 210–16 (1971).
10. EASON, G. On the torsional impulsive loading of an elastic half space. *Q. Jl. Mech. appl. Math.* **17,** 279–92 (1964).
11. EWING, W. M., JARDETZKY, W. S., and PRESS, F. *Elastic waves in layered media.* McGraw-Hill, New York (1957).
12. FREDERICK, J. R. *Ultrasonic engineering.* John Wiley and Sons, New York (1965).
13. FUNG, Y. C. *Foundations of solid mechanics.* Prentice-Hall, New Jersey (1965)
14. GOODIER, J. N. and BISHOP, R. E. D. On critical reflections of elastic waves at free surfaces. *J. appl. Phys.* **23,** 124–6 (1952).
15. ——, JAHSMAN, W. E., and RIPPERGER, E. A. An experimental surface-wave method for recording force-time curves in elastic impacts. *J. appl. Mech.* **26,** 3–7 (1959).
16. GUTENBERG, B. Energy ratio of reflected and refracted seismic waves. *Bull. seism. Soc. Am.* **34,** 85–102 (1944).
17. GUTZWILLER, M. C. The impact of a rigid circular cylinder on an elastic solid. *Proc. R. Soc.* **A255,** 153–91 (1962).
18. HILL, J. L. Torsional-wave propagation from a rigid sphere semi-embedded in an elastic half-space. *J. acoust. Soc. Am.* **40,** No. 2, 376–9 (1966).
19. HUNTER, S. C. Energy absorbed by elastic waves during impact. *J. Mech. Phys. Solids* **5,** 162–71 (1957).
20. JEFFREYS, H. and JEFFREYS, B. S. *Methods of mathematical physics.* Cambridge University Press (1946).
21. KNOPOFF, L. On Rayleigh wave velocities. *Bull. seism. Soc. Am.* **42,** 307–8 (1952).
22. LAMB, H. On the propagation of tremors over the surface of an elastic solid. *Phil. Trans. R. Soc.* **A203,** 1–42 (1904).
23. LANG, H. A. Surface displacements in an elastic half-space. *Z. angew Math. Mech.* **41,** 141–53 (1961).
24. LEWIS, D. and DALLY, J. W. Photoelastic analysis of Rayleigh wave propagation in wedges. *J. geophys. Res.* **75,** 3387–98 (1970).
25. LORD, A. E. (Jr.). Geometric diffraction loss in longitudinal and shear-wave attenuation measurements in an isotropic half-space. *J. acoust. Soc. Am.* **39,** 650–62 (1966).
26. LOVE, A. E. H. *Some problems of geodynamics.* Cambridge University Press (1911 and 1926).
27. MARINO, R. L. and DALLY, J. W. Rayleigh wave propagation along curved boundaries. *Develop. Mech.* **5,** 819–31 (1969).
28. MCLACHLAN, N. W. *Bessel functions for engineers.* Clarendon Press, Oxford (1961).

29. Miklowitz, J. Elastic wave propagation. In *Applied mechanics surveys* (Ed. H. N. Abramson, H. Liebowitz, J. M. Crowley, and S. Juhasz). Spartan Books, Washington D.C. (1966).
30. Miller, G. F. and Pursey, H. The field and radiation impedance of mechanical radiators on the free surface of a semi-infinite isotropic solid. *Proc. R. Soc.* **A223,** 521–41 (1954).
31. —— and Pursey, H. On the partition of energy between elastic waves in a semi-infinite solid. *Proc. R. Soc.* **A233,** 55–69 (1955).
32. Mitra, M. Von. Disturbance produced in an elastic half-space by an impulsive twisting moment applied to an attached rigid circular disc. *Z. angew Math. Mech.* **38,** 40–3 (1958).
33. ——. Disturbance produced in an elastic half-space by impulsive normal pressure. *Proc. Camb. phil. Soc. math. phys. Sci.* **60,** 683–96 (1961).
34. Pekeris, C. L. The seismic surface pulse. *Proc. natn. Acad. Sci. U.S.A.* **41,** 469–80 (1955).
35. ——. The seismic buried pulse. *Proc. natn. Acad. Sci. U.S.A.* **41,** 629–38 (1955).
36. —— and Lifson, H. Motion of the surface of a uniform elastic half-space produced by a buried pulse. *J. Acoust. Soc. Am.* **29,** 1233–8 (1957).
37. Rayleigh, J. W. S. On waves propagated along the plane surface of an elastic solid. *Proc. Lond. math. Soc.* **17,** 4–11 (1887).
38. Reinhardt, H. W. and Dally, J. W. Some characteristics of Rayleigh wave interaction with surface flaws. *Mater. Eval.* **28,** 213–20 (1970).
39. Riley, W. F. and Dally, J. W. A photoelastic analysis of stress wave propagation in a layered model. *Geophysics* **31,** 881–99 (1966).
40. Stoneley, R. Elastic waves at the surface of separation of two solids. *Proc. R. Soc.* **A106,** 416–28 (1924).
41. Tatel, H. E. Note on the nature of a seismogram. II. *J. Geophys. Res.* **59,** 289–94 (1954).
42. Thau, S. A. and Dally, J. W. Subsurface Characteristics of the Rayleigh Wave. *Int. J. Eng. Sci.* **7,** 37–52 (1969).
43. Viktorov, I. A. *Rayleigh and Lamb waves: physical theory and applications* Plenum Press, New York (1967).
44. Woods, R. D. Screening of surface waves in soils. *J. Soil Mech. Founds Div. Am. Soc. civ. Engrs.* July **94,** 951–79 (1968).

Problems

6.1. Verify the statement made earlier (following (6.1.95)) that the extraneous roots of the Rayleigh wave equation correspond to the special cases of plane-wave reflection in a half-space (see, for example, (6.1.52), (6.1.56)).

6.2. Establish whether SH surface waves can exist in a traction-free half-space.

6.3. As indicated in the text, Goodier and Bishop [14] used a limiting process on wave amplitude ratios to obtain the results summarized by Fig. 6.7. From a review of the given reference, obtain the grazing SV wave results by the limiting process.

6.4. Consider the reflection of P, SV, and SH waves from a completely fixed boundary. Determine the amplitude ratios and investigate any special cases of reflection. Include the possibility of waves at grazing incidence.

6.5. Consider a semi-infinite media, $y \geq 0$, and the cases of incident plane harmonic P or SV waves. Let the boundary conditions be those of elastic restraint (that is, $\tau_{yy} = ku_y$, $\tau_{xy} = 0$ at $y = 0$) and obtain the results for the reflected wave amplitude ratios.

6.6. Consider the case of a semi-infinite medium with a slightly wavy boundary. Thus let the medium generally occupy the region $y > 0$, with the exact surface given by $y = \varepsilon f(x)$, where ε is a small parameter. First show that the unit normal vector **n** (directed into the media) has components

$$l = -\varepsilon f'(x)(1+\varepsilon^2 f'^2)^{-\frac{1}{2}}, \qquad m = (1+\varepsilon^2 f'^2)^{-\frac{1}{2}}.$$

Then show that traction-free boundary conditions on the free surface are given by $-\tau_{yy}\varepsilon f' + \tau_{xy} = 0$, $-\tau_{xy}\varepsilon f' + \tau_{yy} = 0$ on $y = \varepsilon f(x)$.

6.7. Consider two parallel line SH wave sources of equal strength, separated by the distance $2a$, in an infinite media. What conditions exist on the mid plane (that is, the plane perpendicular to the plane containing the sources and located midway between the sources) if the sources are (a) in phase and (b) 180° out of phase?

6.8. Consider a half-space excited by a strip surface SH wave source. That is, consider the half-space $y \geq 0$, with the prescribed displacements $u_z(x, 0, t) = 0$, $|x| > a$, and $u_x = u_y = 0$. Obtain the transformed solution and attempt inversion. Are the results for a line source recovered for distances several wavelengths from the source?

6.9. Attempt to verify the expression for u_x as given by (6.3.49). That is, as indicated in the text, apply the results (6.3.48) to (6.3.27).

6.10. Consider a half-space defined by $z \geq 0$, with a rigid hemisphere of radius a embedded at the origin. Let the hemisphere be undergoing torsional oscillations about the z-axis given by $\Omega_0 \exp(i\omega t)$, where Ω_0 is maximum angular displacement. Formulate the problem in terms of spherical coordinates. Note that only one of the displacement equations of motion survives. Solve for the resulting wave propagation.

6.11. Consider the case of two semi-infinite media in contact. Let the boundary conditions be of the transverse slip type. That is,

$$\tau_{yy} = \tau'_{yy}, \qquad \tau_{yx} = \tau_{yz} = \tau'_{yx} = \tau'_{yz} = 0.$$

Solve for reflection–refraction of incident plane harmonic P and SV waves.

6.12. Consider two semi-infinite media, having different material properties, bonded together. Develop the governing theory and equations for the reflection and transmission of SH waves across the boundary between the two media.

6.13. Consider a layered media as defined by Fig. 6.31. Consider the case of plane harmonic SH waves travelling in the lower media and impinging on the interface at some oblique angle. Solve for the resulting reflection–transmission expressions. Are there any special cases of interest involving critical incidence angles and/or material properties?

6.14. As an extension of the previous problem, obtain the reflection–transmission expressions for incident P and incident SV waves.

6.15. Again consider the case of two semi-infinite media in contact (see Fig. 6.30 for coordinate and media definitions). For the case of incident plane harmonic P and SV waves, several patterns of emerging waves are possible, depending on incoming wave-type, incidence angle, and whether $c_1 \gtrless c_1'$. Investigate the behaviour of the reflected–refracted wave system for incidence beyond the critical angle.

Scattering and diffraction of elastic waves

WE now wish to study the interaction of elastic waves with discontinuities or boundaries of more complex shape than that of the half-space of the last chapter. Such problems might arise when waves propagating through an infinite medium encounter cavities, inclusions, or cracks. Such problems, generally denoted as scattering and diffraction problems, have been of long standing interest in acoustics and electro-magnetic theory. Problems of interest in these fields include sonar detection, architectural design, radar applications, and antenna design. Interest in the field of elasticity has been more recent, with problems in ultrasonic testing, dynamic stress concentrations, and blast effects on buried structures representing some areas of application.

The basic nature of the problem is the same for all fields. Namely, a propagating harmonic wave or a pulse (usually considered plane for either case) is considered to encounter a discontinuity, in the form of an inclusion (representing, say, a flaw in the material) or a slit or wedge (representative of a crack). As a consequence of reflection phenomena, such as discussed in Chapter 6 for the half-space, a scattering of elastic waves occurs. In addition, wave diffraction occurs near the edge of the discontinuity. The problem then is to determine the resultant stress and displacement fields with emphasis on near-fields (stress concentrations) or far-field effects (radiation patterns, total energy reflected and transmitted).

The increased complexity of the elastic-reflection phenomena, due to mode conversion effects, yields scattering problems in solids more complex than their acoustic or electro-magnetic counterparts. However, most tools of analysis for the simpler problems, some of which have been only recently developed, can also be applied to elastic scattering.

There is no section allocated for purely experimental studies on scattering and diffraction because of the relative scarcity of results in this area. Durelli and Riley [7] performed a photo-elastic study in which the stress distribution on the boundary of a circular hole in a large plate was determined under dynamic pulse loading. Shea [37] has measured the dynamic stress concentration factor of a circular hole in a plate under passage of a sharp pulse and noted a variation of this factor with frequency. The results were in qualitative agreement with those predicted by Pao [33]. Reference is also made to the experimental study by Wells and Post [49] and by Beebe [4], where photoelasticity was employed to determine the dynamic stress distribution about a running crack. Finally, it should be noted that a number of studies have been done on the reflection and transmission of ultrasonic waves by discontinuities from the standpoint of use in non-destructive testing.

In the following, the first section will be devoted to the cylindrical and spherical cavity problems. The second section will be devoted to a diffraction problem. Discussion of other methods of analysing problems in this area will be given.

7.1. Scattering of waves by cavities

The scattering of elastic waves by cavities or inclusions of simple geometry, such as cylindrical or spherical, represents our first consideration. The coordinate systems for such geometries enable separation of variables to be performed on the wave equation and solutions in terms of infinite series are possible. Solutions of this type are particularly attractive when the incident waves are simple harmonic in contrast, say, to sharp pulses.

7.1.1. *Scattering of SH waves by a cylindrical cavity*

One of the simplest cases of elastic scattering is that of SH waves impinging on a cylindrical cavity, where the plane of polarization is parallel to the cavity axis. If it is recalled that SH waves do not undergo mode conversion at a free surface under these circumstances, then the resultant displacement and stress fields will be governed by a single scalar equation. The problem becomes, consequently, identical to the equivalent acoustical problem.

1. *Governing equations.* Consider a plane harmonic SH wave system propagating in the positive z direction, and impinging on an infinite, cylindrical cavity of radius a as shown in Fig. 7.1. The displacement field for the SH wave system will be given by

$$u_x = u(y, z, t), \qquad u_y = u_z = 0. \tag{7.1.1}$$

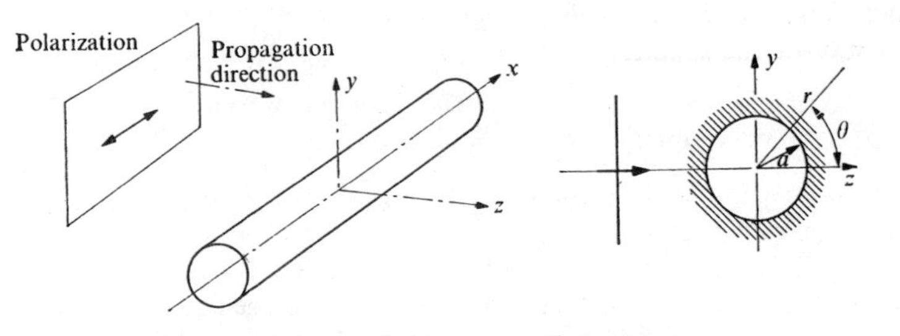

FIG. 7.1. SH waves incident on a cylindrical cavity.

The displacement equations of motion then reduce to

$$\nabla^2 u = \frac{1}{c_2^2}\frac{\partial^2 u}{\partial t^2}, \tag{7.1.2}$$

where ∇^2 is the Laplacian in polar coordinates. The traction-free cavity surface requires the boundary conditions

$$\tau_{rr} = \tau_{r\theta} = \tau_{rx} = 0, \qquad r = a, \tag{7.1.3}$$

where

$$\tau_{rr} = \lambda\left(\frac{\partial u_r}{\partial r}+\frac{1}{r}\frac{\partial u_\theta}{\partial \theta}+\frac{u_r}{r}+\frac{\partial u_x}{\partial x}\right)+2\mu\frac{\partial u_r}{\partial r},$$

$$\tau_{r\theta} = \mu\left(\frac{1}{r}\frac{\partial u_r}{\partial \theta}+\frac{\partial u_\theta}{\partial r}-\frac{u_\theta}{r}\right), \tag{7.1.4}$$

$$\tau_{rx} = \mu\left(\frac{\partial u_x}{\partial r}+\frac{\partial u_r}{\partial x}\right).$$

However, as a consequence of the characteristics of an SH field, we have that

$$u_r = u_\theta = 0, \qquad \frac{\partial}{\partial x} = 0, \tag{7.1.5}$$

so that the boundary conditions reduce to

$$\partial u/\partial r = 0, \qquad r = a. \tag{7.1.6}$$

An incident SH wave propagating in the positive z direction is given by

$$u_i = U_0 \exp\{i(\omega t-\gamma z)\} = U_0 \exp\{i\gamma(c_2 t-z)\}. \tag{7.1.7}$$

When the incident wave strikes the cavity, reflection will occur, setting up a scattered wavefield,

$$u_s = u_s(r, \theta, t). \tag{7.1.8}$$

The total displacement field will then be given as

$$u(r, \theta, t) = u_i+u_s, \tag{7.1.9}$$

where $u(r, \theta, t)$ must satisfy the governing equation and the boundary conditions.

2. *Expansion in infinite series.* The incident wavefield automatically satisfies the wave equation for $\gamma c_2 = \omega$. Let

$$u_s = R\Theta e^{i\omega t} = U_s e^{i\omega t}, \tag{7.1.10}$$

and substitute in (7.1.2), giving

$$R''\Theta+\frac{1}{r}R'\Theta+\frac{1}{r^2}R\Theta'' = -\gamma^2 R\Theta, \qquad \gamma^2 = \omega^2/c_2^2. \tag{7.1.11}$$

This separates to

$$\Theta''+k^2\Theta = 0, \tag{7.1.12}$$

$$R''+\frac{1}{r}R'+\left(\gamma^2-\frac{k^2}{r^2}\right)R = 0. \tag{7.1.13}$$

The solution of (7.1.12) is

$$\Theta = A\cos k\theta+B\sin k\theta. \tag{7.1.14}$$

Symmetry with respect to the x-axis requires $B = 0$ while the requirement that Θ be single-valued (that is, $\Theta(\theta) = \Theta(\theta+2\pi)$) indicates k is an integer, say n.

We recognize (7.1.13) as Bessel's equation of order n having the solutions

$$R = AH_n^{(2)}(\gamma r)+BH_n^{(1)}(\gamma r), \tag{7.1.15}$$

where the Hankel-function solution form has been chosen in anticipation of the convenient exponential representation of the asymptotic behaviour. We now impose the requirement that the scattered wavefield must be outward propagating. This condition will enable us to determine which Hankel function is appropriate. As we have shown previously, for large values of the argument, the Hankel functions have the following approximate expressions,

$z \gg 1$:

$$H_n^{(1)}(z) \sim \left(\frac{2}{\pi z}\right)^{\frac{1}{2}}\exp\left\{i\left(z-\frac{\pi}{4}-\frac{n\pi}{2}\right)\right\}(1-\ldots),$$
$$H_n^{(2)}(z) \sim \left(\frac{2}{\pi z}\right)^{\frac{1}{2}}\exp\left\{-i\left(z-\frac{\pi}{4}-\frac{n\pi}{2}\right)\right\}(1+\ldots). \tag{7.1.16}$$

Recalling that the time dependence is $\exp(i\omega t)$, it is seen that an outward propagating wave must have the form $\exp\{i(\omega t-\gamma r)\}$, so that $H_n^{(2)}$ is appropriate in the present case. Hence let $B = 0$ in (7.1.15). The scattered field thus is given by

$$U_s(r, \theta) = \sum_{n=0}^{\infty} A_n H_n^{(2)}(\gamma r)\cos n\theta, \tag{7.1.17}$$

where A_n must be determined.

We now have solutions which satisfy the wave equation so that only substitution in the boundary condition remains. The form of $U_s(\theta)$ is ideally suited for this, but this is not the case for the incoming wave (7.1.7). What is required is a Bessel-function representation of the plane wave. Writing u_i as

$$u_i = U_i(r, \theta)e^{i\omega t} = U_0 e^{-i\gamma r\cos\theta}e^{i\omega t}, \tag{7.1.18}$$

it is found that†

$$U_i(r, \theta) = U_0\sum_{n=0}^{\infty}\varepsilon_n(-1)^n J_n(\gamma r)\cos n\theta \qquad (\varepsilon_0 = 1,\ \varepsilon_n = 2). \tag{7.1.19}$$

† Pp. 55–7 of Reference [23] or p. 96 of Reference [21].

With (7.1.17), (7.1.18) we have the incident and scattered wavefields similarly represented. Substitution in the boundary condition, where the time dependence has been omitted, gives

$$\frac{\partial U_i}{\partial r}+\frac{\partial U_s}{\partial r} = 0, \qquad r = a. \tag{7.1.20}$$

Using (7.1.17) and (7.1.19), we obtain

$$\sum_{n=0}^{\infty}\left\{U_0\varepsilon_n(-1)^n\frac{\mathrm{d}J_n(\gamma r)}{\mathrm{d}r}+A_n\frac{\mathrm{d}H_n^{(2)}(\gamma r)}{\mathrm{d}r}\right\}\cos n\theta\bigg|_{r=a} = 0. \tag{7.1.21}$$

Solving for A_n, we have

$$A_n = \frac{-\varepsilon_n(-1)^n U_0 J_n'(\gamma a)}{H_n^{(2)\prime}(\gamma a)}. \tag{7.1.22}$$

By use of recursion formula, we have

$$\begin{aligned} J_n'(\gamma a) &= \frac{\gamma}{2}\{J_{n-1}(\gamma a)-J_{n+1}(\gamma a)\}, & J_0'(\gamma a) &= -\gamma J_1(\gamma a),\\ Y_n'(\gamma a) &= \frac{\gamma}{2}\{Y_{n-1}(\gamma a)-Y_{n+1}(\gamma a)\}, & Y_0'(\gamma a) &= -\gamma Y_1(\gamma a),\\ H_n^{(2)\prime}(\gamma a) &= \frac{\gamma}{2}\{H_{n-1}^{(2)}(\gamma a)-H_{n+1}^{(2)}(\gamma a)\}, & H_0^{(2)\prime}(\gamma a) &= -\gamma H_1^{(2)}(\gamma a), \end{aligned} \tag{7.1.23}$$

so that

$$\begin{aligned} A_0 &= -U_0\frac{J_1(\gamma a)}{H_1^{(2)}(\gamma a)},\\ A_n &= -2(-1)^n U_0\frac{J_{n+1}(\gamma a)-J_{n-1}(\gamma a)}{H_{n+1}^{(2)}(\gamma a)-H_{n-1}^{(2)}(\gamma a)}. \end{aligned} \tag{7.1.24}$$

This may be expressed in somewhat simpler form by writing

$$A_0 = U_0\frac{1}{1-\mathrm{i}(Y_1/J_1)}, \qquad A_n = -2(-1)^n U_0\bigg/\left(1-\mathrm{i}\frac{Y_{n+1}-Y_{n-1}}{J_{n+1}-J_{n-1}}\right), \tag{7.1.25}$$

which are of the general form

$$\frac{1}{1-\mathrm{i}q} = \frac{1}{1+q^2}+\frac{\mathrm{i}q}{1+q^2}. \tag{7.1.26}$$

Let $q = \cot\gamma$, so that

$$\frac{1}{1-\mathrm{i}q} = \sin\gamma(\sin\gamma+\mathrm{i}\cos\gamma) = \mathrm{i}\sin\gamma\mathrm{e}^{-\mathrm{i}\gamma}. \tag{7.1.27}$$

Then we may write for A_0, A_n

$$A_0 = -U_0\mathrm{i}\sin\gamma_0\exp(-\mathrm{i}\gamma_0), \qquad A_n = 2(-1)^{n+1}U_0\sin\gamma_n\exp(-\mathrm{i}\gamma_n), \tag{7.1.28}$$

where

$$\cot \gamma_0 = \frac{Y_1(\gamma a)}{J_1(\gamma a)}, \qquad \cot \gamma_n = \frac{Y_{n+1}(\gamma a) - Y_{n-1}(\gamma a)}{J_{n+1}(\gamma a) - J_{n-1}(\gamma a)}. \tag{7.1.29}$$

With this, the scattered solution takes the form

$$u_s(r, \theta, t) = U_0 e^{i\omega t}\left\{-i \sin \gamma_0 \exp(-i\gamma_0) H_0^{(2)}(\gamma r) + \right.$$
$$\left. + 2\sum_{n=1}^{\infty}(-i)^{n+1} \sin \gamma_n \exp(-i\gamma_n) H_n^{(2)}(\gamma r) \cos n\theta\right\}. \tag{7.1.30}$$

We may study the scattered-field configuration by computing the above series. However, by considering the far field, for which $\gamma r \gg 1$, the result may be simplified by using the asymptotic forms (7.1.16). This gives

$\gamma r \gg 1$:

$$u_s(r, \theta, t) = U_0 \sqrt{\left(\frac{2}{\pi \gamma r}\right)} \exp\{i(\omega t - \gamma r)\}\left[-i \sin \gamma_0 \exp\{-i(\gamma_0 - \pi/4)\} + \right.$$
$$\left. + 2\sum_{n=1}^{\infty}(-i)^{n+1} \sin \gamma_n \exp\left\{-i\left(\gamma_n - \frac{2n+1}{4}\pi\right)\right\} \cos n\theta\right]$$
$$= U_0 \sqrt{\left(\frac{2}{\pi \gamma r}\right)} \exp\{i(\omega t - \gamma r)\} \psi_s(\theta). \tag{7.1.31}$$

In the acoustic analogue of this problem, $u_s \sim p_s$, the scattered pressure field. Considerable interest is connected with the scattered power of the field, where†

$$P_s = \frac{2I_0}{\pi \gamma} \int_0^{2\pi} |\psi_s(\theta)|^2 \, d\theta, \tag{7.1.32}$$

and I_0 is a constant (in acoustics, the intensity of the incident wave). A plot of P_s versus γa gives the results shown in Fig. 7.2(a). Of equal interest is to consider the behaviour of $|\psi_s|^2$, appearing in the integrand of P_s. A plot of this versus θ indicates the directional characteristics of the scattering, and is shown in Fig. 7.2(b). The representations are in terms of the wavenumber parameter γa. Recalling that $\gamma = 2\pi/\lambda$, where λ is the wavelength, it is seen that, for $a \simeq \lambda(\gamma a = 1)$, most of scattered energy is back toward the incoming wave (hence 'back scattering'). The interaction of this field, primarily a result of reflection with the incoming wave will produce an irregular radiation field on the forward side of the cylinder. On the other hand, there is little scattered energy to the rear of the cylinder, so that the toal field there is quite

† See Lindsay [21, pp. 91–101] for complete coverage of this problem, including development of expressions for P_s.

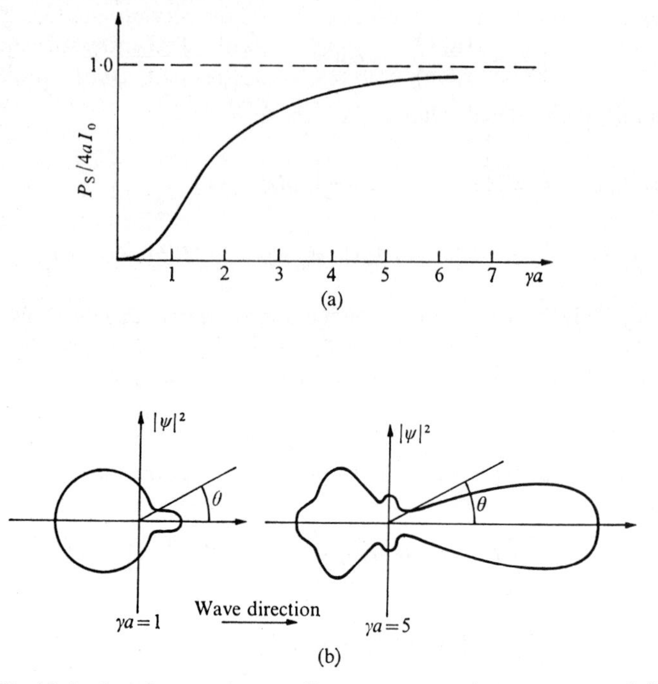

FIG. 7.2. (a) A plot of scattered power P_s versus wavenumber parameter and (b) the directional characteristics of the scattering for $\gamma a = 1, 5$. (After Lindsay [21, Figs. 3.8, 3.9].)

similar to the incident wavefield. The case of $a \gg \lambda(\gamma a = 5)$, or the short-wavelength case, indicates that a great deal of forward scattering has occurred. This field will interfere with the incident field to form a 'shadow zone'. For $\gamma a \gg 1$, we would expect the region behind the cylinder to be relatively undisturbed. However, we recall from optics that diffraction effects (interaction of radiation with the edges of objects) prevent objects from casting sharp shadows, regardless of wavelength. Thus, as γa becomes large, we would expect a shadow zone to be developed, but with indistinct edges. This would be borne out by calculations of $|\psi|^2$, for it would be found that its angular distribution would increase in complexity, containing many lobes and dips, and preventing complete annihilation of the incident field in the shadow region.

7.1.2. *Scattering of compressional waves by a spherical obstacle*

Let us now consider plane harmonic compressional waves impinging on a spherical obstacle. Such a problem is possibly more representative of actual situations of included flaws or 'stress raisers' in materials. The approach followed, however, will be similar to that of the previous section: that is, expansion of the wavefields in infinite series. The increased algebraic complexity of the present problem may be quickly surmised if it is realized that a total of five fields must be so expanded in the general case—the incident field, scattered compressional field, scattered shear field, refracted compression field, and refracted shear field. The latter two fields arise, of course, when the obstacle is taken as an elastic inclusion as opposed to an empty cavity. The coverage of this problem will be based on the work of Pao and Mow [34].

1. *Governing equations.* Consider a plane harmonic compression wave impinging on a spherical obstacle of radius a, as shown in Fig. 7.3. In this

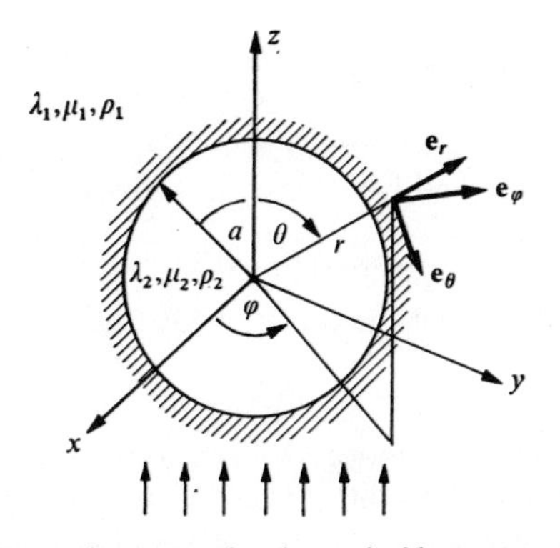

FIG. 7.3. Plane harmonic compressional wave incident on a spherical obstacle.

situation, subscripts 1 are used to identify the infinite medium, while subscripts 2 identify the properties of the spherical obstacle. Because of the axisymmetry of the present problem (this would not be the case if shear waves were impinging), we have that

$$\mathbf{u} = u_r\mathbf{e}_r + u_z\mathbf{e}_z. \tag{7.1.33}$$

In terms of potential functions, we follow the approach used in the half-space analysis (see § 6.3) and write

$$\mathbf{u} = \nabla\Phi + \nabla x\left(\frac{\partial\Psi}{\partial\theta}\mathbf{e}_\phi\right), \tag{7.1.34}$$

where Φ, Ψ satisfy

$$\nabla^2\Phi = \frac{1}{c_\alpha^2}\frac{\partial^2\Phi}{\partial t^2}, \qquad \nabla^2\Psi = \frac{1}{c_\beta^2}\frac{\partial^2\Psi}{\partial t^2}, \tag{7.1.35}$$

and where

$$c_\alpha^2 = (\lambda+2\mu)/\rho, \qquad c_\beta^2 = \mu/\rho. \tag{7.1.36}$$

The stress–displacement equations are

$$\begin{aligned}
\tau_{rr} &= \lambda U + 2\mu\frac{\partial u_r}{\partial r},\\
\tau_{\theta\theta} &= \lambda U + 2\mu\left(\frac{1}{r}\frac{\partial u_\theta}{\partial \theta} + \frac{u_r}{r}\right),\\
\tau_{\phi\phi} &= \lambda U + 2\mu\left(\frac{u_r}{r} + \frac{\cot\theta}{r}u_\theta\right),\\
\tau_{r\theta} &= \mu\left(\frac{1}{r}\frac{\partial u_r}{\partial \theta} + \frac{\partial u_\theta}{\partial r} - \frac{u_\theta}{r}\right),
\end{aligned} \tag{7.1.37}$$

where

$$U = \nabla\cdot\mathbf{u} = \frac{\partial u_r}{\partial r} + \frac{2}{r}u_r + \frac{1}{r}\frac{\partial u_\theta}{\partial \theta} + \frac{\cot\theta}{r}u_\theta. \tag{7.1.38}$$

2. *Incident, scattered, and refracted wavefields.* The incident compression wave propagating in the positive z direction is given by

$$\Phi^{(\mathrm{i})} = \Phi_0 \exp\{\mathrm{i}(\alpha_1 z - \omega t)\}, \qquad \Psi^{(\mathrm{i})} = 0. \tag{7.1.39}$$

In order for $\Phi^{(\mathrm{i})}$ to satisfy the first of (7.1.35) we must have

$$\alpha_1^2 = \omega^2\rho_1/(\lambda_1 + 2\mu_1). \tag{7.1.40}$$

We recall in the simpler scattering problem of the previous section that it became necessary, when substituting in the boundary conditions, to expand the incoming plane wavefield in the 'natural' eigenfunctions of the coordinate system of the problem. In anticipation of this, we expand $\Phi^{(\mathrm{i})}$ in a similar fashion, giving†

$$\Phi^{(\mathrm{i})} = \Phi_0 \sum_{n=0}^{\infty}(2n+1)\mathrm{i}^n j_n(\alpha_1 r)P_n(\cos\theta). \tag{7.1.41}$$

We have omitted the $\exp(-\mathrm{i}\omega t)$ time dependence in this expression, and will do so in the following. The $j_n(\alpha_1 r)$ is the previously encountered spherical Bessel function given as

$$j_n(z) = \left(\frac{\pi}{2z}\right)^{\frac{1}{2}} J_{n+\frac{1}{2}}(z), \tag{7.1.42}$$

and $P_n(\cos\theta)$ are the Legendre polynomials.

Of course (7.1.41) represents but one of the solution forms to the Helmholtz equation, $\nabla^2\Phi + k^2\Phi = 0$, that arises in considering either of (7.1.35), with Φ or Ψ having harmonic time dependence and the spherical Laplacian being given by

$$\nabla^2 = \frac{1}{r^2}\frac{\partial}{\partial r}\left(r^2\frac{\partial}{\partial r}\right) + \frac{1}{r^2\sin\theta}\frac{\partial}{\partial\theta}\left(\sin\theta\frac{\partial}{\partial\theta}\right) + \frac{1}{r^2\sin^2\theta}\frac{\partial^2}{\partial\phi^2}. \tag{7.1.43}$$

† P. 1466 of Reference [27].

Separation of variables leads to

$$\left\{\frac{\partial}{\partial r}\left(r^2\frac{dR}{dr}\right)R^{-1}+\frac{1}{\sin\theta}\frac{\partial}{\partial\theta}\left(\sin\theta\frac{d\Theta}{d\theta}\right)\Theta^{-1}+k^2r^2\right\}\sin^2\theta = -\frac{d^2\tilde{\phi}}{d\phi^2} = \gamma^2, \quad (7.1.44)$$

where $\Phi = R(r)\,\Theta(\theta)\tilde{\phi}(\phi)$ and γ^2 is the first separation constant. Thus

$$\frac{d^2\tilde{\phi}}{d\phi^2}+\gamma^2\tilde{\phi} = 0. \quad (7.1.45)$$

The requirement that $\tilde{\phi}$ be single-valued leads to $\gamma^2 = m^2$, where m is an integer. The second separation of (7.1.44) gives

$$\frac{1}{\sin\theta}\frac{d}{d\theta}\left(\sin\theta\frac{d\Theta}{d\theta}\right)+\left(l^2-\frac{m^2}{\sin^2\theta}\right)\Theta = 0, \quad (7.1.46)$$

$$\frac{d}{dr}\left(r^2\frac{dR}{dr}\right)+\left(k^2-\frac{l^2}{r^2}\right)R = 0. \quad (7.1.47)$$

Now the solutions of the preceding are the Legendre polynomials, where l^2 is constrained to be

$$l^2 = n(n+1), \qquad n \sim \text{integer}, \quad (7.1.48)$$

and

$$\Theta_n = P_n(\cos\theta), \quad (7.1.49)$$

where the first few Legendre polynomials are given by

$$\begin{gathered} P_0(z) = 1, \qquad P_1(z) = z, \qquad P_2(z) = \tfrac{1}{2}(3z^2-1), \\ P_3(z) = \tfrac{1}{2}(5z^3-3z), \qquad P_4(z) = \tfrac{1}{8}(35z^4-30z^2+3),\ldots. \end{gathered} \quad (7.1.50)$$

Finally, (7.1.47) has the solutions

$$R(r) = A\sqrt{\left(\frac{\pi}{2kr}\right)}J_{n+\frac{1}{2}}(kr)+B\sqrt{\left(\frac{\pi}{2kr}\right)}Y_{n+\frac{1}{2}}(kr), \quad (7.1.51)$$

where $J_{n+\frac{1}{2}}$ and $Y_{n+\frac{1}{2}}$ are the half-order Bessel functions. The spherical Bessel and Hankel functions are thus

$$\begin{gathered} j_n(kr) = \sqrt{\left(\frac{\pi}{2kr}\right)}J_{n+\frac{1}{2}}(kr), \qquad n_n(kr) = \sqrt{\left(\frac{\pi}{2kr}\right)}Y_{n+\frac{1}{2}}(kr), \\ h_n^{(1)}(kr) = \sqrt{\left(\frac{\pi}{2kr}\right)}H_{n+\frac{1}{2}}^{(1)}(kr). \end{gathered} \quad (7.1.52)$$

We recall that the spherical Hankel functions have the unique property of being representable by a finite sum; thus

$$h_n^{(1)}(kr) = \frac{e^{ikr}(-i)^{-n+1}}{kr}\sum_{l=0}^{n}\frac{(n+l)!}{l!(n-l)!}\left(\frac{i}{2kr}\right)^l. \quad (7.1.53)$$

Hence (7.1.51) may also be written

$$R(r) = Ah_n^{(1)}(kr)+Bh_n^{(2)}(kr). \tag{7.1.54}$$

Continuing, the scattered wavefield, which will have dilatational and rotational components, will be given by

$$\Phi^{(s)} = \sum_{n=0}^{\infty} A_n h_n(\alpha_1 r)P_n(\cos\theta), \qquad \Psi^{(s)} = \sum_{n=0}^{\infty} B_n h_n(\beta_1 r)P_n(\cos\theta), \tag{7.1.55}$$

where $\beta_1^2 = \omega^2\rho_1/\mu_1$. In the above, $h_n = h_n^{(1)}$, where the spherical Hankel function of the first kind has been selected because it represents waves diverging from the obstacle. The impinging waves will set up the waves within the obstacle. Since this region will include the origin $r = 0$, the resulting solutions must be well behaved there. This leads to the selection of $j_n(z)$, so that

$$\begin{aligned}\Phi^{(r)} &= -\sum_{n=0}^{\infty} C_n j_n(\alpha_2 r)P_n(\cos\theta),\\ \Psi^{(r)} &= -\sum_{n=0}^{\infty} D_n j_n(\beta_2 r)P_n(\cos\theta),\end{aligned} \tag{7.1.56}$$

where the superscript identifies the refracted wave system. The minus sign is for later convenience in algebraic manipulations. The presence of the $r^{-\frac{1}{2}}$ in the definition of $j_n(kr)$ may raise the question of the behaviour of j_n at the origin. It turns out, however, that for small z

$$j_n(z) \to \frac{z^n}{1.3.5...(2n+1)}, \qquad n_n(z) \to \frac{1.1.3.5...(2n-1)}{z^{n+1}}, \tag{7.1.57}$$

so that $j_n(z)$ is still well behaved, while $n_n(z)$ is singular.

3. *Displacements and stresses.* Substitution into the boundary conditions requires the stresses in terms of the potential functions Φ, Ψ and in turn, in terms of the spherical Bessel functions. This data is compiled in the following:

Displacements

$$u_r = \frac{\partial\Phi}{\partial r}+\frac{1}{r}(D_\theta\Psi), \qquad u_\theta = \frac{1}{r}\frac{\partial\Phi}{\partial\theta}-\frac{\partial}{\partial\theta}(D_r\Psi), \tag{7.1.58}$$

where

$$D_r = \frac{1}{r}\frac{\partial}{\partial r}(r), \qquad D_\theta = \frac{1}{\sin\theta}\frac{\partial}{\partial\theta}\left(\sin\theta\frac{\partial}{\partial\theta}\right). \tag{7.1.59}$$

For the incident field, let

$$\Phi = \Phi^{(i)}, \qquad \Psi = 0. \tag{7.1.60}$$

For the scattered field, let

$$\Phi = \Phi^{(s)}, \qquad \Psi = \Psi^{(s)}. \tag{7.1.61}$$

For the refracted field, let

$$\Phi = \Phi^{(r)}, \qquad \Psi = \Psi^{(r)}. \tag{7.1.62}$$

Stresses

$$\tau_{rr} = 2\mu\left\{-\frac{\beta^2}{2}\Phi - \frac{2}{r}\frac{\partial\Phi}{\partial r} - \frac{1}{r^2}D_\theta\Phi + \frac{D_\theta}{r}\left(\frac{\partial\Psi}{\partial r} - \frac{\Psi}{r}\right)\right\}, \tag{7.1.63}$$

$$\tau_{\theta\theta} = 2\mu\left\{-\frac{(\beta^2-2\alpha^2)}{2}\Phi + \frac{1}{r}\frac{\partial\Phi}{\partial r} + \frac{1}{r^2}D_\theta\Phi - \frac{\cot\theta}{r^2}\frac{\partial\Phi}{\partial\theta}\right\} +$$

$$+2\mu\left\{\frac{\cot\theta}{r}\frac{\partial}{\partial\theta}D_r\Psi - \frac{1}{r}D_\theta\frac{\partial\Psi}{\partial r}\right\}, \tag{7.1.64}$$

$$\tau_{\phi\phi} = 2\mu\left\{-\frac{(\beta^2-2\alpha^2)}{2}\Phi + \frac{1}{r}\frac{\partial\Phi}{\partial r} + \frac{\cot\theta}{r^2}\frac{\partial\Phi}{\partial\theta} + \frac{D_\theta\Psi}{r^2} - \frac{\cot\theta}{r}\frac{\partial}{\partial\theta}D_r\Psi\right\}, \tag{7.1.65}$$

$$\tau_{r\theta} = \mu\frac{\partial}{\partial\theta}\left(\frac{2}{r}\frac{\partial\Phi}{\partial r} - \frac{2}{r^2}\Phi + \beta^2\Psi + \frac{2}{r}\frac{\partial\Psi}{\partial r} + \frac{2}{r^2}\Psi + \frac{2}{r^2}D_\theta\Psi\right), \tag{7.1.66}$$

where Φ, Ψ take on the superscripts appropriate to the particular field and $\alpha^2,\beta^2 = \alpha_1^2,\beta_1^2$ when the incident or scattered field is used and $\alpha^2,\beta^2 = \alpha_2^2,\beta_2^2$ for the refracted field.

Introducing the symbols ε_{ij}, which are linear combinations of the spherical harmonics defined later, we have for the displacements and stresses in medium 1;

$$\begin{aligned}
\Phi_1 &= \Phi^{(i)} + \Phi^{(s)}, \qquad \Psi_1 = \Psi^{(s)},\\
u_{r_1} &= \frac{1}{r}\sum_{n=0}^{\infty}(-\Phi_0\varepsilon_1 + A_n\varepsilon_{11} + B_n\varepsilon_{12})P_n,\\
u_{\theta_1} &= \frac{1}{r}\sum_{n=0}^{\infty}(-\Phi_0\varepsilon_2 + A_n\varepsilon_{21} + B_n\varepsilon_{22})\frac{\mathrm{d}P_n}{\mathrm{d}\theta},\\
\tau_{rr_1} &= \frac{2\mu_1}{r}\sum_{n=0}^{\infty}(-\Phi_0\varepsilon_3 + A_n\varepsilon_{31} + B_n\varepsilon_{32})P_n,\\
\tau_{r\theta_1} &= \frac{2\mu_1}{r}\sum_{n=0}^{\infty}(-\Phi_0\varepsilon_4 + A_n\varepsilon_{41} + B_n\varepsilon_{42})\frac{\mathrm{d}P_n}{\mathrm{d}\theta}.
\end{aligned} \tag{7.1.67}$$

For medium 2 we have

$$\begin{aligned}
\Phi_2 &= \Phi^{(r)}, \qquad \Psi_2 = \Psi^{(r)},\\
u_{r_2} &= -\frac{1}{r}\sum_{n=0}^{\infty}(C_n\varepsilon_{13} + D_n\varepsilon_{14})P_n,\\
u_{\theta_2} &= -\frac{1}{r}\sum_{n=0}^{\infty}(C_n\varepsilon_{23} + D_n\varepsilon_{24})\frac{\mathrm{d}P_n}{\mathrm{d}\theta},\\
\tau_{rr_2} &= -\frac{2\mu_2}{r}\sum_{n=0}^{\infty}(C_n\varepsilon_{33} + D_n\varepsilon_{34})P_n,\\
\tau_{r\theta_2} &= -\frac{2\mu_2}{r^2}\sum_{n=0}^{\infty}(C_n\varepsilon_{43} + D_n\varepsilon_{44})\frac{\mathrm{d}P_n}{\mathrm{d}\theta},
\end{aligned} \tag{7.1.68}$$

where $P_n = P_n(\cos\theta)$. The ε_j and ε_{ij} are defined as

$$
\begin{aligned}
\varepsilon_1 &= -\mathrm{i}^n(2n+1)\{nj_n(\alpha_1 r)-\alpha_1 r j_{n+1}(\alpha_1 r)\},\\
\varepsilon_2 &= -\mathrm{i}^n(2n+1)j_n(\alpha_1 r),\\
\varepsilon_3 &= -\mathrm{i}^n(2n+1)\{(n^2-n-\tfrac{1}{2}\beta_1^2 r^2)j_n(\alpha_1 r)+2\alpha_1 r j_{n+1}(\alpha_1 r)\},\\
\varepsilon_4 &= -\mathrm{i}^n(2n+1)\{(n-1)j_n(\alpha_1 r)-\alpha_1 r j_{n+1}(\alpha_1 r)\},\\
\varepsilon_{11} &= nh_n(\alpha_1 r)-\alpha_1 r h_{n+1}(\alpha_1 r),\\
\varepsilon_{21} &= h_n(\alpha_1 r),\\
\varepsilon_{31} &= (n^2-n-\tfrac{1}{2}\beta_1^2 r^2)h_n(\alpha_1 r)+2\alpha_1 r h_{n+1}(\alpha_1 r),\\
\varepsilon_{41} &= (n-1)h_n(\alpha_1 r)-\alpha_1 r h_{n+1}(\alpha_1 r),\\
\varepsilon_{12} &= -n(n+1)h_n(\beta_1 r),\\
\varepsilon_{22} &= -(n+1)h_n(\beta_1 r)+\beta_1 r h_{n+1}(\beta_1 r),\\
\varepsilon_{32} &= -n(n+1)\{(n-1)h_n(\beta_1 r)-\beta_1 r h_{n+1}(\beta_1 r)\},\\
\varepsilon_{42} &= -(n^2-1-\tfrac{1}{2}\beta_1^2 r^2)h_n(\beta_1 r)-\beta_1 r h_{n+1}(\beta_1 r),\\
\varepsilon_{13} &= nj_n(\alpha_2 r)-\alpha_2 r j_{n+1}(\alpha_2 r),\\
\varepsilon_{23} &= j_n(\alpha_2 r),\\
\varepsilon_{33} &= (n^2-n-\tfrac{1}{2}\beta_2^2 r^2)j_n(\alpha_2 r)+2\alpha_2 r j_{n+1}(\alpha_2 r),\\
\varepsilon_{43} &= (n-1)j_n(\alpha_2 r)-\alpha_2 r j_{n+1}(\alpha_2 r),\\
\varepsilon_{14} &= -n(n+1)j_n(\beta_2 r),\\
\varepsilon_{24} &= -(n+1)j_n(\beta_2 r)+\beta_2 r j_{n+1}(\beta_2 r),\\
\varepsilon_{34} &= -n(n+1)\{(n-1)j_n(\beta_2 r)-\beta_2 r j_{n+1}(\beta_2 r)\},\\
\varepsilon_{44} &= -(n^2-1-\tfrac{1}{2}\beta_2^2 r^2)j_n(\beta_2 r)-\beta_2 r j_{n+1}(\beta_2 r).
\end{aligned}
\tag{7.1.69}
$$

4. *Boundary conditions.* We shall consider the case of an elastic inclusion, considered to be bonded to the medium. Continuity of the displacement and stress fields gives

$$
\left.
\begin{aligned}
u_r^{(\mathrm{i})}+u_r^{(\mathrm{s})} &= u_r^{(\mathrm{r})},\\
u_\theta^{(\mathrm{i})}+u_\theta^{(\mathrm{s})} &= u_\theta^{(\mathrm{r})},\\
\tau_{rr}^{(\mathrm{i})}+\tau_{rr}^{(\mathrm{s})} &= \tau_{rr}^{(\mathrm{r})},\\
\tau_{r\theta}^{(\mathrm{i})}+\tau_{r\theta}^{(\mathrm{s})} &= \tau_{r\theta}^{(\mathrm{r})}.
\end{aligned}
\right\}\quad r=a
\tag{7.1.70}
$$

Taking u_r as an example, this gives

$$
\sum_{n=0}^{\infty}(-\Phi_0 E_1+A_n E_{11}+B_n E_{12}+C_n E_{13}+D_n E_{14})P_n = 0,
\tag{7.1.71}
$$

where we have let

$$E_i = \varepsilon_i \big|_{r=a}, \qquad E_{ij} = \epsilon_{ij} \big|_{r=a}. \tag{7.1.72}$$

Vanishing of each term of (7.1.71) gives

$$A_n E_{11} + B_n E_{12} + C_n E_{13} + D_n E_{14} = \Phi_0 E_1. \tag{7.1.73}$$

The remaining three boundary conditions yield similar equations. The results, which are four equations in A_n, B_n, C_n, D_n, may be put in matrix form as

$$\begin{bmatrix} E_{11} & E_{12} & E_{13} & E_{14} \\ E_{21} & E_{22} & E_{23} & E_{24} \\ E_{31} & E_{32} & pE_{33} & pE_{34} \\ E_{41} & E_{42} & pE_{43} & pE_{44} \end{bmatrix} \begin{bmatrix} A_n \\ B_n \\ C_n \\ D_n \end{bmatrix} = \Phi_0 \begin{bmatrix} E_1 \\ E_2 \\ E_3 \\ E_4 \end{bmatrix}, \tag{7.1.74}$$

where $p = \mu_2/\mu_1$. Numerical evaluation of the coefficients presents no particular problem, except in the case of very long wavelengths (small arguments) or short wavelengths (large arguments). For those cases, replacement of the various spherical Bessel functions by their asymptotic limits may be necessary.

5. *Special cases.* The generality of the elastic inclusion problem enables a number of useful special cases to be derived, including the cases of a cavity, fluid inclusion, and rigid sphere, as well as these various types of scatterers in a fluid medium. However, except for the cavity case, the remaining cases must be deduced by limiting processes that contain not a few subtleties. Only results will be presented here.

Cavity: The boundary conditions are

$$\begin{aligned} \tau_{rr}^{(i)} + \tau_{rr}^{(s)} &= 0, \\ \tau_{r\theta}^{(i)} + \tau_{r\theta}^{(s)} &= 0, \end{aligned} \qquad r = a, \tag{7.1.75}$$

giving

$$\begin{bmatrix} E_{31} & E_{32} \\ E_{41} & E_{42} \end{bmatrix} \begin{bmatrix} A_n \\ B_n \end{bmatrix} = \Phi_0 \begin{bmatrix} E_3 \\ E_4 \end{bmatrix}. \tag{7.1.76}$$

Fluid inclusion: The boundary conditions are

$$\begin{aligned} u_r^{(i)} + u_r^{(s)} &= u_r^{(r)}, \\ \tau_{rr}^{(i)} + \tau_{rr}^{(s)} &= \tau_{rr}^{(r)}, \\ \tau_{r\theta}^{(i)} + \tau_{r\theta}^{(s)} &= 0, \end{aligned} \qquad r = a. \tag{7.1.77}$$

These conditions reflect the incapability of an inviscid fluid to transmit shear. Hence $\tau_{r\theta}$ must vanish at $r = a$ while the displacements u_θ may be discontinuous. The resulting equations for the coefficients are

$$\begin{bmatrix} E_{11} & E_{12} & E_{13} \\ E_{31} & E_{32} & E_{33}^{f} \\ E_{41} & E_{42} & 0 \end{bmatrix} \begin{bmatrix} A_n \\ B_n \\ C_n \end{bmatrix} = \Phi_0 \begin{bmatrix} E_1 \\ E_2 \\ E_3 \end{bmatrix}, \tag{7.1.78}$$

where

$$E_{33}^{f} = -\tfrac{1}{2}(\rho_2/\rho_1)\beta_1^2 r^2 j_n(\alpha_2 r). \tag{7.1.79}$$

Rigid sphere: Under the action of the impinging waves, the sphere will translate as a rigid body, so that the boundary conditions become

$$\begin{aligned} u_r^{(i)}+u_r^{(s)} &= U_z \cos\theta, \\ u_\theta^{(i)}+u_\theta^{(s)} &= -U_z \sin\theta, \end{aligned} \qquad r = a, \tag{7.1.80}$$

where

$$m\ddot{U}_z = \iint (\tau_{rr}\cos\theta - \tau_{r\theta}\sin\theta)a^2 \sin\theta \, d\theta \, d\phi, \tag{7.1.81}$$

and $m = 4\pi a^3\rho_2/3$. The coefficient matrix is given as

$$\begin{bmatrix} E_{11} & E_{12} \\ E_{21} & E_{22} \end{bmatrix} \begin{bmatrix} A_n \\ B_n \end{bmatrix} = \Phi_0 \begin{bmatrix} E_1 \\ E_2 \end{bmatrix}, \qquad n \neq 1, \tag{7.1.82}$$

while the E_{ij} take on slightly different values than given by (7.1.69) for $n = 1$.

The work of this section represents a formal solution to the stated problem. Unfortunately, numerical results for the various displacement and stress fields are rather scant. The dynamic stress concentration factor has been found by Pao [33] for the case of a hole in a plate, a problem with many similarities to the cylindrical cavity problem. Results tend to show only a 10–15 per cent increase in stress concentration at the most critical frequencies.

A number of other investigations have been carried out on scattering of harmonic waves by cylindrical and spherical obstacles. Ying and Truell [50] and Einspruch and Truell [8] have investigated the scattering of compressional waves by spherical obstacles and obtained results for the scattered field and the scattering cross-section. Knopoff [16] has also investigated this problem and that of shear wave scattering by obstacles [17] and obtained results for the far-field radiation patterns. The case of waves scattered by a hole in a plate has also been studied by McCoy [22]. Chieng [5] has investigated the scattering of elastic waves by multiple cylindrical cavities in a medium. The motion of a cylindrical rigid inclusion under harmonic P and SV waves has been investigated by Miles [26], while Mow [28, 29] has obtained results for the transient motion of a rigid spherical inclusion. Thau and Pao [45, 46, 47] have considered the case of wave scattering by parabolic cylindrical obstacles. Separation of variables and series expansions still pertain for this geometry, but, since the eigenfunctions of the problem are not orthogonal, large systems of equations must be solved simultaneously. Perturbation techniques are also applied in the references cited to supplement the expansion results in the low-frequency range. The investigations cited are only a partial list of the many studies that have been conducted in this area. Mow and Pao [30] have presented an extensive monograph on this subject. We close by noting, as in so many other areas of wave mechanics, that Rayleigh [35] made some of the earliest significant contributions.

7.2. Diffraction of plane waves

We now wish to consider the interaction of propagating plane waves with obstacles of more complex shape than the cylindrical and spherical obstacles of the last section. In particular, consider the presence of sharp edges in the radiation field. Such situations are also associated with diffraction effects in optics, wherein sharp shadows are precluded.

Now a characteristic of the problems of the last section was that the boundaries were such that coordinate systems for which the wave equation was separable were appropriate. There are, however, only eleven coordinate systems for which the scalar wave equation is separable† and when the vector wave equation governs (as in the general elasticity case) separation is possible in even fewer.‡ With the exception of some special axially symmetric problems, this accounts for the geometric shapes that can be treated by separation of variables. It is evident, therefore, that for those problems where the bounding surface is not separable, or 'not quite separable', some different mathematical tools must be used. This results in a certain degree of escalation in mathematical complexity, an unfortunate but unavoidable by-product of considering a wider class of problems.

For such problems, the use of integral-equation techniques offers powerful methods of solution. Included are the Green's function method and the Wiener–Hopf technique. The Green's function approach, which is widely used, will only be briefly discussed here. The Wiener–Hopf technique, also widely used, will be the method applied to the simplest diffraction problem. In addition to these methods, certain variational techniques also pertain.§

Finally, it should be noted that only scalar diffraction problems will be considered in the following. That is, only a single wave equation will be considered. Of course, this is completely descriptive of the acoustical situation or special elastic-wave cases (for example, SH waves impinging on certain boundaries), but is not adequate for the general elastic case, where a scalar and a vector wave equation govern. However, in many instances, the solution techniques developed for simpler problems can be extended to the more complex cases.

7.2.1. *Discussion of the Green's function approach*

The basic idea of the Green's function method is as follows. Suppose it is desired to calculate a field due to a source distribution; first calculate the effects of each elementary portion of source and call this result $G(\mathbf{r}/\mathbf{r}_0)$, where $\mathbf{r}_0$ represents the element at $\mathbf{r}_0$; then the total field at $\mathbf{r}$ due to the

† See Morse and Feshbach [27, pp. 494–523]. They are (1) rectangular, (2) circular cylindrical, (3) elliptic cylindrical, (4) parabolic cylindrical, (5) spherical, (6) conical, (7) parabolic, (8) prolate spheroidal, (9) oblate spheroidal, (10) ellipsoidal, and (11) paraboloidal.

‡ The first five listed in the preceding footnote.

§ See Morse and Feshbach [27, p. 1513–53] as a start in this field.

source distribution $\rho(\mathbf{r}_0)$ is the integral of $G\rho$ over the whole range of $\mathbf{r}_0$. The function $G(\mathbf{r}/\mathbf{r}_0)$ is the Green's function for the problem.

By considering the inhomogeneous Helmholtz equation as an illustration, these points may be brought out. Thus consider the Helmholtz equation

$$\nabla^2\psi+k^2\psi = -4\eta\rho(\mathbf{r}), \tag{7.2.1}$$

which results from the steady-state wave equation. Replace $\rho(\mathbf{r})$ by the loading or source,

$$\rho(\mathbf{r}) = \delta(\mathbf{r}-\mathbf{r}_0). \tag{7.2.2}$$

Then $G(\mathbf{r}/\mathbf{r}_0)$ will be the solution of

$$\nabla^2G+k^2G = -4\pi\delta(\mathbf{r}-\mathbf{r}_0). \tag{7.2.3}$$

Now by manipulating (7.2.1) and (7.2.3) and applying Gauss' theorem, the following integral equation results,†

$$\psi(\mathbf{r}) = \iiint \rho(\mathbf{r}_0)G(\mathbf{r}/\mathbf{r}_0)\,\mathrm{d}v+4\pi\iint\{G(\mathbf{r}/\mathbf{r}_0)\operatorname{grad}\psi^{\mathrm{s}}-\psi^{\mathrm{s}}\operatorname{grad}G(\mathbf{r}/\mathbf{r}_0)\}\,\mathrm{d}A. \tag{7.2.4}$$

Proper consideration of the surface integral in the above enables the boundary conditions appropriate to the given problem to be incorporated.

We will not attempt to go into further detail on this method. The point is, it is basically an integral-equation technique in which the determination of a special solution (the Green's function) enables a great number of additional solutions to be found. For examples of the application of this method to diffraction problems one is referred to Friedlander [10], Knopoff [15], and Thomas [48].

An alternative to the Green's function approach, called the Kirchhoff integral-equation method, has been used by Friedman and Shaw [11], Banaugh [2, 3], and others. As described by Ko and Karlsson [18], the method is based on integral representations of a potential which satisfies a wave equation. A solution for the scattered potential is then given in terms of the distributions of retarded values of the potential and its derivatives on a closed boundary surface together with volume distributions, in the field, of the sources which appear as inhomogeneous terms of the wave equation. Other applications of this technique have been by Fredricks [9] and Shaw [36].

7.2.2. *The Sommerfeld diffraction problem*

We shall now consider a specific diffraction problem as a vehicle for illustrating the general scope of the analysis that must be associated with waves impinging on irregular (that is, inseparable) surfaces. It shall also be used to illustrate a particular integral equation approach known as the Wiener–Hopf technique. This technique provides an extension to the range of problems that may be treated by integral transforms.

† Morse and Feshbach [27, pp. 804–6].

In the problem at hand, plane harmonic waves are considered to impinge on a rigid, half-plane barrier. In this situation, some of the incoming wavefield will reflect from the flat barrier surface, while some will miss the surface completely. Some of the incoming wavefield will strike near the edge of the barrier, and be diffracted into the shadow region behind the barrier. A rough illustration is given in Fig. 7.4. This is the classic Sommerfield diffraction

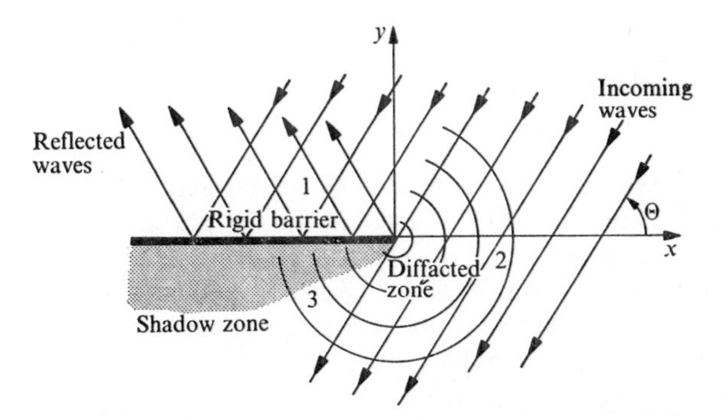

FIG. 7.4. Diffraction of waves by a rigid barrier.

problem, and has been solved by several methods.† The nature of the problem may be one of incident acoustic waves on a rigid barrier, or of incident SH waves on an interior crack. The following analysis of this problem, using the Wiener–Hopf method,‡ is rather involved and uses a number of concepts from complex variable theory. As far as possible, subsidiary discussion, meant to clarify certain points, will be given.

1. *Formulation.* Steady-state plane waves are considered to be propagating in from infinity and striking a rigid barrier. This incident wavefield is given as

$$\phi_i(x, y) = e^{-i(kx\cos\Theta + ky\sin\Theta)} \qquad (0 < \Theta < \pi), \tag{7.2.5}$$

where the time factor $\exp(-i\omega t)$ has been omitted in the preceding and all following expressions. Reflection and diffraction at the barrier will give rise to a diffracted–reflected field $\phi(x, y)$. The total field $\phi_t(x, y)$ will thus be

$$\phi_t(x, y) = \phi(x, y) + \phi_i(x, y), \tag{7.2.6}$$

where ϕ_t satisfies the Helmholtz equation

$$\nabla^2\phi_t + k^2\phi_t = 0, \qquad k^2 = \omega^2/c^2. \tag{7.2.7}$$

Since $\phi_i(x, y)$ satisfies the Helmholtz equation identically, the result (7.2.7) in effect reduces to

$$\nabla^2\phi + k^2\phi = 0. \tag{7.2.8}$$

† See, for example, Stoker [40, pp. 109, 141] or Morse and Feshbach [27, pp. 1383].
‡ See Noble [31] for this and many other diffraction problems solved using this technique.

The incident wave impinging on the barrier yields three regions of wave interaction, labelled 1, 2, and 3 in Fig. 7.4. They consist of the following:
region 1: incident, reflected, and diffracted waves;
region 2: incident and diffracted waves;
region 3: diffracted waves.
The primary boundary condition for the problem is

$$\partial\phi_t/\partial y = 0 \qquad (-\infty < x \leqslant 0, y = 0). \tag{7.2.9}$$

From (7.2.5) and (7.2.6) this gives

$$\partial\phi/\partial y = ik \sin \Theta e^{-ikx\cos\Theta} \qquad (y = 0, -\infty < x \leqslant 0). \tag{7.2.10}$$

(In terms of SH elastic waves, this would correspond to $\partial u/\partial y = 0$; that is, a stress-free boundary or slit.) In addition to this primary condition, which is the obvious one and serves to define the problem, a number of additional conditions, such as continuity or rate of decay, must be imposed on the field quantity ϕ. These conditions will be summarized in the following. The various roles these conditions play will be brought out as the analysis proceeds.

Continuity of ϕ_t, ϕ. We require that $\partial\phi_t/\partial y$ and therefore $\partial\phi/\partial y$ be continuous for $-\infty < x < \infty$ at $y = 0$. It is also required that ϕ_t, and therefore ϕ, be continuous for $0 < x < +\infty$. However, ϕ_t will be discontinuous at $y = 0$, $-\infty < x \leqslant 0$ (that is, on the barrier). This behaviour is illustrated in Fig. 7.5.

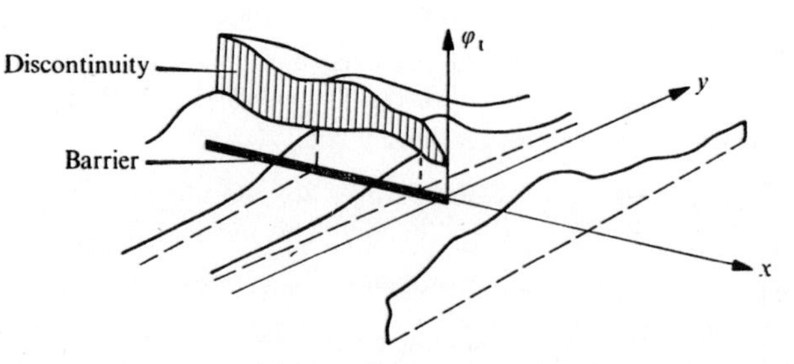

FIG. 7.5. Illustration of the discontinuity in ϕ_t that occurs at the barrier.

Bounds on ϕ. In determining the convergence of Fourier integrals that arise later, it is of value to put upper bounds on the diffracted field ϕ. Consider first regions 1 and 3 of Fig. 7.5. In region 1 as $x \to -\infty$ for any $y > 0$, the total field will be given by the reflected field. Hence, letting k be complex

$$k = k_1 + ik_2, \qquad k_1, k_2 > 0, \tag{7.2.11}$$

we have that $\phi_t \simeq \phi$, and that

$$|\phi| < C_1 \exp(k_2 x \cos \Theta), \qquad y > 0, x \to -\infty. \tag{7.2.12}$$

Similarly, in region 3, as $x \to -\infty$, $y < 0$, which is the shadow region, only a diffracted wave is present. But, as will be shown presently, this will be negligible at large values, so that

$$\phi = \phi_t - \phi_i \to \phi_i, \qquad x \to -\infty, y < 0. \tag{7.2.13}$$

Hence

$$|\phi| < C_2 \exp(k_2 x \cos \Theta), \qquad y < 0, x \to -\infty. \tag{7.2.14}$$

(The constants arising in the above and in the immediately following are arbitrary.)

As $x \to +\infty$, the diffracted field in terms of far-field effects may be considered to be produced by a source placed at the leading edge ($x = 0$) of the barrier. The approximate differential equation for this situation would be

$$\nabla^2\phi + k^2\phi = -4\pi\delta(x)\delta(y), \tag{7.2.15}$$

which has the solution

$$\phi(x, y) = \pi i H_0^{(1)}(kr) \qquad r^2 = x^2 + y^2. \tag{7.2.16}$$

The asymptotic behaviour of the Hankel function has been considered previously and is given by

$$\lim_{r \to \infty} H_0^{(1)}(kr) \sim C_3 r^{-\frac{1}{2}} \exp(-k_2 r)\exp(ik_1 r). \tag{7.2.17}$$

Hence we may state

$$|\phi| < C_4 \exp\{-k_2(x^2+y^2)^{\frac{1}{2}}\}, \qquad y \lessgtr 0, x \to +\infty. \tag{7.2.18}$$

Edge conditions. It would not be surprising if the solution had a singularity at the origin, where the barrier edge is located. Generally some type of physical arguments must be employed. The importance of these conditions, which are presented in the summary following, is in establishing the uniqueness of the final solution.

2. *Summary of equations and boundary conditions.*
(a) The total field ϕ_t is given by

$$\phi_t = \phi_i + \phi, \tag{7.2.19}$$

where

$$\phi_i = e^{-i(kx \cos \Theta + ky \sin \Theta)}, \qquad k = k_1 + ik_2. \tag{7.2.20}$$

Determine ϕ, which satisfies

$$\nabla^2\phi + k^2\phi = 0 \tag{7.2.21}$$

and the following boundary conditions.
(b) $y = 0$, $-\infty < x \leqslant 0$,

$$\partial\phi_t/\partial y = 0, \tag{7.2.22}$$

so that

$$\partial\phi/\partial y = ik \sin \Theta e^{-ikx\cos\Theta}. \tag{7.2.23}$$

Also, $\partial\phi_t/\partial y$, hence $\partial\phi/\partial y$, continuous for $y = 0$, $-\infty < x < \infty$.

(c) $y = 0$, $0 < x < \infty$, ϕ_t, ϕ continuous.

(d) For $y \gtrless 0$

$$\begin{aligned} &\text{(i) } x \to -\infty, && |\phi| < C_1 \exp(k_2 \cos\Theta), \\ &\text{(ii) } x \to +\infty, && |\phi| < C_2 \exp\{-k_2(x^2+y^2)^{\frac{1}{2}}\}. \end{aligned} \tag{7.2.24}$$

(e) At edge

$$\begin{aligned} \partial\phi_t/\partial y &\to C_3 x^{-\frac{1}{2}}, && x \to +0,\ y = 0, \\ \phi_t &\to C_4, && x \to +0,\ y = 0, \\ \phi_t &\to C_5, && x \to -0,\ y = +0, \\ \phi_t &\to C_6, && x \to -0,\ y = -0. \end{aligned} \tag{7.2.25}$$

3. *Application of Fourier transform.* We seek to apply the Fourier transform, defined as

$$F(\alpha, y) = \frac{1}{\sqrt{(2\pi)}} \int_{-\infty}^{\infty} f(x, y) e^{i\alpha x}\, dx, \tag{7.2.26}$$

with inverse

$$f(x, y) = \frac{1}{\sqrt{(2\pi)}} \int_{-\infty}^{\infty} F(\alpha, y) e^{-i\alpha x}\, d\alpha, \tag{7.2.27}$$

where α is complex and given as $\alpha = \sigma + i\tau$. In the present problem, the semi-infinite nature of the barrier and consequent discontinuity in ϕ leads us to break up the Fourier transform as follows:

$$\Phi(\alpha, y) = \Phi_+(\alpha, y) + \Phi_-(\alpha, y) = \frac{1}{\sqrt{(2\pi)}} \int_{-\infty}^{\infty} \phi(x, y) e^{i\alpha x}\, dx, \tag{7.2.28}$$

where

$$\Phi_+(\alpha, y) = \frac{1}{\sqrt{(2\pi)}} \int_{0}^{\infty} \phi e^{i\alpha x}\, dx, \qquad \Phi_-(\alpha, y) = \frac{1}{\sqrt{(2\pi)}} \int_{-\infty}^{0} \phi e^{i\alpha x}\, dx. \tag{7.2.29}$$

It is important to know the region of regularity, or analyticity, of the transformed quantity Φ in the complex α-plane. From the bounds on Φ given in (7.2.24) it may be easily established that

$$\begin{aligned} &\Phi_+ \text{ analytic for } \tau > -k_2, \\ &\Phi_- \text{ analytic for } \tau < k_2 \cos\Theta. \end{aligned} \tag{7.2.30}$$

Thus, as shown in Fig. 7.6, the regions of analyticity of Φ_+, Φ_- overlap so that the entire function $\Phi(\alpha, y)$ is regular in the strip $-k_2 < \tau < k_2 \cos\Theta$.

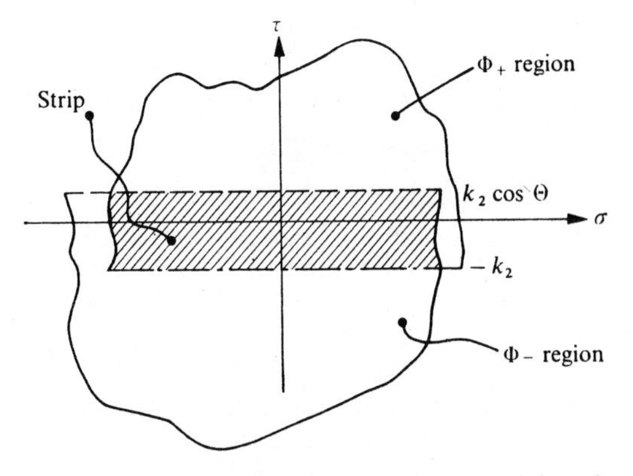

FIG. 7.6. Regions where Φ_+, Φ_- are analytic in the α-plane, and the strip where both are commonly analytic.

This discussion on regions of regularity of the Fourier transform is not peculiar to the Wiener–Hopf development, but pertains to any application of the transform. In the present case the regions are established as follows. Consider Φ_+ and $\alpha = \sigma + i\tau$. Then the first of (7.2.29) becomes

$$\Phi_+ = \int_0^\infty \phi e^{-\tau x}\, dx, \tag{7.2.31}$$

where the $e^{i\sigma x}$ has been dropped since it does not affect the convergence of the integral. As $x \to +\infty$, $|\phi| < \exp(-k_2 x)$, according to the boundary conditions. Hence

$$\Phi_+ \cong \int_0^\infty \exp\{-(\tau + k_2)x\}\, dx, \tag{7.2.32}$$

which exists as long as $\tau + k_2 > 0$, or $\tau > -k_2$. In other words, Φ_+ is analytic in the half-plane $\tau > -k_2$. The condition on Φ_- follows a similar development, except the boundary condition $|\phi| < \exp(-k_2 x \cos \Theta)$ as $x \to -\infty$ is used.

The Fourier transform of the Helmholtz equation then leads to

$$\frac{d^2\Phi(\alpha, y)}{dy^2} - \gamma^2\Phi(\alpha, y) = 0, \qquad \gamma^2 = \alpha^2 - k^2, \tag{7.2.33}$$

where Φ is defined in the sense of (7.2.28).

4. *Transformed solution—with some boundary conditions applied.* The differential equation (7.2.33) is easily solved. However, it must be recognized that $\phi(x, y)$, hence $\Phi(\alpha, y)$, is discontinuous across $y = 0$. Although this discontinuity is only for $x < 0$, we are forced at this point to admit two

solutions, for $y \geq 0$ and $y \leq 0$. Thus we have

$$\Phi(\alpha, y) = \begin{cases} A_1(\alpha)e^{-\gamma y} + B_1(\alpha)e^{\gamma y}, & y \geqslant 0 \\ A_2(\alpha)e^{-\gamma y} + B_2(\alpha)e^{\gamma y}, & y \leqslant 0. \end{cases} \tag{7.2.34}$$

First we will impose a radiation-type condition at infinity, where $y \to \pm\infty$. Thus we want ϕ to decay to zero far from the barrier. At first glance, it would appear that $y \to +\infty$ would require $B_1(\alpha) = 0$ and, as $y \to -\infty$, $A_2(\alpha) = 0$. Although these are the correct choices, the rational selection of the zero coefficients follows a more recondite line of reasoning.

Thus, selection of $B_1 = A_2 = 0$ in the preceding would imply that the real part of γ (which is complex) is greater than zero. This is not an obvious fact and requires closer consideration of γ. Now

$$\gamma = (\alpha^2 - k^2)^{\frac{1}{2}} = (\alpha + k)^{\frac{1}{2}}(\alpha - k)^{\frac{1}{2}}, \tag{7.2.35}$$

where γ is a multi-valued function with branch points at $\pm k$. In order to make these functions single-valued, branch cuts are incorporated in the complex plane. Although the appropriate cuts for the functions encountered in our analysis will be the subject of additional later discussion, it will suffice presently to indicate their selection, as shown in Fig. 7.7. In the figure, branch

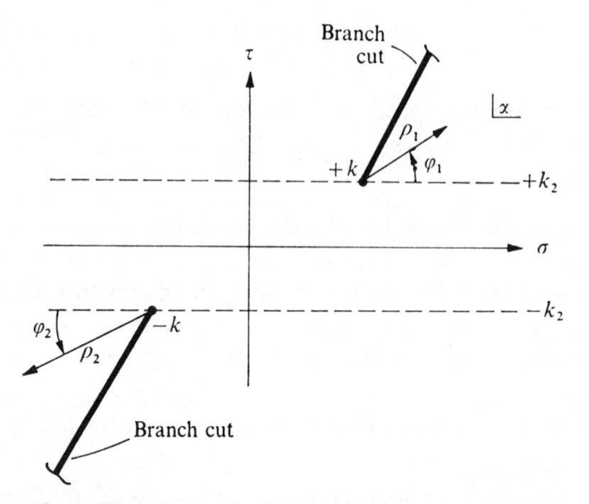

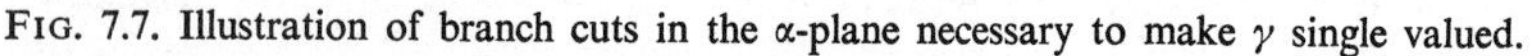
FIG. 7.7. Illustration of branch cuts in the α-plane necessary to make γ single valued.

points are at $\pm k$ and branch cuts extend from these points to $+\infty$ (from $+k$) and $-\infty$ (from $-k$). Using the coordinates ρ_1, ρ_2, ϕ_1, ϕ_2, we have

$$\alpha = k + \rho_1 \exp(i\phi_1), \qquad \alpha = -k + \rho_2 \exp(i\phi_2). \tag{7.2.36}$$

Thus

$$\gamma = \{\rho_1 \exp(i\phi_1)\}^{\frac{1}{2}}\{\rho_2 \exp(i\phi_2)\}^{\frac{1}{2}} = (\rho_1\rho_2)^{\frac{1}{2}} \exp\{\tfrac{1}{2}i(\phi_1 + \phi_2)\}. \tag{7.2.37}$$

Now the relevant statement for the present problem is that for α in the strip $-k_2 < \text{Im}(\alpha) < +k_2$, which encompasses the region of analyticity, then $\text{Re}(\gamma) > 0$. From (7.2.37) this implies $\cos(\phi_1+\phi_2)/2 > 0$. In the figure, $\pm k$ have been located at $\pm k_2$, the maximum permissible distances from the σ-axis. For these cases, with positive ϕ_1, ϕ_2 defined as shown, it should be evident that, as $\sigma \to \pm\infty$, $\cos(\phi_1+\phi_2)/2 \to 0$, and $\cos(\phi_1+\phi_2)/2 > 0$ for lesser values of σ.

With $\text{Re}\,\gamma > 0$ established, we are justified in reducing (7.2.34) to

$$\Phi(\alpha, y) = \begin{cases} A_1(\alpha)e^{-\gamma y}, & y \geqslant 0 \\ B_2(\alpha)e^{\gamma y}, & y \leqslant 0. \end{cases} \tag{7.2.38}$$

From boundary condition (b) regarding the continuity of $\partial\phi/\partial y$, hence $\partial\Phi/\partial y$, for $-\infty < x < \infty$, we have

$$\partial\Phi(\alpha, +0)/\partial y = \partial\Phi(\alpha, -0)/\partial y, \tag{7.2.39}$$

giving

$$-A_1(\alpha) = B_2(\alpha) = -A(\alpha). \tag{7.2.40}$$

Hence

$$\Phi(\alpha, y) = \begin{cases} A(\alpha)e^{-\gamma y}, & y \geqslant 0 \\ -A(\alpha)e^{\gamma y}, & y \leqslant 0. \end{cases} \tag{7.2.41}$$

5. *Introduction of* $\Phi_+(0)$, $\Phi_-(+0)$ *and the remaining boundary conditions.* Up to this point, we have essentially 'followed our nose' in the application of the Fourier transform to a problem with peculiar boundary conditions and, although the application of the Wiener–Hopf technique has been threatened, the analysis till now has not deviated from standard transform theory. However, the need for deviation has arisen in the following fashion. The discontinuity in $\phi(x, y)$ as x goes from > 0 to <0 has forced the definition of $\Phi(\alpha, y)$ in the Φ_+, Φ_- sense, as given by (7.2.29). Thus, having started with the transform and inverse transform defined on the infinite interval

$$(-\infty < x < \infty),$$

we have ended up with transform expressions defined on the semi-infinite intervals, $0 < x < \infty$, $-\infty < x < 0$. The basic idea of the Wiener–Hopf technique is to extend functions defined in one region ($-\infty < x < 0$, say) into a hitherto undefined region ($0 < x < \infty$, say), so that complete functions ($-\infty < x < \infty$) may result. Some of the characteristic means of analysis of implementing this simple concept constitute the apparatus of the Wiener–Hopf technique.

For brevity of notation in the following, we shall replace $\Phi(\alpha, y)$ by $\Phi(\alpha)$, or $\Phi(y)$ when there is no risk of confusion. Thus, for example,

$$\Phi_+(\alpha, 0) \to \Phi_+(0). \tag{7.2.42}$$

We now define

$$\Phi_-(\pm 0) = \lim_{y\to\pm 0} \frac{1}{\sqrt{(2\pi)}} \int_{-\infty}^{0} \phi e^{i\alpha x}\, dx,$$

$$\Phi_+(\pm 0) = \lim_{y\to\pm 0} \frac{1}{\sqrt{(2\pi)}} \int_{0}^{\infty} \phi e^{i\alpha x}\, dx, \tag{7.2.43}$$

$$\Phi'_+(\alpha, y) = \frac{1}{\sqrt{(2\pi)}} \int_{0}^{\infty} \frac{\partial\phi}{\partial y} e^{i\alpha x}\, dx,$$

$$\Phi'_-(\alpha, y) = \frac{1}{\sqrt{(2\pi)}} \int_{-\infty}^{0} \frac{\partial\phi}{\partial y} e^{i\alpha x}\, dx. \tag{7.2.44}$$

The definitions of $\Phi'_+(\pm 0)$, $\Phi'_-(\pm 0)$ are similar to those of (7.2.29).

Before applying the boundary conditions at $y = 0$, we make the following observation: continuity of $\phi(x, y)$, $\partial\phi/\partial y$ for $x > 0$ and continuity of $\partial\phi/\partial y$ for $x < 0$, all at $y = 0$, gives

$$\begin{aligned} \Phi_+(+0) &= \Phi_+(-0) = \Phi_+(0), \\ \Phi'_+(+0) &= \Phi_+(-0) = \Phi'_+(0), \\ \Phi'_-(+0) &= \Phi'_-(-0) = \Phi'_-(0). \end{aligned} \tag{7.2.45}$$

We now enforce the conditions of continuity across $y = 0$ in the following essential statements:

$$\begin{aligned} \Phi_+(0) + \Phi_-(+0) &= A(\alpha), \\ \Phi_+(0) + \Phi_-(-0) &= -A(\alpha), \\ \Phi'_+(0) + \Phi'_-(0) &= -\gamma A(\alpha). \end{aligned} \tag{7.2.46}$$

In interpreting these statements, we should focus attention on the third equation of (7.2.46). In applying the conditions on $\partial\phi/\partial y$ for $x < 0$ (see (b) of (2) summary of equations and boundary conditions), and hence on Φ', we see that only a portion of Φ' is known, namely $\Phi'_-(0)$. The Wiener–Hopf technique continues a function from its known domain ($x < 0$ here) to the entire domain ($x > 0$). This effectively introduces the unknown $\Phi'_+(0)$; the Wiener–Hopf method seeks to determine it by the process of analytic continuation. With $\Phi'_1(0)$ known, $A(\alpha)$ is known.

6. *Reduction to standard Wiener–Hopf equations.* We will be applying the process of analytic continuation to the various functions of (7.2.46). Hence, it will be desirable to work only with functions whose domain of regularity is known. Since $A(\alpha)$ is unknown in all respects, it will be eliminated from the above. Adding the first two equations of (7.2.46), we obtain

$$2\Phi_+(0) = -\Phi_-(+0) - \Phi_-(-0). \tag{7.2.47}$$

Next, subtract the second from the first of (7.2.46) to yield an equation for $A(\alpha)$. Substitute this in the third of (7.2.46) to give

$$\Phi'_+(0)+\Phi'_-(0) = -\frac{\gamma}{2}\{\Phi_-(+0)-\Phi_-(-0)\}. \tag{7.2.48}$$

Now, as has been mentioned, $\Phi'_-(0)$ is known. Thus, from the boundary conditions,

$$\Phi'_-(0) = \frac{1}{\sqrt{(2\pi)}}\int\limits_{-\infty}^{0} e^{i\alpha x}(ik\sin\Theta e^{-ikx\cos\Theta})\,dx,$$

$$= \frac{ik\sin\Theta}{\sqrt{(2\pi)}}\int\limits_{-\infty}^{0} e^{ix(\alpha-k\cos\Theta)}\,dx. \tag{7.2.49}$$

This integral is readily found by contour integration† to give

$$\Phi'_-(0) = \frac{k\sin\Theta}{\sqrt{(2\pi)(\alpha-k\cos\Theta)}}. \tag{7.2.50}$$

Now, from (7.2.47) and (7.2.48) we have the sum and difference of two quantities $\Phi_-(+0)$ and $\Phi_-(-0)$. Define

$$2D_- = \Phi_-(+0)-\Phi_-(-0), \qquad 2S_- = \Phi_-(+0)+\Phi_-(-0). \tag{7.2.51}$$

With these definitions and (7.2.50), we obtain the two equations

$$\begin{aligned} \Phi_+(0) &= -S_-, \\ \Phi'_+(0)+\frac{k\sin\Theta}{\sqrt{(2\pi)(\alpha-k\cos\Theta)}} &= -\gamma D_-. \end{aligned} \tag{7.2.52}$$

These equations have been put in the 'standard' Weiner–Hopf form of

$$R(\alpha)\Phi_+(\alpha)+S(\alpha)\Psi_-(\alpha)+T(\alpha) = 0, \tag{7.2.53}$$

where R, S, T are known functions and Φ_+, Ψ_- are unknown functions, known only to be regular in the upper half-plane (Φ_+) and lower half-plane (Ψ_-), respectively. Thus, in the present case, we have from the first and second equations, respectively, of (7.2.52)

$$\begin{aligned} &R(\alpha) = 1, \quad S(\alpha) = 1, \quad T(\alpha) = 0, \\ &R(\alpha) = 1, \quad S(\alpha) = \gamma, \quad T(\alpha) = k\sin\Theta/\sqrt{(2\pi)(\alpha-k\cos\Theta)}. \end{aligned} \tag{7.2.54}$$

In our case, S_-, D_- are unknown (as are Φ'_+, Φ_+), but regular in $\tau < k_2\cos\Theta$.

† Thus, replace x by $z = x+iy$ and integrate around the contour $-\infty < x < 0$, $0 < iy < +\infty$, and a quarter-circle from $+i\infty$ to $-\infty$.

7. *Outline of the fundamental step.* We are now on the verge of applying the fundamental step of the Wiener–Hopf procedure. This is to find a function $K_+(\alpha)$, and $K_-(\alpha)$ such that

$$R(\alpha)/S(\alpha) = K_+(\alpha)/K_-(\alpha), \tag{7.2.55}$$

where $K_+(\alpha)$ is regular in $\tau > \tau_-$ ($-k_2$ in our case), and $K_-(\alpha)$ is regular in $\tau < \tau_+$ ($k_2 \cos \Theta$ in our case). Then (7.2.53) becomes

$$K_+(\alpha)\Phi_+(\alpha)+K_-(\alpha)\Psi_-(\alpha)+K_-(\alpha)T(\alpha)/S(\alpha) = 0. \tag{7.2.56}$$

Decompose K_-T/S into the form

$$K_-(\alpha)T(\alpha)/S(\alpha) = H_+(\alpha)+H_-(\alpha). \tag{7.2.57}$$

Then (7.2.56) may be written as

$$K_+(\alpha)\Phi_+(\alpha)+H_+(\alpha) = -K_-(\alpha)\Psi_-(\alpha)-H_-(\alpha) = J(\alpha). \tag{7.2.58}$$

By this rearrangement we have defined a new function $J(\alpha)$. Supposedly $J(\alpha)$ is regular only in the strip $\tau_- < \tau < \tau_+$ ($k_2 < \tau < k_2 \cos \Theta$ in our case). However, $K_+\Phi_++H_+$ is regular in the upper half-plane and $K_-\Psi_-+H_-$ is regular in the lower half-plane. Hence, by analytic continuation the function is regular over the whole α-plane.

The next step is to determine $J(\alpha)$, and this is done by applying boundary conditions (in our case, 2(e) (see p. 414)) and a form of Liouville's theorem from complex variables. Once $J(\alpha)$ is known solutions for Φ_+, Ψ_- follow from (7.2.58), so that, in principle, the problem is solved. Thus, in our present diffraction problem, steps 1–6 have been for the purpose of putting everything in the Wiener–Hopf form.

We note that the concept of analytic continuation of functions in the complex plane, referred to several times thus far, is discussed in any book on complex-variable theory.† Basically, the idea is the following: suppose a function $f_1(\alpha)$ is regular in region D_1 of the α-plane and $f_2(\alpha)$ is regular in region D_2 and that D_1, D_2 overlap; suppose $f_1 = f_2$ in the overlap region, then f_2 is said to be the analytic continuation of f_1 into D_2. A standard example illustrating this concept is the function

$$f_1(z) = 1+z+z^2 \ldots = \sum_{n=0}^{\infty} z^n, \qquad |z| < 1. \tag{7.2.59}$$

But, consider

$$f_2(z) = \frac{1}{1-z}, \qquad z \neq 1, \tag{7.2.60}$$

and note that

$$f_1(z) = f_2(z), \qquad |z| < 1. \tag{7.2.61}$$

Then $f_2(z)$ is the analytic continuation of $f_1(z)$ into the remainder of the complex plane $z \neq 1$.

† See, for example, Churchill [6, pp. 188–94] or Morse and Feshbach [27, pp. 389–98].

8. *Factorization of the Wiener–Hopf equations.* Consider the second of (7.2.52), and note that $\gamma = (\alpha-k)^{\frac{1}{2}}(\alpha+k)^{\frac{1}{2}}$, so that we may write

$$\frac{\Phi'_+(0)}{(\alpha+k)^{\frac{1}{2}}}+(\alpha-k)^{\frac{1}{2}}D_-+\frac{k\sin\Theta}{\sqrt{(2\pi)}(\alpha+k)^{\frac{1}{2}}(\alpha-k\cos\Theta)}=0. \qquad (7.2.62)$$

Now, in factoring γ in this way, we define the branches of the factors as follows:

$$\begin{aligned}\sigma &\to +\infty, \qquad (\alpha-k)^{\frac{1}{2}} \to \alpha^{\frac{1}{2}},\\ \sigma &\to -\infty, \qquad (\alpha+k)^{\frac{1}{2}} \to \alpha^{\frac{1}{2}}.\end{aligned} \qquad (-k_2 < \tau < k_2) \qquad (7.2.63)$$

Then the factor $(\alpha+k)^{\frac{1}{2}}$ is regular for $\tau > -k_2$, while $(\alpha-k)^{\frac{1}{2}}$ is regular for $\tau < +k_2$. Consequently, (7.2.62) has been brought to the general stage of the Wiener–Hopf procedure given by (7.2.56).

(*Remark.* Additional discussion on the branches of γ, referred to in the preceding and first referred to with reference to Fig. 7.7 is warranted. The liberty will be taken of reviewing the basic aspects of the subject as presented in complex variable theory.

Thus consider $f(\alpha) = \alpha^{\frac{1}{2}}$, and let $\alpha = \rho\exp(\mathrm{i}\phi)$. This function is multi-valued, since $\rho\exp\{\mathrm{i}(\phi+2n\pi)\} = \rho\exp(\mathrm{i}\phi)$, so that

$$f(\alpha) = \rho^{\frac{1}{2}}\mathrm{e}^{\mathrm{i}\phi/2}, \qquad \rho^{\frac{1}{2}}\mathrm{e}^{\mathrm{i}(\phi/2+\pi)},\ldots.$$

We may make $f(\alpha)$ single-valued by requiring $-\pi < \phi < \pi$, as shown in Fig. 7.8(a). With this restriction, the first branch of $f(\alpha)$ is generated in the $f(\alpha)$-plane (shaded region of Fig. 7.8(b)). This branch may be identified by the fact that $+p$ in the α-plane is taken to $+p^{\frac{1}{2}}$. The location of the branch cut is arbitrary, and the effects of other choices are shown in Fig. 7.8(c) and (d).

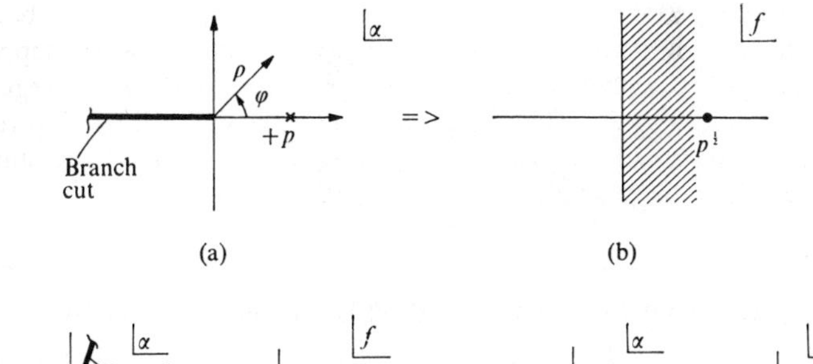

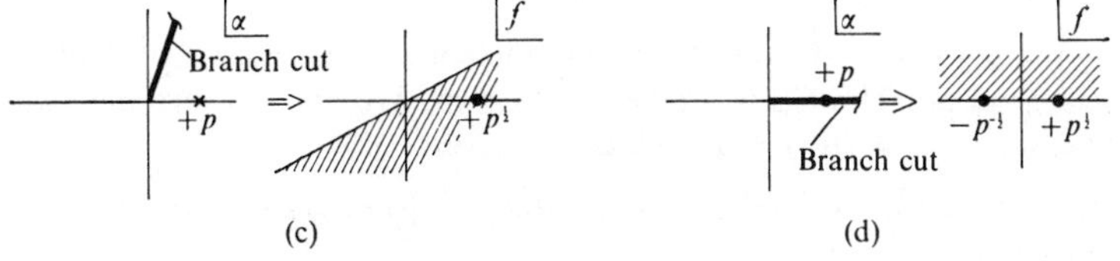

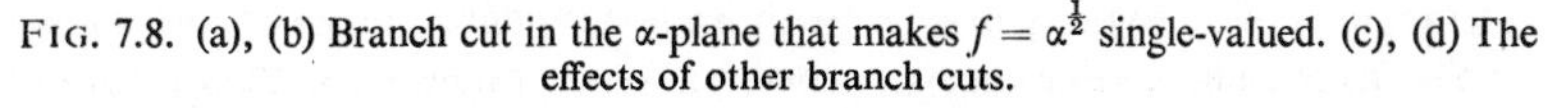

FIG. 7.8. (a), (b) Branch cut in the α-plane that makes $f = \alpha^{\frac{1}{2}}$ single-valued. (c), (d) The effects of other branch cuts.

Now, again referring to (a) in the preceding figure, the second branch of $f(\alpha)$ is obtained by restricting ϕ such that

$$\pi < \phi < 2\pi, \qquad -2\pi < \phi < -\pi,$$

This would generate the region $\tau < 0$ in Fig. 7.8(b) with similar restrictions on ϕ generating the other branches of Fig. 7.8(c) and (d).

To extend this, consider the two functions

$$\chi = (\alpha)^{\frac{1}{2}}, \qquad \psi = (-\alpha)^{\frac{1}{2}}.$$

Consider the first two branches of each of these functions, where the branch cut, common for both, has been taken in the upper half α-plane, as in Fig. 7.8(a). Select the branches so that

(1) if p = positive real number, $\chi = +p^{\frac{1}{2}}$;
(2) if $\alpha = -p$, $\psi = +p^{\frac{1}{2}}$.

This means selecting the first branch of χ and the second branch of ψ, as may be seen in the illustrations of Fig. 7.9(a) and (b). Thus the first branch

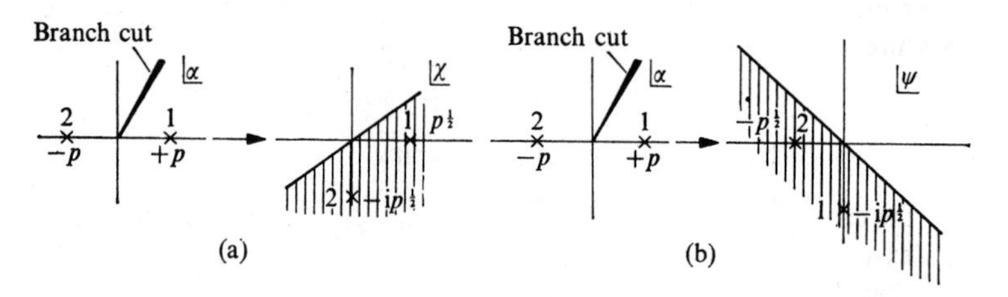

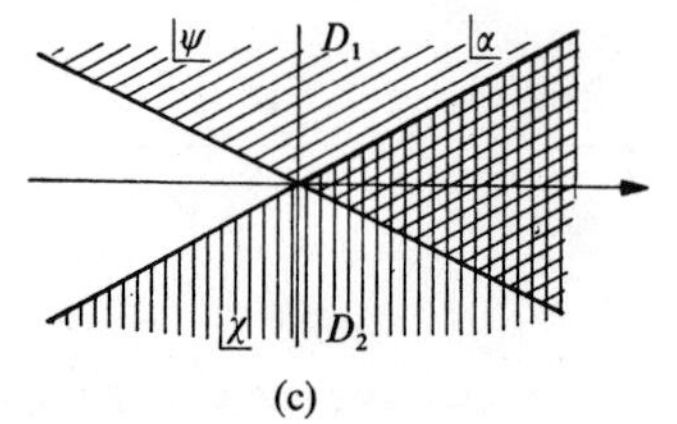

FIG. 7.9. (a), (b) Branches of χ, ψ. (c) The common zone of analyticity.

of χ takes $+p$ to $+p^{\frac{1}{2}}$, while the first branch of ψ takes $-p$ to $-p^{\frac{1}{2}}$. However, the second branch of ψ will take $-p$ to $+p^{\frac{1}{2}}$. We now observe that, along the real axis ($\tau = 0$), $\chi = \mathrm{i}\psi$. Thus, $+p$ gives

$$\chi = p^{\frac{1}{2}}, \qquad \psi = -\mathrm{i}p^{\frac{1}{2}}, \qquad \mathrm{i}\psi = +p^{\frac{1}{2}}.$$

The regions of analyticity of χ, ψ, with branches defined as given, are shown in Fig. 7.9(c). The double-cross-hatched region represents the common zone of of analyticity of χ and ψ. However, $\chi = i\psi$ along the real axis. Then, by a basic theorem of analytic continuation, χ is the continuation of ψ from D_2 to D_1 (and vice versa).

Although this review has been rather lengthy, it provides the necessary background for (7.2.63). Thus, we have factored γ as

$$\gamma = \chi_1\chi_2, \qquad \chi_1 = (\alpha-k)^{\frac{1}{2}}, \qquad \chi_2 = (\alpha+k)^{\frac{1}{2}}.$$

The cuts for χ_1, χ_2 have been shown previously in Fig. 7.7. With these definitions, it should be apparent that the first branches of χ_1, χ_2 are appropriate for (7.2.62).)

Hence, to continue, we see that (7.2.62) is in a form such that $\Phi_+'/(\alpha+k)^{\frac{1}{2}}$ is regular for $\tau > -k_2$, while $(\alpha-k)^{\frac{1}{2}}D_-$ is regular for $\tau < +k_2$.

We now consider the third term of the right-hand side of (7.2.62), with the idea of decomposing it to the form $H_+(\alpha)$, $H_-(\alpha)$ given in the Wiener–Hopf procedure outline step (7.2.57). In particular, we first note that this term, because of the pole at $\alpha = k\cos\Theta$ and the branch point of $(\alpha+k)^{\frac{1}{2}}$, is regular only in the strip $-k_2 < \tau < k_2\cos\Theta$. But, following the previous procedure, we wish to resolve

$$\frac{k\sin\Theta}{\sqrt{(2\pi)}(\alpha+k)^{\frac{1}{2}}(\alpha-k\cos\Theta)} = H_+(\alpha)+H_-(\alpha). \tag{7.2.64}$$

As in the case of much previous work in this section, the resolution relies on complex-variable theory.

The basic theorem on which the resolution relies is stated approximately as follows:

THEOREM *Let $f(\alpha)$ be analytic in the strip $\tau_- < \tau < \tau_+$ and bounded as $|\sigma| \to \infty$. Then for $\tau_- < c < \tau < d < \tau_+$, we have*

$$f(\alpha) = f_+(\alpha)+f_-(\alpha),$$

$$f_+(\alpha) = \frac{1}{2\pi i}\int_{-\infty+ic}^{-\infty+ic}\frac{f(\zeta)}{\zeta-\alpha}\,d\zeta, \qquad f_-(\alpha) = -\frac{1}{2\pi i}\int_{-\infty+id}^{+\infty+id}\frac{f(\zeta)}{\zeta-\alpha}\,d\zeta.$$

where $f_+(\alpha)$ is regular for all $\tau > \tau_-$, $f_-(\alpha)$ is regular for all $\tau < \tau_+$.

To implement the preceding theorem in the present case, we select contours and cuts as shown in Fig. 7.10. The point α is taken in the strip

$$-k_2 < \tau < k_2\cos\Theta.$$

Consider the case of $f_+(\alpha)$.

$$\frac{1}{\sqrt{(2\pi)}}\int_{-\infty-ik_2}^{+\infty-ik_2}\frac{k\sin\Theta\,d\zeta}{(\zeta+k)^{\frac{1}{2}}(\zeta-k\cos\Theta)(\zeta-\alpha)}+\int_{\Gamma+} = 2\pi i\sum \text{Res.} \tag{7.2.65}$$

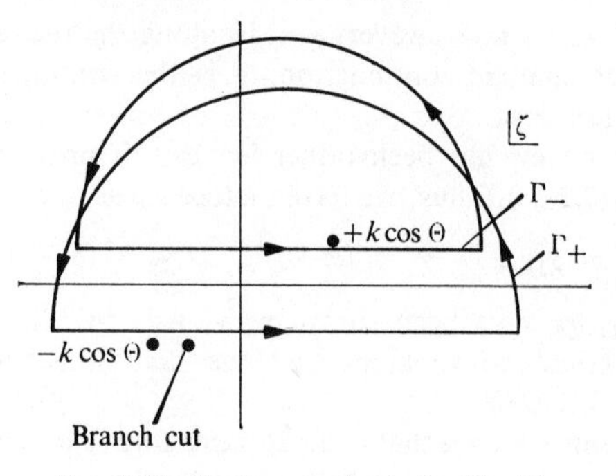

FIG. 7.10. Contours and cuts for H_+, H_-.

The integral about Γ_+ vanishes for the contour radius approaching infinity. Simple poles exist at $\zeta = \alpha$, $k\cos\Theta$, which have residues of

$$\text{Res} = \frac{k\sin\Theta}{(\alpha+k)^{\frac{1}{2}}(\alpha-k\cos\Theta)},\ \frac{k\sin\Theta}{(k\cos\Theta+k)^{\frac{1}{2}}(k\cos\Theta-\alpha)}, \tag{7.2.66}$$

respectively. Then $H_+(\alpha)$ becomes

$$H_+(\alpha) = \frac{k\sin\Theta}{\sqrt{(2\pi)}(\alpha-k\cos\Theta)}\left\{\frac{1}{(\alpha+k)^{\frac{1}{2}}}-\frac{1}{(k+k\cos\Theta)^{\frac{1}{2}}}\right\}. \tag{7.2.67}$$

In a similar fashion $H_-(\alpha)$ may be evaluated, where only a singularity $\zeta = k\cos\Theta$ is contained in the integral. The results are

$$\frac{k\sin\Theta}{\sqrt{(2\pi)}(\alpha+k)^{\frac{1}{2}}(\alpha-k\cos\Theta)} = \frac{k\sin\Theta}{\sqrt{(2\pi)}(\alpha-k\cos\Theta)}\left\{\frac{1}{(\alpha+k)^{\frac{1}{2}}}-\frac{1}{(k+k\cos\Theta)^{\frac{1}{2}}}\right\}+$$
$$+\frac{k\sin\Theta}{\sqrt{(2\pi)}(k+k\cos\Theta)^{\frac{1}{2}}(\alpha-k\cos\Theta)}$$
$$= H_+(\alpha)+H_-(\alpha). \tag{7.2.68}$$

With this resolution at hand, we now rearrange (7.2.62) according to the Wiener–Hopf form given in (7.2.58),

$$\frac{\Phi'_+(0)}{(\alpha+k)^{\frac{1}{2}}}+H_+(\alpha) = -(\alpha-k)^{\frac{1}{2}}D_- - H_-(\alpha) = J(\alpha). \tag{7.2.69}$$

9. *Determination of* $J(\alpha)$. The form (7.2.69) defines a function regular in $\tau > -k_2$, $\tau < k_2\cos\Theta$ and in the overlap strip—hence it is regular in the whole α-plane. Now the exact form of $J(\alpha)$ is found from an extended form of Liouville's theorem, which is approximately stated as follows: *If* $f(\alpha)$ *is a function such that* $|f(\alpha)| \leqslant M$ *for all* α, M *being a constant, then* $f(\alpha)$ *is a constant*. The extension of this theorem that pertains in the present case is as

follows: *If $f(\alpha)$ is a function such that $|f(\alpha)| \leqslant M|\alpha|^p$ as $|\alpha| \to \infty$, where M, p are constants, then $f(\alpha)$ is a polynomial of degree less than or equal to p.*

The application of this extended theorem in the present case used boundary condition (e) of the originally stated problem. These were conditions placed on the edge. Thus, having arrived at this stage of the analysis, it would be apparent that additional conditions of a particular type would need be invoked before the solution could proceed. This accounts for the rather peculiar form of these conditions.

The detailed application of these conditions will be foregone here, and the results stated. It can be shown that $J(\alpha)$ must be regular in the α-plane and tend to zero as $|\alpha| \to \infty$. Then, with this conclusion, the direct application of Liouville's theorem indicates that

$$J(\alpha) = 0. \tag{7.2.70}$$

10. *The Wiener–Hopf solution.* Having 'determined' $J(\alpha)$ in the last section, we now have from (4.2.69)

$$\begin{aligned} \Phi'_+(0) &= -(\alpha+k)^{\frac{1}{2}}H_+(\alpha), \\ D_- &= -(\alpha-k)^{\frac{1}{2}}H_-(\alpha). \end{aligned} \tag{7.2.71}$$

Then from the third eqn of (7.2.46) we have

$$\begin{aligned} A(\alpha) &= -\frac{1}{\gamma}\left\{\Phi'_+(0)+\frac{k\sin\Theta}{\sqrt{(2\pi)}(\alpha-k\cos\Theta)}\right\}, \\ &= -\frac{1}{\sqrt{(2\pi)}}\frac{k\sin\Theta}{(k+k\cos\Theta)^{\frac{1}{2}}(\alpha-k)^{\frac{1}{2}}(\alpha-k\cos\Theta)}. \end{aligned} \tag{7.2.72}$$

The Fourier inversion formula must now be applied to (7.2.43). The usual Fourier inversion is given as

$$f(x) = \frac{1}{\sqrt{(2\pi)}}\int_{-\infty}^{\infty} F(\alpha)e^{-i\alpha x}\,d\alpha. \tag{7.2.73}$$

However, for $F(\alpha)$ analytic in the strip $\tau_- < \tau < \tau_+$ we may, without loss of generality, shift the contour slightly above or below the $\tau = 0$ axis assumed in the above form, to give

$$f(x) = \frac{1}{\sqrt{(2\pi)}}\int_{i\tau-\infty}^{i\tau+\infty} F(\alpha)e^{-i\alpha x}\,d\alpha. \tag{7.2.74}$$

Using the above form of the Fourier inversion, the Weiner–Hopf solution for the total field is given by

$$\phi_+(x, y) = e^{-ik(x\cos\Theta+y\sin\Theta)} \mp \frac{1}{2\pi}(k-k\cos\Theta)^{\frac{1}{2}}\int_{-\infty+ia}^{+\infty+ia}\frac{e^{-i\alpha x\mp\gamma y}}{(\alpha-k)^{\frac{1}{2}}(\alpha-k\cos\Theta)}\,d\alpha, \tag{7.2.75}$$

where the upper sign is for $y \geqslant 0$, the lower sign for $y \leqslant 0$.

11. *The final solution.* The result (7.2.75) may be considered to be the Wiener–Hopf solution. It may appear somewhat of an anticlimax to have brought forth, after so much labour, such an apparently incomplete solution. However, this is not actually the case, since the integral resulting is 'well known'. Carrying out the details again requires complex-variable theory of a fairly heavy calibre, but the results are in terms of Fresnel integrals, defined as

$$F(v) = \int_{v}^{\infty} \exp(\mathrm{i}u^2)\,\mathrm{d}u. \tag{7.2.76}$$

The properties of these integrals have been extensively tabulated.† The results are

$$\phi_t(x, y) = \mathrm{e}^{-\mathrm{i}k(x\cos\Theta + y\sin\Theta)} \mp \frac{I}{2\pi}, \tag{7.2.77}$$

where

$$I = 2\sqrt{\pi}\mathrm{e}^{\pi\mathrm{i}/4}[-\mathrm{e}^{-\mathrm{i}kr\cos(\theta-\Theta)}F\{\sqrt{(2kr)}\cos\tfrac{1}{2}(\theta-\Theta)\}+$$
$$+\mathrm{e}^{-\mathrm{i}kr\cos(\theta+\Theta)}F\{\sqrt{(2kr)}\cos\tfrac{1}{2}(\theta+\Theta)\}], \qquad -\pi \leqslant \theta \leqslant \pi. \tag{7.2.78}$$

For details on completing the above integration, as well as for more detailed discussion of various points of the Wiener–Hopf procedure, the reader is referred to Noble [31].

12. *Discussion.* As an illustration of diffraction theory, where the boundary conditions are such as to remove separation of variables as an approach, the Sommerfeld diffraction problem has been considered. As an illustration of the method of solution, the Wiener–Hopf technique has been developed. We should first note that the problem considered was the simplest possible case. Complications due to additional edges or due to finite thicknesses could be added singly or in combination. Also, as has been pointed out, the simplest elastic-wave case, equivalent to the acoustic problem, was considered. Hence, if the analysis appeared formidable in the present problem, the difficulties inherent in more complex geometries are easily imagined.

As was mentioned before formal analysis was undertaken, an integral equation approach to such diffraction problems is appropriate. Possible approaches are the Green's function, Fourier transforms, or Fredholm integral equations. Although the Wiener–Hopf technique was presented in the context of the Fourier approach, the basic principles involving analytic continuation may be used in the other two methods.

In any event, the analysis of the present problem was rather intricate, and it is open to question whether the Green's function or Fredholm equation approach might not have been simpler. For example, Stoker‡ solves the

† See Abramowitz and Stegun [1].
‡ Pp. 109–33, 141–5 of Reference [40].

Sommerfeld problem by two techniques, one of them the Wiener–Hopf method. In summarizing the two approaches, he refers to the latter as having the air of a *tour de force* in function theory, amongst other things. In actuality, this method should be regarded as another tool of analysis, having limitations as well as points of strength. Many of the intricate points of analysing by this method have their analogues in the other approaches, for example.

We note that the technical literature of diffraction problems is quite voluminous and no effort is made to review the many contributions here. We note that Sih and coworkers [20, 38, 39] have considered P and S waves impinging on cracks of finite size and have employed or extended the Wiener–Hopf method in these studies. Thau and coworkers (see, for example, [41, 42, 43, and 44]) have investigated a large number of diffraction problems, including obstacles in a half-space and elastic obstacles. The Wiener–Hopf procedure arises in a number of these studies.

7.2.3. *Geometric acoustics*

Having established the formidability of elastic scattering and diffraction problems in the previous sections, the natural question arises whether new approaches can be developed to these problems that will generate useful information without some of the mathematical difficulties associated with present methods. Since a considerable amount of work has been done in acoustics and optics, it is logical to see if new methods have evolved in these areas for problems in scattering and diffraction, in the hope that they might be applied to elastic scattering.

As a matter of fact, it appears that certain methods have been devised in the field of acoustics for handling complex diffraction problems. These methods, which are approximate in nature, have been modelled on methods first developed in optics and are generally denoted geometrical acoustics. They represent extensions of geometrical optics.

As in the case of geometric optics, geometric acoustics is based on the postulate that fields propagate along rays. However, in optics it is well known that ray theory does not account for the 'bending' of light around corners, in other words diffraction regions. To circumvent this, the acoustic development introduces the concept of diffracted rays, in addition to the usual optic-type rays. These rays are capable of travelling in part on the surface of an object, hence enabling the diffraction effect to be produced. Furthermore, the theory assigns a field value at each point along the ray (including amplitude, phase, and polarization). The total field at a point is taken to be the sum of all rays passing through that point. Hence, by using the theory of geometrical acoustics, an explicit expression may be obtained for the field produced at any point when a wave hits a smooth, convex object.

The principal developments in this area have been made by J. B. Keller. References [12], [13], [14], and [19] represent only a very partial list of developments in this field by Keller and his co-workers. Other developments have been made by Zauderer [51, 52]. The book by Friedlander [10] is devoted to this aspect of considering diffraction problems. Miklowitz and coworkers [24, 25, 32] have applied this technique to elastic scattering problems.

References

1. ABRAMOWITZ, M. and STEGUN, I. A. *Handbook of mathematical functions: with formulas, graphs, and mathematical tables.* Dover Publications, New York (1965).
2. BANAUGH, R. P. *Scattering of acoustic and elastic waves by surfaces of arbitrary shape.* Univ. of Calif. Rad. Lab. Report No. UCRL-6779 (1962).
3. ——. Application of integral representations of displacement potentials in elastodynamics. *Bull. seism. Soc. Am.* **54,** 1073–86 (1964).
4. BEEBE, W. M. *An experimental investigation of dynamic crack propagation in plastic and metals.* Tech. Rep. AFML-TR-66-249, Air Force Material Laboratory (Nov. 1966).
5. CHIENG, S. L. Multiple scattering of elastic waves by parallel cylinders. *J. appl. Mech.* **36,** 523–7 (1969).
6. CHURCHILL, R. V. *Introduction to complex variables and applications.* McGraw-Hill, New York (1948).
7. DURELLI, A. J. and RILEY, W. F. Stress distribution on the boundary of a circular hole in a large plate during passage of a stress pulse of long duration. *J. appl. Mech.* **28,** 245–51 (1961).
8. EINSPRUCH, N. G. and TRUELL, R. Scattering of a plane longitudinal wave by a spherical fluid obstacle in an elastic medium. *J. acoust. Soc. Am.* **32,** 214–20 (1960).
9. FREDRICKS, R. W. Diffraction of an elastic pulse in a loaded half-space. *J. acoust. Soc. Am.* **33,** 17–22 (1961).
10. FRIEDLANDER, F. G. *Sound Pulses.* Cambridge University Press (1958).
11. FRIEDMAN, M. B. and SHAW, R. P. Diffraction of pulses by cylindrical obstacles of arbitrary cross section. *J. appl. Mech.* **29,** 40 (1962).
12. KELLER, J. B. Geometrical acoustics, I. The theory of weak shock waves. *J. appl. Phys.* **25,** 938–47 (1954).
13. ——. Geometrical theory of diffraction. *J. opt. Soc. Am.* **52,** 116–30 (1962).
14. —— and KARAL, F. C. (Jr.). Geometrical theory of elastic surface-wave excitation and propagation. *J. acoust. Soc. Am.* **36,** 32–40 (1964).
15. KNOPOFF, L. Diffraction of elastic waves. *J. acoust. Soc. Am.* **28,** 2, 217–29 (1956).
16. ——. Scattering of compression waves by spherical obstacles. *Geophysics* **24,** 30–9 (1959).
17. ——. Scattering of shear waves by spherical obstacles. *Geophysics,* **24,** 209–19 (1959).
18. KO, W. L. and KARLSSON, T. Application of Kirchhoff's integral equation formulation to an elastic wave scattering problem. *J. appl. Mech.* **34,** 921–30 (1967).

19. LEVY, B. R. and KELLER, J. B. Diffraction by a smooth object. *Commun. pure appl. Math.* **12,** 159–209 (1959).
20. LOEBER, J. F. and SIH, G. C. Diffraction of antiplane shear waves by a finite crack. *J. acoust. Soc. Am.* **44,** 90–8 (1968).
21. LINDSAY, R. B. *Mechanical radiation.* McGraw-Hill, New York (1960).
22. MCCOY, J. J. Effects of non-propagating plate waves on dynamical stress concentrations. *Int. J. Solids Struct.* **4,** 355–70 (1968).
23. MCLACHLAN, N. W. *Bessel functions for engineers.* Clarendon Press, Oxford (1961).
24. MIKLOWITZ, J. Pulse propagation in a viscoelastic solid with geometric dispersion. In *Stress waves in anelastic solids,* pp. 255–76. Springer-Verlag, Berlin (1964).
25. ——. Scattering of a plane elastic compressional pulse by a cylindrical cavity. *Proc. XIth Int. Congr. appl. Mech.* Springer-Verlag, Berlin (1966).
26. MILES, J. W. Motion of a rigid cylinder due to a plane elastic wave. *J. acoust. Soc. Am.* **32,** 1656–9 (1960).
27. MORSE, P. and FESHBACH, H. *Methods of theoretical physics* Vols. 1 and 2. McGraw-Hill, New York (1953).
28. MOW, C. C. Transient response of a rigid spherical inclusion in an elastic medium. *J. appl. Mech.* **32,** 637 (1965).
29. ——. On the transient motion of a rigid spherical inclusion in an elastic medium and its inverse problem. *J. appl. Mech.* **33,** 807 (1966).
30. —— and PAO, Y. H. *The diffraction of elastic waves and dynamic stress concentrations.* RAND Rep. R-482-PR (April 1971).
31. NOBLE, B. *Methods based on the Wiener–Hopf technique.* Pergamon Press, New York (1958).
32. NORWOOD, F. R. and MIKLOWITZ, J. Diffraction of transient elastic waves by a spherical cavity. *J. appl. Mech.* **34,** 735–44 (1967).
33. PAO, Y. H. Dynamical stress concentration in an elastic plate. *J. appl. Mech.* **29,** 299–305 (1962).
34. —— and MOW, C. C. Scattering of plane compressional waves by a spherical obstacle. *J. appl. Phys.* **34,** 493–9 (1963).
35. RAYLEIGH, J. W. S. *The theory of sound* Vols. I and II. Dover Publications, New York (1945).
36. SHAW, R. P. Retarded potential approach to the scattering of elastic pulses by rigid obstacles of arbitrary shape. *J. acoust. Soc. Am.* **44,** 745–8 (1968).
37. SHEA, R. Dynamic stress-concentration Factors. *Exp. Mech.* **21,** 20–4 (1964).
38. SIH, G. C. and LOEBER, J. F. Torsional vibration of an elastic solid containing a penny-shaped crack. *J. acoust. Soc. Am.* **44,** 1237–45 (1968).
39. —— ——. Wave propagation in an elastic solid with a line of discontinuity or finite crack. *Q. appl. Math.* **27,** 193–213 (1969).
40. STOKER, J. J. *Water waves.* Wiley–Interscience, New York (1957).
41. THAU, S. A. Dynamic reactions along a rigid-smooth wall in an elastic half-space with a moving boundary load. *Int. J. Solids Struct.* **4,** 1–13 (1968).
42. ——. Motion of a finite rigid strip in an elastic half-space subjected to blast wave loading. *Int. J. Solids Struct.* **7,** 193–211 (1971).

43. THAU, S. A. and LU, T.-H. Dynamic stress concentration at a cylindrical inclusion in an elastic medium with an arbitrarily stiff bond. *Int. J. Mech. Sci.* **11,** 677–88 (1969).
44. —— ——. Diffraction of transient horizontal shear waves by a finite rigid ribbon. *Int. J. Eng. Sci.* **8,** 857–74 (1970).
45. —— and PAO, Y. H. Diffractions of horizontal shear waves by a parabolic cylinder and dynamic stress concentrations. *J. appl. Mech.* **33,** 785–92 (1966).
46. —— ——. Stress-intensification near a semi-infinite rigid-smooth strip due to diffraction of elastic waves. *J. appl. Mech.* **34,** 119–26 (1967).
47. —— ——. Wave function expansions and perturbation method for the diffraction of elastic waves by a parabolic cylinder. *J. appl. Mech.* **34,** 915–20 (1967).
48. THOMAS, D. P. Electromagnetic diffraction by two coaxial discs. *Proc. Camb. phil. Soc. Math. Phys. Sci.* **60,** 621–34 (1964).
49. WELLS, A. A. and POST, D. *The dynamic stress distribution surrounding a running crack—a photoelastic analysis.* Nav. Res. Laboratory, NRL-4935, Washington D.C. (April 1957).
50. YING, C. F. and TRUELL, R. Scattering of a plane longitudinal wave by a spherical obstacle in an isotropically elastic solid. *J. appl. Phys.* **27,** 1086–97 (1956).
51. ZAUDERER, E. Wave propagation around a convex cylinder. *J. Math. Mech.* **13,** 171–86 (1964).
52. ——. Wave propagation around a smooth object. *J. Math. Mech.* **13,** 187–200 (1964).

Problems

7.1. Formulate and solve the problem of scattering of plane harmonic SH waves incident on a rigid immovable cylinder. That is, let the boundary conditions be given by $u = 0$ at $r = a$.

7.2. Attempt a general formulation and solution of the problem of plane, harmonic SH waves incident on an elastic cylinder. Then, using a procedure similar to that illustrated in the text for the spherical scattering problem, attempt to recover various special cases such as the cavity, the rigid, immovable cylinder (see Problem 7.1), and the rigid cylinder capable of translation.

7.3. Consider the results for scattering of SH waves by a cylindrical cavity. Assume $\gamma a \gg 1$, and obtain an approximate expression for the scattered power for this case of Rayleigh scattering.

7.4. Consider the results given for scattering of waves by a spherical cavity (7.1.76). Verify the reduction of the general results (7.1.74) to the simpler form. Assume $\alpha_1 a \gg 1$ (Rayleigh scattering) and obtain approximate expressions for the coefficient matrix and the scattered wavefields.

7.5. Attempt to trace through the analysis of scattering of SH waves by a slit, where the edges of the slit are now considered to be clamped (that is, $u = 0$, $-\infty < x \leq 0$, $y = 0$) instead of traction-free as in the text.

8 Wave propagation in plates and rods

WE NOW consider the propagation of waves in plates and rods (and cylindrical shells) as governed by the exact equations of elasticity. This represents the first instance where geometries, previously studied using strength-of-material theories, are again considered. We recall that the simple theories, such as for longitudinal waves in rods or flexural waves in beams, were restricted to low frequencies owing to kinematical limitations. It is to be expected, therefore, that the exact theories of plates, rods, and shells will find greatest application at high frequencies and for transient loading conditions.

The frequency equations for waves in infinite rods and plates have been known for some time. Thus, Pochhammer (1876), and Chree (1889) developed the results for the rod and Rayleigh (1889) and Lamb (1889) presented the results for the plate. Very little additional work occurred in this area for many years, however. In 1948, Davies [11] studied, both theoretically and experimentally, a number of aspects of longitudinal waves in rods. As part of that study, the first few branches of the Pochhammer–Chree frequency equation for the rod were developed and the propagation characteristics of a transient according to the exact theory were considered. Since Davies' investigation, many aspects have been thoroughly investigated. Most attention has been given to understanding the complete frequency spectrum of the plate and the rod, to the analysis of transient disturbances, and to the development and application of approximate theories for plates, rods, and shells.

Our study of these geometries will be in the order just mentioned. Thus the propagation of waves in plates will be investigated first, with the development of the Rayleigh–Lamb frequency equation and the spectrum being the objective. The development of the Pochhammer–Chree equation for the rod will then follow, with the complete spectrum for that geometry also presented. The forced motion of plates and rods, both harmonic and transient, will be studied. Finally, the development of approximate theories for the various geometries will be considered.

8.1. Continuous waves in a plate

As previously mentioned, the early developments in wave propagation in plates were by Rayleigh and Lamb. The Rayleigh–Lamb theory pertains to the propagation of continuous, straight crested waves in a plate, infinite in extent and having traction-free surfaces. Plane-strain conditions apply. Our study will develop the frequency equation for this theory and present the complete frequency spectrum. The considerations here are somewhat complicated, however, and it will be advisable, before considering the Rayleigh–Lamb case, to study the propagation of SH waves in a plate. Recalling the simpler circumstances pertaining to this case in our previous studies in infinite and semi-infinite media and in wave scattering, we expect and will find a simpler theory governing these waves in a plate. Boundary conditions other than traction-free ones will also be considered.

8.1.1. *SH waves in a plate*

In considering waves in a plate, which has two boundary surfaces, two sources of complexity arise. First, multiple reflections of waves between the boundary surfaces occur. Secondly, mode conversion of P and SV waves occurs. By restricting to the SH wave case, the second complication is removed, and we may restrict our attention to the first area.

Consider, then, a plate of infinite extent in the z direction and of thickness $2b$, as shown in Fig. 8.1. The coordinate system is selected with the x, z-plane

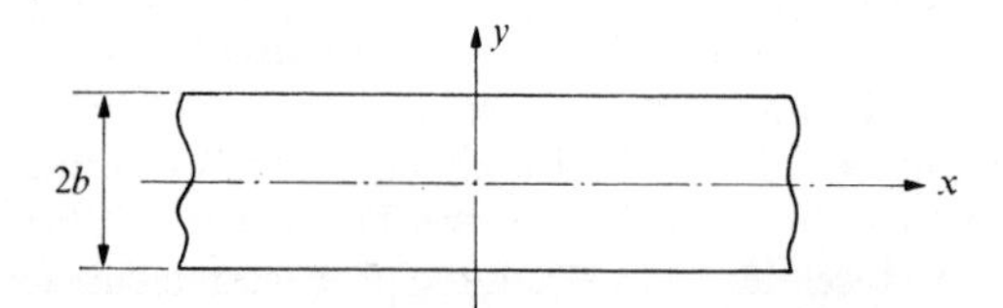

FIG. 8.1. Coordinate system for a plate of thickness $2b$.

coinciding with the middle surface of the plate, with y positive upward. For SH waves, we have the governing equations

$$\nabla^2 u_z = \frac{1}{c_2^2}\frac{\partial^2 u_z}{\partial t^2}, \tag{8.1.1}$$

where $u_z = u_z(x, y, t)$. We immediately consider solutions of the wave equation given by

$$u_z = h(y)e^{i(\xi x - \omega t)}. \tag{8.1.2}$$

Substitution in the wave equation gives

$$\frac{d^2 h}{dy^2} + \beta^2 h = 0, \qquad \beta^2 = \frac{\omega^2}{c_2^2} - \xi^2, \tag{8.1.3}$$

which leads to

$$u_z = (A_1 \sin \beta y + A_2 \cos \beta y)e^{i(\xi x - \omega t)}. \tag{8.1.4}$$

This solution form is most convenient to the problem at hand. However, by way of interpretation, we note that an alternative solution form could as well be

$$u_z = A_1' e^{i(\xi x - \beta y - \omega t)} + A_2' e^{i(\xi x + \beta y - \omega t)}, \tag{8.1.5}$$

which explicitly brings out the plane-wave nature of the motion.

The boundary conditions for the problem are $\tau_{yy} = \tau_{xy} = \tau_{zy} = 0$ at $y = \pm b$. With the SH wave restriction, only the last condition on τ_{zy} is non-trivial, being given by

$$\partial u_z/\partial y = 0, \qquad y = \pm b. \tag{8.1.6}$$

Applying these conditions to the solution (8.1.4) gives

$$\begin{aligned} A_1 \cos \beta b - A_2 \sin \beta b &= 0, \\ A_1 \cos \beta b + A_2 \sin \beta b &= 0, \end{aligned} \tag{8.1.7}$$

from which results the frequency equation

$$\cos \beta b \sin \beta b = 0. \tag{8.1.8}$$

This equation is satisfied by

$$\beta b = n\pi/2 \qquad (n = 0, 1, 2, 3, 4, \ldots). \tag{8.1.9}$$

These results show that harmonic SH waves may propagate only under special conditions, as given by (8.1.9). Thus, given a frequency ω, the resulting wavenumber ξ will be given by $\xi = (\omega^2/c_2^2 - \beta^2)^{\frac{1}{2}}$, where β is given by (8.1.9).

Let us consider the motion in more detail. We first note that the displacement solution (8.1.4) involves a symmetric (A_2) and antisymmetric (A_1) motion with respect to the $y = 0$ mid surface. Now suppose that the frequency equation is satisfied by $\cos \beta b = 0$. From (8.1.7) we have that $A_2 = 0$ must hold so that the motion is antisymmetric, given by

$$u_z = A_1 \sin \beta_n y e^{i(\xi x - \omega t)}, \tag{8.1.10}$$

where, from $\cos \beta b = 0$, we have that $n = 1, 3, 5, \ldots$. Similarly, if $\sin \beta b = 0$ in (8.1.8), we have from (8.1.7) that $A_1 = 0$, and the resulting symmetric motion is given by

$$u_z = A_2 \cos \beta_n y e^{i(\xi x - \omega t)}, \tag{8.1.11}$$

where $n = 0, 2, 4, \ldots$. The resulting form of the y variation of displacement across the plate thickness is shown for the first few modes in Fig. 8.2(a) and (b).

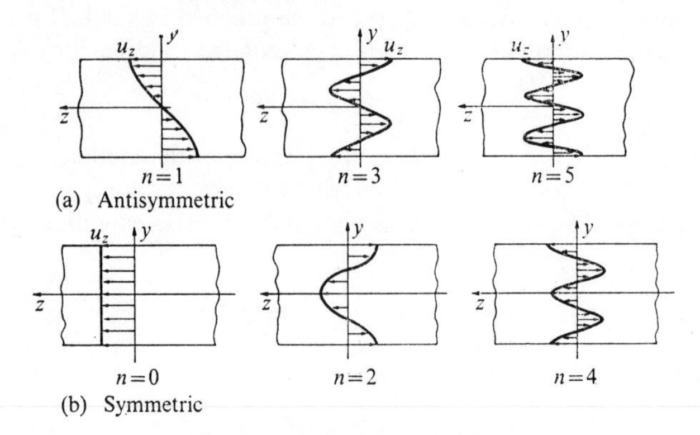

FIG. 8.2. The y-variation in displacement for the (a) first three antisymmetric, and (b) first three symmetric SH plate modes.

The frequency spectrum for SH waves results from (8.1.9). Thus we write $\beta^2 b^2 = (n\pi/2)^2$, and using the definition of β^2 obtain

$$\frac{\omega^2 b^2}{c_2^2} = \left(\frac{n\pi}{2}\right)^2 + \xi^2 b^2. \tag{8.1.12}$$

Letting

$$\Omega = 2b\omega/\pi c_2, \qquad \bar{\xi} = 2b\xi/\pi, \tag{8.1.13}$$

we have

$$\Omega^2 = n^2 + \bar{\xi}^2, \tag{8.1.14}$$

or

$$\bar{\xi} = \pm(\Omega^2 - n^2)^{\frac{1}{2}}, \tag{8.1.15}$$

where n odd gives the antisymmetric and n even the symmetric roots. Using $\omega = \xi c$ or, in non-dimensional form, $\Omega = \bar{\xi}\bar{c}$, where $\bar{c} = c/c_2$ we may obtain the dispersion curve equations. The group velocity, given by $c_g = d\omega/d\xi$ or, in non-dimensional form, by $\bar{c}_g = d\Omega/d\bar{\xi}$ also follows from (8.1.15).

Examining (8.1.15), we see that $n = 0$ gives the result that $\bar{\xi} = \Omega$ or, in dimensional form, $\omega = \xi c_2$. Thus we see that the first mode is non-dispersive, whereas all other modes propagate dispersively. Next, we see that, for $\Omega > n$, $\bar{\xi}$ is real and the spectrum consists of a family of hyperbolas ($n = 1, 2, 3, \ldots$). At $\bar{\xi} = 0$, $\Omega = n$ and the cutoff frequencies for the various modes result. For $\Omega < n$, $\bar{\xi}$ is imaginary and the spectrum consists of a family of circles ($n = 1, 2, 3 \ldots$). The resulting spectrum, showing just the positive real and positive imaginary branches for positive Ω is shown in Fig. 8.3. The solid lines correspond to the symmetric (n even) modes, the dashed lines are the antisymmetric modes (n odd). The interpretation of the imaginary wavenumbers,

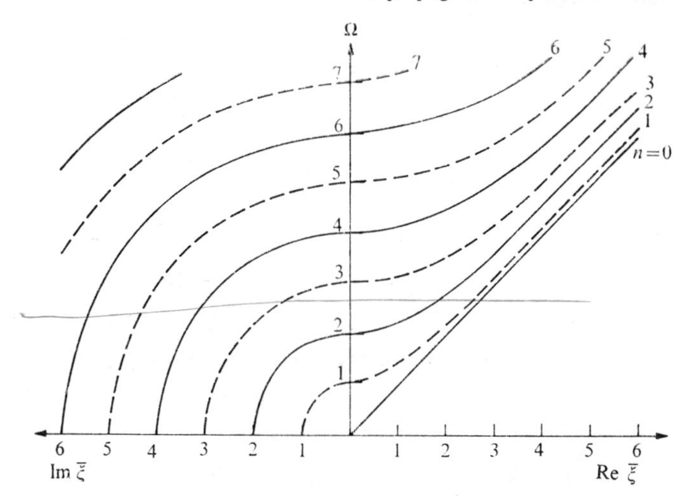

FIG. 8.3. Frequency spectrum for SH waves in a plate. Solid lines are the symmetric modes, dashed lines are the antisymmetric modes.

arising from (8.1.15) and shown in the figure, is quite familiar at this stage. Thus, for $\bar{\xi} = \pm i\bar{\xi}'$, we have

$$u_z = (A_1 \sin \beta y + A_2 \cos \beta y)\exp(\pm\bar{\xi}'x)\exp(-i\omega t), \qquad (8.1.16)$$

corresponding to a non-propagating, spatially varying disturbance. The importance of such solutions in forced-motion and wave-reflection problems has been well established from previous studies. Finally, we note that for a given frequency there are only a finite number of propagating SH modes (that is, the line Ω = constant cuts only a finite number of branches of the spectrum). This suggests it would not be possible to form an arbitrary stress distribution by a Fourier superposition of propagating modes. However, if the imaginary branches are included, an infinite mode set is obtained and the formation of an arbitrary stress distribution becomes possible.

8.1.2. *Waves in a plate with mixed boundary conditions*

As our next step on the road to the Rayleigh–Lamb equation for plates, we consider the case of a plate having mixed boundary conditions. In contrast to the previous section, we now admit P and SV waves in the plate and consider conditions of plane strain. The resulting wave behaviour, although more complicated than the SH wave case, will be less complicated than the case of traction-free surfaces. While the case of mixed boundary conditions is not of great practical importance, the results obtained here will be found to play a very important role in obtaining the Rayleigh–Lamb spectrum.

We again consider the plate geometry of Fig. 8.1. If conditions of plane strain hold in the z direction, we have that $u_z = \partial/\partial z = 0$ and

$$u_x = \frac{\partial \Phi}{\partial x} + \frac{\partial H_z}{\partial y}, \tag{8.1.17}$$

$$u_y = \frac{\partial \Phi}{\partial y} - \frac{\partial H_z}{\partial x}, \tag{8.1.18}$$

where $u_x = u_x(x, y, t)$, $u_y = u_y(x, y, t)$. Further, we have

$$\nabla^2 \Phi = \frac{1}{c_1^2}\frac{\partial^2 \Phi}{\partial t^2}, \qquad \nabla^2 H_z = \frac{1}{c_2^2}\frac{\partial^2 H_z}{\partial t^2}. \tag{8.1.19}$$

The case of mixed boundary conditions considered here is expressed by

$$u_y = \tau_{xy} = \tau_{zy} = 0, \qquad y = \pm b. \tag{8.1.20}$$

We consider solutions to (8.1.19) of the form

$$\Phi = f(y)\mathrm{e}^{\mathrm{i}(\xi x - \omega t)}, \qquad H_z = \mathrm{i}h_z(y)\mathrm{e}^{\mathrm{i}(\xi x - \omega t)}, \tag{8.1.21}$$

where the factor i has been inserted in H_z for later convenience. Substitution in the wave equations gives for f, h_z

$$\frac{\mathrm{d}^2 f}{\mathrm{d}y^2} + \alpha^2 f = 0, \qquad \frac{\mathrm{d}^2 h_z}{\mathrm{d}y^2} + \beta^2 h_z = 0, \tag{8.1.22}$$

where

$$\alpha^2 = \omega^2/c_1^2 - \xi^2, \qquad \beta^2 = \omega^2/c_2^2 - \xi^2. \tag{8.1.23}$$

We obtain

$$f = A \sin \alpha y + B \cos \alpha y, \qquad h_z = C \sin \beta y + D \cos \beta y. \tag{8.1.24}$$

The resulting potentials and displacements are

$$\Phi = (A \sin \alpha y + B \cos \alpha y)\mathrm{e}^{\mathrm{i}(\xi x - \omega t)}, \tag{8.1.25}$$

$$H_z = \mathrm{i}(C \sin \beta y + D \cos \beta y)\mathrm{e}^{\mathrm{i}(\xi x - \omega t)}, \tag{8.1.26}$$

$$u_x = \mathrm{i}\{\xi(A \sin \alpha y + B \cos \alpha y) + \beta(C \cos \beta y - D \sin \beta y)\}\mathrm{e}^{\mathrm{i}(\xi x - \omega t)}, \tag{8.1.27}$$

$$u_y = \{\alpha(A \cos \alpha y - B \sin \alpha y) + \xi(C \sin \beta y + D \cos \beta y)\}\mathrm{e}^{\mathrm{i}(\xi x - \omega t)}. \tag{8.1.28}$$

The stresses in terms of the potentials have been previously given by (6.1.9)–(6.1.12) in the study of the half-space. Writing the stresses so that the Φ and H_z contributions are separated, we obtain

$$\tau_{xx} = \mu[\{2\alpha^2 - k^2(\xi^2 + \alpha^2)\}(A \sin \alpha y + B \cos \alpha y) - \\ -2\xi\beta(C \cos \beta y - D \sin \beta y)]\mathrm{e}^{\mathrm{i}(\xi x - \omega t)}, \tag{8.1.29}$$

$$\tau_{yy} = \mu[\{2\xi^2 - k^2(\xi^2 + \alpha^2)\}(A \sin \alpha y + B \cos \alpha y) + \\ +2\beta\xi(C \cos \beta y - D \sin \beta y)]\mathrm{e}^{\mathrm{i}(\xi x - \omega t)}, \tag{8.1.30}$$

$$\tau_{xy} = \mathrm{i}\mu\{2\alpha\xi(A \cos \alpha y - B \sin \alpha y) - \\ -(\beta^2 - \xi^2)(C \sin \beta y + D \cos \beta y)\}\mathrm{e}^{\mathrm{i}(\xi x - \omega t)}. \tag{8.1.31}$$

For the plane-strain case at hand, τ_{zz} is obtainable from τ_{xx} and τ_{yy}, while $\tau_{zx} = \tau_{zy} = 0$.

We apply the solutions for u_y, τ_{xy} to the boundary conditions (8.1.20) and obtain

$$\begin{aligned}
\alpha(A\cos\alpha b - B\sin\alpha b) + \xi(C\sin\beta b + D\cos\beta b) &= 0,\\
\alpha(A\cos\alpha b + B\sin\alpha b) - \xi(C\sin\beta b - D\cos\beta b) &= 0,\\
2\alpha\xi(A\cos\alpha b - B\sin\alpha b) - (\beta^2-\xi^2)(C\sin\beta b + D\cos\beta b) &= 0,\\
2\alpha\xi(A\cos\alpha b + B\sin\alpha b) + (\beta^2-\xi^2)(C\sin\beta b - D\cos\beta b) &= 0. \qquad (8.1.32)
\end{aligned}$$

Adding and subtracting the first two and the second two equations gives

$$\begin{aligned}
\alpha A\cos\alpha b + \xi D\cos\beta b &= 0,\\
\alpha B\sin\alpha b - \xi C\sin\beta b &= 0,\\
2\alpha\xi A\cos\alpha b - (\beta^2-\xi^2)D\cos\beta b &= 0,\\
2\alpha\xi B\sin\alpha b + (\beta^2-\xi^2)C\sin\beta b &= 0, \qquad (8.1.33)
\end{aligned}$$

which re-group to

$$\begin{bmatrix} \alpha\cos\alpha b & \xi\cos\beta b \\ 2\alpha\xi\cos\alpha b & -(\beta^2-\xi^2)\cos\beta b \end{bmatrix}\begin{bmatrix} A \\ D \end{bmatrix} = 0, \qquad (8.1.34)$$

$$\begin{bmatrix} \alpha\sin\alpha b & -\xi\sin\beta b \\ 2\alpha\xi\sin\alpha b & (\beta^2-\xi^2)\sin\beta b \end{bmatrix}\begin{bmatrix} B \\ C \end{bmatrix} = 0. \qquad (8.1.35)$$

Before expanding the determinant of coefficients of the preceding results, we note that the displacements (8.1.27) and (8.1.28) contain symmetric and antisymmetric components. Thus, for u_x we see that the B and C terms give symmetric displacements with respect to $y = 0$, while the A and D terms give antisymmetric displacements. For u_y, the B and C terms and A and D terms again give symmetric and antisymmetric displacements, respectively. The general form of the various components are shown in Fig. 8.4.

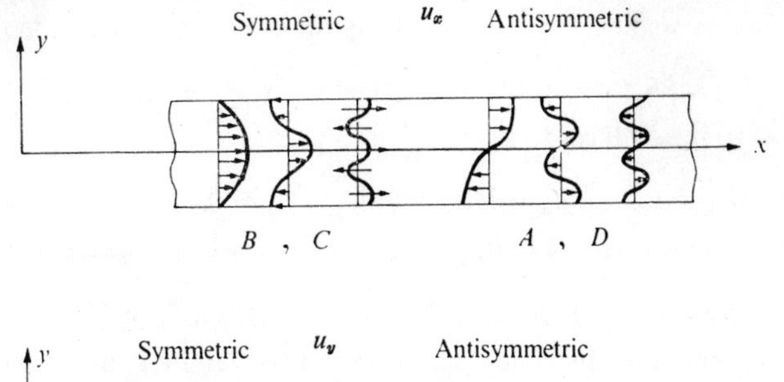

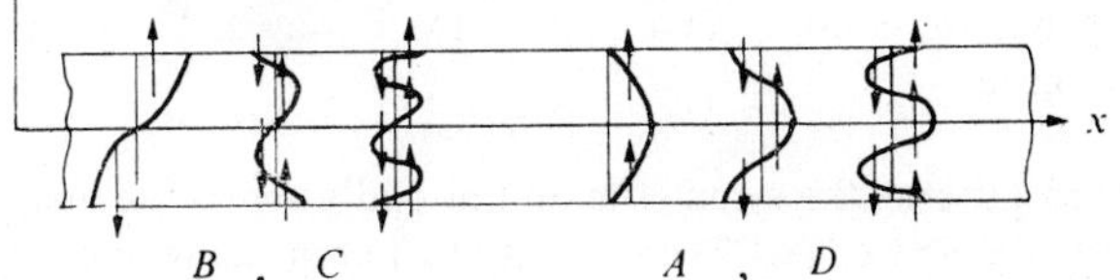

FIG. 8.4. Symmetric and antisymmetric components of the u_x, u_y displacements.

Continuing, we see that the result (8.1.34) corresponds to the antisymmetric modes. Thus, expanding the determinant, we obtain

$$\alpha(\beta^2+\xi^2)\cos \alpha b \cos \beta b = 0. \tag{8.1.36}$$

This frequency equation will be satisfied by

$$\alpha = 0, \qquad \alpha b = m\pi/2, \qquad \beta b = n\pi/2 \qquad (m, n = 1, 3, 5, \ldots). \tag{8.1.37}$$

If $\alpha = 0$ or $\alpha b = m\pi/2$, we have from (8.1.34) that $D = 0$, so that

$$u_x = \mathrm{i}\xi A \sin \alpha_m y \mathrm{e}^{\mathrm{i}(\xi x - \omega t)}, \qquad u_y = \alpha_m A \cos \alpha_m y \mathrm{e}^{\mathrm{i}(\xi x - \omega t)}. \tag{8.1.38}$$

If $\beta b = n\pi/2$, then $A = 0$, and we have

$$u_x = -\mathrm{i}\beta D \sin \beta_n y \mathrm{e}^{\mathrm{i}(\xi x - \omega t)}, \qquad u_y = \xi D \cos \beta_n y \mathrm{e}^{\mathrm{i}(\xi x - \omega t)}. \tag{8.1.39}$$

In the case of (8.1.38), we have antisymmetric modes resulting from the reflection of P waves within the plate, while (8.1.39) results from the reflection of SV waves. It is possible for these modes to be uncoupled from one another owing to the boundary conditions. Thus, recall in the study of the half-space that it was found that P and SV waves reflect without mode conversion from mixed boundary constraints of the present type.

In a similar way, we have that the result (8.1.35) governs symmetric modes in the plate. The frequency equation is

$$\alpha(\beta^2+\xi^2)\sin \alpha b \sin \beta b = 0. \tag{8.1.40}$$

This will be satisfied by

$$\alpha = 0, \qquad \alpha b = m\pi, \qquad \beta b = n\pi \qquad (m = 0, \quad m, n = 1, 2, 3, \ldots). \tag{8.1.41}$$

For $\alpha b = m\pi$, we have that $C = 0$, giving the symmetric P waves in the plate

$$u_x = \mathrm{i}\xi B \cos \alpha_m y \mathrm{e}^{\mathrm{i}(\xi x - \omega t)}, \qquad u_y = -\alpha_m B \sin \alpha_m y \mathrm{e}^{\mathrm{i}(\xi x - \omega t)}. \tag{8.1.42}$$

For $m = 0$, this reduces to

$$u_x = \mathrm{i}\xi B \mathrm{e}^{\mathrm{i}(\xi x - \omega t)}, \qquad u_y = 0. \tag{8.1.43}$$

For $\beta b = n\pi$, we have $B = 0$, giving the symmetric SV waves in the plate

$$u_x = \mathrm{i}\beta_n C \cos \beta_n y \mathrm{e}^{\mathrm{i}(\xi x - \omega t)}, \qquad u_y = \xi C \sin \beta_n y \mathrm{e}^{\mathrm{i}(\xi x - \omega t)}. \tag{8.1.44}$$

In obtaining the frequency spectrum for waves in a mixed boundary condition plate, we consider first that of the P waves. From (8.1.37) and (8.1.41), we have

$$\alpha b = m\pi/2 \qquad (m = 0, 1, 2, 3, \ldots), \tag{8.1.45}$$

where m even governs the symmetric and m odd governs the antisymmetric waves. Using the first of (8.1.23), this may be put in the form

$$\Omega^2 = k^2(m^2+\bar{\xi}^2) \qquad (m = 0, 1, 2, \ldots), \tag{8.1.46}$$

where now

$$\Omega = 2b\omega/\pi c_2, \qquad \bar{\xi} = 2b\xi/\pi. \tag{8.1.47}$$

In a similar fashion, for the SV waves we have from (8.1.37) and (8.1.41)

$$\beta b = n\pi/2 \qquad (n = 1, 2, 3 \ldots), \tag{8.1.48}$$

where n even and odd governs, respectively, the symmetric and antisymmetric SV waves. Using the second of (8.1.23), this may be put in the form

$$\Omega^2 = (n^2 + \bar{\xi}^2) \qquad (n = 1, 2, 3, \ldots). \tag{8.1.49}$$

In order to plot the spectrum, Poisson's ratio must be specified, since it enters into (8.1.46) through k^2.

The curves for the real branches of the spectrum are seen to be hyperbolas. The cutoff frequencies, given by $\bar{\xi} \to 0$, are $\Omega = km, n$ for the P and SV waves. For $\Omega < km$, the wavenumbers become imaginary. Replacing $\bar{\xi}$ by $i\bar{\xi}$ in (8.1.46), it is seen that the P wave branches of the spectrum are ellipses. For $\Omega < n$, the wavenumbers of the respective SV waves become imaginary and, it is seen from (8.1.49), in the form of circles. The resulting spectrum, plotted for $\nu = 0{\cdot}31$, so that $k^2 = 1{\cdot}91$, is shown in Fig. 8.5.

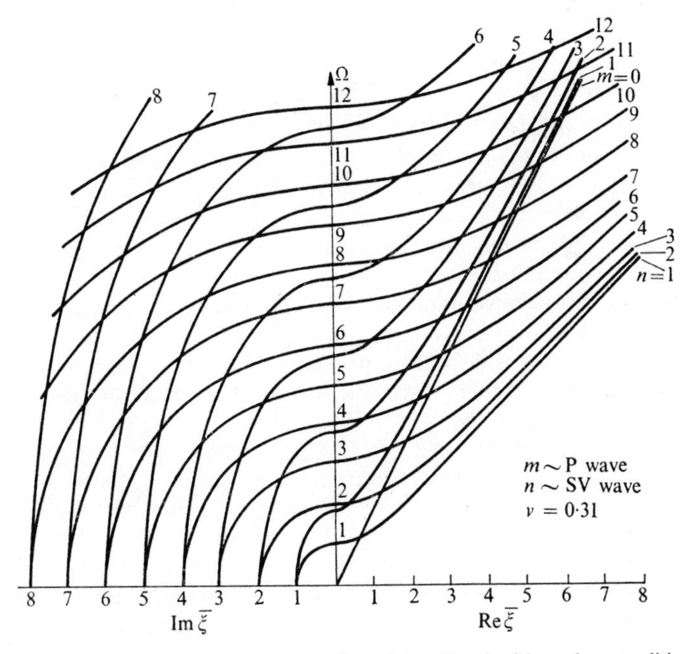

FIG. 8.5. Frequency spectrum for waves in a plate with mixed boundary conditions.

For the spectrum, we see that the first symmetric P wave ($m = 0$) is non-dispersive, whereas all other modes are dispersive. For large wavenumber ($\xi \to \infty$), it is seen from (8.1.46) that P wave branches are asymptotic to $\Omega = k\xi$. In dimensional form, this gives $\omega = \xi c_1$. Thus, at high frequency and short wavelength, the P wave modes propagate at the dilatational wave speed. Similarly, for the SV waves, $\xi \to \infty$ gives the asymptote $\Omega = \xi$ or $\omega = \xi c_2$. Thus the SV-wave modes propagate at the shear-wave velocity at high frequencies and short wavelengths. An illustration of the first few P and SV modes is shown in Fig. 8.6.

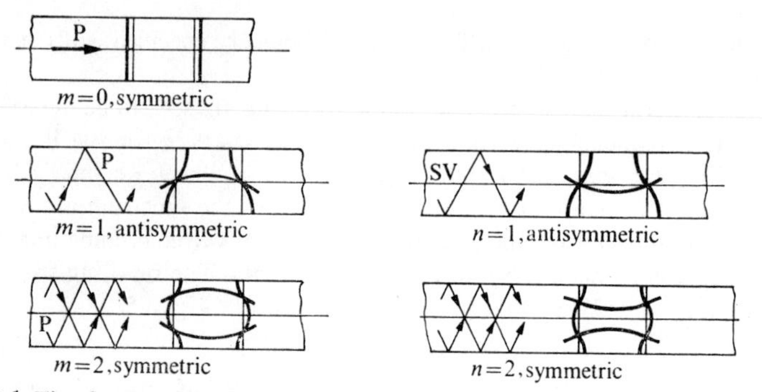

FIG. 8.6. First few P and SV modes in a mixed boundary condition plate. (After Mindlin [63, Fig. 17].)

8.1.3. *The Rayleigh–Lamb frequency equation for the plate*

We now consider the case of waves in a plate having traction-free boundaries. This is the case of greatest practical interest and is the classical case first studied by Rayleigh and Lamb. In our development, we can make direct use of many of the equations given in the previous study of waves in a mixed boundary-condition plate.

Thus, again consider waves of plane strain propagating in the x direction in a plate of thickness $2b$ (see Fig. 8.1) having traction-free boundaries. The governing equations for displacements, potential functions, and stresses, as given in the previous section by eqn (8.1.25)–(8.1.31), still hold. The boundary conditions are now given by

$$\tau_{yy} = \tau_{xy} = \tau_{zy} = 0, \quad y = \pm b, \tag{8.1.50}$$

where the last condition on τ_{zy} is satisfied identically. We now resolve our considerations into the cases of symmetric and antisymmetric waves.

Consider first the case of symmetric waves. From our previous considerations, we know the symmetric displacements to be given by

$$u_x = \mathrm{i}(B\xi \cos \alpha y + C\beta \cos \beta y)\mathrm{e}^{\mathrm{i}\varphi}, \tag{8.1.51}$$

$$u_y = (-B\alpha \sin \alpha y + C\xi \sin \beta y)\mathrm{e}^{\mathrm{i}\varphi}, \tag{8.1.52}$$

where we have let $A = D = 0$ in (8.1.27) and (8.1.28). The phase factor in the preceding equations is simply $\psi = \xi x - \omega t$. We apply the boundary conditions to eqns (8.1.30) and (8.1.31) for τ_{yy}, τ_{xy}, letting $A = D = 0$, and obtain

$$\begin{aligned}(\xi^2-\beta^2)B\cos\alpha b+2\xi\beta C\cos\beta b &= 0,\\ \pm\mathrm{i}\{-2\xi\alpha B\sin\alpha b+(\xi^2-\beta^2)C\sin\beta b\} &= 0.\end{aligned} \tag{8.1.53}$$

Thus the four boundary conditions on $y = \pm b$ reduce to two unique equations in B and C. Equating the determinant of coefficients to zero, we obtain from (8.1.53) the frequency equation

$$\frac{\tan\beta b}{\tan\alpha b} = -\frac{4\alpha\beta\xi^2}{(\xi^2-\beta^2)^2}. \tag{8.1.54}$$

This is the Rayleigh–Lamb frequency equation for the propagation of symmetric waves in a plate. From (8.1.53) we also obtain the amplitude ratios

$$\frac{B}{C} = -\frac{2\xi\beta\cos\beta b}{(\xi^2-\beta^2)\cos\alpha b} = \frac{(\xi^2-\beta^2)\sin\beta b}{2\xi\alpha\sin\alpha b}. \tag{8.1.55}$$

Now consider the case of the antisymmetric modes, given by the displacements

$$u_x = \mathrm{i}(\xi A\sin\alpha y-\beta D\sin\beta y)\mathrm{e}^{\mathrm{i}\psi}, \tag{8.1.56}$$

$$u_y = (\alpha A\cos\alpha y+\xi D\cos\beta y)\mathrm{e}^{\mathrm{i}\psi}. \tag{8.1.57}$$

Substituting the expressions for the stresses τ_{yy}, τ_{xy} in the boundary conditions, where $B = C = 0$ in (8.1.30) and (8.1.31), we obtain

$$\begin{aligned}\pm\{(\xi^2-\beta^2)A\sin\alpha b-2\beta\xi D\sin\beta b\} &= 0,\\ 2\alpha\xi A\cos\alpha b-(\beta^2-\xi^2)D\cos\beta b &= 0.\end{aligned} \tag{8.1.58}$$

This gives the Rayleigh–Lamb frequency equation for antisymmetric waves in a plate,

$$\frac{\tan\beta b}{\tan\alpha b} = -\frac{(\xi^2-\beta^2)^2}{4\alpha\beta\xi^2}, \tag{8.1.59}$$

and the amplitude ratios

$$\frac{A}{D} = \frac{2\xi\beta\sin\beta b}{(\xi^2-\beta^2)\sin\alpha b} = -\frac{(\xi^2-\beta^2)\cos\beta b}{2\xi\alpha\cos\alpha b}. \tag{8.1.60}$$

We may combine the frequency equations for symmetric and antisymmetric waves into a single equation given by

$$F(\alpha,\beta,\xi) = \frac{\tan\beta b}{\tan\alpha b}+\left\{\frac{4\alpha\beta\xi^2}{(\xi^2-\beta^2)^2}\right\}^{\pm 1} = 0, \qquad \begin{cases}+1 = \text{symmetric}\\ -1 = \text{antisymmetric}\end{cases} \tag{8.1.61}$$

where, we recall,

$$\alpha^2 = \omega^2/c_1^2-\xi^2, \qquad \beta^2 = \omega^2/c_2^2-\xi^2. \tag{8.1.62}$$

Hence the problem is: given the frequency ω, determine the wavenumbers satisfying the Rayleigh–Lamb equation and, from the relationship $\omega = \xi c$, establish the propagation velocity of the waves. Although the frequency equation was derived long ago and is fairly simple in appearance, a complete understanding of the spectrum, including the behaviour of the higher modes and complex branches, has come about only comparatively recently.

8.1.4. *The general frequency equation for a plate*

Our objective is to study in detail the Rayleigh–Lamb equation. Before doing so, a more general development of the frequency equations for the various types of waves in a plate will be presented. The analysis will recover the symmetric and antisymmetric waves of the Rayleigh–Lamb case as well as the previously-studied SH wave case. The results will show that it is not necessary to consider the various wave-types independently, but that they resolve themselves in the natural course of analysis. The work is based on that of Meeker and Meitzler [48].

Thus, again consider straight-crested waves propagating in a plate in the positive x direction, where the governing equations are

$$\mathbf{u} = \nabla\Phi + \nabla\times\mathbf{H}, \qquad \nabla.\mathbf{H} = 0,$$
$$\nabla^2\Phi = \frac{1}{c_1^2}\frac{\partial^2\Phi}{\partial t^2}, \qquad \nabla^2\mathbf{H} = \frac{1}{c_2^2}\frac{\partial^2\mathbf{H}}{\partial t^2}. \tag{8.1.63}$$

Since we wish to generalize somewhat from plane strain, all three components of the vector potential must be retained. The displacements are

$$u_x = \frac{\partial\Phi}{\partial x} + \frac{\partial H_z}{\partial y},$$
$$u_y = \frac{\partial\Phi}{\partial y} - \frac{\partial H_z}{\partial x}, \tag{8.1.64}$$
$$u_z = -\frac{\partial H_x}{\partial y} + \frac{\partial H_y}{\partial x},$$

where variations in the z direction have been excluded, so that $\partial/\partial z = 0$.

As before, we consider solutions of the general form

$$\Phi = f(y)e^{i(\xi x - \omega t)}, \qquad H_x = h_x(y)e^{i(\xi x - \omega t)},$$
$$H_y = h_y(y)e^{i(\xi x - \omega t)}, \qquad H_z = h_z(y)e^{i(\xi x - \omega t)}, \tag{8.1.65}$$

where, in following the notation of Reference [48], we have not introduced additional factors of i as was done previously in (8.1.21). Substitution in the differential eqns (8.1.63) gives

$$\begin{aligned}\Phi &= (A \cos \alpha y + B \sin \alpha y)e^{i(\xi x - \omega t)},\\ H_x &= (C \cos \beta y + D \sin \beta y)e^{i(\xi x - \omega t)},\\ H_y &= (E \cos \beta y + F \sin \beta y)e^{i(\xi x - \omega t)},\\ H_z &= (G \cos \beta y + H \sin \beta y)e^{i(\xi x - \omega t)}.\end{aligned} \tag{8.1.66}$$

The displacements are then given by

$$\begin{aligned}u_x &= \{i\xi(A \cos \alpha y + B \sin \alpha y) + \beta(-G \sin \beta y + H \cos \beta y)\}e^{i(\xi x - \omega t)},\\ u_y &= \{\alpha(-A \sin \alpha y + B \cos \alpha y) - i\xi(G \cos \beta y + H \sin \beta y)\}e^{i(\xi x - \omega t)},\\ u_z &= \{-\beta(-C \sin \beta y + D \cos \beta y) + i\xi(E \cos \beta y + F \sin \beta y)\}e^{i(\xi x - \omega t)}.\end{aligned} \tag{8.1.67}$$

The boundary conditions are as previously given by (8.1.50), where we write the stresses in the form

$$\begin{aligned}\tau_{yy} &= (\lambda + 2\mu)\frac{\partial u_y}{\partial y} + \lambda\frac{\partial u_x}{\partial x},\\ \tau_{yx} &= \mu\left(\frac{\partial u_y}{\partial x} + \frac{\partial u_x}{\partial y}\right),\\ \tau_{yz} &= \mu\frac{\partial u_z}{\partial y}.\end{aligned} \tag{8.1.68}$$

The boundary conditions yield six equations in the eight unknown constants $A, B, \ldots, H$. The remaining two equations result from the divergence condition on $\mathbf{H}$, given by

$$\frac{\partial H_x}{\partial x} + \frac{\partial H_y}{\partial y} = 0. \tag{8.1.69}$$

If the results for H_x, H_y are substituted in the above, and real and imaginary parts equated to zero, two equations result which would permit either C or D and E or F to be eliminated from the boundary condition equations. Proceeding in an alternative manner as in Reference [48], we evaluate (8.1.69) at $y = \pm b$ to generate two additional equations. The resulting boundary condition and divergence condition equations are

$$\begin{aligned}&\{(\lambda + 2\mu)\alpha^2 + \lambda\xi^2\}(A \cos \alpha b + B \sin \alpha b) + 2i\mu\xi\beta(-G \sin \beta b + H \cos \beta b) = 0,\\ &\{(\lambda + 2\mu)\alpha^2 + \lambda\xi^2\}(A \cos \alpha b - B \sin \alpha b) + 2i\mu\xi\beta(G \sin \beta b + H \cos \beta b) = 0,\\ &\beta^2(C \cos \beta b + D \sin \beta b) + i\xi\beta(-E \sin \beta b + F \cos \beta b) = 0,\\ &\beta^2(C \cos \beta b - D \sin \beta b) + i\xi\beta(E \sin \beta b + F \cos \beta b) = 0,\\ &2i\xi\alpha(-A \sin \alpha b + B \cos \alpha b) + (\xi^2 - \beta^2)(G \cos \beta b + H \sin \beta b) = 0,\\ &2i\xi\alpha(A \sin \alpha b + B \cos \alpha b) + (\xi^2 - \beta^2)(G \cos \beta b - H \sin \beta b) = 0,\\ &\beta(-E \sin \beta b + F \cos \beta b) + i\xi(C \cos \beta b + D \sin \beta b) = 0,\\ &\beta(E \sin \beta b + F \cos \beta b) + i\xi(C \cos \beta b - D \sin \beta b) = 0.\end{aligned} \tag{8.1.70}$$

This constitutes a system of eight homogeneous equations in the constants $A, B, \ldots, H$. A necessary and sufficient condition for the existence of a solution is that the determinant of coefficients must vanish. This determinant is given by

$$\begin{vmatrix} c\cos\alpha b & c\sin\alpha b & 0 & 0 & -f\sin\beta b & f\cos\beta b & 0 & 0 \\ c\cos\alpha b & -c\sin\alpha b & 0 & 0 & f\sin\beta b & f\cos\beta b & 0 & 0 \\ 0 & 0 & -h\sin\beta b & h\cos\beta b & 0 & 0 & \beta^2\cos\beta b & \beta^2\sin\beta b \\ 0 & 0 & h\sin\beta b & h\cos\beta b & 0 & 0 & \beta^2\cos\beta b & -\beta^2\sin\beta b \\ -d\sin\alpha b & d\cos\alpha b & 0 & 0 & g\cos\beta b & g\sin\beta b & 0 & 0 \\ d\sin\alpha b & d\cos\alpha b & 0 & 0 & g\cos\beta b & -g\sin\beta b & 0 & 0 \\ 0 & 0 & -\beta\sin\beta b & \beta\cos\beta b & 0 & 0 & \mathrm{i}\xi\cos\beta b & \mathrm{i}\xi\sin\beta b \\ 0 & 0 & \beta\sin\beta b & \beta\cos\beta b & 0 & 0 & \mathrm{i}\xi\cos\beta b & -\mathrm{i}\xi\sin\beta b \end{vmatrix} = 0, \tag{8.1.71}$$

where c, d, f, g, h are defined by

$$c = \{(\lambda+2\mu)\alpha^2+\lambda\xi^2\},$$
$$d = 2\mathrm{i}\xi\alpha, \quad f = 2\mathrm{i}\mu\xi\beta,$$
$$g = \xi^2-\beta^2, \quad h = \mathrm{i}\xi\beta. \tag{8.1.72}$$

The eight columns of the determinant are associated, respectively, with the constants A, B, E, F, G, H, C, D. By adding and subtracting rows and columns of the coefficient determinant, it is possible to recast it to the form

$$\begin{matrix} B & G & E & D & A & H & C & F \end{matrix}$$
$$\begin{vmatrix} \text{---} & \text{---} & 0 & 0 & 0 & 0 & 0 & 0 \\ \text{---} & \text{---} & 0 & 0 & 0 & 0 & 0 & 0 \\ 0 & 0 & \text{---} & \text{---} & 0 & 0 & 0 & 0 \\ 0 & 0 & \text{---} & \text{---} & 0 & 0 & 0 & 0 \\ 0 & 0 & 0 & 0 & \text{---} & \text{---} & 0 & 0 \\ 0 & 0 & 0 & 0 & \text{---} & \text{---} & 0 & 0 \\ 0 & 0 & 0 & 0 & 0 & 0 & \text{---} & \text{---} \\ 0 & 0 & 0 & 0 & 0 & 0 & \text{---} & \text{---} \end{vmatrix} = 0, \tag{8.1.73}$$

where the constants are positioned with the appropriate columns. A series of manipulations that takes (8.1.71) to the form (8.1.73) is as follows (use R and C to indicate 'row' and 'column,' with R1 being the top row, C1 being the left column):

(1) add R1 to R2, R3 to R4, R5 to R6, R7 to R8;
(2) subtract R2 from R1, R4 from R3, R6 from R5, R8 from R7;
(3) interchange R2, R6, R4, R7, leaving R1, R3, R5, R8 in place;
(4) interchange C2 to C1, C1 to C5, C5 to C2; interchange C4, C8; leave C3, C6, C7 in place.

The determinant (8.1.73) may be expanded to the product of four subdeterminants,

$$\begin{vmatrix} \mathrm{i}\xi\cos\beta b & \beta\cos\beta b \\ \beta^2\cos\beta b & h\cos\beta b \end{vmatrix} \times \begin{vmatrix} -\beta\sin\beta b & \mathrm{i}\xi\sin\beta b \\ h\sin\beta b & \beta^2\sin\beta b \end{vmatrix} \times \begin{vmatrix} c\cos\alpha b & f\cos\beta b \\ -d\sin\alpha b & g\sin\beta b \end{vmatrix} \times$$

$$\times \begin{vmatrix} g\cos\beta b & d\cos\alpha b \\ f\sin\beta b & c\sin\alpha b \end{vmatrix} = 0. \quad (8.1.74)$$

The coefficients associated with the subdeterminants are

$$|C, F| \times |E, D| \times |A, H| \times |B, G| = 0. \quad (8.1.75)$$

Various solutions to (8.1.74) are possible for various non-zero combinations of the constants. Thus

Solution I: $A, B, D, E, G, H = 0$, $C, F \neq 0$ gives

$$u_x = u_y = 0, \quad (8.1.76)$$
$$u_z = (C\beta + \mathrm{i}F\xi)\sin\beta y \exp\{\mathrm{i}(\xi x - \omega t)\}.$$

Solution II: $A, B, C, F, G, H = 0$, $D, E \neq 0$ gives

$$u_x = u_y = 0, \quad (8.1.77)$$
$$u_z = (-\beta D + \mathrm{i}\xi E)\cos\beta y \exp\{\mathrm{i}(\xi x - \omega t)\}.$$

Solution III: $B, C, D, E, F, G = 0$, $A, H \neq 0$ gives

$$u_x = (\mathrm{i}\xi A\cos\alpha y + \beta H\cos\beta y)\exp\{\mathrm{i}(\xi x - \omega t)\}, \quad (8.1.78)$$
$$u_y = -(\alpha A\sin\alpha y + \xi H\sin\beta y)\exp\{\mathrm{i}(\xi x - \omega t)\},$$
$$u_z = 0.$$

Solution IV: $A, C, D, E, F, H = 0$, $B, G \neq 0$ gives

$$u_x = (\mathrm{i}\xi B\sin\alpha y - \beta G\sin\beta y)\exp\{\mathrm{i}(\xi x - \omega t)\},$$
$$u_y = (\alpha B\cos\alpha y - \mathrm{i}\xi G\cos\beta y)\exp\{\mathrm{i}(\xi x - \omega t)\}, \quad (8.1.79)$$
$$u_z = 0.$$

Allowing for difference in coefficient notation and the inclusion of i in basic definitions, it may be seen that (8.1.78) and (8.1.79) correspond to the previously derived Rayleigh–Lamb results (8.1.54) and (8.1.59). We have, in addition, obtained the cases of antisymmetric (8.1.76) and symmetric (8.1.77) SH modes in the more general derivation.

The frequency equations are obtained in the usual way, by expanding the various determinants. These give the following.

Solution I: $C, F \neq 0$. The determinant in question is

$$\begin{vmatrix} \mathrm{i}\xi \cos \beta b & \beta \cos \beta b \\ \beta^2 \cos \beta b & \mathrm{i}\xi\beta \cos \beta b \end{vmatrix} = 0, \tag{8.1.80}$$

giving

$$\beta(\xi^2+\beta^2)\cos^2 \beta b = 0. \tag{8.1.81}$$

Solution II: $D, E \neq 0$. We have

$$\begin{vmatrix} -\beta \sin \beta b & \mathrm{i}\xi \sin \beta b \\ -\mathrm{i}\xi\beta \sin \beta b & \beta^2 \sin \beta b \end{vmatrix} = 0, \tag{8.1.82}$$

giving

$$\beta(\xi^2+\beta^2)\sin^2 \beta b = 0. \tag{8.1.83}$$

Solutions III, IV: The results for these cases are precisely those derived previously for the Rayleigh–Lamb case.

8.1.5. *Analysis of the Rayleigh–Lamb equation*

In our investigation of SH waves in plates and P and SV waves in mixed boundary condition plates, we found the situation considerably simpler than for the Rayleigh–Lamb case. Thus, for SH waves, only a single wave-type could exist, and the frequency equation was quite simple and permitted analytical solution. For the case of mixed boundary conditions, two types of waves were possible, but they were uncoupled from one another. Further, the frequency equations were again quite simple and permitted analytical solution. For the Rayleigh–Lamb case, none of these situations hold. Both P and SV waves exist for any given mode owing to mode conversion at the traction-free surfaces. Further, the frequency equation does not permit simple analytical solution. We shall proceed through several steps in analysing this case.

1. *Physical interpretation of the Rayleigh–Lamb modes.* The displacements in the plate are given by (8.1.27) and (8.1.28). The nature of the symmetric and antisymmetric displacement components has been previously illustrated by Fig. 8.4. Now, representing $\sin \alpha y$, $\cos \alpha y$, $\sin \beta y$, and $\cos \beta y$ in exponential form, we see that it is possible to express the displacement u_x as

$$u_x = \mathrm{i}\left[\frac{A\xi}{2\mathrm{i}}\left\{\left(1+\mathrm{i}\frac{B}{A}\right)\mathrm{e}^{\mathrm{i}(\xi x+\alpha y-\omega t)}+\left(1-\mathrm{i}\frac{B}{A}\right)\mathrm{e}^{\mathrm{i}(\xi x-\alpha y-\omega t)}\right\}+\right.$$
$$\left.+\frac{C\beta}{2}\left\{\left(1+\mathrm{i}\frac{D}{C}\right)\mathrm{e}^{\mathrm{i}(\xi x-\beta y-\omega t)}+\left(1-\mathrm{i}\frac{D}{C}\right)\mathrm{e}^{\mathrm{i}(\xi x-\beta y-\omega t)}\right\}\right], \tag{8.1.84}$$

with a similar expression holding for u_y. For a given frequency, the amplitude ratios B/A, D/C would be fixed. This explicitly shows the plane-wave nature of the disturbance.

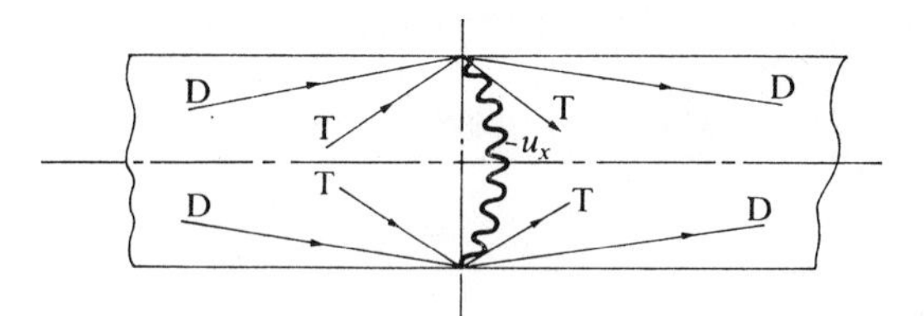

FIG. 8.7. Variation of the symmetric displacement u_x resulting from the reflection and interference of dilatational and shear waves. D = dilatational, T = transverse. (After Redwood [80, Fig. 5.1].)

Figure 8.7 shows how we may consider the modes of propagation in a plate using the plane-wave representation. Consider a pair of dilatational waves to be incident at such an angle that their reflection and interference produces u_x having a symmetric one-half cycle variation across y. Consider a pair of shear waves incident at a smaller angle. Since they are at the same frequency, their reflection and interference would produce a greater number of cyclic variations. The superposition of the two effects yields the results shown for u_x.

2. *Various regions of the Rayleigh–Lamb equation.* We recall that

$$\begin{aligned} \alpha^2 &= \omega^2/c_1^2 - \xi^2 = \xi^2(c^2/c_1^2 - 1), \\ \beta^2 &= \omega^2/c_2^2 - \xi^2 = \xi^2(c^2/c_2^2 - 1). \end{aligned} \tag{8.1.85}$$

Hence, depending on whether $\xi^2 \gtreqless \omega^2/c_1^2,\ \omega^2/c_2^2$ or whether $c^2 \gtreqless c_1^2,\ c_2^2$, we may have α, β being real, zero, or imaginary. Then the frequency equation is correspondingly altered as follows.

Region I, $\xi > \omega/c_2$. It follows also that $\xi > \omega/c_1$ and that $c < c_2, c_1$. Then we replace α, β in the frequency equation by $i\alpha'$, $i\beta'$, where $\alpha'^2 = -\alpha^2$, $\beta'^2 = -\beta^2$, and obtain

$$\frac{\tanh \beta' b}{\tanh \alpha' b} = \left\{\frac{4\alpha'\beta'\xi^2}{(\xi^2 - \beta'^2)^2}\right\}^{\pm 1}. \tag{8.1.86}$$

Region II, $\omega/c_2 > \xi > \omega/c_1$. It follows that $c_2 < c < c_1$, and we replace α by $i\alpha'$ in the frequency equation, giving

$$\frac{\tan \beta b}{\tanh \alpha' b} = \pm\left\{\frac{4\alpha'\beta\xi^2}{(\xi^2 - \beta^2)^2}\right\}^{\pm 1}. \tag{8.1.87}$$

Region III, $\xi < \omega/c_1$. It follows that $c > c_1$, and the frequency equation has the previously derived form (8.1.61).

3. *Reduction to thin-plate results.* Let us consider the case when the transverse wavelength with respect to the thickness is quite large, so that $2\pi/\beta$, $2\pi/\alpha \gg b$. The analysis of regions I and II yield the results of interest.

Region I. Consider both symmetric and antisymmetric cases. For the former, there are no roots. For the antisymmetric case, we expand the hyperbolic tangent

$$\tanh x = x(1-\tfrac{1}{3}x^2+\ldots). \tag{8.1.88}$$

Retaining the first two terms, the frequency equation is reduced to

$$\frac{\beta'(1-\frac{1}{3}\beta'^2b^2)}{\alpha'(1-\frac{1}{3}\alpha'^2b^2)} = \frac{(\xi^2+\beta'^2)^2}{4\xi^2\alpha'\beta'}. \tag{8.1.89}$$

Put this in the form

$$-(\xi^2-\beta'^2) = \tfrac{4}{3}\xi^2\beta'^4b^2-\tfrac{1}{3}\alpha'^2b^2(\xi^2+\beta'^2)^2. \tag{8.1.90}$$

Discarding terms of higher order than $(c/c_2)^4$, we obtain

$$\frac{c}{c_2} = 2\xi b\left\{\frac{1}{3}\left(1-\frac{1}{k^2}\right)\right\}^{\frac{1}{2}}. \tag{8.1.91}$$

Using $c_1^2 = k^2c_2^2$, $k^2 = 2(1-\nu)/(1-2\nu)$, and $c_2^2 = \mu/\rho$, this becomes

$$c = \xi b\left\{\frac{E}{3\rho(1-\nu^2)}\right\}^{\frac{1}{2}}. \tag{8.1.92}$$

This result, with the linear dependence of c on ξ, agrees with that derived from classical plate theory.† It pertains, of course, to the flexural vibration case, and represents only a single vibrational mode in a limited frequency range in the over-all frequency spectrum.

Region II: The antisymmetric case has no roots. The symmetric case becomes

$$\frac{\beta}{\alpha'} = \frac{4\xi^2\alpha'\beta}{(\xi^2-\beta^2)^2}, \tag{8.1.93}$$

giving

$$\frac{c}{c_2} = 2\left(1-\frac{1}{k^2}\right)^{\frac{1}{2}}, \tag{8.1.94}$$

or

$$c = \left\{\frac{E}{\rho(1-\nu^2)}\right\}^{\frac{1}{2}}. \tag{8.1.95}$$

This is the thin-plate or plane-stress analogue of the bar velocity $c_0 = (E/\rho)^{\frac{1}{2}}$ of longitudinal rod theory.

4. *Lamé modes.* A special class of exact solutions, called the Lamé modes but evidently first identified by Lamb in 1917, can be obtained by considering the special case of $\xi = \beta$. Then from the definition of α^2 and β^2 we have that $\alpha^2 = \omega^2/c_1^2-\beta^2$, $2\xi^2 = \omega^2/c_2^2$. The roots for this case are in region II. The frequency equation reduces to:

$$\begin{aligned} &\text{symmetric: } \tan\beta b \to \infty, \quad \beta = n\pi/2b \quad (n = 1, 3, \ldots),\\ &\text{antisymmetric: } \tan\beta b = 0, \quad \beta = n\pi/2b \quad (n = 0, 2, 4, \ldots). \end{aligned} \tag{8.1.96}$$

† See Chapter 4, eqn (4.2.39).

The frequency is given by

$$\omega = \sqrt{2}\xi c_2 = \pi n c_2/\sqrt{2}b \qquad (n = 0, 1, 2, \ldots). \tag{8.1.97}$$

These modes will be examined in more detail in our analysis of bounded plates.

5. *Cutoff frequencies.* The cutoff frequencies for the various plate modes will be obtained by considering $\xi \to 0$. For this limiting value, the Rayleigh–Lamb equation reduces to:

$$\begin{aligned} &\text{symmetric: } \sin \beta b \cos \alpha b = 0, \\ &\text{antisymmetric: } \sin \alpha b \cos \beta b = 0. \end{aligned} \tag{8.1.98}$$

Symmetric case (thickness modes):

$$\begin{aligned} &\cos \alpha b = 0, \qquad \alpha b = \pi p/2 \qquad (p = 1, 3, 5, \ldots), \\ &\sin \beta b = 0, \qquad \beta b = \pi q/2 \qquad (q = 0, 2, 4, \ldots). \end{aligned} \tag{8.1.99}$$

From (8.1.51) and (8.1.52), we see that the displacements are now given by

$$u_x = \mathrm{i}\beta C \cos \beta y \mathrm{e}^{-\mathrm{i}\omega t}, \qquad u_y = -\alpha B \sin \alpha y \mathrm{e}^{-\mathrm{i}\omega t}. \tag{8.1.100}$$

For the case of $\cos \alpha b = 0$, we see from the boundary condition eqns (8.1.53) that $C = 0$, so that

$$u_x = 0, \qquad u_y = -\alpha B \sin \alpha y \mathrm{e}^{-\mathrm{i}\omega t}, \qquad \alpha = \pi p/2b. \tag{8.1.101}$$

Such a mode is called a 'thickness-stretch' mode. For the case of $\sin \beta b = 0$, we have $B = 0$ and

$$u_x = \mathrm{i}\beta C \cos \beta y \mathrm{e}^{-\mathrm{i}\omega t}, \qquad u_y = 0, \qquad \beta = \pi q/2b. \tag{8.1.102}$$

These are called the 'thickness-shear' modes.

Antisymmetric case: From (8.1.98) we have

$$\begin{aligned} &\sin \alpha b = 0, \qquad \alpha b = \pi p/2 \qquad (p = 0, 2, 4, \ldots), \\ &\cos \beta b = 0, \qquad \beta b = \pi q/2 \qquad (q = 2, 4, 6, \ldots). \end{aligned} \tag{8.1.103}$$

From (8.1.56) and (8.1.57) the displacements are

$$u_x = -\mathrm{i}\beta D \sin \beta y \mathrm{e}^{-\mathrm{i}\omega t}, \qquad u_y = \alpha A \cos \alpha y \mathrm{e}^{-\mathrm{i}\omega t}, \tag{8.1.104}$$

with the boundary condition equations being given by (8.1.58). For $\sin \alpha\beta = 0$, we obtain $D = 0$ and

$$u_x = 0, \qquad u_y = \alpha A \cos \alpha y \mathrm{e}^{-\mathrm{i}\omega t}, \qquad \alpha = \pi p/2b. \tag{8.1.105}$$

For $\cos \beta b = 0$, we have $A = 0$ and

$$u_x = -\mathrm{i}\beta D \sin \beta y \mathrm{e}^{-\mathrm{i}\omega t}, \qquad u_y = 0, \qquad \beta = \pi q/2b. \tag{8.1.106}$$

6. *The short-wavelength limit.* Some information on the asymptotic behaviour of the branches is obtainable by letting $\xi \to \infty$. For region I, we have

$\tanh \beta' b / \tanh \alpha' b \to 1$ as $\xi \to \infty$, so that the frequency equation reduces to

$$4\xi^2\alpha'\beta' = (\xi^2 + \beta'^2)^2, \tag{8.1.107}$$

for the symmetric and antisymmetric cases. This is merely the Rayleigh surface wave equation. The Rayleigh results enter here since, for such small wavelengths, the finite-thickness plate appears as a semi-infinite media. Hence vibrational energy is transmitted mainly along the surface of the plate.

For region II, we have, as $\xi \to \infty$, that

$$\frac{\{4(\alpha'/\xi)(\beta/\xi)\}^{\pm 1}}{(1-\beta^2/\xi^2)^2} \to 0. \tag{8.1.108}$$

Consider first the symmetric case (positive exponent) and note that

$$\frac{\alpha'}{\xi} = \left(1 - \frac{c^2}{c_1^2}\right)^{\frac{1}{2}}, \qquad \frac{\beta}{\xi} = \left(\frac{c^2}{c_2^2} - 1\right)^{\frac{1}{2}}. \tag{8.1.109}$$

As $\xi \to \infty$, the product $(\alpha'/\xi)(\beta/\xi) \to 0$ and the question is whether $c \to c_1$ or c_2 to give this result. Since $1/c_1 < 1/c_2$, we conclude that β/ξ will approach zero the fastest. Hence, as $\xi \to \infty$, $c \to \sqrt{2}c_2$. The same results hold for the antisymmetric case (negative exponent), as can be shown by writing the right-hand side of (8.1.87) as

$$\frac{(2 - c^2/c_2^2)^2}{(1 - c^2/c_1^2)^{\frac{1}{2}}(c^2/c_2^2 - 1)^{\frac{1}{2}}} \to 0. \tag{8.1.110}$$

Finally, for region III, there are no roots in the short-wavelength limit.

7. *The grid of bounds.* The analysis of the Rayleigh–Lamb equation thus far has extracted some information on the behaviour of the branches of the frequency spectrum and dispersion curves for plate waves. This information is rather limited, however, and the details of the spectrum are far from complete. However, Mindlin† has developed a technique for constructing a grid of bounding curves which define the spectrum exactly at selected points and approximately (within the bounds) in the intervening regions. We proceed to develop the basic aspects of the technique.

We first recall the analysis of waves in a plate under mixed boundary conditions, presented in the previous section. The boundary conditions were given by $u_y = \tau_{xy} = 0$, $y = \pm b$. The frequency equations for this case were given by (8.1.36) and (8.1.40). The frequency spectrum was given by Fig. 8.6.

We now consider the case of a plate having *elastically restrained* boundaries. This is specified by the boundary conditions

$$\tau_{yy} = \mp e u_y, \qquad \tau_{xy} = 0, \qquad y = \pm b, \tag{8.1.111}$$

† Mindlin [63, pp. 210–14] gives a thorough exposition of the technique and interpretation of the spectrum.

where e is the elastic modulus of the foundation. By letting $e \to 0, \infty$ the cases of traction-free and mixed boundary conditions are recovered. From the previously given expressions for displacements and stresses in a plate (8.1.27), (8.1.28), (8.1.30), and (8.1.31), the preceding boundary conditions give:

Symmetric case ($A = D = 0$):

$$\mu\{B(\xi^2-\beta^2)\cos \alpha b+2C\xi\beta \cos \beta b\} = e(B\alpha \sin \alpha b-C\xi \sin \beta b),$$
$$2B\xi\alpha \sin \alpha b-C(\xi^2-\beta^2)\sin \beta b = 0. \tag{8.1.112}$$

Antisymmetric case ($B = C = 0$):

$$\mu\{A(\xi^2-\beta^2)\sin \alpha b-2D\xi\beta \sin \beta b\} = -e(A\alpha \cos \alpha b+D\xi \cos \beta b),$$
$$2A\xi\alpha \cos \alpha b+D(\xi^2-\beta^2)\cos \beta b = 0. \tag{8.1.113}$$

The frequency equations for symmetric and antisymmetric waves in an elastically restrained plate are obtainable from the determinants of coefficients.

Now, from the standpoint of complexity, the frequency equations resulting from (8.1.112) and (8.1.113) are considerably more complex than the traction-free Rayleigh–Lamb equation or the rigid-restraint case. However, it is to be expected that these equations would furnish a smooth transition of the frequency spectrum from the case of the rigid boundary ($e \to \infty$), for which the simple results (8.1.36) and (8.1.40) are available, to the traction-free case ($e \to 0$). While conceptually this may be obvious enough, the usefulness of this approach lies in certain special solutions to (8.1.112) and (8.1.113) and their relation to the mixed boundary condition case.

Consider, then, the frequency spectrum for the case of mixed boundary conditions, or rigid restraint, as given in Fig. 8.5. Isolate a portion of the spectrum where the branches of the P and SV waves intersect. A typical region is shown in Fig. 8.8(a). We recall that the m integers refer to the dilatational modes and the n integers refer to the equivoluminal modes. Furthermore, the even integers (both m and n) are symmetric modes and the odd integers are

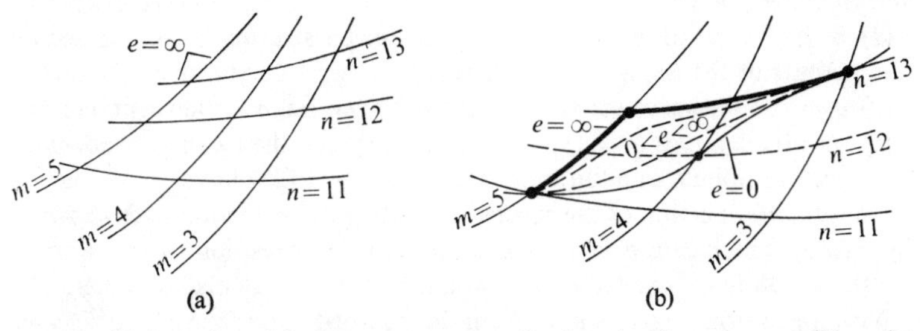

FIG. 8.8. (a) A portion of the frequency spectrum for the mixed boundary condition plate and (b) transition of a portion of the spectrum from $e = 0$ to $e \to \infty$ conditions. (Based on Mindlin [63, Fig. 20].)

antisymmetric modes. Also, the fact that these branches are for the rigid boundary is indicated by $e = \infty$.

The essential stages in determining the transition of the spectrum from rigid to traction-free conditions now arise. Considering first the antisymmetric case, we refer to (8.1.113) and note that these are satisfied by

$$\cos \alpha b = \cos \beta b = 0, \tag{8.1.114}$$

$$A/D = \pm 2\xi\beta/(\xi^2 - \beta^2). \tag{8.1.115}$$

Now (8.1.114) is identical to the antisymmetric mode case for mixed boundary conditions given by (8.1.36). For this type of condition we recall that $\cos \alpha b = 0$, $\alpha b = m\pi/2$ (m odd) represents the dilatational modes and $\cos \beta b = 0$, ($\beta b = n\pi/2$, n odd) represents the shear modes. The point here is that $\cos \alpha b = \cos \beta b = 0$ represents the intersection points of the branches $m = 1, 3, 5, 7, \ldots$ and $n = 1, 3, 5, 7, \ldots$. The significance of this is that, regardless of boundary constraint (that is, $0 \leqslant e \leqslant \infty$), certain common points of the frequency spectrum have been located. In the case of the limited region shown in Fig. 8.8(a), these would be the two points at $m = 5$, $n = 11$ and $m = 3$, $n = 13$.

The next factor that enables us to further structure the spectrum is the special case of $e = 0$. Thus, for this case, we see from (8.1.113) that

$$\sin \alpha b = \sin \beta b = 0, \tag{8.1.116}$$

$$A/D = \pm(\xi^2 - \beta^2)/2\xi\alpha, \tag{8.1.117}$$

represent solutions. Now (8.1.116) are the frequency equations for the previously encountered case of mixed boundary conditions, symmetric modes given by (8.1.40). These branches are the ($m, n =$ even) curves of the spectrum of Fig. 8.5. Hence their simultaneous satisfaction by (8.1.116) represents the intersection points of the (m, n even)-curves. Referring to Fig. 8.8(a) this would be the point $m = 4$, $n = 12$ for the region shown.

With these results, we have the following information regarding the antisymmetric modes:

(1) $0 \leq e \leq \infty$; all curves of the spectrum pass through the intersection points of the ($m, n =$ odd)-curves;

(2) $e = \infty$; the spectrum is represented by the ($m, n =$ odd)-curves;

(3) $e = 0$ all curves of the spectrum pass through the ($m, n =$ even)-intersection points, in addition to those points of (2) above.

The nature of a portion of the spectrum meeting these conditions is shown in Fig. 8.8(b). The location of a specific point in the transition from $e = \infty$ to $e = 0$, excepting those cited above, would have to be calculated numerically.

With the information available from the grid-of-bounds considerations plus that previously available on cutoff frequencies, Lame' modes, and high-frequency limits, a fairly detailed construction of the spectrum becomes

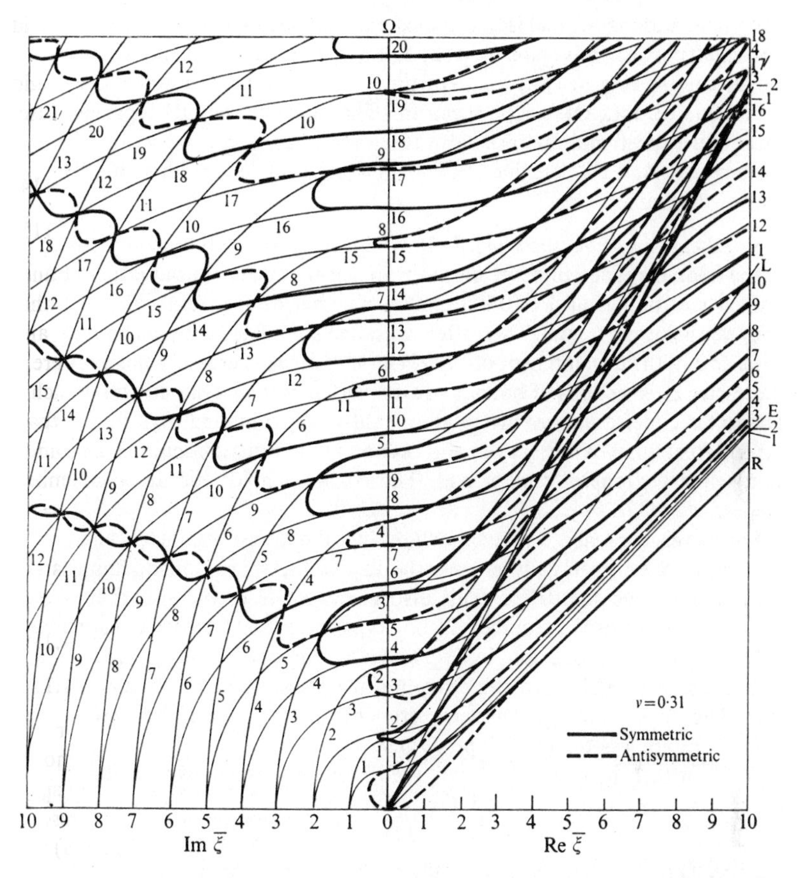

FIG. 8.9. Frequency spectrum for the Rayleigh–Lamb equation. (After Mindlin [63, Fig. 19].)

possible. Additional information on the curvature of the branches at the cutoff frequency (not developed here) is used. The resulting spectrum as given by Mindlin [63] for the case of Poisson's ratio of $\nu = 0{\cdot}31$ is shown in Fig. 8.9. The heavy solid lines of the spectrum correspond to the symmetric modes, the dashed lines to the antisymmetric modes. The thin solid lines are the grid of bounds provided by the frequency spectrum for the mixed boundary condition plate. The integers associated with the various modes are indicated in the imaginary plane. The normalizations of wavenumber and frequency are the same as used previously. Thus

$$\bar{\xi} = 2b\xi/\pi, \qquad \Omega = 2b\omega/\pi c_2. \tag{8.1.118}$$

The integers along the real Ω-axis indicate the mode number, with the integers on the left corresponding to $p = 1, 2, 3, \ldots$ (thickness stretch) and on the right corresponding to $q = 1, 2, 3, \ldots$ (thickness shear). Finally, we note the four rays OR, OE, OL, OD in the spectrum. The rays OE and OD represent the shear wave speed and dilatational wave speed lines, respectively, and divide regions I, II, and III. The line OR is the Rayleigh wave speed line, and OL is that of the Lamé modes.

8. *Discussion of the spectrum.* Following very closely Mindlin's discussion of the frequency spectrum, we start at zero wavenumber. The curvature of the branch and bound at $\bar{\xi} = 0$ determines whether a given branch starts out above or below its bound. Thereafter, a branch is confined between bounds, as described in the development of (7) *The grid of bounds*, crossing them only at successive intersections of bounds m even with n even and m odd with n odd. The antisymmetric modes are predominantly equivoluminal ($|A/D| < 1$) at the intersections m even, n even and predominantly dilatational ($|A/D| > 1$) at the intersections m odd, n odd. The converse holds for the symmetric modes.

For increasing real $\bar{\xi}$, every branch, except the lowest symmetric and antisymmetric ones, eventually crosses the ray OD. As mentioned, this is the dilatational wave speed line. This also corresponds to the situation when $\alpha = 0$, and gives rise to Goodier–Bishop type waves that were investigated for the half-space (see, for example, Fig. 6.12). Further, OD is the bound $m = 0$, so that the symmetric branches cross it at its intersections with the bounds n even. The slopes of all the symmetric branches are the same at these crossover points and are less than the slope of OD.

With further increase of $\bar{\xi}$, all the branches, except the lowest antisymmetric one, cross the ray OL at its intersections with the bounds $n = 1, 2, 3, \ldots$, to which the branches are tangent. At these points, the mode is purely equivoluminal and is composed of SV waves reflecting at 45° angles of incidence (these are the Lamé modes, to be discussed in more detail later). Since the modes of these points are purely equivoluminal, these are the only points that remain fixed as Poisson's ratio is varied.

After crossing OL, all the real branches except the lowest symmetric and antisymmetric ones approach the ray OE, representing the shear-wave velocity. This is in accord with the results obtained in (6) *The short-wavelength limit.* Only the lowest symmetric branch intersects OE, resulting in a Goodier–Bishop wave at that point. Thereafter, the lowest symmetric and antisymmetric branches approach the ray OR asymptotically, corresponding to the Rayleigh surface wave velocity.

9. *Complex branches.* Our considerations thus far have been for real, positive frequency and real or imaginary wavenumbers. It turns out that branches of the frequency spectrum having complex wavenumber also exist.

Referring to our basic definitions (8.1.23) for α, β we see if ξ is real, then α, β may be both real, both imaginary, or α imaginary and β real. If ξ is imaginary, α, β will be real, and if ξ is complex, in general, α, β will both be complex. The resulting spatial behaviour of the waves for some of these possibilities are as follows:

ξ real; α, β real:

$$\mathrm{e}^{\mathrm{i}(\xi x \pm \alpha y - \omega t)}, \qquad \mathrm{e}^{\mathrm{i}(\xi x \pm \beta y - \omega t)}. \tag{8.1.119}$$

ξ real; α, β imaginary:

$$\mathrm{e}^{\pm \alpha y}\mathrm{e}^{\mathrm{i}(\xi x - \omega t)}, \qquad \mathrm{e}^{\pm \beta y}\mathrm{e}^{\mathrm{i}(\xi x - \omega t)}. \tag{8.1.120}$$

ξ imaginary; α, β real:

$$\mathrm{e}^{\pm \xi x}\mathrm{e}^{\mathrm{i}(\pm \alpha y - \omega t)}, \qquad \mathrm{e}^{\pm \xi x}\mathrm{e}^{\mathrm{i}(\pm \beta y - \omega t)}. \tag{8.1.121}$$

ξ complex; α, β complex:

Let

$$\xi = a + ib, \qquad \alpha = c + \mathrm{i}d, \qquad \beta = e + \mathrm{i}f, \tag{8.1.122}$$

$$\mathrm{e}^{\pm bx}\mathrm{e}^{\pm dy}\mathrm{e}^{\mathrm{i}(bx + cy - \omega t)}, \qquad \mathrm{e}^{\pm bx}\mathrm{e}^{\pm fy}\mathrm{e}^{\mathrm{i}(bx + ey - \omega t)}. \tag{8.1.123}$$

We have encountered real and complex wavenumbers previously,† and realize that such wavenumbers play a role in forced vibration and reflection problems.

In considering complex branches of the spectrum, we confine our attention to the first three symmetric modes of the spectrum for which there is a single, complex branch. The representation of the spectrum in Fig. 8.9 is for positive real and positive imaginary wavenumber. The spectrum is actually symmetric with respect to the real and imaginary planes owing to the occurrence of the wavenumber as ξ^2. Figure 8.10 is a three-dimensional representation of a portion of the spectrum containing the first three symmetric modes. A complex branch is seen originating from a minimum point on the second longitudinal mode $L(2)$. Other complex branches originate from other such minima of the spectrum. From the complete spectrum of Fig. 8.9, it is seen that such minima occur rather rarely.

The portion of the $L(2)$ branch between the cutoff frequency and the minimum was an early puzzling feature of the spectrum, since a negative group velocity was indicated. The existence of complex branches, while of significance, particularly in regard to the question of Fourier synthesis of pulses, did not in itself resolve the question of negative group velocity. However, Folk [16] showed that the proper path of integration for a Fourier pulse analysis must be such that the slope $\mathrm{d}\omega/\mathrm{d}\xi$ (or $\mathrm{d}\Omega/\mathrm{d}\bar{\xi}$) is always of the same sign. This requires interpreting the spectrum in terms of just the solid or just the dashed lines in Fig. 8.12.

† See § 3.3 for the case of complex wavenumber.

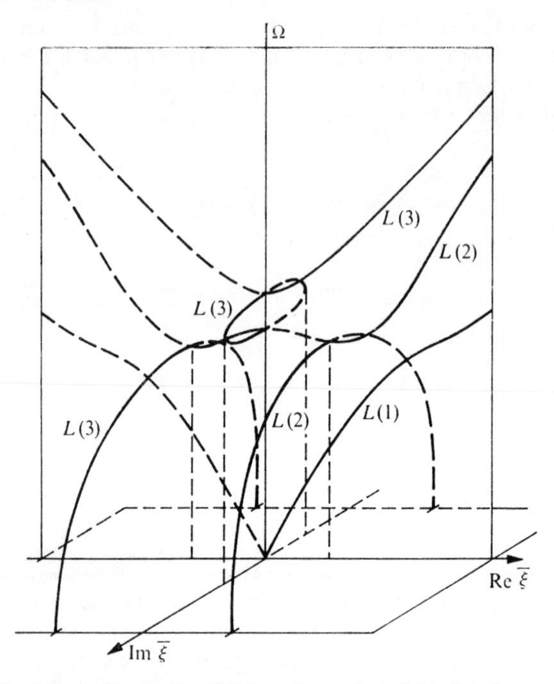

FIG. 8.10. Three-dimensional representation of a portion of the plate frequency spectrum for the first three longitudinal modes. (After Mindlin [63, Fig. 23].)

10. *Dispersion curves.* Information on the phase velocity is obtainable by directly considering the Rayleigh–Lamb equation in terms of c and ξ, or by computing the results from the roots of the frequency spectrum using the relation $\omega = \xi c$. The results for the first few modes are shown in Fig. 8.11 for a Poisson's ratio of $\nu = \frac{1}{3}$. The normalized phase velocity is given by $\bar{c} = c/c_2$. Again, the solid curves are the symmetric, the dashed curves are the antisymmetric modes. The light solid lines represent grid-of-bounds curves.

11. *Status.* We close this section on the Rayleigh–Lamb equation by quoting Mindlin's† evaluation of the understanding in this field.

Lamb [39] in 1917, studied the lowest symmetric and antisymmetric modes. He also identified the cut-off modes, the Lamé [40] modes and certain aspects of the high-frequency spectrum. Bounds α = constant, β = constant were employed by Holden [26], in 1951, to construct a portion of the spectrum of symmetric modes for real wavenumbers and this method was considerably elaborated and extended to include the antisymmetric modes and imaginary phase velocities by Onoe [74]. Analogous bounds were employed by Mindlin [60], in 1951, to construct the branches of a similar transcendental equation, which appears in the theory of

† P. 215 of Reference [63].

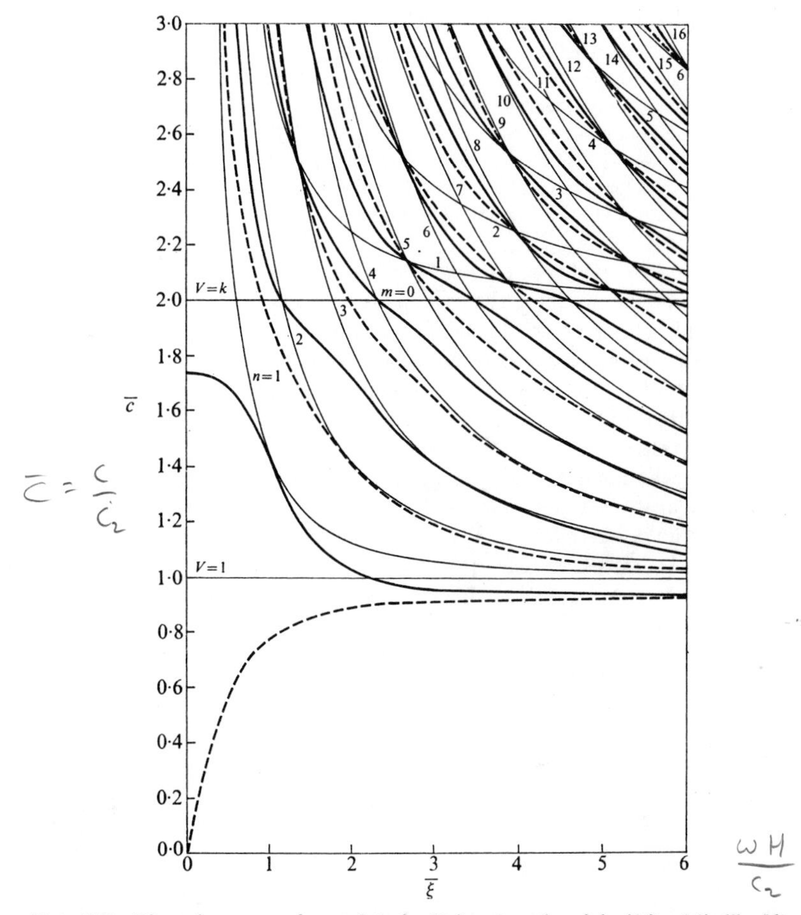

FIG. 8.11. Dispersion curves for a plate for Poisson's ratio of $\frac{1}{3}$. (After Mindlin [63, Fig. 18].)

vibrations of crystal plates, and this method was applied later to (8.1.61). The intricate behavior of the branches in the neighborhood of zero wavenumber and the important role of Poisson's ratio were studied by Onoe [74], Mindlin [36], and Mindlin and Onoe [67]. The imaginary loop that connects the second and third symmetric branches, in a certain range of Poisson's ratio, had been brought to light previously in computations by Aggarwal and Shaw [2], in 1954, and a computation of the family of imaginary branches has been performed by Lyon [44]. The existence of modes at the cut-off frequencies, other than the modes identified by Lamb, was established by Mindlin [62]. The phenomenon of the occasional appearance of phase and group velocities of opposite sign was encountered and explained by Tolstoy and Usdin [87], and the criteria for the existence of this anomaly in any branch were given by Mindlin [61]. The possibility of resolving the waves in the plate into pairs of reflecting P and SV waves was noticed by Harrison [24]. Complex wavenumbers and phase velocities associated with real frequencies were found by Mindlin and Medick [66] in an approximate theory of vibrations of plates. The existence and important aspects of the behaviour of the complex branches of (8.1.61) were established by Onoe [67, pp. 14–17]. At this writing, the understanding of the roots of the equation appears to be reasonably complete except for the behaviour of the higher complex branches when both the real and imaginary parts of the wavenumber are large.

8.1.6. *Circular crested waves in a plate*

We have now considered the case of straight-crested waves in a plate in some detail. However, problems of axisymmetric loading will result in circularly crested waves, and the immediate question is whether the frequency equation is different for such waves. Goodman [21] investigated this aspect and found that, in fact, the Rayleigh–Lamb equation is again generated. We will review briefly the basic aspects of his analysis.

We use cylindrical coordinates r, θ, z with the z coordinate measured perpendicular to the mid plane of the plate. We assume axisymmetric conditions hold, so that $u_\theta = \partial/\partial\theta = 0$. The plate is of thickness $2b$, with traction-free surfaces, so that

$$\tau_{zz} = \tau_{zr} = \tau_{z\theta} = 0, \qquad z = \pm b. \tag{8.1.124}$$

For the axisymmetric case, $\tau_{z\theta} = 0$ always. The resulting form of the displacement equations of motion for this situation is

$$(\lambda+\mu)\frac{\partial\Delta}{\partial r}+\mu\nabla^2 u_r-\mu\frac{u_r}{r^2} = \rho\frac{\partial^2 u_r}{\partial t^2}, \tag{8.1.125}$$

$$(\lambda+\mu)\frac{\partial\Delta}{\partial z}+\mu\nabla^2 u_z = \rho\frac{\partial^2 u_z}{\partial t^2}, \tag{8.1.126}$$

where Δ is the dilatation, given by

$$\Delta = \frac{\partial u_r}{\partial r}+\frac{u_r}{r}+\frac{\partial u_z}{\partial z}. \tag{8.1.127}$$

By analogy with the statical solution of the problem, assume a solution for the dilatation to be given by

$$\Delta = A\frac{\mu}{\lambda+\mu}J_0(\xi r)e^{i\omega t}\begin{cases}\cosh \alpha z\\ \sinh \alpha z\end{cases}, \tag{8.1.128}$$

where A, ξ, α are constants, The cosh, sinh yield, respectively, symmetrical and antisymmetrical motions relative to the mid plane. For a dilatational solution of the preceding form, and referring to (8.1.127), assume the displacements to be of the form

$$u_r = f_1(z)J_1(\xi r)e^{i\omega t}, \qquad u_z = f_2(z)J_0(\xi r)e^{i\omega t}. \tag{8.1.129}$$

Substituting these and (8.1.128) in the definition of the dilatation gives the requirement that

$$\xi f_1+\frac{\mathrm{d}f_2}{\mathrm{d}z} = A\frac{\mu}{\lambda+\mu}\begin{cases}\cosh \alpha z\\ \sinh \alpha z\end{cases}. \tag{8.1.130}$$

Substituting (8.1.128) and (8.1.129) in the equations of motion gives, for the symmetric case of $\cosh \alpha z$,

$$\begin{aligned}\frac{\mathrm{d}^2 f_1}{\mathrm{d}z^2}-\beta^2 f_1 &= A\xi \cosh \alpha z,\\ \frac{\mathrm{d}^2 f_2}{\mathrm{d}z^2}-\beta^2 f_2 &= -A\alpha \sinh \alpha z,\end{aligned} \tag{8.1.131}$$

where $\beta^2 = \xi^2-\omega^2/c_2^2$. For the antisymmetric case, where $\sinh \alpha z$ is used in (8.1.128), the $\cosh \alpha z$, $\sinh \alpha z$ terms are interchanged in (8.1.131). The solutions of (8.1.131), with constants adjusted so as to satisfy the conditions (8.1.130), give

$$\begin{aligned}u_r &= J_1(\xi r)\left\{B \sinh \beta z-B'\frac{\beta}{\xi} \cosh \beta z+A\frac{\xi}{\alpha^2-\beta^2} \cosh \alpha z\right\}\mathrm{e}^{i\omega t},\\ u_z &= J_0(\xi r)\left\{B' \sinh \beta z-B\frac{\xi}{\beta} \cosh \beta z-A\frac{\alpha}{\alpha^2-\beta^2} \sinh \alpha z\right\}\mathrm{e}^{i\omega t},\end{aligned} \tag{8.1.132}$$

where $\alpha^2 = \xi^2-\omega^2/c_1^2$. For the antisymmetric case, it is again only necessary to interchange $\sinh \alpha z$, $\cosh \alpha z$.

The stresses τ_{zz}, τ_{zr} are given for the symmetric case by

$$\tau_{zz} = \mu J_0(\xi r)\left\{2B'\beta \cosh \beta z-2B\xi \sinh \beta z-A\frac{\xi^2+\beta^2}{\alpha^2-\beta^2} \cosh \alpha z\right\}\mathrm{e}^{i\omega t}, \tag{8.1.133}$$

$$\tau_{rz} = \mu J_1(\xi r)\left\{\frac{2\alpha\xi}{\alpha^2-\beta^2}A \sinh \alpha z+B\frac{\xi^2+\beta^2}{\beta} \cosh \beta z-B'\frac{\xi^2+\beta^2}{\beta} \sinh \beta z\right\}\mathrm{e}^{i\omega t}. \tag{8.1.134}$$

In order for the stresses to vanish on $y = \pm b$, the constant B and the determinant of coefficients resulting from (8.1.33) and (8.1.34) must vanish. The resulting frequency equation is

$$\frac{\tanh \beta b}{\tanh \alpha b} = \frac{4\alpha\beta\xi^2}{(\xi^2-\beta^2)^2}. \tag{8.1.135}$$

This we recognize as the Rayleigh–Lamb equation for symmetric waves in a plate in 'region I' of the spectrum.† Considering antisymmetric waves would similarly yield the antisymmetric Rayleigh–Lamb equation.

† See the previous section, eqns (8.1.85)–(8.1.87).

While the objective of reviewing a portion of the Goodman analysis has now been accomplished, namely showing that the Rayleigh–Lamb equation also governs circular-crested waves in a plate, we shall continue a little further. The displacements are now given by

$$u_r = AJ_1(\xi r)\left(\frac{\xi}{\alpha^2-\beta^2}\right)\left(\cosh \alpha z - \frac{\xi^2+\beta^2}{2\xi^2}\frac{\cosh \alpha b}{\cosh \beta b}\cosh \beta z\right)e^{i\omega t},$$
$$u_z = -AJ_0(\xi r)\left(\frac{\alpha}{\alpha^2-\beta^2}\right)\left(\sinh \alpha z - \frac{2\xi^2}{\xi^2+\beta^2}\frac{\sinh \alpha b}{\sinh \beta b}\sinh \beta z\right)e^{i\omega t}, \quad (8.1.136)$$

for the symmetric case. Thus, although the frequency–wavenumber relationship holds whether the waves are straight or circularly crested, the displacements (and stresses) vary according to Bessel functions rather than trigonometric functions, in so far as the radial coordinate is concerned. For large values of r, we have that†

$$J_0(\xi r) \to \frac{\sin \xi r + \cos \xi r}{\sqrt{(\pi \xi r)}}, \qquad J_1(\xi r) \to \frac{\sin \xi r - \cos \xi r}{\sqrt{(\pi \xi r)}}. \quad (8.1.137)$$

Thus far from the origin the motion becomes periodic in r. Actually, 'far' occurs rather rapidly, within four or five zeros of the Bessel function. As r becomes very large, the straight crested behaviour is the limit of circular-crested waves.

8.1.7. *Bounded plates—SH and Lamé modes*

Having considered the propagation of continuous waves in infinite plates, a natural next question is whether the natural frequencies of finite plates can be determined in analogy to the situation in simple plate theory. Generally, the answer to this is negative. Thus the problem of the natural frequencies of a rectangular parallelopiped plate of arbitrary dimensions, say $2a$, $2b$, $2c$, and having traction-free surfaces is presently intractable. Nevertheless, considerations of such problems are relevant and necessary in many practical situations. In this section we will review the simple case of SH modes and the Lamé modes first mentioned in the analysis of the Rayleigh–Lamb equation.

Consider first the case of SH modes, and consider a plate having x and y dimensions of $2a$ and $2b$, as shown in Fig. 8.12(a) and to be infinite in extent in the z direction. For SH waves we have the governing equation given by $\nabla^2 u_z = (\partial^2 u_z/\partial t^2)/c_2^2$, and the non-trivial boundary conditions given by

$$x = \pm a, \qquad \tau_{xz} = \frac{\partial u_z}{\partial x} = 0,$$
$$y = \pm b, \qquad \tau_{zy} = \frac{\partial u_z}{\partial y} = 0. \quad (8.1.138)$$

† McLachlan [46].

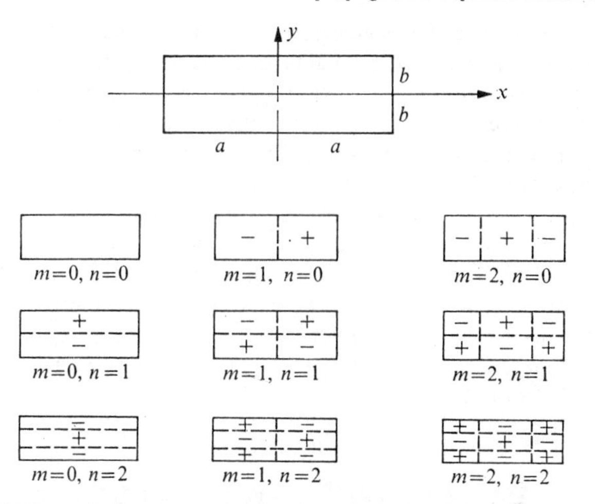

FIG. 8.12. (a) Plate finite in the x, y directions but of infinite extent in the z direction and (b) illustration of the first few SH modes.

The solution to this problem is quite simple, being quite like the case for vibrations of rectangular membranes, except that the present boundary conditions are on the derivative along the boundary. Thus, considering a solution of the form $u_z = X(x)Y(y)\exp(-i\omega t)$, we obtain

$$X_m = A_m \sin \xi_m x + B_m \cos \xi_m x, \qquad \xi_m = m\pi/2a,$$
$$Y_n = C_n \sin \alpha_n y + D_n \cos \alpha_n y, \qquad \alpha_n = n\pi/2b, \tag{8.1.139}$$

where $m, n = 1, 2, 3 \ldots$ and $\alpha_n^2 = \omega_{mn}^2/c_2^2 - \xi_m^2$. The values $m, n = 0$ are also roots but give only rigid-body translation.

For the results (8.1.139) we have modes symmetric with respect to x given by $A_m = 0$, m even, antisymmetric in x by $B_m = 0$, m odd. Modes symmetric in y are given by $C_n = 0$, n even and antisymmetric in y by $D_n = 0$, n odd. The natural frequencies are given by

$$\omega_{mn} = \pi\left\{\left(\frac{m}{2a}\right)^2 + \left(\frac{n}{2b}\right)^2\right\}^{\frac{1}{2}}. \tag{8.1.140}$$

The first few mode shapes are shown in Fig. 8.12(b). Now, our 'bounded' plate in this analysis is only partly so, being infinite in the z direction. The SH solutions satisfying traction-free boundary conditions on $x = \pm a$, $y = \pm b$ would not satisfy such conditions on any surface $z =$ constant. However, if the dimensions of the finite plate were such that $2c \gg 2a$, $2b$ and SH wave type of excitation was given, the results from the present analysis would probably closely predict most modes. However, quite possibly extraneous modes due to the finiteness of the plate would be noted also.

We now consider the Lamé modes. Recall from the analysis of the Rayleigh–Lamb equation (p. 448) that the Lamé modes represented special solutions to the frequency equation given when $\xi = \beta$. It was found (see (8.1.96)) that $\beta = n\pi/2b$, where n odd was the symmetric case, n even was the antisymmetric case. The displacements and stresses are given by (8.1.27)–(8.1.31). Recall, in general, that for the symmetric modes in a plate, we have $A = D = 0$ in the displacement–stress expressions, while for the antisymmetric modes, we have $B = C = 0$. For the special case of the symmetric Lamé modes, we refer to (8.1.53), the boundary condition equations. For $\xi = \beta$ and $\tan \beta b = \infty$, so that $\cos \beta b = 0$, we have that $B = 0$ must occur. For the antisymmetric Lamé modes refer to (8.1.58), where again $\xi = \beta$ and $\sin \beta b = 0$, so that we must have $A = 0$. The resulting potentials, displacements, and stresses are:

Symmetric:

$$\Phi = 0, \qquad H_z = \mathrm{i}C \sin \beta y \mathrm{e}^{\mathrm{i}\psi}, \tag{8.1.141}$$

$$u_x = \mathrm{i}\beta C \cos \beta y \mathrm{e}^{\mathrm{i}\psi}, \qquad u_y = \xi C \sin \beta y \mathrm{e}^{\mathrm{i}\psi}, \tag{8.1.142}$$

$$\tau_{xx} = -2\mu\xi\beta C \cos \beta y \mathrm{e}^{\mathrm{i}\psi},$$

$$\tau_{yy} = 2\mu\xi\beta C \cos \beta y \mathrm{e}^{\mathrm{i}\psi}, \tag{8.1.143}$$

$$\tau_{xy} = 0.$$

Antisymmetric:

$$\Phi = 0, \qquad H_z = \mathrm{i}D \cos \beta y \mathrm{e}^{\mathrm{i}\psi}, \tag{8.1.144}$$

$$u_x = -\mathrm{i}\beta D \sin \beta y \mathrm{e}^{\mathrm{i}\psi}, \qquad u_y = \xi D \cos \beta y \mathrm{e}^{\mathrm{i}\psi}, \tag{8.1.145}$$

$$\tau_{xx} = 2\mu\xi\beta D \sin \beta y \mathrm{e}^{\mathrm{i}\psi},$$

$$\tau_{yy} = 2\mu\xi\beta D \sin \beta y \mathrm{e}^{\mathrm{i}\psi}, \tag{8.1.146}$$

$$\tau_{xy} = 0,$$

where $\psi = \xi x - \omega t$.

A quite meaningful physical interpretation of these results is possible by using exponential representations for $\sin \beta y$, $\cos \beta y$. Thus, for the symmetric u_x displacement, we may write

$$u_x = \mathrm{i}\frac{\beta C}{2}\{\mathrm{e}^{\mathrm{i}(\xi x + \beta y - \omega t)} + \mathrm{e}^{\mathrm{i}(\xi x - \beta y - \omega t)}\}. \tag{8.1.147}$$

Since $\beta = \xi$, it is seen that the result is in terms of plane waves at 45° with respect to the boundaries. We also note that dilatational effects are completely absent ($\Phi = 0$) in this mode and, finally, recall that the case of SV waves incident at 45° was found to be a special case of wave reflection, where no conversion occurs and only SV waves are reflected. We thus conclude that the Lamé modes result from a propagating train of SV waves at 45°. The general

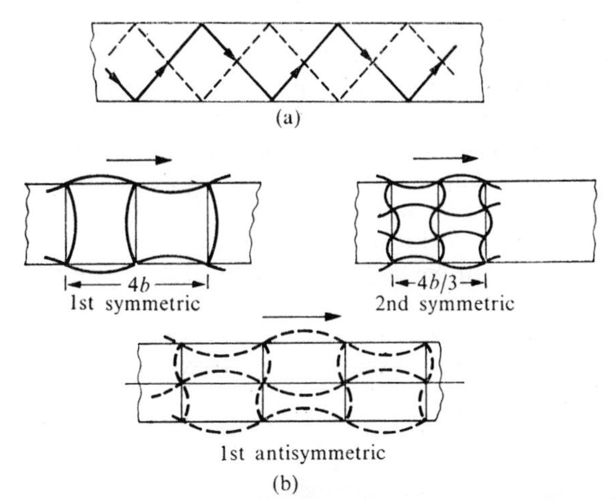

FIG. 8.13. Formation of (a) symmetric Lamé modes by SV waves at 45°, and (b) the first few Lamé modes.

situation for symmetric modes is shown in Fig. 8.13(a), while the first two symmetric modes and the first antisymmetric mode are shown in Fig. 8.13(b).

Our considerations thus far, while interesting, do not seem to relate to bounded plates. Now, a plate bounded by, say, $x = \pm a$ would have to satisfy conditions of $\tau_{xx} = \tau_{xy} = 0$ on these surfaces. We note that $\tau_{xy} = 0$ for the Lamé modes. Further, we note for, say, the symmetric case that $\mathrm{Re}\tau_{xx} = 0$ for $\cos(\xi x - \omega t) = 0$. These represent propagating planes of zero τ_{xx} stress. To achieve the standing-wave case necessary for a bounded plate, merely add an oppositely propagating wave system. This would be obtained, for the symmetric case, directly from (8.1.27)–(8.1.31), where ψ is replaced by $\psi' = -\xi x - \omega t$. For the stress τ_{xx}, this would give

$$\begin{aligned}\tau_{xx} &= -2\mu\xi\beta C \cos \beta y e^{i(\xi x - \omega t)} - 2\mu\xi\beta C \cos \beta y e^{-i(\xi x + \omega t)} \\ &= -4\mu\xi\beta C \cos \beta y \cos \xi x e^{-i\omega t}.\end{aligned} \tag{8.1.148}$$

For the symmetric Lamé modes, $\beta = \xi = n\pi/2b$, n odd. From (8.1.148) we see that

$$\tau_{xx} = 0, \qquad \frac{n\pi x}{2b} = \frac{p\pi}{2}, \qquad p \text{ odd}, \tag{8.1.149}$$

so that $x = pb/n$ give locations of zero τ_{xx} stress. Since $\tau_{xy} = 0$ already, these establish the dimensions of bounded plates capable of free vibration in the Lamé modes. Again, as for the SH wave case, the z dimension of the plate would have to be large compared to the x and y dimensions in order for the Lamé modes to be excited. Mindlin [63, pp. 220–6] has given an extensive discussion of other bounded-plate cases, obtained for mixed boundary conditions and other special situations.

8.2. Waves in circular rods and cylindrical shells

We now extend our considerations to wave effects in solid, circular cylinders and cylindrical shells. The situation here is analogous to that in plates, in that these geometries have been previously studied in the context of strength-of-material theories. It was found that longitudinal and torsional waves were governed by the wave equation and only a single, non-dispersive mode of propagation was possible in each case. The propagation of flexural waves was governed by Bernoulli–Euler theory and such waves propagated dispersively although, again, only a single mode of propagation was possible. Refinements of these simple equations resulted, in the case of rods, by including lateral-inertia effects, and in the case of beams, by considering shear and rotary-inertia effects. Waves in shells were studied using only the membrane theory.

The coverage in this area will be rather brief compared to that of plates. There are many similarities between the two situations which justify briefer coverage. Thus, the SH and longitudinal modes of the plate have their analogues in torsional and longitudinal rod modes. The antisymmetric plate modes are analogous to the flexural rod modes. The major objective will be to present frequency-spectrum and dispersion-curve information.

It is possible, in studying torsional, longitudinal, and flexural waves in rods and shells, to consider each type of motion separately and derive frequency equations for each. This was done, for example, in the case of plates where SH and plane-strain modes were considered. However, as was also done in the case of plates, it is possible to develop a general frequency equation for the problem and to then resolve it into the various modes. This was done by Gazis [20] in the analysis of cylindrical shells, for example. Meeker and Meitzler [48], following the approach of Gazis, developed the Pochhammer–Chree equations for a solid cylinder in this manner. The approach taken in this section will follow that of the cited references.

8.2.1. *The frequency equation for the solid rod*

Consider a solid, circular, cylindrical rod as shown in Fig. 8.14. As in our previous treatment of plate problems, we shall formulate the displacements

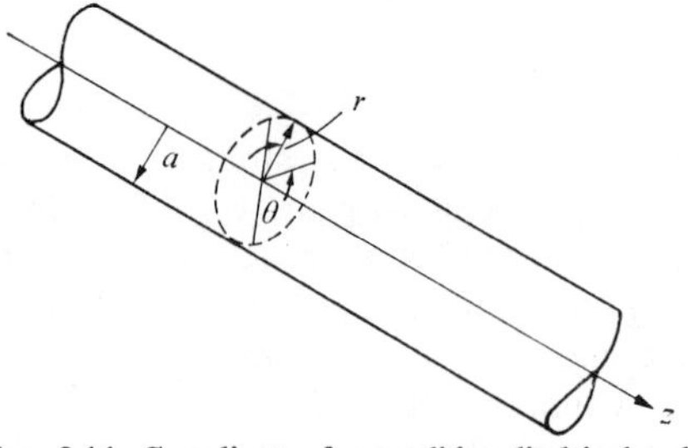

FIG. 8.14. Coordinates for a solid, cylindrical rod.

and stresses in terms of potential functions Φ and $\mathbf{H}$. Cylindrical coordinates are appropriate here, and we write for the displacements

$$\mathbf{u} = \nabla\Phi+\nabla\times\mathbf{H}, \qquad \nabla.\mathbf{H} = F(\mathbf{r}, t). \tag{8.2.1}$$

An alternate condition on the divergence has been chosen in (8.2.1), where $F(\mathbf{r}, t)$ is an arbitrary function. As pointed out by Gazis, the choice of $F(\mathbf{r}, t)$ is arbitrary due to the gauge invariance of the Helmholtz resolution.† The use of this condition will come somewhat later. The scalar components of $\mathbf{u}$ are given by

$$\begin{aligned} u_r &= \frac{\partial\Phi}{\partial r}+\frac{1}{r}\frac{\partial H_z}{\partial\theta}-\frac{\partial H_\theta}{\partial z}, \\ u_\theta &= \frac{1}{r}\frac{\partial\Phi}{\partial\theta}+\frac{\partial H_r}{\partial z}-\frac{\partial H_z}{\partial r}, \\ u_z &= \frac{\partial\Phi}{\partial z}+\frac{1}{r}\frac{\partial}{\partial r}(rH_\theta)-\frac{1}{r}\frac{\partial H_r}{\partial\theta}. \end{aligned} \tag{8.2.2}$$

The potentials Φ and $\mathbf{H}$ satisfy the scalar and vector wave equations,

$$\nabla^2\Phi = \frac{1}{c_1^2}\frac{\partial^2\Phi}{\partial t^2}, \qquad \nabla^2\mathbf{H} = \frac{1}{c_2^2}\frac{\partial^2\mathbf{H}}{\partial t^2}, \tag{8.2.3}$$

where

$$\nabla^2\Phi = \frac{\partial^2\Phi}{\partial r^2}+\frac{1}{r}\frac{\partial\Phi}{\partial r}+\frac{1}{r^2}\frac{\partial^2\Phi}{\partial\theta^2}+\frac{\partial^2\Phi}{\partial z^2}, \tag{8.2.4}$$

$$\nabla^2\mathbf{H} = \left(\nabla^2 H_r-\frac{H_r}{r^2}-\frac{2}{r^2}\frac{\partial H_\theta}{\partial\theta}\right)\mathbf{e}_r+\left(\nabla^2 H_\theta-\frac{H_\theta}{r^2}+\frac{2}{r^2}\frac{\partial H_r}{\partial\theta}\right)\mathbf{e}_\theta+\nabla^2 H_z\mathbf{e}_z. \tag{8.2.5}$$

The stresses are given by Hooke's law, $\tau_{ij} = \lambda\,\Delta\delta_{ij}+2\mu e_{ij}$, where

$$\begin{aligned} e_{rr} &= \frac{\partial u_r}{\partial r}, \qquad e_{\theta\theta} = \frac{1}{r}\frac{\partial u_\theta}{\partial\theta}+\frac{u_r}{r}, \qquad e_{zz} = \frac{\partial u_z}{\partial z}, \\ e_{r\theta} &= \frac{1}{2}\left(\frac{1}{r}\frac{\partial u_r}{\partial\theta}+\frac{\partial u_\theta}{\partial r}-\frac{u_\theta}{r}\right), \qquad e_{rz} = \frac{1}{2}\left(\frac{\partial u_z}{\partial r}+\frac{\partial u_r}{\partial z}\right), \\ e_{\theta z} &= \frac{1}{2}\left(\frac{\partial u_\theta}{\partial z}+\frac{1}{r}\frac{\partial u_z}{\partial\theta}\right). \end{aligned} \tag{8.2.6}$$

The boundary conditions for the problem will be given by

$$\tau_{rr} = \tau_{r\theta} = \tau_{rz} = 0, \quad r = a. \tag{8.2.7}$$

We now consider the conditions under which harmonic waves may propagate in a cylinder. Thus consider Φ, H_r, etc. to be of the general form

$$\Phi = f(r)\Theta_\phi(\theta)\mathrm{e}^{\mathrm{i}(\xi z-\omega t)}, \qquad H_r = h_r(r)\Theta_r(\theta)\mathrm{e}^{\mathrm{i}(\xi z-\omega t)},\ldots. \tag{8.2.8}$$

† See, for example, Morse and Feshbach [70, p. 207].

When these are substituted in the scalar and vector wave equations, sine and cosine solutions result for the θ-dependence. Thus, for Φ we have

$$f''\Theta_\phi+\frac{1}{r}f'\Theta_\phi+\frac{1}{r^2}f\Theta_\phi''-\xi^2 f\Theta_\phi = -\frac{\omega^2}{c_1^2}f\Theta_\phi, \tag{8.2.9}$$

giving

$$r^2\frac{f''}{f}+r\frac{f'}{f}-\left(\xi^2-\frac{\omega^2}{c_1^2}\right)r^2 = -\Theta_\phi''/\Theta_\phi = k^2. \tag{8.2.10}$$

Thus,

$$\Theta_\phi = A\sin k\theta+B\cos k\theta. \tag{8.2.11}$$

Single-valuedness requirements on Θ_ϕ make $k=n$, an integer. Similar solutions also hold for Θ_r, Θ_θ, Θ_z. Furthermore, later requirements on the nature of the θ dependence for the longitudinal, torsional, and flexural modes would lead us to discard either sine or cosine terms in the various Θ results. The resulting expressions for Φ, H_r, H_θ, H_z become

$$\begin{aligned} \Phi &= f(r)\cos n\theta e^{i(\xi z-\omega t)}, \\ H_r &= h_r(r)\sin n\theta e^{i(\xi z-\omega t)}, \\ H_\theta &= h_\theta(r)\cos n\theta e^{i(\xi z-\omega t)}, \\ H_z &= h_z(r)\sin n\theta e^{i(\xi z-\omega t)}. \end{aligned} \tag{8.2.12}$$

We now proceed with the determination of the r dependence for the four functions. Starting with Φ, we have from (8.2.11) that

$$\frac{d^2f}{dr^2}+\frac{1}{r}\frac{df}{dr}+\left(\alpha^2-\frac{n^2}{r^2}\right)f = 0, \tag{8.2.13}$$

where $\alpha^2 = \omega^2/c_1^2-\xi^2$. This is Bessel's equation of order n having the solution

$$f(r) = AJ_n(\alpha r), \tag{8.2.14}$$

where the second solution $Y_n(\alpha r)$ has been discarded because of its singular behaviour at the origin. The equation that results for h_z is similar to (8.2.13) with α^2 replaced by β^2. The solution is

$$h_z(r) = B_3J_n(\beta r), \tag{8.2.15}$$

where $\beta^2 = \omega^2/c_2^2-\xi^2$. The remaining two equations in $h_r(r)$, $h_\theta(r)$ will be coupled, as study of the vector Laplacian (8.2.5) will reveal. The resulting equations are

$$\begin{aligned} &\frac{d^2h_r}{dr^2}+\frac{1}{r}\frac{dh_r}{dr}+\frac{1}{r^2}(-n^2h_r+2nh_\theta-h_r)-\xi^2h_r+\frac{\omega^2}{c_2^2}h_r = 0, \\ &\frac{d^2h_\theta}{dr^2}+\frac{1}{r}\frac{dh_\theta}{dr}+\frac{1}{r^2}(-n^2h_\theta+2nh_r-h_\theta)-\xi^2h_\theta+\frac{\omega^2}{c_2^2}h_\theta = 0. \end{aligned} \tag{8.2.16}$$

These equations may be solved simultaneously for h_r, h_θ. Thus subtract the first from the second to give

$$\left\{\frac{d^2}{dr^2}+\frac{1}{r}\frac{d}{dr}+\beta^2-\frac{(n+1)^2}{r^2}\right\}(h_r-h_\theta) = 0. \tag{8.2.17}$$

This has the solution

$$h_r-h_\theta = 2B_2J_{n+1}(\beta r). \tag{8.2.18}$$

Add the two equations of (8.2.16) to give

$$\left\{\frac{d^2}{dr^2}+\frac{1}{r}\frac{d}{dr}+\beta^2-\frac{(n-1)^2}{r^2}\right\}(h_r+h_\theta) = 0. \tag{8.2.19}$$

This has the solution,

$$h_r+h_\theta = 2B_1J_{n-1}(\beta r). \tag{8.2.20}$$

Adding and subtracting (8.2.18) and (8.2.20), we obtain

$$\begin{aligned} h_r &= B_1J_{n-1}(\beta r)+B_2J_{n+1}(\beta r),\\ h_\theta &= B_1J_{n-1}(\beta r)-B_2J_{n+1}(\beta r). \end{aligned} \tag{8.2.21}$$

There are four constants associated with the components of displacement, with three boundary conditions to be applied. The property of gauge invariance can now be used to eliminate one of the constants, without loss of generality. Setting $B_1 = 0$, which results in $h_r(r) = -h_\theta(r)$, the resulting displacements and some of the stresses are

$$\begin{aligned} u_r &= \{f'+(n/r)h_z+\xi h_r\}\cos n\theta e^{i(\xi z-\omega t)},\\ u_\theta &= \{-(n/r)f+\xi h_r-h_z'\}\sin n\theta e^{i(\xi z-\omega t)},\\ u_z &= \{-\xi f-h_r'-(n+1)h_r/r\}\cos n\theta e^{i(\xi z-\omega t)}, \end{aligned} \tag{8.2.22}$$

$$\begin{aligned} \tau_{rr} &= \left[-\lambda(\alpha^2+\xi^2)f+2\mu\left\{f''+\frac{n}{r}\left(h_z'-\frac{h_z}{r}\right)+\xi h_r'\right\}\right]\cos n\theta e^{i(\xi z-\omega t)},\\ \tau_{r\theta} &= \mu\left[-\frac{2n}{r}\left(f'-\frac{f}{r}\right)-(2h_z''-\beta^2h_z)-\xi\left(\frac{n+1}{r}h_r-h_r'\right)\right]\sin n\theta e^{i(\xi z-\omega t)},\\ \tau_{rz} &= \mu\left[-2\xi f'-\frac{n}{r}\left\{h_r'+\left(\frac{n+1}{r}-\beta^2+\xi^2\right)h_r\right\}-\frac{n\xi}{r}h_z\right]\cos n\theta e^{i(\xi z-\omega t)}. \end{aligned} \tag{8.2.23}$$

The procedure to obtain the frequency equation is to substitute the results for f, h_r, h_θ, h_z in (8.2.23) evaluated at $r = a$. The resulting determinant of coefficients, which yields the frequency equation, is

$$|a_{ij}| = 0 \qquad (i, j = 1, 2, 3), \tag{8.2.24}$$

where

$$\begin{aligned}
a_{11} &= \left\{\frac{\lambda(\alpha^2+\xi^2)(\alpha a)^2}{2\mu\alpha^2}+(\alpha a)^2-n^2\right\}J_n(\alpha a)+\alpha a J_n'(\alpha a),\\
a_{12} &= \{n^2-(\beta a)^2\}J_n(\beta a)-\beta a J_n'(\beta a),\\
a_{13} &= 2n\{\beta a J_n'(\beta a)-J_n(\beta a)\},\\
a_{21} &= n\{\alpha a J_n'(\alpha a)-J_n(\alpha a)\},\\
a_{22} &= -n\{\beta a J_n'(\beta a)-J_n(\beta a)\},\\
a_{23} &= -\{2n^2-(\beta a)^2\}J_n(\beta a)+2\beta a J_n'(\beta a),\\
a_{31} &= -\alpha a J_n'(\alpha a),\\
a_{32} &= -\frac{\beta^2-\xi^2}{2\xi^2}\beta a J_n'(\beta a),\\
a_{33} &= nJ_n(\beta a).
\end{aligned} \tag{8.2.25}$$

8.2.2. *Torsional, longitudinal, and flexural modes in a rod*

The result (8.2.24) is a general frequency equation in the same sense as (8.1.61), obtained for the plate. As in the case of the plate, various special modes, such as torsional, longitudinal, and flexural result from the general case.

1. *Torsional modes*. The family of torsional modes results when only the u_θ displacement is assumed to exist. Such a displacement field is obtained if only $H_z \neq 0$ is assumed, resulting in

$$H_z = B_3 J_0(\beta r)e^{i(\xi z-\omega t)}, \tag{8.2.26}$$

and

$$u_\theta = BJ_1(\beta r)e^{i(\xi z-\omega t)}. \tag{8.2.27}$$

We have replaced $-\beta B_3$ that results from differentiating $J_0(\beta r)$ by B in (8.2.27). The frequency equation for the torsional modes may be obtained by using the boundary condition $\tau_{r\theta} = 0$ and noting from (8.2.23) that this is merely

$$r\frac{\partial}{\partial r}\left(\frac{u_\theta}{r}\right) = 0, \qquad r = a, \tag{8.2.28}$$

or by setting $n = 0$ in (8.2.24). From the latter procedure, (8.2.24) reduces to the term a_{23} and its cofactor matrix. The remaining elements are zero. We thus have,

$$\begin{vmatrix} a_{11}' & a_{12}' \\ a_{31}' & a_{32}' \end{vmatrix} a_{23}' = 0, \tag{8.2.29}$$

where

$$\begin{aligned}
a_{11}' &= \left\{\frac{\lambda(\alpha^2+\xi^2)(\alpha a)^2}{2\mu\alpha^2}+(\alpha a)^2\right\}J_0(\alpha a)+\alpha a J_0'(\alpha a),\\
a_{12}' &= -(\beta a)^2 J_0(\beta a)-\beta a J_0'(\beta a),\\
a_{31}' &= -\alpha a J_0'(\alpha a), \qquad a_{32}' = -\frac{\beta^2-\xi^2}{2\xi^2}\beta a J_0'(\beta a),\\
a_{23}' &= (\beta a)^2 J_0(\beta a)+2\beta a J_0'(\beta a).
\end{aligned} \tag{8.2.30}$$

The present case of torsional waves is governed by $a'_{23} = 0$, giving

$$\beta a J_0(\beta a) = 2J_1(\beta a), \tag{8.2.31}$$

as the frequency equation for torsional waves. This frequency equation would result if (8.2.28) were applied directly. Some of the roots of the torsional wave frequency equation are

$$\beta_1 a = 5{\cdot}136, \quad \beta_2 a = 8{\cdot}417, \quad \beta_3 a = 11{\cdot}62, \dots . \tag{8.2.32}$$

Given a root, say $\beta_p a$, the resulting frequency–wavenumber relation is

$$(\beta_p a)^2 = (\omega a/c_2)^2 - (\xi a)^2. \tag{8.2.33}$$

If this spectrum is plotted, it will be found to be quite similar to the SH modes of a plate (see Fig. 8.3). It is seen that the torsional modes governed by (8.2.33) propagate dispersively.

There is an additional solution of the frequency eqn (8.2.31) given by $\beta = 0$. Upon examining the governing equation for $h_z(r)$ (see comments leading to (8.2.15)), it is found that we have $\mathrm{d}^2h_z/\mathrm{d}r^2 + r^{-1}\,\mathrm{d}h_z/\mathrm{d}r = 0$, having a solution $h_z = A + B\ln r$. This does not yield an acceptable displacement field. However, if one examines the displacement equations of motion under the conditions of the present problem, where $u_\theta = u_\theta(r, z)$, $u_r = u_z = 0$, the only non-trivial equation is

$$\frac{\partial^2 u_\theta}{\partial r^2} + \frac{1}{r}\frac{\partial u_\theta}{\partial r} - \frac{u_\theta}{r^2} + \frac{\partial^2 u_\theta}{\partial z^2} = \frac{1}{c_2^2}\frac{\partial^2 u_\theta}{\partial t^2}. \tag{8.2.34}$$

Considering a solution $u_\theta = U(r)\exp\{\mathrm{i}(\xi z - \omega t)\}$ gives

$$\frac{\mathrm{d}^2 U}{\mathrm{d}r^2} + \frac{1}{r}\frac{\mathrm{d}U}{\mathrm{d}r} + \left(\beta^2 - \frac{1}{r^2}\right)U = 0. \tag{8.2.35}$$

For $\beta = 0$, the resulting solution is

$$U = \frac{A}{r} + Br. \tag{8.2.36}$$

The singular behaviour at $r = 0$ requires $A = 0$. Thus, for $\beta = 0$, we have a displacement field given by

$$u_\theta = Br\mathrm{e}^{\mathrm{i}(\xi z - \omega t)}, \tag{8.2.37}$$

where $\xi = \omega/c_2$. This is the lowest mode of propagation of torsional waves and it is this mode of propagation that is described by the 'strength-of-materials' approach. It represents the exceptional case where elasticity and strength of materials yield the same results. This mode propagates non-dispersively.

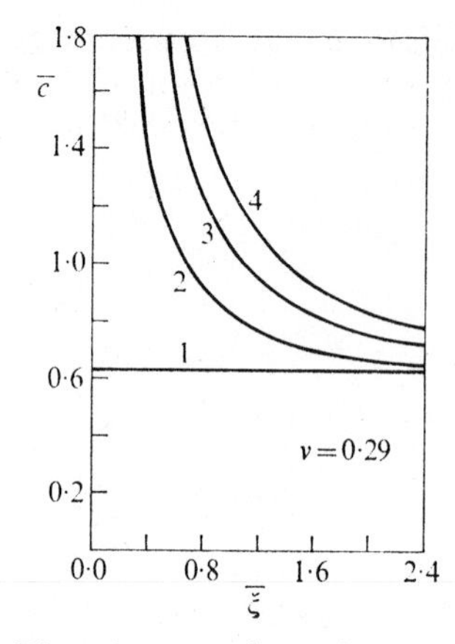

FIG. 8.15. Dispersion curves for torsional waves in a rod.

The dispersion curves for the first four torsional modes ($\beta a = 0,\ 5{\cdot}136,\ 8{\cdot}417,\ 11{\cdot}62$) are plotted in Fig. 8.15, where

$$\bar{c} = c/c_1, \qquad \bar{\xi} = a\xi/2\pi. \tag{8.2.38}$$

As previously mentioned, the frequency spectrum has the same shape as for SH waves in a plate. In fact, the torsional modes for a rod are considered the analogue of the SH plate modes.

2. *The longitudinal modes.* The reduction of the general frequency equation for the case of $n = 0$ led to the result (8.2.29), for which the term $a'_{23} = 0$ yielded the torsional modes. It turns out that the frequency equation for the longitudinal modes is given by the cofactor matrix. We proceed to develop this case in more detail.

Consider the propagation when $u_\theta = \partial/\partial\theta = 0$. From (8.2.2) we have

$$u_r = \frac{\partial\Phi}{\partial r} - \frac{\partial H_\theta}{\partial z}, \qquad u_z = \frac{\partial\Phi}{\partial z} + \frac{1}{r}\frac{\partial}{\partial r}(rH_\theta), \tag{8.2.39}$$

so that determination of Φ and H_θ shall be sufficient to prescribe the motion. The solutions for Φ and H_θ have been given previously by (8.2.12), where $n = 0$ in the $\cos n\theta$ dependence. We thus have that

$$\Phi = AJ_0(\alpha r)e^{i(\xi z - \omega t)}, \qquad H_\theta = -B_2 J_1(\beta r)e^{i(\xi z - \omega t)}. \tag{8.2.40}$$

Substitution of the above in the non-trivial boundary conditions, which are $\tau_{rr} = \tau_{rz} = 0$, $r = a$, yields the frequency equation. This was given in (8.2.29) as a cofactor matrix. It expands to give

$$\frac{2\alpha}{a}(\beta^2+\xi^2)J_1(\alpha a)J_1(\beta a) - (\beta^2-\xi^2)^2 J_0(\alpha a)J_1(\beta a) - \\ -4\xi^2\alpha\beta J_1(\alpha a)J_0(\beta a) = 0. \tag{8.2.41}$$

This result is referred to as the 'Pochhammer' frequency equation for the longitudinal modes. It was first published in 1876, but because of its complexity, detailed calculations of the roots did not appear until much later (the 1940s).

The displacements for this mode are given by

$$u_r = B_2\left\{-\frac{A}{B_2}\alpha J_1(\alpha r)+\mathrm{i}\xi J_1(\beta r)\right\}\mathrm{e}^{\mathrm{i}(\xi z-\omega t)},$$
$$u_z = B_2\left\{\frac{A}{B_2}\mathrm{i}\xi J_0(\alpha r)-\beta J_0(\beta r)\right\}\mathrm{e}^{\mathrm{i}(\xi z-\omega t)}, \tag{8.2.42}$$

where

$$\frac{A}{B_2} = -\left(\frac{\beta}{\alpha}\right)^2\frac{\beta^2-\xi^2}{2\xi^2}\frac{J_1(\beta a)}{J_1(\alpha a)}. \tag{8.2.43}$$

The interpretation of the displacement fields as being a resultant of dilatational and shear waves that holds for plates also holds here. Thus, in the case of u_z the $J_0(\alpha r)$ term of (8.2.42) may be interpreted as the longitudinal component of a set of plane dilatational waves whose normals form a conical surface, where the axis of the cone is the z-axis. The $J_0(\beta r)$ term of (8.2.42) is the longitudinal component of a set of transverse waves. Figure 8.7, first drawn for the case of plates, also applies for this interpretation if the normals shown in that figure are interpreted as the normals of two sets of diametrically opposite waves. As before, it is the transverse component that produces the rapid variation of u_z.

A plot of the first three modes of the dispersion curve is shown in Fig. 8.16, where the dimensionless velocity and wavenumber are given by

$$\bar{c} = c/c_0, \qquad \bar{\xi} = a\xi/2\pi, \tag{8.2.44}$$

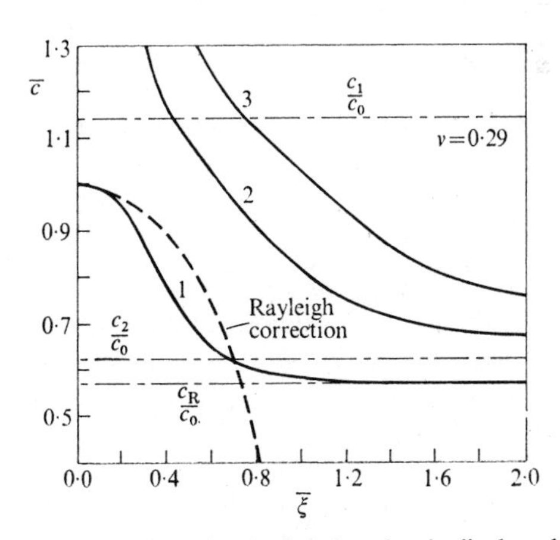

FIG. 8.16. Dispersion curves for the first three longitudinal modes in a rod.

and c_0 is the classical bar velocity, $c_0 = \sqrt{(E/\rho)}$. It is of interest to observe the behaviour of the first mode. As $\xi \to 0$, the phase velocity approaches the classical bar velocity. However, the deviation from c_0 increases for increasing ξ, emphasizing the diminishing accuracy of classical rod theory for high frequency. This aspect, of course, was also brought out in Chapter 2. As ξ increases to large values, we note that the wave velocity approaches that of Rayleigh waves. This phenomena, also encountered in plates, indicates that at higher and higher frequencies the disturbance is mainly confined to the surface.

Returning to the low-frequency behaviour of the first mode, some additional comments may be made. If J_0 and J_1 are expanded in a power series as

$$\begin{aligned} J_0(z) &= 1-\tfrac{1}{4}(z)^2+\tfrac{1}{64}(z^4)+\ldots, \\ J_1(z) &= \tfrac{1}{2}(z)-\tfrac{1}{16}(z)^3+\ldots, \end{aligned} \tag{8.2.45}$$

and $J_0(z) \simeq 1$, $J_1(z) \simeq z/2$ are used as approximations, then the insertion of (8.2.45) in (8.2.41) will yield the results $\omega^2/\xi^2 = E/\rho$, which is the bar velocity. If the next higher-order terms are retained in the expansion, a dispersive relationship is obtained and corresponds to the results obtained when lateral-inertia effects are included in the strength-of-materials approach. This was studied in Chapter 2 as the Love theory for longitudinal waves in rods. It is also designated as the Rayleigh correction, and is shown as the dashed line in the preceding figure.

The complete frequency spectrum for longitudinal waves in a rod has been obtained by Onoe, McNiven, and Mindlin [75]. The methods used closely follow those used for constructing the plate spectrum, such as establishing cutoff frequencies, asymptotic behaviour, and, most importantly, a grid of boundary curves. A three-dimensional illustration of the spectrum, showing real, imaginary, and complex branches, is shown in Fig. 8.17. Additional illustrations of just the real and imaginary branches, and the detailed behaviour of the lowest modes are also included in the paper by Onoe *et al.* The non-dimensional frequency and wavenumber parameters are given by

$$\Omega = \omega a/\delta c_2, \qquad \bar{\xi} = \xi a, \tag{8.2.46}$$

where δ is the lowest non-zero root of $J_1(\delta) = 0$.

A particularly interesting feature of the spectrum is the behaviour of the higher complex branches. We note that the lowest complex branch emanates from a minimum of the second longitudinal mode, just as in plate theory. Most of the higher complex branches, however, emanate from minima of the imaginary branches, interconnecting the various branches, or dropping to the $\Omega = 0$ plane. This behaviour was not shown for the plate spectrum, but it is quite reasonable to expect similar behaviour.

3. *The flexural modes.* The general characteristic eqn (8.2.24) was found to yield the torsional and longitudinal frequency equations for $n = 0$. For these

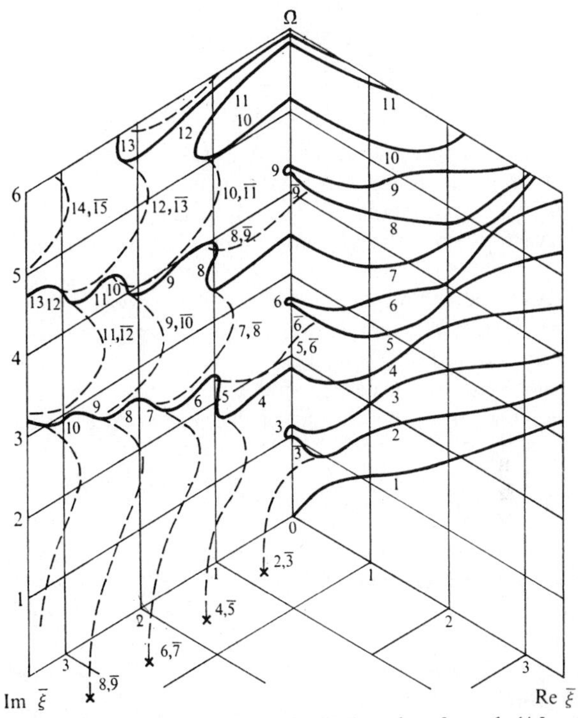

FIG. 8.17. Frequency spectrum for the longitudinal modes of a rod. (After Onoe *et al.* [75, Fig. 6].)

cases one or two of the displacement components were found to vanish. In order to investigate the propagation of flexural waves, all displacement components will exist. The case of $n = 1$ corresponds to the lowest-order family of flexural modes. The displacements are given from (8.2.22) and the frequency equations, more complicated than those for the longitudinal and torsional modes and even more so than the frequency equation for flexural modes in a plate, is obtained by expanding (8.2.24). Hudson [27] has carried out calculations for some of the lowest branches of flexural modes and Pao and Mindlin [76, 77] have investigated them in some detail. As given by Reference [77], the resulting Pochhammer frequency equation is

$$J_1(\bar{\alpha})J_1^2(\bar{\beta})(f_1\mathscr{J}_\beta^2+f_2\mathscr{J}_\alpha\mathscr{J}_\beta+f_3\mathscr{J}_\beta+f_4\mathscr{J}_\alpha+f_5) = 0, \tag{8.2.47}$$

where

$$\begin{aligned} f_1 &= 2(\bar{\beta}^2-\bar{\xi}^2)^2, \quad f_2 = 2\bar{\beta}^2(5\bar{\xi}^2+\bar{\beta}^2), \\ f_3 &= \bar{\beta}^6-10\bar{\beta}^4-2\bar{\beta}^4\bar{\xi}^2+2\bar{\beta}^2\bar{\xi}^2+\bar{\beta}^2\bar{\xi}^4-4\bar{\xi}^4, \\ f_4 &= 2\bar{\beta}^2(2\bar{\beta}^2\bar{\xi}^2-\bar{\beta}^2-9\bar{\xi}^2), \\ f_5 &= \bar{\beta}^2(-\bar{\beta}^4+8\bar{\beta}^2-2\bar{\beta}^2\bar{\xi}^2+8\bar{\xi}^2-\bar{\xi}^4), \end{aligned} \tag{8.2.48}$$

and where

$$\bar{\alpha} = \alpha a, \quad \bar{\beta} = \beta a, \qquad \bar{\xi} = \xi a, \quad \Omega = \omega a/c_2,$$
$$\mathscr{J}_x = xJ_0(x)/J_1(x). \tag{8.2.49}$$

Using the techniques previously used for obtaining the plate spectrum and the spectrum for the longitudinal modes for a rod, Pao and Mindlin obtained a grid of boundary curves for the flexural modes. The considerations are somewhat more complicated in the flexural-mode case, since it is found that the boundary curves do not resolve into simple dilatational and shear modes. The resulting spectrum, showing the real and imaginary branches, is shown in Fig. 8.18. It is to be noted that some of the imaginary branches drop to the imaginary $\bar{\xi}$-axis. Complex branches interconnect the gaps in the imaginary branches or drop to the $\Omega = 0$ plane.

Higher-order flexural modes are generated by considering $n \geqslant 2$. Frequency spectra having the general appearance of Fig. 8.18 arise. Evidently, the possibility of coupling between certain higher-order flexural modes and the lowest (or lower) longitudinal modes exists, and is suggested by some of the branches of the individual spectra crossing.

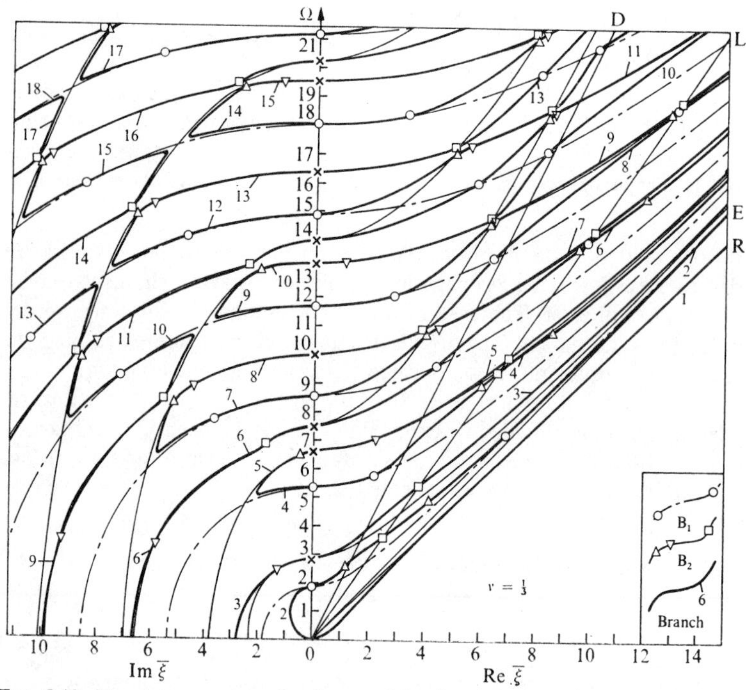

FIG. 8.18. Frequency spectrum for the $n = 1$ family of flexural modes. (After Pao [76, Fig. 4].)

8.2.3. *Waves in cylindrical shells*

The analysis of wave propagation in circular, cylindrical shells, according to elasticity theory, has been done by Gazis [19, 20]. We recall that the analysis of waves in rods was based on the procedure used by Gazis. Hence, the development of the frequency equations for shells may proceed quite straightforwardly. Somewhat greater complexity of expression results from the two sets of boundary conditions at inner and outer free surfaces. The analysis follows that of Gazis [20].

The cylindrical shell geometry has an inner radius a and outer radius b, with cylindrical coordinates as shown earlier in Fig. 8.14. The governing equations used for the rod study, (8.2.1)–(8.2.5) still hold. The boundary conditions are now given by

$$\tau_{rr} = \tau_{r\theta} = \tau_{rz} = 0, \qquad r = a, b, \tag{8.2.50}$$

where a and b are internal and external radii, respectively. The solution forms (8.2.12) still hold. However, Gazis uses a slightly different form, given by

$$\begin{aligned} \Phi &= f(r)\cos n\theta \cos(\omega t+\xi z), \\ H_r &= h_r(r)\sin n\theta \sin(\omega t+\xi z), \\ H_\theta &= h_\theta(r)\cos n\theta \sin(\omega t+\xi z), \\ H_z &= h_3(r)\sin n\theta \cos(\omega t+\xi z), \end{aligned} \tag{8.2.51}$$

where $h_3(r)$ replaces the previously used $h_z(r)$. Bessel equations of various order and argument again govern f, h_r, h_θ, h_3 and are given by (8.2.14) (which also holds for h_3 with α^2 replaced by β^2) and (8.2.21). The solutions are given by

$$\begin{aligned} f &= AZ_n(\alpha_1 r)+BW_n(\alpha_1 r), \\ h_3 &= A_3Z_n(\beta_1 r)+B_3W_n(\beta_1 r), \\ 2h_1 &= h_r-h_\theta = 2A_1Z_{n+1}(\beta_1 r)+2B_1W_{n+1}(\beta_1 r), \\ 2h_2 &= h_r+h_\theta = 2A_2Z_{n-1}(\beta_1 r)+2B_2W_{n-1}(\beta_1 r). \end{aligned} \tag{8.2.52}$$

To interpret this solution, we first note that the Bessel equations will have solutions J and Y if α^2, $\beta^2 > 0$ and solutions I, K for α^2, $\beta^2 < 0$, where I and K are the modified Bessel functions. Furthermore, both solutions to Bessel's equation apply in the present shell analysis, since singular behaviour at the origin (used to discard the solution Y_n in the rod analysis) is no longer a consideration. Thus, in (8.2.52) Z_n corresponds to J_n if α^2 or β^2 is greater than zero, and to I_n if they are less than zero. Similarly, W_n corresponds to Y_n or K_n. Also, $\alpha_1 = |\alpha|$, $\beta_1 = |\beta|$, where as before

$$\alpha^2 = \omega^2/c_1^2-\xi^2, \qquad \beta^2 = \omega^2/c_2^2-\xi^2. \tag{8.2.53}$$

The property of gauge invariance is used to eliminate two of the integration constants of (8.2.52). Owing to this property, any one of the three potentials h_1, h_2, h_3 may be set to zero without loss of generality. Setting $h_2 = 0$ gives

$h_r = -h_\theta = h_1$. The resulting displacements and stresses are then given by

$$\begin{aligned} u_r &= \{f'+(n/r)h_3+\xi h_1\}\cos n\theta \cos(\omega t+\xi z), \\ u_\theta &= \{-(n/r)f+\xi h_1-h_3'\}\sin n\theta \cos(\omega t+\xi z), \\ u_z &= \{-\xi f-h_1'-(n+1)h_1/r\}\cos n\theta \sin(\omega t+\xi z), \end{aligned} \tag{8.2.54}$$

$$\begin{aligned} \tau_{rr} &= \left[-\lambda(\alpha^2+\xi^2)f+2\mu\left\{f''+\frac{n}{r}\left(h_3'-\frac{h_3}{r}\right)+\xi h_1'\right\}\right]\times \\ &\qquad\qquad \times \cos n\theta \cos(\omega t+\xi z), \\ \tau_{r\theta} &= \mu\left[-\frac{2n}{r}\left(f'-\frac{f}{r}\right)-(2h_3''-\beta^2 h_3)-\xi\left(\frac{n+1}{r}h_1-h_1'\right)\right]\times \\ &\qquad\qquad \times \sin n\theta \cos(\omega t+\xi z), \\ \tau_{rr} &= \mu\left[-2\xi f'-\frac{n}{r}\left\{h_1'+\left(\frac{n+1}{r}-\beta^2+\xi^2\right)h_1\right\}-\frac{n\xi}{r}h_3\right]\times \\ &\qquad\qquad \times \cos n\theta \sin(\omega t+\xi z). \end{aligned} \tag{8.2.55}$$

The frequency equation results from substituting (8.2.55) in the boundary conditions (8.2.50). The six constants A, B, A_1, B_1, A_2, B_2 appear in each of the six boundary condition equations. The resulting determinant of coefficients is given by

$$|c_{ij}| = 0 \qquad (i, j = 1, \ldots, 6). \tag{8.2.56}$$

The elements of the first three rows are given by

$$\begin{aligned} c_{11} &= \{2n(n-1)-(\beta^2-\xi^2)a^2\}Z_n(\alpha_1 a)+2\lambda_1\alpha_1 a Z_{n+1}(\alpha_1 a), \\ c_{12} &= 2\xi\beta_1 a^2 Z_n(\beta_1 a)-2\xi a(n+1)Z_{n+1}(\beta_1 a), \\ c_{13} &= -2n(n-1)Z_n(\beta_1 a)+2\lambda_2 n\beta_1 a Z_{n+1}(\beta_1 a), \\ c_{14} &= \{2n(n-1)-(\beta^2-\xi^2)a^2\}W_n(\alpha_1 a)+2\alpha_1 a W_{n+1}(\alpha_1 a), \\ c_{15} &= 2\lambda_2\xi\beta_1 a^2 W_n(\beta_1 a)-2(n+1)\xi a W_{n+1}(\beta_1 a), \\ c_{16} &= -2n(n-1)W_n(\beta_1 a)+2n\beta_1 a W_{n+1}(\beta_1 a), \\ c_{21} &= 2n(n-1)Z_n(\alpha_1 a)-2\lambda_1 n\alpha_1 a Z_{n+1}(\alpha_1 a), \\ c_{22} &= -\xi\beta_1 a^2 Z_n(\beta_1 a)+2\xi a(n+1)Z_{n+1}(\beta_1 a), \\ c_{23} &= -\{2n(n-1)-\beta^2 a^2\}Z_n(\beta_1 a)-2\lambda_2\beta_1 a Z_{n+1}(\beta_1 a), \\ c_{24} &= 2n(n-1)W_n(\alpha_1 a)-2n\alpha_1 a W_{n+1}(\alpha_1 a), \\ c_{25} &= -\lambda_2\xi\beta_1 a^2 W_n(\beta_1 a)+2\xi a(n+1)W_{n+1}(\beta_1 a), \\ c_{26} &= -\{2n(n-1)-\beta^2 a^2\}W_n(\beta_1 a)-2\beta_1 a W_{n+1}(\beta_1 a), \\ c_{31} &= 2n\xi\alpha_1 Z_n(\alpha_1 a)-2\lambda_1\xi\alpha_1 a^2 Z_{n+1}(\alpha_1 a), \\ c_{32} &= n\beta_1 a Z_n(\beta_1 a)-(\beta^2-\xi^2)a^2 Z_{n+1}(\beta_1 a), \\ c_{33} &= -n\xi a Z_n(\beta_1 a), \\ c_{34} &= 2n\xi a W_n(\alpha_1 a)-2\xi\alpha_1 a^2 W_{n+1}(\alpha_1 a), \\ c_{35} &= \lambda_2 n\beta_1 a W_n(\beta_1 a)-(\beta^2-\xi^2)a^2 W_{n+1}(\beta_1 a), \\ c_{36} &= -n\xi a W_n(\beta_1 a), \end{aligned} \tag{8.2.57}$$

with the remaining three rows obtained from the first three by replacing a by b. The parameters λ_1 and λ_2 are $+1$ if the Bessel functions J and Y are used and -1 when I and K are used. The frequency equation, of course, is most complicated. However, Gazis obtains a number of special cases of interest, as well as numerical results. Some of these are considered in the following.

1. *Motion independent of z*. Motion independent of z occurs when $\xi = 0$. This gives the cutoff frequencies for the shell. Under these conditions, the determinant (8.2.56) breaks into the product of subdeterminants,

$$D_1 D_2 = 0, \tag{8.2.58}$$

where

$$D_1 = \begin{vmatrix} c_{11} & c_{13} & c_{14} & c_{16} \\ c_{21} & c_{23} & c_{24} & c_{26} \\ c_{41} & c_{43} & c_{44} & c_{46} \\ c_{51} & c_{53} & c_{54} & c_{56} \end{vmatrix}, \qquad D_2 = \begin{vmatrix} c_{32} & c_{35} \\ c_{62} & c_{65} \end{vmatrix}, \tag{8.2.59}$$

and the c_{ij} are given by (8.2.57) with $\xi = 0$.

The case of $D_1 = 0$ corresponds to the case of plane-strain motion extensively analysed by Gazis in a separate paper [19]. In considering this case, suppose first that motion is also independent of θ. This is given by $n = 0$ in the solutions. It is found under these conditions of axially symmetric vibrations that the extensional and shear modes uncouple. The results presented by Gazis for this case are shown in Fig. 8.19(a) and (b). The parameter h is the

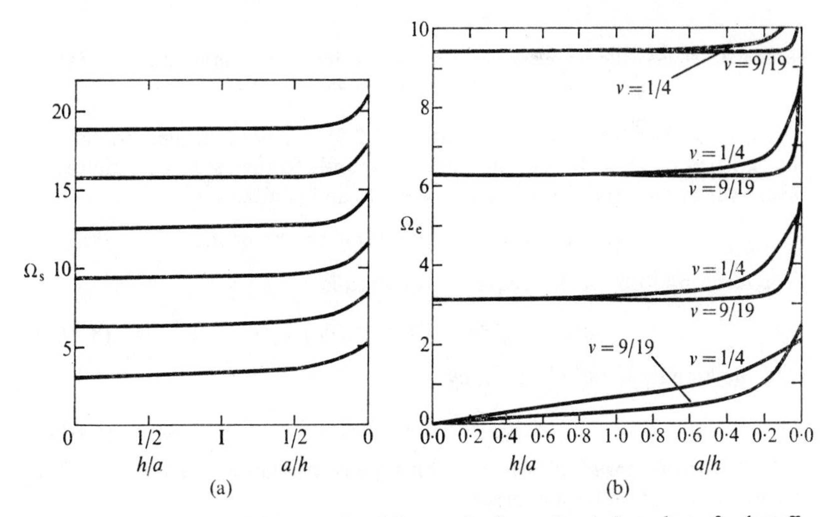

FIG. 8.19. (a) Shear and (b) extensional frequencies for motion independent of z (cutoff frequencies) and also axially symmetric. (After Gazis [19, Figs. 2, 3].)

thickness of the shell, $h = b-a$. The abscissae are divided into the two regions $0 \leq h/a \leq 1$ and $1 \geq a/h > 0$. For h/a small, the behaviour approaches that of a plate, while for a/h small, the behaviour approaches that of the solid cylinder. The non-dimensional frequencies are

$$\Omega_e = \omega h/c_1, \qquad \Omega_s = \omega h/c_2. \tag{8.2.60}$$

The extensional frequencies are seen to depend on Poisson's ratio. Continuing the case of plane-strain vibrations, consider now the non-axially symmetric case ($n \neq 0$). Shown in Fig. 8.20 are the results for $n = 2$ for $\nu = \frac{1}{3}$.

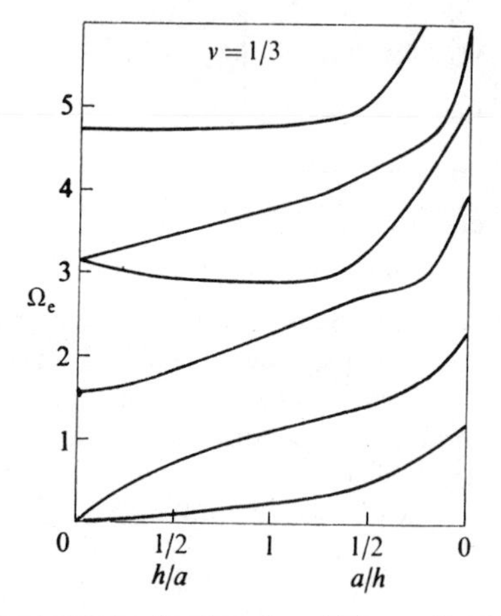

FIG. 8.20. Non-axially symmetric vibrations ($n = 2$) for motion independent of z. (After Gazis [19, Figs. 7, 8].)

Now consider the frequency equation (8.2.58) for z-independent motion to be satisfied by $D_2 = 0$. This motion corresponds to the case of longitudinal shear where only the u_z displacement occurs and is given by

$$u_z = \{A_1\beta J_n(\beta r)+B_1\beta Y_n(\beta r)\}\cos n\theta \sin \omega t. \tag{8.2.61}$$

The resulting form of the frequency equation is

$$J_n'(\beta a)Y_n'(\beta a)-J_n'(\beta b)Y_n'(\beta a) = 0, \tag{8.2.62}$$

and the amplitude ratio is given by

$$A_1/B_1 = -Y_1'(\beta a)/J_1'(\beta a). \tag{8.2.63}$$

2. *Motion independent of* θ. For this type of motion we have $n = 0$. The frequency equation degenerates to

$$D_3D_4 = 0, \tag{8.2.64}$$

where

$$D_3 = \begin{vmatrix} c_{11} & c_{12} & c_{14} & c_{15} \\ c_{31} & c_{32} & c_{34} & c_{35} \\ c_{41} & c_{42} & c_{44} & c_{45} \\ c_{61} & c_{62} & c_{63} & c_{65} \end{vmatrix}, \qquad D_4 = \begin{vmatrix} c_{23} & c_{26} \\ c_{53} & c_{56} \end{vmatrix}. \tag{8.2.65}$$

Considering first $D_3 = 0$, this is found to represent the longitudinal modes involving only the displacements u_r, u_z. As is the case for plates and solid rods, these modes involve both dilatational and equivoluminal waves through the potentials f and h_1. A special class of solutions corresponding to equivoluminal Lamé-type modes is found to exist for the special case of

$$\beta^2 = \xi^2 > 0.$$

The case of torsional modes, where $f = h_1 = 0$, occurs for $D_4 = 0$. The resulting displacements involve only u_θ and the frequency equation is given by

$$J_2(\beta a) Y_2(\beta b) - J_2(\beta b) Y_2(\beta a) = 0. \tag{8.2.66}$$

This equation is the same that results for the axially symmetric shear modes in (1) *Motion independent of z* (p. 477). As in the case of the solid rod, the lowest torsional mode is non-dispersive.

3. *Additional numerical results.* In Reference [20], Gazis presents numerical results for a variety of cases of waves in cylinders. In Fig. 8.21 additional results for the cutoff frequencies are presented for the case of $n = 1$ (see

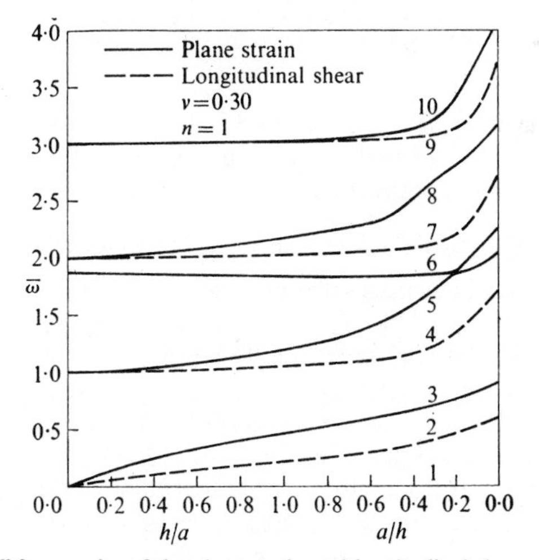

FIG. 8.21. Cutoff frequencies of the plane-strain and longitudinal-shear modes for $n = 1$. (After Gazis [20, Fig. 1].)

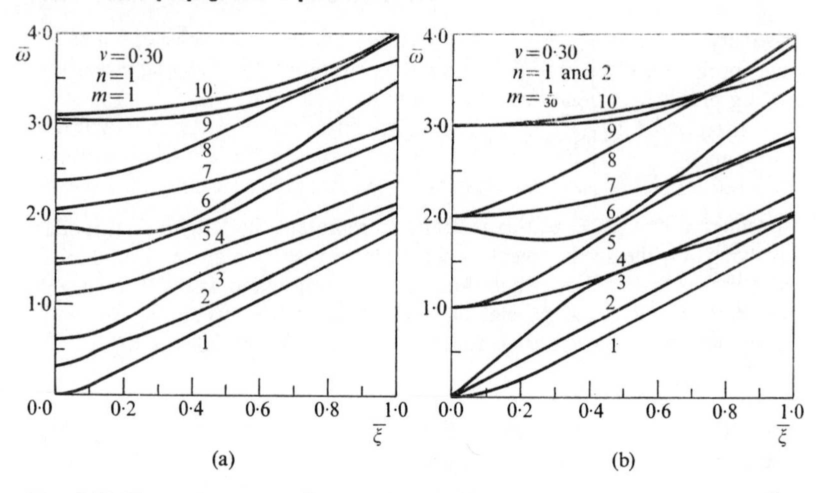

FIG. 8.22. Frequency spectrum for waves in a shell for $n = 1$ and (a) $m = 1$, (b) $m = \frac{1}{30}$, $n = 1$ and 2. (After Gazis [20, Figs. 3–6].)

Fig. 8.19 for the case of $n = 0$). The non-dimensional frequency $\bar{\omega}$ and wavenumber are now given by

$$\bar{\omega} = \omega h/\pi c_2, \qquad \bar{\xi} = h\xi/2\pi. \tag{8.2.67}$$

The relation of $\bar{\omega}$ to $\bar{\xi}$ depends, in the case of the shell, on the ratio of thickness to radius, as well as Poisson's ratio. Results are shown in Fig. 8.22 for the frequency spectrum for two different values of m, where

$$m = h/R, \tag{8.2.68}$$

and R is the mean radius. The case of $n = 1$ is that of the lowest flexural mode. For the case of a very thin shell, $m = 1/30$, the results for $n = 1$ and 2 are indistinguishable for the scale shown.

8.3. Approximate theories for waves in plates, rods, and shells

Let us review our knowledge of wave motion in plates, rods, and shells. As a result of our studies of §§ 8.1, 8.2, we realize these geometries are capable of infinitely many modes of vibration. This aspect was summarized in the multi-branched frequency spectra of the various geometries. The extensive analysis and results presented for the Rayleigh–Lamb and Pochhammer–Chree spectrum and results presented for the case of cylindrical shells suggests that the propagation of continuous waves according to the exact theory of elasticity is fairly well understood. However, the solution of free- and forced-vibration problems has become rather difficult since the cost of applying exact theory has been to escalate the complexity of the governing equations and boundary conditions. This is brought out by the relative sparseness of solutions for the free vibrations of finite solids or the difficulty of solving transient-loading problems for semi-infinite plates and rods.

At the other extreme, we have the strength-of-material theories of rods, plates, and shells, studied in the first few chapters. The simple wave equation described the longitudinal and torsional motion of rods, and the flexural motion of beams and plates was described by single, fourth-order partial differential equations. In the context of exact theory, these simpler theories contained only a single degree of freedom and approximated the lowest branch of the various exact spectra only over a limited range. However, the simple governing equations of the strength-of-material theories made solution of free- and forced-vibration problems comparatively easy, although the solutions were limited in frequency range.

The question is whether other theories can be developed, valid for higher frequencies than the simple theories, containing some of the features of exact theories and yet still yielding tractable governing equations. The Love and Timoshenko theories for rods represent instances of such developments. In the former, the basic kinematical assumptions of classical longitudinal rod theory were retained and only an additional lateral-inertia effect was considered. In the latter, the kinematical assumptions on the deformation of Bernoulli–Euler theory of beams were refined to include shear-deformation and rotary-inertia effects. Such developments, although *ad hoc* in nature, served to yield higher-order strength-of-material theories.

Over the years, many developments have occurred in this field, with various approaches used. Green [22] surveyed the development of dispersion relations for bars and divided the development of approximate theories into two classes. In one class, approximations are made in the equations of motion and Green listed contributions by Rayleigh [79], Love [43], Timoshenko [86], Prescott [78], Mindlin and Herrmann [64], Volterra [89], Bishop [6], and Kynch [37]. In the second class, solutions of the exact equations which only approximately satisfy the boundary conditions are considered and contributions by Chree [7], Morse [71], and Kynch and Green [38] were reviewed.

Although many and varied developments in approximate theories have occurred, the approach that has become prevalent was first applied to the case of plates by Mindlin [59] and to the case of rods by Mindlin and Herrman [64]. A large number of subsequent contributions by Mindlin and his coworkers and many others have placed the development of approximate theories for rods, plates, and shells on a very systematic basis. Much of the work in this area was based on the analysis of the vibration of piezoelectric plates used as elements in electronic systems. Thus, much of the literature pertains to coupled electro-mechanical or anisotropic elastic elements. The coverage here will be restricted to the isotropic, elastic case.

The approach will be as follows. First the basic early development by Mindlin [59] will be set forth, since it contains many of the basic ingredients of all later approximate theories, yet can be definitely related to the Timoshenko beam theory. The more modern work on plates, by Mindlin and Medick [66], will then be reviewed. The case of rods will be covered, and the work of Mindlin and Herrmann [64] and Mindlin and McNiven [65] reviewed, with these two papers serving as the rod counterparts of the two previously-cited plate papers. Finally, discussion of developments in the area of shell studies will cover work by Herrmann and Mirsky [25], McNiven, Shah, and Sackman [47], and others.

8.3.1. *An approximate theory for plate flexural waves*

The first development of an approximate theory will be for flexural waves in plates and will follow Mindlin's first contribution. The essential feature of this development and those that follow for other approximate theories is the integration of the exact equations of elasticity across the plate thickness. Other features include the restriction of deformation to finite degrees of freedom and the introduction of adjustment coefficients. These last are generalizations of the method used in the development of Timoshenko beam theory.

1. *Plate stresses and stress–strain relations.* Consider a differential element of plate of thickness h, having coordinates as shown in Fig. 8.23, and

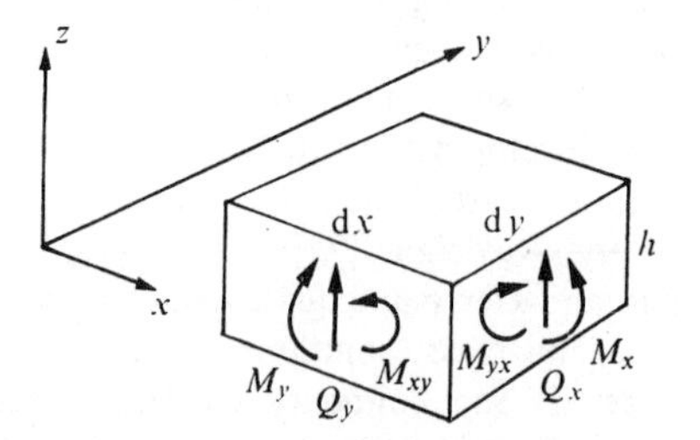

FIG. 8.23. An element of plate subjected to forces and moments.

subjected to bending and twisting moments and shear forces as shown.

The plate stresses M_x, M_y, M_{xy}, Q_x, Q_y are defined by

$$(M_x, M_y, M_{yx}) = \int_{-h/2}^{h/2} (\tau_{xx}, \tau_{yy}, \tau_{xy}) z \, dz,$$
$$(Q_x, Q_y) = \int_{-h/2}^{h/2} (\tau_{xz}, \tau_{yz}) \, dz, \tag{8.3.1}$$

which are the same as the definitions of classical plate theory.

The stress–strain equations of elasticity, given by the general Hooke's law are $\tau_{ij} = \lambda\,\Delta\delta_{ij} + 2\mu\varepsilon_{ij}$ where $\Delta = \varepsilon_{kk}$. The usual procedure in elasticity is to use the six stress–strain equations in conjunction with the strain–displacement relations to reduce the three equations of motion (in the six stress components) to equations in the displacement components. In the present case, there are only five plate-stress components, and these will be defined in terms of five plate strain components. Plate displacements will be introduced later. Start by writing the stress–strain equation for τ_{zz} in the form

$$\varepsilon_{zz} = -\frac{\lambda}{\lambda+2\mu}(\varepsilon_{xx}+\varepsilon_{yy})+(\lambda+2\mu)^{-1}\tau_{zz}. \tag{8.3.2}$$

Now eliminate ε_{zz} from the stress–strain equations for ε_{xx}, ε_{yy}, giving the five equations

$$\begin{aligned}
\tau_{xx} &= \frac{4\mu(\lambda+\mu)}{\lambda+2\mu}\varepsilon_{xx}+\frac{2\mu\lambda}{\lambda+2\mu}\varepsilon_{yy}+\frac{\lambda}{\lambda+2\mu}\tau_{zz},\\
\tau_{yy} &= \frac{2\mu\lambda}{\lambda+2\mu}\varepsilon_{xx}+\frac{4\mu(\lambda+\mu)}{\lambda+2\mu}\varepsilon_{yy}+\frac{\lambda}{\lambda+2\mu}\tau_{zz},\\
\tau_{xy} &= 2\mu\varepsilon_{xy}, \qquad \tau_{yz} = 2\mu\varepsilon_{yz}, \qquad \tau_{zx} = 2\mu\varepsilon_{zx}.
\end{aligned} \tag{8.3.3}$$

Using relations between the elastic constants, these may be rewritten as

$$\begin{aligned}
\tau_{xx} &= \frac{E}{1-\nu^2}(\varepsilon_{xx}+\nu\varepsilon_{yy})+\frac{\nu}{1-\nu}\tau_{zz},\\
\tau_{yy} &= \frac{E}{1-\nu^2}(\varepsilon_{yy}+\nu\varepsilon_{xx})+\frac{\nu}{1-\nu}\tau_{zz},\\
\tau_{xy} &= 2\mu\varepsilon_{xy}, \qquad \tau_{yz} = 2\mu\varepsilon_{yz}, \qquad \tau_{zx} = 2\mu\varepsilon_{zx}.
\end{aligned} \tag{8.3.4}$$

The plate stress–strain equations are obtained by integrating the preceding over the thickness of the plate. Thus, for M_x we have

$$M_x = \int_{-h/2}^{h/2} \tau_{xx} z\,dz = \frac{E}{1-\nu^2}\int_{-h/2}^{h/2}(\varepsilon_{xx}+\nu\varepsilon_{yy})z\,dz+\frac{\nu}{1-\nu}\int_{-h/2}^{h/2}\tau_{zz}z\,dz. \tag{8.3.5}$$

The integral contribution of the stress τ_{zz} is now neglected. Upon defining the plate strain components

$$\begin{aligned}
(\Gamma_x, \Gamma_y, \Gamma_{xy}) &= \frac{12}{h^3}\int_{-h/2}^{h/2}(\varepsilon_{xx}, \varepsilon_{yy}, \varepsilon_{xy})z\,dz,\\
(\Gamma_{xz}, \Gamma_{yz}) &= \frac{1}{h}\int_{-h/2}^{h/2}(\varepsilon_{xz}, \varepsilon_{yz})\,dz,
\end{aligned} \tag{8.3.6}$$

we obtain for M_x and M_y

$$M_x = D(\Gamma_x + \nu\Gamma_y), \qquad M_y = D(\Gamma_y + \nu\Gamma_x), \tag{8.3.7}$$

where D is the usual plate stiffness given as

$$D = \frac{Eh^3}{12(1-\nu^2)}. \tag{8.3.8}$$

For M_{xy}, we have

$$M_{xy} = 2\mu \int_{-h/2}^{h/2} \varepsilon_{xy} z \, dz = \frac{E}{1+\nu} \int_{-h/2}^{h/2} \varepsilon_{xy} z \, dz, \tag{8.3.9}$$

giving

$$M_{xy} = (1-\nu) D \Gamma_{xy}. \tag{8.3.10}$$

The plate shear forces are treated somewhat differently than in classical plate theory, which neglects shear effects. Here, a Timoshenko-like shear effect is introduced by writing

$$Q_x = 2\mu' \int \varepsilon_{xz} \, dz, \tag{8.3.11}$$

where $\mu' = \kappa^2 \mu$. Thus

$$Q_x = \mu' h \Gamma_{xz}. \tag{8.3.12}$$

The motivation behind the introduction of the modified shear modulus μ' is similar to that of beam theory. Thus if we denote $\bar{\varepsilon}_{xz}$ as the exact shear strain, then the exact shear force $\bar{Q}_x$ is

$$\bar{Q}_x = 2\mu \int_{-h/2}^{h/2} \bar{\varepsilon}_{xz} \, dz. \tag{8.3.13}$$

However, as a consequence of neglecting certain stresses and because of kinematical assumptions that will be made, we have an inexact strain ε_{xz} that gives an inexact shear force Q_x, where $Q_x \neq \bar{Q}_x$. We introduce the shear coefficient κ^2, leading to the modified shear modulus μ' such that Q_x, $\bar{Q}_x$ are brought into agreement. We may then summarize our plate stress–strain relations as

$$\begin{aligned} M_x &= D(\Gamma_x + \nu\Gamma_y), \qquad M_y = D(\Gamma_y + \nu\Gamma_x), \\ M_{xy} &= (1-\nu) D \Gamma_{xy}, \\ Q_x &= \mu' h \Gamma_{xz}, \qquad Q_y = \mu' h \Gamma_{yz}. \end{aligned} \tag{8.3.14}$$

Two comments are in order. First, (8.3.14) could have been obtained by setting $\tau_{zz} = 0$ at the outset. However, adopting the present procedure reveals that a weighted average of τ_{zz} was neglected and not τ_{zz} itself. Second, the determination of κ^2 in the approximate plate theory will be handled quite differently than in previous beam or plate developments, which used elasticity analysis to establish the shear coefficient.

2. *Kinematics and plate stress–displacement relations*. We have the strain–displacement equations of elasticity given by $\varepsilon_{ij} = (u_{i,j} + u_{j,i})/2$. Then,

integrating the strains across the plate thickness as suggested by the plate strain definitions (8.3.6), gives

$$\begin{aligned}\int_{-h/2}^{h/2} (\varepsilon_{xx}, \varepsilon_{yy}, \varepsilon_{xy}) z \, dz &= \int_{-h/2}^{h/2} \left(\frac{\partial u}{\partial x}, \frac{\partial v}{\partial y}, \frac{\partial u}{\partial y}+\frac{\partial v}{\partial x}\right) z \, dz, \\ \int_{-h/2}^{h/2} (\varepsilon_{xz}, \varepsilon_{yz}) \, dz &= \int_{-h/2}^{h/2} \left(\frac{\partial u}{\partial z}+\frac{\partial w}{\partial x}, \frac{\partial v}{\partial z}+\frac{\partial w}{\partial y}\right) dz.\end{aligned} \tag{8.3.15}$$

This last expression explicitly brings into the development the displacements. The important assumptions on the kinematics of deformation are now made by considering u, v, w to be given by

$$u(x, y, z, t) = z\psi_x(x, y, t), \quad v(x, y, z, t) = z\psi_y(x, y, t), \quad w(x, y, z, t) = \bar{w}(x, y, t). \tag{8.3.16}$$

Several general remarks are in order here. First, the w displacements are independent of z, so that thickness modes (stretch and shear), possible in the exact theory, are ruled out in this approximation. Second, the assumed u, v, w displacement field represents the two-dimensional generalization of the Timoshenko assumptions. If shear deformations were to be neglected, as in Bernoulli–Euler beam theory or classical plate theory, we would have $\partial w/\partial x = -\psi_x$, $\partial w/\partial y = -\psi_y$, and

$$u(x, y, z, t) = -z \, \partial w/\partial x, \quad v(x, y, z, t) = -z \, \partial w/\partial y.$$

Further, if it were desired to develop a theory capable of predicting higher-order deformations, more refined kinematical assumptions would be required. A systematic way of doing this would be to expand u, v, w as power series in the thickness coordinate. Thus, using the general indicial form given below as a guide,

$$u_j(x, y, z, t) = \sum_{n=0}^{\infty} z^n u_j^{(n)}(x, y, t), \tag{8.3.17}$$

we would have

$$\begin{aligned} u &= u^{(0)}+zu^{(1)}+z^2u^{(2)}+\dots, \\ v &= v^{(0)}+zv^{(1)}+z^2v^{(2)}+\dots, \\ w &= w^{(0)}+zw^{(1)}+z^2w^{(2)}+\dots . \end{aligned} \tag{8.3.18}$$

Thus, in the present case, we have only retained the $u^{(1)}$, $v^{(1)}$, $w^{(0)}$ terms of the expansion. In neglecting the $u^{(0)}$, $v^{(0)}$ terms, we have, in effect, neglected extensional vibrations in the x, y directions. The remaining terms have been discarded as higher order in the present development, but could be retained in a more exact analysis. Other expansions could be used. In particular, an expansion in Legendre polynomials in the thickness coordinates will be shown later. Finally, it is seen how various vibrational modes may be selectively studied by suppressing or retaining various terms of the expansion.

Returning to the development, the assumed deformations (8.3.16) are inserted into the right-hand side of the strain–displacement integrals (8.3.15). For Γ_x, we have for the various steps

$$\Gamma_x = \frac{12}{h^3}\int_{-h/2}^{h/2} \varepsilon_x z\,dz = \frac{12}{h^3}\int_{-h/2}^{h/2} \frac{\partial u}{\partial x} z\,dz$$
$$= \frac{\partial \psi_x}{\partial x}\frac{12}{h^3}\int_{-h/2}^{h/2} z^2\,dz = \frac{\partial \psi_x}{\partial x}. \tag{8.3.19}$$

Similarly, for Γ_{xz},

$$\Gamma_{xz} = \frac{1}{h}\int_{-h/2}^{h/2} \varepsilon_{xz}\,dz = \frac{1}{h}\int_{-h/2}^{h/2}\left(\frac{\partial u}{\partial z}+\frac{\partial w}{\partial x}\right)dz$$
$$= \left(\psi_x+\frac{\partial \bar{w}}{\partial x}\right)\frac{1}{h}\int_{-h/2}^{h/2} dz = \psi_x+\frac{\partial \bar{w}}{\partial x}. \tag{8.3.20}$$

We thus have for the plate strain–displacement relations,

$$\Gamma_x = \frac{\partial \psi_x}{\partial x}, \qquad \Gamma_y = \frac{\partial \psi_y}{\partial y}, \qquad \Gamma_{xy} = \frac{\partial \psi_y}{\partial x}+\frac{\partial \psi_x}{\partial y},$$
$$\Gamma_{xz} = \psi_x+\frac{\partial \bar{w}}{\partial x}, \qquad \Gamma_{yz} = \psi_y+\frac{\partial \bar{w}}{\partial y}. \tag{8.3.21}$$

The plate stress–displacement relations are

$$M_x = D\left(\frac{\partial \psi_x}{\partial x}+\nu\frac{\partial \psi_y}{\partial y}\right), \qquad M_y = D\left(\frac{\partial \psi_y}{\partial y}+\nu\frac{\partial \psi_x}{\partial x}\right),$$
$$M_{xy} = \frac{1-\nu}{2}D\left(\frac{\partial \psi_y}{\partial x}+\frac{\partial \psi_x}{\partial y}\right), \tag{8.3.22}$$
$$Q_x = \kappa^2 Gh\left(\frac{\partial \bar{w}}{\partial x}+\psi_x\right), \qquad Q_y = \kappa^2 Gh\left(\frac{\partial \bar{w}}{\partial y}+\psi_y\right).$$

The result for classical plate theory is recovered from the above by letting $\psi_x = -\partial\bar{w}/\partial x$, $\psi_y = -\partial\bar{w}/\partial y$. Then $M_x = -D(\partial^2\bar{w}/\partial x^2+\nu\partial^2\bar{w}/\partial y^2)$, the classical plate result, with M_y and M_{xy} following in a similar fashion. Of course, $Q_x = Q_y = 0$ then occurs in (8.3.32). This anomaly is the well-known result of neglecting shear deformations, yet retaining shear forces in the equilibrium equations. The shear forces are in fact determined from the equilibrium equations in classical theory.

3. *Equations of motion.* Consider now the stress equations of motion of three-dimensional elasticity theory, given as

$$\begin{aligned}\frac{\partial\tau_{xx}}{\partial x}+\frac{\partial\tau_{xy}}{\partial y}+\frac{\partial\tau_{xz}}{\partial z} &= \rho\ddot{u},\\ \frac{\partial\tau_{yx}}{\partial x}+\frac{\partial\tau_{yy}}{\partial y}+\frac{\partial\tau_{yz}}{\partial z} &= \rho\ddot{v},\\ \frac{\partial\tau_{zx}}{\partial x}+\frac{\partial\tau_{zy}}{\partial y}+\frac{\partial\tau_{zz}}{\partial z} &= \rho\ddot{w}.\end{aligned} \tag{8.3.23}$$

To convert these to the plate stress equations of motion first multiply the first two of (8.3.23) by z and integrate across the thickness. Exhibiting the operations only for the first case, we have

$$\frac{\partial}{\partial x}\int_{-h/2}^{h/2}\tau_{xx}z\,dz+\frac{\partial}{\partial y}\int_{-h/2}^{h/2}\tau_{xy}z\,dz+\int_{-h/2}^{h/2}\frac{\partial\tau_{xz}}{\partial z}z\,dz = \rho\int_{-h/2}^{h/2}\ddot{u}z\,dz. \tag{8.3.24}$$

The first two integrals of the preceding are M_x, M_{xy} respectively. The third expression is integrated by parts, giving

$$\int_{-h/2}^{h/2}\frac{\partial\tau_{xz}}{\partial z}z\,dz = [z\tau_{xz}]_{-h/2}^{h/2}-\int_{-h/2}^{h/2}\tau_{xz}\,dz = -Q_x, \tag{8.3.25}$$

where $z\tau_{xz}$ is zero at $\pm h/2$, since τ_{xz} vanishes on the boundaries. The right-hand side of (8.3.24) gives

$$\rho\int_{-h/2}^{h/2}\ddot{u}z\,dz = \rho\ddot{\psi}_x\int_{-h/2}^{h/2}z^2\,dz = \frac{\rho h^3\ddot{\psi}_x}{12}. \tag{8.3.26}$$

The second of (8.3.23) is handled in the same way. The third equilibrium equation is integrated directly. Thus

$$\frac{\partial}{\partial x}\int_{-h/2}^{h/2}\tau_{zx}\,dz+\frac{\partial}{\partial y}\int_{-h/2}^{h/2}\tau_{zy}\,dz+\int_{-h/2}^{h/2}\frac{\partial\tau_{zz}}{\partial z}\,dz = \rho\int_{-h/2}^{h/2}\ddot{w}\,dz. \tag{8.3.27}$$

The first two integrals yield Q_x and Q_y respectively. The third integral gives

$$\int_{-h/2}^{h/2}\frac{\partial\tau_{zz}}{\partial z}\,dz = [\tau_{zz}]_{-h/2}^{h/2} = -q_1-(-q_2) = q_2-q_1 = q. \tag{8.3.28}$$

The right-hand side of (8.3.28) represents the external loading on the plate

The plate stress equations of motion thus become

$$\frac{\partial M_x}{\partial x}+\frac{\partial M_{xy}}{\partial y}-Q_x = \frac{\rho h^3}{12}\frac{\partial^2 \psi_x}{\partial t^2},$$
$$\frac{\partial M_{xy}}{\partial x}+\frac{\partial M_y}{\partial y}-Q_y = \frac{\rho h^3}{12}\frac{\partial^2 \psi_y}{\partial t^2}, \tag{8.3.29}$$
$$\frac{\partial Q_x}{\partial x}+\frac{\partial Q_y}{\partial y}+q = \rho h\frac{\partial^2 \bar{w}}{\partial t^2}.$$

When the plate stress–displacement eqns (8.3.22) are substituted in the preceding, we obtain the equations of motion

$$\frac{D}{2}\left\{(1-\nu)\nabla^2\psi_x+(1+\nu)\frac{\partial \Phi}{\partial x}\right\}-\kappa^2\mu h\left(\psi_x+\frac{\partial \bar{w}}{\partial x}\right) = \frac{\rho h^3}{12}\frac{\partial^2 \psi_x}{\partial t^2},$$
$$\frac{D}{2}\left\{(1-\nu)\nabla^2\psi_y+(1+\nu)\frac{\partial \Phi}{\partial y}\right\}-\kappa^2\mu h\left(\psi_y+\frac{\partial \bar{w}}{\partial y}\right) = \frac{\rho h^3}{12}\frac{\partial^2 \psi_y}{\partial t^2}, \tag{8.3.30}$$
$$\kappa^2\mu h(\nabla^2\bar{w}+\Phi)+q = \rho h(\partial^2\bar{w}/\partial t^2),$$

where

$$\Phi = \frac{\partial \psi_x}{\partial x}+\frac{\partial \psi_y}{\partial y}. \tag{8.3.31}$$

4. *Kinetic and potential energy.* If our aim was only to derive the governing differential equations for a higher-order plate theory, the task would be complete with the results (8.3.30). Development of energy expressions for the plate is also necessary, since this quantity is an important parameter in its own right, and also lays down the groundwork for establishing the proper boundary conditions. Proceeding, we have the strain energy function of elasticity given by

$$2W = \tau_{xx}\varepsilon_{xx}+\tau_{yy}\varepsilon_{yy}+\tau_{zz}\varepsilon_{zz}+\tau_{xy}\varepsilon_{xy}+\tau_{yz}\varepsilon_{yz}+\tau_{zx}\varepsilon_{zx}$$
$$= \tau_{xx}\frac{\partial u}{\partial x}+\tau_{yy}\frac{\partial v}{\partial y}+\tau_{zz}\frac{\partial w}{\partial z}+\tau_{xy}\left(\frac{\partial v}{\partial x}+\frac{\partial u}{\partial y}\right)+$$
$$+\tau_{yz}\left(\frac{\partial w}{\partial y}+\frac{\partial v}{\partial z}\right)+\tau_{zx}\left(\frac{\partial w}{\partial x}+\frac{\partial u}{\partial z}\right). \tag{8.3.32}$$

We substitute the expressions for u, v, w, given by (8.3.16) into (8.3.32) and integrate over the plate thickness to give

$$2\bar{W} = M_x\frac{\partial \psi_x}{\partial x}+M_y\frac{\partial \psi_y}{\partial y}+M_{yx}\left(\frac{\partial \psi_y}{\partial x}+\frac{\partial \psi_x}{\partial y}\right)+$$
$$+Q_x\left(\frac{\partial \bar{w}}{\partial x}+\psi_x\right)+Q_y\left(\frac{\partial \bar{w}}{\partial y}+\psi_y\right)$$
$$= M_x\Gamma_x+M_y\Gamma_y+M_{yx}\Gamma_{yx}+Q_x\Gamma_{xz}+Q_y\Gamma_{yz}. \tag{8.3.33}$$

This represents the plate strain energy per unit area. Using the plate stress–strain relations (8.3.14) in the preceding, we obtain

$$4\overline{W} = D(1+\nu)(\Gamma_x+\Gamma_y)^2+D(1-\nu)\{(\Gamma_x-\Gamma_y)^2+\Gamma_{yx}^2\}+ \\ +2\kappa^2\mu h(\Gamma_{yz}^2+\Gamma_{xz}^2). \quad (8.3.34)$$

We note that $\overline{W}$ is positive definite if $E(1+\nu)$, $E(1-\nu)$ are required to be positive. Further, we note that

$$\frac{\partial\overline{W}}{\partial\Gamma_x} = M_x, \quad \frac{\partial\overline{W}}{\partial\Gamma_y} = M_y, \quad \frac{\partial\overline{W}}{\partial\Gamma_{yx}} = M_{yx}, \quad \frac{\partial\overline{W}}{\partial\Gamma_{zx}} = Q_x, \quad (8.3.35)$$

$$\partial\overline{W}/\partial\Gamma_{zy} = Q_y.$$

The kinetic energy per unit volume is given as $\rho(\dot{u}^2+\dot{v}^2+\dot{w}^2)/2$. Using (8.3.16) and integrating over the thickness gives

$$T = \frac{\rho h^3}{24}\left\{\left(\frac{\partial\psi_x}{\partial t}\right)^2+\left(\frac{\partial\psi_y}{\partial t}\right)^2\right\}+\frac{\rho h}{2}\left(\frac{\partial\bar{w}}{\partial t}\right)^2 \quad (8.3.36)$$

as the kinetic energy per unit area for the plate. The kinetic and potential energies in the plate at time t are given by

$$\bar{T} = \iint\left\{\frac{\rho h^3}{24}(\dot{\psi}_x^2+\dot{\psi}_y^2)+\frac{\rho h}{2}\dot{\bar{w}}^2\right\}\,\mathrm{d}x\,\mathrm{d}y, \qquad \bar{V} = \iint\overline{W}\,\mathrm{d}x\,\mathrm{d}y. \quad (8.3.37)$$

The total energy at time t is the sum of $\bar{T}$ and $\bar{V}$ and may be written as

$$\bar{T}+\bar{V} = \int_{t_0}^{t}\mathrm{d}t\iint\frac{\partial T}{\partial t}\,\mathrm{d}x\,\mathrm{d}y+\int_{t_0}^{t}\mathrm{d}t\iint\frac{\partial\overline{W}}{\partial t}\,\mathrm{d}x\,\mathrm{d}y+\bar{T}_0+\bar{V}_0, \quad (8.3.38)$$

where $\bar{T}_0$, $\bar{V}_0$ represent the values of $\bar{T}$, $\bar{V}$ at t_0.

The time derivatives of (8.3.36) and (8.3.33) are given by

$$\frac{\partial T}{\partial t} = \frac{\rho h^3}{12}(\dot{\psi}_x\ddot{\psi}_x+\dot{\psi}_y\ddot{\psi}_y)+\rho h\dot{\bar{w}}\ddot{\bar{w}}, \quad (8.3.39)$$

$$\frac{\partial\overline{W}}{\partial t} = \frac{\partial\overline{W}}{\partial\Gamma_x}\frac{\partial\Gamma_x}{\partial t}+\frac{\partial\overline{W}}{\partial\Gamma_y}\frac{\partial\Gamma_y}{\partial t}+\frac{\partial\overline{W}}{\partial\Gamma_{xy}}\frac{\partial\Gamma_{xy}}{\partial t}+\frac{\partial\overline{W}}{\partial\Gamma_{zx}}\frac{\partial\Gamma_{zx}}{\partial t}+\frac{\partial\overline{W}}{\partial\Gamma_{zy}}\frac{\partial\Gamma_{zy}}{\partial t}. \quad (8.3.40)$$

By using relations (8.3.35) and the expressions (8.3.21) in (8.3.40), we obtain

$$\frac{\partial\overline{W}}{\partial t} = \left(M_x\frac{\partial}{\partial x}+M_{yx}\frac{\partial}{\partial y}+Q_x\right)\frac{\partial\psi_x}{\partial t}+\left(M_{yx}\frac{\partial}{\partial x}+M_y\frac{\partial}{\partial y}+Q_y\right)\frac{\partial\psi_y}{\partial t}+ \\ +\left(Q_x\frac{\partial}{\partial x}+Q_y\frac{\partial}{\partial y}\right)\frac{\partial\overline{W}}{\partial t}. \quad (8.3.41)$$

By integrating (8.3.41) by parts, as contained in (8.3.38), the contributions of the external forces to the total energy may be found. As an example, we have

$$\iint \left(M_x \frac{\partial \dot{\psi}_x}{\partial x} + M_{yx} \frac{\partial \dot{\psi}_x}{\partial y} \right) \mathrm{d}x\, \mathrm{d}y = \iint \left\{ \frac{\partial}{\partial x}(M_x \dot{\psi}_x) + \frac{\partial}{\partial y}(M_{yx} \dot{\psi}_x) \right\} \mathrm{d}x\, \mathrm{d}y - \\ - \iint \left(\frac{\partial M_x}{\partial x} + \frac{M_{yx}}{\partial y} \right) \dot{\psi}_x\, \mathrm{d}x\, \mathrm{d}y. \quad (8.3.42)$$

We apply Green's theorem in evaluating the first integral on thé right-hand side of the above, to obtain

$$\iint \left(M_x \frac{\partial \dot{\psi}_x}{\partial x} + M_{yx} \frac{\partial \dot{\psi}_x}{\partial y} \right) \mathrm{d}x\, \mathrm{d}y = \oint \dot{\psi}_x (M_x l + M_{yx} m)\, \mathrm{d}s - \\ - \iint \left(\frac{\partial M_x}{\partial x} + \frac{\partial M_{yx}}{\partial y} \right) \dot{\psi}_x\, \mathrm{d}x\, \mathrm{d}y, \quad (8.3.43)$$

where l, m are direction cosines of the normal to the boundary. We thus obtain

$$\iint \frac{\partial \overline{W}}{\partial t}\, \mathrm{d}x\, \mathrm{d}y = \oint \left\{ \dot{\psi}_x (M_x l + M_{yx} m) + \dot{\psi}_y (M_{yx} l + M_y m) + \dot{w}(Q_x l + Q_y m) \right\} \mathrm{d}s - \\ - \iint \left\{ \dot{\psi}_x \left(\frac{\partial M_x}{\partial x} + \frac{\partial M_{yx}}{\partial y} - Q_x \right) + \right. \\ \left. + \dot{\psi}_y \left(\frac{\partial M_{yx}}{\partial x} + \frac{\partial M_y}{\partial y} - Q_y \right) + \dot{w} \left(\frac{\partial Q_x}{\partial x} + \frac{\partial Q_y}{\partial y} \right) \right\} \mathrm{d}x\, \mathrm{d}y. \quad (8.3.44)$$

in terms of coordinates $\bar{\nu}$ and s measured normal to and along the boundary, the line integral above becomes

$$\oint (\dot{\psi}_{\bar{\nu}} M_{\bar{\nu}} + \dot{\psi}_s M_{\bar{\nu}s} + \dot{w} Q_{\bar{\nu}})\, \mathrm{d}s. \quad (8.3.45)$$

With these results, (8.3.38) becomes

$$\overline{T} + \overline{V} = \int_{t_0}^{t} \mathrm{d}t \oint (\dot{\psi}_{\bar{\nu}} M_{\bar{\nu}} + \dot{\psi}_s M_{\bar{\nu}s} + \dot{w} Q_{\bar{\nu}})\, \mathrm{d}s + \\ + \int_{t_0}^{t} \mathrm{d}t \iint \left\{ \dot{\psi}_x \left(\frac{\rho h^3}{12} \ddot{\psi}_x - \frac{\partial M_x}{\partial x} - \frac{\partial M_{yx}}{\partial y} + Q_x \right) + \right. \\ + \dot{\psi}_y \left(\frac{\rho h^3}{12} \ddot{\psi}_y - \frac{\partial M_{yx}}{\partial x} - \frac{\partial M_y}{\partial y} + Q_y \right) + \\ \left. + \dot{w} \left(\rho h \ddot{w} - \frac{\partial Q_x}{\partial x} - \frac{\partial Q_y}{\partial y} \right) \right\} \mathrm{d}x\, \mathrm{d}y + \overline{T}_0 + \overline{V}_0. \quad (8.3.46)$$

The expressions in the parenthesis of the area integral are the plate equations of motion. We presume these are satisfied, so that the preceding reduces to

$$\bar{T}+\bar{V} = \int_{t_0}^{t} dt \oint \left(\frac{\partial \psi_{\bar{\nu}}}{\partial t} M_{\bar{\nu}} + \frac{\partial \psi_s}{\partial t} M_{\bar{\nu}s} + \frac{\partial \bar{w}}{\partial t} Q_{\bar{\nu}} \right) ds +$$
$$+ \int_{t_0}^{t} dt \iint q \frac{\partial \bar{w}}{\partial t} \, dx \, dy + \bar{T}_0 + \bar{V}_0. \quad (8.3.47)$$

This result is a statement that the total energy at time t is equal to the sum of the energy at time t_0 and the work done by the external forces along the edge of the plate and over the surface in the time interval $(t-t_0)$.

5. *Boundary conditions.* The establishment of boundary and initial conditions follows from uniqueness-of-solutions arguments. This argument is used in establishing boundary conditions for the general elasticity case, and it only needs slight rephrasing here to apply to the present form of the total energy. Thus, consider two sets of plate displacements and surface loads. If the components of each set satisfy the equations of motion so will their differences. Then the energies and the plate stress components calculated from the differences will satisfy an equation of the same form as (8.3.47). If the right-hand side of (8.3.47) vanishes, $\bar{T}$ and $\bar{V}$ vanish separately, since they are both positive. If $\bar{T}$ vanishes, the kinetic energy per unit volume vanishes, since it is positive, and hence plate velocities vanish. If $\bar{V}$ vanishes $\overline{W}$ vanishes, since it is positive, and if $\overline{W}$ vanishes so do the plate strain and stress components. Hence, if the right-hand side of (8.3.47) vanishes, the two systems must be identical, except possibly for a rigid-body displacement.

We now return to the single solution to establish the boundary conditions (those conditions necessary for vanishing of the right-hand side of (8.3.47)). It is required that one each of the three pairs of quantities

$$\dot{\psi}_{\bar{\nu}} M_{\bar{\nu}}, \qquad \dot{\psi}_s M_{\bar{\nu}s}, \qquad \dot{\bar{w}} Q_{\bar{\nu}} \quad (8.3.48)$$

be specified on the edge of the plate. Throughout the plate, either q or $\bar{w}$ must be given as must be the initial values of ψ_x, ψ_y, $\bar{w}$, and their time derivatives.

It is of interest to compare the boundary conditions (8.3.48) to those of classical plate theory. For illustration, consider the plate edge given by $x =$ constant. Thus we have

$$\dot{\psi}_x M_x, \qquad \dot{\psi}_y M_{xy}, \qquad \dot{\bar{w}} Q_x = 0. \quad (8.3.49)$$

Now, $\psi_y = Q_y/\kappa^2\mu h - \partial\bar{w}/\partial y$, from (8.3.22), and in classical theory this reduces to $\psi_y = -\partial\bar{w}/\partial y$. Thus

$$\dot{\psi}_y M_{xy} \to -\frac{\partial^2 \bar{w}}{\partial t \, \partial y} M_{xy}. \quad (8.3.50)$$

However,

$$-\int \frac{\partial^2 \bar{w}}{\partial t\, \partial y} M_{xy}\, dy = \int \frac{\partial \bar{w}}{\partial t} \frac{\partial M_{xy}}{\partial y}\, dy. \tag{8.3.51}$$

Hence the boundary conditions reduce to

$$-\frac{\partial \dot{\bar{w}}}{\partial x} M_x, \qquad \frac{\partial \bar{w}}{\partial t}\left(\frac{\partial M_{xy}}{\partial y} + Q_x\right) = 0, \tag{8.3.52}$$

which are the Kirchhoff conditions of classical plate theory.

5. *Propagation of harmonic waves.* The differential equations of motion may be reduced to a single equation for $\bar{w}$ by differentiating the first and second of (8.3.30) and adding the results to give

$$\left(D\nabla^2 - \mu' h - \frac{\rho h^3}{12} \frac{\partial^2}{\partial t^2}\right)\Phi = \mu' h \nabla^2 \bar{w}. \tag{8.3.53}$$

Next eliminate Φ between (8.3.53) and the third equation of (8.3.30), to give

$$\left(\nabla^2 - \frac{\rho}{\mu'} \frac{\partial^2}{\partial t^2}\right)\left(D\nabla^2 - \frac{\rho h^3}{12} \frac{\partial^2}{\partial t^2}\right)\bar{w} + \rho h \frac{\partial^2 \bar{w}}{\partial t^2} = \left(1 - \frac{D\nabla^2}{\mu' h} + \frac{\rho h^2}{12\mu'} \frac{\partial^2}{\partial t^2}\right) q. \tag{8.3.54}$$

We have already studied the dispersion relations for exact plate theory. Now consider the conditions under which straight-crested harmonic waves may propagate. It is sufficient to let

$$\bar{w} = \cos \xi(x - ct), \tag{8.3.55}$$

and substitute in (8.3.54) with $q = 0$. The resulting dispersion relation is

$$\frac{h^2 \xi^2}{12}\left(1 - \frac{c^2}{\kappa^2 c_2^2}\right)\left(\frac{c_p^2}{c^2} - 1\right) = 1, \tag{8.3.56}$$

where c_2 is the shear-wave velocity and

$$c_p = \left(\frac{E}{\rho(1-\nu^2)}\right)^{\frac{1}{2}}, \tag{8.3.57}$$

which is the plate analogue of the bar velocity $(E/\rho)^{\frac{1}{2}}$. Equation (8.3.56) may be put in the form

$$\left(\frac{c^2}{c_2^2}\right)^2 - \left(\kappa^2 + \frac{c_p^2}{c_2^2} - \frac{12\kappa^2}{h^2 \xi^2}\right)\left(\frac{c^2}{c_2^2}\right) + \kappa^2 \frac{c_p^2}{c_2^2} = 0. \tag{8.3.58}$$

There will thus be two roots for c^2. Now consider the behaviour as $\xi \to \infty$ (that is, high frequency). From the last we obtain $c^2 = \kappa^2 c_2^2$. However, according to exact theory, this velocity should approach the Rayleigh surface wave velocity (see Fig. 8.16, for example). This gives a method for adjusting

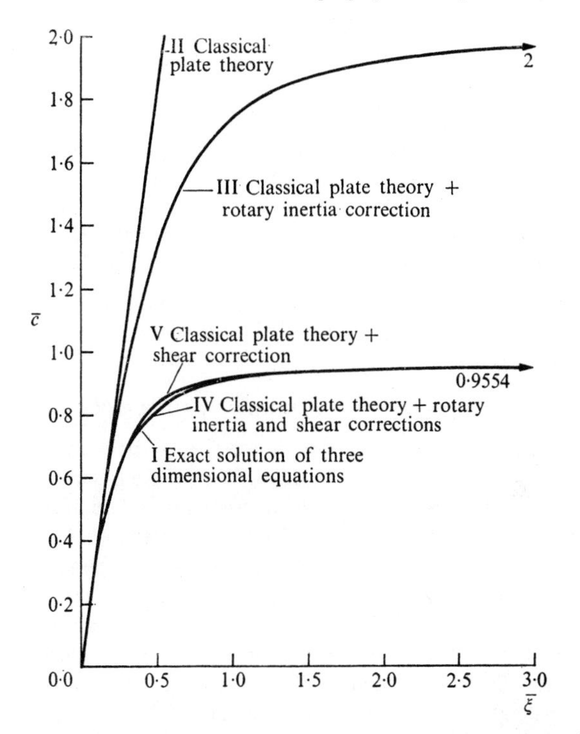

FIG. 8.24. The dispersion curve resulting from Mindlin's approximate theory for flexural waves in a plate (Curve IV) and results from other theories. (From Mindlin [59, Fig. 1].)

the constant κ^2. Thus, in order for $c \to c_R$ (the Rayleigh velocity) as ξ becomes large, we must have $\kappa = c_R/c_2$. This ratio has been plotted in Fig. 6.8 for various Poisson's ratio.

The resulting dispersion curve is shown as curve IV in Fig. 8.24 for the case of $\nu = \frac{1}{2}$, for which $\kappa = 0{\cdot}9554$. Also shown in the figure are the dispersion curves for the lowest flexure mode of exact theory (I), classical plate theory (II), classical theory with a rotary-inertia correction (III), and classical plate theory with shear correction only (V). The curves are quite similar to those previously displayed in the study of Timoshenko beam theory (see Fig. 3.13). For the first mode of vibration, it is not possible to distinguish the difference between I and IV. The dimensionless velocity and wavenumber of the figure are $\bar{c} = c/c_s$ and $\bar{\xi} = \xi h/2\pi$.

We may investigate the conditions under which pure thickness-shear vibrations may occur by letting

$$\psi_y = \bar{w} = 0, \qquad \psi_x = e^{i\omega t}. \tag{8.3.59}$$

By letting $\xi = 0$ we are, in effect, investigating the cutoff frequency. Substituting in the equations of motion (8.3.30) with $q = 0$ yields

$$\omega_c = 2\sqrt{3}(\kappa c_2/h). \tag{8.3.60}$$

From exact theory, the value for the cutoff frequency of the first antisymmetric mode of thickness-shear vibration is $\omega = \pi c_2/h$. To make these two results identical requires $\kappa^2 = \pi^2/12$. This value of κ^2 differs from the previously obtained value of $\kappa^2 = c_R^2/c_2^2$. The present value has been obtained by adjusting the low-frequency behaviour and the other the high-frequency behaviour of the approximate theory. Which technique is better may depend on the problem at hand in terms of frequency range or the primary vibration mode.

8.3.2. *An approximate theory for extensional waves in plates*

The development of another approximate theory for waves in plates that differs in several respects from the study of the previous section will now be presented. First, the description of a different type of motion, extensional (longitudinal) instead of flexural, will be considered. As discussed in the previous section, this is conveniently done in approximate theories by retaining and omitting various terms in the expansions for the displacements. Secondly, the energy equation will be used as the starting point, instead of directly integrating the equations of motion across the thickness. Thirdly, the displacements will be expanded in Legendre polynomials of the thickness coordinate instead of the thickness coordinate itself. In addition, the deformation kinematics are more elaborate than for the flexural theory previously considered, and a larger number of adjustment parameters enter. The procedures for determining these parameters are accordingly more intricate than before. The theory presented here is that developed by Mindlin and Medick [66].

1. *Expansion of the displacements.* To enable indicial notation to be used, let the plate coordinates be x_1, x_2, x_3, with x_2 being the thickness coordinate and the x_1, x_3-plane being the mid plane of the plate. The plate thickness is $2b$. Assume that the displacements u_1, u_2, u_3 may be expanded in terms of the Legendre polynomials P_n as

$$u_j = \sum_{n=0}^{\infty} P_n(\eta) u_j^{(n)}(x_1, x_3, t) \qquad (j = 1, 2, 3), \tag{8.3.61}$$

where $\eta = x_2/b$ and the first few polynomials are

$$P_0(\eta) = 1, \qquad P_1(\eta) = \eta, \qquad P_2(\eta) = (3\eta^2 - 1)/2, \qquad P_3(\eta) = (5\eta^3 - 3\eta)/2,$$

$$P_n(\eta) = \frac{1}{2^n n!} \frac{d^n(\eta^2 - 1)^n}{d\eta^n}. \tag{8.3.62}$$

The reason for expanding in $P(\eta)$ instead of η itself (as, for example, suggested in the discussion leading to (8.3.17)) was given by Mindlin and Medick as follows. Expansions in η of second and higher powers lead to awkward mathematical forms owing to lack of orthogonality of the terms of a power series. Although similarly awkward forms occur using the Legendre polynomials, they generally do not occur until third-order terms are reached. Since the present theory will be at most third order, this will not seriously affect the work.

2. *Stress equations of motion.* We start with the variational equations of motion as obtained from Hamilton's principle, given by

$$\int_V (\tau_{ij,i}-\rho\ddot{u}_j)\,\delta u_j\,\mathrm{d}V = 0. \tag{8.3.63}$$

Now

$$\begin{aligned}\ddot{u}_j &= \sum_{n=0}^{\infty} P_n(\eta)\ddot{u}_j^{(n)}(x_1, x_3, t),\\ \delta u_j &= \sum_{n=0}^{\infty} P_n(\eta)\,\delta u_j^{(n)}(x_1, x_3, t).\end{aligned} \tag{8.3.64}$$

Substitution of the above in the equations of motion gives two integrals to be evaluated:

$$\begin{aligned}I_1 &= \int_V \tau_{ij,i}\left(\sum_{n=0}^{\infty} P_n(\eta)\,\delta u_j^{(n)}\right)\mathrm{d}V,\\ I_2 &= \rho\int_V \left(\sum_{n=0}^{\infty} P_m(\eta)\ddot{u}_j^{(m)}\right)\left(\sum_{n=0}^{\infty} P_n(\eta)\,\delta u_j^{(n)}\right)\mathrm{d}V.\end{aligned} \tag{8.3.65}$$

Considering first I_1, we have

$$I_1 = b\int_A \mathrm{d}A\left\{\sum_{m=0}^{\infty} \delta u_j^{(n)}(x_1, x_3, t)\int_{-1}^{1} \tau_{ij,i}P_n(\eta)\,\mathrm{d}\eta\right\}. \tag{8.3.66}$$

Now, using conventional notation on the derivatives, it is evident that

$$\int_{-1}^{1} \frac{\partial\tau_{1j}}{\partial x_1}P_n(\eta)\,\mathrm{d}\eta = \frac{\partial}{\partial x_1}\int_{-1}^{1} \tau_{1j}P_n(\eta)\,\mathrm{d}\eta \tag{8.3.67}$$

and

$$\int_{-1}^{1} \frac{\partial\tau_{3j}}{\partial x_3}P_n(\eta)\,\mathrm{d}\eta = \frac{\partial}{\partial x_3}\int_{-1}^{1} \tau_{3j}P_n(\eta)\,\mathrm{d}\eta. \tag{8.3.68}$$

The integrals are analogous to the plate stresses that have arisen in earlier approximate theories. For the partial derivatives with respect to x_2, we have

$$\int_{-1}^{1} \frac{\partial \tau_{2j}}{\partial x_2} P_n(\eta)\, d\eta = \int_{-1}^{1} \frac{\partial}{\partial x_2}(\tau_{2j} P_n(\eta))\, d\eta - \int_{-1}^{1} \tau_{2j} \frac{dP_n(\eta)}{dx_2}\, d\eta$$

$$= \left[\frac{1}{b}\tau_{2j} P_n(\eta)\right]_{-1}^{1} - \frac{1}{b}\int_{-1}^{1} \tau_{2j} \frac{dP_n(\eta)}{d\eta}\, d\eta. \qquad (8.3.69)$$

Now it is known that

$$\frac{dP_n(\eta)}{d\eta} = \sum_{m=1,3,\cdots}^{n} D_{mn} P_{n-m}(\eta), \qquad D_{mn} = 2(n-m)+1. \qquad (8.3.70)$$

Thus

$$\int_{-1}^{1} \tau_{2j} \frac{dP_n}{d\eta}\, d\eta = \sum_{m=1,3,\cdots}^{n} D_{mn} \int_{-1}^{1} \tau_{2j} P_{n-m}(\eta)\, d\eta. \qquad (8.3.71)$$

We now define the plate stresses as

$$\tau_{ij}^{(n)} = \int_{-1}^{1} \tau_{ij} P_n\, d\eta. \qquad (8.3.72)$$

The integral I_1 then becomes

$$I_1 = b\int_A dA\Bigg(\sum_{n=0}^{\infty} \delta u_1^{(n)}\left\{\frac{\partial \tau_{11}^{(n)}}{\partial x_1} + \frac{\partial \tau_{31}^{(n)}}{\partial x_3} - \frac{1}{b}\left(\sum_{m=1,3,\cdots}^{n} D_{mn}\tau_{21}^{(n-m)}\right) + \frac{1}{b}[\tau_{21}P_n(\eta)]_{-1}^{1}\right\} +$$

$$+\sum_{n=0}^{\infty} \delta u_2^{(n)}\left\{\frac{\partial \tau_{12}^{(n)}}{\partial x_1} + \frac{\partial \tau_{32}^{(n)}}{\partial x_3} - \frac{1}{b}\left(\sum_{m=1,3,\cdots}^{\infty} D_{mn}\tau_{22}^{(n-m)}\right) + \frac{1}{b}[\tau_{22}P_n(\eta)]_{-1}^{1}\right\} +$$

$$+\sum_{n=0}^{\infty} \delta u_3^{(n)}\left\{\frac{\partial \tau_{13}^{(n)}}{\partial x_1} + \frac{\partial \tau_{33}^{(n)}}{\partial x_3} - \frac{1}{b}\left(\sum_{m=1,3,\cdots}^{\infty} D_{mn}\tau_{23}^{(n-m)}\right) + \frac{1}{b}[\tau_{23}P_n(\eta)]_{-1}^{1}\right\}\Bigg). \qquad (8.3.73)$$

Now consider I_2. We have

$$I_2 = \rho b\int_A dA\left\{\int_{-1}^{1}\left(\sum_{m=0}^{\infty} P_m(\eta)\ddot{u}_j^{(m)}\right)\left(\sum_{n=0}^{\infty} P_n(\eta)\, \delta u_j^{(n)}\right) d\eta\right\},$$

$$= \rho b\int_A dA\left\{\sum_{m=0}^{\infty}\sum_{n=0}^{\infty} \ddot{u}_j^{(m)}\, \delta u_j^{(n)} \int_{-1}^{1} P_m(\eta)P_n(\eta)\, d\eta\right\}. \qquad (8.3.74)$$

Now

$$\int_{-1}^{1} P_m(\eta)P_n(\eta)\, d\eta = \begin{cases} 0, & m \neq n \\ C_n, & m = n \end{cases}, \qquad C_n = 2/(2n+1). \qquad (8.3.75)$$

Then

$$I_2 = \rho b \int_A \mathrm{d}A \left(\sum_{n=0}^{\infty} C_n \ddot{u}_j^{(n)} \, \delta u_j^{(n)} \right). \tag{8.3.76}$$

We now define

$$F_j^{(n)} = [\tau_{2j} P_n(\eta)]_{-1}^{1}. \tag{8.3.77}$$

Then I_1, I_2 combine to give

$$\begin{aligned}
I_1 - I_2 &= \int_V (\tau_{ij,i} - \rho \ddot{u}_j) \, \delta u_j \, \mathrm{d}V \\
&= b \int_A \mathrm{d}A \Bigg[\sum_{n=0}^{\infty} \delta u_1^{(n)} \left\{ \frac{\partial \tau_{11}^{(n)}}{\partial x_1} + \frac{\partial \tau_{31}^{(n)}}{\partial x_3} - \frac{1}{b} \left(\sum_{m=1,3,\ldots}^{n} D_{mn} \tau_{21}^{(n-m)} \right) + \right. \\
&\qquad \left. + F_1^{(n)} - \rho C_n \ddot{u}_1^{(n)} \right\} + \\
&\quad + \sum_{n=0}^{\infty} \delta u_2^{(n)} \left\{ \frac{\partial \tau_{12}^{(n)}}{\partial x_1} + \frac{\partial \tau_{32}^{(n)}}{\partial x_3} - \frac{1}{b} \left(\sum_{m=1,3,\ldots}^{\infty} D_{mn} \tau_{22}^{(n-m)} \right) + F_2^{(n)} - \rho C_n \ddot{u}_2^{(n)} \right\} + \\
&\quad + \sum_{m=0}^{\infty} \delta u_3^{(m)} \{\ldots\} \Bigg] = 0.
\end{aligned} \tag{8.3.78}$$

Using index notation this reduces to

$$I_1 - I_2 = \int_A \mathrm{d}A \left(\sum_{n=0}^{\infty} \delta u_j^{(n)} \left\{ b \tau_{1j,1}^{(n)} + b \tau_{3j,3}^{(n)} - \sum_{m=1,3,\ldots}^{n} D_{mn} \tau_{2j}^{(n-m)} + F_j^{(n)} - \rho C_n \ddot{u}_j^{(n)} \right\} \right) = 0. \tag{8.3.79}$$

We thus arrive at the stress equations of motion

$$b \tau_{1j,1}^{(n)} + b \tau_{3j,3}^{(n)} - \sum_{m=1,3,\ldots}^{n} D_{mn} \tau_{2j}^{(n-m)} + F_j^{(n)} = \rho C_n \ddot{u}_j^{(n)} \qquad (j = 1, 2, 3)$$
$$(n = 0, 1, 2, \ldots). \tag{8.3.80}$$

3. *The plate strains.* The plate stress–strain relations must now be derived. Starting with the definition of infinitesimal strain $\varepsilon_{ij} = (u_{i,j} + u_{j,i})/2$, we obtain

$$2\varepsilon_{ij} = \sum_{n=0}^{\infty} \{P_n(\eta) u_i^{(n)}\}_{,j} + \sum_{n=0}^{\infty} \{P_n(\eta) u_j^{(n)}\}_{,i}. \tag{8.3.81}$$

Now

$$\begin{aligned}
\sum_{n=0}^{\infty} \{P_n(\eta) u_i^{(n)}\}_{,1} &= \sum_{n=0}^{\infty} P_n(\eta) u_{i,1}^{(n)}, \\
\sum_{n=0}^{\infty} \{P_n(\eta) u_i^{(n)}\}_{,3} &= \sum_{n=0}^{\infty} P_n(\eta) u_{i,3}^{(n)}, \\
\sum_{n=0}^{\infty} \{P_n(\eta) u_i^{(n)}\}_{,2} &= \sum_{n=0}^{\infty} \frac{\mathrm{d}P_n(\eta)}{\mathrm{d}x_2} u_i^{(n)} = \frac{1}{b} \sum_{n=0}^{\infty} \left(\sum_{m=1,3}^{n} D_{mn} P_{n-m}(\eta) \right) u_i^{(n)}.
\end{aligned} \tag{8.3.82}$$

Then the strains ε_{11}, ε_{33}, ε_{13} have the form

$$2\varepsilon_{ij}=\sum_{n=0}^{\infty}P_n(\eta)(u_{i,j}^{(n)}+u_{j,i}^{(n)}) \qquad (i,j\neq 2). \tag{8.3.83}$$

The strains ε_{12}, ε_{32} have the form

$$2\varepsilon_{i2}=\frac{1}{b}\sum_{n=0}^{\infty}\left\{\sum_{m=1,3}^{n}D_{mn}P_{n-m}(\eta)\right\}u_i^{(n)}+\sum_{n=0}^{\infty}P_n(\eta)u_{2,i}^{(n)} \qquad (i\neq 2). \tag{8.3.84}$$

The strain ε_{22} is given by

$$2\varepsilon_{22}=\frac{2}{b}\sum_{n=0}^{\infty}\left\{\sum_{m=1,3}^{\infty}D_{mn}P_{n-m}(\eta)\right\}u_2^{(n)}. \tag{8.3.85}$$

The general expression for ε_{ij} that combines the various aspects of (8.3.81)–(8.3.85) is then

$$2\varepsilon_{ij}=\sum_{n=0}^{\infty}\sum_{m=1,3}^{\infty}\{(u_{i,j}^{(n)}+u_{j,i}^{(n)})P_n+(\delta_{2j}u_i^{(n)}+\delta_{2i}u_j^{(n)})b^{-1}D_{mn}P_{n-m}\}. \tag{8.3.86}$$

We are seeking plate stress–strain relations, where plate stresses of some order n, $\tau_{ij}^{(n)}$ will be related to plate strains of similar order. Thus it is desired, effectively, to have ε_{ij} expanded in the form

$$\varepsilon_{ij}=\sum_{n=0}^{\infty}P_n(\eta)\varepsilon_{ij}^{(n)}. \tag{8.3.87}$$

Although the first bracketed term of (8.3.86) is of this form, some manipulation must be done to bring the entire expression into this form. According to Mindlin and Medick, considering the double sum as a triangular array and interchanging the order of summation of columns and rows brings about the desired result, with

$$2\varepsilon_{ij}^{(n)}=u_{i,j}^{(n)}+u_{j,i}^{(n)}+\frac{(2n+1)}{b}\sum_{m=1,3}^{\infty}(\delta_{2j}u_i^{(m+n)}+\delta_{2i}u_j^{(m+n)}). \tag{8.3.88}$$

4. *The plate stress–strain relations.* We now have the conventional stresses related to the plate strains by

$$\tau_{ij}=c_{ijkl}\varepsilon_{kl}=c_{ijkl}\sum_{n=0}^{\infty}P_n(\eta)\varepsilon_{kl}^{(n)}. \tag{8.3.89}$$

These may be substituted into the defining equation for the plate stresses (8.3.72) to give

$$\begin{aligned}\tau_{ij}^{(n)}&=\int_{-1}^{1}P_n(\eta)c_{ijkl}\sum_{m=0}^{\infty}P_m\varepsilon_{kl}^{(m)}\,d\eta\\&=c_{ijkl}\sum_{m=0}^{\infty}\varepsilon_{kl}^{(m)}\int_{-1}^{1}P_n(\eta)P_m(\eta)\,d\eta\\&=C_nc_{ijkl}\varepsilon_{kl}^{(n)},\end{aligned} \tag{8.3.90}$$

where the orthogonality condition (8.3.75) has been employed.

We use the reduced index notation given by

$$\tau_1 = \tau_{11}, \quad \tau_2 = \tau_{22}, \quad \tau_3 = \tau_{33}, \quad \tau_4 = \tau_{23}, \quad \tau_5 = \tau_{31}, \quad \tau_6 = \tau_{12},$$

$$\varepsilon_1 = \varepsilon_{11}, \quad \varepsilon_2 = \varepsilon_{22}, \quad \varepsilon_3 = \varepsilon_{33}, \quad \varepsilon_4 = 2\varepsilon_{23}, \quad \varepsilon_5 = 2\varepsilon_{31}, \quad \varepsilon_6 = 2\varepsilon_{12}. \tag{8.3.91}$$

Then

$$\tau_p^{(n)} = C_n c_{pq} \varepsilon_q^{(n)} \qquad (p, q = 1, 2). \tag{8.3.92}$$

5. *Strain energy.* To complete the definition of the important quantities of elastodynamics in terms of plate stresses and strains, consider the strain energy, given by

$$2U = c_{ijkl}\varepsilon_{ij}\varepsilon_{kl} = c_{pq}\varepsilon_p\varepsilon_q. \tag{8.3.93}$$

We define the plate strain energy density by

$$\bar{U} = \int_{-1}^{1} U \, d\eta. \tag{8.3.94}$$

Then

$$\bar{U} = \frac{1}{2}\int_{-1}^{1} c_{pq}\left(\sum_{n=0}^{\infty} P_n(\eta)\varepsilon_p^{(n)}\right)\left(\sum_{m=0}^{\infty} P_m(\eta)\varepsilon_q^{(m)}\right) d\eta,$$

$$= \tfrac{1}{2}c_{pq}\left(\sum_{n=0}^{\infty}\sum_{m=0}^{\infty}\varepsilon_p^{(n)}\varepsilon_q^{(m)}\int_{-1}^{1} P_n(\eta)P_m(\eta)\, d\eta\right),$$

$$= \tfrac{1}{2}c_{pq}\sum_{n=0}^{\infty} C_n \varepsilon_p^{(n)}\varepsilon_q^{(n)}. \tag{8.3.95}$$

Also this equals

$$\bar{U} = \tfrac{1}{2}\sum_{n=0}^{\infty} \tau_q^{(n)}\varepsilon_q^{(n)}. \tag{8.3.96}$$

Further, we see from (8.3.95) that $\partial \bar{U}/\partial \varepsilon_p^{(n)} = c_{pq}C_n\varepsilon_q^{(n)}$, so that

$$\tau_p^{(n)} = \partial \bar{U}/\partial \varepsilon_p^{(n)}. \tag{8.3.97}$$

Finally, we have the kinetic energy, given for exact theory by

$$K = \tfrac{1}{2}\rho \dot{u}_j \dot{u}_j. \tag{8.3.98}$$

For the plate we define $\bar{K}$ by

$$\bar{K} = \int_{-1}^{1} K \, d\eta = \frac{1}{2}\int_{-1}^{1} \rho\left(\sum_{m=0}^{\infty} P_m(\eta)\dot{u}_j^{(m)}\right)\left(\sum_{n=0}^{\infty} P_n(\eta)\dot{u}_j^{(n)}\right) d\eta,$$

$$= \tfrac{1}{2}\rho\sum_{n=0}^{\infty} C_n \dot{u}_j^{(n)}\dot{u}_j^{(n)}. \tag{8.3.99}$$

6. *The equations for extensional vibrations.* The developments thus far have been quite general. If considerations are now restricted only to extensional (that is, longitudinal or symmetric) motions of the plate, only certain terms of the expansion may be retained. Now the first few terms of u_1, u_2, u_3 are

$$u_1 = P_0 u_1^{(0)} + P_1 u_1^{(1)} + P_2 u_1^{(2)} + \ldots,$$

$$u_2 = P_0 u_2^{(0)} + P_1 u_2^{(1)} + P_2 u_2^{(2)} + \ldots,$$

$$u_3 = P_0 u_3^{(0)} + P_1 u_3^{(1)} + P_2 u_3^{(2)} + \ldots. \tag{8.3.100}$$

Observing the forms of $P_0, P_1, P_2, \ldots$, the forms of the thickness displacement variation for the first few modes are shown in Fig. 8.25. For $j = 1, 3$ the

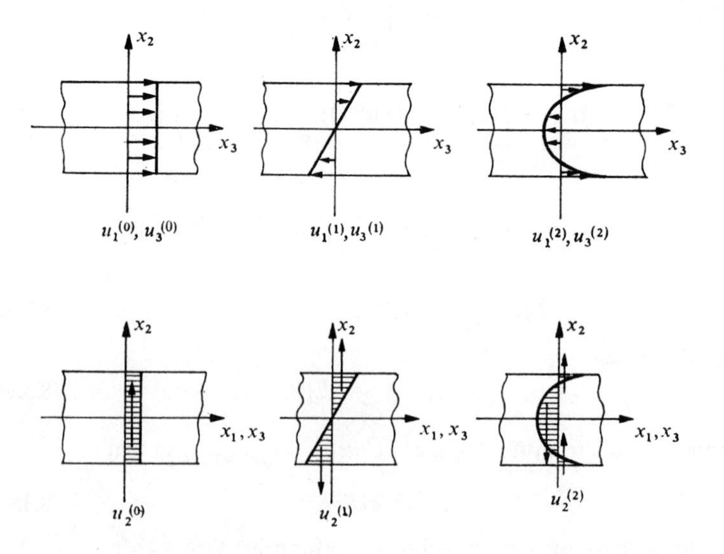

FIG. 8.25. Thickness variation of the first few displacement modes $u_j^{(n)}$.

symmetric modes of $u_j^{(n)}$ are $u_1^{(0)}$, $u_2^{(0)}$, $u_1^{(2)}$, $u_3^{(2)}$, $u_1^{(4)}$, $u_2^{(4)}$, ... For $j = 2$, the symmetric modes are $u_2^{(1)}$, $u_2^{(3)}$, $u_2^{(5)}$, ... Thus, only n = even terms are retained in $u_1^{(n)}$, $u_3^{(n)}$ and n = odd terms in $u_2^{(n)}$. A general rule covering the selection of the symmetric displacement modes is that $n+j$ = odd. The extension of this rule to cover the appropriate stresses $\tau_{ij}^{(n)}$ and strains $\varepsilon_{ij}^{(n)}$ is that $i+j+n$ = even. On this basis, the governing equations for extensional vibrations may be summarized. For the stress equations of motion, we have

$$\frac{\partial\tau_1^{(0)}}{\partial x_1}+\frac{\partial\tau_5^{(0)}}{\partial x_3}+\frac{F_1^{(0)}}{b}=2\rho\frac{\partial^2u_1^{(0)}}{\partial t^2},$$

$$\frac{\partial\tau_5^{(0)}}{\partial x_1}+\frac{\partial\tau_3^{(0)}}{\partial x_3}+\frac{F_3}{b}=2\rho\frac{\partial^2u_3^{(0)}}{\partial t^2},$$

$$\frac{\partial\tau_6^{(1)}}{\partial x_1}+\frac{\partial\tau_4^{(1)}}{\partial x_3}-\frac{\tau_2^{(0)}}{b}+\frac{F_2^{(1)}}{b}=\frac{2\rho}{3}\frac{\partial^2u_2^{(1)}}{\partial t^2},$$

$$\frac{\partial\tau_1^{(2)}}{\partial x_1}+\frac{\partial\tau_5^{(2)}}{\partial x_3}-\frac{3\tau_6^{(1)}}{b}+\frac{F_1^{(2)}}{b}=\frac{2\rho}{5}\frac{\partial^2u_1^{(2)}}{\partial t^2},$$

$$\frac{\partial\tau_5^{(2)}}{\partial x_1}+\frac{\partial\tau_3^{(2)}}{\partial x_3}-\frac{3\tau_4^{(1)}}{b}+\frac{F_3^{(2)}}{b}=\frac{2\rho}{5}\frac{\partial^2u_3^{(2)}}{\partial t^2},$$

$$\frac{\partial\tau_6^{(3)}}{\partial x_1}+\frac{\partial\tau_4^{(3)}}{\partial x_3}-\frac{5\tau_2^{(2)}}{b}-\frac{\tau_2^{(0)}}{b}+\frac{F_2^{(3)}}{b}=\frac{2\rho}{7}\frac{\partial^2u_2^{(3)}}{\partial t^2}. \qquad (8.3.101)$$

The strains are

$$\varepsilon_1^{(0)}=\partial u_1^{(0)}/\partial x_1,$$

$$\varepsilon_2^{(0)}=(u_2^{(1)}+u_2^{(3)}+\ldots)/b,$$

$$\varepsilon_3^{(0)}=\partial u_3^{(0)}/\partial x_3,$$

$$\varepsilon_5^{(0)}=\partial u_3^{(0)}/\partial x_1+\partial u_1^{(0)}/\partial x_3,$$

$$\varepsilon_4^{(1)}=\partial u_2^{(1)}/\partial x_3+3(u_3^{(2)}+u_3^{(4)}+\ldots)/b,$$

$$\varepsilon_6^{(1)}=\partial u_2^{(1)}/\partial x_1+3(u_1^{(2)}+u_1^{(4)}+\ldots)/b,$$

$$\varepsilon_1^{(2)}=\partial u_1^{(2)}/\partial x_1,$$

$$\varepsilon_2^{(2)}=5(u_2^{(3)}+u_2^{(5)}+\ldots)/b,$$

$$\varepsilon_3^{(2)}=\partial u_3^{(2)}/\partial x_3,$$

$$\varepsilon_5^{(2)}=\partial u_3^{(2)}/\partial x_1+\partial u_1^{(2)}/\partial x_3,$$

$$\varepsilon_4^{(3)}=\partial u_2^{(3)}/\partial x_3+7(u_3^{(4)}+u_3^{(6)}+\ldots)/b,$$

$$\varepsilon_6^{(3)}=\partial u_2^{(3)}/\partial x_1+7(u_1^{(4)}+u_1^{(6)}+\ldots)/b. \qquad (8.3.102)$$

The stress–strain relations are given by

$$
\begin{aligned}
\tau_1^{(0)} &= 2\{(\lambda+2\mu)\varepsilon_1^{(0)}+\lambda(\varepsilon_2^{(0)}+\varepsilon_3^{(0)})\},\\
\tau_2^{(0)} &= 2\{(\lambda+2\mu)\varepsilon_2^{(0)}+\lambda(\varepsilon_3^{(0)}+\varepsilon_1^{(0)})\},\\
\tau_3^{(0)} &= 2\{(\lambda+2\mu)\varepsilon_3^{(0)}+\lambda(\varepsilon_1^{(0)}+\varepsilon_2^{(0)})\},\\
\tau_5^{(0)} &= 2\mu\varepsilon_5^{(0)},\\
\tau_4^{(1)} &= 2\mu\varepsilon_4^{(1)}/3,\\
\tau_6^{(1)} &= 2\mu\varepsilon_6^{(1)}/3,\\
\tau_1^{(2)} &= 2\{(\lambda+2\mu)\varepsilon_1^{(2)}+\lambda(\varepsilon_2^{(2)}+\varepsilon_3^{(2)})\}/5,\\
\tau_2^{(2)} &= 2\{(\lambda+2\mu)\varepsilon_2^{(2)}+\lambda(\varepsilon^{(2)}+\varepsilon_1^{(2)})\}/5,\\
\tau_3^{(2)} &= 2\{(\lambda+2\mu)\varepsilon_3^{(2)}+\lambda(\varepsilon_1^{(2)}+\varepsilon_3^{(2)})\}/5,\\
\tau_5^{(2)} &= 2\mu\varepsilon_5^{(2)}/5,\\
\tau_4^{(3)} &= 2\mu\varepsilon_4^{(3)}/7,\\
\tau_6^{(3)} &= 2\mu\varepsilon_6^{(3)}/7.
\end{aligned}
\tag{8.3.103}
$$

The strain energy and kinetic energy densities are given by

$$
\begin{aligned}
2\bar{U} = {} & \tau_1^{(0)}\varepsilon_1^{(0)}+\tau_3^{(0)}\varepsilon_3^{(0)}+\tau_5^{(0)}\varepsilon_5^{(0)}+\tau_2^{(0)}\varepsilon_2^{(0)}+\tau_4^{(1)}\varepsilon_4^{(1)}+\tau_6^{(1)}\varepsilon_6^{(1)}+\\
& +\tau_1^{(2)}\varepsilon_1^{(2)}+\tau_3^{(2)}\varepsilon_3^{(2)}+\tau_5^{(2)}\varepsilon_5^{(2)}+\tau_2^{(2)}\varepsilon_2^{(2)}+\tau_4^{(3)}\varepsilon_4^{(3)}+\tau_6^{(3)}\varepsilon_6^{(3)}+\ldots,
\end{aligned}
\tag{8.3.104}
$$

$$
\bar{K} = \rho(\dot{u}_1^{(0)}\dot{u}_1^{(0)}+\dot{u}_3^{(0)}\dot{u}_3^{(0)}+\tfrac{1}{3}\dot{u}_2^{(1)}\dot{u}_2^{(1)}+\tfrac{1}{5}\dot{u}_1^{(2)}\dot{u}_1^{(2)}+\tfrac{1}{5}\dot{u}_3^{(2)}\dot{u}_3^{(2)}+\tfrac{1}{7}\dot{u}_2^{(3)}\dot{u}_2^{(3)}+\ldots).
\tag{8.3.105}
$$

7. *Truncation of the series.* We now retain only $u_1^{(0)}$, $u_1^{(2)}$, $u_3^{(0)}$, $u_3^{(2)}$, $u_2^{(1)}$, $u_2^{(3)}$. All other $u_j^{(n)}$ we set equal to zero. These modes have been shown in the previous illustrations. We are retaining two degrees of freedom in the form of the u_1, u_3 displacements and two degrees of freedom in the u_2 displacement. From Mindlin and Medick:

The terms $u_1^{(0)}$, $u_3^{(0)}$ are the amplitudes of uniform distributions of displacements which occur in low-frequency extensional and shear motions in the plane of the plate: $u_2^{(1)}$ is the amplitude of a linear distribution of displacement which is an approximation to the exact sinusoidal distribution in the lowest, symmetric, thickness-stretch mode; $u_1^{(2)}$ and $u_3^{(2)}$ are the amplitudes of quadratic distributions of displacements which are approximations to the sinusoidal distributions in the lowest symmetric, thickness-shear mode, and the face-shear mode of the same order.

Some slight complications enter at this stage. We first note that $u_2^{(3)}$ has been retained, representing a third-order displacement term. However, we will only be interested in strains of order two, $\varepsilon_i^{(2)}$ or less. However, terms third-order in $u_2^{(n)}$ contribute to these strains (see $\varepsilon_2^{(2)}$ of (8.3.102)). Several stresses in turn contain $\varepsilon_2^{(2)}$. Thus, a second-order theory in strains may contain terms of yet higher order in displacement. The particular strain and stress expressions containing contributions from $u_2^{(3)}$ are

$$\varepsilon_2^{(0)} = (u_2^{(1)}+u_2^{(3)})/b, \qquad \varepsilon_2^{(2)} = 5u_2^{(3)}/b, \tag{8.3.106}$$

$$\begin{aligned} \tau_1^{(2)} &= 2\{(\lambda+2\mu)\varepsilon_1^{(2)}+\lambda(\varepsilon_2^{(2)}+\varepsilon_3^{(2)})\}/5, \\ \tau_2^{(2)} &= 2\{(\lambda+2\mu)\varepsilon_2^{(2)}+\lambda(\varepsilon_3^{(2)}+\varepsilon_1^{(2)})\}/5, \\ \tau_3^{(2)} &= 2\{(\lambda+2\mu)\varepsilon_3^{(2)}+\lambda(\varepsilon_1^{(2)}+\varepsilon_2^{(2)})\}/5. \end{aligned} \tag{8.3.107}$$

In addition, $\tau_2^{(0)}$ contains a contribution from $\varepsilon_2^{(0)}$. It should be noted that $u_2^{(3)}$ is not the sole contributor to this latter strain, whereas it is for the strain $\varepsilon_2^{(2)}$.

Now, it is recognized that $\varepsilon_2^{(2)}$ expansions may result from $\tau_1^{(2)}$, $\tau_3^{(2)}$ stresses due to the Poisson effect. As it now stands, however, coupling with the $u_2^{(3)}$ mode occurs as a result of such strains. The procedure taken is to neglect $\tau_2^{(2)}$, the stress associated with the designated expansion and contraction. With $\tau_2^{(2)} = 0$, we have from the second of (8.3.107) that

$$\varepsilon_2^{(2)} = -\frac{\lambda}{\lambda+2\mu}(\varepsilon_3^{(2)}+\varepsilon_1^{(2)}). \tag{8.3.108}$$

This relates the strain of interest to strains containing only the second-order displacement modes and replaces the definition of $\varepsilon_2^{(2)}$ in terms of $u_2^{(3)}$. The remaining stresses are

$$\begin{aligned} \tau_1^{(2)} &= 2E'(\varepsilon_1^{(2)}+\nu\varepsilon_3^{(2)})/5, \\ \tau_3^{(2)} &= 2E'(\varepsilon_3^{(2)}+\nu\varepsilon_1^{(2)})/5, \end{aligned} \tag{8.3.109}$$

where

$$\nu = \frac{\lambda}{2(\lambda+\mu)}, \qquad E' = \frac{4\mu(\lambda+\mu)}{\lambda+2\mu} = \frac{E}{1-\nu^2}. \tag{8.3.110}$$

Thus the contribution of a strain mode due, in reality, to a yet higher-order displacement mode is approximated by lower-order strain and displacement modes. Finally, the term $u_2^{(3)}$ is dropped from the strain $\varepsilon_2^{(0)}$ and thus from the stress $\tau_2^{(0)}$.

It is recognized, of course, that an approximate theory having limited degrees of freedom has been developed. Such a theory obviously cannot reproduce the higher branches of the exact frequency spectrum. In all probability, it will also not agree with the lowest branches in all detail, since the omission of higher polynomial terms will adversely affect the agreement of the lower modes as well as omit the higher modes. By introducing a number of parameters analogous to the Timoshenko shear coefficients, it is possible to retain some freedom of adjustment of the resulting dispersion curves. Mindlin and Medick state that the incorrect distributions of displacement affect the frequencies mainly through the thickness strains and velocities. They let

$$\begin{aligned} \varepsilon_2^{(0)} &= \kappa_1\varepsilon_2^{(0)}, & \dot{u}_2^{(1)} &= \kappa_3\dot{u}_2^{(1)}, \\ \varepsilon_4^{(1)} &= \kappa_2\varepsilon_4^{(1)}, & \dot{u}_1^{(2)} &= \kappa_4\dot{u}_1^{(2)}, \\ \varepsilon_6^{(1)} &= \kappa_2\varepsilon_6^{(1)}, & \dot{u}_3^{(2)} &= \kappa_4\dot{u}_3^{(2)}. \end{aligned} \tag{8.3.111}$$

Thus, four coefficients have been introduced and are available for later adjustment.

With the restriction to extensional motion, truncation of the series to include second-order effects, including the manipulations to remove coupling to $u_2^{(3)}$, and the introduction of the adjustment coefficients, the complete equations for the second-order approximate theory are now summarized. The kinetic and strain energy densities are now given by

$$K^{(2)} = \rho\{\dot{u}_1^{(0)}\dot{u}_1^{(0)}+\dot{u}_3^{(0)}\dot{u}_3^{(0)}+\tfrac{1}{3}\kappa_3^2\dot{u}_2^{(1)}\dot{u}_2^{(1)}+\tfrac{1}{5}\kappa_4^2(\dot{u}_1^{(2)}\dot{u}_1^{(2)}+\dot{u}_3^{(2)}\dot{u}_3^{(2)})\}, \tag{8.3.112}$$

$$\begin{aligned} \bar{U}^{(2)} = {} & (\lambda+2\mu)(\varepsilon_1^{(0)}\varepsilon_1^{(0)}+\kappa_1^2\varepsilon_2^{(0)}\varepsilon_2^{(0)}+\varepsilon_3^{(0)}\varepsilon_3^{(0)})+ \\ & +2\lambda(\kappa_1\varepsilon_2^{(0)}\varepsilon_3^{(0)}+\varepsilon_3^{(0)}\varepsilon_1^{(0)}+\kappa_1\varepsilon_1^{(0)}\varepsilon_2^{(0)})+ \\ & +\mu\varepsilon_5^{(0)}\varepsilon_5^{(0)}+\tfrac{1}{3}\mu\kappa_2^2(\varepsilon_4^{(1)}\varepsilon_4^{(1)}+\varepsilon_6^{(1)}\varepsilon_6^{(1)})+ \\ & +\tfrac{1}{5}E'(\varepsilon_1^{(2)}\varepsilon_1^{(2)}+\varepsilon_3^{(2)}\varepsilon_3^{(2)}+2\nu\varepsilon_1^{(2)}\varepsilon_3^{(2)})+\tfrac{1}{5}\mu\varepsilon_5^{(2)}\varepsilon_5^{(2)}. \end{aligned} \tag{8.3.113}$$

The stress–strain relations are

$$\begin{aligned} \tau_1^{(0)} &= 2\{(\lambda+2\mu)\varepsilon_1^{(0)}+\lambda(\kappa_2\varepsilon_2^{(0)}+\varepsilon_1^{(0)})\}, \\ \tau_2^{(0)} &= 2\{\kappa_1^2(\lambda+2\mu)\varepsilon_2^{(0)}+\lambda\kappa_1(\varepsilon_3^{(0)}+\varepsilon_1^{(0)})\}, \\ \tau_3^{(0)} &= 2\{(\lambda+2\mu)\varepsilon_3^{(0)}+\lambda(\varepsilon_1^{(0)}+\kappa_1\varepsilon_2^{(0)})\}, \\ \tau_5^{(0)} &= 2\mu\varepsilon_5^{(0)}, \\ \tau_4^{(1)} &= \tfrac{2}{3}\mu\kappa_2^2\varepsilon_4^{(1)}, \\ \tau_6^{(1)} &= \tfrac{2}{3}\mu\kappa_2^2\varepsilon_6^{(1)}, \\ \tau_1^{(2)} &= \tfrac{2}{5}E'(\varepsilon_1^{(2)}+\nu\varepsilon_3^{(2)}), \\ \tau_3^{(2)} &= \tfrac{2}{5}E'(\varepsilon^{(2)}+\nu\varepsilon_1^{(2)}), \\ \tau_5^{(2)} &= \tfrac{2}{5}\mu\varepsilon_5^{(2)}. \end{aligned} \tag{8.3.114}$$

Strain displacements:

$$
\begin{aligned}
\varepsilon_1^{(0)} &= \frac{\partial u_1^{(0)}}{\partial x_1}, \\
\varepsilon_2^{(0)} &= \frac{u_2^{(1)}}{b}, \\
\varepsilon_3^{(0)} &= \frac{\partial u_3^{(0)}}{\partial x_3}, \\
\varepsilon_5^{(0)} &= \frac{\partial u_3^{(0)}}{\partial x_1} + \frac{\partial u_1^{(0)}}{\partial x_3}, \\
\varepsilon_4^{(1)} &= \frac{3u_3^{(2)}}{b} + \frac{\partial u_2^{(1)}}{\partial x_3}, \\
\varepsilon_6^{(1)} &= \frac{\partial u_2^{(1)}}{\partial x_1} + \frac{3u_1^{(2)}}{b}, \\
\varepsilon_1^{(2)} &= \frac{\partial u_1^{(2)}}{\partial x_1}, \\
\varepsilon_3^{(2)} &= \frac{\partial u_3^{(2)}}{\partial x_3}, \\
\varepsilon_5^{(2)} &= \frac{\partial u_3^{(2)}}{\partial x_1} + \frac{\partial u_1^{(2)}}{\partial x_3}.
\end{aligned}
\tag{8.3.115}
$$

Stress equations of motion:

$$
\begin{aligned}
\frac{\partial \tau_1^{(0)}}{\partial x_1} + \frac{\partial \tau_5^{(0)}}{\partial x_3} + \frac{F_1^{(0)}}{b} &= 2\rho \frac{\partial^2 u_1^{(0)}}{\partial t^2}, \\
\frac{\partial \tau_5^{(0)}}{\partial x_1} + \frac{\partial \tau_3^{(0)}}{\partial x_3} + \frac{F_3^{(0)}}{b} &= 2\rho \frac{\partial^2 u_3^{(0)}}{\partial t^2}, \\
\frac{\partial \tau_6^{(1)}}{\partial x_1} + \frac{\partial \tau_4^{(1)}}{\partial x_3} - \frac{\tau_2^{(0)}}{b} + \frac{F_2^{(1)}}{b} &= \frac{2\rho\kappa_3^2}{3} \frac{\partial^2 u_2^{(1)}}{\partial t^2}, \\
\frac{\partial \tau_1^{(2)}}{\partial x_1} + \frac{\partial \tau_5^{(2)}}{\partial x_3} - \frac{3\tau_6^{(1)}}{b} + \frac{F_1^{(2)}}{b} &= \frac{2\rho\kappa_4^2}{5} \frac{\partial^2 u_1^{(2)}}{\partial t^2}, \\
\frac{\partial \tau_5^{(2)}}{\partial x_1} + \frac{\partial \tau_3^{(2)}}{\partial x_3} - \frac{3\tau_4^{(1)}}{b} + \frac{F_3^{(2)}}{b} &= \frac{2\rho\kappa_4^2}{5} \frac{\partial^2 u_3^{(2)}}{\partial t^2}.
\end{aligned}
\tag{8.3.116}
$$

The displacement equations of motion are

$$\mu\,\nabla^2 u_1^{(0)}+(\lambda+\mu)\frac{\partial e_0}{\partial x_1}+\frac{\lambda\kappa_1}{b}\frac{\partial u_2^{(1)}}{\partial x_1}+\frac{F_1^{(0)}}{2b}=\rho\frac{\partial^2 u_1^{(0)}}{\partial t^2},$$

$$\mu\,\nabla^2 u_3^{(0)}+(\lambda+\mu)\frac{\partial e_0}{\partial x_3}+\frac{\lambda\kappa_1}{b}\frac{\partial u_2^{(1)}}{\partial x_3}+\frac{F_3^{(0)}}{2b}=\rho\frac{\partial^2 u_3^{(0)}}{\partial t^2},$$

$$\mu\kappa_2^2\,\nabla^2 u_2^{(1)}-\frac{3\lambda\kappa_1 e_0}{b}-\frac{3\kappa_1^2(\lambda+2\mu)u_2^{(1)}}{b^2}+\frac{3\mu\kappa_2^2 e_2}{b}+\frac{3F_2^{(1)}}{2b}=\rho\kappa_3^2\frac{\partial^2 u_2^{(1)}}{\partial t^2},$$

(8.3.117)

$$\frac{E'}{2}\left\{(1-\nu)\,\nabla^2 u_1^{(2)}+(1+\nu)\frac{\partial e_2}{\partial x_1}\right\}-\frac{5\mu\kappa_2^2}{b}\left(\frac{\partial u_2^{(1)}}{\partial x_1}+\frac{3u_1^{(2)}}{b}\right)+\frac{5F_1^{(2)}}{2b}=\rho\kappa_4^2\frac{\partial^2 u_1^{(2)}}{\partial t^2},$$

$$\frac{E'}{2}\left\{(1-\nu)\,\nabla^2 u_3^{(2)}+(1+\nu)\frac{\partial e_2}{\partial x_3}\right\}-\frac{5\mu\kappa_2^2}{b}\left(\frac{\partial u_2^{(1)}}{\partial x_3}+\frac{3u_3^{(2)}}{b}\right)+\frac{5F_3^{(2)}}{2b}=\rho\kappa_4^2\frac{\partial^2 u_3^{(2)}}{\partial t^2},$$

where

$$\nabla^2=\frac{\partial^2}{\partial x_1^2}+\frac{\partial^2}{\partial x_3^2},\qquad e_0=\frac{\partial u_1^{(0)}}{\partial x_1}+\frac{\partial u_3^{(0)}}{\partial x_3},\qquad e_2=\frac{\partial u_1^{(2)}}{\partial x_1}+\frac{\partial u_3^{(2)}}{\partial x_3}. \quad (8.3.118)$$

8. *Propagation of harmonic waves.* Consider straight-crested waves propagating in the x_1 direction, so that $\partial/\partial x_3 = 0$. The plate displacement equations of motion reduce to

$$\mu\frac{\partial^2 u_1^{(0)}}{\partial x_1^2}+(\lambda+\mu)\frac{\partial e_0}{\partial x_1}+\frac{\lambda\kappa_1}{b}\frac{\partial u_2^{(1)}}{\partial x_1}=\rho\ddot{u}_1^{(0)},$$

$$\mu\frac{\partial^2 u_3^{(0)}}{\partial x_1^2}=\rho\ddot{u}_3^{(0)},$$

$$\mu\kappa_2^2\frac{\partial^2 u_2^{(1)}}{\partial x_1^2}-\frac{3\lambda\kappa_1 e_0}{b}-\frac{3\kappa_1^2(\lambda+2\mu)}{b^2}u_2^{(1)}+\frac{3\mu\kappa_2^2 e_2}{b}=\rho\kappa_3^2\ddot{u}_2^{(1)}, \quad (8.3.119)$$

$$\frac{E'}{2}\left\{(1-\nu)\frac{\partial^2 u_1^{(2)}}{\partial x_1^2}+(1+\nu)\frac{\partial e_2}{\partial x_1}\right\}-\frac{5\mu\kappa_2^2}{b}\left(\frac{\partial u_2^{(1)}}{\partial x_1}+\frac{3u_1^{(2)}}{b}\right)=\rho\kappa_4^2\ddot{u}_1^{(2)},$$

$$\frac{E'(1-\nu)}{2}\frac{\partial^2 u_3^{(2)}}{\partial x_1^2}-\frac{15\mu\kappa_2^2}{b^2}u_3^{(2)}=\rho\kappa_4^2\ddot{u}_3^{(2)},$$

where

$$e_0=\frac{\partial u_1^{(0)}}{\partial x_1},\qquad e_2=\frac{\partial u_1^{(2)}}{\partial x_1}. \quad (8.3.120)$$

The five equations may be represented symbolically as

$$\begin{aligned} f_1\{u_1^{(0)}, u_2^{(1)}\} &= \rho\ddot{u}_1^{(0)}, \\ f_2\{u_3^{(0)}\} &= \rho\ddot{u}_3^{(0)}, \\ f_3\{u_2^{(1)}, u_1^{(0)}, u_1^{(2)}\} &= \rho\ddot{u}_2^{(1)}, \\ f_4\{u_1^{(2)}, u_2^{(1)}\} &= \rho\ddot{u}_1^{(2)}, \\ f_5\{u_3^{(2)}\} &= \rho\ddot{u}_3^{(2)}. \end{aligned} \tag{8.3.121}$$

We see that f_1, f_3, f_4 couple $u_1^{(0)}$, $u_1^{(2)}$, $u_2^{(1)}$, while f_2 and f_5 are independent in $u_3^{(0)}$, $u_3^{(2)}$ respectively. The latter two are identified with the face-shear modes.

Considering the three coupled equations, let

$$\begin{aligned} u_1^{(0)} &= A \sin \xi x_1 e^{i\omega t}, \\ u_2^{(1)} &= B \cos \xi x_1 e^{i\omega t}, \\ u_1^{(2)} &= C \sin \xi x_1 e^{i\omega t}. \end{aligned} \tag{8.3.122}$$

Substitution in the first, third, and fourth of (8.3.121) gives the frequency equation

$$|a_{ij}| = 0 \qquad (i, j = 1, 2, 3), \tag{8.3.123}$$

where

$$\begin{aligned} a_{11} &= k^2z^2-\Omega^2, \\ a_{22} &= \kappa_2^2 z^3/3+4\kappa_1^2k^2/\pi^2-\kappa_3^2\Omega^2/3, \\ a_{33} &= E'z^2/5\mu+12\kappa_2^2/\pi^2-\kappa_4^2\Omega^2/5, \\ a_{12} &= 2\kappa_1(k^2-2)z/\pi = a_{21}, \\ a_{23} &= -2\kappa_2^2 z/\pi = a_{32}, \\ a_{13} &= a_{31} = 0, \end{aligned} \tag{8.3.124}$$

and

$$\begin{gathered} z = \frac{2\xi b}{\pi}, \qquad \Omega = \frac{\omega}{\omega_s}, \qquad \omega_s = \frac{\pi(\mu/\rho)^{\frac{1}{2}}}{2b}, \\ k^2 = (\lambda+2\mu)/\mu = 2(1-\nu)/(1-2\nu). \end{gathered} \tag{8.3.125}$$

In addition, three amplitude ratios are obtained

$$A_i : B_i : C_i = 1 : a_i/z_i : a_i/b_i, \qquad i = 1, 2, 3, \tag{8.3.126}$$

where

$$\begin{aligned} a_i &= \pi(\Omega^2-k^2z_i^2)/2\kappa_1(k^2-2), \\ b_i &= \frac{3}{2\pi}\left(\frac{\pi\kappa_4^2\Omega^2}{15\kappa_2^2}-4\right)-\frac{2\pi(k^2-1)z_i^2}{5\kappa_2^2k^2}. \end{aligned} \tag{8.3.127}$$

9. *Determination of the* κ_i. It should be recalled in the flexural wave theory of the previous section that the adjustment coefficient was obtained by

matching the short-wavelength behaviour ($\xi \to \infty$) to that of exact theory. An alternate procedure was to match the behaviour at the cutoff frequency ($\xi \to 0$). Mindlin and Medick describe the considerations in obtaining the κ_i and, in particular, the advantage in matching the approximate to the exact spectrum at $\xi = 0$ as follows.

> The relation between Ω and z [in eqn (8.3.123)] should match, as closely as possible, the corresponding relation obtained from the three-dimensional equations. The match is improved, within the framework of the present approximation, by choosing appropriate values for the coefficients κ_i; but since the κ_i are constants, a perfect match can be made only at one value of z for each of them. Now, large enough z corresponds to frequencies high enough to enter the range of modes that have not been included in the approximate equations. In a plate vibrating at such high frequencies under practical (i.e., not mixed) edge conditions, the high modes would, in general, couple with the ones of lower order. Thus the applicability of the approximate equations to bounded plates is limited to frequencies below the lowest frequency of the lowest neglected mode. There is, then, little advantage to be gained in matching the approximate and exact solutions at short wavelengths (large z) at the expense of a good match at long wavelengths. In fact, we go to the extreme and do all of the matching in the neighborhood of $z = 0$ primarily because of the intricate behavior of the exact solution at long wavelengths and also because this choice results in a reasonably good match out to as short wavelengths as the frequency limitation permits.

For matching purposes, the cutoff frequencies are obtained as $\xi \to 0$ ($z \to 0$, where z is the non-dimensionalized wavenumber). The three modes at $z \to 0$ are extensional, thickness stretch, and thickness shear. Now two of the cutoff frequencies are

$$\Omega^2 = 12\kappa_1^2 k^2/\pi^2\kappa_3^2, \qquad 60\kappa_2^2/\pi^2\kappa_4^2. \tag{8.3.128}$$

By adjusting the ratios κ_1^2/κ_3^2, κ_2^2/κ_4^2, the intercepts may be brought into proper adjustment. By equating slopes of the second two branches at their cutoff frequencies, the other two relations to establish the coefficients are found. Exact agreement in magnitude and curvature of the frequency spectrum at the cutoff frequencies is obtained. This is at some sacrifice in accuracy of the face shear modes, however.

10. *The frequency spectrum.* The frequency equation relating Ω to z is a cubic in Ω^2. Application of Decartes' rule of signs shows that there are three real positive, two real positive and one negative or one real positive and two complex conjugate roots. It is also found that Poisson's ratio greatly influences the spectrum, with the critical ranges defined by $\nu \gtreqless \frac{1}{3}$. The case of $\nu < \frac{1}{3}$ will be discussed first, where reference is made to Fig. 8.26.

The three branches shown are the simple longitudinal $L(1)$, the thickness stretch $L(2)$ and the thickness shear $L(3)$. Consider Ω to take on the values indicated by the lines (1), (2), and so forth. (1) For $\Omega \geqslant 2$, the three positive real roots $L(1)$, $L(2)$, $L(3)$ result. However, since these roots result from z_1^2, z_2^2,

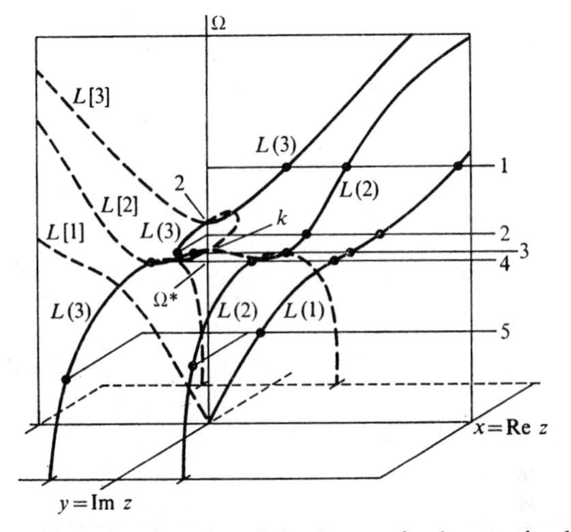

FIG. 8.26. The frequency spectrum ($\nu < \frac{1}{3}$) for the second-order extensional plate theory.

z_3^2, there are also roots $-L(1)$, $-L(2)$, $-L(3)$ in the negative x-plane. (2) For $k \leqslant \Omega < 2$, there are still two positive real roots $L(1)$, $L(2)$. However, $L(3)$ becomes imaginary. Reflections of the real and imaginary roots still occur. (3) For $\Omega^* < \Omega < k$, there are again three real roots. However, the root $L(3)$ has become negative, so that there are two real positive and one real negative. It is the branch $L[3]$, previously the reflection of $L(3)$, that becomes real and positive. (4) As $\Omega \to \Omega^*$, the roots $L[3]$, $L(2)$ approach a minimum. So, at the same time, do $L[2]$, $L(3)$ in the negative x-plane. (5) For $\Omega < \Omega^*$, the roots $L(2)$, $L[3]$ become complex conjugates of one another, given by $x \pm iy$. This also occurs for the branches $L[3]$, $L(2)$. There continues to be the real positive root $L(1)$, with its reflection $L[1]$.

For $\nu < \frac{1}{3}$, the thickness-shear mode has a higher frequency than the thickness-stretch mode. As $\nu \to \frac{1}{3}$, the cutoff frequencies approach one another and the imaginary loop shrinks to a point. For $\nu > \frac{1}{3}$, the two branches are interchanged. The spectrums for the cases $\nu \gtrless \frac{1}{3}$ are shown in Fig. 8.27. Also shown as dashed lines in the figures are the branches of the spectrum as obtained from exact theory. The agreement is seen to be remarkably close for the frequency range shown.

It should be noted that as ν continues to increase beyond $\nu = \frac{1}{3}$, the cutoff frequency of the thickness-stretch mode continues up the Ω-axis. However, there are restrictions on how high up we may go, since the second thickness-shear mode will be encountered at $\Omega = 4$. Even before this value is reached, coupling with this mode should be important. However, it is not included in the present approximate theory. Consequently, ν is limited to values less than

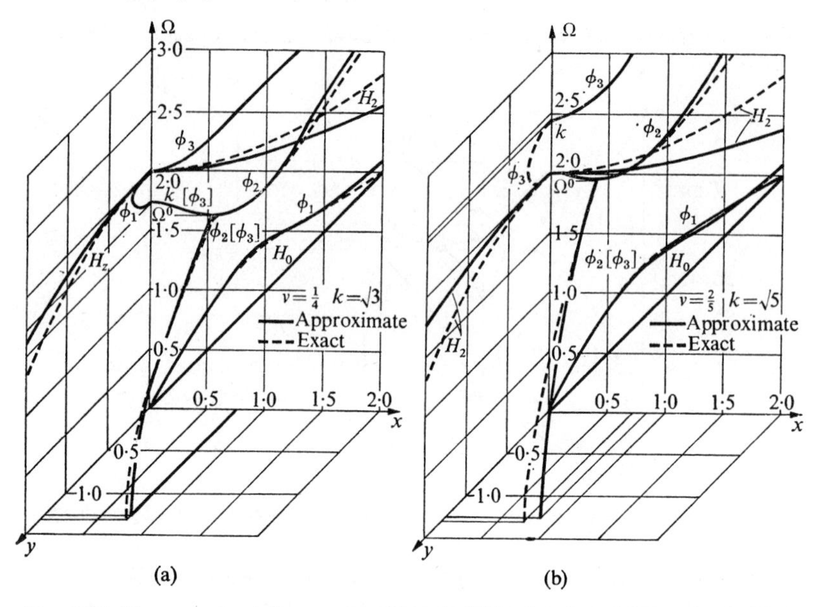

FIG. 8.27. The approximate frequency spectrum (solid lines) and exact frequency spectrum (dashed lines) for the cases of (a) $\nu < \frac{1}{3}$ and (b) $\nu > \frac{1}{3}$. (After Mindlin and Medick [66, Fig. 4].)

about $\frac{7}{16}$. Finally, as $z \to \infty$, the limiting phase velocities are found to be

$$c = \frac{\kappa_2}{\kappa_3}\left(\frac{\mu}{\rho}\right)^{\frac{1}{2}}, \quad \left(\frac{\lambda+2\mu}{\rho}\right)^{\frac{1}{2}}, \quad \frac{1}{\kappa_4}\left\{\frac{E}{\rho(1-\nu^2)}\right\}^{\frac{1}{2}}. \tag{8.3.129}$$

According to exact theory, the first of these should be the Rayleigh velocity, and the second and third the shear (equivoluminal) wave velocity.

8.3.3. *Approximate theories for longitudinal waves in rods*

A number of approximate theories for rods have appeared, as suggested in the survey by Green [22]. However, the developments presented here are again by Mindlin and his coworkers. The techniques are quite similar to those reviewed for plates, so the presentation will be rather brief. Both developments are for axially symmetric waves.

1. *Mindlin–Herrmann theory.* The development of an approximate theory for longitudinal waves in rods was presented by Mindlin and Herrmann [64] in 1950, appearing at about the same time as the work by Mindlin on an approximate theory for flexural waves in plates. Consider, then, a circular rod of radius a having a traction-free outer surface and having cylindrical co-ordinates as shown in Fig. 8.14. Starting from energy considerations, we have

that the change in internal energy in a volume V and in a time interval $t-t_0$ is given by

$$\Delta U = \int_{t_0}^{t} dt \int_{v} \frac{\partial}{\partial t}(T+V)\, dv, \tag{8.3.130}$$

where T and V are the kinetic and strain energy densities. For the former

$$T = \tfrac{1}{2}\rho(\dot{u}_r^2+\dot{u}_\theta^2+\dot{u}_z^2), \tag{8.3.131}$$

while for the latter we have that $V = V(\varepsilon_{rr}, \varepsilon_{\theta\theta}, \varepsilon_{zz}, \varepsilon_{r\theta}, \varepsilon_{rz}, \varepsilon_{\theta z})$, in general, and that

$$\partial V/\partial t = \tau_{rr}\dot{\varepsilon}_{rr}+\tau_{\theta\theta}\dot{\varepsilon}_{\theta\theta}+\ldots+\tau_{\theta z}\dot{\varepsilon}_{\theta z}. \tag{8.3.132}$$

The strains are related to the displacements in cylindrical coordinates by

$$\varepsilon_{rr} = \frac{\partial u_r}{\partial r}, \qquad \varepsilon_{\theta\theta} = \frac{1}{r}\frac{\partial u_\theta}{\partial \theta}+\frac{u_r}{r}, \qquad \varepsilon_{zz} = \frac{\partial u_z}{\partial z},$$

$$\varepsilon_{r\theta} = \left(\frac{1}{r}\frac{\partial u_r}{\partial \theta}+\frac{\partial u_\theta}{\partial r}-\frac{u_\theta}{r}\right), \qquad \varepsilon_{rz} = \left(\frac{\partial u_z}{\partial r}+\frac{\partial u_r}{\partial z}\right), \tag{8.3.133}$$

$$\varepsilon_{\theta z} = \left(\frac{\partial u_\theta}{\partial z}+\frac{1}{r}\frac{\partial u_z}{\partial \theta}\right).$$

The following kinematics of deformation are now assumed:

$$u_r = \frac{r}{a}u(z, t), \qquad u_\theta = 0, \qquad u_z = w(z, t). \tag{8.3.134}$$

Thus, two modes of deformation are admitted as illustrated in Fig. 8.28. With

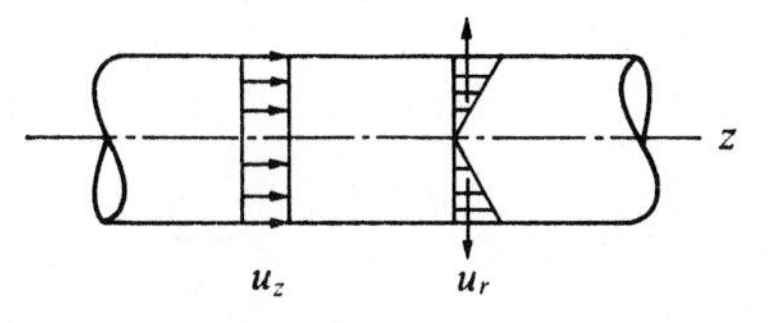

FIG. 8.28. Kinematics of the assumed deformations of u_r, u_z in a rod.

the assumed kinematics we have

$$\frac{\partial T}{\partial t} = \rho(\dot{u}_r\ddot{u}_r+\dot{u}_z\ddot{u}_z) = \rho\left(\frac{r^2}{a^2}\dot{u}\ddot{u}+\dot{w}\ddot{w}\right), \tag{8.3.135}$$

and

$$\frac{\partial V}{\partial t} = \frac{\tau_{rr}}{a}\dot{u}+\frac{\tau_{\theta\theta}}{a}\dot{u}+\tau_{zz}\frac{\partial \dot{w}}{\partial z}+\tau_{rz}\frac{r}{a}\frac{\partial \dot{u}}{\partial z}, \tag{8.3.136}$$

where $\varepsilon_{rr} = u/a$, $\varepsilon_{\theta\theta} = u/a$, $\varepsilon_{zz} = \partial w/\partial z$, $\varepsilon_{r\theta} = 0$, $\varepsilon_{rz} = (r/a)\partial u/\partial z$, $\varepsilon_{\theta z} = 0$.

Substituting these results in the volume integral of (8.3.130), simply called I, gives

$$I = 2\pi\int_b^c dz\int_0^a\left\{\rho\left(\frac{r^2}{a^2}\dot{u}\ddot{u}+\dot{w}\ddot{w}\right)+\frac{\tau_{rr}}{a}\dot{u}+\frac{\tau_{\theta\theta}}{a}\dot{u}+\tau_{zz}\frac{\partial\dot{w}}{\partial z}+\tau_{rz}\frac{r}{a}\frac{\partial\dot{u}}{\partial z}\right\}r\,dr. \quad (8.3.137)$$

Again noting that u, w are independent of r, we define the rod stresses

$$P_r = \int_0^a \tau_{rr}r\,dr, \qquad P_\theta = \int_0^a \tau_{\theta\theta}r\,dr,$$

$$P_z = \int_0^a \tau_{zz}r\,dr, \qquad Q = \int_0^a \frac{\tau_{zr}r^2}{a}\,dr. \quad (8.3.138)$$

Then the integral I becomes

$$I = 2\pi\int_b^c dz\left\{\rho\left(\frac{a^2}{4}\dot{u}\ddot{u}+\frac{a^2}{2}\dot{w}\ddot{w}\right)+\frac{P_r}{a}\dot{u}+\frac{P_\theta}{a}\dot{u}+P_z\frac{\partial\dot{w}}{\partial z}+Q\frac{\partial\dot{u}}{\partial z}\right\}. \quad (8.3.139)$$

The last two terms of (8.3.139) may be integrated by parts. Carrying this out and re-grouping the various terms, the change in internal energy is then given by

$$\Delta U = 2\pi\int_{t_0}^t dt\int_b^c\left\{\left(\rho\frac{a^2}{4}\ddot{u}+\frac{P_r+P_\theta}{a}-\frac{\partial Q}{\partial z}\right)\dot{u}+\left(\rho\frac{a^2}{2}\ddot{w}-\frac{\partial P_z}{\partial z}\right)\dot{w}\right\}dz+$$

$$+2\pi\int_{t_0}^t dt[Q\dot{u}+P_z\dot{w}]_b^c. \quad (8.3.140)$$

With this result in hand, consider now the stress equations of motion of elasticity, given by

$$\frac{\partial\tau_{rr}}{\partial r}+\frac{\partial\tau_{zr}}{\partial z}+\frac{\tau_{rr}-\tau_{\theta\theta}}{r} = \rho\frac{\partial^2 u_r}{\partial t^2},$$

$$\frac{\partial\tau_{zr}}{\partial r}+\frac{\partial\tau_{zz}}{\partial z}+\frac{1}{r}\tau_{zr} = \rho\frac{\partial^2 u_z}{\partial t^2} \quad (8.3.141)$$

for the present case of axial symmetry. To convert these to the rod equations, multiply the first by r^2, the second by r, and integrate over the cross-section. Thus, for the first case, we have

$$\int_0^a\left\{r^2\frac{\partial\tau_{rr}}{\partial r}+r^2\frac{\partial\tau_{zr}}{\partial z}+r(\tau_{rr}-\tau_{\theta\theta})\right\}dr = \frac{\rho}{a}\int_0^a r^3\ddot{u}\,dr. \quad (8.3.142)$$

Carrying out the integration gives

$$\frac{\partial Q}{\partial z}-\frac{P_\theta+P_r}{a}+aR = \rho\frac{a^2}{4}\frac{\partial^2 u}{\partial t^2}, \tag{8.3.143}$$

where $R = \tau_{rr}|_{r=a}$. For the second equation of motion, there results

$$\frac{\partial P_z}{\partial z}+aZ = \rho\frac{a^2}{2}\frac{\partial^2 w}{\partial t^2}, \tag{8.3.144}$$

where $Z = \tau_{zr}|_{r=a}$. These are the rod stress equations of motion.

The boundary conditions and initial conditions follow next. Thus using (8.3.143) and (8.3.144) in (8.3.140) gives

$$\Delta U = 2\pi a \int_{t_0}^{t} \mathrm{d}t \int_{b}^{c} (R\dot{u}+Z\dot{w})\,\mathrm{d}z+2\pi \int_{t_0}^{t} (Q\dot{u}+P_z\dot{w})\,\mathrm{d}t. \tag{8.3.145}$$

This is simply the statement of work–energy which says that the change of internal energy is equal to the work done on the system in the time interval t_0 to t. By applying uniqueness-of-solutions considerations, similar to those used in the approximate plate development, the boundary and initial conditions follow:

(a) along the rod, one each of the pairs of quantities $R\dot{u}$, $Z\dot{w}$ must be given;
(b) at the ends of the rod, one each of the two pairs of quantities $Q\dot{u}$, $P_z\dot{w}$ must be given;
(c) the initial displacements and velocities must be given.

The rod stress–displacement relations are now required. These are obtained in the usual fashion, starting with the generalized Hooke's law. In cylindrical coordinates for axisymmetry, these are

$$\begin{aligned}
\tau_{rr} &= \lambda\Delta+2\mu\varepsilon_{rr},\\
\tau_{\theta\theta} &= \lambda\Delta+2\mu\varepsilon_{\theta\theta},\\
\tau_{zz} &= \lambda\Delta+2\mu\varepsilon_{zz},\\
\tau_{zr} &= \mu\varepsilon_{zr},\\
\Delta &= \varepsilon_{rr}+\varepsilon_{\theta\theta}+\varepsilon_{zz}.
\end{aligned} \tag{8.3.146}$$

Substituting the assumed displacements in the strain definitions, the resulting strains in Hooke's law, and the resulting stresses in the rod stresses, and carrying out the integrations gives the rod stress–displacement relations. To enable the approximate theory to be adjusted for better agreement with exact theory, the constants κ, κ_1 are introduced. The results are

$$P_r = P_\theta = \frac{\kappa_1^2}{2}\left\{2a(\lambda+\mu)u+a^2\lambda\frac{\partial w}{\partial z}\right\},$$

$$P_z = \frac{1}{2}\left\{2a\lambda u+a^2(\lambda+2\mu)\frac{\partial w}{\partial z}\right\}, \tag{8.3.147}$$

$$Q = \frac{\kappa^2 a^2\mu}{4}\frac{\partial u}{\partial z}.$$

When these relations are substituted in the rod stress equations of motion, we obtain

$$a^2\kappa^2\mu\frac{\partial^2 u}{\partial z^2}-8\kappa_1^2(\lambda+\mu)u-4a\kappa_1^2\lambda\frac{\partial w}{\partial z} = \rho a^2\frac{\partial^2 u}{\partial t^2},$$

$$2a\lambda\frac{\partial u}{\partial z}+a^2(\lambda+2\mu)\frac{\partial^2 w}{\partial z^2} = \rho a^2\frac{\partial^2 w}{\partial t^2}. \tag{8.3.148}$$

The effects of body forces and surface loads have been omitted from the above equations. The first equation is mainly associated with radial shear and inertia. The second equation is similar to classical rod theory, except for the added $2a\lambda(\partial u/\partial z)$ term. Note also that the resulting wave velocity is that of dilatational waves instead of the bar velocity.

Now consider the propagation of harmonic waves, letting $R = Z = 0$ and

$$u = Ae^{i\xi(z-ct)}, \qquad w = Be^{i\xi(z-ct)}. \tag{8.3.149}$$

Substituting these expressions in the displacement equations of motion and evaluating the resulting determinant of coefficients gives the dispersion relation

$$\left\{\frac{\kappa^2\mu}{\rho}+\frac{8\kappa_1^2(\lambda+\mu)}{a^2\kappa^2\rho}-c^2\right\}\left(\frac{\lambda+2\mu}{\rho}-c^2\right)-\frac{8\kappa_1^2\lambda^2}{a^2\kappa^2\rho^2} = 0. \tag{8.3.150}$$

Consider now the long- and short-wavelength limits. As $\xi \to 0$, the limiting phase velocities are

$$c^2 = E/\rho = c_0^2, \qquad c^2 = \infty. \tag{8.3.151}$$

Both agree with exact results. The first, it is noted, is the classical bar velocity. As $\xi \to \infty$, the results are

$$c^2 = \kappa^2\frac{\mu}{\rho} = \kappa^2 c_2^2, \qquad c^2 = \frac{\lambda+2\mu}{\rho} = c_1^2. \tag{8.3.152}$$

In exact theory, the first of these is the Rayleigh velocity c_R^2. In order to bring the short-wavelength limit into agreement, it is necessary to select κ^2 as $\kappa^2 = c_R^2/c_2^2$. The second limit in the exact theory is $c_2^2 = \mu/\rho$. This discrepancy cannot be corrected in this present form of the theory. Selection of κ_1 to obtain better agreement of approximate and exact theory is based on certain

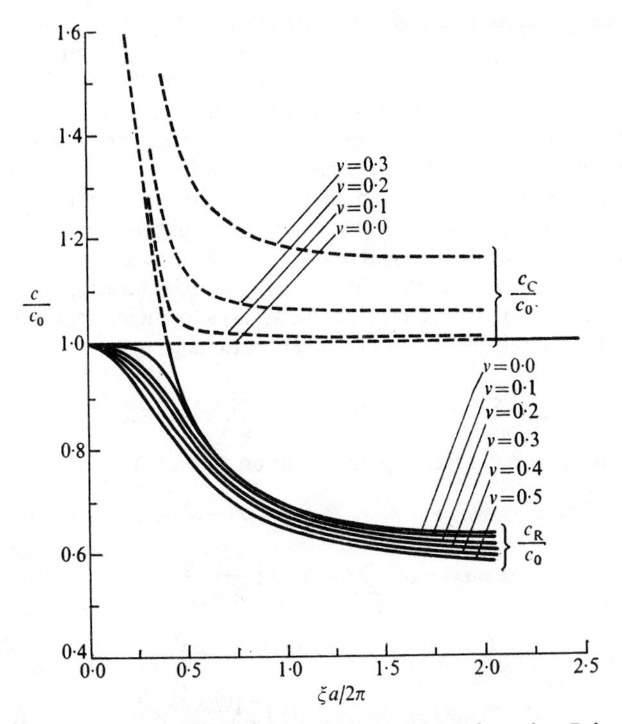

FIG. 8.29. Dispersion curves for Mindlin–Herrmann rod theory for various Poisson's ratio. (After Mindlin and Herrmann [64, Fig. 2].)

observations by Hudson [29] on the behaviour of the first-mode curves of exact theory for various values of Poisson's ratio. He observed that the curves are all tangent at a common point, given by $a/\lambda = 0{\cdot}293$, $c/c_2 = \sqrt{2}$. Adjusting the approximate theory curves for a similar point of tangency gives

$$\kappa_1^2 = 0{\cdot}422(2-\kappa^2) \tag{8.3.153}$$

The resulting dispersion curves for the two-mode rod theory are shown in Fig. 8.29 for various values of Poisson's ratio. The value of κ^2 has been selected according to $\kappa^2 = c_R^2/c_2^2$, while κ_1^2 has been taken as unity. It should be noted that the modes uncouple for $\nu = 0$.

2. *Mindlin–McNiven theory.* Similar to the refinements that followed Mindlin's first theory for plates, additional refinements followed the work on rods by Mindlin and Herrmann. The theory presented by Mindlin and McNiven [65] will now be briefly considered. Recall that Mindlin–Herrmann theory (p. 510) was a two-mode theory. However, considering the exact frequency spectrum for the rod, it must be appreciated that the cutoff

frequencies of the second and third symmetric rod modes can interchange positions, depending on Poisson's ratio. This was brought out in detail by the Mindlin–Medick theory of plates. Thus, one of the objectives of the Mindlin–McNiven development was to include a third degree of freedom or mode. In addition the adjustment coefficients, determined from the high frequency behaviour and other arguments in the Mindlin–Herrmann development, were obtained in the refined rod theory in the manner used by Mindlin and Medick; that is by adjusting the long-wavelength behaviour to match the exact spectrum. Only the essential results of the theory will be presented here.

Again consider a rod as shown in Fig. 8.14. For axisymmetry, $u_\theta = \partial/\partial\theta =$ and $u_r = u_r(r, z, t)$, $u_z = u_z(r, z, t)$. Now, expand these displacements in terms of Jacobi polynomials in the radial coordinate

$$u_r = \sum_{n=0}^{\infty} \mathscr{U}_n(\alpha)u_n(z, t), \qquad u_z = \sum_{n=0}^{\infty} \mathscr{W}_n(\alpha)w_n(z, t), \tag{8.3.154}$$

where $\alpha = r/a$ and the Jacobi polynomials are given by†

$$\mathscr{U}_0(\alpha) = \alpha, \qquad \mathscr{U}_1(\alpha) = \alpha - \tfrac{3}{2}\alpha^3, \; \dots,$$

$$\mathscr{U}_n(\alpha) = \alpha + \sum_{k=1}^{n} (-1)^k \binom{n}{k} \frac{(n+2)_k}{(k+1)!} \alpha^{2k+1},$$

$$\mathscr{W}_0(\alpha) = 1, \qquad \mathscr{W}_1(\alpha) = 1 - 2\alpha^2, \; \dots, \tag{8.3.155}$$

$$\mathscr{W}_n(\alpha) = 1 + \sum_{k=1}^{n} (-1)^k \binom{n}{k} \frac{(n+1)_k}{k!} \alpha^{2k},$$

where

$$\binom{n}{k} = \frac{n(n-1)(n-2)\dots(n-k+1)}{k!},$$

$$(\beta)_k = \beta(\beta+1)(\beta+2)\dots(\beta+k-1). \tag{8.3.156}$$

Now, the objective of using Jacobi polynomials in the expansion instead of the radial coordinate itself is quite similar to that of using Legendre polynomials instead of the thickness coordinate in the Mindlin–Medick plate theory. That is, these polynomials possess convenient orthogonality properties that simplify many of the integrations to be performed. These properties are given by

$$4(n+1)^3 \int_0^1 \alpha \mathscr{U}_m \mathscr{U}_n \, d\alpha = \begin{cases} 0, & m \neq n \\ 1, & m = n, \end{cases}$$

$$2(2n+1) \int_0^1 \alpha \mathscr{W}_m \mathscr{W}_n \, d\alpha = \begin{cases} 0, & m \neq n \\ 1, & m = n. \end{cases} \tag{8.3.157}$$

† See, for example, Reference [45].

The derivative property is

$$\frac{\mathrm{d}\mathscr{W}_n}{\mathrm{d}\alpha} = \sum_{k=0}^{n-1} C_{nk}\mathscr{U}_k, \qquad C_{nk} = 4(k+1)^2(-1)^{n+k}. \tag{8.3.158}$$

The rod stress equations of motion are obtained by substituting the assumed displacements into the variational equations of motion and integrating across the cross-section. The results are

$$\begin{aligned} \frac{\mathrm{d}Q_n}{\mathrm{d}z} - \frac{P_n}{a} + \frac{(-1)^n}{n+1}\frac{R}{a} &= \frac{\rho}{4(n+1)^3}\frac{\partial^2 u_n}{\partial t^2}, \\ \frac{\mathrm{d}F_n}{\mathrm{d}z} - \sum_{k=0}^{n-1} C_{nk}\frac{Q_k}{a} + (-1)^n\frac{Z}{a} &= \frac{\rho}{2(2n+1)}\frac{\partial^2 w_n}{\partial t^2}, \end{aligned} \tag{8.3.159}$$

where the rod stresses are given by

$$\begin{aligned} P_n &= \int_0^1 \left(\tau_{\theta\theta}\mathscr{U}_n + \tau_{rr}\alpha\frac{\mathrm{d}\mathscr{U}_n}{\mathrm{d}\alpha}\right)\mathrm{d}\alpha, \\ Q_n &= \int_0^1 \tau_{rz}\mathscr{U}_n\alpha\,\mathrm{d}\alpha, \\ F_n &= \int_0^1 \tau_{zz}\mathscr{W}_n\alpha\,\mathrm{d}\alpha, \end{aligned} \tag{8.3.160}$$

and

$$R = \tau_{rr}\big|_{r=a}, \qquad Z = \tau_{rz}\big|_{r=a}. \tag{8.3.161}$$

The strains are given by

$$\begin{aligned} \varepsilon_{rr} &= \frac{\partial u_r}{\partial r} = \sum_{n=0}^{\infty} \frac{\partial \mathscr{U}_n}{\partial \alpha}\frac{u_n}{a}, \\ \varepsilon_{\theta\theta} &= \frac{u_r}{r} = \sum_{n=0}^{\infty} \frac{\mathscr{U}_n}{\alpha}\frac{u_n}{a}, \\ \varepsilon_{zz} &= \frac{\partial u_z}{\partial z} = \sum_{n=0}^{\infty} \mathscr{W}_n\frac{\partial w_n}{\partial z}, \\ \varepsilon_{rz} &= \frac{\partial u_r}{\partial z} + \frac{\partial u_z}{\partial r} = \sum_{n=0}^{\infty}\left(\mathscr{U}_n\frac{\partial u_n}{\partial z} + \frac{\partial \mathscr{W}_n}{\partial \alpha}\frac{w_n}{a}\right). \end{aligned} \tag{8.3.162}$$

The rod stress–strain relations are obtained by substituting (8.3.162) into the usual stress–strain relations

$$\begin{aligned}\tau_{rr} &= \partial U/\partial \varepsilon_{rr} = \lambda\Delta + 2\mu\varepsilon_{rr},\\ \tau_{\theta\theta} &= \partial U/\partial \varepsilon_{\theta\theta} = \lambda\Delta + 2\mu\varepsilon_{\theta\theta},\\ \tau_{zz} &= \partial U/\partial \varepsilon_{zz} = \lambda\Delta_{zz} + 2\mu\varepsilon_{zz},\\ \tau_{rz} &= \partial U/\partial \varepsilon_{rz} = \mu\varepsilon_{rz},\end{aligned} \tag{8.3.163}$$

where $\Delta = \varepsilon_{rr} + \varepsilon_{\theta\theta} + \varepsilon_{zz}$ and U is the strain energy, and then substituting the resulting stresses into the definitions (8.3.160). A few of the terms are

$$\begin{aligned}P_0 &= 2(\lambda+\mu)\frac{u_0}{a} + \lambda\frac{\partial w_0}{\partial z} + \dots,\\ Q_0 &= \frac{\mu}{4}\left(\frac{\partial u_0}{\partial z} - \frac{4w_1}{a}\right) + \dots,\\ F_0 &= \frac{\lambda+2\mu}{2}\frac{\partial w_0}{\partial z} + \lambda\frac{u_0}{a} + \dots,\\ F_1 &= \frac{\lambda+2\mu}{6}\frac{\partial w_1}{\partial z} + \dots .\end{aligned} \tag{8.3.164}$$

A second-order approximate theory is desired. To this end, retain only the terms associated with the u_0, w_0, and w_1 displacements. With this restriction, the rod displacements, stresses, and strains are redefined as

$$u = u_0, \qquad w = w_0, \qquad \psi = w_1, \tag{8.3.165}$$

$$\begin{aligned}P_r &= P_0 = \int_0^1 (\tau_{rr} + \tau_{\theta\theta})\alpha \, d\alpha,\\ P_z &= P_0 = \int_0^1 \tau_{zz}\alpha \, d\alpha,\\ P_\psi &= F_1 = \int_0^1 \tau_{zz}(1-2\alpha^2)\alpha \, d\alpha,\\ P_{rz} &= Q_0 = \int_0^1 \tau_{rz}\alpha^2 \, d\alpha,\end{aligned} \tag{8.3.166}$$

$$\begin{aligned}\Gamma_r &= u/a, & \Gamma_z &= \partial w/\partial z,\\ \Gamma_\psi &= \partial \psi/\partial z, & \Gamma_{rz} &= \partial u/\partial z - 4\psi/a.\end{aligned} \tag{8.3.167}$$

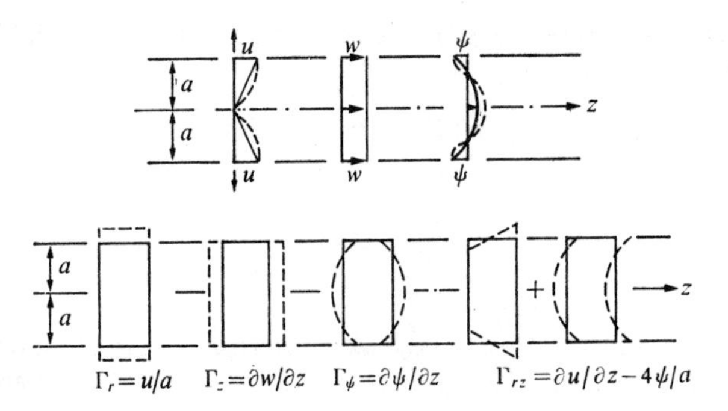

FIG. 8.30. Displacement and strain components for a second-order longitudinal rod theory. (After Mindlin and McNiven [65, Fig. 1].)

The displacement and strain components are shown in Fig. 8.30. In order to partially compensate for the effects of the omitted higher-order terms on the remaining modes, the adjustment coefficients κ_1, κ_2, κ_3, κ_4 are introduced, where Γ_r, Γ_{rz}, $\dot{u}$, ψ are replaced by $\kappa_1\Gamma_r$, $\kappa_2\Gamma_{rz}$, $\kappa_3\dot{u}$, and $\kappa_4\dot{\psi}$.

The resulting strain and kinetic energy densities for the second-order theory with the adjustment parameters incorporated are given by

$$\begin{aligned} U_2 &= \tfrac{1}{2}\{2(\lambda+\mu)\kappa_1^2\Gamma_r^2+\tfrac{1}{2}(\lambda+2\mu)\Gamma_z^2+2\lambda\kappa_1\Gamma_r\Gamma_z+ \\ &\quad +\tfrac{1}{4}\mu\kappa_2^2\Gamma_{rz}^2+\tfrac{1}{6}(\lambda+2\mu)\Gamma_\psi^2\}, \\ K_2 &= \frac{\rho}{2}(\tfrac{1}{4}\kappa_3^2\dot{u}^2+\tfrac{1}{2}\dot{w}^2+\tfrac{1}{6}\kappa_4^2\dot{\psi}^2). \end{aligned} \tag{8.3.168}$$

The second-order stress–strain–displacement relations are given by

$$\begin{aligned} P_r &= \partial U_2/\partial\Gamma_r = 2(\lambda+\mu)\kappa_1^2\Gamma_r+\lambda\kappa_1\Gamma_z, \\ &= \{2(\lambda+\mu)\kappa_1^2 u+\lambda\kappa_1 w'\}a^{-1}, \\ P_z &= \partial U_2/\partial\Gamma_z = \tfrac{1}{2}(\lambda+2\mu)\Gamma_z+\lambda\kappa_1\Gamma_r, \\ &= \{\tfrac{1}{2}(\lambda+2\mu)w'+\lambda\kappa_1 u\}a^{-1}, \\ P_{rz} &= \partial U_2/\partial\Gamma_{rz} = \tfrac{1}{4}\mu\kappa_2^2\Gamma_{rz}, \\ &= \tfrac{1}{4}\mu\kappa_2^2(u'-4\psi)a^{-1}, \\ P_\psi &= \partial U_2/\partial\Gamma_\psi = \tfrac{1}{6}(\lambda+2\mu)\Gamma_\psi, \\ &= \tfrac{1}{6}(\lambda+2\mu)\psi' a^{-1} \end{aligned} \tag{8.3.169}$$

The stress equations of motion are given by

$$\frac{\partial P_{rz}}{\partial z}-P_r+R = \tfrac{1}{4}\rho a\kappa_3^2\ddot{u},$$

$$\frac{\partial P_z}{\partial z}+Z = \tfrac{1}{2}\rho a\ddot{w}, \tag{8.3.170}$$

$$\frac{\partial P_\psi}{\partial z}+4P_{rz}-Z = \tfrac{1}{6}\rho a\kappa_4^2\ddot{\psi}.$$

Finally, the displacement equations of motion are

$$(u''-4\psi')-8(\lambda+\mu)\kappa_1^2u-4\lambda\kappa_1w'+4aR = \rho a^2\kappa_3^2\ddot{u},$$

$$(\lambda+2\mu)w''+2\lambda\kappa_1u'+2aZ = \rho a^2\ddot{w}, \tag{8.3.171}$$

$$(\lambda+2\mu)\psi''+6\mu\kappa_2^2(u'-4\psi)-6aZ = \rho a^2\kappa_4^2\ddot{\psi}.$$

For propagation of harmonic waves, let

$$u = A\cos\gamma z e^{i\omega t}, \qquad w = B\sin\gamma z e^{i\omega t}, \qquad \psi = C\sin\gamma z e^{i\omega t} \tag{8.3.172}$$

and substitute in the displacement equations of motion (8.3.171) with $R = Z = 0$. The frequency equation is obtained from the resulting determinant of coefficients, given by

$$\begin{vmatrix} a_{11} & a_{12} & a_{13} \\ a_{12} & a_{22} & 0 \\ a_{13} & 0 & a_{33} \end{vmatrix} = 0, \tag{8.3.173}$$

where

$$a_{11} = \kappa_2^2\delta^2\zeta^2+8(k^2-1)\kappa_1^2-\kappa_3^2\delta^2\Omega^2,$$

$$a_{22} = 2\delta^2(k^2\zeta^2-\Omega^2),$$

$$a_{33} = 6(k^2\delta^2\zeta^2+24\kappa_2^2-\kappa_4^2\delta^2\Omega^2), \tag{8.3.174}$$

$$a_{12} = 4(k^2-2)\kappa_1\delta\zeta,$$

$$a_{13} = 12\kappa_2^2\,\delta\zeta,$$

and where

$$\Omega = \omega/\omega_s, \qquad \omega_s = \delta c_2/a, \qquad \zeta = \gamma a/\delta,$$

$$k^2 = c_1^2/c_2^2,$$

and where δ is the lowest non-zero root of $J_1(\delta) = 0$. The adjustment coefficients are obtained by matching the approximate spectrum to the exact spectrum at $\zeta = 0$. The procedures given by Mindlin–McNiven result in the ordinates, slopes, and curvatures of the three branches of the approximate spectrum being identical to those of the exact spectrum at $\zeta = 0$. The resulting branches are very similar to the first three branches of Fig. 8.17 for $\xi < 2$. The deviations between approximate and exact theories is slight and occurs only at smaller wavelengths. As Poisson's ratio is increased, the cutoff frequencies of the second and third branches, corresponding to the radial mode and axial shear mode, approach one another, becoming equal at $\nu = 0{\cdot}2833$. For larger values of Poisson's ratio, the radial mode is above the axial shear mode. The behaviour is quite similar to that shown in Fig. 8.27 for plates.

8.3.4. *Approximate theories for waves in shells*

An extensive discussion was previously given on the many and varied approximate theories for vibrations of shells.† The lowest-order theory, membrane theory, was used in the analysis of waves in a cylindrical shell in the earlier chapter. Bending theories, such as by Love [43] and others represent a higher-order shell theory that has found wide application in shell vibration problems. Rather little has appeared on using these theories in transient wave-propagation problems, however. Quite possibly this is due to the inherent limitations imposed by the shear-rigidity assumption, limitations that were also noted in the case of Bernoulli–Euler beam theory and classical plate theory. A number of developments in yet-higher-order shell theories, containing shear deformations, rotary inertia and other deformations, have appeared, and it is to this area that our attention will be directed, with only cylindrical shells being considered.

Several developments in higher-order shell theories appeared within a short period of time. Mirsky and Herrmann [69] included shear effects in both the axial and circumferential direction and rotary-inertia effects in the study of non-axially symmetric waves in a cylindrical shell. This work was preceded by the study of axially symmetric motions [68], including shear effects and transverse normal stress, and a yet earlier study by Herrmann and Mirsky [25] for the axially symmetric case with shear effects included. Cooper and Naghdi [9] presented a theory including shear effects and rotary inertia for non-axially symmetric motions. They showed reduction of their equations to those of Love and Donnell upon selectively omitting various higher-order effects. This work was preceded by the study of the axially symmetric case by Naghdi and Cooper [72]. Lin and Morgan [42] also developed the equations for axially symmetric motions including shear and rotary-inertia effects.

† Chapter 4, § 4.3.

Yu [90] reduced the Herrmann–Mirsky results to Donnell-type equations. A most useful survey of these developments, including comparison of the dispersion curves and frequency spectra of the theories was given by Greenspon [23]. Also included was discussion and comparison of results for bending and membrane theories. McNiven, Shah, and Sackman [47] have presented a theory for axially symmetric waves that includes the first radial mode, as well as longitudinal and shear modes. The development is quite similar to that of Mindlin and McNiven [65] for rods.

In the following, the basic aspects of the Mirsky–Herrmann and Cooper–Naghdi theories will be presented, with the main emphasis on giving dispersion curves and frequency spectra. The comparisons of the various theories by Greenspon will be reviewed followed by the results from McNiven *et al.*

1. *Mirsky–Herrmann theory* [68]. Consider an infinite, cylindrical shell of mean radius R and thickness h. Let x be the axial coordinate, θ be the polar angle, and z be the radial distance from the middle surface of the shell, measured positively outwards. The displacement kinematics are assumed to be given by

$$\begin{aligned}\bar{u}_x(x,\theta,z,t) &= u(x,\theta,t)+z\psi_x(x,\theta,t),\\ \bar{u}_\theta(x,\theta,z,t) &= v(x,\theta,t)+z\psi_\theta(x,\theta,t),\\ \bar{u}_z(x,\theta,z,t) &= w(x,\theta,t).\end{aligned} \tag{8.3.175}$$

It is seen that the ψ_z, ψ_θ terms bring in the axial and circumferential shear effects.

The basic steps of the theory involve, first, integration of the strain energy density expression across the shell thickness. This leads to the shell stress expressions

$$\begin{aligned}
N_{xx} &= \int_{-h/2}^{h/2} \sigma_{xx}(1+z/R)\,dz, & M_{x\theta} &= \int_{-h/2}^{h/2} \sigma_{x\theta}(1+z/R)\,dz,\\
M_{xx} &= \int_{-h/2}^{h/2} \sigma_{xx}(1+z/R)z\,dz, & N_{\theta x} &= \int_{-h/2}^{h/2} \sigma_{x\theta}\,dz,\\
N_{\theta\theta} &= \int_{-h/2}^{h/2} \sigma_{\theta\theta}\,dz, & M_{\theta x} &= \int_{-h/2}^{h/2} \sigma_{x\theta}z\,dz,\\
M_{\theta\theta} &= \int_{-h/2}^{h/2} \sigma_{\theta\theta}z\,dz, & Q_x &= \int_{-h/2}^{h/2} \sigma_{xz}(1+z/R)\,dz,\\
N_{x\theta} &= \int_{-h/2}^{h/2} \sigma_{x\theta}(1+z/R)\,dz, & Q_\theta &= \int_{-h/2}^{h/2} \sigma_{z\theta}\,dz.
\end{aligned} \tag{8.3.176}$$

The kinetic energy expression is then developed. Using the two energy expressions in conjunction with Hamilton's principle leads to the shell stress equations of motion. The shell stress–strain–displacements are developed from Hooke's law. Somewhat paralleling the method used by Mindlin [59] in obtaining the approximate plate theory, the effects of the stress σ_{zz} are dropped. Furthermore, the adjustment coefficients κ_x, κ_θ are introduced in conjunction with the strains ε_{xz}, $\varepsilon_{\theta z}$. The resulting stress-displacement equations are given by

$$N_{xx} = E_p\frac{\partial u}{\partial x}+\left(\frac{D}{R}\right)\frac{\partial \psi_x}{\partial x}+\left(\frac{\nu E_p}{R}\right)\left(w+\frac{\partial v}{\partial \theta}\right),$$

$$N_{\theta x} = Gh\frac{\partial v}{\partial x}+\left(\frac{Gh}{R}\right)\left(1+\frac{Gh}{R}\right)\frac{\partial u}{\partial \theta}-\left(\frac{GI}{R^2}\right)\frac{\partial \psi_x}{\partial \theta},$$

$$M_{xx} = \left(\frac{D}{R}\right)\left(\frac{\partial u}{\partial x}+R\frac{\partial \psi_x}{\partial x}+\nu\frac{\partial \psi_\theta}{\partial \theta}\right),$$

$$M_{\theta x} = GI\left(\frac{\partial \psi_\theta}{\partial x}+R^{-2}\frac{\partial u}{\partial \theta}+R^{-1}\frac{\partial \psi_x}{\partial \theta}\right),$$

$$N_{x\theta} = G\left(h\frac{\partial v}{\partial x}+\left(\frac{I}{R}\right)\frac{\partial \psi_\theta}{\partial x}+\left(\frac{h}{R}\right)\frac{\partial u}{\partial \theta}\right),$$

$$N_{\theta\theta} = \left(\frac{E_p}{R}+\frac{D}{R^3}\right)\left(w+\frac{\partial v}{\partial \theta}\right)-\left(\frac{D}{R^2}\right)\frac{\partial \psi_\theta}{\partial \theta}+\nu E_p\frac{\partial u}{\partial x},$$

$$M_{x\theta} = \left(\frac{GI}{R}\right)\left(\frac{\partial v}{\partial x}+R\frac{\partial \psi_\theta}{\partial x}+\frac{\partial \psi_x}{\partial \theta}\right),$$

$$M_{\theta\theta} = \left(\frac{D}{R}\right)\frac{\partial \psi_\theta}{\partial \theta}-\left(\frac{D}{R^2}\right)\left(w+\frac{\partial v}{\partial \theta}\right)+D\nu\frac{\partial \psi_x}{\partial x},$$

$$Q_x = \kappa_x^2 Gh\left(\psi_x+\frac{\partial w}{\partial x}\right),$$

$$Q_\theta = \left(\kappa_\theta^2\frac{Gh}{R}\right)\left(1+\frac{h^2}{12R^2}\right)\left(\frac{\partial w}{\partial \theta}-v+R\psi_\theta\right), \tag{8.3.177}$$

where G is the shear modulus, $E_p = Eh/(1-\nu^2)$, $I = h^3/12$, and

$$D = Eh^3/12(1-\nu^2).$$

The resulting five displacement equations of motion, where body forces are omitted, are given by

$$\left\{E_p\frac{\partial^2}{\partial x^2}+\left(\frac{Gh}{R^2}\right)\left(1+\frac{I}{hR^2}\right)\frac{\partial^2}{\partial\theta^2}-\rho h\frac{\partial^2}{\partial t^2}\right\}u+$$
$$+\left\{\left(\frac{D}{R}\right)\frac{\partial^2}{\partial x^2}-\left(\frac{GI}{R^3}\right)\frac{\partial^2}{\partial\theta^2}-\left(\frac{\rho I}{R}\right)\frac{\partial^2}{\partial t^2}\right\}\psi_x+$$
$$+\left\{\left(\frac{E_p(1+\nu)}{2R}\right)\frac{\partial^2}{\partial x\,\partial\theta}\right\}v+\left\{\left(\frac{\nu E_p}{R}\right)\frac{\partial}{\partial x}\right\}w=0,$$

$$\left\{\left(\frac{D}{R}\right)\frac{\partial^2}{\partial x^2}-\left(\frac{GI}{R^3}\right)\frac{\partial^2}{\partial\theta^2}-\left(\frac{\rho I}{R}\right)\frac{\partial^2}{\partial t^2}\right\}u+$$
$$+\left\{D\frac{\partial^2}{\partial x^2}+\left(\frac{GI}{R^2}\right)\frac{\partial^2}{\partial\theta^2}-\kappa_x^2Gh-\rho I\frac{\partial^2}{\partial t^2}\right\}\psi_x+$$
$$+\left\{\left(\frac{D(1+\nu)}{2R}\right)\frac{\partial^2}{\partial x\,\partial\theta}\right\}\psi_\theta+\left(-\kappa_x^2Gh\frac{\partial}{\partial x}\right)w=0,$$

$$\left\{\left(\frac{E_p(1+\nu)}{2R}\right)\frac{\partial^2}{\partial x\,\partial\theta}\right\}u+$$
$$+\left\{Gh\frac{\partial^2}{\partial x^2}+\left(\frac{E_p}{R^2}+\frac{D}{R^4}\right)\frac{\partial^2}{\partial\theta^2}-\left(\kappa_\theta^2\frac{G}{R^2}\right)\left(h+\frac{I}{R^2}\right)-\rho h\frac{\partial^2}{\partial t^2}\right\}v+$$
$$+\left\{\left(\frac{GI}{R}\right)\frac{\partial^2}{\partial x^2}-\left(\frac{D}{R^3}\right)\frac{\partial^2}{\partial\theta^2}+\left(\kappa_\theta^2\frac{G}{R}\right)\left(h+\frac{I}{R^2}\right)-\left(\frac{\rho I}{R}\right)\frac{\partial^2}{\partial t^2}\right\}\psi_\theta+$$
$$+\left\{\frac{E_p}{R^2}+\frac{D}{R^4}+\left(\kappa_\theta^2\frac{G}{R^2}\right)\left(h+\frac{I}{R^2}\right)\right\}\frac{\partial w}{\partial\theta}=0,$$

$$\left\{\left(\frac{D(1+\nu)}{2R}\right)\frac{\partial^2}{\partial x\,\partial\theta}\right\}\psi_x+$$
$$+\left\{\left(\frac{GI}{R}\right)\frac{\partial^2}{\partial x^2}-\left(\frac{D}{R^3}\right)\frac{\partial^2}{\partial\theta^2}+\left(\kappa_\theta^2\frac{G}{R}\right)\left(h+\frac{I}{R^2}\right)-\left(\frac{\rho I}{R}\right)\frac{\partial^2}{\partial t^2}\right\}v+$$
$$+\left\{GI\frac{\partial^2}{\partial x^2}+\left(\frac{D}{R^2}\right)\frac{\partial^2}{\partial\theta^2}-\kappa_\theta^2G\left(h+\frac{I}{R^2}\right)-\rho I\frac{\partial^2}{\partial t^2}\right\}\psi_\theta+$$
$$+\left\{-\frac{D}{R^3}-\left(\kappa_\theta^2\frac{G}{R}\right)\left(h+\frac{I}{R^2}\right)\right\}\frac{\partial w}{\partial\theta}=0,$$

$$\left\{-\left(\frac{E_p\nu}{R}\right)\frac{\partial}{\partial x}\right\}u+\left\{\kappa_x^2Gh\frac{\partial}{\partial x}\right\}\psi_x+\left\{-\left(\kappa_\theta^2\frac{G}{R^2}\right)\left(h+\frac{I}{R^2}\right)-\left(\frac{E_p}{R^2}+\frac{D}{R^4}\right)\right\}\frac{\partial v}{\partial\theta}+$$
$$+\left\{\frac{D}{R^3}+\left(\kappa_\theta^2\frac{G}{R}\right)\left(h+\frac{I}{R^2}\right)\right\}\frac{\partial\psi_\theta}{\partial\theta}+$$
$$+\left\{\kappa_x^2Gh\frac{\partial^2}{\partial x^2}+\left(\kappa_\theta^2\frac{G}{R^2}\right)\left(h+\frac{I}{R^2}\right)\frac{\partial^2}{\partial\theta^2}-\left(\frac{E_p}{R^2}+\frac{D}{R^4}\right)-\rho h\frac{\partial^2}{\partial t^2}\right\}w=0. \quad (8.3.178)$$

Wave propagation is studied by considering solutions

$$\begin{aligned} u(x, \theta, t) &= U e^{i(\omega t - \alpha x)} \cos n\theta, \\ \psi_x(x, \theta, t) &= \psi e^{i(\omega t - \alpha x)} \cos n\theta, \\ v(x, \theta, t) &= V e^{i(\omega t - \alpha x)} \sin n\theta, \\ \psi_\theta(x, \theta, t) &= \Phi e^{i(\omega t - \alpha x)} \sin n\theta, \\ w(x, \theta, t) &= W e^{i(\omega t - \alpha x)} \cos n\theta. \end{aligned} \tag{8.3.179}$$

Substitution in the equations of motion gives the frequency-equation determinant

$$\begin{vmatrix} A & B & C & 0 & D \\ B & E & 0 & F & G \\ C & 0 & H & J & K \\ 0 & F & J & L & M \\ D & G & K & M & W \end{vmatrix} = 0, \tag{8.3.180}$$

where

$$\begin{aligned} A &= 4\pi^2\delta^2(s^2-2N)-m^2n^2(1+\tfrac{1}{12}m^2), \\ B &= \tfrac{1}{3}m\pi^2\delta^2(s^2-2N)+\tfrac{1}{12}m^3n^2, \\ C &= 2N\pi mn\delta(1+\nu)i, \\ D &= 4N\pi\nu m\delta i, \\ E &= \tfrac{1}{3}\pi^2\delta^2(s^2-2N)-\tfrac{1}{12}m^2n^2-\kappa_x^2, \\ F &= \tfrac{1}{6}Nmn\pi\delta(1+\nu)i, \\ G &= -2\kappa_x^2\pi\delta i, \\ H &= m^2(\kappa_\theta^2+2Nn^2)(1+\tfrac{1}{12}m^2)-4\pi^2\delta^2(s^2-1), \\ J &= -\tfrac{1}{3}m\pi^2\delta^2(s^2-1)-\kappa_\theta^2 m(1+\tfrac{1}{12}m^2)-\tfrac{1}{6}Nm^2n^2, \\ K &= m^2n(1+\tfrac{1}{12}m^2)(\kappa_\theta^2+2N), \\ L &= \kappa_\theta^2(1+\tfrac{1}{12}m^2)+\tfrac{1}{6}Nm^2n^2-\tfrac{1}{3}\pi^2\delta^2(s^2-1), \\ M &= -\kappa_\theta^2 mn(1+\tfrac{1}{12}m^2)-\tfrac{1}{6}Nm^3n, \\ W &= -4\pi^2\delta^2(s^2-\kappa_x^2)+m^2(1+\tfrac{1}{12}m^2)(\kappa_\theta^2n^2+2N), \end{aligned} \tag{8.3.181}$$

where

$$m = h/R, \quad N = 1/(1-\nu), \quad s = c/c_2, \quad \delta = h/L, \tag{8.3.182}$$

and L is the wavelength, given by $2\pi/\alpha$. For the special case of axial symmetry, it is shown by the authors that the determinant degenerates into the product of two subdeterminants corresponding to pure torsional modes and to longitudinal modes. The adjustment coefficients κ_x, κ_θ are obtained by comparing the cutoff frequencies of the thickness shear modes in the axial and circumferential directions to those of exact theory.

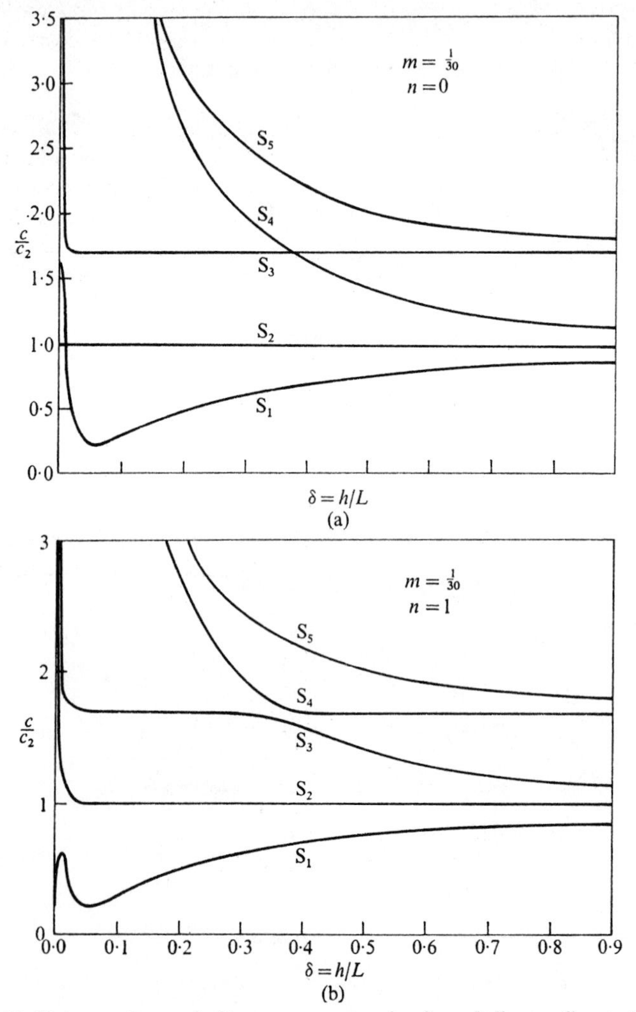

FIG. 8.31. Data on phase velocity versus wavenumber in a shell according to Mirsky–Herrmann theory for (a) $m = h/R = \frac{1}{30}$, $n = 0$; (b) $m = h/R = \frac{1}{30}$, $n = 1$. (After Herrmann and Mirsky [68, Figs. 1, 2].)

Results are presented by Mirsky and Herrmann for the axial phase velocity versus wavenumber for a number of situations, of which only two are shown in Fig. 8.31. The phase velocity is non-dimensionalized by the shear velocity c_2, while the wavenumber parameters δ is that defined by (8.3.182). Poisson's ratio is $\nu = 0{\cdot}3$ in all cases. In Fig. 8.31(a), the axisymmetric case $(n = 0)$ is shown for a rather thin shell $(h/R = \frac{1}{30})$. The non-dispersive S_2 mode is associated with purely torsional motion, S_4 is the second torsional mode. The mode S_1 is longitudinal. Flexural motion $(n = 1)$ is described by Fig. 8.31(b). All of the modes are now coupled.

2. *Naghdi–Cooper theory* [9, 72]. Again consider a cylindrical shell of thickness h and axial coordinate x. In this development, the circumferential coordinate is s, measured at the mid surface of the shell (this would be given by $s = R\theta$ in the Mirsky–Herrmann theory) and ζ is the coordinate directed radially outward from the mid surface (corresponding to z in the Mirsky–Herrmann development). The displacement kinematics are assumed to be

$$\begin{aligned} U_x &= u_x(x, s, t) + \zeta\beta_x(x, s, t), \\ U_s &= u_s(x, s, t) + \zeta\beta_s(x, s, t), \\ U_\zeta &= w(x, s, t), \end{aligned} \tag{8.3.183}$$

which are the same form as (8.3.175).

Cooper and Naghdi have developed the governing equations based on a variational theorem by Reissner [81]. The resulting stress equations of motion are found to be

$$\begin{aligned} \frac{\partial N_x}{\partial x} + \frac{\partial N_{sx}}{\partial s} &= \rho h \frac{\partial^2 u_x}{\partial t^2} + \frac{\rho h^3}{12a}\frac{\partial^2 \beta_x}{\partial t^2}, \\ \frac{\partial N_{xs}}{\partial x} + \frac{\partial N_s}{\partial s} + \frac{V_s}{a} &= \rho h \frac{\partial^2 u_s}{\partial t^2} + \frac{\rho h^3}{12a}\frac{\partial^2 \beta_s}{\partial t^2}, \\ \frac{\partial V_x}{\partial x} + \frac{\partial V_s}{\partial s} - \frac{N_s}{a} &= \rho h \frac{\partial^2 w}{\partial t^2}, \\ \frac{\partial M_x}{\partial x} + \frac{\partial M_{sx}}{\partial s} - V_x &= \rho \frac{h^3}{12}\left(\frac{\partial^2 \beta_x}{\partial t^2} + \frac{1}{a}\frac{\partial^2 u_x}{\partial t^2}\right), \\ \frac{\partial M_{xs}}{\partial x} + \frac{\partial M_s}{\partial s} - V_s &= \rho \frac{h^3}{12}\left(\frac{\partial^2 \beta_s}{\partial t^2} + \frac{1}{a}\frac{\partial^2 u_s}{\partial t^2}\right), \end{aligned} \tag{8.3.184}$$

where the shell stress–displacement equations are given by

$$
\begin{aligned}
N_x &= C\left[\left\{\frac{\partial u_x}{\partial x}+\nu\left(\frac{\partial u_s}{\partial s}+\frac{w}{a}\right)\right\}+\frac{h^2}{12a}\frac{\partial \beta_x}{\partial x}\right],\\
N_s &= C\left[\left\{\left(\frac{\partial u_s}{\partial s}+\frac{w}{a}\right)+\nu\frac{\partial u_x}{\partial x}\right\}-\frac{h^2}{12a}\frac{\partial \beta_s}{\partial s}\right],\\
N_{xs} &= \frac{(1-\nu)}{2}C\left\{\left(\frac{\partial u_x}{\partial s}+\frac{\partial u_s}{\partial x}\right)+\frac{h^2}{12a}\frac{\partial \beta_s}{\partial x}\right\},\\
N_{sx} &= \frac{(1-\nu)}{2}C\left\{\left(\frac{\partial u_x}{\partial s}+\frac{\partial u_s}{\partial x}\right)+\frac{h^2}{12a}\left(\frac{1}{a}\frac{\partial u_x}{\partial s}-\frac{\partial \beta_x}{\partial s}\right)\right\},\\
M_x &= D\left\{\left(\frac{\partial \beta_x}{\partial x}+\nu\frac{\partial \beta_s}{\partial s}\right)+\frac{1}{a}\frac{\partial u_x}{\partial x}\right\},\\
M_s &= D\left\{\left(\frac{\partial \beta_s}{\partial s}+\nu\frac{\partial \beta_x}{\partial x}\right)-\frac{1}{a}\left(\frac{\partial u_s}{\partial s}+\frac{w}{a}\right)\right\},\\
M_{xs} &= \frac{(1-\nu)}{2}D\left\{\left(\frac{\partial \beta_x}{\partial s}+\frac{\partial \beta_s}{\partial x}\right)+\frac{1}{a}\frac{\partial u_s}{\partial x}\right\},\\
M_{sx} &= \frac{(1-\nu)}{2}D\left\{\left(\frac{\partial \beta_x}{\partial s}+\frac{\partial \beta_s}{\partial x}\right)-\frac{1}{a}\frac{\partial u_x}{\partial s}\right\},\\
V_x &= \kappa G h\left(\frac{\partial w}{\partial x}+\beta_x\right),\\
V_s &= \kappa G h\left\{\frac{\partial w}{\partial s}-\left(\frac{u_s}{a}-\beta_s\right)\right\},
\end{aligned}
\tag{8.3.185}
$$

where a is the mean radius of the shell, ρ, E, G, ν are the material constants, $D = Eh^3/12(1-\nu^2)$ and $C = Eh/(1-\nu^2)$. Only a single adjustment coefficient is present in the theory. The resulting displacement equations of motion, referred to as System (I) by the authors, are given by

$$
\begin{aligned}
&\frac{\partial^2 u_x}{\partial x^2}+\frac{(1-\nu)}{2}\left(1+\frac{h^2}{12a^2}\right)\frac{\partial^2 u_x}{\partial s^2}+\frac{(1+\nu)}{2}\frac{\partial^2 u_s}{\partial x\,\partial s}+\frac{\nu}{a}\frac{\partial w}{\partial x}+\\
&\qquad+\frac{h^2}{12a}\left\{\frac{\partial^2 \beta_x}{\partial x^2}-\frac{(1-\nu)}{2}\frac{\partial^2 \beta_x}{\partial s^2}\right\}-\gamma^2\frac{\partial^2}{\partial t^2}\left(u_x+\frac{h^2}{12a}\beta_x\right)=0,\\
&\frac{\partial^2 u_s}{\partial s^2}+\frac{(1-\nu)}{2}\frac{\partial^2 u_s}{\partial x^2}+\frac{(1+\nu)}{2}\frac{\partial^2 u_x}{\partial x\,\partial s}+\frac{1}{a}\frac{\partial w}{\partial s}+\frac{\kappa_1}{a}\left(\beta_s-\frac{u_s}{a}+\frac{\partial w}{\partial s}\right)-\\
&\qquad-\frac{h^2}{12a}\left\{\frac{\partial^2 \beta_s}{\partial s^2}-\frac{(1-\nu)}{2}\frac{\partial^2 \beta_s}{\partial x^2}\right\}-\gamma^2\frac{\partial^2}{\partial t^2}\left(u_s+\frac{h^2}{12a}\beta_s\right)=0,\\
&\kappa_1\left(\nabla^2 w+\frac{\partial \beta_x}{\partial x}+\frac{\partial \beta_s}{\partial s}-\frac{1}{a}\frac{\partial u_s}{\partial s}\right)-\frac{1}{a}\left(\frac{\partial u_s}{\partial s}+\frac{w}{a}+\nu\frac{\partial u_x}{\partial x}\right)+\frac{h^2}{12a^2}\frac{\partial \beta_s}{\partial s}-\gamma^2\frac{\partial^2 w}{\partial t^2}=0,
\end{aligned}
\tag{8.3.186}
$$

$$\frac{h^2}{12}\left\{\frac{\partial^2\beta_x}{\partial x^2}+\frac{(1-\nu)}{2}\frac{\partial^2\beta_x}{\partial s^2}+\frac{(1+\nu)}{2}\frac{\partial^2\beta_s}{\partial x\,\partial s}\right\}-$$

$$-\kappa_1\left(\frac{\partial w}{\partial x}+\beta_x\right)+\frac{h^2}{12a}\left\{\frac{\partial^2 u_x}{\partial x^2}-\frac{(1-\nu)}{2}\frac{\partial^2 u_x}{\partial s^2}\right\}-\gamma^2\frac{h^2}{12}\frac{\partial^2}{\partial t^2}\left(\beta_x+\frac{u_x}{a}\right)=0,$$

$$\frac{h^2}{12}\left\{\frac{\partial^2\beta_s}{\partial s^2}+\frac{(1-\nu)}{2}\frac{\partial^2\beta_s}{\partial x^2}+\frac{(1+\nu)}{2}\frac{\partial^2\beta_x}{\partial x\,\partial s}\right\}-\kappa_1\left(\frac{\partial w}{\partial s}+\beta_s-\frac{u_s}{a}\right)-$$

$$-\frac{h^2}{12a}\left\{\frac{\partial^2 u_s}{\partial s^2}-\frac{(1-\nu)}{2}\frac{\partial^2 u_s}{\partial x^2}+\frac{1}{a}\frac{\partial w}{\partial s}\right\}-\gamma^2\frac{h^2}{12}\frac{\partial^2}{\partial t^2}\left(\beta_s+\frac{u_s}{a}\right)=0,$$

where

$$\nabla^2=\frac{\partial^2}{\partial x^2}+\frac{\partial^2}{\partial s^2},\qquad \gamma^2=\frac{(1-\nu^2)\rho}{E},\qquad \kappa_1=\frac{(1-\nu)\kappa}{2}. \tag{8.3.187}$$

When the effects of shear deformation and rotary inertia are neglected in the development of the System (I) equations, Love's equations for the shell are recovered.

An alternate set of shell equations is then presented by the authors, called System (II), with the intent that, when shear effects and rotary inertia are neglected, the more convenient shell equations of Donnell will be recovered. The simplification procedure involves neglecting certain displacement contributions, certain middle surface strains, and certain terms in h^2/a. The resulting shell stress–displacement equations are

$$N_x=C\left\{\frac{\partial u_x}{\partial x}+\nu\left(\frac{\partial u_s}{\partial s}+\frac{w}{a}\right)\right\},\qquad N_s=C\left\{\left(\frac{\partial u_s}{\partial s}+\frac{w}{a}\right)+\nu\frac{\partial u_x}{\partial x}\right\},$$

$$N_{xs}=N_{sx}=Gh\left(\frac{\partial u_x}{\partial s}+\frac{\partial u_s}{\partial x}\right),$$

$$M_x=D\left(\frac{\partial\beta_x}{\partial x}+\nu\frac{\partial\beta_s}{\partial s}\right),\qquad M_s=D\left(\frac{\partial\beta_s}{\partial s}+\nu\frac{\partial\beta_x}{\partial x}\right), \tag{8.3.188}$$

$$M_{xs}=M_{sx}=\frac{(1-\nu)}{2}D\left(\frac{\partial\beta_x}{\partial s}+\frac{\partial\beta_s}{\partial x}\right),$$

$$V_x=\kappa Gh\left(\frac{\partial w}{\partial x}+\beta x\right),\qquad V_s=\kappa Gh\left(\frac{\partial w}{\partial s}+\beta_s\right).$$

Certain additional minor adjustments are made in the displacement equations of motion (8.3.186) before substituting the preceding to give the System (II) equations

$$\frac{\partial^2 u_x}{\partial x^2}+\frac{(1-\nu)}{2}\frac{\partial^2 u_x}{\partial s^2}+\frac{(1+\nu)}{2}\frac{\partial^2 u_s}{\partial x\,\partial s}+\frac{\nu}{a}\frac{\partial w}{\partial x}-\gamma^2\frac{\partial^2 u_x}{\partial t^2}=0,$$

$$\frac{\partial^2 u_s}{\partial s^2}+\frac{(1-\nu)}{2}\frac{\partial^2 u_s}{\partial x^2}+\frac{(1+\nu)}{2}\frac{\partial^2 u_x}{\partial x\,\partial s}+\frac{1}{a}\frac{\partial w}{\partial s}-\gamma^2\frac{\partial^2 u_s}{\partial t^2}=0,$$

$$\kappa_1\left(\nabla^2 w+\frac{\partial\beta_x}{\partial x}+\frac{\partial\beta_s}{\partial s}\right)-\frac{1}{a}\left(\frac{\partial u_s}{\partial s}+\frac{w}{a}+\nu\frac{\partial u_x}{\partial x}\right)-\gamma^2\frac{\partial^2 w}{\partial t^2}=0,$$

$$\frac{h^2}{12}\left\{\frac{\partial^2\beta_x}{\partial x^2}+\frac{(1-\nu)}{2}\frac{\partial^2\beta_x}{\partial s^2}+\frac{(1+\nu)}{2}\frac{\partial^2\beta_s}{\partial x\,\partial s}\right\}-\kappa_1\left(\frac{\partial w}{\partial x}+\beta_x\right)-\gamma^2\frac{h^2}{12}\frac{\partial^2\beta_x}{\partial t^2}=0,$$

$$\frac{h^2}{12}\left\{\frac{\partial^2\beta_s}{\partial s^2}+\frac{(1-\nu)}{2}\frac{\partial^2\beta_s}{\partial x^2}+\frac{(1+\nu)}{2}\frac{\partial^2\beta_x}{\partial x\,\partial s}\right\}-\kappa_1\left(\frac{\partial w}{\partial s}+\beta_s\right)-\gamma^2\frac{h^2}{12}\frac{\partial^2\beta_s}{\partial t^2}=0. \tag{8.3.189}$$

The case of axially symmetric torsion-free wave propagation is considered in Reference [72]. Dispersion relations are obtained by considering propagating harmonic waves

$$\begin{bmatrix} u_x \\ \beta_x \\ w \end{bmatrix}=\begin{bmatrix} A' \\ B' \\ C' \end{bmatrix}\exp\{\mathrm{i}(mx-pt)\}, \tag{8.3.190}$$

and substituting in the various equations of motion, appropriately simplified by letting $u_s=\beta_s=\partial/\partial s=0$. Dispersion equations are obtained for the System (I), System (II), and the Donnell equations, called System (III). The value of the adjustment coefficient is $\kappa_1=\frac{5}{6}$ and is, according to Naghdi and Cooper, 'a natural consequence of the consistent assumptions for the stresses and displacements employed in Reissner's variational theorem'. The resulting dispersion curves for the axially symmetric motion as given by the three systems of equations mentioned are shown in Fig. 8.32 for a Poisson's ratio of $\nu=0{\cdot}3$ and a single value of the ratio h/a. The non-dimensional velocity is given by c/c_{p}, where c_{p} is the longitudinal thin plate velocity,

$$c_{\mathrm{p}}=\{E/\rho(1-\nu^2)^{\frac{1}{2}}\}.$$

The wavelength is λ. Using the System (I) and (II) equations, Cooper and Naghdi [9] also studied the propagation of torsional waves and of non-axially symmetric waves.

3. *Greenspon's comparisons.* As previously mentioned, Greenspon [23] made a comparison of the results for several theories of waves in cylindrical shells. The theories included not only those just reviewed, but results from exact theory, membrane theory, and several bending theories. The discussion pertains to theories and contribution by Rayleigh [79] and Baron and

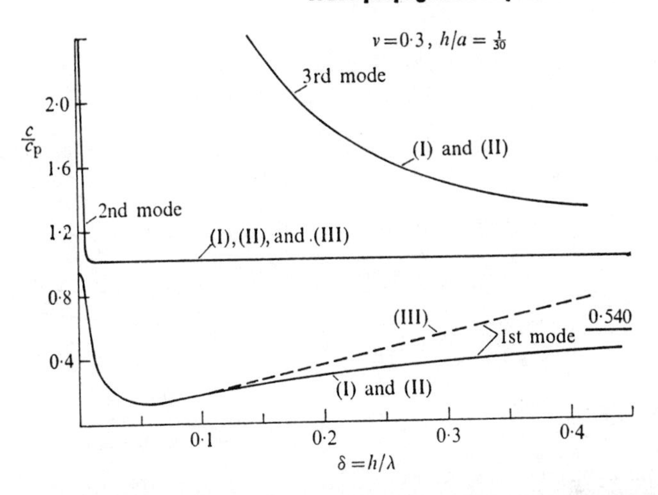

FIG. 8.32. Dispersion curves for axially symmetric waves in a shell according to Naghdi–Cooper theory for $h/a = \frac{1}{30}$. (After Naghdi and Cooper [72, Fig. 1(a)].)

Bleich [4] in the field of membrane shells, Love [43], Flugge [15], Arnold and Warburton [3], and Kennard [34] in the field of shell-bending theories, and the works of Mirsky and Herrmann [68], [69], Lin and Morgan [42], Yu [90], and Cooper and Naghdi [9, 72] for thick-shell (shear-deformation) theories.

The results of the comparisons are presented as dispersion curves and frequency spectra for various values of n, for a Poisson's ratio of $\nu = 0{\cdot}3$. The dimensionless velocity, frequency, and wavenumbers are respectively

$$\bar{c} = c/c_2, \qquad \Omega = \omega a_0/c_p, \qquad \beta = 2\pi a_0/\lambda, \tag{8.3.191}$$

where a_0 is the outside radius of the shell and λ is the wavelength. Also defined is the thickness ratio α, where

$$\alpha = a_i/a_0, \tag{8.3.192}$$

and a_i is the inner radius of the shell. A portion of the results of the comparison are given in Fig. 8.33. The rather elaborate code for identifying the various theories and contributions is given in conjunction with the figure.

4. *The theory of McNiven et al.* [47]. The thick-shell theories put forth by Mirsky and Herrmann, Naghdi and Cooper, and others contain shear-deformation effects. By invoking more elaborate assumptions on the deformation kinematics, higher-order shell theories are obtainable. McNiven *et al.* have presented a higher-order approximate theory for axisymmetric motions in cylindrical shells. The development is quite similar in spirit to that of Mindlin and Medick [66] for plates and Mindlin and McNiven [65] for rods. It contains the first three axisymmetric modes,

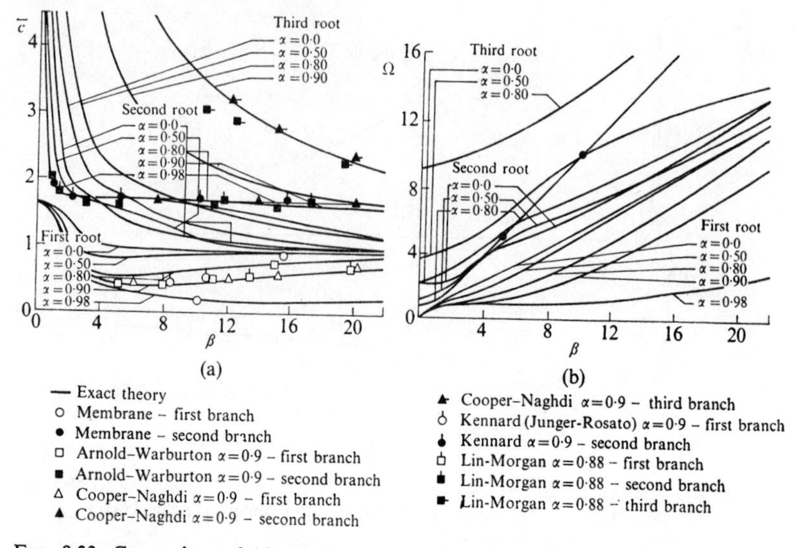

FIG. 8.33. Comparison of (a) dispersion curves and (b) frequency spectra for waves in shells according to various theories for $n = 0$ (axially symmetric case). (After Greenspon [23, Fig. 1].)

comprising the longitudinal, first radial, and first axial shear modes. Adjustment coefficients (four in number) are introduced and evaluated by matching the frequency spectrum of the approximate theory to that of the exact theory at the cutoff frequencies.

The real branches of the resulting frequency spectrum are shown in Fig. 8.34 for various cases. The non-dimensional frequency Ω and wave-number ζ are given by

$$\Omega = \omega h/\pi c_2, \qquad \zeta = h\xi/2\pi, \tag{8.3.193}$$

where h is the shell thickness. The parameter a^* is the ratio of outer to inner radius, $a^* = b/a$. Thus the results are for various thickness–curvature ratios and for two values of Poisson's ratio. The approximate theory is indicated by the dashed lines. The first three modes of exact theory are shown by the solid lines, the fourth mode of exact theory by dot–dashed line. The behaviour of the imaginary and complex branches were also determined by McNiven *et al.* [47].

8.4. Forced motion of plates and rods

The objective in this section is to consider a limited number of problems in the forced motion of elastic plates and rods. The exact analysis of a simple problem in SH waves in plates and a qualitative analysis of longitudinal waves

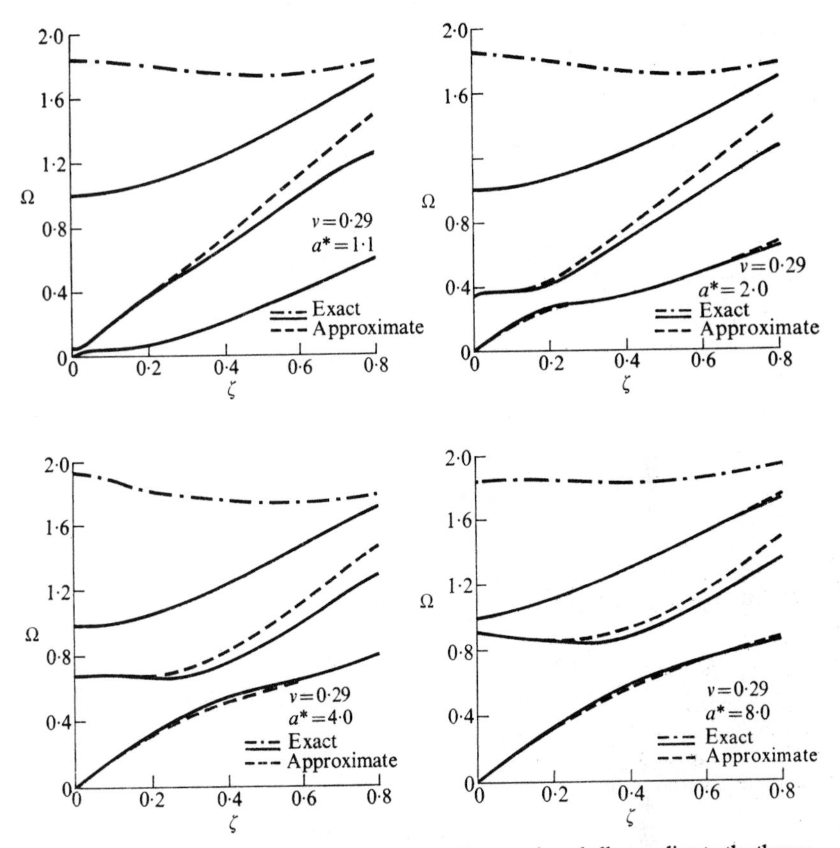

FIG. 8.34. Frequency spectra for axially symmetric waves in a shell according to the theory of McNiven *et al.* for various curvatures and Poisson's ratio = 0·29. (After McNiven *et al.* [47, Part I, Fig. 3].)

in rods will enable a general understanding of wave propagation in these structures to be obtained. The rather large class of problems involving the mathematical analysis of transient pulse propagation in plates and rods will then be discussed. However, analysis in this area, generally involving extensions of the saddle-point method presented in an earlier chapter, is rather complex. Attention will be confined mainly to presenting the results, in terms of stress waveforms, obtained by many contributors in this area.

8.4.1. *SH waves in a plate*

The case of SH waves has served well in several past situations to illustrate basic aspects of propagation or analysis, unobscured by the complicating effects due to mode conversion occurring from P and SV wave interactions with a surface. This will again be the case. Thus, consider a semi-infinite plate, defined by $y = \pm b$, $x \geqslant 0$, traction-free on the surfaces $y = \pm b$ and subjected to the end stress $\tau_{zx}(0, y, t)$. Such a loading will excite only SH waves in this geometry. Two basic problems will be considered, one involving harmonic loading and the second involving transient loading and homogeneous initial conditions.

1. *Harmonic SH waves.* The case of pure SH waves, polarized in the z direction, we know to be governed by the scalar wave equation in the u_z displacement

$$\nabla^2 u_z(x, y, t) = \frac{1}{c_2^2}\frac{\partial^2 u_z(x, y, t)}{\partial t^2}. \tag{8.4.1}$$

The boundary conditions for the problem are

$$\tau_{yz} = 0, \qquad y = \pm b, \qquad \tau_{xz}(0, y, t) = \tau(y)e^{-i\omega t}. \tag{8.4.2}$$

Also $\tau_{yy} = \tau_{xy} = 0$ on $y = \pm b$, but these are identically satisfied for SH waves.

Consider a solution

$$u_z(x, y, t) = U(x, y)e^{-i\omega t}, \tag{8.4.3}$$

and obtain, upon substituting in (8.4.1),

$$\nabla^2 U + k^2 U = 0, \qquad k^2 = \omega^2/c_2^2. \tag{8.4.4}$$

A separation of variables $U = X(x)Y(y)$ used in the preceding gives

$$dX^2/dx^2 + k^2 XY = -d^2Y/dy^2 = \alpha^2. \tag{8.4.5}$$

For the y variable there results

$$Y = A\sin\alpha y + B\cos\alpha y. \tag{8.4.6}$$

The first boundary condition of (8.4.2) reduces simply to $dY/dy = 0$, $y = \pm b$. The solution (8.4.6) thus becomes

$$Y_n = A_n \sin\alpha_n y + B_n\cos\alpha_n y, \quad \alpha_n = n\pi/2b \quad (n = 0, 1, 2, \ldots). \tag{8.4.7}$$

The x variation becomes

$$d^2X_n/dx^2 + \xi_n^2 X_n = 0, \qquad \xi_n^2 = k^2 - \alpha_n^2, \tag{8.4.8}$$

which has the solution, for $\xi_n^2 > 0$, of

$$X_n = C_n \exp(i\xi_n x) + D_n \exp(-i\xi_n x). \tag{8.4.9}$$

For $\xi_n^2 < 0$, we have

$$X_n = C_n' \exp(-\xi_n' x) + D_n' \exp(\xi_n' x). \tag{8.4.10}$$

We set $D_n = D_n' = 0$ since the D_n term violates the radiation condition for

$\xi_n^2 > 0$ and D_n' gives exponentially increasing displacements for $\xi_n^2 < 0$. The complete solution may then be written as

$$u_z(x, y, t) = \frac{B_0}{2}\exp\{i(\xi_0 x - \omega t)\} + \\ + \sum_{n=1}^{N}(A_n \sin \alpha_n y + B_n \cos \alpha_n y)\exp\{i(\xi_n x - \omega t)\} + \\ + \sum_{N+1}^{\infty}(A_n' \sin \alpha_n y + B_n' \cos \alpha_n y)\exp(-\xi_n' x)\exp(-i\omega t), \quad (8.4.11)$$

where

$$n < N, \quad \xi_n^2 > 0, \\ n \geq N, \quad \xi_n^2 < 0. \quad (8.4.12)$$

The A_n, B_n, A_n', B_n' are determined from the boundary conditions at $x = 0$, given by the second of (8.4.2). Thus, we have that $\tau_{xz}(x, y, t) = (\mu/2)\partial u_z/\partial x$, leading to

$$\tau(y) = \frac{\mu}{2}\left\{i\frac{B_0 \xi_0}{2} + i\sum_{n=1}^{N}\xi_n(A_n \sin \alpha_n y + B_n \cos \alpha_n y) - \\ - \sum_{N+1}^{\infty}\xi_n'(A_n' \sin \alpha_n y + B_n' \cos \alpha_n y)\right\}. \quad (8.4.13)$$

Multiplying both sides of (8.4.13) by $\sin \alpha_n y$ or $\cos \alpha_n y$ and integrating over the interval $-b \leqslant y \leqslant b$ gives

$$A_n = \frac{2}{i\mu b\xi_n}\int_{-b}^{b}\tau(y)\sin \alpha_n y \, dy,$$

$$A_n' = -\frac{2}{\mu b\xi_n'}\int_{-b}^{b}\tau(y)\sin \alpha_n y \, dy,$$

$$B_n = \frac{2}{i\mu b\xi_n}\int_{-b}^{b}\tau(y)\cos \alpha_n y \, dy, \quad (8.4.14)$$

$$B_n' = -\frac{2}{\mu b\xi_n'}\int_{-b}^{b}\tau(y)\cos \alpha_n y \, dy.$$

Let us now interpret the solution (8.4.13), with coefficients given by (8.4.14). This can best be done by also referring to the frequency spectrum for SH waves in a plate, given by Fig. 8.3. Each branch of the spectrum corresponds to a specific mode of propagation. Suppose that the driving frequency ω is such that

$$3 < \Omega < 4. \tag{8.4.15}$$

We recall that $\Omega = 2b\omega/\pi c_2$ is the non-dimensional frequency used in the figure. It is seen for this frequency range that four propagating modes ($n = 0, 1, 2, 3$) are capable of being exciting. All of the non-propagating modes (those with imaginary wavenumber) for $n \geqslant 4$ are capable of being excited. In the solution (8.4.13), the A_n, B_n are associated with the propagating modes, the A'_n, B'_n with the non-propagating modes. The degree to which a given mode is excited will depend on the amplitude coefficients which will in turn depend on the stress distribution $\tau(y)$. We note that the A'_n, B'_n are spatially decaying modes, so that, far from the end of the plate, their effect would not be felt† and only the propagating modes would contribute.

Suppose, as a simple example, that

$$\tau(0, y, t) = \tau_0 e^{-i\omega t}. \tag{8.4.16}$$

Then $A_n = B_n = A'_n = B'_n = 0$, $n \neq 0$ and we have $B_0 = 2\tau_0/i\mu b\xi_0$, so that

$$u_z(x, y, t) = \frac{\tau_0}{i\mu b\xi_0} \exp\{i(\xi_0 x - \omega t)\}, \tag{8.4.17}$$

where $\xi_0^2 = \omega^2/c_2^2$. For this particular example, only the lowest mode, which happens to be non-dispersive, is excited.

As a second simple example, suppose that

$$\tau_{xz}(0, y, t) = \tau(y)e^{-i\omega t}, \tag{8.4.18}$$

where $\tau(y)$ is symmetric with respect to y and ω is such that $2 < \Omega < 3$. Owing to the symmetry, $A_n = A'_n = 0$. Furthermore, only B_0, B_2 will be propagating for the given frequency range. Thus

$$u_z(x, y, t) = \frac{B_0}{2} \exp\{i(\xi_0 x - \omega t)\} + B_2 \cos \alpha_2 y \exp\{i(\xi_2 x - \omega t)\} +$$

$$+ \sum_{n=4,6,\ldots}^{\infty} B'_n \cos \alpha_n y \exp(-\xi'_n x)\exp(-i\omega t). \tag{8.4.19}$$

2. *Transient SH waves*. Suppose now that a transient disturbance is applied to the end of the plate such that

$$\tau_{xz}(0, y, t) = g(y)h(t),$$

$$u_z(x, y, 0) = \partial u_z(x, y, 0)/\partial t = 0. \tag{8.4.20}$$

Applying the Laplace transform to the governing eqn (8.4.1) gives

$$\nabla^2 \bar{u}_z(x, y, s) - \frac{s^2}{c_2^2} \bar{u}_z(x, y, s) = 0. \tag{8.4.21}$$

† This is somewhat too general a statement. Thus, if $\Omega \to 4$ (for example), the decay constant becomes quite small ($\xi'_n \sim 0$) and the decay distance can become appreciable. See, for example, Torvik [88].

Again using separation of variables $\bar{u}_z = \bar{X}(x, s)Y(y)$ gives

$$Y_n(y) = A_n \sin \alpha_n y + B_n \cos \alpha_n y, \quad \alpha_n = n\pi/2b, \tag{8.4.22}$$

and, for the governing equation for $\bar{X}(x, s)$,

$$\frac{d^2\bar{X}}{dx^2} - \xi_n^2 \bar{X} = 0, \qquad \xi_n^2 = \frac{s^2}{c_2^2} + \alpha_n^2. \tag{8.4.23}$$

This has the solution

$$\bar{X} = C \exp(\xi_n x) + D \exp(-\xi_n x). \tag{8.4.24}$$

We require $C = 0$ on the basis of the increasing exponential behaviour associated with that term. The solution may now be written as

$$\bar{u}_z(x, y, s) = \frac{B_0}{2} \exp(-\xi_0 x) + \sum_{n=1}^{\infty} (A_n \sin \alpha_n y + B_n \cos \alpha_n y) \exp(-\xi_n x). \tag{8.4.25}$$

The solution coefficients may be determined from the boundary conditions of (8.4.20). Thus we have

$$\bar{\tau}_{xz}(0, y, s) = \frac{\mu}{2} \frac{\partial \bar{u}_z(0, y, s)}{\partial x} = g(y)\bar{h}(s). \tag{8.4.26}$$

Substituting (8.4.25) in the last gives

$$g(y)\bar{h}(s) = -\frac{\mu}{2}\left\{\frac{\xi_0 B_0}{2} + \sum_{n=1}^{\infty} \xi_n (A_n \sin \alpha_n y + B_n \cos \alpha_n y)\right\}. \tag{8.4.27}$$

From this, the coefficients A_n, B_n are found to be

$$A_n = -\frac{2\bar{h}(s)}{\mu \xi_n} \int_{-b}^{b} g(y) \sin \alpha_n y \, dy,$$

$$B_n = -\frac{2\bar{h}(s)}{\mu \xi_n} \int_{-b}^{b} g(y) \cos \alpha_n y \, dy. \tag{8.4.28}$$

Designating the integrals as

$$a_n = \int_{-b}^{b} g(y) \sin \alpha_n y \, dy, \qquad b_n = \int_{-b}^{b} g(y) \cos \alpha_n y \, dy, \tag{8.4.29}$$

we have the transformed solution given by

$$\bar{u}_z(x, y, s) = -\frac{2\bar{h}(s)}{\mu}\left\{\frac{b_0 \exp(-\xi_0 x)}{2\xi_0} + \sum_{n=1}^{\infty} (a_n \sin \alpha_n y + b_n \cos \alpha_n y) \frac{\exp(-\xi_n x)}{\xi_n}\right\}. \tag{8.4.30}$$

The Laplace inversion is obtained from the typical term

$$\mathscr{L}^{-1}\left\{\frac{\bar{h}(s)\exp(-\xi_n x)}{\xi_n}\right\}. \tag{8.4.31}$$

We note from Laplace transform tables† that

$$f_n(t) = \mathscr{L}^{-1}\left\{\frac{\exp(-\xi_n x)}{\xi_n}\right\} = \begin{cases} 0, & 0 < t < x/c_2 \\ c_2 J_0\{\alpha_n c_2(t^2 - x^2/c_2^2)^{\frac{1}{2}}\}, & x/c_2 < t < \infty. \end{cases} \tag{8.4.32}$$

For an arbitrary time variation $h(t)$, the final inversion would have to be expressed in terms of the convolution integral. Thus

$$u_z(x, y, t) = -\frac{2}{\mu}\left\{\frac{b_0}{2}\int_0^t h(t-\tau)f_0(\tau)\,d\tau + \right.$$

$$\left. + \sum_{n=1}^{\infty}(a_n \sin \alpha_n y + b_n \cos \alpha_n y)\int_0^t h(t-\tau)f_n(\tau)\,d\tau\right\}. \tag{8.4.33}$$

A simple evaluation of this integral is possible only for special values of $h(t)$. Thus, suppose $h(t) = \delta(t)$. Then we have

$$u_z(x, y, t) = -\frac{2}{\mu}\left\{\frac{b_0}{2}f_0(t) + \sum_{n=1}^{\infty}(a_n \sin \alpha_n y + b_n \cos \alpha_n y)f_n(t)\right\}. \tag{8.4.34}$$

The main objective in this analysis of SH wave propagation is to show how the nature of the propagation in a plate or rod can be affected by two aspects of the input: first, the spatial distribution of the exciting stress determines the degree to which various modes may or may not be excited and, second, the time variation of the excitation determines which of the excited modes may be propagating and which non-propagating.

8.4.2. *Pulse propagation in an infinite rod*

While the theoretical analysis of pulse propagation in rods and plates is generally quite complicated, by using stationary-phase and group-velocity concepts, it is possible to obtain a qualitative understanding of the action. We now consider the case of an initial pulse applied to an infinitely long rod, and closely follow the original arguments of Davies [11] for this case.

1. *Basic stationary-phase argument.* Assuming that at $t = 0$ the initial pulse consists of a Fourier superposition of continuous harmonic waves then, for $t > 0$, the dispersion of the various frequency components will distort the

† See, for example, Churchill [8, Appendix 3].

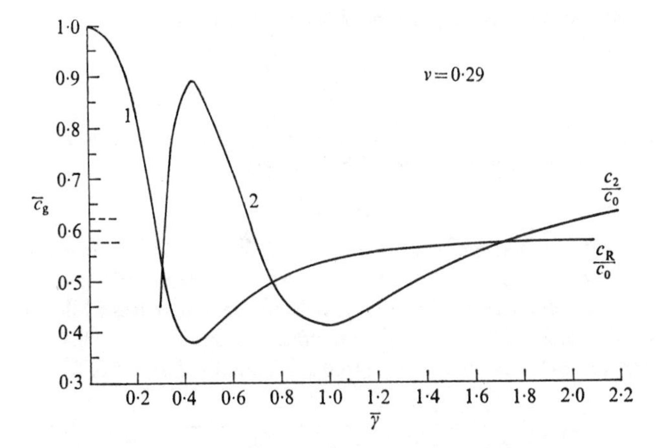

FIG. 8.35. Group velocity versus wavenumber for the first two longitudinal modes of a circular rod. (After Davies [11, Fig. 13].)

pulse, as has been discussed in earlier sections. According to stationary-phase arguments, the dominant components of the disturbance at a point x and time t will satisfy the condition that the phase ϕ is stationary, where

$$\phi = \gamma x - \omega t. \tag{8.4.35}$$

By differentiating, this may be shown to be equivalent to† $c_g(\omega) = x/t$. Thus for a given value of x and at a given time t, the frequency of the disturbance will be such that $c_g(\omega) = x/t$.

2. *Construction of period, c_g curves.* In order to determine the variation in the disturbance with time at a given location, it is useful to plot the period of the disturbance as a function of x/t. The procedure for obtaining this plot will be as follows:

(a) for a given $t/x = 1/c_g$, determine the value of γ corresponding to c_g;
(b) determine the value of phase velocity c, corresponding to γ;
(c) the period T, will be given by $T = 2\pi/\gamma c$.

Now, the dispersion curves for the first three rod modes were given earlier by Fig. 8.16. The group-velocity curves for the first two modes are given in Fig. 8.35. The non-dimensional velocities and wavenumbers are given by

$$\bar{c}_g = c_g/c_0, \qquad \bar{c} = c/c_0, \qquad \bar{\gamma} = a\gamma/2\pi, \tag{8.4.36}$$

† See Chapter 1, § 6.

where c_0 is the bar velocity, $c_0 = (E/\rho)^{\frac{1}{2}}$. Further, we introduce the non-dimensional distance $\bar{x}$, time $\bar{t}$, and period $\bar{T}$, where

$$\bar{x} = x/a, \qquad \bar{t} = t/T_a, \qquad \bar{T} = T/T_a, \tag{8.4.37}$$

where a is the bar radius and $T_a = a/c_0$. With these, the basic relationship between group velocity, distance, and time becomes

$$\bar{c}_g \doteq \bar{x}/\bar{t}. \tag{8.4.38}$$

Considering the first mode, it is seen that $0{\cdot}38 < \bar{c}_g < 1{\cdot}0$. It follows that $1{\cdot}0 < \bar{t}/\bar{x} < 2{\cdot}64$. The wavenumber $\bar{\gamma}$ is zero at $\bar{c}_g = 1{\cdot}0$, which gives a period of infinity. As $\bar{c}_g$ decreases from 1·0, the wavenumber increases while the phase velocity $\bar{c}$ decreases. The period thus decreases from infinity. This and other aspects yet to be discussed are presented in Fig. 8.36 as a plot of $\bar{T}$ versus $\bar{t}/\bar{x}$.

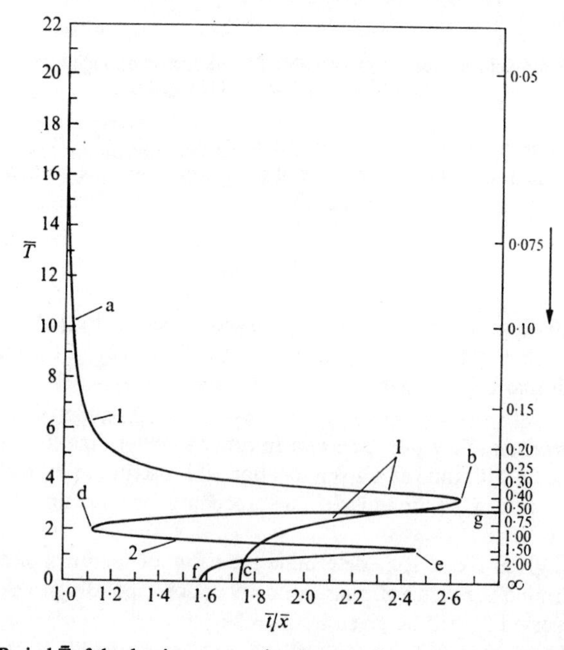

FIG. 8.36. Period $\bar{T}$ of the dominant group in a bar at a distance $\bar{x}$ and time $\bar{t}$. (After Davies [11, Fig. 15].)

Thus the aspect of the period decreasing from infinity is indicated by a of curve 1. Continuing, the group velocity of the first mode decreases to a minimum, given by $\bar{c}_g = 0{\cdot}38$ at $\bar{\gamma} = 0{\cdot}43$. This yields a period value of $\bar{T} = 3{\cdot}2$ and a maximum value of $\bar{t}/\bar{x} = 2{\cdot}64$ (point (b) of Fig. 8.36). Referring back to the group-velocity curve, it is seen that as $\bar{\gamma}$ increases beyond the minimum point ($\bar{\gamma} = 0{\cdot}43$), that the group velocity approaches a limiting value of c_R/c_0, where c_R is the Rayleigh velocity. Since $\bar{\gamma} \to \infty$ in this limit, the period approaches zero, as given by point (c) in Fig. 8.36.

Now consider the second mode. As seen from Fig. 8.35, the $\bar{c}_g$ curve has a maximum value at $\bar{c}_g = 0{\cdot}89$ (where $\bar{\gamma} = 0{\cdot}44$), a local minimum at $\bar{c}_g = 0{\cdot}42$ (where $\bar{\gamma} = 1{\cdot}0$) and approaches a limiting value of c_2/c_0. Although not completely drawn in, the $\bar{c}_g$ curve is approaching zero in the vicinity of $\bar{\gamma} = 0{\cdot}2$. This corresponds to the local minimum of the second branch of the frequency spectrum for the rod. In Fig. 8.35, the maximum point of the $\bar{c}_g$ curve gives point d on curve 2, the local minimum gives point (e). The limiting value as $\bar{\gamma} \to \infty$ is given by point (f). Finally, as $\bar{c}_g \to 0$, the wave-number and phase velocity approach definite limits, giving a limiting value to the period, indicated by point (g) of Fig. 8.36.

3. *Propagation of a sharp pulse.* Let us now use Fig. 8.35 to describe, in general terms, the disturbance at a point x due to a disturbance at the origin at $t = 0$. Suppose the disturbance is in the form of an infinite magnitude, infinitely short-duration pulse (a Dirac delta function). The Fourier spectrum of such a pulse has equal amplitudes for all Fourier components. If dispersion were absent, the c_g versus γ and c versus γ curves would simply be horizontal straight lines. The resulting T versus t/x, or $\bar{T}$ versus $\bar{t}/\bar{x}$ curve would be a vertical straight line located at $\bar{t} = 1$. This would indicate that waves of all periods would arrive simultaneously at a point, thereby reconstructing the delta pulse.

Consider initially only the time response in the first mode at a point x. The first disturbance is felt at $\bar{t}/\bar{x} = 1$ (that is, $t = x/c_0$) and consists of long-period waves travelling at the bar velocity c_0. With increasing time $\bar{t}$, the period of the disturbance decreases, indicating an increase in frequency. This change in frequency is at first rapid, since movement is along the nearly vertical portion of the curve, and then less rapid, as the $\bar{T}$ versus $\bar{t}/\bar{x}$ curve flattens out.

For $\bar{t}/\bar{x} \geqslant 1{\cdot}73$ it is seen that $\bar{T}$ versus $\bar{t}$ is double-valued. Thus, at $\bar{t} = 1{\cdot}73$ the nature of the signal arriving at x becomes more complex. Since the lower branch of the $\bar{T}$ curve approaches $\bar{t} = 1{\cdot}73$ nearly vertically, this indicates that nearly all of the waves having a period of $\bar{T} < 0{\cdot}5$ arrive simultaneously, thus superposing a high-frequency pulse on the lower-frequency upper-branch contribution. As $\bar{t}$ increases beyond 1·73, the signal contains two frequency contributions which, at first, are quite different but, as $\bar{t}/\bar{x}$ increases toward 2·64, approach the same frequency. At $\bar{t}/\bar{x} = 2{\cdot}64$, the disturbance due to the first mode suddenly ceases, with the period of the terminating signal given by $\bar{T} = 3{\cdot}2$. Thus, the total duration of the signal (due to the first-mode contribution) is $\bar{t}/\bar{x} = 1{\cdot}64$. In terms of absolute time, $t = \bar{t}x/c_0$. Hence, for larger values of x, it is seen that the total duration increases.

An attempt has been made to sketch the approximate character of the vibration in Fig. 8.37. The contributions from the upper and lower branches of the first mode are shown separately and superimposed to represent the total signal. The main aspect to be noted is the frequency behaviour, since amplitude contributions are uncertain. The main points to be observed are the arrival of the high-frequency contribution at $\bar{t}/\bar{x} = 1{\cdot}73$ and, as $\bar{t}/\bar{x} \to 2{\cdot}64$, the appearance of a 'beating'-type disturbance as the frequencies of the two branches approach the same value.

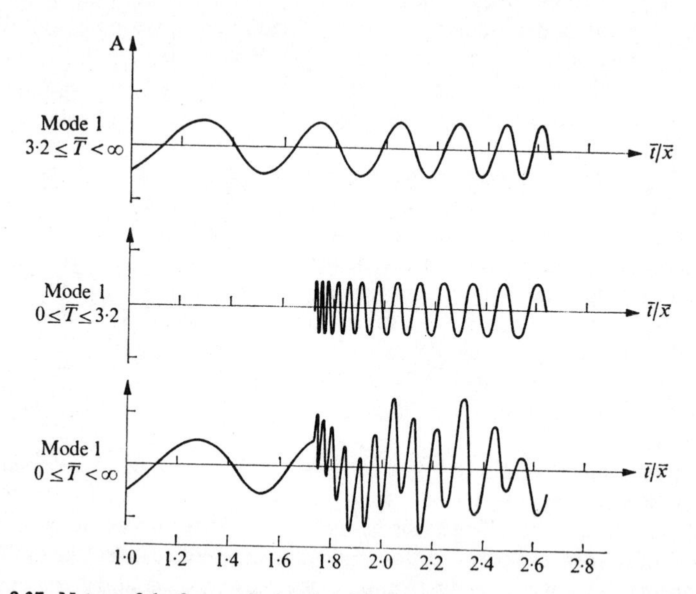

FIG. 8.37. Nature of the first mode signal arriving at a given location x for changing time.

Conceptually, there is no difficulty in adding the second-mode contribution to the above. The contribution would start at $\bar{t}/\bar{x} = 1{\cdot}12$ with the disturbance having a period $\bar{T} = 2$. For $\bar{t} > 1{\cdot}12$, the second mode would have, initially, two frequency components. At $\bar{t}/\bar{x} = 1{\cdot}57$ (point f) the third branch of the second mode enters in, so that, for $\bar{t} > 1{\cdot}57$, three frequency components contribute. As $\bar{t} \to 2{\cdot}44$ (point e) two of the components converge and terminate, but, unlike the first mode, signals from the second mode continues to arrive owing to the branch of the second mode that approaches the limiting value $\bar{T} = 2{\cdot}8$. This, we recall, was a result of the zero group velocity occurring for this mode. No effort is made here to sketch the signal behaviour. Obviously, the combined signal from the first and second modes would be quite complicated.

8.4.3. *Transient compressional waves in semi-infinite rods and plates*

Over the years, the problem of a transient load applied to the end of a semi-infinite rod has been extensively analysed. The geometry and loading situation is of considerable practical interest in experimental investigations of dynamic material properties and in measuring impact forces. Furthermore, geometry and loading are sufficiently simple to make theoretical analysis of the wave propagation, while complicated, still possible. Usually, the end of the rod or plate is considered to be subjected to a step loading in stress or to a step change in velocity. The solution to this problem according to elementary rod theory is well known; that is, the step pulse propagates undistorted with the bar velocity c_0. On theoretical grounds and by experimental observation, distortion of the step pulse with distance travelled is expected and has been observed for such a severe transient. The problem, then, is the theoretical analysis of the situation.

The usual procedure has been to apply integral transforms to the exact equations of elasticity. Usually, the inversion integrals have been evaluated by the saddle-point method, so that the results were valid only for the far field. Complications usually arise in this phase of the analysis owing to the coincidence of the saddle points with poles of the integrand under certain conditions. In many cases, only the contributions of the first mode were included in the inversion. The integral-transform technique has also been applied to various approximate rod theories, with inversion being performed by the saddle-point method. Numerical methods have been used in select cases to obtain near-field data. The results from a number of investigators will be reviewed briefly in the following.

1. *Skalak's solution for impact of two semi-infinite rods.* The case of longitudinal impact of two semi-infinite rods was first considered by Skalak [85], using the situation shown in Fig. 8.38. The approach used was both

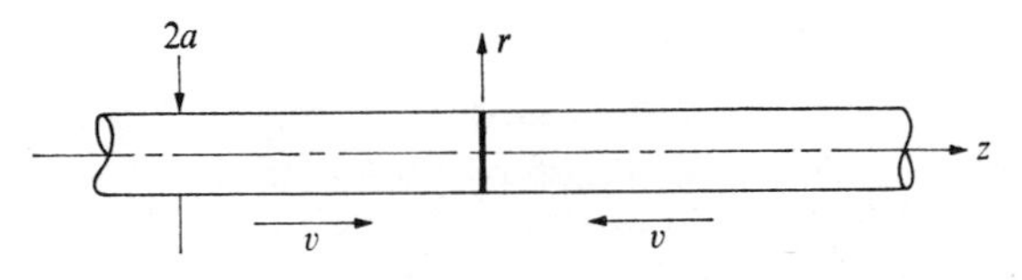

FIG. 8.38. Longitudinal impact of two semi-infinite rods.

unusual and ingenious and involved the superposition of two solutions. The first involves the impact of two semi-infinite rods, as shown in Fig. 8.38, but which are also completely constrained against lateral motion. Such a situation is quite like that of two semi-infinite media impacting, where plane-strain conditions prevail in the radial direction. Such an impact yields a displacement wave situation given by

$$u_z(z, t) = \begin{cases} -vt, & z > c_1 t \\ -\dfrac{z}{c_1}v, & 0 \leqslant z < c_1 t, \end{cases} \tag{8.4.39}$$

where $z > 0$. Such a wave, due to the radial constraint, leads to radial stresses of

$$\tau_{rr}(a, z, t) = \lambda\, \partial u_z/\partial z. \tag{8.4.40}$$

The second part consists of solving the problem of an infinite rod, initially at rest, but subjected to a traction on the surface of the bar. The traction consists of the radial stress (8.4.40), but opposite in sign. Thus, at time $t = 0$, a radial stress, directed radially outward, begins propagating from the origin in the z direction. This system of surface tractions, when superimposed on those of (8.4.40), given by the first part of the problem, exactly cancel and produce the traction-free lateral surfaces of the two colliding semi-infinite rods.

Solution of the second part is, of course, the difficult part. The problem is specified by the displacement equations of motion

$$\begin{aligned} (\lambda+2\mu)\frac{\partial \Delta}{\partial r} + 2\mu\frac{\partial \Omega}{\partial z} &= \rho\frac{\partial^2 u_r}{\partial t^2}, \\ (\lambda+2\mu)\frac{\partial \Delta}{\partial z} - \frac{2\mu}{r}\frac{\partial}{\partial r}(r\Omega) &= \rho\frac{\partial^2 u_z}{\partial t^2}, \end{aligned} \tag{8.4.41}$$

where

$$\Delta = \frac{1}{r}\frac{\partial(ru_r)}{\partial r} + \frac{\partial u_z}{\partial z}, \qquad \Omega = \frac{1}{2}\left(\frac{\partial u_r}{\partial z} - \frac{\partial u_z}{\partial r}\right), \tag{8.4.42}$$

and the boundary conditions

$$\tau_{rr}|_{r=a} = R(z, t) = \begin{cases} \dfrac{\lambda v}{c_1}, & -c_1 t < z < c_1 t \\ 0, & |z| > c_1 t \end{cases}, \qquad \tau_{rz}|_{r=a} = 0. \tag{8.4.43}$$

Skalak uses a Fourier–Laplace transform pair defined by

$$\bar{g}(\gamma, p) = \frac{1}{4\pi^2}\int_{-\infty}^{\infty} e^{-i\gamma z}\, dz \int_0^{\infty} g(z, t) e^{-ipt}\, dt, \tag{8.4.44}$$

with inverse

$$g(z, t) = \int_{-\infty}^{\infty} e^{i\gamma z}\, d\gamma \int_{-\infty-i\alpha}^{\infty-i\alpha} \bar{g}(\gamma, p) e^{ipt}\, dp. \tag{8.4.45}$$

Applying the transforms to the governing eqns (8.4.41) and (8.4.42) and eliminating $\bar{u}_r$, $\bar{u}_z$ gives two equations in $\bar{\Delta}$, $\bar{\Omega}$

$$\frac{\partial^2\bar{\Delta}}{\partial r^2}+\frac{1}{r}\frac{\partial\bar{\Delta}}{\partial r}+h^2\bar{\Delta}=0,$$
$$\frac{\partial^2\bar{\Omega}}{\partial r^2}+\frac{1}{r}\frac{\partial\bar{\Omega}}{\partial r}+\left(k^2-\frac{1}{r^2}\right)\bar{\Omega}=0, \tag{8.4.46}$$

where

$$h^2=\frac{\rho p^2}{\lambda+2\mu}-\gamma^2, \qquad k^2=\frac{\rho p^2}{\mu}-\gamma^2. \tag{8.4.47}$$

The solutions to (8.4.46) are then

$$\bar{\Delta}=BJ_0(hr), \qquad \bar{\Omega}=DJ_1(kr). \tag{8.4.48}$$

These last solutions are substituted for $\bar{\Delta}$, $\bar{\Omega}$ in the transformed equations of motion, and the results solved for $\bar{u}_r$, $\bar{u}_z$, giving

$$\bar{u}_r=A\frac{\partial}{\partial r}J_0(hr)+C\gamma J_1(kr),$$
$$\bar{u}_z=\mathrm{i}\gamma AJ_0(hr)+\frac{\mathrm{i}C}{r}\frac{\partial}{\partial r}\{rJ_1(kr)\}. \tag{8.4.49}$$

The transformed stresses are then obtained, using (8.4.49) and substituted in the transformed boundary conditions and the resulting two equations solved for A and C, with the results

$$A=\frac{\bar{R}}{F}\left(2\gamma^2-\frac{\rho p^2}{\mu}\right)J_1(ka), \qquad C=-\frac{2\bar{R}\gamma}{F}\frac{\partial J_0(ha)}{\partial a}, \tag{8.4.50}$$

where $\bar{R}$ is the transform of the load function and

$$F=\left\{2\mu\frac{\partial^2J_0(ha)}{\partial a^2}-\frac{p^2\rho\lambda}{\lambda+2\mu}J_0(ha)\right\}\left(2\gamma^2-\rho\frac{p^2}{\mu}\right)J_1(ka)-2\mu\gamma\frac{\partial J_1(ka)}{\partial a}2\gamma\frac{\partial J_0(ha)}{\partial a}. \tag{8.4.51}$$

This last is the Pochhammer frequency equation for axially symmetric waves in an infinite rod. Applying the inversion gives for $\partial u_z/\partial z$

$$\frac{\partial u_z(r,z,t)}{\partial z}=-\int_{-\infty}^{\infty}\int_{-\infty-\mathrm{i}\alpha}^{\infty-\mathrm{i}\alpha}\bar{R}\{\gamma^2(2\gamma^2-\rho p^2/\mu)J_1(ka)J_0(hr)+$$
$$+2\gamma^2hkJ_1(ha)J_0(kr)\}\frac{\mathrm{e}^{\mathrm{i}(\gamma z+pt)}}{F}\,\mathrm{d}\gamma\,\mathrm{d}p, \tag{8.4.52}$$

where, for the present case,

$$\bar{R}=\frac{\lambda v}{2\pi c_1^2(\gamma^2-p^2/c_1^2)}. \tag{8.4.53}$$

The inversion with respect to p is carried out first using contour integration in the complex p-plane. The contour consists of the line $p = i\alpha$ and the semicircle closed in the upper p-plane. The contribution from the latter path is zero so that the p-integral is $2\pi i \sum$ Res of the enclosed poles. The poles are given by $F = 0$, the roots of the Pochhammer equation. The result is given by

$$\frac{\partial u_z}{\partial z} = \frac{vi}{\pi}\int_0^\infty \sum_n \left[\frac{-\gamma^2\lambda\left(2\gamma^2 - \frac{\rho p^2}{\mu}\right)J_1(ka)J_0(hr) - \gamma^2 2\lambda hkJ_1(ha)J_0(ka)}{c_1^2\left(\gamma^2 - \frac{p^2}{c_1^2}\right)\frac{\partial F}{\partial p}} \times e^{i(\gamma z + pt)}\right]_{p=p_n} d\gamma, \quad (8.4.54)$$

where $p = p_n$ corresponds to the branches of the Pochhammer–Chree frequency spectrum, given by $p_n = p_n(\gamma)$.

The evaluation of the results (8.4.54), which are exact to this stage, is in general not possible. However, by considering t large, the saddle-point method of approximately evaluating the integral may be used. The results are given by

$$u_z' = \frac{v}{c_0}\left\{\frac{1}{\pi}\int_0^\infty \frac{\sin(\alpha''\eta + \frac{1}{3}\eta^3)}{\eta}\,d\eta + \frac{1}{\pi}\int_0^\infty \frac{\sin(\alpha'\eta + \frac{1}{3}\eta^3)}{\eta}\,d\eta\right\}, \quad (8.4.55)$$

where

$$\alpha'' = \frac{z''}{(3dt)^{\frac{1}{3}}}, \qquad \alpha' = \frac{z'}{(3dt)^{\frac{1}{3}}},$$
$$z'' = -z - c_0 t, \qquad z' = z - c_0 t. \quad (8.4.56)$$

The integrals appearing in the result are related to the Airy integral†

$$(Ai)(\alpha) = \frac{1}{\pi}\int_0^\infty \cos(\alpha\eta + \tfrac{1}{3}\eta^3)\,d\eta. \quad (8.4.57)$$

Thus it becomes possible to express the results as

$$\frac{\partial u_z}{\partial z} = \frac{v}{c_8}\left\{\frac{1}{6} + \int_0^{\alpha''} (Ai)(\alpha)\,d\alpha + \frac{1}{6} + \int_0^{\alpha'} (Ai)(\alpha)\,d\alpha\right\}. \quad (8.4.58)$$

The numerical compilation of (8.4.58) follows from tables of Airy integrals.

A plot of the resulting waveform for the strain $\varepsilon_z = \partial u_z/\partial z$ is shown in Fig. 8.39. The pulse shape is seen to be highly oscillatory. The rectangular pulse shown in the figure represents the input pulse as well as the propagated pulse as predicted by elementary longitudinal rod theory. Note that

† See, for example, p. 447 of [1].

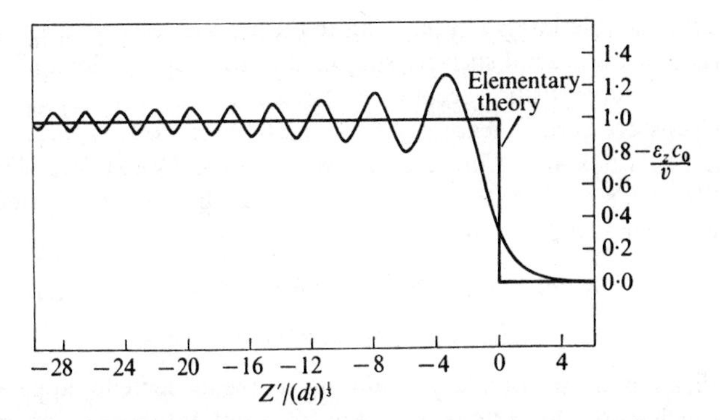

FIG. 8.39. Predicted strain wave $\varepsilon_z = \partial u_z/\partial z$ resulting from the impact of two semi-infinite rods. (After Skalak [85, Fig. 5].)

$z' = z - c_0 t = 0$ corresponds to the arrival time of the pulse according to classical theory. The Skalak analysis shows the rise of the pulse to begin at an earlier time, but with maximum amplitude not being attained until a later time than predicted by simple rod theory. As will be shown in § 8.5, the first few oscillations of the predicted pulse are in excellent agreement with experimental observations.

As pointed out by Skalak, the information essential to the preceding theoretical development is the first two terms of the expansion of the phase velocity for the first mode about $\gamma = 0$. Consequently, it is to be expected that approximate rod theories which duplicate the first mode near $\gamma = 0$ will yield results identical to the analysis that stems from exact elasticity and the Pochhammer equation. The Love theory for the rod, which incorporates lateral-inertia effects into the classical rod theory, does closely approximate the first mode for a limited region (see Fig. 2.27). Davies' analysis of a step pulse using Love's theory predicted oscillatory behaviour of the pulse (see Fig. 2.29) and, as checked by Skalak, the results of Fig. 2.29 and Fig. 8.39 agree quite closely when appropriate scale adjustments are made. The conclusion reached is that, despite the fact that many higher-order stress effects are present in the exact theory of waves in rods, the predominant effect causing deviation of the far-field pulse shape from that predicted by elementary theory is that of lateral inertia.

2. *Analysis by Folk et al. of the step pressure pulse on a rod.* As previously mentioned, many investigations have appeared on transient waves in rods, of which that of Skalak was the first to use exact theory in the analysis. The case of a step pressure pulse applied to the end of a rod was analysed by Folk, Fox, Shook, and Curtis [17], with other contributions on the flexural pulse case and experimental results being given by Fox and Curtis [18], DeVault and Curtis [12], and Curtis [10]. The method of analysis used in all cases was the direct

application of double-integral transforms to the problem of a semi-infinite rod having traction-free lateral surfaces subjected to the step end load. This is in contrast to the superposition technique of Skalak.

Let us consider certain features of the analysis of Folk *et al.* [17]. The governing equations are as previously given by (8.4.41) and (8.4.42) in the review of Skalak's work, with the initial conditions and boundary conditions on the lateral surface given by

$$u_r = u_z = \partial u_r/\partial t = \partial u_z/\partial t = 0, \tag{8.4.59}$$

and

$$\tau_{rr}(z, a, t) = \tau_{rz}(z, a, t) = 0. \tag{8.4.60}$$

The specification of the boundary conditions on the loaded end, apparently a simple enough matter for a step pressure pulse, is not, it turns out, completely obvious. The problem arises in the application of the integral transforms. As stated by Folk *et al.*:

The heart of the problem is to choose transforms which will not only provide solutions to the differential equations, but will ask for the right initial and boundary information. The Fourier transform is suitable for excluding t, but not z. If one attempts to exclude z by applying the Fourier transformation [to eqns (8.4.41), (8.4.42)] it is found that four pieces of information are needed at the end of the bar, whereas the behavior of the bar is completely determined when two conditions are specified. This difficulty can be avoided by excluding z through the proper use of sine and cosine transformations. The method of solution then follows standard procedure.

With sine, cosine, and exponential Fourier transforms defined respectively by

$$\begin{aligned} f^{\mathrm{S}}(\gamma, r, t) &= \int_0^\infty f(z, r, t)\sin(\gamma z)\,\mathrm{d}z, \\ f^{\mathrm{C}}(\gamma, r, t) &= \int_0^\infty f(z, r, t)\cos(\gamma z)\,\mathrm{d}z, \\ f^{\mathrm{F}}(\gamma, r, \omega) &= \int_0^\infty f(\gamma, r, t)\mathrm{e}^{\mathrm{i}\omega t}\,\mathrm{d}t, \end{aligned} \tag{8.4.61}$$

it results that the sine transform applied to the first of (8.4.41) and (8.4.42) and the cosine transform applied to the second of these equations ask for the specification of τ_{zz} and u_r at $z = 0$. These are specified to be

$$\tau_{zz}(r, 0, t) = -P_0 H\langle t\rangle, \qquad u_r(r, 0, t) = 0. \tag{8.4.62}$$

These are, of course, mixed boundary conditions. It turns out that no combination of transforms asks for pure stress or pure displacement conditions. This feature, it should be noted, arises in all of the exact analyses of transients in rods and plates by integral-transform methods. A difficulty of the mixed conditions is that, while the condition on τ_{zz} is realistic, that on the radial displacement is not in accord with the physical conditions of most experiments. It turns out that this approximation does not seriously affect the predicted far-field response.

Application of the proper transforms gives as the transformed solution

$$E^{\mathrm{SF}} = \overset{\mathrm{SF}}{\varepsilon}_{\theta\theta} + \overset{\mathrm{SF}}{\varepsilon}_{zz} = \frac{\overset{\mathrm{SF}}{u_r}}{r} - \gamma \overset{\mathrm{CF}}{u_z}$$

$$= \frac{(\lambda+2\mu)A}{\rho\omega^2}\left\{\frac{hJ_1(hr)}{r} + \gamma^2 J_0(hr)\right\} + \frac{2\mu\gamma B}{\rho\omega^2}\left\{\frac{J_1(kr)}{r} - kJ_0(kr)\right\} + \frac{\mathrm{i}P_0\gamma}{(\lambda+2\mu)\omega h^2}, \tag{8.4.63}$$

where

$$A(\gamma,\omega) = \frac{\mathrm{i}P_0\rho\lambda\gamma(k^2-\gamma^2)\omega J_1(ka)}{(\lambda+2\mu)^2\mu h^2\Phi(\gamma,\omega)},$$
$$B(\gamma,\omega) = \frac{\mathrm{i}P_0\rho\lambda\gamma^2\omega J_1(ha)}{(\lambda+2\mu)\mu^2 h\Phi(\gamma,\omega)}, \tag{8.4.64}$$

and

$$\Phi(\gamma,\omega) = \frac{2h}{a}(k^2+\gamma^2)J_1(ha)J_1(ka) - $$
$$-(k^2-\gamma^2)^2 J_0(ha)J_1(ka) - 4\gamma^2 hkJ_1(ha)J_0(ka). \tag{8.4.65}$$

The function $\Phi(\gamma, \omega) = 0$ is the Pochhammer equation. The particular form of the solution (8.4.63) was chosen so that the results could be compared directly with experimentally measured parameters.

The inverse sine transformation with respect to the wavenumber γ is first carried out. Since E^{SF} is an odd function in γ, the inverse transform is given by

$$E^{\mathrm{F}} = -\frac{\mathrm{i}}{\pi}\int_{-\infty}^{\infty} E^{\mathrm{SF}} \mathrm{e}^{\,\mathrm{i}\gamma z}\,\mathrm{d}\gamma. \tag{8.4.66}$$

This integral is evaluated by the Cauchy residue theorem by using a contour along the real γ-axis and a semicircle in the upper half-plane. The latter contribution vanishes for a circle radius approaching infinity, so that the integral (8.4.66) is given by $2\pi\mathrm{i}\sum$ Res of the poles enclosed by the contour.

It is at this stage that Folk *et al.* developed the arguments for including or excluding poles from the contour that directly relates to the interpretation of the portions of the branches of the frequency spectrum for rods (and plates) that indicate negative group velocity.† Thus, the position of a pole in the γ-plane depends on ω and is given by $\gamma_q = \gamma_q(\omega)$. This symbolic relationship represents nothing more than the branches of the Pochhammer–Chree

† For example, the region to the left of the local minimum or the second symmetrical branch of Fig. 8.9 and Fig. 8.10 for the plate case.

frequency spectrum, as shown in Fig. 8.17. Thus, for a given value ω_0 (or Ω_0 for the figure), an $\omega = \omega_0$ plane would be pierced at an infinite number of points by the branches of the spectrum, each representing a pole location for $\omega = \omega_0$. Many of the poles would be along the real and imaginary γ-axes, but some would be complex, owing to the complex branches of the spectrum.

Consider the poles on the real γ-axis. The question is: which poles should be included and which poles should be excluded from the contour. The answer to this is given by anticipating the inverse Fourier transform on the frequency ω. The path used in that transform will not be along the real ω-axis, but along an axis arbitrarily close to and parallel with the real axis, but still slightly above it. Then the ω used in calculating the poles of γ should have a small but positive imaginary component $\mathrm{i}\alpha$. Expanding $\gamma_q(\omega)$ about a frequency ω_0 gives

$$\gamma_q(\omega) = \gamma_q(\omega_0) + \frac{\mathrm{d}\gamma_q}{\mathrm{d}\omega}\Delta\omega + \dots \tag{8.4.67}$$

Substitution of $\mathrm{i}\alpha$ for $\Delta\omega$ shows that γ_q will have a small but non-zero imaginary component whose sign is positive if $\mathrm{d}\gamma_q/\mathrm{d}\omega$ is positive. Hence only those poles on the real axis are included which meet this criterion. With this, the inverse sine transformed result is given by

$$E^{\mathrm{F}} = \frac{\mathrm{i}P_0(1-\nu)}{E} \sum_q F(r, \gamma_q, \omega)\exp(\mathrm{i}\gamma_q z), \tag{8.4.68}$$

where

$$F(r, \gamma_q, \omega) = \left(\frac{4\nu(1+\nu)}{(1-\nu)^2}\frac{\gamma}{\omega h}\left[\frac{(k^2-\gamma^2)J_1(ka)}{h}\left\{\frac{hJ_1(hr)}{r} + \gamma^2 J_0(hr)\right\} + \right.\right.$$
$$\left.\left. + 2\gamma^2 J_1(ha)\left\{\frac{J_1(kr)}{r} - kJ_0(kr)\right\}\right]\frac{1}{\partial\Phi/\partial\gamma}\right)_{\gamma=\gamma_q(\omega)}. \tag{8.4.69}$$

The inverse transform of E^{F} then gives

$$E(z, r, t) = \frac{1}{2\pi}\int_{-\infty+\mathrm{i}\alpha}^{\infty+\mathrm{i}\alpha} E^{\mathrm{F}} \mathrm{e}^{-\mathrm{i}\omega t}\,\mathrm{d}\omega$$

$$= -\frac{P_0(1-\nu)}{E}\sum_q \frac{1}{2\pi\mathrm{i}}\int_{-\infty+\mathrm{i}\alpha}^{\infty+\mathrm{i}\alpha} F(r, \gamma_q, \omega)\exp\{\mathrm{i}(\gamma_q z - \omega t)\}\,\mathrm{d}\omega. \tag{8.4.70}$$

Each integral in the result (8.4.70) corresponds to the contribution of an individual mode of the Pochhammer spectrum for the axisymmetric case.

The saddle-point method is used to evaluate the integrals appearing in (8.4.70) for locations far from the end of the rod. Referring back to the development of this method in Chapter 6 (§ 2), we see that the saddle points will be obtained from $df(\omega)/d\omega = 0$, where we write for the exponential in (8.4.70)

$$\exp\{i(\gamma_q z-\omega t)\} = \exp\{zf(\omega)\}, \qquad f(\omega) = i\left(\frac{\gamma_q}{z}-\frac{\omega}{z}t\right). \tag{8.4.71}$$

Expansion about the saddle point $\bar{\omega}$ gives

$$F(r, \gamma_q, \omega) \cong F(r, \bar{\gamma}_q, \bar{\omega}),$$

$$\gamma_q z-\omega t \cong \gamma_q z-\bar{\omega}t+\frac{z}{2}\frac{d^2\gamma_q}{d\omega^2}(\omega-\bar{\omega})^2, \tag{8.4.72}$$

which holds as long as no poles or zeros are located near $\bar{\omega}$. The results are given by

$$E(z, r, t) = -\frac{P_0(1-\nu)}{E}\left\{1+\sum_q A_q \sin\left(\bar{\gamma}_q z-\bar{\omega}t\pm\frac{\pi}{4}\right)\right\}, \tag{8.4.73}$$

where

$$A_q = 2F(r, \bar{\gamma}_q, \bar{\omega})\left(2\pi z\left|\frac{d^2\gamma_q}{d\omega^2}\right|\right)^{-\frac{1}{2}}. \tag{8.4.74}$$

This expression predicts that near a time $t = z/c_g$, there should appear oscillations having a period $2\pi/\bar{\omega}$, a wavelength $2\pi/\bar{\gamma}_q$ and an amplitude A_q.

The approximation used in obtaining the results (8.4.73) breaks down when $d^2\gamma_q/d\omega^2 \to 0$, or when there are poles and zeros of $F(r, \gamma_q, \omega)$ near $\bar{\omega}$. Furthermore, for the first mode, F has a pole coinciding with the saddle point at $\bar{\omega} = 0$ at the maximum group velocity. It becomes necessary to use a modification of the usual saddle-point method, where a third-order term is included. This technique is applied to the evaluation of the first-mode contribution near maximum group velocity. This gives the head of the pulse behaviour as

$$E(r, z, t) \cong \frac{P_0(1-\nu)}{F}\left\{\frac{1}{3}+\int_0^B (\mathrm{Ai})(-B)\, dB\right\}, \tag{8.4.75}$$

where

$$B = \left(t-\frac{z}{c_0}\right)\left(\frac{4c_0^3}{3\nu^2 a^2 z}\right)^{\frac{1}{3}}, \tag{8.4.76}$$

and where $(\mathrm{Ai})(x)$ is the Airy integral. Now, at $t = z/c_0$, the integral of (8.4.75) is zero, so that the pulse has reached $\frac{1}{3}$ of its final value. The behaviour of the pulse is oscillatory and is the same as shown in Fig. 8.39 for Skalak's results. A 27 per cent overshoot occurs on the first oscillation.

The case of a step pressure pulse applied on a semicircular area of the end of the rod has been analysed by DeVault and Curtis [12]. Such a loading produces flexural as well as longitudinal disturbances. While the analysis is more complicated than for the uniform pressure pulse, the analysis follows generally the same steps just outlined.

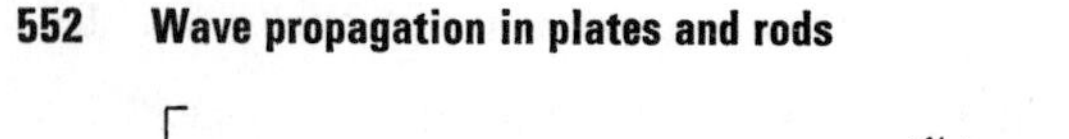

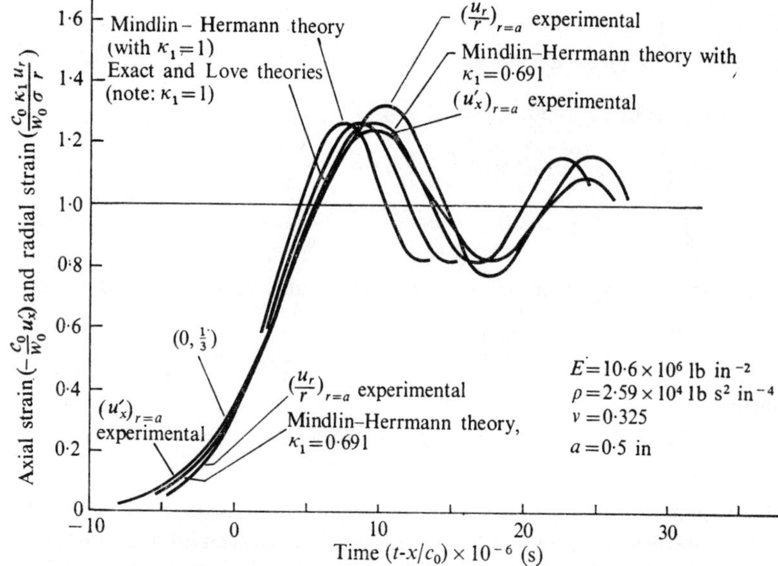

FIG. 8.40. Head-of-the-pulse behaviour as predicted by Mindlin–Herrmann, Love, and exact theories. (After Miklowitz [53, Fig. 4].)

3. *Other results for compressional waves in a rod.* A number of other results have been obtained for compressional waves in a rod. In several cases, approximate equations of motion for the rod have been used. Miklowitz [51] used the Mindlin–Herrmann theory for the rod and considered the case of a step pressure pulse loading and obtained a formal solution. The integral solution was numerically evaluated by Miklowitz [52] for the near-field response for various Poisson's ratio. Anomalous stress discontinuities were noted for arrival times given by x/c_1, corresponding to major wave contributions travelling at the dilatational velocity.† In a later contribution, Miklowitz [53] used the stationary-phase method, also partially covered in Reference [52], in conjunction with Mindlin–Herrmann theory to obtain the far-field response. Results were obtained for two values of the correction factor κ_1 appearing in the Mindlin–Herrmann theory and are shown in Fig. 8.40. Also shown are the results from the Skalak analysis using exact theory and the Davies analysis using Love's theory. Experimental results are also shown.

Kaul and McCoy [33] applied the approximate equations for the rod developed by Mindlin and McNiven [65] to the case of a step pressure pulse

† This aspect was previously commented on in the analysis of longitudinal waves in membrane shells in § 4.3.

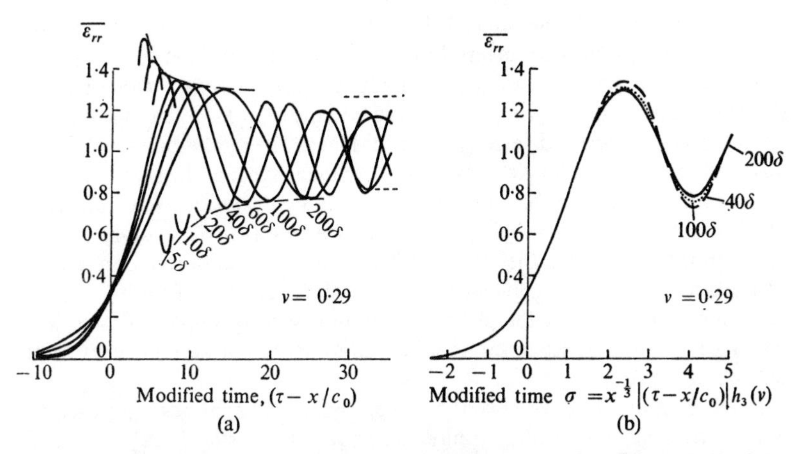

FIG. 8.41. Radial strain $\bar{\varepsilon}_{rr}$ as a function of the modified time (a) $\tau - x/c_0$ and (b) of the modified time σ. (After Kaul and McCoy [33, Figs. 4, 5].)

applied to a semi-infinite rod. Integral-transform methods were used and the steepest-descent or saddle-point method applied to obtain the far-field, head-of-the-pulse response. The solution is in terms of the Airy function. Results are shown in Fig. 8.41 for the radial strain as a function of time, where

$$\varepsilon_{rr} = \frac{u(x,\tau)}{a}, \qquad \bar{\varepsilon}_{rr} = \frac{k_1 E}{\nu P_0}\varepsilon_{rr}, \tag{8.4.77}$$

and $u(x, t)$ is the displacement function associated with the kinematical assumptions of the theory. Thus

$$\begin{aligned} u_x(r,x,\tau) &= w(x,\tau)+(1-2r^2/a^2)\psi(x,\tau), \\ u_r(r,x,\tau) &= (r/a)u(x,\tau). \end{aligned} \tag{8.4.78}$$

The parameters x and τ are non-dimensional distance along the rod and time, defined by

$$x = \delta(z/a), \quad \tau = (\delta t/a)c_2, \quad J_1(\delta) = 0, \quad \delta = 3{\cdot}8317. \tag{8.4.79}$$

The parameter k_1 in (8.4.77) is an adjustment coefficient of the approximate theory. For $\nu = 0{\cdot}29$, $k_1^2 = 0{\cdot}7739$. The amplitude of the applied stress is $P_0/2$. In Fig. 8.41 (a), the evolution of the head of the pulse at various distances from the end is shown. In Fig. 8.41 (b), the modified time is

$$\sigma = \frac{\{(\tau - x/c_0)\}h_3(\nu)}{x^{\frac{1}{3}}}, \qquad h_3(\nu) = c_0\left(\frac{2k_1}{\sqrt{3\nu\delta k_3}}\right)^{\frac{2}{3}}, \qquad k_3^2 = 0{\cdot}9754, \tag{8.4.80}$$

which is similar to the base used to present the data of Skalak. As pointed out by Kaul and McCoy and earlier by Fox and Curtis [18], it becomes evident from the representation of Fig. 8.41 (b) that for different values of x, the initial portion of the curve is the same. This leads to the prediction that the time of the initial rise should vary inversely as the cube of the distance of travel.

Returning to results obtained by analysis using the exact equations of elasticity, Jones and Norwood [32] considered the semi-infinite rod subjected to both pressure step and velocity transient loads. The saddle-point technique was used to obtain the far-field response. A unique aspect of the work was that cross-section warping effects on the first-mode contribution were retained. The basic Airy integral solution obtained by previous investigations is contained in the result, with the warping effects adding second-order corrections that are of decreasing significance for increasing distances of travel. A comparison of the results of the pressure step and velocity impact showed less than 1 per cent difference in the solutions at a distance of 20 diameters from the end of the rod.

In all of the analyses reviewed thus far, asymptotic methods were used to obtain far-field, head-of-the-pulse information. Bertholf [5] obtained numerical solutions to the exact elasticity equations for the case of axisymmetric waves in semi-infinite and finite rods. The case of a step pulse was considered, with the case of free-end and rigid-lubricated conditions existing at the opposite end being considered. Results are shown in Fig. 8.42 for the axial strain in a one-diameter long cylindrical bar at various times for the free-end case.

To interpret the data of the figure, first note that the abscissa coordinate z/d is distance along the rod, where d is the rod diameter and z is the axial

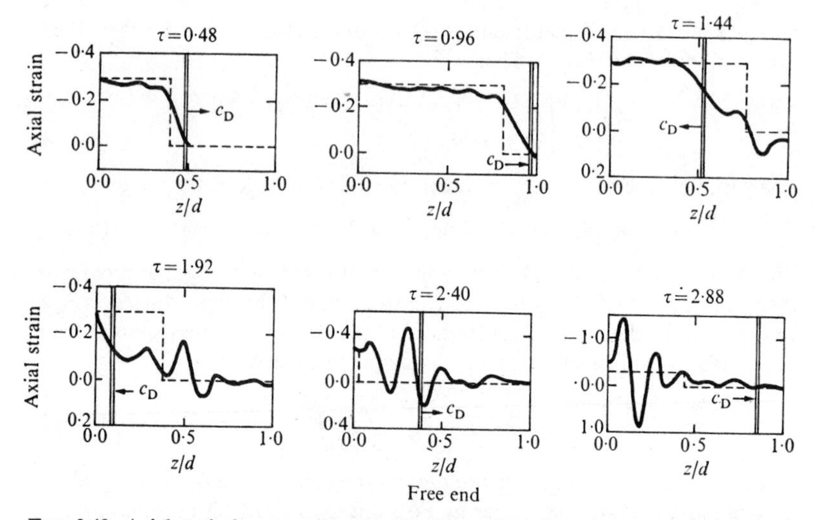

FIG. 8.42. Axial strain in a one-diameter long bar subjected to a step pressure pulse for various values of non-dimensional time τ. (After Bertholf [5, Figs. 12, 13].)

coordinate. The loaded end is at $z/d = 0$. The non-dimensional time τ is given by

$$\tau = tc_1/d. \tag{8.4.81}$$

The results were computed for a specific rod material (24S-T aluminum) for a specific impact stress. For our considerations, it is sufficient to consider the axial-strain ordinate as some relative value. Finally, the dashed line in the figures represents the wave system that would exist according to simple, one-dimensional rod theory. The vertical line $c_D (c_D = c_1)$ represents the position of a dilatational wavefront initiated at the instant of impact. As would be expected, the strain wave behaviour is considerably more intricate than the predictions of simple theory. Other results are given by Bertholf for the case of a two-diameter length rod, and the pulse behaviour in a semi-infinite rod.

Returning to the case of the semi-infinite rod, Kennedy and Jones [35] presented results having several features of interest. In all analyses reviewed thus far, the velocity or pressure step pulses applied to the end of the rod have been assumed to be uniformly distributed over the end. Kennedy and Jones considered the case of radial variation in the applied step pressure pulse, as shown in Fig. 8.43. The stress applied to the end of the rod is prescribed by

$$\tau_{zz}(0, r, t) = P(r)H\langle t\rangle, \tag{8.4.82}$$

where

$$P(r) = P_0(p+1)\{1-(r/a)^2\}^p, \quad p \geq 0. \tag{8.4.83}$$

Thus, the parameter p defines the nature of the radial, axisymmetric stress distribution. The value of $p = 0$ gives the case of a uniform pressure distribution considered in all previous studies. The second feature of interest in the

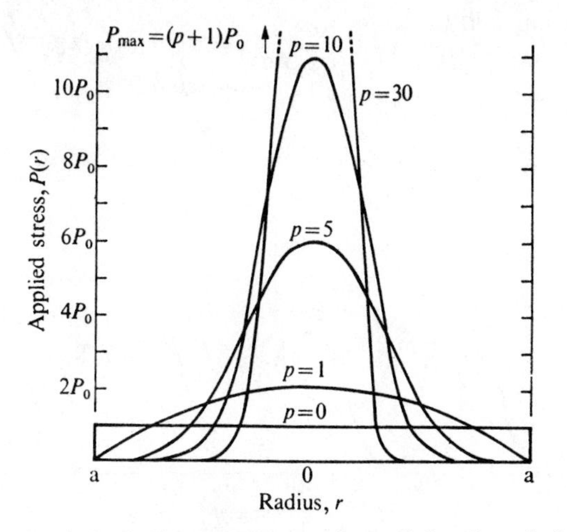

FIG. 8.43. Radial variation in the step pressure pulse applied to the end of a semi-infinite rod. (After Kennedy and Jones [35, Fig. 1].)

Kennedy–Jones analysis is that both near-field and far-field results are obtained. The numerical techniques developed by Bertholf [5] are applied to the near-field analysis, while the saddle-point asymptotic method is used for the far-field analysis.

Results were obtained for the specific impact situation of a steel bar of radius $a = 1{\cdot}27$ cm subjected to an end loading for which $P_0 = 1{\cdot}0$ kbar (1 bar $\simeq$ 14·4 lb in^{-2}). The behaviour of the strain $E = \varepsilon_{zz} + \varepsilon_{\theta\theta}$ at the surface of the bar is shown in Fig. 8.44. It is sufficient here to interpret the strain ordinate scale arbitrarily. The abscissa z/a corresponds to various positions along the rod. The results are for three values of the loading parameter $p = 0$, 10, 30. The solid lines in the plots are the results from numerical analysis, the dashed lines are the results from asymptotic analysis. A first observation to be made for the far field, $z/a = 40$, is that numerical and analytical results are in fair agreement and, furthermore, that there is very

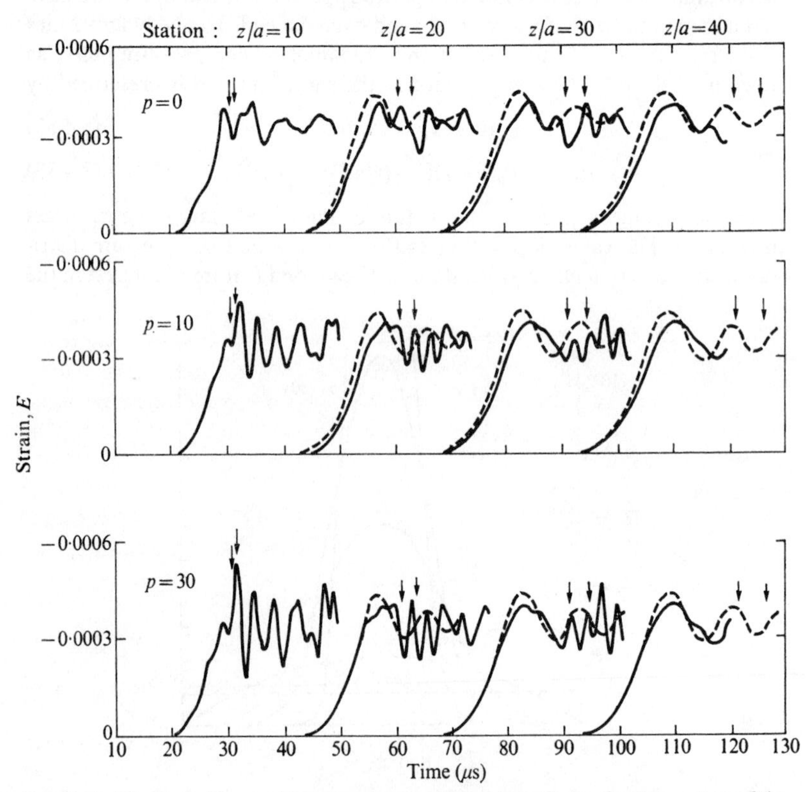

FIG. 8.44. The strain $E = \varepsilon_{zz} + \varepsilon_{\theta\theta}$ at various locations along the bar for three values of the loading parameter p. (After Kennedy and Jones [35, Fig. 4].)

little difference in the response for $p = 0, 10, 30$. This last merely indicates a type of dynamic St. Venant's principle, where the response far from the load is insensitive to variations in the load distribution. Now consider the very near-field response ($z/a = 10$) and the differences between the $p = 0, 10, 30$ behaviour. For the more severely concentrated load ($p = 30$), it is to be expected that higher rod modes will be more strongly excited than for a uniform distribution ($p = 0$). This is indicated by the greater presence of high-frequency signal for $p = 30$ compared to $p = 0$, with the case of $p = 10$ being intermediate. The small vertical arrows in the figure indicate the arrival time of second- and third-mode contributions. Numerous other results were given by Kennedy and Jones, including the behaviour of E along the axis, along a radius of $r/a = 0{\cdot}5$ and the stress behaviour.

4. *Transient waves in plates.* A number of analyses have appeared on transient waves in plates as governed by the exact equations of elasticity or higher-order plate theories. In the main, methods of analysis are quite similar to those employed in analysis of rod problems. Thus, integral-transform techniques are employed and far-field results obtained by asymptotic methods. Because of the many similarities of the techniques reviewed in the last few sections, the coverage here of this rather large subject will be rather brief.

The case of a semi-infinite rectangular bar subjected to a longitudinal step pulse has been considered by Jones and Ellis [30, 31]. The situation is the plate analogue of the semi-infinite rod problem extensively discussed in the last few sections. The bar is defined by $y = \pm a, z = \pm b, x > 0$ where $b \ll a$. Loading occurs at the $x = 0$ face. Plane-stress conditions were assumed to prevail in the z direction, with the exact elasticity equations being applied to the x, y coordinates. The case of plane-strain conditions in the z direction is easily obtained, of course, by changing the elastic constants. The method of analysis closely follows that first used by Folk *et al.* [17], except that higher-order cross-section warping effects are retained in the far-field evaluation. This aspect has been previously mentioned in conjunction with the rod analysis of Jones and Norwood [32] and derived from the work of Jones and Ellis. In addition, second-mode contributions are included in the results. The results obtained by Jones and Ellis for a specific impact situation at a location along the axis of the bar is shown in Fig. 8.45. The ordinate is the strain

$$E(\varepsilon_{xx}+\varepsilon_{yy})/P_0,$$

where E is Young's modulus and P_0 is the applied pressure. The influence of the various modes and correction terms is shown by the various curves.

Several contributions on compressional and flexural waves in plates have been made by Miklowitz and his coworkers. Thus Miklowitz [56] considered an elastic plate subjected to a pair of concentrated loads, as shown in

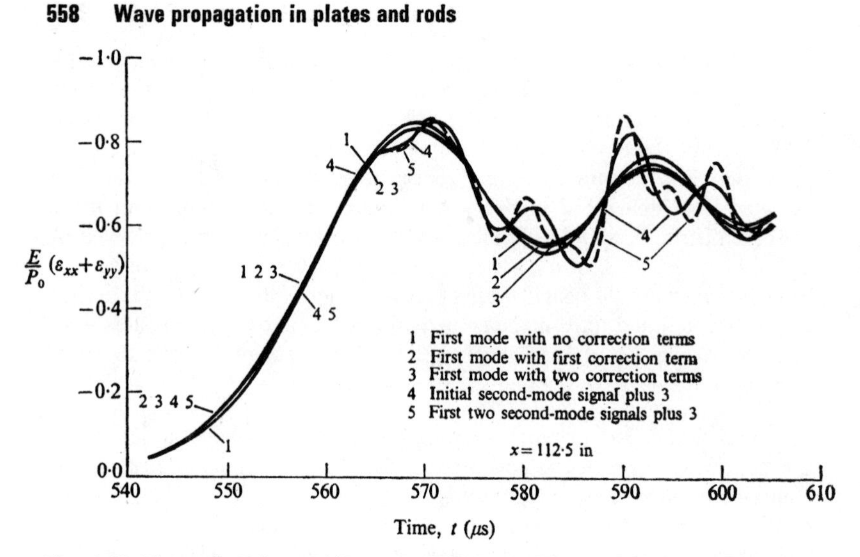

FIG. 8.45. The strain $E(\varepsilon_{xx}+\varepsilon_{yy})/P_0$ versus time at a specific location along the axis of a rectangular bar subjected to a step pressure pulse. (After Jones and Ellis [31, Fig. 9].)

Fig. 8.46(a). The symmetry of the loading with respect to the mid surface of the plate enables the problem to be reduced to the situation shown in Fig. 8.46(b) of a plate on a rigid, lubricated half-space. The method of stationary phase is used to obtain the far-field horizontal and vertical displacements. The results are shown in Fig. 8.47 for the former component, where

$$\bar{u}_\rho = \frac{\pi\mu}{2P_0} u_\rho(\rho, \xi, \tau) \times 10^{-3}, \tag{8.4.84}$$

and where ρ, ξ, τ are dimensionless radial, axial, and time coordinates, given by

$$\rho = r/H, \quad \xi = z/H, \quad \tau = c_2 t/H, \tag{8.4.85}$$

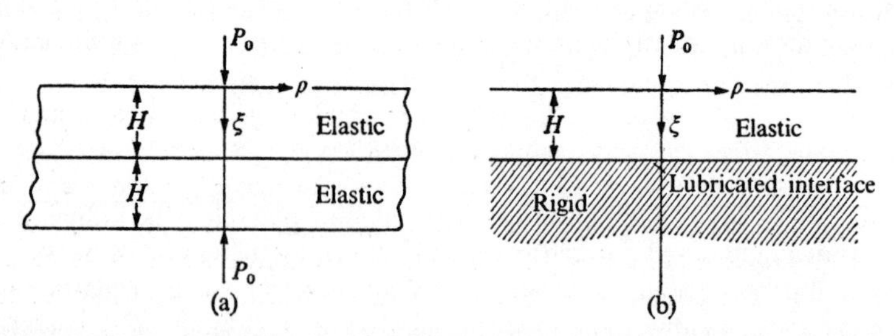

FIG. 8.46. (a) An elastic plate of thickness $2H$ subjected to a pair of concentrated loads and (b) the equivalent case of a plate on a rigid half-space. (After Miklowitz [56, Figs. 1, 2].)

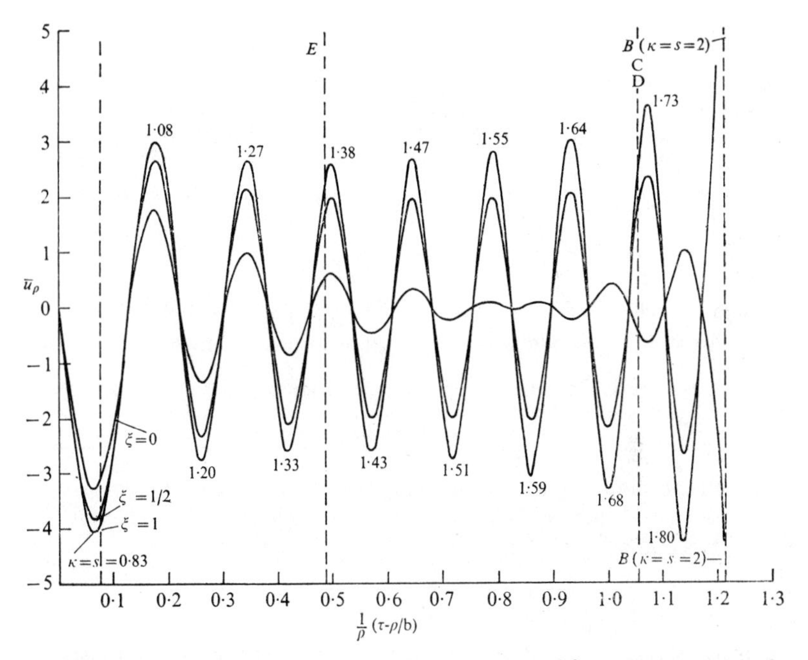

FIG. 8.47. Horizontal displacement $\bar{u}_\rho$ at the station $\rho = 20$ and $\xi = 0, \frac{1}{2}, 1$ as a result of a step load applied to the plate. (After Miklowitz [56, Fig. 4].)

and b is the velocity ratio

$$b = c_p/c_2 = \{E/\rho(1-\nu^2)\}^{\frac{1}{2}}/c_2. \tag{8.4.86}$$

The displacements are shown for three values of depth, $\xi = 0, \frac{1}{2}, 1$ at the station $\rho = 20$. Only the first plate-mode contribution is included in the analysis. The highly oscillatory nature of the response is to be noted. Also, the apparent increasing amplitude with the passage of time is to be noted. However, considerations related to arrival time of other mode contributions suggests that only the region from zero to the dashed line at E is sufficiently accurate.

Miklowitz [54] gave the solution to the case of a hole punched in a stretched elastic plate. However, only the plane-stress plate equations were used. In a later work, Scott and Miklowitz [82] considered the case of a radial step displacement applied to the wall of a circular, cylindrical cavity. The case of a radial step pressure pulse was considered in a later contribution by the same authors [83]. The Laplace and extended Hankel transforms were applied to the exact elasticity equations, and the method of stationary phase used to obtain the far-field response. The time behaviour of the radial and vertical

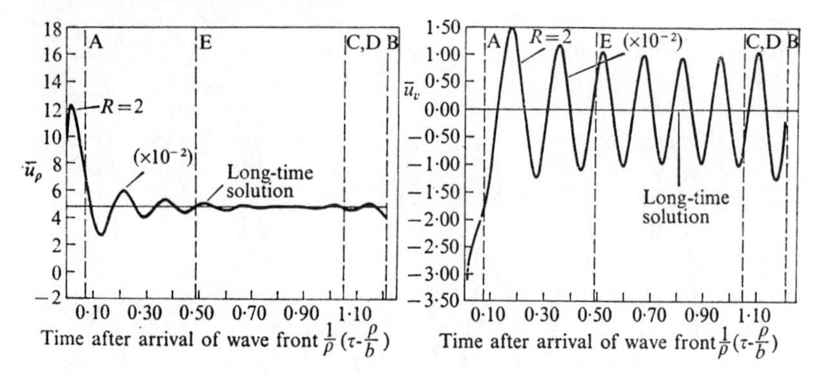

FIG. 8.48. Horizontal and vertical displacement components $\bar{u}_\rho$, $\bar{u}_v$ in a plate containing a circular hole subjected to a step radial displacement for $R = 2{\cdot}0$ at a station $\rho = 20$, $\xi = 1$. (After Scott and Miklowitz [82, Fig. 2].)

displacement components is shown in Fig. 8.48 for the step displacement case for a particular ratio of hole radius to plate half-thickness. The various parameters appearing in the figure are defined as follows:

$$\bar{u}_\rho = \frac{\pi(1-2\nu)}{8\nu u_0} u_\rho, \qquad \bar{u}_v = \frac{\pi(1-2\nu)}{8\nu u_0} u_v, \qquad R = \frac{a}{H}, \tag{8.4.87}$$

where $u_\rho = u_\rho(\rho, \tau, \xi)$, $u_v = u_v(\rho, \tau, \xi)$ are the radial and vertical displacements in terms of the dimensionless variables defined by (8.4.87). The time after arrival of the wavefront $(\tau-\rho/b)/\rho$ is the same as for Fig. 8.47. The observation point is at $\rho = 20$ and $\xi = 1$, corresponding to a point on the surface of the plate.

As in the case of the work in Reference [56], it is not possible to obtain head-of-the-pulse data from the approximate solution. This behaviour is associated with the group-velocity minimum. The difficulty is that the integrals which arise in extending the stationary phase to the point of a group-velocity minimum have not been computed. This is in contrast to the case of the rod and plate subjected to longitudinal impact, where the Airy integral results. Other considerations put forth by Scott and Miklowitz, based on period-arrival time considerations of the second and third modes and on the higher-frequency components of the first mode suggest that region 0–E is the strongest part of the results.

The analysis of flexural waves in a plate governed by the Mindlin plate equations was put forth by Miklowitz [55]. The case of a concentrated step-function load was considered using Laplace transform theory. The contours in the complex plane involved in the inversion are quite similar to those shown in Chapter 3 (Fig. 3.17) for the Timoshenko beam analysis. Results were presented for the moment and shear responses $M_r(r, t)$, $M_\theta(r, t)$, and $Q_r(r, t)$

for step and rectangular pulse function input at various stations. Scott and Miklowitz [84] have presented formal solutions, using Laplace and finite Fourier transforms of a general class of non-axisymmetric, transient plate wave-propagation problems. A particular case of a normal, half-ring load has been evaluated. Finally, Miklowitz [57] has presented a very extensive review of transient wave analysis in plates. A discussion of various methods of solution is given.

8.5. Experimental studies on waves in rods and plates

The main objective of this section will be to present experimental results illustrating the dispersive characteristics of rods and plates. This will include results illustrating multiple reflection of pulses and dispersion of sharp pulses as well as some studies on mode coupling.

8.5.1. *Multiple reflections within a waveguide*

It is known, of course, that dispersion is caused by the multiple reflections and mode conversions within a waveguide. It is of interest to present some of the results explicitly showing the early stages of this process. In a study by Kolsky [36], a sharp pulse was initiated on one side of a short, cylindrical slab of steel and a wave detector placed on the opposite side, as shown in Fig. 8.49(a). The initial portion of the wave system sensed by the detector is shown in Fig. 8.49(b). It is possible to relate the general character of the disturbance to waves travelling specific paths. Thus, in Fig. 8.49(b), disturbance 1 arrives by the direct path 1 and travels at the dilatational wave speed. The next signal arrival (2) arrives by path 2 and is a P–P wave. Signal 3 arriving by path 3 is a P–S wave. Signals 4, 5, and 6 are, respectively P–P–P, S–S, and P–P–S waves. The increased number of reflections makes the signal increasingly difficult to decipher after that.

A portion of a study by Evans, Hadley, Eisler, and Silverman [13] also provides insight into the early stages of wave development. A diagram of the experimental arrangement used for making the measurements is shown in Fig. 8.50(a). The model was a 2·15 in. (5·46 cm) thick specimen of plastic. The nature of the pulses received at six locations on the surface (each 1 cm apart) are shown in Fig. 8.50(b). The first signal that arrives at the detector is L, the directly travelling longitudinal wave. The next arrival sensed by the detector used by Evans *et al.* is called S and is the surface wave. The next arrival is the strong signal R_1 caused by a P–P wave reflected from the bottom of the plate. For the more distant receiver locations this signal actually precedes the surface wave arrival. The next signal C_1 results from both P–S and S–P waves reflected from the bottom of the plate. The signal peak R_2 results from a P–P–P–P wave system while C_2 results from a combination of P–P–P–S, P–P–S–P, P–S–P–P, and S–P–P–P wave systems.

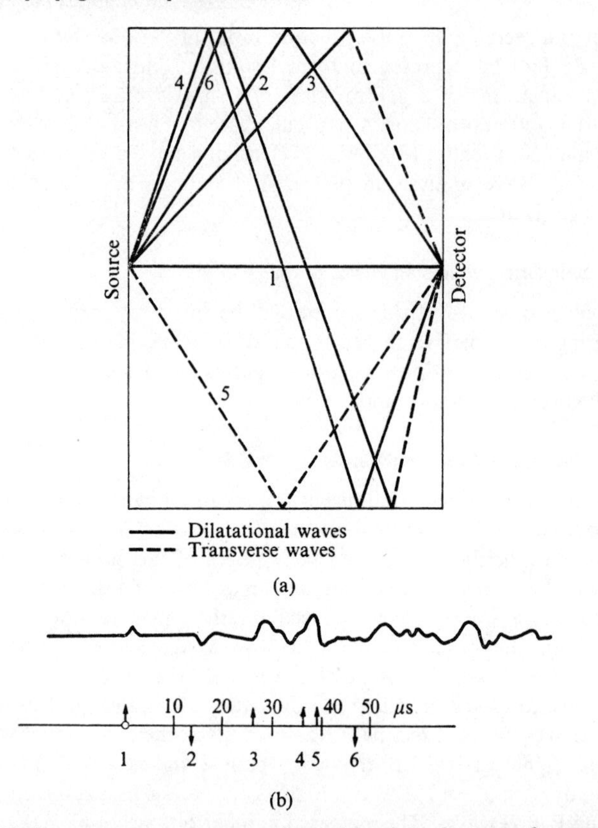

FIG. 8.49. (a) Source–detector arrangement for studying the early stages of wave reflection in a slab, and (b) the resulting signal at the detector. (After Kolsky [36].)

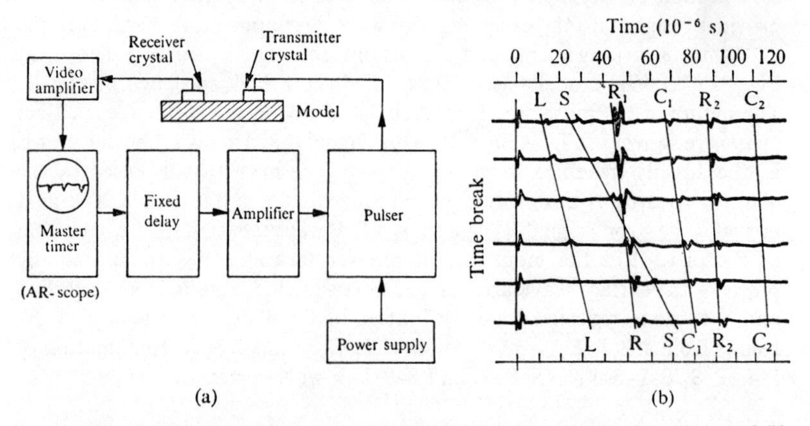

FIG. 8.50. (a) Experimental arrangement for observing pulse reflections in a plate, and (b) pulse forms observed at six stations. (After Evans *et al.* [13, Figs. 1, 3].)

8.5.2. *Dispersion of a sharp pulse in a cylindrical rod*

Experiments were conducted by Oliver [73] on the dispersion of sharp pulses in an elastic rod. An electrically pulsed piezoelectric transducer was used to induce a short duration (about 20 μs) pulse into a 19 ft, (5.8 m), 1 in. (2·54 cm) diameter steel rod. A second transducer, placed at various locations along the rod, including the end, was used to detect the propagating waves. Figure 8.51(a) is a block diagram of the experimental apparatus.

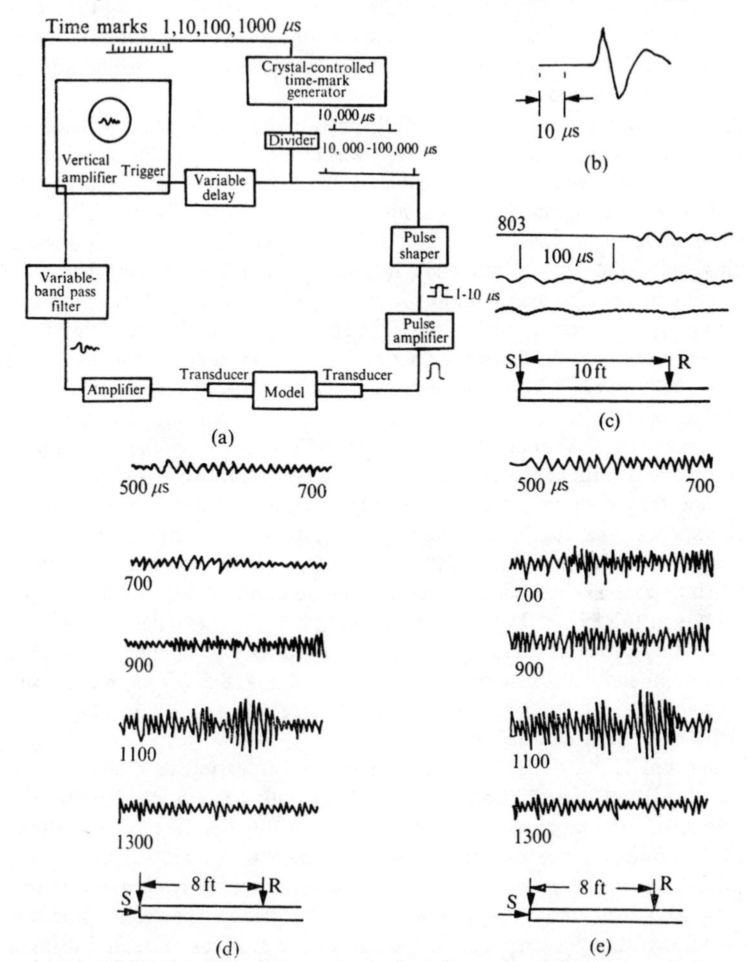

FIG. 8.51. (a) Block diagram of apparatus used in wide-band pulse studies on waves in rods; (b) acoustic pulse applied to the rod; (c) flexural disturbance induced by transversely pulsing the rod; (d) dispersed first longitudinal mode in a rod; (e) first and second longitudinal modes. (After Oliver [73, Figs. 2, 3, 5, 7, 9].)

Shown in Fig. 8.51(b) is the waveform of the acoustic pulse induced into the rod. By changing the orientation of the driving transducer, it was possible to induce antisymmetric or symmetric waves into the rod. For example, Fig. 8.51(c) shows the detected waveform 10 ft. (3·05 m) from the source when the pulse was imparted transversely to the rod, and represents a flexural disturbance similar in its early stages to those presented in § 3.6 on flexural waves. By longitudinally pulsing the rod, symmetrical waves could be induced. Two examples of the response are shown in Fig. 8.51(d) and (e). In the case of (d), the band pass of the filter was set to optimally detect the first longitudinal mode while in (e) the high-frequency cutoff was raised to include the second symmetric mode.

A very qualitative discussion on the dispersion of a sharp pulse, considering only the first mode, was given in § 8.4. It was pointed out that the initial signal would be of low frequency, with frequency increasing and an abrupt arrival of high-frequency components. Referring to Fig. 8.51(d) and following the discussion of Oliver, it is seen that at about 490 μs the train of long-period waves begins, with the signal frequency increasing to about 760 μs. The low-frequency waves arriving at about 750 μs are pointed out by Oliver as due to some antisymmetrical contributions. The short-period waves, resulting from the lower branch of the first-mode period-arrival time curve of Fig. 8.36 (and shown qualitatively in Fig. 8.37), arrive at about 760 μs. From this time on the disturbance consists of two parts, one a wave of increasing period, one of decreasing period. At about 900 μs, one period is about twice the other and at 1000 μs one is $\frac{3}{2}$ times the other. As time increases the two signals approach the same frequency and a distinct beating action is noted, as from about 1150 μs to 1250 μs. At the time corresponding to a minimum group velocity, the largest contribution occurs (at about 1250 μs). This is usually designated the 'Airy' phase. Theoretically, the first-mode contribution should sharply terminate at this point. The continuing signal noted in the detector output is explained by Oliver as resulting from a continued leakage of energy down the rod from an end resonance mode. In the case of Fig. 8.51(e), both first- and second-mode contributions are present. The second-mode contribution is initially noted at about 580 μs.

Hsieh and Kolsky [28] have considered the dispersion of a sharp pulse (about 2 μs) both theoretically and experimentally. The applied pulse was represented, for analytical purposes, by an error function of the noted duration. A Fourier superposition of 64 harmonics was used to predict the response, where the propagation velocity of each frequency component was determined from the Pochhammer–Chree frequency spectrum. The experimentally obtained response of the end of the bar was measured using a capacitance gauge. The results are shown in Fig. 8.52. The dashed line represents the end displacement that would have resulted from the error function pulse if dispersion were absent. The solid line is the result predicted by the

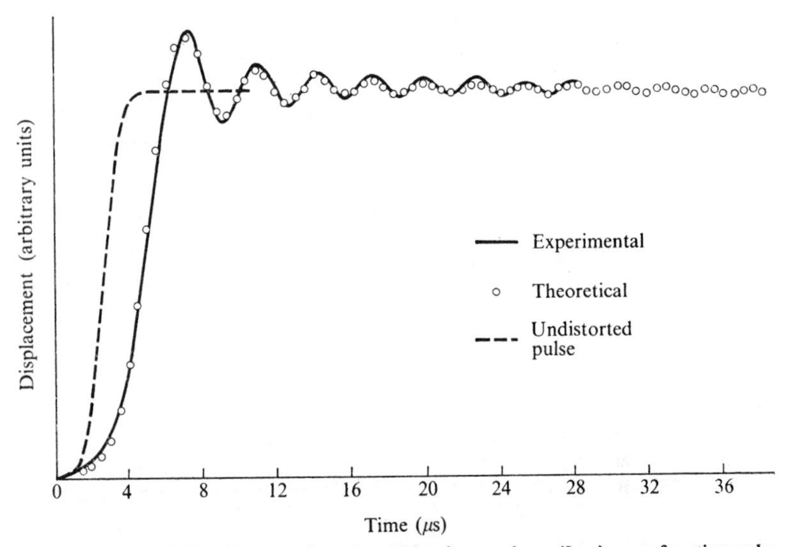

FIG. 8.52. The end displacement of a rod resulting from a sharp (2 μs) error-function pulse as predicted by Fourier superposition, simple theory, and as measured by experiment. (After Hsieh and Kolsky [28].)

Fourier superposition and the small circles give the experimentally measured displacements.

8.5.3. *Experimental results for step pulses*

A number of experimental investigations have been conducted in which a longitudinal step load was applied to the end of a rod or plate and the subsequent propagation measured. Miklowitz and Nisewanger [58] in 1956 used a shock tube to produce a 300 lb in^{-2} ($20{\cdot}7 \times 10^5$ N m^{-2}) step pressure pulse in a 1 in. (2·54 cm) diameter aluminum rod. Condenser microphone and strain-gauge transucers were used to measure the wave propagation at various locations along the rod. The microphone transducer was sensitive to the radial displacement of the rod surface while the strain gauges were oriented to be sensitive to the axial strain. The resulting displacement and strain records are shown in Fig. 8.53(a) and (b). It should be noted that the experimental boundary conditions are given by a step longitudinal stress and zero end shear stress. The mathematical boundary conditions for this problem, it will be recalled, are step longitudinal stress and zero radial displacement.

Intrepreting the records, it is apparent that the response at any station is quite different from the applied step function. For the locations close to the loaded end, it is to be noted that there is a considerable amount of high-frequency activity, particularly in the strain records. This is evidence of the

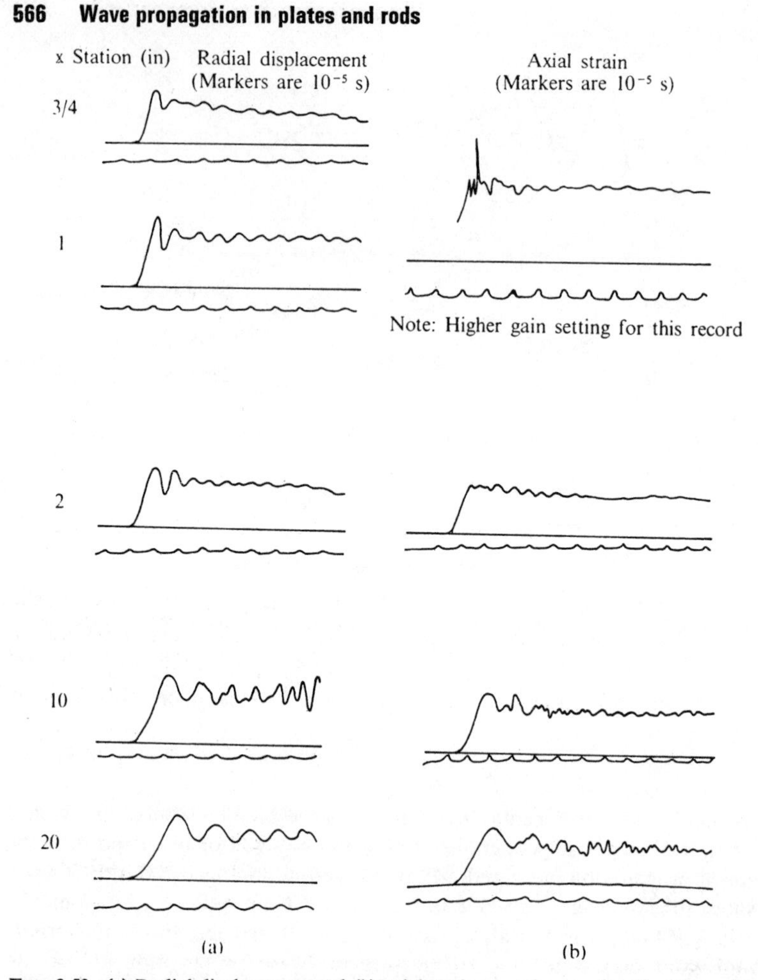

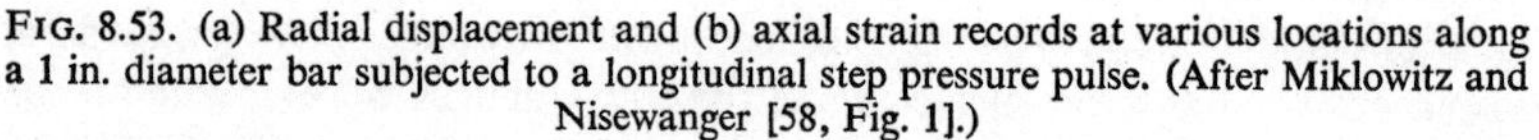

Fig. 8.53. (a) Radial displacement and (b) axial strain records at various locations along a 1 in. diameter bar subjected to a longitudinal step pressure pulse. (After Miklowitz and Nisewanger [58, Fig. 1].)

presence of higher-mode activity in the pulse. Considering the records from the more remote locations ($x = 20$ in. $= 51$ cm, say), it is seen that the head-of-the-pulse consists of the low-frequency, Airy function response predicted by exact theory and the various approximate theories for the rod such as the Love theory or the Mindlin–Herrmann theory.

Fox and Curtis [18, 10] have presented experimental results on longitudinal and bending waves in a rod subjected to step loading. A shock tube apparatus was used to produce the pressure pulse in a 1·5 in. (3·81 cm) diameter

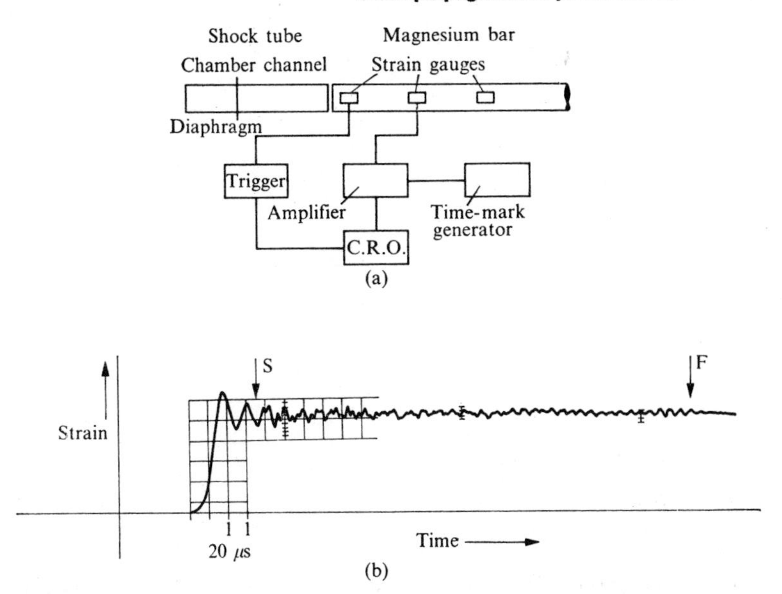

FIG. 8.54. (a) Apparatus used for the production and detection of stress waves, and (b) strain record at a distance of $z = 1{\cdot}51$ m from the end of a rod subjected to a longitudinal step pressure load. (After Fox and Curtis [18, Figs. 1, 2].)

magnesium bar. Piezoelectric strain gauges were used to detect the surface strains at several locations along the bar. The general experimental arrangement is shown in Fig. 8.54(a). The shock tube produced a 45 lb in^{-2} ($3{\cdot}14 \times 10^5$ N m^{-2}) step pulse in the bar that had less than a 1 μs rise time. A typical strain record for a longitudinal step pulse loading is shown in Fig. 8.54(b).

It is seen that the head-of-the-pulse has the Airy function behaviour predicted analytically for the far-field response. After the first one or two oscillations, the pulse becomes quite irregular in shape, as marked by the arrow S in the figure. On the basis of period-arrival time considerations, such as predicted in the Davies analysis of § 8.4, this behaviour is clearly associated with the arrival of second-mode contributions. As time passes, these contributions become of less importance and the signal becomes more regular. As the signal approaches the arrow F, the amplitude actually increases somewhat, and a type of beating action is perceptible. This again corresponds to the group-velocity minimum of the first mode and represents what was identified as the Airy phase in the review of Oliver's work earlier in this section. It should be noted that the signal terminates rather abruptly at this point.

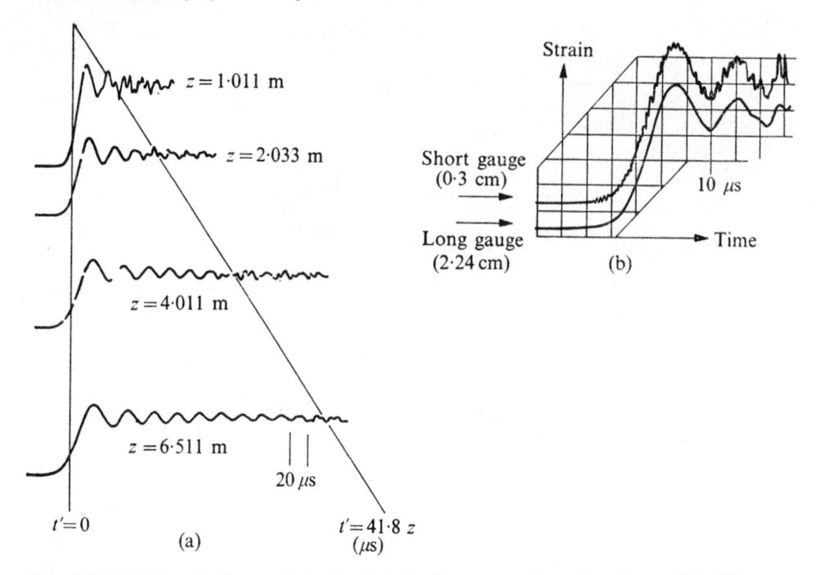

FIG. 8.55. (a) Onset of second mode contributions at various locations; (b) difference in strain signals as detected by short and long strain gauges. (After Curtis [10, Figs. 8, 9].)

Several other measurements of interest are presented in the work of Fox and Curtis. The strain record at various locations along the bar is shown in Fig. 8.55(a), showing the onset of the second-mode vibrations. A particularly interesting result is shown in Fig. 8.55(b) which brings out the influence of gauge size on the strain measurements. Using the short strain gauge, a large amount of high frequency is indicated that does not show on the records from larger gauges at the same location. By constructing frequency–arrival-time charts for the first six bar modes, Curtis showed on a qualitative basis that these higher modes appeared to account for the high-frequency activity.

Finally, mention should be made of the tests reported by Curtis [10] in which both longitudinal and flexural waves were induced in the rod using the shock tube. The technique used was to mask off the end of the rod so that the step pulse acted only on a semicircular end region, yielding a net moment on the rod as well as a net longitudunal force. Figure 8.56(a) shows the combined longitudinal–flexural response for a particular gauge location. The time scale of the record is greatly compressed over that of Fig. 8.54(b), so that about one-third of the longitudinal pulse of Fig. 8.54(b) is compressed into the first 200 μs shown in Fig. 8.56(a). The initiation of the flexural portion of the disturbance is marked by the arrow in the latter figure. The general behaviour of the flexural wave is seen to be quite similar to that previously shown in § 3.6 for the far-field response to impulsive moments. Figure 8.56(b) is a frequency

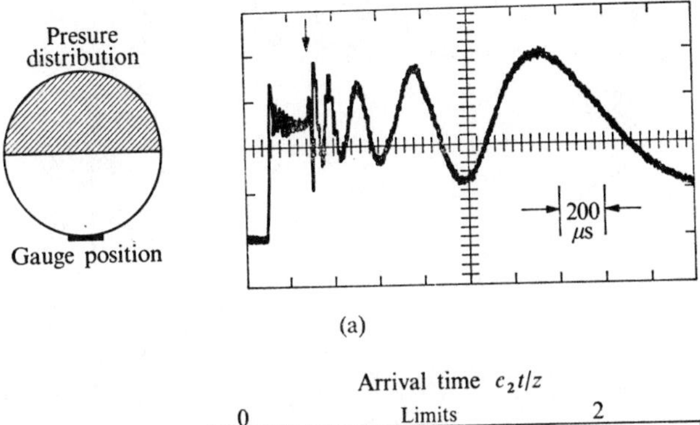

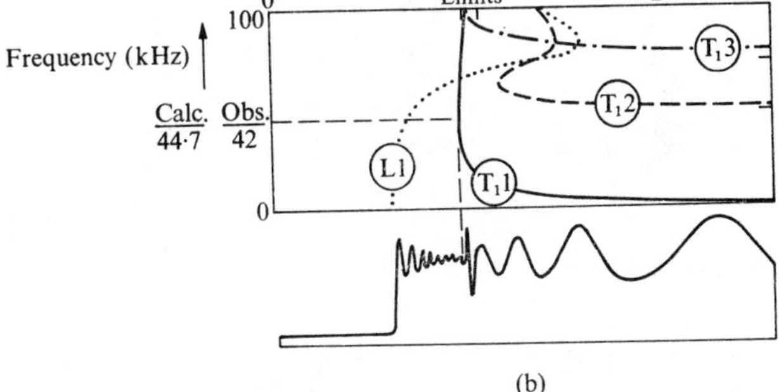

FIG. 8.56. (a) Combined longitudinal–flexural response at a gauge location of $z = 164$ cm from the end subjected to a step pressure pulse over a semicircular region; (b) frequency–arrival-time plot showing the flexural disturbance to be associated with the first transverse mode. (After Curtis [10, Figs. 11, 12].)

arrival-time plot showing the onset of the disturbance is clearly associated with the first antisymmetric, or flexural, mode.

In addition to the work described here, mention should be made of the results by Jones and Ellis [31], who were interested in the measurement of longitudinal waves propagating in a long, rectangular strip. A shock tube was used to step-pulse load the end of the strip quite in the manner shown for the circular rod in Fig. 8.54(a). The strips were of aluminum, 130 in. (3·3 m) in length, had a nominal depth of 1·5 in. (3·81 cm), and were of three different thicknesses, 0·064 in., 0·126 in., 0·252 in. (0·163 cm, 0·320 cm, 0·640 cm). Both condenser-microphone and strain-gauge records were taken at two different locations along the bar. Figure 8.57 shows the response from the microphone and strain-gauge pick-ups at the two locations for the three strip thicknesses. The microphone response at $x = 112·5$ in. $= 2·86$ m has the familiar Airy function form. A considerable amount of high-frequency

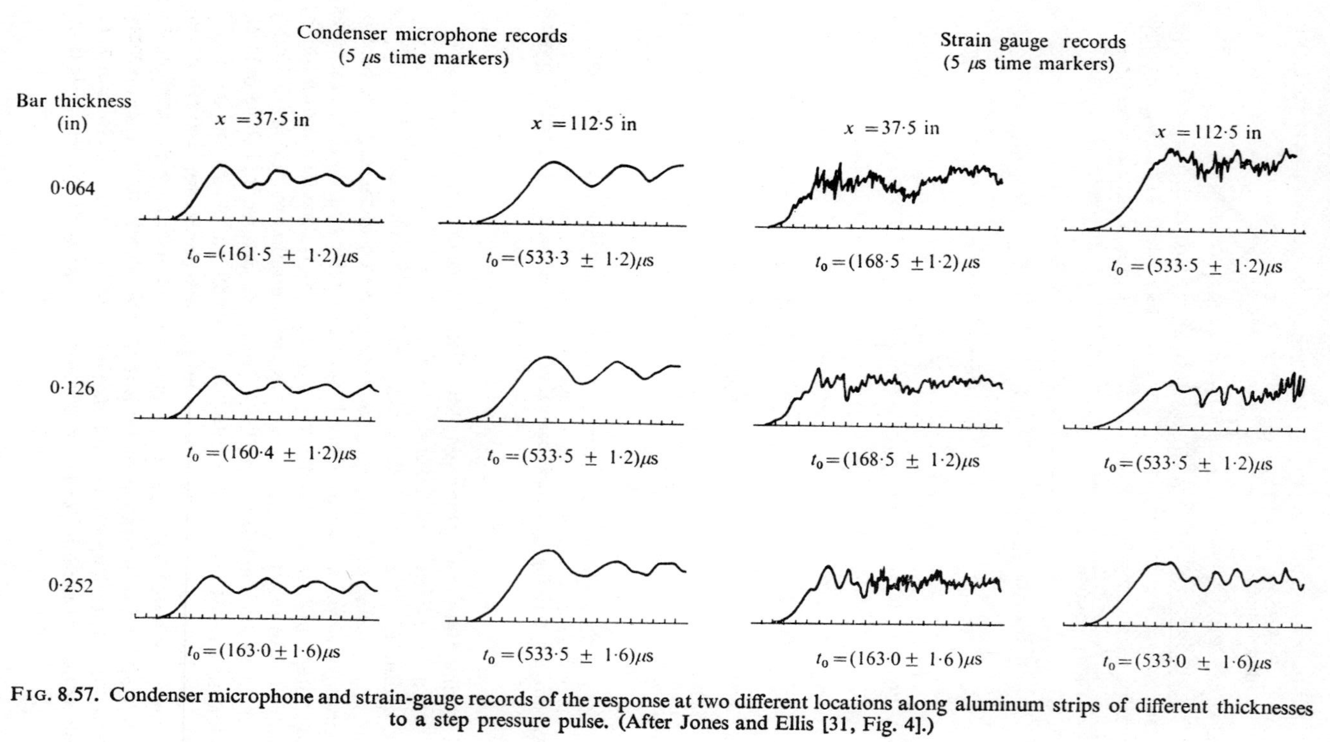

FIG. 8.57. Condenser microphone and strain-gauge records of the response at two different locations along aluminum strips of different thicknesses to a step pressure pulse. (After Jones and Ellis [31, Fig. 4].)

activity is to be noted in the strain-gauge responses, particularly at the closer location ($x = 37{\cdot}5$ in. $= 0{\cdot}95$ m) and for the thinner strips.

8.5.4. *Other studies of waves in cylindrical rods and shells*

Meitzler [49] has reported on the interesting phenomenon of mode coupling that can occur in longitudinal wave propagation in wires at critical frequencies. The basic aspects of the experimental equipment used in the investigation is shown in Fig. 8.58(a). A narrow-band acoustic pulse was applied to the end of

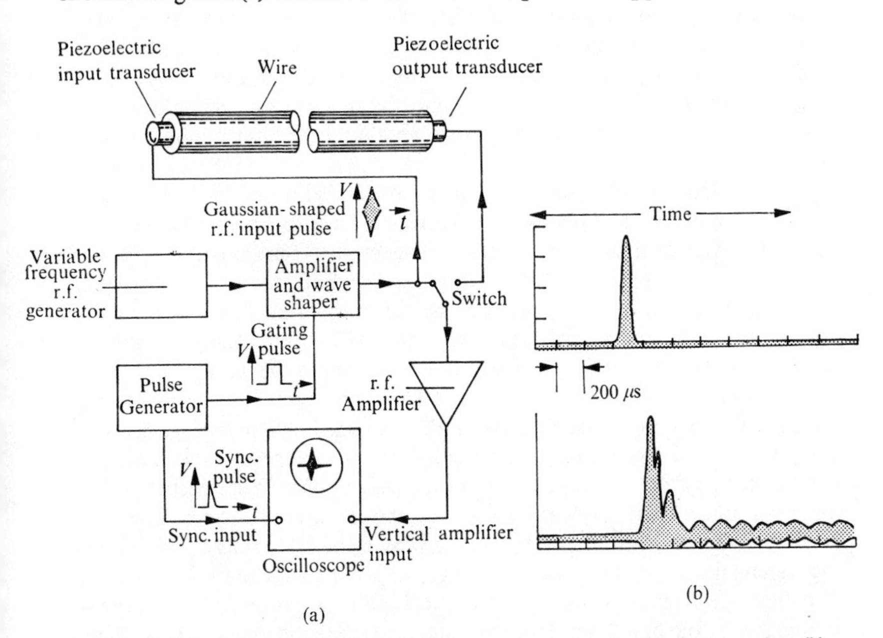

FIG. 8.58. (a) Schematic of apparatus used to apply a narrow-band pulse to a wire; (b) resulting wave transmission in an aluminum wire having a 0·178 cm diameter and a 365·8 cm length. Input pulse duration, 50 μs. Top figure is output at $f = 1{\cdot}280$ MHz, lower figure is output at $f = 1{\cdot}404$ MHz. (After Meitzler [49, Figs. 1, 2].)

the wire by a piezoelectric transducer and detected by the same means at the other end. The carrier frequency of the pulse was varied from 0·5 MHz to 4·0 MHz and a Gaussian pulse modulation was used having a duration of about 50 μs. A typical set of observations is shown in Fig. 8.58(b).

Using the apparatus described, a first-mode longitudinal pulse was initiated in the wire. The expected behaviour was for the modulated pulse to propagate at the group velocity of the carrier frequency, maintaining its Gaussian shape. In general, this is what occurred and the upper figure of Fig.8 .58(b) is typical of the output (and input) pulse at most frequencies. However, at certain critical frequencies the output was considerably distorted, such as in the lower figure of Fig. 8.58(b). Other examples are presented by Meitzler. It was found that the extent of the pulse distortion was dependent on the wire material and wire length. Critical frequencies for pulse distortion were also noted for other modes. The explanation given by Meitzler for the observed distortion was the phenomenon of mode coupling. By this was meant that, at critical frequencies, the initial energy launched into the wire in one mode would, during propagation, excite another rod mode. The resulting partition of the propagating energy into two modes considerably distorted the pulse. The explanation for the occurrence of this phenomenon was the presence of surface imperfections of the wire. Meeker and Meitzler [48] briefly review this phenomenon. Lange [41] has also contributed in this area, showing that the conversion of energy from the first longitudinal to first flexural mode can be observed in the long-wavelength limit.

Meitzler [50] has also reported on the phenomenon of backward wave transmission in rods and plates. The theoretical basis for this phenomenon is associated with a small region of the frequency spectrum having group and phase velocity of opposite signs. Referring to Fig. 8.10, such a region occurs, for positive real wavenumber, between the minimum of the $L(2)$ mode and the cutoff frequency, and is the dashed segment of the $L(2)$ branch. A similar segment exists for negative real wavenumber, where the group velocity is positive and the phase velocity is negative. The experimental arrangement used for investigating this phenomenon is the same as shown in Fig. 8.58(a). Figure 8.59(a) shows experimental results for the change in pulse delay for various carrier frequencies of the input Gaussian pulse. What occurs is that an input pulse excites several modes of the 212 cm, 0·144 cm diameter, isoelastic wire. As the carrier frequency is varied, the delay time of the various modes change or are expected to change, in accord with the predictions of Pochhammer–Chree theory. A theoretical plot of group velocity versus frequency for the first few rod modes is shown by the solid lines of Fig. 8.59(b). The modes $L(0, 1)$, $L(0, 2)$, $L(0, 3)$ refer to the first three, longitudinal modes while $F(1, 1)$, identified in Fig. 8.59(a), is the lowest flexural mode. Shown on Fig. 8.59(a) as solid lines are the theoretical frequency–delay-time curves predicted by Pochhammer–Chree theory. Thus the various peaks in the experimental response are fairly clearly associated with specific modes.

In interpreting Fig. 8.59(a), it is first noted that the first flexural mode $F(1, 1)$ is excited to a slight extent, as is evident in the top four photographs. The first longitudinal mode $L(0, 1)$ is strongly evident in all photographs. Consider now the $L(0, 2)$ and $L(0, 3)$ modes, which are identified as the $L(2)$ and $L(3)$ modes in the frequency spectrum of Fig. 8.10. Starting with the $L(0, 2)$ mode of Fig. 8.59(a) at high frequency (the bottom photograph) and decreasing frequency, it is seen that the group velocity becomes lower for increasing frequency. This is seen to be consistent with the $L(2)$ mode of the

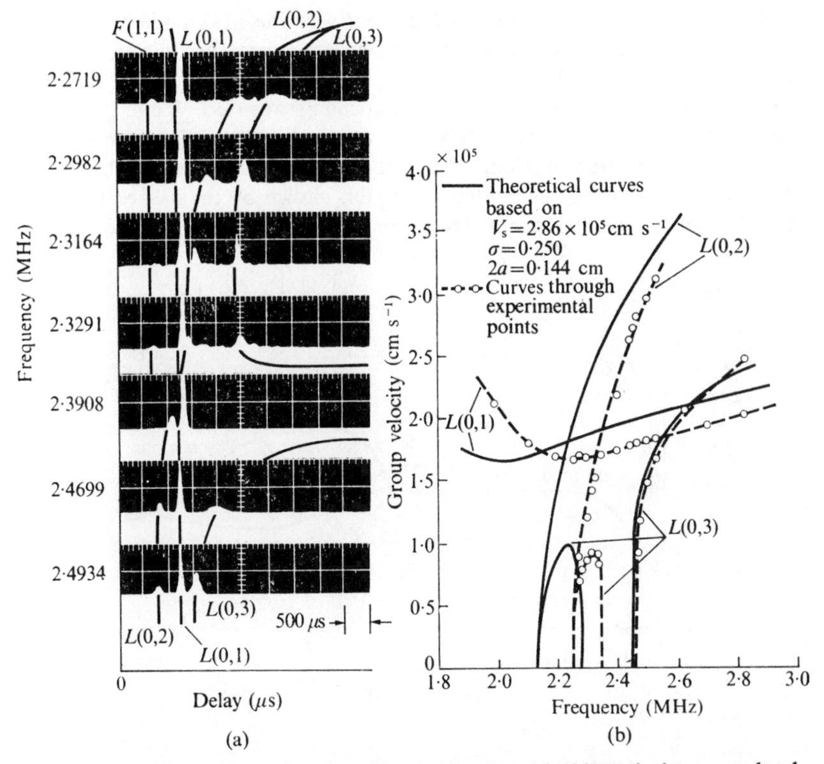

FIG. 8.59. (a) Variations in delay of the various modes excited by a single, narrow-band Gaussian pulse in an elastic wire; (b) theoretical and experimental curves of group velocity versus frequency. (After Meitzler [50, Figs. 4, 5].)

frequency spectrum. As frequency is decreased, the $L(2)$ mode approaches a local minimum, at which a cutoff frequency occurs. For the material and geometry used in the experiment, this is about at 2·2 MHz, or slightly above that of the top photograph of Fig. 8.59(a). It is seen in the latter figure that the $L(0, 2)$ pulse shape is becoming quite spread out at higher frequency, consistent with the increased dispersiveness as the minimum of the $L(2)$ mode is approached.

Considering now the $L(0, 3)$ mode of Fig. 8.59(a), it is seen for decreasing frequency in the lower two photographs that the delay is increasing, as is the amount of dispersion. At a frequency of 2·3908 MHz, the mode is completely absent. This behaviour is consistent with the $L(3)$ mode of the dispersion curve as it approaches its cutoff frequency. Now, for a further decrease in frequency, the $L(0, 3)$ mode is again present (top four photographs) with variation in delay and, again, increasing dispersiveness as frequency decreases to 2·2719 MHz. This behaviour is consistent with the $L(3)$ mode of the dispersion curve in the region of negative real wavenumber, where group velocity is positive but phase velocity negative. Again, it corresponds, in Fig. 8.10, to the short portion of the $L(3)$ branch occurring between $\bar{\xi} = 0$ and the local minimum occurring at a negative value of $\bar{\xi}$, and is the region of so-called backward wave transmission. Using the experimental results, Meitzler developed the group-velocity–frequency plots shown by the dashed lines in Fig. 8.59(b). Some difference existed between the experimental and theoretical

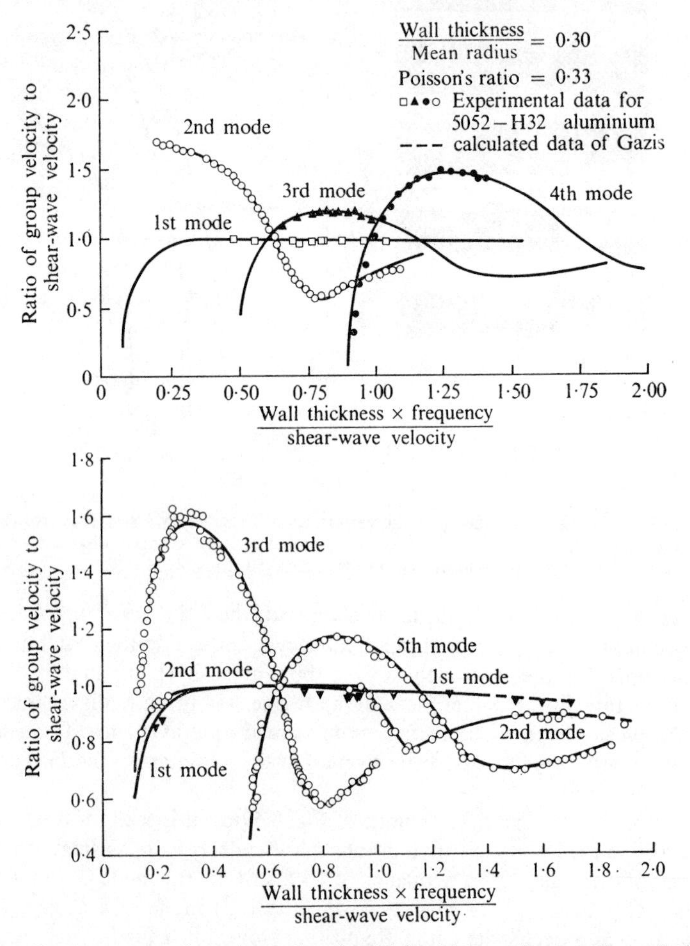

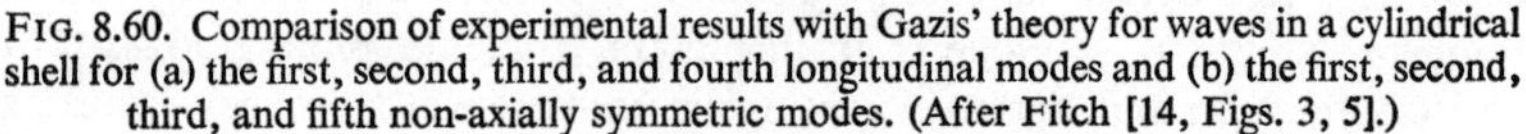

FIG. 8.60. Comparison of experimental results with Gazis' theory for waves in a cylindrical shell for (a) the first, second, third, and fourth longitudinal modes and (b) the first, second, third, and fifth non-axially symmetric modes. (After Fitch [14, Figs. 3, 5].)

results, as may be seen. Meitzler hypothesized, and put forth supporting arguments to show, that elastic anisotropy of the material was a cause of these discrepancies.

Experimental studies on the propagation of waves in cylindrical shells have been reported by Fitch [14]. The apparatus used for these tests was again essentially that shown in Fig. 8.54(a), except that a ring-shaped transducer was used to initiate the pulse into the cylindrical test specimen. The test specimen was a 6 ft (1·83 m) long hollow cylinder of 5052-H32 aluminum alloy. The outer diameter was $\frac{3}{8}$ in. (0·95 cm) and the wall thickness was 0·050 in. (1·27 mm). Narrow-band pulses were propagated in the shell and the group velocity of various modes were measured. The experimental results were compared to the theoretical predictions of the theory put forth by Gazis [20]. The results for the first few symmetric and non-symmetric modes are shown in Fig. 8.60. The comparison between theory and experiment was found to be extremely close. A comparison was also made by Fitch between the theoretical and experimental results for the second longitudinal mode in the shell and the theoretical results for the second longitudinal mode in a strip, with again excellent agreement being found.

References

1. ABRAMOWITZ, M. and STEGUN, I. A. *Handbook of mathematical functions: with formulas, graphs, and mathematical tables.* Dover Publications, New York (1965).
2. AGGRAWAL, R. R. and SHAW, E. A. G. Axially symmetric vibrations of a finite isotropic disk-IV. *J. acoust. Soc. Am.* **26,** 341 (1954).
3. ARNOLD, R. N. and WARBURTON, G. B. Flexural vibrations of the walls of thin cylindrical shells having freely supported ends. *Proc. R. Soc.* **A197,** 238–56 (1949).
4. BARON, M. L. and BLEICH, H. H. Tables of frequencies and modes of free vibration of infinitely long thin cylindrical shells. *J. appl. Mech.* **21,** 178–84 (1954).
5. BERTHOLF, L. D. Numerical solution for two-dimensional elastic wave propagation in finite bars. *J. appl. Mech.* **34,** 725–34 (1967).
6. BISHOP, R. E. D. Longitudinal waves in Beams. *Aeronant. Q.* **3,** 280 (1952).
7. CHREE, C. The equations of an isotropic elastic solid in polar and cylindrical coordinates, their solutions and applications. *Trans. Camb. phil. Soc. Math. Phys. Sci.* **14,** 250 (1889).
8. CHURCHILL, R. V. *Operational mathematics.* McGraw-Hill, New York (1958).
9. COOPER, R. M. and NAGHDI, P. M. Propagation of nonaxially symmetric waves in elastic cylindrical shells. *J. acoust. Soc. Am.* **29,** 1365–72 (1957).
10. CURTIS, C. W. Propagation of an elastic strain pulse in a semi-infinite bar, in *Stress wave propagation in materials,* (Ed. N. Davids) p. 15–43. Interscience, New York (1960).
11. DAVIES, R. M. A critical study of the Hopkinson pressure bar. *Phil. Trans. R. Soc.* **A240,** 375–457 (1948).

12. DeVault, G. P. and Curtis, C. W. Elastic cylinder with free lateral surface and mixed time-dependent end conditions. *J. acoust. Soc. Am.* **34,** 421–32 (1962).
13. Evans, J. F., Hadley, C. F., Eisler, J. D., and Silverman, D. A three-dimensional seismic wave model with both electrical and visual observation of waves. *Geophysics* **19,** 220–36 (1954).
14. Fitch, A. H. Observation of elastic-pulse propagation in axially symmetric and nonaxially symmetric longitudinal modes of hollow cylinders. *J. acoust. Soc. Am.* **35,** 706–8 (1962).
15. Flügge, W. *Statik und Dynamik der Schalen*, pp. 115 and 230. Springer-Verlag, Berlin (1934).
16. Folk, R. T. *Time dependent boundary value problems in elasticity.* Ph.D. Dissertation, Lehigh University, Bethlehem, Pennsylvania (1958).
17. ——, Fox, G., Shook, C. A., and Curtis, C. W. Elastic strain produced by sudden application of pressure to one end of a cylindrical bar—I. Theory. *J. acoust. Soc. Am.* **30,** 552–58 (1958).
18. Fox, G. and Curtis, C. W. Elastic strain produced by sudden application of pressure to one end of a cylindrical bar—II. Experimental observations. *J. acoust. Soc. Am.* **30,** 559–63 (1958).
19. Gazis, D. C. Exact analysis of the plane-strain vibrations of thick-walled hollow cylinders. *J. acoust. Soc. Am.* **30,** 786–94 (1958).
20. ——. Three-dimensional investigation of the propagation of waves in hollow circular cylinders—I. Analytical foundation II. Numerical results. *J. acoust. Soc. Am.* **31,** 568–78 (1959).
21. Goodman, L. E. Circular-crested vibrations of an elastic solid bounded by two parallel planes. *Proc. Ist Natn. Congr. appl. Mech.* pp. 65–73. ASME, New York (1952).
22. Green, W. A. Dispersion relations for elastic waves in bars, in *Progress in Solid Mechanics*, Vol. I, (Eds. I. N. Sneddon and R. Hill), Chap. 5. North-Holland Publishing Company, Amsterdam (1960).
23. Greenspon, J. E. Vibrations of a thick-walled cylindrical shell—comparison of the exact theory with approximate theories. *J. acoust. Soc. Am.* **32,** 571–8 (1960).
24. Harrison, M. *The propagation of elastic waves in a plate.* Report No. 872, the David W. Taylor model basin (1954).
25. Herrmann, G. and Mirsky, I. Three-dimensional and shell-theory analysis of axially symmetric motions of cylinders. *J. appl. Mech.* **23,** 563–8 (1956).
26. Holden, A. N. Longitudinal modes of elastic waves in isotropic cylinders and slabs. *Bell Sys. tech. J.* **30,** 956–69 (1951).
27. Hudson, G. E. Dispersion of elastic waves in solid circular cylinders. *Phys. Rev.* **63,** 46–51 (1943).
28. Hsieh, D. Y. and Kolsky, H. An experimental study of pulse propagation in elastic cylinders. *Proc. phys. Soc.* **71,** 608–12 (1958).
29. Hudson, G. E. Dispersion of elastic waves in solid circular cylinders. *Phys. Rev.* **63,** 46 (1943).
30. Jones, O. E. and Ellis, A. T. Longitudinal strain pulse propagation in wide rectangular bars, Part I—Theoretical considerations. *J. appl. Mech.* **30,** 51–60 (1963).

31. Jones, O. E. and Ellis, A. T. Longitudinal strain pulse propagation in wide rectangular bars, Part II—Experimental observations and comparisons with theory. *J. appl. Mech.* **30,** 61–9 (1963).
32. —— and Norwood, F. R. Axially symmetric cross-sectional strain and stress distributions in suddenly loaded cylindrical elastic bars. *J. appl. Mech.* **34,** 718–24 (1967).
33. Kaul, R. K. and McCoy, J. J. Propagation of axisymmetric waves in a circular semi-infinite elastic rod. *J. acoust. Soc. Am.* **36,** 653–60 (1964).
34. Kennard, E. H. The new approach to shell theory: circular cylinders. *J. appl. Mech.* **75,** 33–40 (1953).
35. Kennedy, L. W. and Jones, O. E. Longitudinal wave propagation in a circular bar loaded suddenly by a radially distributed end stress. *J. appl. Mech.* **36,** 470–8 (1969).
36. Kolsky, H. The propagation of longitudinal elastic waves along cylindrical bars. *Phil. Mag.* **45,** 712–26 (1954).
37. Kynch, G. J. *Br. J. appl. Phys.* **8,** 64 (1957).
38. —— and Green, W. A. *Q. Jl. Mech. appl. Math.* **10,** 63 (1957).
39. Lamb, H. On waves in an elastic plate. *Proc. R. Soc.* **A93,** 114 (1917).
40. Lamé, G. *Leçons sur la théorie mathématique de l'elasticité des corps solides* (2nd edn). Gauthier-Villars, Paris (1866).
41. Lange, J. N Mode conversion in the long-wavelength limit. *J. acoust. Soc. Am.* **41,** 1449–52 (1967).
42. Lin, T. C. and Morgan, G. W. A study of axisymmetric vibrations of cylindrical shells as affected by rotatory inertia and transverse shear. *J. appl. Mech.* **23,** 255–61 (1956).
43. Love, A. E. H. *A treatise on the mathematical theory of elasticity.* Dover Publications, New York (1944).
44. Lyon, R. H. Response of an elastic plate to localized driving forces. *J. acoust. Soc. Am.* **27,** 259 (1955).
45. Magnus, W. and Oberhettinger, F. *Special functions of mathematical physics.* Chelsea Publishing Company, New York (1949).
46. McLachlan, N. W. *Bessel functions for engineers.* Clarendon Press, Oxford (1961).
47. McNiven, H. D., Shah, A. H., and Sackman, J. L. Axially symmetric waves in hollow, elastic rods. Parts I and II. *J. acoust. Soc. Am.* **40,** 784–92 and 1073–6 (1966).
48. Meeker, T. R. and Meitzler, A. H. Guided wave propagation in elongated cylinders and plates, in *Physical acoustics,* (Ed. W. P. Mason) Vol. 1, Part A, Chap. 2. Academic Press, New York (1964).
49. Meitzler, A. H. Mode coupling occurring in the propagation of elastic pulses in wires. *J. acoust. Soc. Am.* **33,** 435–45 (1961).
50. ——. Backward-wave transmission of stress pulses in elastic cylinders and plates. *J. acoust. Soc. Am.* **38,** 835–42 (1965).
51. Miklowitz, J. Travelling compressional waves in an elastic rod according to the more exact one-dimensional theory. *Proc. IInd U.S. natn. Congr. appl. Mech.* pp. 179–186. ASME, New York (1955).

52. MIKLOWITZ, J. The propagation of compressional waves in a dispersive elastic rod, Part I—Results from the theory. *J. appl. Mech.* **24,** 231–9 (1957).
53. ——. On the use of approximate theories of an elastic rod in problems of longitudinal impact. *Proc. IIIrd U.S. natn. Congr. appl. Mech.* pp. 215–24. ASME, New York (1958).
54. ——. Plane-stress unloading waves emanating from a suddenly punched hole in a stretched elastic plate. *J. appl. Mech.* **27,** 165–7 (1960).
55. ——. Flexural stress waves in an infinite elastic plate due to a suddenly applied concentrated transverse load. *J. appl. Mech.* **27,** 681–9 (1960).
56. ——. Transient compressional waves in an infinite elastic plate or elastic layer overlying a rigid half-space. *J. appl. Mech.* **29,** 53–60 (1962).
57. ——. Transient wave propagation in elastic rods and plates. *J. geophys. Res.* **68,** 1190–2 (1963).
58. —— and NISEWANGER, C. R. The propagation of compressional waves in a dispersive elastic rod, Part II—Experimental results and comparison with theory. *J. appl. Mech.* **24,** 240–4 (1957).
59. MINDLIN, R. D. Influence of rotatory inertia and shear on flexural motions of isotropic, elastic plates. *J. appl. Mech.* **18,** 31 (1951).
60. ——. The thickness shear and flexural vibrations of crystal plates. *J. appl. Phys.* **22,** 316 (1951).
61. ——. *An introduction to the mathematical theory of vibrations of elastic plates.* U.S. Army Signal Corps Engineering Laboratories, Fort Monmouth, New Jersey (1955).
62. ——. Vibrations of an infinite elastic plate at its cut-off frequencies. *Proc. IIIrd U.S. natn. Congr. appl. Mech.* p. 225 (1958).
63. ——. Waves and vibrations in isotropic, elastic plates. In *Structural Mechanics* (Eds. J. N. Goodier and N. Hoff). pp. 199–232 (1960).
64. —— and HERRMANN, G. A one-dimensional theory of compressional waves in an elastic rod. *Proc. Ist U.S. natn. Congr. appl. Mech.* pp. 187–91 (1950).
65. —— and MCNIVEN, H. D. Axially symmetric waves in elastic rods. *J. appl. Mech.* **27,** 145–51 (1960).
66. —— and MEDICK, M. A. Extensional vibrations of elastic plates. *J. appl. Mech.* **26,** 561–9 (1959).
67. —— and ONOE, M. Mathematical theory of vibrations of elastic plates. *Proceedings of the XIth Annual Symposium on Frequency Control* pp. 17–40. U. S. Army Signal Corps Engineering Laboratories, Fort Monmouth, New Jersey (1957).
68. MIRSKY, I. and HERRMANN, G. Nonaxially symmetric motions of cylindrical shells. *J. acoust. Soc. Am.* **29,** 1116–23 (1957).
69. —— ——. Axially symmetric motions of thick cylindrical shells. *J. appl. Mech.* **25,** 97–102 (1958).
70. MORSE, P. and FESHBACH, H. *Methods of theoretical physics* Vol. I and II. McGraw-Hill, New York (1953).
71. MORSE, R. W. Velocity of compressional waves in rods of rectangular cross-section. *J. acoust. Soc. Am.* **22,** 219 (1950).
72. NAGHDI, P. M. and COOPER, R. M. Propagation of elastic waves in cylindrical shells, including the effects of transverse shear and rotatory inertia. *J. acoust. Soc. Am.* **28,** 56–63 (1956).

73. Oliver, J. Elastic wave dispersion in a cylindrical rod by a wide-band short-duration pulse technique. *J. acoust. Soc. Am.* **29**, 189–94 (1957).
74. Onoe, M. A study of the branches of the velocity-dispersion equations of elastic plates and rods. *Report Joint Commitee on Ultrasonics of the Institute of Electrical Communication Engineers and the Acoustical Society of Japan* (1955).
75. ——, McNiven, H. D., and Mindlin, R. D. Dispersion of axially symmetric waves in elastic rods. *J. appl. Mech.* **29**, 729–34 (1962).
76. Pao, Y.-H. The dispersion of flexural waves in an elastic circular cylinder, Part II. *J. appl. Mech.* **29**, 61–4 (1962).
77. —— and Mindlin, R. D. Dispersion of flexural waves in an elastic, circular cylinder. *J. appl. Mech.* **27**, 513–20 (1960).
78. Prescott, J. Elastic waves and vibrations of thin rods. *Phil. Mag.* **33**, 703 (1942).
79. Rayleigh, J. W. S. *The theory of sound*, Vol. I and II. Dover Publications, New York (1945).
80. Redwood, M. *Mechanical Waveguides*. Pergamon Press, New York (1960).
81. Reissner, E. *J. math. Phys.* **29**, 90–5 (1950).
82. Scott, R. A. and Miklowitz, J. Transient compressional waves in an infinite elastic plate with a circular cylindrical cavity. *J. appl. Mech.* **31**, 627–34 (1964).
83. —— ——. Transient compressional waves in an infinite elastic plate generated by a time-dependent radial body force. *J. appl. Mech.* **32**, 706–8 (1965).
84. —— ——. Transient non-axisymmetric wave propagation in an infinite isotropic elastic plate. *Int. J. Solids Struct.* **5**, 65–79 (1969).
85. Skalak, R. Longitudinal impact of a semi-infinite circular elastic bar. *J. appl. Mech.* **34**, 59–64 (1957).
86. Timoshenko, S. P. On the correction for shear of the differential equation for transverse vibrations of prismatic bars. *Phil. Mag.* Ser. 6, **41**, 744 (1921).
87. Tolstoy, I. and Usdin, E. Wave propagation in elastic plates: low and high mode dispersion. *J. acoust. Soc. Am.* **29**, 37–42 (1957).
88. Torvik, J. Reflection of wave trains in semi-infinite plates. *J. acoust. Soc. Am.* **41**, 346–53 (1967).
89. Volterra, E. A one-dimensional theory of wave-propagation in elastic rods based on the 'method of internal constraints'. *Ing.-Arch.* **23**, 410 (1955).
90. Yu, Y.-Y. Vibrations of thin cylindrical shells analysed by means of Donnell-type equations. *J. Aerospace Sci.* **25**, 699–715 (1958).

Problems

8.1. Sketch the group velocity curves for the SH wave modes of a plate.

8.2. In addition to the frequency spectrum (ω versus ξ) and dispersion curves (c versus ξ), wave propagation data is sometimes presented in terms of c versus ω. Sketch this set of curves for the SH wave modes in a plate.

8.3. Consider the propagation of SH waves in a plate of thickness $2b$, where fixed boundary conditions govern at $y = \pm b$. Obtain the frequency equation and compare the resulting frequency spectrum to that of SH waves in a traction-free plate.

8.4. Consider the case of SH waves propagating in a plate with elastically restrained boundaries. Thus, at $y = \pm b$, we have $\tau_{yz} = ku_z$. Sketch the frequency spectrum. Illustrate the transition between fixed boundaries (see Problem 8.3) and free boundaries (Fig. 8.3) as k varies from infinity to zero.

8.5. Consider the propagation of SH waves in a symmetrical, three-layered plate. Using coordinates of Fig. 8.1, let the plate be defined by the mid layer $y = \pm b$, with attached layers at $y = +b$ and $y = -b$ each of thickness a. Assume the layers are of the same material and have a shear velocity less than the mid layer. Derive the frequency equation for propagation of waves in the positive x direction. Determine what simplifications occur in the frequency equation when $\gamma b \gg 1$ or $\gamma b \ll 1$.

8.6. Wave propagation in plates has been presented in terms of plane-strain conditions. Express the governing equations and solution forms for plane stress conditions. Thus, referring to Fig. 8.1, assume the thickness of the plate in the z direction to be finite ($z = \pm a$) and small compared to longitudinal wavelengths. Only changes in elastic constants should be necessary. What restrictions will be placed on the number of modes of the Rayleigh–Lamb spectrum that can be considered in describing wave propagation in such a system?

8.7. Using the basic relationship that $d\omega/d\gamma = c_g$, sketch the approximate form of the group-velocity curves for the first six modes shown for the plate in Fig. 8.9 for $\mathrm{Re}\,\xi \leqslant 3$. Note regions of negative group velocity for certain of the modes, and recall the discussion given to this aspect in conjunction with Fig. 8.10.

8.8. Consider a semi-infinite plate having traction-free lateral surfaces on $y = \pm b$ and having a stress-free edge at $x = 0$. Investigate the reflection of incident longitudinal plane waves from the boundary. Thus, obtain reflection coefficient ratios for the various wave components.

8.9. Consider the vibrations of a bounded plate, where the dimensions are $2a$ and $2b$ in the x, y directions, respectively, while plane-strain conditions prevail in the z direction. Establish sets of mixed boundary conditions on $x = \pm a$, $y = \pm b$ that enable the problem of the free vibrations of such a plate to be solved.

8.10. Attempt to discover if the analogue of the Lamé mode in plates also exists for circular rods. Thus, do special solutions to the Pochhammer–Chree equation exist for $\xi = \beta$?

8.11. Starting with the Pochhammer–Chree equation, attempt to recover the 'thin rod' results by assuming that $2\pi/\beta$, $2\pi/\alpha \gg a$. Thus, see if results analogous to 8.1.92, 8.1.95 for the plate can be obtained for the rod.

8.12. Suppose one has a thin, cylindrical rod of diameter d, length l, such that the longitudinal resonance is adequately described by simple classical rod theory. Give a qualitative sketch of the expected change in resonant frequency that would occur if (a) the length is held constant and the diameter is increased and (b) the diameter is held constant and the length is increased.

8.13. Consider the propagation of pure torsional waves in the hollow cylinder described by the boundary conditions (8.2.50). Using the displacement equations of motion directly, obtain the frequency equation for the propagation of torsional waves. The result should, of course, agree with (8.2.66) obtained by the general, potential function approach.

8.14. Attempt to derive the frequency equation for pure torsional waves in a composite rod. The rod is defined by an inner cylinder of radius a attached to an outer shell of inner radius a, outer radius b. Assume the shear-wave velocity of the inner cylinder is greater than that of the shell.

8.15. Consider the development of an approximate theory for longitudinal waves in a plate, assuming displacements of the form

$$u(x, y, t) = u^{(0)}(x, t) + y^2 u^{(2)}(x, t)$$

$$v(x, y, t) = y v^{(1)}(x, t) + y^3 v^{(3)}(x, t)$$

which restrict the motion to symmetric only. Sketch the various displacement modes for these displacements. Derive the plate stress equations of motion and boundary conditions for the assumed displacement forms. Derive the plate stress–displacement relations. Simplify the development by letting

$$u^{(2)} = v^{(3)} = 0,$$

and give the resulting stress equations of motion, stress–displacement relations, and boundary conditions. Obtain the plate displacement equations of motion. Obtain the frequency equation and draw the frequency spectrum, using the non-dimensionalized frequency and wavenumbers $\Omega = 2b\omega/\pi c_2$, $\bar{\xi} = 2b\xi/\pi$. Determine the cutoff frequencies of the plate, comparing the values obtained with those of exact theory for $\nu = 0{\cdot}31$.

8.16. Consider the Mindlin approximate plate equations, given by (8.3.30) and the problem of reflection of flexural waves from the edge of a semiinfinite plate, $z = \pm h/2$, $y \geqslant 0$. First consider plane harmonic waves to be at normal incidence to the boundary $y = 0$ and obtain the reflected wave amplitude ratios. Now suppose plane waves arrive at oblique incidence. Determine the reflected wave system, including expressions for amplitude ratios.

8.17. Consider the case of longitudinal waves in a stepped rod, as governed by Mindlin–Herrmann theory. Let the rod be defined by $r = a$, $z \leqslant 0$, $r = b$, $z > 0$, where $b > a$. The material properties are the same on either side of the step at $z = 0$. Consider a harmonic, longitudinal wave to be propagating toward the step in the smaller rod. Determine the expressions for the reflected–transmitted wave systems, including amplitude ratios, if possible. Can any statements be made regarding the magnitude of the discontinuity relative to wavelength and/or rod radii? Are comparisons to the reflection–transmission results for the classical rod, given in Chapter 1, possible?

8.18. Consider the propagation of torsional waves in a semi-infinite cylinder of radius a. First consider the displacement $u_\theta(r, \theta, 0, t) = U_0 \exp(i\omega t)$ applied at $z = 0$, and attempt to solve for the wave propagation. What would be the nature of the radial distribution of displacement necessary to excite and propagate only the first torsional mode? Now attempt to solve the case of an applied displacement $u_0 H\langle t\rangle$ applied at $z = 0$. The Laplace transform is suggested. In all of the above work, it is suggested that the displacement equations of motion be used directly.

Appendix A: The elasticity equations

THE objective here is to review briefly the basic equations for an elastic continuum, including the concepts of strain and stress and the development of the constitutive relations. All equations will be referred to Cartesian coordinates, enabling the intricacies of tensor calculus to be avoided. The results for cylindrical and spherical coordinates will merely be summarized.

A.1. Notation

The use of index notation, summation conventions, and certain symbols enables the equations of a continuum to be developed and displayed with remarkable brevity. We first delineate all variables by numerical indices instead of by individual letters or alphabetic subscripts. Thus x_1, x_2, x_3 instead of x, y, z or x_x, x_y, x_z. Using index notation, we may write the equation

$$u = a_1x_1+a_2x_2+a_3x_3. \tag{A.1.1}$$

Using a conventional summation symbol, this may be written as

$$u = \sum_{i=1}^{3} a_i x_i. \tag{A.1.2}$$

We now introduce the summation convention wherein a repeated index denotes summation over the range of the index. Thus, (A.1.1) simply reduces to

$$u = a_i x_i \qquad (i = 1, 2, 3). \tag{A.1.3}$$

A repeated index may be changed to a different repeating index. Thus

$$a_i x_i = a_j x_j. \tag{A.1.4}$$

More than one double index may appear. Thus the equation

$$\begin{aligned} v = {} & a_{11}x_1x_1+a_{12}x_1x_2+a_{13}x_1x_3+ \\ & +a_{21}x_2x_1+a_{22}x_2x_2+a_{23}x_2x_3+ \\ & +a_{31}x_3x_1+a_{32}x_3x_2+a_{33}x_3x_3 \end{aligned} \tag{A.1.5}$$

becomes

$$v = a_{ij}x_i x_j \qquad (i, j = 1, 2, 3). \tag{A.1.6}$$

One indicial equation may represent several equations in extended notation. Thus

$$w_i = a_{ij}x_j \qquad (i, j = 1, 2, 3). \tag{A.1.7}$$

is equivalent to

$$\begin{aligned} w_1 &= a_{11}x_1+a_{12}x_2+a_{13}x_3, \\ w_2 &= a_{21}x_1+a_{22}x_2+a_{23}x_3, \\ w_3 &= a_{31}x_1+a_{32}x_2+a_{33}x_3. \end{aligned} \tag{A.1.8}$$

Two special symbols find wide use in indicial representation of equations. The first is the Kronecker delta defined as

$$\delta_{ij} = \begin{cases} +1, & i = j \\ 0, & i \neq j. \end{cases} \tag{A.1.9}$$

The second is the permutation symbol defined as

$$e_{ijk} = \begin{cases} +1, & ijk \text{ even permutation of } 1, 2, 3 \\ -1, & ijk \text{ odd permutation of } 1, 2, 3 \\ 0, & \text{any two indices equal.} \end{cases} \tag{A.1.10}$$

Thus we have $\delta_{11} = \delta_{22} = \delta_{33} = 1$, $\delta_{12} = 0$, $\delta_{23} = 0$, etc., and

$$e_{123} = e_{231} = e_{312} = 1, \qquad e_{213} = e_{321} = e_{132} = -1,$$
$$e_{112} = 0, \qquad e_{223} = 0, \text{ etc.}$$

Finally, we note the derivative notation where differentiation with respect to a variable will be indicated by a comma followed by an index. Thus

$$u_{,j} = \frac{\partial u}{\partial x_j}, \qquad v_{i,j} = \frac{\partial v_i}{\partial x_j}. \tag{A.1.11}$$

A number of the common vector operations may be easily written in index notation. A few of these are summarized in the following with their corresponding index notation form given alongside.

$$\begin{aligned} \mathbf{a} \cdot \mathbf{b} &\sim a_i b_i, \\ \mathbf{a} \times \mathbf{b} &\sim e_{ijk} a_k b_j, \\ \nabla\phi &\sim \phi_{,i}, \\ \nabla \cdot \mathbf{A} &\sim A_{i,i}, \\ \nabla \times \mathbf{A} &\sim e_{ijk} A_{k,j}, \\ \nabla \cdot \nabla\phi = \nabla^2\phi &\sim \phi_{,ii}. \end{aligned} \tag{A.1.12}$$

A.2. Strain

Consider a continuous medium of volume V and surface S that undergoes deformation. Before deformation, point P_0 is located by the position vector

X_i and P_1, a neighbouring point of P_0 is located by the vector dX_i from P_0. After deformation, P_0 goes into P_0' and is located by the vector x_i and P_1 goes into P_1' and is located by the vector dx_i relative to P_0'. The displacement of P_0 to P_0' is measured by the vector u_i. The displacement of P_1 to P_1' is measured by $\hat{u}_i$. The final volume and surface of the deformed body is V' and S' respectively. These various quantities are shown in Fig. A.1.

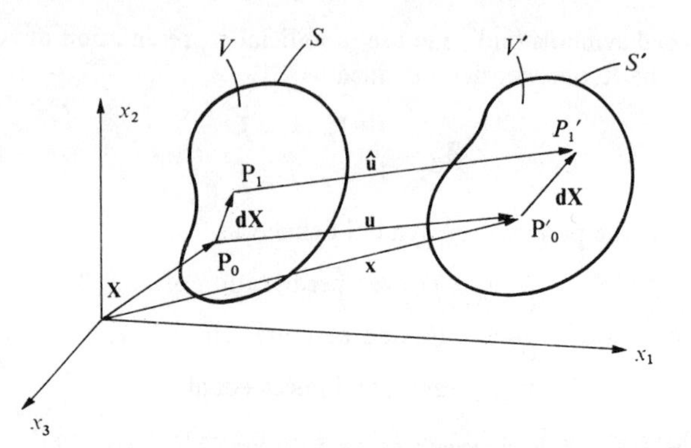

FIG. A.1. Deformation of a continuum of volume V into the volume V'.

The relationships between the various vector quantities are

$$x_i = X_i + u_i, \qquad u_i + dx_i = \hat{u}_i + dX_i. \tag{A.2.1}$$

But, from the first equation we have that $dx_i = dX_i + du_i$. Substituting in the second equation we have

$$\hat{u}_i = u_i + du_i. \tag{A.2.2}$$

To first order, we may express du_i as

$$du_i = u_{i,j}\, dx_j, \tag{A.2.3}$$

which may be put in the form

$$du_i = \tfrac{1}{2}(u_{j,i} + u_{j,i})\, dx_i + \tfrac{1}{2}(u_{i,j} - u_{j,i})\, dx_i. \tag{A.2.4}$$

We then define the infinitesimal strain and rotation tensors respectively as

$$e_{ij} = \tfrac{1}{2}(u_{i,j} + u_{j,i}), \qquad \omega_{ij} = \tfrac{1}{2}(u_{i,j} - u_{j,i}). \tag{A.2.5}$$

The result (A.2.4) emphasizes that the kinematics of an arbitrary neighbouring point of P_0 is governed by the local strain-gradient field $u_{i,j}$ and that the motion is a combination of local distortion effects e_{ij} and also local rigid-body rotation effects ω_{ij}.

A.3. Stress

Consider a continuum of volume V and surface S that is acted upon by various forces as shown in Fig. A.2(a). As a result of these forces, tractive

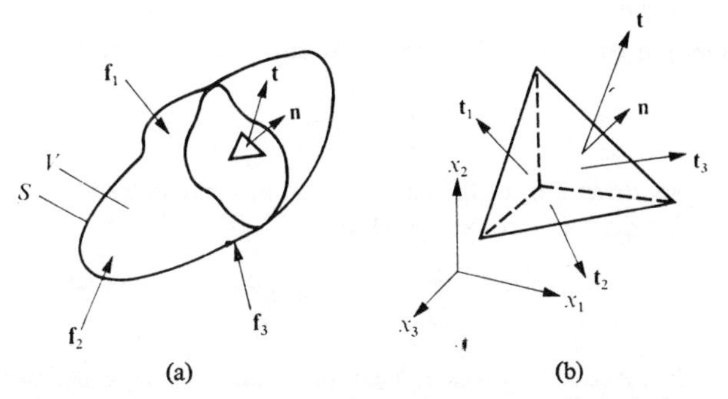

FIG. A.2. (a) A continuum subjected to forces, and (b) a trihedral element of that continuum.

forces will act on an arbitrary surface element within the body, as shown in Fig. A.2(b). The traction vector is given by

$$\mathbf{t} = t_j \mathbf{i}_j, \tag{A.3.1}$$

where the traction components t_j serve to define the stress tensor τ_{ij} by

$$t_i = \tau_{ij} \mathbf{n}_j. \tag{A.3.2}$$

In Cartesian extended notation these equations take the form

$$\begin{aligned} t_x &= \tau_{xx} l + \tau_{xy} m + \tau_{xz} n, \\ t_y &= \tau_{yx} l + \tau_{yy} m + \tau_{yz} n, \\ t_z &= \tau_{zx} l + \tau_{zy} m + \tau_{zz} n. \end{aligned} \tag{A.3.3}$$

A.4. Conservation equations

Certain fundamental axioms are essential to the construction of a continuum theory. Although the goal of this brief review is limited to the equations of infinitesimal, isotropic elasticity, development along certain fundamental lines is helpful in appreciating the basic unity of the theory and for observing the means by which effects of electrical or thermal phenomena may be included.

The axioms accepted as true on the basis of our experience in the physical world are:

conservation of mass, conservation of momentum, conservation of moment of momentum, and conservation of energy.

Should the inclusion of electrical or thermal effects be sought, axioms pertaining to the principle of entropy, conservation of charge, Faraday's law, and Ampere's law would need be appended. Our concern will only be with the mechanical effects, however. The mathematical statements of the foregoing principles will now be given.

A.4.1. *Conservation of mass*

Consider the volume V of surface S. At any instant, the total mass is given by

$$m = \int_V \rho \, \mathrm{d}V. \tag{A.4.1}$$

The principle, or axiom, of the conservation of mass states that time rate of change of this mass is zero or $\mathrm{D}m/\mathrm{D}t = 0$. Thus

$$\frac{\mathrm{D}}{\mathrm{D}t} \int_V \rho \, \mathrm{d}V = 0, \tag{A.4.2}$$

where ρ is the mass density. A statement of the last form is sometimes referred to as global conservation of mass.

(*Remark.* A few words on the nature of the time derivative are warranted. Consider the time variation of the quantity $F(x_1, x_2, x_3, t)$ with time. We may take two viewpoints: (1) focusing our attention on a point in space and determine the time rate of change of F at that point; (2) focusing our attention on a material point, following this point as time passes, and determining the time rate of change of F at this material point. Regardless of viewpoint, however, we have that

$$\frac{\mathrm{D}F}{\mathrm{D}t} = \left.\frac{\partial F}{\partial t}\right|_{x_i = \text{constant}} + \left.\frac{\partial F}{\partial x_i}\right|_{t = \text{constant}} \dot{x}_i. \tag{A.4.3}$$

Now, by viewpoint (1), $\dot{x}_i = 0$ in the above so that

$$\mathrm{D}F/\mathrm{D}t = \partial F/\partial t. \tag{A.4.4}$$

This is known as the spatial or Eulerian description. From viewpoint (2), however, the extended derivative (A.4.3) must be used and such a form is called the material or Lagrangian description. In our treatment of elasticity, the former will generally be used, so that distinction between total and partial derivatives disappears.)

The differential equation form of (A.4.2), sometimes referred to as the local conservation of mass equation, is obtained by taking the derivative inside the integral. Thus

$$\int_V \frac{\mathrm{D}\rho}{\mathrm{D}t} \, \mathrm{d}V = \int_V \left(\frac{\mathrm{D}\rho}{\mathrm{D}t} + \rho \frac{\mathrm{D}(\mathrm{d}V)}{\mathrm{D}t} \right)$$

$$= \int_V \left(\frac{\mathrm{D}\rho}{\mathrm{D}t} + \rho \frac{\partial \dot{x}_i}{\partial x_i} \right) \mathrm{d}V = 0. \tag{A.4.5}$$

Taken in the Lagrangian sense, we note that $\frac{\mathrm{D}\rho}{\mathrm{D}t} = \frac{\partial \rho}{\partial t} + \frac{\partial \rho}{\partial x_j}\frac{\partial x_j}{\partial t}$, so that (A.4.5) becomes

$$\int_V \left\{\frac{\partial \rho}{\partial t} + \frac{\partial}{\partial x_i}(\rho \dot{x}_i)\right\} \mathrm{d}V = 0, \tag{A.4.6}$$

or, since this must hold for arbitrary V,

$$\frac{\partial \rho}{\partial t} + (\rho v_{i,i}) = 0. \tag{A.4.7}$$

This latter form of the conservation of mass, or continuity equation, finds application most frequently in fluid dynamics.

A.4.2. *Conservation of momentum*

Consider the volume V of surface S. Associated with every point of V will be a velocity v_i, while acting on the body will be surface tractions t_i and body forces f_i. Conservation of momentum states that the time rate of change of momentum is equal to the total force acting on the body. Taken in the global sense, this statement has the form

$$\frac{\mathrm{D}}{\mathrm{D}t}\int_V v_i \rho \, \mathrm{d}V = \int_S t_i \, \mathrm{d}S + \int_V f_i \rho \, \mathrm{d}V. \tag{A.4.8}$$

This expression may also be put in differential equation, or local form, by using Gauss's theorem.

Green–Gauss theorem: In vector form, this is

$$\int_V \nabla . \mathbf{u} \, \mathrm{d}V = \int_S \mathbf{u} . \mathbf{n} \, \mathrm{d}S,$$

or, in Cartesian tensor form,

$$\int_V u_{j,j} \, \mathrm{d}V = \int_S u_j n_j \, \mathrm{d}S.$$

Thus upon letting $t_i = \tau_{ji} n_j$, (A.4.8) becomes

$$\int_V (\rho \ddot{u}_i - \tau_{ji,j} - \rho f_i) \, \mathrm{d}V = 0 \tag{A.4.9}$$

or, since V is arbitrary,

$$\tau_{ji,j} + \rho f_i = \rho \ddot{u}_i. \tag{A.4.10}$$

A.4.3. *Conservation of moment of momentum*

This principle states that the time rate of change of moment of momentum is equal to the sum of moments on the body. In the global sense, and using vector form, this is

$$\frac{\mathrm{D}}{\mathrm{D}t}\int_V \mathbf{r}\times\mathbf{v}\rho\,\mathrm{d}V = \int_S \mathbf{r}\times\mathbf{t}\,\mathrm{d}S + \int_V \mathbf{r}\times\mathbf{f}\rho\,\mathrm{d}V. \tag{A.4.11}$$

In tensor form, this is

$$\int_V e_{ijk}\ddot{u}_k x_j \rho\,\mathrm{d}V = \int_S e_{ijk}t_k x_j\,\mathrm{d}S + \int_V e_{ijk}f_k x_j \rho\,\mathrm{d}V. \tag{A.4.12}$$

This may be put in differential form, or the local sense, by using Gauss's theorem. Thus

$$\begin{aligned}\int_S e_{ijk}t_k x_j\,\mathrm{d}S &= \int_S e_{ijk}x_j(\tau_{lk}n_l)\,\mathrm{d}S \\ &= \int_V (e_{ijk}x_j\tau_{lk})_{,l}\,\mathrm{d}V.\end{aligned} \tag{A.4.13}$$

Carrying out the differentiations gives

$$\int_S e_{ijk}t_k x_j\,\mathrm{d}S = \int_V (e_{ijk}\tau_{jk} + e_{ijk}x_j\tau_{lk,l})\,\mathrm{d}V. \tag{A.4.14}$$

Inserting (A.4.14) in (A.4.12), and using the conservation of momentum results, gives

$$\int_V (e_{ijk}\tau_{jk})\,\mathrm{d}V = 0 \tag{A.4.15}$$

or

$$e_{ijk}\tau_{jk} = 0. \tag{A.4.16}$$

This is equivalent to

$$\tau_{ij} = \tau_{ji}, \qquad i \neq j. \tag{A.4.17}$$

Thus symmetry of the stress tensor results from balance of moment of momentum. Thus reduces the unknown stresses from nine to six.

A.4.4. *Conservation of energy*

Conservation of energy states that the time rate of change of kinetic energy and internal energy is equal to the work done upon the body by the external forces per unit time and the sum of all other energies per unit time.

The kinetic $\mathscr{K}$ is defined as

$$\mathscr{K} = \frac{1}{2}\int_V v_i v_i \rho\,\mathrm{d}V. \tag{A.4.18}$$

The internal energy $\mathscr{E}$ is defined as

$$\mathscr{E} = \int_V \varepsilon\rho\,\mathrm{d}V. \tag{A.4.19}$$

where ε is the internal energy per unit mass. The work done by the external forces is

$$W = \int_S t_i v_i\,\mathrm{d}S + \int_V f_i v_i \rho\,\mathrm{d}V, \tag{A.4.20}$$

so that conservation of energy takes the form

$$\dot{\mathscr{K}}+\dot{\mathscr{E}} = W, \tag{A.4.21}$$

or

$$\frac{\mathrm{D}}{\mathrm{D}t}\int_V (\tfrac{1}{2}v_i v_i+\varepsilon)\rho \, \mathrm{d}V = \int_S t_i v_i \, \mathrm{d}S + \int_V f_i v_i \rho \, \mathrm{d}V. \tag{A.4.22}$$

Eqn (A.4.22) may be put in differential form by converting the surface integral to a volume integral as follows:

$$\int_V (\dot{u}_i\ddot{u}_i+\dot{\varepsilon})\rho \, \mathrm{d}V = \int_V (\tau_{ji}\dot{u}_{i,j}) \, \mathrm{d}V + \int_V f_i \dot{u}_i \rho \, \mathrm{d}V, \tag{A.4.23}$$

or

$$\int_V \{(\rho\ddot{u}_i - \tau_{ji,j} - \rho f_i)\dot{u}_i + (\rho\dot{\varepsilon} - \tau_{ji}\dot{u}_{i,j})\} \, \mathrm{d}V = 0. \tag{A.4.24}$$

The first parenthesis is the momentum equation (A.4.10). Hence the above reduces to

$$\rho\dot{\varepsilon} = \tau_{ji}\dot{u}_{i,j}. \tag{A.4.25}$$

Finally noting that

$$\dot{u}_{i,j} = \dot{\varepsilon}_{ji} + \dot{\omega}_{ij} \tag{A.4.26}$$

and that the product of a symmetric tensor (τ_{ji}) and an antisymmetric tensor (ω_{ij}) is zero, we have

$$\rho\dot{\varepsilon} = \tau_{ji}\dot{\varepsilon}_{ji}. \tag{A.4.27}$$

To summarize, the fundamental conservation equations have given

$$\begin{aligned}
&\text{mass:} && \partial\rho/\partial t + (\rho\dot{u}_{i,i}) = 0, \\
&\text{momentum} && \tau_{ji,j} + \rho f_i = \rho\ddot{u}_i, \\
&\text{moment of momentum} && \tau_{ij} = \tau_{ji}, \qquad (i \neq j), \\
&\text{energy} && \rho\dot{\varepsilon} = \tau_{ji}\dot{\varepsilon}_{ij}.
\end{aligned} \tag{A.4.28}$$

The above is a system of eight equations in thirteen unknowns ($\rho = 1$, $u_i = 3$, $\tau_{ij} = 9$), while f_i is assumed given and ε_{ij} is expressible in terms of u_i. The nature of the medium is determined by ε. Clearly the eight equations are inadequate for a unique determination of the thirteen unknowns.

Hence, the foregoing must be supplemented by additional equations. The need for added equations is also clear from purely physical considerations; the preceding equations are valid for any type of media, so that no differentiation has been made between, say, fluids, and solids. However, two different material bodies having the same geometry and mass distribution and subjected to identical external forces respond differently. This difference is a function of their intrinsic material properties, and it is these properties that must be incorporated into the continuum model.

A.5. Constitutive equations

The function of constitutive equations is to relate states of deformation with states of traction. In the case of ideally elastic bodies, two methods were developed, one by Green (1841) and one by Cauchy (1829). Brief presentations of both will be given.

A.5.1. *Green's method*

The essence of the formulation is that the body is perfectly elastic, so that there are no dissipative mechanisms, and the constitutive equations must be derivable from an internal energy function, which is a function of the strain. Thus define $U = \rho\varepsilon$, so that

$$U = U(\varepsilon_{ij}). \tag{A.5.1}$$

Then

$$\dot{U} = \frac{\partial U}{\partial \varepsilon_{ij}}\dot{\varepsilon}_{ij}, \tag{A.5.2}$$

so that from (A.4.27) we have

$$\tau_{ij} = \partial U/\partial \varepsilon_{ij}. \tag{A.5.3}$$

The precise functional form of $U(\varepsilon_{ij})$ is now the question. By expanding U in a power series of ε_{ij}, and retaining only the quadratic terms, we have

$$U \cong a_{ijkl}\varepsilon_{ij}\varepsilon_{kl}. \tag{A.5.4}$$

Discarding the linear terms of the expansion effectively postulates zero initial stress, while discarding the higher powers is in line with the assumption of small strains. From (A.5.3) we thus have

$$\tau_{ij} = a_{ijkl}\varepsilon_{kl}. \tag{A.5.5}$$

A.5.2. *Cauchy's method*

The essence of Cauchy's method is the assumption of a direct functional relationship between stress and strain of the form

$$\tau_{ij} = \tau_{ij}(\varepsilon_{ij}). \tag{A.5.6}$$

Expanding $\tau_{ij}(\varepsilon_{ij})$ in a power series and discarding the constant term (no initial stress) and the higher-order terms (infinitesimal elasticity) gives

$$\tau_{ij} = a_{ijkl}\varepsilon_{kl}. \tag{A.5.7}$$

Apparently, both Green's and Cauchy's method have led to the same results for τ_{ij}. However, such is not the case. To show this the 81 constants a_{ijkl} appearing in either result may be reduced to 36 by observing the symmetry with respect to ij and kl.† Thus both results may be put temporarily in the form

$$\tau_i = c_{ij}\varepsilon_j, \tag{A.5.8}$$

† See, for example, Sokolnikoff [52, pp. 59–60].

where

$$\tau_1 = \tau_{11}, \qquad \tau_2 = \tau_{22}, \qquad \tau_3 = \tau_{33},$$
$$\tau_4 = \tau_{23}, \qquad \tau_5 = \tau_{31}, \qquad \tau_6 = \tau_{12}, \tag{A.5.9}$$

and similarly for the ε_{ij}. However, from (A.5.3) and (A.5.4) of Green's method, which now take the form

$$\tau_i = \partial U/\partial \varepsilon_i, \qquad U = c_{ij}\varepsilon_i\varepsilon_j, \tag{A.5.10}$$

we are led to conclude that $c_{ij} = c_{ji}$, thereby reducing the elastic constants from 36 to 21. No such conclusion may be drawn, without further restrictive assumptions, in Cauchy's method. To reduce Cauchy's results to Green's, we assume the existence of a strain energy function, taking the form

$$U = \tfrac{1}{2}c_{ij}\varepsilon_i\varepsilon_j, \tag{A.5.11}$$

with the property that

$$\tau_i = \partial U/\partial \varepsilon_i. \tag{A.5.12}$$

This reduces the number of constants for the general anisotropic linear case to 21, in agreement with Green's results.

A.5.3. *Isotropic elastic solid*

By assuming homogeneity and isotropy, the number of constants reduces from 21 to 2.† Reverting to double subscript notation for the stresses, this takes the form

$$\tau_{ij} = \lambda\varepsilon_{kk}\delta_{ij} + 2\mu\varepsilon_{ij}, \tag{A.5.13}$$

where λ, μ are known as the Lamé constants. The latter is the material shear modulus. Under special loadings, such as simple tension or pure shear, the resulting relations between stress and strain are of such form as to make it convenient to define additional elastic constants, such as Young's modulus, Poisson's ratio, and the bulk modulus. These are merely combinations of λ and μ. Thus

$$E = \mu(3\lambda+2\mu)/(\lambda+\mu) = \text{Young's modulus},$$
$$\nu = \tfrac{1}{2}\lambda/(\lambda+\mu) = \text{Poisson's ratio}, \tag{A.5.14}$$
$$K = \lambda+\tfrac{2}{3}\mu = \text{bulk modulus}.$$

The inverse strain–stress equations are

$$\varepsilon_{ij} = \frac{(1+\nu)}{E}\tau_{ij} - \frac{\nu}{E}\Theta\delta_{ij}, \qquad \Theta = \tau_{11}+\tau_{22}+\tau_{33}. \tag{A.5.15}$$

If we go back to the system of conservation equations (A.4.27), we see that the addition of eqns (A.5.13), for the isotropic case, make the equations

† See, for example, Sokolnikoff [52, pp. 62–6].

of a continuum determinate. Thus we have

(1) mass and momentum give four equations in ten unknowns (ρ, u_i, $\tau_{ij} = 10$). The moment of momentum results have already reduced the stresses from nine to six;

(2) the energy equation has generated the stress–strain results, giving six equations;

(3) generally, density is taken as specified, so that the conservation of mass equation may be omitted, giving a total of nine equations in nine unknowns.

Of couse, it should be apparent that the resulting constitutive equations have introduced constants (ranging from 2 in number for the isotropic case to 21 for the general anisotropic case). It is here that the vital role of experiment becomes apparent, whereby critical tests are performed, the results related to theoretical models, and the constants determined.

A.6. Solution uniqueness and boundary conditions

The equations for the case of linear, isotropic elasticity are

$$\begin{aligned} \tau_{ij,j}+\rho f_i &= \rho \ddot{u}_i, \\ \tau_{ij} &= \lambda\varepsilon_{kk}\delta_{ij}+2\mu\varepsilon_{ij}, \\ \varepsilon_{ij} &= \tfrac{1}{2}(u_{i,j}+u_{j,i}), \\ \omega_{ij} &= \tfrac{1}{2}(u_{i,j}-u_{j,i}). \end{aligned} \tag{A.6.1}$$

The momentum equation (the first of (A.6.1)) may be expressed in terms of displacements by utilizing the stress–strain and strain–displacement relations of (A.6.1) giving

$$(\lambda+\mu)u_{j,ji}+\mu u_{i,jj}+\rho f_i = \rho \ddot{u}_i, \tag{A.6.2}$$

or, in vector form

$$(\lambda+\mu)\nabla\nabla\cdot\mathbf{u}+\mu\nabla^2\mathbf{u}+\rho\mathbf{f} = \rho\ddot{\mathbf{u}}. \tag{A.6.3}$$

A fundamental question pertaining to the preceding is whether a solution that satisfies the equations as well as certain boundary and initial conditions is a unique solution. Phrased somewhat differently, we may ask 'what boundary and initial conditions must be imposed on the above equations to insure that a resulting solution, satisfying equations and conditions, is a unique solution?' Thus the questions of proper boundary conditions and uniqueness of solutions are inseparable.

A.6.1. *Uniqueness*

The question of uniqueness is approached by supposing two solutions to the same problem may exist and then showing that such an assumption leads to a contradiction.

Consider two unique solutions exist, given by u_i', u_i'', τ_{ij}', τ_{ij}'', ε_{ij}', ε_{ij}''. If these are indeed solutions, then the linear combination formed by taking the difference of the two solutions is also a solution. We thus form the difference system, defined as

$$u_i = u_i' - u_i'', \qquad \tau_{ij} = \tau_{ij}' - \tau_{ij}'',$$
$$\varepsilon_{ij} = \varepsilon_{ij}' - \varepsilon_{ij}'', \qquad f_i = f_i' - f_i'', \qquad t_i = t_i' - t_i''. \tag{A.6.4}$$

Conservation of energy still holds for the difference system so from (A.4.22), we have

$$\frac{\mathrm{D}}{\mathrm{D}t}\int_V (\tfrac{1}{2}\rho\dot{u}_i\dot{u}_i + U)\,\mathrm{d}V = \int_S t_i\dot{u}_i\,\mathrm{d}S + \int_V f_i\dot{u}_i\rho\,\mathrm{d}V. \tag{A.6.5}$$

This, we recall, is a statement that the time rate of change of kinetic and potential energy equals the work done per unit time. Carrying out an integration with respect to time on the preceding gives

$$\int_V [\tfrac{1}{2}\rho\dot{u}_i\dot{u}_i + U]_{t_0}^{t}\,\mathrm{d}V = \int_{t_0}^{t}\mathrm{d}t\int_S t_i\dot{u}_i\,\mathrm{d}S + \int_{t_0}^{t}\mathrm{d}t\int_V f_i\dot{u}_i\,\mathrm{d}V. \tag{A.6.6}$$

It is seen that the first integral of the right-hand side contains the surface tractions t_i and time derivatives of displacements with the specification of these values at an initial time t_0 implied by the lower limit of integration. Likewise, the second integral contains body forces f_i and velocities $\dot{u}_i$.

Now, the conditions on the boundary S are prescribed by combinations of t_i and u_i, while interior forces are given by f_i. The point is that starting from identical conditions on S and in V, two different solutions have arisen. If conditions are the same, it follows that the right-hand side must be zero, since $t_i = t_i' - t_i''$, $f_i = f_i' - f_i''$ and both t_i', t_i'', etc. are constrained to be the same. Consequently

$$\int_V [\tfrac{1}{2}\rho\dot{u}_i\dot{u}_i + U]_{t_0}^{t}\,\mathrm{d}V = 0. \tag{A.6.7}$$

Using definitions (A.4.18), (A.4.19), this is of the form

$$\mathscr{K} + \mathscr{E} = \mathscr{K}_0 + \mathscr{E}_0. \tag{A.6.8}$$

However, $\mathscr{K}_0$, $\mathscr{E}_0$ are based on the initial velocities $\dot{u}_i$ and displacements u_i of the difference system, and these are zero. Hence

$$\mathscr{K} + \mathscr{E} = 0. \tag{A.6.9}$$

Furthermore, both $\mathscr{K}$ and $\mathscr{E}$ are positive definite†, indicating that

$$\mathscr{K} = \mathscr{E} = 0. \tag{A.6.10}$$

Hence, the two supposedly different solutions must, in fact, be identical, so that only one solution exists.

† The positive definiteness of the strain energy can be used to place theoretical restrictions on the range of the elastic constants. For example, it may be shown that $\mu > 0$, $3\lambda + 2\mu > 0$ must hold for the elastic case.

A.6.2. *Boundary conditions*

Additional remarks are now warranted on the boundary and initial conditions. Return now to the single solution u_i, τ_{ij}, ε_{ij}, etc. and note the previously appearing integrals

$$\int_{t_0}^{t} \mathrm{d}t \int_S t_i \dot{u}_i \,\mathrm{d}S, \qquad \int_{t_0}^{t} \mathrm{d}t \int_V f_i \dot{u}_i \rho \,\mathrm{d}V. \tag{A.6.11}$$

Consider the first integral, and the surface shown in Fig. A.3. Then

$$\begin{aligned} t_i u_i &= t_n u_n + t_s u_s + t_t u_t \\ &= \tau_{nn} u_n + \tau_{ns} u_s + \tau_{nb} u_b. \end{aligned} \tag{A.6.12}$$

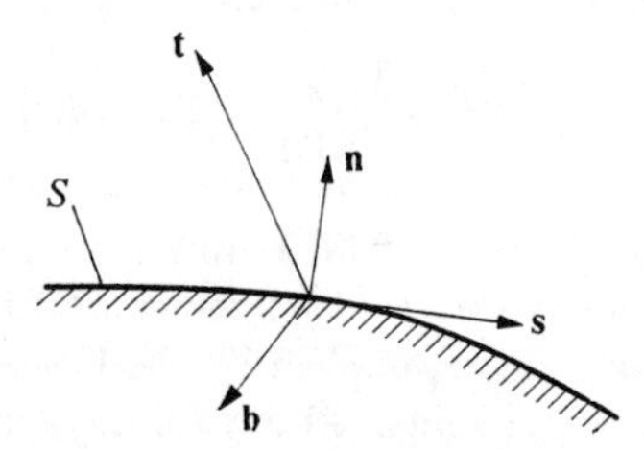

FIG. A.3. Base-vector triad at the surface of a continuum.

The disappearance of the above surface integral in the previous proof was based on specifying surface conditions and from (A.6.12), we see these conditions on S must be

$$\tau_{nn}, \qquad \tau_{ns}, \qquad \tau_{nb} \quad \text{or} \quad u_i, \tag{A.6.13}$$

or a proper mix of τ_{ij}, u_i. In addition, the specification of τ_{ij} or u_i at $t = t_0$ is required.

In the case of the second integral of (A.6.11), we see the specification of f_i is required, and the values of u_i, $\dot{u}_i$ at $t = t_0$. Thus we must have

(1) in V, at each point f_i and initial values of u_i and $\dot{u}_i$;

(2) on S, one member of each of the three products

$$\tau_{nn} u_n, \qquad \tau_{ns} u_s, \qquad \tau_{nb} u_b$$

as well as initial values of these quantities.

Thus the boundary conditions have resulted as sufficient conditions for uniqueness of solution.

A.7. Other continua

The fundamental unity of the study of continua, such as solids, liquids, and gases, is emphasized by the applicable equations of balance. The constitutive equations are then the distinguishing characteristics between various materials. In our work, we shall be concerned mainly with homogeneous isotropic solids, characterized by (A.6.1). It is of interest to remark on some other directions of study in solid and fluid mechanics, keeping in mind the universal applicability of the equations of balance, and noting the way in which constitutive relations are introduced.

1. *Anisotropy*. Elastic materials that are fundamentally anisotropic are crystals. Eqn (A.5.5) would still hold. However, depending on the degree of anisotropy, the number of elastic constants could range from 3 to 21.

2. *Piezoelectric effects*: It is well known that certain crystals and poled ceramics develop an electrical charge when strained and vice versa. This coupled electro-mechanical effect considerably complicates formulation of the continuum equations. Equations of mass, momentum, and moment of mementum would not be affected, however.

We would find the energy equation changed, since the work done by electric fields on charges would have to be incorporated in this balance equation. In addition, as was mentioned in § A.4, the number of fundamental axioms would have to be broadened to conservation of charge, Faraday's law, and Ampere's law. These axioms uncoupled from solid mechanics would lead to the well-known Maxwell equations, a set of field equations nearly as complicated as those of elasticity. The electro-mechanical coupling thus creates problems which, in general, are exceptionally difficult.

3. *Thermal effects*: Heat effects may be included in the energy balance equation. However, an axiom regarding entropy must also be added to the four equations of balance. Although heat effects will be neglected in our work, severe temperature gradients can cause interaction effects between heat and mechanical vibration, leading to thermoelastic waves.

4. *Fluids (gases and liquids)*. Study of hydrodynamics, gasdynamics, supersonic, and hypersonic flow, etc. are all studies of continua characterized by particular constitutive equations. Depending on compressibility, viscosity, and heat, these equations take on varying degrees of complexity, just as do their solid analogues.

For example, the constitutive equation of incompressible, inviscid liquids, or of ideal gases, are simpler than those of isotropic solids. In all three cases, these represent the simplest models of a type of continua. As additional effects are added to solids, liquids, and gases, the resulting constitutive equations take their most complex form in the case of solids. As a consequence, considerable progress has been made in, say, the study of the propagation of shock waves in fluids, whereas less work has taken place in this area of solids.

5. *Viscoelastic effects*. In the case of solids, it is often found that stresses are significantly affected by strain rates as well as by the strains alone. The

resulting constitutive equation may then take a form $\tau = \tau(\varepsilon, \dot{\varepsilon}, \ddot{\varepsilon}, \ldots)$, with corresponding complexity in subsequent analysis.

6. *Plasticity effects.* Again using solids as an example, it is observed, at sufficient strain amplitudes, that stresses and strains are no longer related in a linear, reversible fashion. The resulting non-linear continuum problem is, of course, exceedingly complex. Nevertheless, plastic wave propagation is an area of mechanics which is being actively studied. It sometimes occurs that experimental results are not adequately explained by amplitude effects (that is, plasticity) alone, but that strain rates also play a role. Thus a visco-plastic theory must be devised.

7. *Microstructure effects.* Continuum models are based on smoothly varying properties, such as density and moduli. However, at smaller and smaller dimensions, approaching grain sizes in metals, the fundamentally heterogeneous nature of solids begins to be of significance. Such might be the case if extremely high-frequency vibrations are occuring. In an effort to incorporate microstructure effects into a continuum theory, much work has been done in the fields of microelasticity, couple-stress theory and multipole continuum mechanics.

8. *Non-linear effects.* The effects of large strains in generating a non-linear set of equations was mentioned in the case of plasticity. In the case of all the foregoing effects, there is usually a linearized set of equations. However these equations have usually resulted from imposing restrictions on an original, fundamentally non-linear set of equations (for example, in isotropic elasticity, small deformations are assumed). Hence, in all cases, at sufficiently severe temperatures or electric fields or strain rates, the various problems can become non-linear.

A.8. Additional energy considerations

We now wish to review the application of work–energy concepts in contrast to equilibrium concepts, to elastodynamics. Interest in energy concepts may be justified on at least three counts. First, the energy stored in a body is an important quantity in its own right. Certain material failure theories, for example, are based on the idea of failure occurring when the strain energy reaches a critical value. Fracture theories also have critical energy levels as their basis. Thus, the Griffith fracture criteria is based on the relation between strain energy and free surface energy during crack growth. Secondly, methods have been developed for determining the deflection of elastic bodies that are based on energy aspects (for example, Castigliano's theorem). Thirdly, numerous methods have been developed for approximate solution of elasto-dynamics problems that are based on energy concepts. In fact, activity and emphasis in this regard has often been so great as to suggest that the only reason for energy consideration was to generate approximate solutions. However, the first considerations are also of considerable importance.

In many of our applications, we will use energy principles to develop approximate theories governing the motion of rods and plates (in contrast to approximate solutions). Developments starting from the law of conservation of energy stated as 'the rate of increase of energy in a body is equal to the rate at which work is done by the external forces' start with

$$\int_V (\dot{T}+\dot{V})\,\mathrm{d}V = \int_S t_i\dot{u}_i\,\mathrm{d}S + \int_V \rho f_i\dot{u}_i\,\mathrm{d}V, \tag{A.8.1}$$

where

$$T = \tfrac{1}{2}\rho\dot{u}_i\dot{u}_i, \qquad V = V(e_{ij}). \tag{A.8.2}$$

Then

$$\dot{T} = \rho\ddot{u}_i\dot{u}_i, \qquad \dot{V} = \frac{\partial V}{\partial e_{ij}}\dot{e}_{ij} = \tfrac{1}{2}\tau_{ij}(\dot{u}_{i,j}+\dot{u}_{j,i}). \tag{A.8.3}$$

Now

$$\begin{aligned}\int_V \dot{V}\,\mathrm{d}V &= \tfrac{1}{2}\int_V \tau_{ij}(\dot{u}_{i,j}+\dot{u}_{j,i})\,\mathrm{d}V\\ &= \tfrac{1}{2}\int_V (\tau_{ij}\dot{u}_i)_{,j}\,\mathrm{d}V - \tfrac{1}{2}\int_V \tau_{ij,j}\dot{u}_i\,\mathrm{d}V + \\ &\quad + \tfrac{1}{2}\int_V (\tau_{ij}\dot{u}_j)_{,i}\,\mathrm{d}V - \tfrac{1}{2}\int_V \tau_{ij,i}\dot{u}_j\,\mathrm{d}V.\end{aligned} \tag{A.8.4}$$

By the divergence theorem,

$$\int_V (\tau_{ij}\dot{u}_{i,j})\,\mathrm{d}V = \int_S \tau_{ij}\dot{u}_i n_j\,\mathrm{d}S, \qquad \int_V (\tau_{ij}\dot{u}_{j,i})\,\mathrm{d}V = \int_S \tau_{ij}\dot{u}_j n_i\,\mathrm{d}S. \tag{A.8.5}$$

Then (A.8.4) becomes

$$\begin{aligned}\int_V \dot{V}\,\mathrm{d}V &= \tfrac{1}{2}\int_S (\tau_{ij}\dot{u}_i n_j + \tau_{ij}\dot{u}_j n_i)\,\mathrm{d}S - \tfrac{1}{2}\int_V (\tau_{ij,j}\dot{u}_i + \tau_{ij,i}\dot{u}_j)\,\mathrm{d}V\\ &= \int_S \tau_{ij}\dot{u}_i n_j\,\mathrm{d}S - \int_V \tau_{ij,j}\dot{u}_i\,\mathrm{d}V.\end{aligned} \tag{A.8.6}$$

Since $t_i = \tau_{ij}n_j$, the above reduces to

$$\int_V \dot{V}\,\mathrm{d}V = \int_S t_i\dot{u}_i\,\mathrm{d}S - \int_V \tau_{ij,j}\dot{u}_i\,\mathrm{d}V. \tag{A.8.7}$$

We thus have (A.8.1), reducing to

$$\int_V (\rho\ddot{u}_i - \tau_{ij,j} - \rho f_i)\dot{u}_i\,\mathrm{d}V = 0. \tag{A.8.8}$$

Many of the developments will stem from *Hamilton's principle*. By way of background, Langhaar [35, pp. 234–5] points out that Newton's equations of motion refer only to a single mass particle. The analysis of the motion of a finite body by Newtonian methods involves subjecting the individual particles to Newton's laws and integrating the effects over the entire body. The Lagrange equations of motion are a generalization of this process and determine the motion of a body with finite degrees of freedom. Which ever equations are used, the general dynamical problem involves specifying the system location and velocity at some time t_0 and solving for the subsequent motion.

Hamilton (1805–65) formulated the problem of dynamics in a somewhat different way by considering the location of the system to be specified at two different times t_0 and t_1, with the intervening motion to be determined. Using the principle of virtual work, D'Alembert's principle, and applying variational calculus, Hamilton derived a general formulation of the equations of mechanics that determines the motion of a system with finite or infinite degrees of freedom. The mathematical statement of his principle is

$$\delta \int_{t_0}^{t_1} (\mathscr{E}+W)\,\mathrm{d}t = 0, \tag{A.8.9}$$

where $\mathscr{E}$ is the total energy of the system, W is the total work (exclusive of work done by inertial forces), and δ is in the sense of the calculus of variations. The above equation states that the motion of a system from t_0 to t_1 is such that the total of the system kinetic energy and system work is an extremum.

In our applications, (A.8.9) takes the somewhat more specific form†

$$\delta \int_{t_0}^{t_1} (\tilde{T}-\tilde{V})\,\mathrm{d}t + \int_{t_0}^{t} \delta W\,\mathrm{d}t = 0, \tag{A.8.10}$$

where

$$\tilde{T} = \int_V T\,\mathrm{d}V, \qquad \tilde{V} = \int_V V\,\mathrm{d}V \tag{A.8.11}$$

and

$$\delta W = \int_S t_i\,\delta u_i\,\mathrm{d}S + \int_V \rho f_i \delta u_i\,\mathrm{d}V, \tag{A.8.12}$$

and T and V have been previously defined by (A.8.2). Carrying out the variation on the kinetic energy we obtain

$$\delta \int_{t_0}^{t_1} \tilde{T}\,\mathrm{d}t = \int_{t_0}^{t_1} \mathrm{d}t \int_V \frac{\partial T}{\partial \dot{u}_i}\,\delta \dot{u}_i\,\mathrm{d}V. \tag{A.8.13}$$

Now

$$\delta \dot{u}_i = \frac{\partial}{\partial t}(\delta u_i), \qquad \delta u_{i,j} = (\delta u_i)_{,j}. \tag{A.8.14}$$

† P. 166 of Reference [36].

So we have

$$\delta\int_t \tilde{T}\,\mathrm{d}t = \int_V \left[\int_t \frac{\partial T}{\partial \dot{u}_i}\frac{\partial}{\partial t}(\delta u_i)\,\mathrm{d}t\right]\mathrm{d}V$$

$$= \left[\int_V \frac{\partial T}{\partial \dot{u}_i}\,\delta u_i\,\mathrm{d}V\right]_{t_0}^{t_1} - \int_t \mathrm{d}t\left\{\int_V \frac{\mathrm{d}}{\mathrm{d}t}\left(\frac{\partial T}{\partial \dot{u}_i}\right)\delta u_i\,\mathrm{d}V\right\}. \qquad \text{(A.8.15)}$$

For the strain energy term we have that

$$\delta\int_{t_0}^{t_1} \tilde{V}\,\mathrm{d}t = \int_{t_0}^{t_1}\mathrm{d}t\int_V \frac{\partial V}{\partial \varepsilon_{ij}}\,\delta\varepsilon_{ij}\,\mathrm{d}V. \qquad \text{(A.8.16)}$$

Now

$$\delta\varepsilon_{ij} = \tfrac{1}{2}\delta(u_{i,j}+u_{j,i}) = \left(\frac{\delta u_i}{2}\right)_{,j}+\left(\frac{\delta u_j}{2}\right)_{,i}. \qquad \text{(A.8.17)}$$

Hence

$$\delta\int_{t_0}^{t_1} \tilde{V}\,\mathrm{d}t = \int_t \mathrm{d}t\int_V \frac{1}{2}\frac{\partial V}{\partial \varepsilon_{ij}}\{(\delta u_i)_{,j}+(\delta u_j)_{,i}\}\,\mathrm{d}V$$

$$= \int_t \mathrm{d}t\left[\frac{1}{2}\int_V \left\{\left(\frac{\partial V}{\partial \varepsilon_{ij}}\,\delta u_i\right)_{,j}+\left(\frac{\partial V}{\partial \varepsilon_{ij}}\,\delta u_j\right)_{,i}\right\}\mathrm{d}V - \right.$$

$$\left. -\frac{1}{2}\int_V \left\{\left(\frac{\partial V}{\partial \varepsilon_{ij}}\right)_{,j}\delta u_i+\left(\frac{\partial V}{\partial \varepsilon_{ij}}\right)_{,i}\delta u_j\right\}\mathrm{d}V\right] \qquad \text{(A.8.18)}$$

We apply the divergence theorem to the first integral, giving

$$\delta\int_{t_0}^{t_1} \tilde{V}\,\mathrm{d}t = \int_t \mathrm{d}t\left[\frac{1}{2}\int \left(\frac{\partial V}{\partial \varepsilon_{ij}}\,\delta u_i n_j+\frac{\partial V}{\partial \varepsilon_{ij}}\,\delta u_j n_i\right)\mathrm{d}S - \right.$$

$$\left. -\frac{1}{2}\int_V \left\{\left(\frac{\partial V}{\partial \varepsilon_{ij}}\right)_{,j}\delta u_i+\left(\frac{\partial V}{\partial \varepsilon_{ij}}\right)_{,i}\delta u_j\right\}\mathrm{d}V\right]. \qquad \text{(A.8.19)}$$

For the second integral of (A.8.10), we have

$$\int_{t_0}^{t_1} \delta W\,\mathrm{d}t = \int_{t_0}^{t_1}\mathrm{d}t\left\{\int_S t_i\delta u_i\,\mathrm{d}S + \int_V \rho f_i\delta u_i\,\mathrm{d}V\right\}. \qquad \text{(A.8.20)}$$

Thus Hamilton's principle gives

$$\int_{t_0}^{t_1}\mathrm{d}t\int_V\left\{-\frac{d}{\mathrm{d}t}\left(\frac{\partial T}{\partial \dot{u}_i}\right)+\frac{1}{2}\left(\frac{\partial V}{\partial \varepsilon_{ij}}\right)_{,j}+\frac{1}{2}\left(\frac{\partial V}{\partial \varepsilon_{ji}}\right)_{,j}+\rho f_i\right\}\delta u_i\,\mathrm{d}V -$$

$$-\int_{t_0}^{t_1}\mathrm{d}t\int_S\left(\frac{1}{2}\frac{\partial V}{\partial \varepsilon_{ij}}n_j+\frac{1}{2}\frac{\partial V}{\partial \varepsilon_{ji}}n_j-t_i\right)\delta u_i\,\mathrm{d}S-\left[\int_V \frac{\partial T}{\partial \dot{u}_i}\,\delta u_i\,\mathrm{d}V\right]_{t_0}^{t_1} \equiv 0. \qquad \text{(A.8.21)}$$

This is the general result. If we recognize

$$\tau_{ij} = \frac{\partial V}{\partial \varepsilon_{ij}}, \qquad \tau_{ij} = \tau_{ji}, \qquad \frac{\partial T}{\partial \dot{u}_i} = \rho \dot{u}_i, \tag{A.8.22}$$

then we obtain

$$\int_{t_0}^{t_1} \mathrm{d}t \int_V (-\rho \ddot{u}_i + \tau_{ij,j} + \rho f_i)\delta u_i \,\mathrm{d}V - \int_{t_0}^{t} \mathrm{d}t \int_S (\tau_{ij} n_j - t_i)\delta u_i \,\mathrm{d}S - \left[\int_V \rho \dot{u}_i \delta u_i \,\mathrm{d}V\right]_{t_0}^{t_1} = 0. \tag{A.8.23}$$

Since $\tau_{ij} n_j = t_i$, the surface integral vanishes identically. At t_0, t_1, $\delta u_i = 0$ so that the last integral vanishes. We thus obtain

$$\int_{t_0}^{t} \mathrm{d}t \int_V (\tau_{ij,j} + \rho f_i - \rho \ddot{u}_i)\, \delta u_i \,\mathrm{d}V \equiv 0. \tag{A.8.24}$$

A.9. Elasticity equations in curvilinear coordinates

The elasticity equations for rectangular Cartesian coordinates have been summarized in indicial form by (A.6.1), with the displacement equations presented by (A.6.2). These equations, as well as various mathematical operators such as gradient and divergence, are required in terms of cylindrical and spherical coordinates in several places of the book. These are summarized in the following.

A.9.1. *Cylindrical coordinates*

A cylindrical coordinate system r, θ, z is shown in Fig. A.4. The elasticity equations are as follows:

$$\begin{aligned} e_{rr} &= \frac{\partial u_r}{\partial r}, \qquad e_{\theta\theta} = \frac{1}{r}\frac{\partial u_\theta}{\partial \theta} + \frac{u_r}{r}, \\ e_{zz} &= \frac{\partial u_z}{\partial z}, \qquad e_{r\theta} = \frac{1}{2}\left(\frac{1}{r}\frac{\partial u_r}{\partial \theta} + \frac{\partial u_\theta}{\partial r} - \frac{u_\theta}{r}\right), \\ e_{rz} &= \frac{1}{2}\left(\frac{\partial u_z}{\partial r} + \frac{\partial u_r}{\partial z}\right), \qquad e_{\theta z} = \frac{1}{2}\left(\frac{\partial u_\theta}{\partial z} + \frac{1}{r}\frac{\partial u_z}{\partial \theta}\right), \end{aligned} \tag{A.9.1}$$

$$\begin{aligned} &\frac{\partial \tau_{rr}}{\partial r} + \frac{1}{r}\frac{\partial \tau_{r\theta}}{\partial \theta} + \frac{\partial \tau_{rz}}{\partial z} + \frac{\tau_{rr} - \tau_{\theta\theta}}{r} + \rho f_r = \rho \ddot{u}_r, \\ &\frac{\partial \tau_{r\theta}}{\partial r} + \frac{1}{r}\frac{\partial \tau_{\theta\theta}}{\partial \theta} + \frac{\partial \tau_{\theta z}}{\partial z} + \frac{2}{r}\tau_{r\theta} + \rho f_\theta = \rho \ddot{u}_\theta, \\ &\frac{\partial \tau_{rz}}{\partial r} + \frac{1}{r}\frac{\partial \tau_{\theta z}}{\partial \theta} + \frac{\partial \tau_{zz}}{\partial z} + \frac{1}{r}\tau_{rz} + \rho f_z = \rho \ddot{u}_z. \end{aligned} \tag{A.9.2}$$

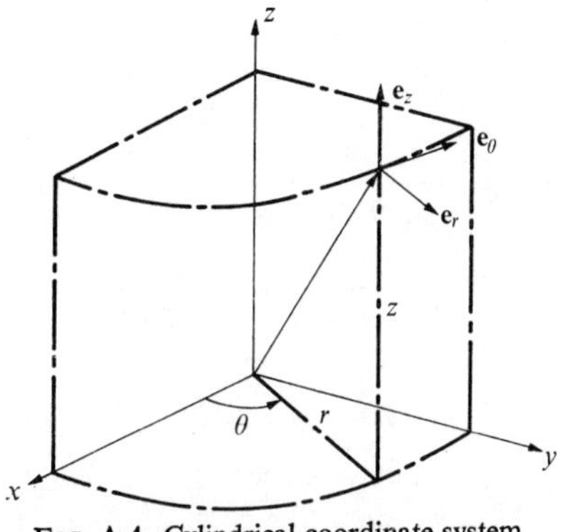

FIG. A.4. Cylindrical coordinate system.

The common operations on scalar and vector fields are the following:

$$
\begin{aligned}
\operatorname{grad} \phi &= \frac{\partial \phi}{\partial r}\mathbf{e}_r + \frac{1}{r}\frac{\partial \phi}{\partial \theta}\mathbf{e}_\theta + \frac{\partial \phi}{\partial z}\mathbf{e}_z, \\
\operatorname{div} \mathbf{A} &= \frac{1}{r}\frac{\partial}{\partial r}(rA_r) + \frac{1}{r}\frac{\partial A_\theta}{\partial \theta} + \frac{\partial A_z}{\partial z}, \\
\operatorname{curl} \mathbf{A} &= \left(\frac{1}{r}\frac{\partial A_z}{\partial \theta} - \frac{\partial A_\theta}{\partial z}\right)\mathbf{e}_r + \left(\frac{\partial A_r}{\partial z} - \frac{\partial A_z}{\partial r}\right)\mathbf{e}_\theta + \left\{\frac{1}{r}\frac{\partial}{\partial r}(rA_\theta) - \frac{1}{r}\frac{\partial A_r}{\partial \theta}\right\}\mathbf{e}_z, \qquad \text{(A.9.3)} \\
\nabla^2 \phi &= \frac{\partial^2 \phi}{\partial r^2} + \frac{1}{r}\frac{\partial \phi}{\partial r} + \frac{1}{r^2}\frac{\partial^2 \phi}{\partial \theta^2} + \frac{\partial^2 \phi}{\partial z^2}, \\
\nabla^2 \mathbf{A} &= \left(\nabla^2 A_r - \frac{A_r}{r^2} - \frac{2}{r^2}\frac{\partial A_\theta}{\partial \theta}\right)\mathbf{e}_r + \left(\nabla^2 A_\theta - \frac{A_\theta}{r^2} + \frac{2}{r^2}\frac{\partial A_r}{\partial \theta}\right)\mathbf{e}_\theta + \nabla^2 A_z \mathbf{e}_z.
\end{aligned}
$$

A.9.2. *Spherical coordinates*

A spherical coordinate system is shown in Fig. A.5. The elasticity equations

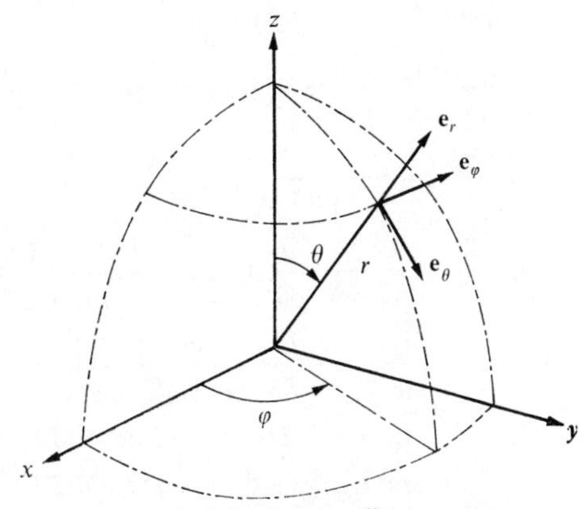

FIG. A.5. Spherical coordinate system.

are as follows:

$$e_{rr} = \frac{\partial u_r}{\partial r}, \qquad e_{\theta\theta} = \frac{1}{r}\frac{\partial u_\theta}{\partial \theta} + \frac{u_r}{r},$$

$$e_{\phi\phi} = \frac{1}{r \sin\theta}\frac{\partial u_\phi}{\partial \phi} + \frac{u_r}{r} + u_\theta \frac{\cot\theta}{r},$$

$$e_{r\phi} = \frac{1}{2}\left(\frac{1}{r \sin\theta}\frac{\partial u_r}{\partial \phi} - \frac{u_\phi}{r} + \frac{\partial u_\phi}{\partial r}\right), \tag{A.9.4}$$

$$e_{r\theta} = \frac{1}{2}\left(\frac{1}{r}\frac{\partial u_r}{\partial \theta} - \frac{u_\theta}{r} + \frac{\partial u_\theta}{\partial r}\right),$$

$$e_{\phi\theta} = \frac{1}{2}\left(\frac{1}{r}\frac{\partial u_\phi}{\partial \theta} - \frac{u_\phi \cot\theta}{r} + \frac{1}{r \sin\theta}\frac{\partial u_\theta}{\partial \phi}\right),$$

$$\frac{\partial \tau_{rr}}{\partial r} + \frac{1}{r\sin\theta}\frac{\partial \tau_{r\phi}}{\partial \phi} + \frac{1}{r}\frac{\partial \tau_{r\theta}}{\partial \theta} + \frac{(2\tau_{rr} - \tau_{\phi\phi} - \tau_{\theta\theta} + \tau_{r\theta}\cot\theta)}{r} + \rho f_r = \rho \ddot{u}_r,$$

$$\frac{\partial \tau_{r\phi}}{\partial r} + \frac{1}{r\sin\theta}\frac{\partial \tau_{\phi\phi}}{\partial \phi} + \frac{1}{r}\frac{\partial \tau_{\phi\theta}}{\partial \theta} + \frac{(3\tau_{r\phi} + 2\tau_{\phi\theta}\cot\theta)}{r} + \rho f_\phi = \rho \ddot{u}_\phi, \tag{A.9.5}$$

$$\frac{\partial \tau_{r\phi}}{\partial r} + \frac{1}{r\sin\theta}\frac{\partial \tau_{\phi\theta}}{\partial \phi} + \frac{1}{r}\frac{\partial \tau_{\theta\theta}}{\partial \theta} + \frac{\{3\tau_{r\theta} + (\tau_{\theta\theta} - \tau_{\phi\phi})\cot\theta\}}{r} + \rho f_\theta = \rho \ddot{u}_\theta.$$

The common operations on scalar and vector fields are

$$\text{grad}\,\Phi = \frac{\partial \Phi}{\partial r}\mathbf{e}_r + \frac{1}{r}\frac{\partial \Phi}{\partial \theta}\mathbf{e}_\theta + \frac{1}{r\sin\theta}\frac{\partial \Phi}{\partial \phi}\mathbf{e}_\phi,$$

$$\text{div}\,\mathbf{A} = \frac{1}{r^2}\frac{\partial}{\partial r}(r^2 A_r) + \frac{1}{r\sin\theta}\frac{\partial}{\partial \theta}(A_\theta \sin\theta) + \frac{1}{r\sin\theta}\frac{\partial A_\phi}{\partial \phi},$$

$$\text{curl}\,\mathbf{A} = \frac{1}{r\sin\theta}\left\{\frac{\partial}{\partial \theta}(A_\phi \sin\theta) - \frac{\partial A_\theta}{\partial \phi}\right\}\mathbf{e}_r + \left\{\frac{1}{r\sin\theta}\frac{\partial A_r}{\partial \phi} - \frac{1}{r}\frac{\partial}{\partial r}(rA_\phi)\right\}\mathbf{e}_\theta + \frac{1}{r}\left\{\frac{\partial}{\partial r}(rA_\theta) - \frac{\partial A_r}{\partial \theta}\right\}\mathbf{e}_\phi,$$

$$\nabla^2\Phi = \frac{1}{r^2}\frac{\partial}{\partial r}\left(r^2\frac{\partial \Phi}{\partial r}\right) + \frac{1}{r^2\sin\theta}\frac{\partial}{\partial \theta}\left(\sin\theta\frac{\partial \Phi}{\partial \theta}\right) + \frac{1}{r^2\sin^2\theta}\frac{\partial^2 \Phi}{\partial \phi^2},$$

$$\begin{aligned}\nabla^2\mathbf{A} = {} & \left\{\nabla^2 A_r - \frac{2}{r^2}A_r - \frac{2}{r^2\sin\theta}\frac{\partial}{\partial \theta}(\sin\theta A_\theta) - \frac{2}{r^2\sin\theta}\frac{\partial A_\phi}{\partial \phi}\right\}\mathbf{e}_r + \\ & + \left\{\nabla^2 A_\theta - \frac{A_\theta}{r^2\sin^2\theta} + \frac{2}{r^2}\frac{\partial A_r}{\partial \theta} - \frac{2}{r^2}\frac{\cos\theta}{\sin^2\theta}\frac{\partial A_\phi}{\partial \phi}\right\}\mathbf{e}_\theta + \\ & + \left\{\nabla^2 A_\phi - \frac{A_\phi}{r^2\sin^2\theta} + \frac{2}{r^2\sin\theta}\frac{\partial A_r}{\partial \phi} + \frac{2}{r^2}\frac{\cos\theta}{\sin^2\theta}\frac{\partial A_\theta}{\partial \phi}\right\}\mathbf{e}_\phi.\end{aligned} \tag{A.9.6}$$

Appendix B. Integral transforms

B.1. General

In mathematics, the concept of a transformation is quite general, and most of the familiar mathematical operations such as multiplication, differentiation and integration can be thought of as transformations of functions under some set of rules. For example, the operation of differentiation is a transformation of the functions $F(t)$ to the functions $F'(t)$ where the transformation operator is represented by D, and

$$\mathrm{D}\{F(t)\} = F'(t). \tag{B.1.1}$$

If we think of some function space as associated with $F(t)$, and another space associated with $F'(t)$, then the transformation D takes points from one space to another as illustrated schematically in Fig. 8.1. It could be said that p' is

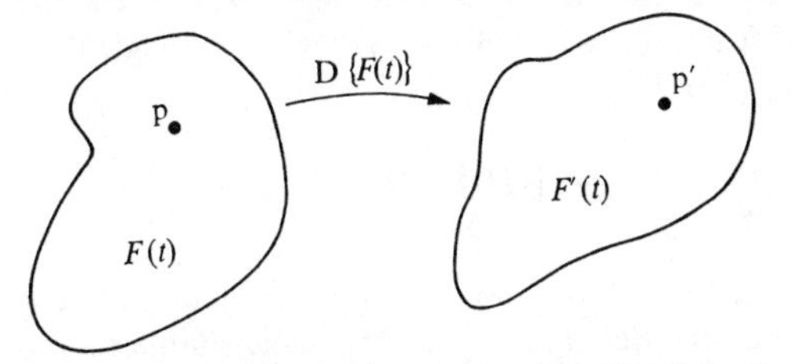

FIG. B.1. Illustration of the transformation of $F(t)$ by D.

the 'image' of point p in $F(t)$ space. Now a transformation that returns p' back to p would be spoken of as an inverse transformation. In terms of D as a derivative, we see that integration would be such an inverse transformation.

In our study of Laplace, Fourier, or other integral transforms, we shall be concerned with transformations of the type $T\{F(t)\}$, where

$$T\{F(t)\} = \int_a^b K(t, s)F(t)\,\mathrm{d}t. \tag{B.1.2}$$

The function $K(t, s)$ will be called the kernel of the transformation and $T\{F(t)\}$ will be an integral transformation of the function $F(t)$ with respect to the kernel $K(t, s)$. The various integral transforms will be characterized by their kernels and their limits a and b.

It will be found that, with certain kernels, the application of integral transforms to ordinary differential equations will yield algebraic expressions in the transformed variable. Partial differential equations will also undergo simplification. By treating the transformed problem, solutions may be found a great deal more easily. However, these solutions are in the transformed space, and the problem of inverting these solutions remains. Obtaining the inverse transformation will usually turn out to be the most formidable part of the entire problem.

B.2. Laplace transforms

B.2.1. *Definition*

If a function $F(t)$, defined for all positive values of the variable t, is multiplied by $\exp(-st)$ and integrated with respect to t from zero to infinity, a new function $f(s)$ of the parameter s is obtained. Thus

$$f(s) = \int_0^\infty e^{-st}F(t)\,dt. \tag{B.2.1}$$

The new function $f(s)$ is called the Laplace transform of $F(t)$. In the context of our previous discussion, we see that $\exp(-st)$ is the kernel of the transformation and $f(s)$ is the image of $F(t)$. We will abbreviate this operation with the notation $\mathscr{L}\{F\}$, where

$$\mathscr{L}\{F(t)\} = \int_0^\infty e^{-st}F(t)\,dt. \tag{B.2.2}$$

Although we shall not make use of this property immediately, it should be noted that s may be a complex quantity. Some simple examples of Laplace transforms of functions follow:

(1) $F(t) = 1,\ t > 0$.

$$\mathscr{L}(1) = \int_0^\infty e^{-st}\,dt = \left[-\frac{1}{s}e^{-st}\right]_0^\infty. \tag{B.2.3}$$

If $s > 0$, then

$$\mathscr{L}(1) = 1/s. \tag{B.2.4}$$

(2) $F(t) = e^{kt},\ t > 0,\ k$ constant.

$$\mathscr{L}(e^{kt}) = \left[\frac{1}{k-s}e^{-(s-k)t}\right]_0^\infty = \frac{1}{s-k}, \qquad s > k. \tag{B.2.5}$$

(3) $F(t) = t$

$$\mathscr{L}(t) = 1/s^2. \tag{B.2.6}$$

B.2.2. *Transforms of derivatives*

The transform of $F'(t)$ may be found through integration by parts. Thus

$$\mathscr{L}\{F'(t)\} = \int_0^\infty e^{-st}F'(t)\,dt$$

$$= [e^{-st}F(t)]_0^\infty + s\int_0^\infty e^{-st}F(t)\,dt. \qquad \text{(B.2.7)}$$

If $F(t)$ is of order $\exp(\alpha t)$ or less as t becomes large and if $s > \alpha$, it follows that

$$\mathscr{L}\{F'(t)\} = sf(s) - F(0). \qquad \text{(B.2.8)}$$

In a similar fashion we may show that

$$\mathscr{L}\{F''(t)\} = s^2f(s) - sF(0) - F'(0), \qquad \text{(B.2.9)}$$

or, in general,

$$\mathscr{L}\{F^{(n)}(t)\} = s^nf(s) - s^{n-1}F(0) - s^{n-2}F'(0) - \\ - \ldots - F^{(n-1)}(0). \qquad \text{(B.2.10)}$$

We see that the Laplace transform reduces $F^{(n)}(t)$ to an algebraic expression in s and includes the initial conditions of the problem (if t represents time) directly in the transformed expression.

B.2.3. *The inverse transform*

The problem of inverting a transformed solution is one of determining a function $F(t)$, given $f(s)$. Assuming that an inversion can be carried out, we indicate the process by the symbol $\mathscr{L}^{-1}$, where

$$\mathscr{L}^{-1}\{f(s)\} = F(t). \qquad \text{(B.2.11)}$$

Thus, $\mathscr{L}^{-1}\{f(s)\}$ denotes a function $F(t)$ whose Laplace transform is $f(s)$. As an example, suppose it is determined that

$$f(s) = 1/(s-k). \qquad \text{(B.2.12)}$$

But, we have already shown that

$$\mathscr{L}(e^{kt}) = 1/(s-k), \qquad \text{(B.2.13)}$$

so that

$$\mathscr{L}^{-1}\{f(s)\} = e^{kt}. \qquad \text{(B.2.14)}$$

As another example, it may be readily shown that, if $F(t) = \sin kt$, then

$$\mathscr{L}(\sin kt) = k/(s^2+k^2). \qquad \text{(B.2.15)}$$

Hence, if a solution in transformed space yields the result

$$f(s) = 1/(s^2+k^2), \qquad \text{(B.2.16)}$$

we then know that,

$$\mathscr{L}^{-1}\left(\frac{1}{s^2+k^2}\right) = \frac{1}{k}\sin kt = F(t). \tag{B.2.17}$$

A brief table of Laplace transforms, for reference purposes, is included at the end of this Appendix. More extensive tables are included in Churchill [7]. The Bateman manuscript project [15] has exhaustive listings of Laplace and other integral transforms. However, the solutions of problems often yield results not in the tables, and the inversion of such cases may require other considerations.

B.2.4. *Partial fractions*

As was stated in the last section, transformed solutions often yield results not tabulated. However, before we resort to extremely complicated inversion processes, every effort should be made to reset the transformed solution into combinations of already tabulated functions. As an example of this, it is often found that transformed solutions are in the form of quotients of polynomials. By the use of partial fractions, these may be decomposed into simpler fractions. For instance, suppose

$$f(s) = \frac{s+1}{s(s+2)}. \tag{B.2.18}$$

From the theory of partial fractions we know this may be expressed as

$$f(s) = \frac{A}{s}+\frac{B}{s+2}, \tag{B.2.19}$$

from which it is found that $A = B = \frac{1}{2}$. Hence

$$f(s) = \frac{1}{2s}+\frac{1}{2(s+2)}. \tag{B.2.20}$$

Then

$$\mathscr{L}^{-1}\{f(s)\} = \mathscr{L}^{-1}\left\{\frac{1}{2s}+\frac{1}{2(s+2)}\right\}$$

$$= \tfrac{1}{2}\mathscr{L}^{-1}\left\{\frac{1}{s}\right\}+\tfrac{1}{2}\mathscr{L}^{-1}\left\{\frac{1}{(s+2)}\right\}. \tag{B.2.21}$$

Now, $\mathscr{L}^{-1}(1/s)$, $\mathscr{L}^{-1}\{1/(s+2)\}$ may be easily found from tables, giving

$$\mathscr{L}^{-1}\{f(s)\} = \tfrac{1}{2}+\tfrac{1}{2}e^{-2t}. \tag{B.2.22}$$

Reference should be made to any basic calculus or algebra book for the details of partial-fraction theory and the techniques of handling more difficult cases.

B.2.5. *Solutions of ordinary differential equations*

The use of Laplace transforms in solving ordinary differential equations with constant coefficients will be illustrated with a simple example. Consider the familiar differential equation

$$\frac{d^2y(t)}{dt^2}+k^2y(t) = 0, \tag{B.2.23}$$

where the initial conditions are

$$y(0) = y_0, \qquad \dot{y}(0) = \dot{y}_0. \tag{B.2.24}$$

Applying the Laplace transform we have,

$$\mathscr{L}\{\ddot{y}(t)\}+k^2\mathscr{L}\{y(t)\} = 0. \tag{B.2.25}$$

We recall that

$$\mathscr{L}\{\ddot{y}(t)\} = s^2\bar{y}(s)-sy(0)-\dot{y}(0), \tag{B.2.26}$$

where we have used $\bar{y}$ to denote the transformed dependent variable. Then, from the initial conditions, we have for the transformed equation

$$s^2\bar{y}(s)-sy_0-\dot{y}_0+k^2\bar{y}(s) = 0. \tag{B.2.27}$$

We solve this simple algebraic equation for $\bar{y}(s)$,

$$\bar{y}(s) = y_0\frac{s}{s^2+k^2}+\frac{\dot{y}_0}{k}\frac{k}{s^2+k^2}, \tag{B.2.28}$$

and see that $\mathscr{L}^{-1}\{\bar{y}(s)\}$ will give $y(t)$, the solution to the problem. From tables, we may find directly that

$$y(t) = \mathscr{L}^{-1}\{\bar{y}(s)\} = y_0 \cos kt+\frac{\dot{y}_0}{k} \sin kt. \tag{B.2.29}$$

As a second example, consider the simultaneous differential equations and initial conditions

$$\frac{d^2Y(t)}{dt^2}-\frac{d^2Z(t)}{dt^2}+\frac{dZ(t)}{dt}-Y(t) = e^t-2,$$

$$2\frac{d^2Y(t)}{dt^2}-\frac{d^2Z(t)}{dt}-2\frac{dY(t)}{dt}+Z(t) = -t, \tag{B.2.30}$$

$$Y(0) = \dot{Y}(0) = Z(0) = \dot{Z}(0) = 0. \tag{B.2.31}$$

Taking the Laplace transform of (B.2.30) we have

$$s^2\bar{y}(s)-s^2\bar{z}(s)+s\bar{z}(s)-\bar{y}(s) = \frac{1}{s-1}-\frac{2}{s},$$

$$2s^2\bar{y}(s)-s^2\bar{z}(s)-2s\bar{y}(s)+\bar{z}(s) = -1/s^2. \tag{B.2.32}$$

Eliminating $\bar{z}(s)$ from the above, we obtain

$$(s^2-2s-1)\bar{y}(s) = \frac{s^2-2s-1}{s(s-1)^2}. \tag{B.2.33}$$

By the use of partial fractions, we find

$$\bar{y}(s) = \frac{1}{s(s-1)^2} = \frac{a_1}{s}+\frac{a_2}{s-1}+\frac{a_3}{(s-1)^2}$$

$$= \frac{1}{s}-\frac{1}{s-1}+\frac{1}{(s-1)^2}. \tag{B.2.34}$$

From tables, the inverse transforms are found to be

$$y(t) = 1-e^t+te^t. \tag{B.2.35}$$

In a similar manner $\bar{z}(s)$ is found to be

$$\bar{z}(s) = \frac{2s-1}{s^2(s-1)^2} = -\frac{1}{s^2}+\frac{1}{(s-1)^2}, \tag{B.2.36}$$

with the inverse given by

$$z(t) = -t+te^t. \tag{B.2.37}$$

B.2.6. *Convolution*

Another very useful method of inverting certain types of transformed solutions will now be shown. Suppose $F(t)$, $G(t)$ are two functions whose transforms are $f(s)$ and $g(s)$. That is

$$f(s) = \mathscr{L}\{F(t)\}, \qquad g(s) = \mathscr{L}\{G(t)\}. \tag{B.2.38}$$

Before we can obtain our desired result, we need a preliminary development called the translation theorem (also called the shift theorem).

THEOREM: *If* $f(s) = \mathscr{L}\{F(t)\}$, *then*

$$e^{-bs}f(s) = \mathscr{L}\{F_b(t)\}, \tag{B.2.39}$$

where

$$F_b(t) = \begin{cases} 0, & 0 < t < b \\ F(t-b), & t > b. \end{cases} \tag{B.2.40}$$

This is easily shown. Thus

$$f(s) = \int_0^\infty e^{-st}F(t)\,dt, \tag{B.2.41}$$

$$e^{-bs}f(s) = \int_0^\infty e^{-s(t+b)}F(t)\,dt, \tag{B.2.42}$$

where b is a constant, assumed greater than zero. Let $t+b = \tau$, so that the last integral becomes

$$e^{-bs}f(s) = \int_b^\infty e^{-s\tau}F(\tau-b)\,d\tau$$

$$= \int_0^\infty e^{-s\tau}F(\tau-b)\,d\tau, \tag{B.2.43}$$

since $t > 0$ is assumed. Hence, if we define $F_b(t)$ as indicated in the theorem, we have proved the desired preliminary result.

Continuing, we apply the translation theorem to $g(s)$, changing notation slightly. Thus

$$e^{-s\tau}g(s) = \mathscr{L}\{G(t-\tau)\}$$

$$= \int_0^\infty e^{-st}G(t-\tau)\,dt, \tag{B.2.44}$$

where it is understood that $G(t) = 0$, $t < 0$. Then we may write

$$f(s)g(s) = g(s)\int_0^\infty F(\tau)e^{-s\tau}\,d\tau$$

$$= \int_0^\infty F(\tau)e^{-s\tau}g(s)\,d\tau. \tag{B.2.45}$$

Incorporating the integral representation of $g(s)$ in the above, we obtain

$$f(s)g(s) = \int_0^\infty F(\tau)\int_0^\infty e^{-st}G(t-\tau)\,dt\,d\tau. \tag{B.2.46}$$

This may be expressed as the limit

$$f(s)g(s) = \lim_{T\to\infty}\int_0^T\int_0^\infty F(\tau)e^{-st}G(t-\tau)\,dt\,d\tau$$

$$= \lim_{T\to\infty}\int_0^\infty e^{-st}\int_0^T F(\tau)G(t-\tau)\,d\tau\,dt. \tag{B.2.47}$$

The above is written as

$$f(s)g(s) = \lim_{T\to\infty}\{I_1(T)+I_2(T)\}, \tag{B.2.48}$$

where

$$I_1(T) = \int_0^T e^{-st}\int_0^T F(\tau)G(t-\tau)\,d\tau\,dt \tag{B.2.49}$$

and

$$I_2(T) = \int_T^\infty e^{-st}\int_0^T F(\tau)G(t-\tau)\,d\tau\,dt. \tag{B.2.50}$$

As $T \to \infty$, $I_2(T) \to 0$. We also recall that $G(t-\tau) = 0$ for $\tau > t$, so that the upper limit of the integration with respect to τ in (B.2.49), and (B.2.50) is replaced by t. Then we have

$$f(s)g(s) = \int_0^\infty e^{-st} \int_0^t F(\tau)G(t-\tau)\,d\tau\,dt. \qquad \text{(B.2.51)}$$

Defining

$$F(t) * G(t) = \int_0^t F(\tau)G(t-\tau)\,d\tau, \qquad \text{(B.2.52)}$$

we have

$$f(s)g(s) = \mathscr{L}\{F(t) * G(t)\} \qquad \text{(B.2.53)}$$

or

$$F(t) * G(t) = \mathscr{L}^{-1}\{f(s)g(s)\}. \qquad \text{(B.2.54)}$$

This is the convolution theorem (sometimes called the '*Faltung* theorem'). We see that the operation gives the inverse transform of the product of two transforms directly in terms of the original functions. As an example,

$$\mathscr{L}^{-1}\left(\frac{1}{s^2}\frac{1}{s-a}\right) = t * e^{at} = \int_0^t \tau e^{a(t-\tau)}\,d\tau$$

$$= e^{at}\int_0^t \tau e^{-a\tau}\,d\tau = \frac{1}{a^2}(e^{at}-at-1). \qquad \text{(B.2.55)}$$

B.2.7. *The inversion integral*

The transformed solutions of many problems may yield results not contained in tables of Laplace transforms. What is needed, therefore, is a so-called 'first-principle' method of inverting a transformed solution that can be applied regardless of the availability of tabulated values. To develop the needed inversion process, we must extend our theory of the Laplace transformation by letting s represent a complex variable. An extension of the Cauchy integral formula provides the desired result.

From complex variables Cauchy's integral formula states that if $f(z)$ is an analytic function inside the closed curve C and if z_0 is a point within C, then

$$f(z_0) = \frac{1}{2\pi i}\int_C \frac{f(z)\,dz}{z-z_0}. \qquad \text{(B.2.56)}$$

An extension of this formula is needed for the present work.

THEOREM. *Let* $f(z)$ *be analytic for* $\text{Re}(z) \geq \gamma$, *where* γ *is a real constant greater than zero. Then, if* $\text{Re}(z_0) > \gamma$,

$$f(z_0) = -\frac{1}{2\pi i}\lim_{\beta\to\infty}\int_{\gamma-i\beta}^{\gamma+i\beta} \frac{f(z)\,dz}{z-z_0}. \qquad \text{(B.2.57)}$$

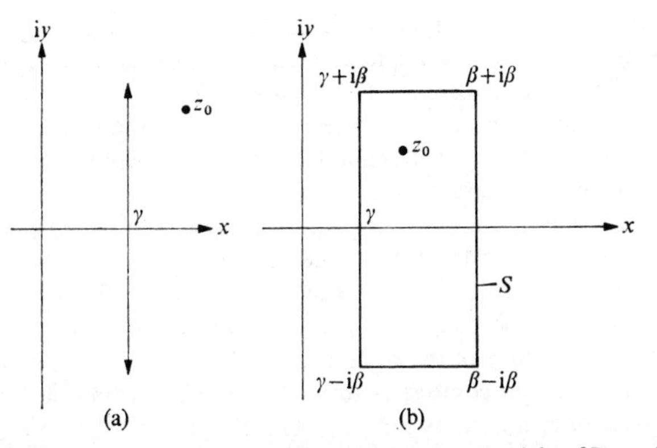

FIG. B.2. (a) The point z_0 in the complex plane located to the right of Re γ, (b) contour enclosing z_0.

This may be illustrated as shown in Fig. B.2(a). The proof of this theorem is somewhat involved, but it consists essentially of choosing a rectangular contour C that encloses z_0, as shown in Fig. B.2(b) and letting $\beta \to \infty$. By certain limiting arguments and inequalities, it is shown that the contribution from the path S vanishes, where S consists of the horizontal and right-hand vertical portions of the path.

We replace z by s in the previous theorem, so that

$$f(s) = \frac{1}{2\pi i} \lim_{\beta\to\infty} \int_{\gamma-i\beta}^{\gamma+i\beta} \frac{f(z)\,dz}{s-z}, \tag{B.2.58}$$

where s is complex and given by $s = x+iy$. Recall $f(s)$ is assumed analytic in the half-plane $\text{Re}(s) > \gamma$. We now formally apply the inverse Laplace transform to the functions on either side of this equation, giving

$$F(t) = \mathscr{L}^{-1}\{f(s)\} = \frac{1}{2\pi i}\mathscr{L}^{-1}\left\{\lim_{\beta\to\infty} \int_{\gamma-i\beta}^{\gamma+i\beta} \frac{f(z)}{s-z}\,dz\right\}$$

$$= \frac{1}{2\pi i} \lim_{\beta\to\infty} \int_{\gamma-i\beta}^{\gamma+i\beta} f(z)\mathscr{L}^{-1}\left(\frac{1}{s-z}\right) dz. \tag{B.2.59}$$

But

$$\mathscr{L}^{-1}\{1/(s-z)\} = e^{zt}, \tag{B.2.60}$$

so that

$$F(t) = \frac{1}{2\pi i} \lim_{\beta\to\infty} \int_{\gamma-i\beta}^{\gamma+i\beta} e^{st} f(s)\,ds, \tag{B.2.61}$$

where, in the above, the dummy variable z has been replaced by s. This is the inversion integral of the Laplace transform. The path of integration is often called the 'Bromwich contour', since Bromwich first devised this method of handling certain integrals that arose in operational mathematics.

The basic concept that will underlie the actual evaluation of inverse Laplace transforms by use of the inversion integral is the application of contour integration in the complex plane. The details of the calculation will depend on the nature of the transformed function $f(s)$. It is usually found that $f(s)$ is either: (1) a single-valued function of s with a finite or an infinite number of poles in the complex s-plane, or (2) a function of s having a number of branch points in the s-plane, as well as a number of poles. In the case of (1) a complete contour is formed from the Bromwich integral by adding a circle of radius R, as shown in Fig. B.3. Enclosed within the entire

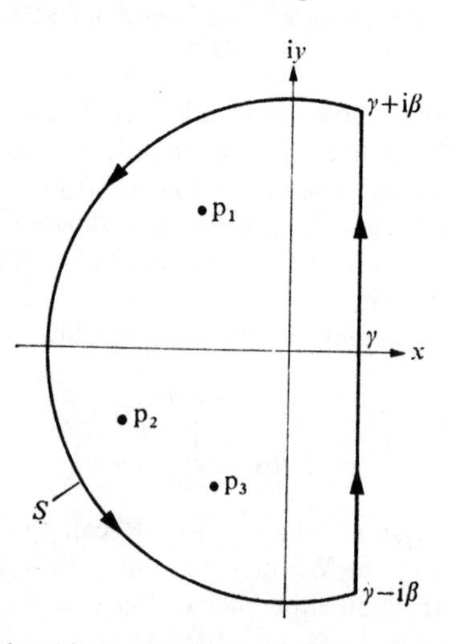

FIG. B.3. Poles enclosed by the Bromwich contour closed in the left-hand plane.

contour C will be a number of poles. From the theory of residues, we know that

$$\int_C f(z)\,dz = 2\pi i \sum_1^n (\text{Res}), \tag{B.2.62}$$

so that in our case we have

$$\frac{1}{2\pi i}\int_S e^{st} f(s)\,ds + \frac{1}{2\pi i}\int_{\gamma-i\beta}^{\gamma+i\beta} e^{st} f(s)\,ds = \sum_1^n (\text{Res}). \tag{B.2.63}$$

The essential point of the method is to then show by proper application of inequalities and limits that

$$\lim_{R\to\infty} \int_S e^{st} f(s)\, ds \to 0. \tag{B.2.64}$$

We are then left only with the Bromwich integral which is the inverse Laplace transform, so that we have

$$\mathscr{L}^{-1}\{f(s)\} = \sum_{1}^{n} (\text{Res}). \tag{B.2.65}$$

Ideas similar to those outlined above also apply in the case of (2). However, the function $f(s)$ is no longer single-valued because of the existence of branch points, so that closing the contour as shown in the previous figure would lead to difficulties. The procedure is still to complete the contour, but to deform the contour so as to envelop the branch cuts that are made to make $f(s)$ a single-valued function.

As an example, suppose

$$f(s) = e^{-\sqrt{s}}/s. \tag{B.2.66}$$

Now $\sqrt{s}$ is not single-valued, and a branch point exists at the origin, in this case. The branch cut necessary to make $\sqrt{s}$ single-valued may be, in theory, of arbitrary orientation. However, a convenient selection is along the negative real axis. We thus have the situation shown in Fig. B.4(a). The contour is completed as shown in Fig. B.4(b). We then have

$$\int_{\gamma-i\beta}^{\gamma+i\beta} + \int_{S_1} + \int_{\Gamma_1} + \int_{\Gamma_2} + \int_{\Gamma_3} + \int_{S_2} = 0, \tag{B.2.67}$$

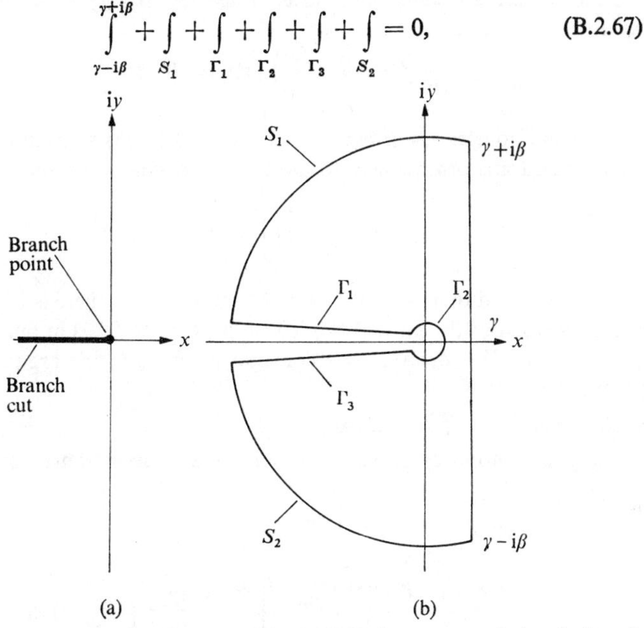

FIG. B.4. (a) Branch point located at the origin; (b) keyhole contour enclosing the branch point.

where it is assumed that no other poles have been enclosed within the contour. Then, again by proper application of inequalities and limits, it must be shown that

$$\int_{S_1} = \int_{S_2} = 0. \tag{B.2.68}$$

It remains then to evaluate $\int_{\Gamma_1}, \int_{\Gamma_2}, \int_{\Gamma_3}$. The second integral will be found to give essentially the residue at the origin, while the first and last integrals will be real integrals (integration is along the real axis). Thus the problem of finding the inverse transform will be reduced to evaluating

$$\mathscr{L}^{-1}\{f(s)\} = -\int_{\Gamma_1} - \int_{\Gamma_2} - \int_{\Gamma_3}. \tag{B.2.69}$$

B.3. Fourier transforms

B.3.1. *Definition*

The Fourier transform of a function $F(t)$, defined for all t, is given by

$$f(\gamma) = \frac{1}{\sqrt{(2\pi)}}\int_{-\infty}^{\infty} e^{i\gamma t}F(t)\,dt, \tag{B.3.1}$$

where $\exp(i\gamma t)$ is the kernel of the transformation. The above is sometimes referred to as the *exponential Fourier transform.* If $F(t)$ is an odd function, the above reduces to

$$f(\gamma) = \sqrt{\left(\frac{2}{\pi}\right)}\int_{0}^{\infty} \sin t\gamma F(t)\,dt, \tag{B.3.2}$$

which is the Fourier *sine transform.* Likewise, if $F(t)$ is an even function, the exponential transform reduces to the Fourier *cosine transform*

$$f(\gamma) = \sqrt{\left(\frac{2}{\pi}\right)}\int_{0}^{\infty} \cos t\gamma F(t)\,dt. \tag{B.3.3}$$

It might be noted that, as in the case of Fourier series, there is not unanimity in the selection of the normalizing constant. It is $1/\sqrt{(2\pi)}$ in our case, but it is sometimes $1/2\pi$, or even as a constant included in the arguments of the exponential.

B.3.2. *Transforms of derivatives*

Using integration by parts, we have for the Fourier transform of dF/dt,

$$\begin{aligned} \mathscr{F}(F') &= \frac{1}{\sqrt{(2\pi)}}\int_{-\infty}^{\infty} e^{i\gamma t}F'(t)\,dt \\ &= \left[\frac{1}{\sqrt{(2\pi)}}e^{i\gamma t}F(t)\right]_{-\infty}^{\infty} - \frac{i\gamma}{\sqrt{(2\pi)}}\int_{-\infty}^{\infty} e^{i\gamma t}F(t)\,dt \\ &= -i\gamma f(\gamma) \end{aligned} \tag{B.3.4}$$

We note that $e^{i\gamma t}F(t) \to 0$, as $t \to \pm\infty$ has been assumed, which places restrictions on the behaviour of $F(t)$ for large t. In general, we may show that

$$\mathscr{F}\{F^{(n)}(t)\} = (-i\gamma)^n f(\gamma). \tag{B.3.5}$$

The Fourier transform of derivatives places conditions on the function at $\pm\infty$, whereas the Laplace transform places one of the conditions at zero, the other at ∞. In either case, certain boundary conditions have been effectively placed on the function. The Fourier sine and cosine transforms have zero as their lower limits, we recall, and should incorporate the initial conditions in a manner similar to the Laplace transform. Thus we have

$$f_c^{(n)}(\gamma) = \sqrt{\left(\frac{2}{\pi}\right)} \int_0^\infty \cos t\gamma F^{(n)}(t)\, dt, \tag{B.3.6}$$

$$f_s^{(n)}(\gamma) = \sqrt{\left(\frac{2}{\pi}\right)} \int_0^\infty \sin t\gamma F^{(n)}(t)\, dt, \tag{B.3.7}$$

as the cosine and sine transforms of d^nF/dt^n. Integration by parts of $f_c^{(n)}(\gamma)$ gives

$$f_c^{(n)} = \sqrt{\left(\frac{2}{\pi}\right)} [F^{(n-1)}(t) \cos t\gamma]_0^\infty + \gamma \sqrt{\left(\frac{2}{\pi}\right)} \int_0^\infty \sin t\gamma F^{(n-1)}(t)\, dt. \tag{B.3.8}$$

We assume

$$\lim_{t\to 0} \sqrt{\left(\frac{2}{\pi}\right)} F^{(n-1)}(t) = a_{n-1}, \qquad \lim_{t\to\infty} F^{(n-1)}(t) = 0, \tag{B.3.9}$$

so that

$$f_c^{(n)} = -a_{n-1} + \gamma f_s^{(n-1)}. \tag{B.3.10}$$

However, integrating $f_s^{(n)}(\gamma)$ by parts, we obtain

$$f_s^{(n)} = -\gamma f_c^{(n-1)}. \tag{B.3.11}$$

From this and the previous equation we have

$$f_c^{(n)} = -a_{n-1} - \gamma^2 f_c^{(n-2)}. \tag{B.3.12}$$

By using these results as a type of recursion formula, we obtain

$$f_c^{(2n)} = -\sum_{r=0}^{n-1} (-1)^r a_{2n-2r-1} \gamma^{2r} + (-1)^n \gamma^{2n} f_c(\gamma), \tag{B.3.13}$$

$$f_c^{(2n+1)} = -\sum_{r=0}^{n} (-1)^r a_{2n-2r} \gamma^{2r} + (-1)^n \gamma^{2n+1} f_s(\gamma). \tag{B.3.14}$$

Similarly, for the sine transforms of derivatives, we obtain

$$f_s^{(2n)} = -\sum_{r=1}^{n} (-1)^r \gamma^{2r-1} a_{2n-2r} + (-1)^{n+1} \gamma^{2n} f_s, \tag{B.3.15}$$

$$f_c^{(2n+1)} = -\sum_{r=1}^{n} (-1)^r \gamma^{2r-1} a_{2n-2r+1} + (-1)^{n+1} \gamma^{2n+1} f_c. \tag{B.3.16}$$

B.3.3. *The inverse transform*

Given a function $f(\gamma)$, the inverse transformation will give the function $F(t)$. Thus we indicate this symbolically by

$$\mathscr{F}^{-1}\{f(\gamma)\} = F(t). \tag{B.3.17}$$

The problem is particularly simple if $f(\gamma)$ is recognized as the transform of a known function $F(t)$. The procedure for handling this case, including the use of partial fractions, has been covered in the section on Laplace transforms. Tables of Fourier transforms are available for such use.

The case frequently arises, however, in which $f(\gamma)$ is not tabulated or cannot be broken into simpler functions which are tabulated. In such cases, a formal inversion process must be used. Even an abbreviated version of the process for Fourier transforms, such as was given for the Laplace transforms, will not be given here. Instead, we merely present a formal definition of the inverse transformation.

Inverse exponential transforms: If $f(\gamma)$ is the Fourier transform of $F(t)$ given by

$$f(\gamma) = \frac{1}{\sqrt{(2\pi)}}\int_{-\infty}^{\infty} e^{i\gamma t}F(t)\,dt, \tag{B.3.18}$$

then $F(t)$ is given by

$$F(t) = \frac{1}{\sqrt{(2\pi)}}\int_{-\infty}^{\infty} e^{-i\gamma t}f(\gamma)\,d\gamma. \tag{B.3.19}$$

Inverse sine, cosine transforms: If $f_s(\gamma)$ is the sine transform

$$f_s(\gamma) = \sqrt{\left(\frac{2}{\pi}\right)}\int_{0}^{\infty} \sin t\gamma F(t)\,dt, \tag{B.3.20}$$

then

$$F(t) = \sqrt{\left(\frac{2}{\pi}\right)}\int_{0}^{\infty} \sin t\gamma f(\gamma)\,d\gamma. \tag{B.3.21}$$

Similarly, if $f_c(\gamma)$ is the cosine transform

$$f_c(\gamma) = \sqrt{\left(\frac{2}{\pi}\right)}\int_{0}^{\infty} \cos t\gamma F(t)\,dt, \tag{B.3.22}$$

then

$$F(t) = \sqrt{\left(\frac{2}{\pi}\right)}\int_{0}^{\infty} \cos t\gamma f(\gamma)\,d\gamma. \tag{B.3.23}$$

A transform whose inverse is identical in form, such as the sine and cosine transforms, is said to be symmetrical. The exponential transform, it is seen, is not quite symmetrical. As in the case of Laplace transforms, γ may be a complex variable, so that the inverse transform may involve contour integration in the complex plane.

B.3.4. *Convolution*

Consider $f(\gamma)$, $g(\gamma)$ to be the Fourier transformations of $F(t)$, $G(t)$ respectively. Then, as in the case of the Laplace transform, there is a convolution theorem which enables the inverse transform of the product $f(\gamma)g(\gamma)$ to be expressed as a convolution-type integral. That is,

$$\mathscr{F}^{-1}\{f(\gamma)g(\gamma)\} = F(t) * G(t). \tag{B.3.24}$$

The actual form of the integral is found as follows:

$$\begin{aligned}\mathscr{F}^{-1}\{f(\gamma)g(\gamma)\} &= \frac{1}{\sqrt{(2\pi)}}\int_{-\infty}^{\infty} f(\gamma)g(\gamma)e^{-i\gamma t}\,d\gamma \\ &= \frac{1}{2\pi}\int_{-\infty}^{\infty} f(\gamma)e^{-i\gamma t}\,d\gamma \int_{-\infty}^{\infty} G(\eta)e^{i\eta\gamma}\,d\eta \\ &= \frac{1}{2\pi}\int_{-\infty}^{\infty} G(\eta)\,d\eta \int_{-\infty}^{\infty} f(\gamma)e^{-i\gamma(t-\eta)}\,d\gamma.\end{aligned} \tag{B.3.25}$$

But

$$F(t-\eta) = \frac{1}{\sqrt{(2\pi)}}\int_{-\infty}^{\infty} f(\gamma)e^{-i\gamma(t-\eta)}\,d\gamma, \tag{B.3.26}$$

so that

$$\begin{aligned}\mathscr{F}^{-1}\{f(\gamma)g(\gamma)\} &= \frac{1}{\sqrt{(2\pi)}}\int_{-\infty}^{\infty} G(\eta)F(t-\eta)\,d\eta \\ &= F * G.\end{aligned} \tag{B.3.27}$$

Thus

$$F * G = \frac{1}{\sqrt{(2\pi)}}\int_{-\infty}^{\infty} G(\eta)F(t-\eta)\,d\eta, \tag{B.3.28}$$

is the convolution, or *Faltung* integral form for the Fourier transform.

B.3.5. *Finite Fourier transforms*

The Fourier-integral transform concepts for extended intervals may be used in problems involving a finite interval. The development starts with Fourier series considerations. The Fourier-series representation of a function periodic in the interval $-l < x < l$ is

$$F(x) = \frac{a_0}{2} + \sum_{n=1}^{\infty} \left(a_n \cos \frac{n\pi}{l} x + b_n \sin \frac{n\pi}{l} x \right), \tag{B.3.29}$$

where

$$a_n = \frac{1}{l} \int_{-l}^{l} F(u) \cos \frac{n\pi}{l} u \, du, \tag{B.3.30}$$

$$b_n = \frac{1}{l} \int_{-l}^{l} F(u) \sin \frac{n\pi}{l} u \, du. \tag{B.3.31}$$

We recall that if $F(x)$ is periodic in $0 < x < l$, or odd in $-l < x < l$, then the representation is

$$F(x) = \frac{2}{l} \sum_{n=1}^{\infty} \sin \frac{n\pi}{l} x \int_0^l F(u) \sin \frac{n\pi}{l} u \, du. \tag{B.3.32}$$

If $F(x)$ is periodic in $0 < x < l$ or even in $-l < x < l$, then

$$F(x) = \frac{1}{l} \int_0^l F(u) \, du + \frac{2}{l} \sum_{n=1}^{\infty} \cos \frac{n\pi x}{l} \int_0^l F(u) \cos \frac{n\pi}{l} u \, du. \tag{B.3.33}$$

We introduce the idea of a *finite Fourier transform* by defining

$$f_c(n) = \frac{1}{l} \int_0^l F(x) \cos \frac{n\pi}{l} x \, dx, \tag{B.3.34}$$

$$f_s(n) = \frac{1}{l} \int_0^l F(x) \sin \frac{n\pi}{l} x \, dx, \tag{B.3.35}$$

as the finite Fourier cosine and sine transforms respectively, where $F(x)$ is defined on the interval $0 < x < l$. Then we define the following inversions:

Finite inverse transform: If $f_c(n)$ is the finite Fourier cosine transform of $F(x)$ in $0 < x < l$, then

$$F(x) = \frac{f_c(0)}{l} + 2 \sum_{n=1}^{\infty} f_c(n) \cos \frac{n\pi}{l} x. \tag{B.3.36}$$

If $f_s(n)$ is the finite Fourier sine transform of $F(x)$, then

$$F(x) = 2 \sum_{n=1}^{\infty} f_s(n) \sin \frac{n\pi}{l} x. \tag{B.3.37}$$

In the above, the series representation plays the role of the inverse transform. Now, the use of finite Fourier transforms does not solve problems which are incapable of solution by the direct application of Fourier series. However, there will be occasion when this approach to a problem will have advantages over the direct series expansions of the various functions.

B.3.6. *The Fourier integral*

Again consider the Fourier series representation of $F(x)$, defined in $-l < x < l$ as

$$F(x) = \frac{a_0}{2} + \sum_{n=1}^{\infty}\left(a_n \cos\frac{n\pi}{l}x + b_n \sin\frac{n\pi}{l}x\right), \tag{B.3.38}$$

where a_n, b_n have been previously defined by (B.3.30) and (B.3.31). If the a_n, b_n definitions are introduced directly into the above, we have

$$F(x) = \frac{1}{2l}\int_{-l}^{l} F(u)\,du +$$

$$+\frac{1}{l}\sum_{n=1}^{\infty}\left\{\int_{-l}^{l} F(u)\cos\frac{n\pi}{l}u \cos\frac{n\pi}{l}x\,dx + \int_{-l}^{l} F(u)\sin\frac{n\pi}{l}u \sin\frac{n\pi}{l}x\,dx\right.$$

$$= \frac{1}{2l}\int_{-l}^{l} F(u)\,du + \frac{1}{l}\sum_{n=1}^{\infty}\int_{-l}^{l} F(u)\cos\frac{n\pi}{l}(u-x)\,du. \tag{B.3.39}$$

Denote the ratio $n\pi/l$ as α. As $l \to \infty$ (so that $\pi/l \to \delta\alpha$) the first term becomes zero and the series becomes

$$F(x) = \frac{1}{\pi}\sum_{n=1}^{\infty}\delta\alpha \int_{-\pi/\delta\alpha}^{\pi/\delta\alpha} F(u)\cos\{n\delta\alpha(u-x)\}\,du$$

$$= \frac{1}{\pi}\int_{-\pi/\delta\alpha}^{\pi/\delta\alpha} F(u)\,du \sum_{n=1}^{\infty}\cos\{(u-x)n\delta\alpha\}\,\delta\alpha. \tag{B.3.40}$$

The limiting value of a series is an integral under the proper conditions, so that

$$\sum_{i=1}^{\infty} F(x_i)\,\delta x_i \to \int_{0}^{\infty} F(x)\,dx. \tag{B.3.41}$$

Now $n\delta\alpha$ is analogous to x_i and $\delta\alpha$ to δx_i, so that we have for $F(x)$,

$$F(x) = \frac{1}{\pi}\int_{-\infty}^{\infty} F(u)\,\mathrm{d}u \int_{0}^{\infty} \cos\{(u-x)\alpha\}\,\mathrm{d}\alpha. \tag{B.3.42}$$

Hence the *Fourier integral* representation of the function $F(x)$ is

$$F(x) = \frac{1}{\pi}\int_{0}^{\infty} \mathrm{d}\alpha \int_{-\infty}^{\infty} F(u)\cos\{(u-x)\alpha\}\,\mathrm{d}u. \tag{B.3.43}$$

It may be easily shown that if $F(x)$ is defined in $0 < x < \infty$, then

$$F(x) = \frac{2}{\pi}\int_{0}^{\infty} \mathrm{d}\alpha \int_{0}^{\infty} F(u)\cos x\alpha \cos u\alpha\,\mathrm{d}u. \tag{B.3.44}$$

B.4. Hankel transforms

B.4.1. *Definitions*

The Hankel transform of order ν of the function $f(x)$ is defined as

$$\bar{f}(\xi) = \int_{0}^{\infty} xf(x)J_\nu(\xi x)\,\mathrm{d}x. \tag{B.4.1}$$

The inverse transform is given by

$$f(x) = \int_{0}^{\infty} \xi\bar{f}(\xi)J_\nu(x\xi)\,\mathrm{d}\xi. \tag{B.4.2}$$

Thus complete symmetry exists in the Hankel transform and its inverse.

As a simple example of a Hankel transform, suppose

$$f(x) = \begin{cases} 1, & 0 < x < a \\ 0, & x > a. \end{cases} \tag{B.4.3}$$

Using the zero-order Hankel transform gives

$$\bar{f}(\xi) = \int_{0}^{a} xJ_0(\xi x)\,\mathrm{d}x. \tag{B.4.4}$$

From the relation

$$\frac{\mathrm{d}}{\mathrm{d}z}\{z^n J_n(z)\} = z^n J_{n-1}(z), \tag{B.4.5}$$

one obtains

$$\bar{f}(\xi) = \frac{1}{\xi}\int_0^a \frac{\mathrm{d}}{\mathrm{d}x}\{xJ_1(\xi x)\}\,\mathrm{d}x = \left[\frac{x}{\xi}J_1(\xi x)\right]_0^a.$$

$$= \frac{a}{\xi}J_1(\xi a). \tag{B.4.6}$$

B.4.2. *Transforms of derivatives and Parseval's theorem*

The Hankel transform of the derivative $\mathrm{d}f/\mathrm{d}x$ is given by

$$\bar{f}_\nu'(\xi) = \int_0^\infty x\frac{\mathrm{d}f}{\mathrm{d}x}J_\nu(\xi x)\,\mathrm{d}x. \tag{B.4.7}$$

Using integration by parts and the derivative and recursion formula for Bessel functions, one obtains

$$\bar{f}_\nu'(\xi) = -\xi\left\{\frac{\nu+1}{2\nu}\bar{f}_{\nu-1}(\xi) - \frac{\nu-1}{2\nu}\bar{f}_{\nu+1}(\xi)\right\}. \tag{B.4.8}$$

Formulas for higher derivatives are obtained by repeated application of (B.4.8).

The Hankel transform finds particular application to problems involving polar and cylindrical coordinate systems. Considerable simplification occurs, for example, to the operator $f''+f'/r-\nu^2f/r^2$. Thus it may be shown that

$$\int_0^\infty r\left(\frac{\mathrm{d}^2f}{\mathrm{d}r^2}+\frac{1}{r}\frac{\mathrm{d}f}{\mathrm{d}r}-\frac{\nu^2}{r^2}f\right)J_\nu(\xi r)\,\mathrm{d}r = -\xi^2\bar{f}_\nu(\xi). \tag{B.4.9}$$

Finally, it should be noted that a simple convolution theorem does not exist for Hankel transforms. This is because there is no simple expression for the product $J_\nu(\xi x)J_\nu(\xi y)$ as there is, say, for $\exp(\mathrm{i}\xi x)\exp(\mathrm{i}\xi y)$ that arises in Fourier transforms. However, there is a simple Parseval theorem which states that

$$\int_0^\infty xf(x)g(x)\,\mathrm{d}x = \int_0^\infty \xi\bar{f}(\xi)\bar{g}(\xi)\,\mathrm{d}\xi, \tag{B.4.10}$$

where $\bar{f}(\xi)$, $\bar{g}(\xi)$ are the Hankel transforms of $f(x)$, $g(x)$.

B.5. Tables of transforms

A few brief tables of Laplace and Fourier transforms are included here for use in analyses and problems presented in the book. More extensive tables are presented in the textbooks by Churchill [7] and Sneddon [51]. The tables in Erdelyi [15] represent some of the most complete listings.

TABLE B.1
Laplace transforms

$$F(t) = \frac{1}{2\pi i} \lim_{\beta\to\infty} \int_{\gamma-i\beta}^{\gamma+i\beta} f(s)e^{st}\,ds, \qquad f(s) = \int_0^\infty F(t)e^{-st}\,dt$$

$F(t)$	$f(s)$
1. $\int_0^t F_1(t-\tau)F_2(\tau)\,d\tau$	$f_1(s)f_2(s)$
2. $F^{(n)}(t)$	$s^n f(s)-s^{n-1}F(0)-\ldots-F^{(n-1)}(0)$
3. $e^{at}F(t)$	$f(s-a)$
4. 1	$\frac{1}{s}$
5. $\frac{t^{n-1}}{(n-1)!}$	$\frac{1}{s^n}$ $(n = 1, 2, \ldots)$
6. $\frac{1}{\sqrt{(\pi t)}}$	$\frac{1}{\sqrt{s}}$
7. e^{at}	$\frac{1}{s-a}$
8. $\frac{1}{(n-1)!}t^{n-1}e^{at}$	$\frac{1}{(s-a)^n}$ $(n = 1, 2, \ldots)$
9. $\frac{1}{a-b}(e^{at}-e^{bt})$	$\frac{1}{(s-a)(s-b)}$
10. $\frac{1}{a-b}(ae^{at}-be^{bt})$	$\frac{s}{(s-a)(s-b)}$
11. $\frac{1}{a}\sin at$	$\frac{1}{s^2+a^2}$
12. $\cos at$	$\frac{s}{s^2+a^2}$
13. $\frac{1}{a}\sinh at$	$\frac{1}{s^2-a^2}$
14. $\cosh at$	$\frac{s}{s^2-a^2}$
15. $\frac{1}{b}e^{at}\sin bt$	$\frac{1}{(s-a)^2+b^2}$
16. $e^{at}\cos bt$	$\frac{s-a}{(s-a)^2+b^2}$
17. $J_0(at)$	$\frac{1}{\sqrt{(s^2+a^2)}}$
18. $H\langle t-k\rangle$	$\frac{e^{-ks}}{s}$

TABLE B.2
Fourier transforms

$$f(x) = \frac{1}{\sqrt{(2\pi)}} \int_{-\infty}^{\infty} F(\xi) e^{-i\xi x}\, d\xi, \qquad F(\xi) = \frac{1}{\sqrt{(2\pi)}} \int_{-\infty}^{\infty} f(x) e^{i\xi x}\, dx.$$

$f(x)$	$F(\xi)$
$\dfrac{\sin(\alpha x)}{x}$	$\begin{cases} \left(\dfrac{\pi}{2}\right)^{\frac{1}{2}} & \lvert\xi\rvert < a \\ 0 & \lvert\xi\rvert > a \end{cases}$
$\left.\begin{matrix} e^{i\omega x} & p < x < q \\ 0 & x < p,\quad x > q \end{matrix}\right\}$	$\dfrac{i}{(2\pi)^{\frac{1}{2}}} \dfrac{e^{ip(\omega+\xi)} - e^{iq(\omega+\xi)}}{\xi}$
$\left.\begin{matrix} e^{-cx+i\omega x} & x > 0 \\ 0 & x < 0 \end{matrix}\right\}$	$\dfrac{i}{(2\pi)^{\frac{1}{2}}(\omega+\xi+ic)}$
$\exp(-px^2), \quad R(p) > 0$	$(2p)^{-\frac{1}{2}} \exp(-\xi^2/4p)$
$\cos(px^2)$	$(2p)^{-\frac{1}{2}} \cos\left(\dfrac{\xi^2}{4p} - \frac{1}{4}\pi\right)$
$\sin(px^2)$	$(2p)^{-\frac{1}{2}} \sin\left(\dfrac{\xi^2}{4p} + \frac{1}{4}\pi\right)$
$\dfrac{1}{\lvert x\rvert}$	$\dfrac{1}{\lvert\xi\rvert}$
$\dfrac{e^{-a\lvert x\rvert}}{\lvert x\rvert^{\frac{1}{2}}}$	$\dfrac{\{(a^2+\xi^2)^{\frac{1}{2}}+a\}^{\frac{1}{2}}}{(a^2+\xi^2)^{\frac{1}{2}}}$
$\left.\begin{matrix} (a^2-x^2)^{-\frac{1}{2}} & \lvert x\rvert < a \\ 0 & \lvert x\rvert > a \end{matrix}\right\}$	$(\frac{1}{2}\pi)^{\frac{1}{2}} J_0(a\xi)$
$\dfrac{\sin\{b(a^2+x^2)^{\frac{1}{2}}\}}{(a^2+x^2)^{\frac{1}{2}}}$	$\begin{cases} 0 & \lvert\xi\rvert > b \\ (\frac{1}{2}\pi)^{\frac{1}{2}} J_0\{a(b^2-\xi^2)^{\frac{1}{2}}\} & \lvert\xi\rvert < b \end{cases}$

TABLE B.3
Fourier cosine transforms

$$f(x) = \sqrt{\left(\frac{2}{\pi}\right)} \int_0^\infty F_c(\xi) \cos \xi x \, d\xi, \qquad F_c(\xi) = \sqrt{\left(\frac{2}{\pi}\right)} \int_0^\infty f(x) \cos \xi x \, dx$$

$f(x)$	$F_c(\xi)$
$\left.\begin{matrix} 1, & 0 < x < a \\ 0, & x > a \end{matrix}\right\}$	$\left(\frac{2}{\pi}\right)^{\frac{1}{2}} \frac{\sin(\xi a)}{\xi}$
e^{-x}	$\left(\frac{2}{\pi}\right)^{\frac{1}{2}} \frac{1}{1+\xi^2}$
$\operatorname{sech}(\pi x)$	$\frac{1}{1+\xi^4}$
$\exp(-x^2)$	$\exp(-\xi^2)$
$\cos(\frac{1}{2}x^2)$	$\frac{1}{\sqrt{2}}\{\cos(\frac{1}{2}\xi^2)+\sin(\frac{1}{2}\xi^2)\}$
$\sin(\frac{1}{2}x^2)$	$\frac{1}{\sqrt{2}}\{\cos(\frac{1}{2}\xi^2)-\sin(\frac{1}{2}\xi^2)\}$

TABLE B.4
Fourier sine transforms

$$f(x) = \sqrt{\left(\frac{2}{\pi}\right)} \int_0^\infty F_s(\xi) \sin \xi x \, d\xi, \qquad F_c(\xi) = \sqrt{\left(\frac{2}{\pi}\right)} \int_0^\infty f(x) \sin \xi x \, dx$$

$f(x)$	$F_s(\xi)$
e^{-x}	$\left(\frac{2}{\pi}\right)^{\frac{1}{2}} \frac{1}{1+\xi^2}$
$x \exp(-x^2/2)$	$\exp(-\xi^2/2)$
$x^n e^{-px}$	$\frac{2^{n+\frac{1}{2}} p^n n! \, \xi}{\pi^{\frac{1}{2}}(p^2+\xi^2)^{n+1}}$
$\cos(ax^2)$	$-a^{-\frac{1}{2}}\left\{\cos\left(\frac{\xi^2}{4a}\right)\operatorname{Si}\left(\frac{\xi}{\sqrt{(2\pi a)}}\right) - \sin\left(\frac{\xi^2}{4a}\right)\operatorname{Ci}\left(\frac{\xi}{\sqrt{(2\pi a)}}\right)\right\}$
$x^{-\frac{1}{2}} \exp(-ax^{\frac{1}{2}})$	$\xi^{-\frac{1}{2}}\{\cos(2a\xi)^{\frac{1}{2}} - \sin(2a\xi)^{\frac{1}{2}}\}$
$\left.\begin{matrix} 0, & 0 < x < a \\ (x^2-a^2)^{-\frac{1}{2}}, & x > a \end{matrix}\right\}$	$\left(\frac{\pi}{2}\right)^{\frac{1}{2}} J_0(a\xi)$

TABLE B.5

Finite Fourier cosine transforms

$$f(x) = \frac{1}{a} f_c(0) + \frac{2}{a} \sum_{n=1}^{\infty} f_c(n) \cos \frac{n\pi x}{a}, \qquad f_c(n) = \int_0^a f(x) \cos \frac{n\pi x}{a} \, dx$$

$f(x)$	$f_c(n)$
1	$\begin{cases} a, & n = 0 \\ 0, & n = 1, 2, 3, \ldots \end{cases}$
$\left.\begin{matrix} 1, & 0 < x < \frac{1}{2}a \\ -1, & \frac{1}{2}a < x < a \end{matrix}\right\}$	$\begin{cases} 0, & n = 0 \\ \frac{2a}{\pi n} \sin(\frac{1}{2}n\pi) & n = 1, 2, \ldots \end{cases}$
x	$\begin{cases} \frac{1}{2}a^2, & n = 0 \\ \left(\frac{a}{\pi n}\right)^2 \{(-1)^n - 1\}, & n = 1, 2, \ldots \end{cases}$
x^2	$\begin{cases} \frac{1}{3}a^3, & n = 0 \\ \frac{2a^3}{\pi^2 n^2} (-1)^n, & n = 1, 2, \ldots \end{cases}$
$\left(1 - \frac{x}{a}\right)^2$	$\begin{cases} \frac{1}{3}a, & n = 0 \\ \frac{2a}{\pi^2 n^2}, & n = 1, 2, \ldots \end{cases}$

TABLE B.6

Finite Fourier sine transforms

$$f(x) = \frac{2}{a} \sum_{n=1}^{\infty} f_s(n) \sin \frac{n\pi x}{a}, \qquad f_s(n) = \int_0^a f(x) \sin \frac{n\pi x}{a} \, dx$$

$f(x)$	$f_s(n)$
1	$\frac{a}{\pi n} \{1 + (-1)^{n+1}\}$
x	$(-1)^{n+1} \frac{a^2}{\pi n}$
$1 - \frac{x}{a}$	$\frac{a}{\pi n}$
$\left.\begin{matrix} x, & 0 \le x \le \frac{1}{2}a \\ a - x, & \frac{1}{2}a \le x \le a \end{matrix}\right\}$	$\frac{2a^2}{\pi^2 n^2} \sin(\frac{1}{2}n\pi)$
x^2	$\frac{a^3(-1)^{n-1}}{\pi n} - \frac{2a^3\{1 - (-1)^n\}}{\pi^3 n^3}$
x^3	$(-1)^n \frac{a^4}{\pi^3} \left(\frac{6}{n^3} - \frac{\pi^2}{n}\right)$
$x(a^2 - x^2)$	$(-1)^{n+1} \frac{6a^4}{\pi^3 n^3}$

TABLE B.7
Hankel transforms

$$f(x) = \int_0^\infty \xi \bar{f}(\xi) J_\nu(\xi x)\,\mathrm{d}\xi, \qquad \bar{f}(\xi) = \int_0^\infty x f(x) J_\nu(\xi x)\,\mathrm{d}x$$

$f(x)$	ν	$\bar{f}(\xi)$
$\begin{cases} x^\nu, & 0 < x < a \\ 0, & x > a \end{cases}$	> -1	$\dfrac{a^{\nu+1}}{\xi} J_{\nu+1}(\xi a)$
$\begin{cases} 1, & 0 < x < a \\ 0, & x > a \end{cases}$	0	$\dfrac{a}{\xi} J_1(a\xi)$
$\begin{cases} (a^2-x^2), & 0 < x < a, \\ 0, & x > a \end{cases}$	0	$\dfrac{4a}{\xi^3} J_1(\xi a) - \dfrac{2a^2}{\xi^2} J_0(\xi a)$
$x^\nu \exp(-px^2)$	> -1	$\dfrac{\xi^\nu}{(2p)^{\nu+1}} \exp(-\xi^2/4p)$
$\dfrac{e^{-px}}{x}$	0	$(\xi^2+p^2)^{-\frac{1}{2}}$
e^{-px}	0	$p(\xi^2+p^2)^{-\frac{3}{2}}$
$x^{-2}e^{-px}$	1	$\dfrac{(\xi^2+p^2)^{\frac{1}{2}}-p}{\xi}$
$\dfrac{e^{-px}}{x}$	1	$\dfrac{1}{\xi} - \dfrac{p}{\xi(\xi^2+p^2)^{\frac{1}{2}}}$
e^{-px}	1	$\xi(\xi^2+p^2)^{-\frac{3}{2}}$
$\dfrac{a}{(a^2+x^2)^{\frac{3}{2}}}$	0	$e^{-a\xi}$
$\dfrac{\sin(ax)}{x}$	0	$\begin{cases} 0, & \xi > a \\ (a^2-\xi^2)^{-\frac{1}{2}}, & 0 < \xi < a \end{cases}$
$\dfrac{\sin(ax)}{x}$	1	$\begin{cases} \dfrac{a}{\xi(\xi^2-a^2)^{\frac{1}{2}}}, & \xi > a \\ 0 & \xi < a \end{cases}$
$\dfrac{\sin(x)}{x^2}$	0	$\begin{cases} \sin^{-1}\left(\dfrac{1}{\xi}\right), & \xi > 1 \\ \frac{1}{2}\pi, & \xi < 1 \end{cases}$

TABLE B.8
Fourier spectra of pulses

Description	$p(t)$	Pulse	Spectrum	$\overline{f_c}(\omega)$†
Rectangular pulse	$A,\ \lvert t\rvert<\frac{\tau}{2}$ $0,\ \lvert t\rvert>\frac{\tau}{2}$	p, A, $-\frac{\tau}{2}$, $\frac{\tau}{2}$, t	$\frac{\tau}{2\pi}$, ω, $\frac{2\pi}{\tau}\ \frac{4\pi}{\tau}\ \frac{6\pi}{\tau}\ldots\left(\frac{2n\pi}{\tau}\right)$	$\frac{\sin\omega\tau/2}{\pi\omega}$
Triangle	$\frac{-2A}{\tau}t+A, 0<t<\frac{\tau}{2}$ $\frac{2A}{\tau}t+A, -\frac{\tau}{2}<t<0$ $0, \lvert t\rvert>\frac{\tau}{2}$	p, A, $-\frac{\tau}{2}$, $\frac{\tau}{2}$, t	$\frac{4\pi}{\tau}\ \frac{8\pi}{\tau}\ldots\left(\frac{4n\pi}{\tau}\right)$, ω	$\frac{2}{\pi\tau\omega^2}\ (1-\cos\omega\tau/2)$
Half sine	$A\cos\pi t/\tau, \lvert t\rvert<\frac{\tau}{2}$ $0, \lvert t\rvert>\frac{\tau}{2}$	p, A, $-\frac{\tau}{2}$, $\frac{\tau}{2}$, t	$0{\cdot}79\tau/\pi^2$, $0{\cdot}33\tau/\pi^2$, $\cdots(2n-1)\frac{\pi}{\tau}$, $\frac{\tau}{\pi^2}$, $\frac{\pi}{\tau}\ \frac{2\pi}{\tau}\ \frac{3\pi}{\tau}\ \frac{5\pi}{\tau}$, ω	$\frac{\cos\omega\tau/2}{\tau\left(\frac{\pi^2}{\tau^2}-\omega^2\right)}$
Sine²	$A\cos^2\frac{\pi t}{\tau}, \lvert t\rvert>\frac{\tau}{2}$ $0, \lvert t\rvert>\frac{\tau}{2}$	p, A, $-\frac{\tau}{2}$, $\frac{\tau}{2}$, t	$0{\cdot}21\tau/\pi$, $\frac{\tau}{8\pi}$, $\frac{\tau}{4\pi}$, $\ldots\frac{2n\pi}{\tau}$, $\frac{\pi}{\tau}\ \frac{2\pi}{\tau}\ \frac{4\pi}{\tau}\ \frac{6\pi}{\tau}$, ω	$\frac{2\pi}{\tau^2\omega\left(\frac{4\pi^2}{\tau^2}-\omega^2\right)}\ \sin\frac{\omega\tau}{2}$
Gaussian	$A\exp(-4t^2/\tau^2)$	p, A, A/e, $-\frac{\tau}{2}$, $\frac{\tau}{2}$, t	$\frac{\tau}{4\sqrt{\pi}}$, $\frac{\tau}{4\sqrt{\pi e}}$, $4/\tau$, ω	$\frac{\tau}{4\sqrt{\pi}}\exp(-\omega^2\tau^2/16)$
Modulated sine	$h(t)\cos\omega_0\tau$ $h(t)\sim$even	p, t, τ	f_c, $\frac{1}{2}h_c(\omega-\omega_0)$, $\frac{1}{2}h_c(\omega+\omega_0)$, $-\omega_0$, ω_0, ω	$f_c(\omega)=(\sqrt{\frac{2}{\pi}})\int_0^\infty h\cos\omega_0 t$ $\times\cos\omega t\,dt$ $=\frac{1}{2}h_c(\omega+\omega_0)-$ $-\frac{1}{2}h_c(\omega-\omega_0)$
Exponential decay	$A\exp(-t/\tau)t>0$ $0, t<0$	p, A, $1/A$, τ, t	$\frac{\tau}{2\pi}$, $\lvert f/A\rvert$, $0{\cdot}707\left(\frac{\tau}{2\pi}\right)$, $-1/\tau$, ω	$f(\omega)/A=\frac{1}{2\pi(1/\tau-i\omega)}$ Note phase dependence.
Rule of thumb		p, A, $A/2$, $\Delta\tau$, t	H, $H/2$, $\Delta\omega$, ω	$\Delta\tau\ \Delta\omega\cong2\pi$ or $\Delta\tau\ \Delta f\cong1$

† $\overline{f_c}(\omega)=\sqrt{(2\pi)}f_c(\omega)/A$

B.7. Fourier spectra of pulses

A mechanical pulse applied to a system is prescribed by its spatial and time distributions. The time variation is frequently replaced by a frequency description in the form of the Fourier spectrum or frequency spectrum of the pulse. The frequency characterization may arise in the course of mathematical analysis due to Fourier transform operations, or merely through a preference for frequency domain instead of time-domain pulse characterization.

The Fourier spectra of a number of common pulse shapes are shown in Table B.8. They are obtained simply by taking the Fourier transform of the pulse $p(t)$. Thus

$$f(\omega) = \frac{1}{\sqrt{(2\pi)}}\int_{-\infty}^{\infty} p(t)e^{i\omega t}\,dt. \tag{B.7.1}$$

If $p(t)$ is an even function, as most of the tabulated cases are, the Fourier cosine transform is appropriate,

$$f_c(\omega) = \sqrt{\left(\frac{2}{\pi}\right)}\int_{0}^{\infty} p(t)\cos\omega t\,dt. \tag{B.7.2}$$

Appendix C: Experimental methods in stress waves

THE object here is to survey a number of the common methods for producing and detecting stress waves and vibrations in structures and elements. The material is meant to provide an elementary background to references made to experimental work in the various chapters of the book. Two topics are considered, the first being a survey of methods for producing transient motion in a solid, the second pertaining to methods for detecting stress waves in a solid. Methods of recording transient and steady signals are only briefly discussed, and no attempt is made to consider the system characteristics of the many detection–recording schemes.

For those interested in the detailed aspects of experimental mechanics, there are many useful references. The *Handbook of experimental stress analysis* by Hetenyi [21] provides a broad coverage of many techniques in both static and dynamic measurements. More recent texts in the area are those of Dove and Adams [13] and Dally and Riley [10]. Both cover photo-elastic methods and strain-gauge techniques, with the former also giving coverage of motion measurement. The *Handbook of shock and vibration*, edited by Harris and Crede [20], has a number of chapters devoted to experimental methods. The characteristics of measurement systems, particularly those used in acoustics and steady vibrations of structures are considered in the texts by Keast [26], Magrab and Bloomquist [37], and Doeblin [12]. Several of the manufacturers of vibration equipment publish short monographs on selected topics in vibration measurement.

A number of survey papers on experimental methods in stress waves and vibrations have appeared. In the area of stress waves, Kolsky has prepared several articles [29, 30, 33]. Hillier [22, 23] has surveyed techniques for determining dynamic material properties. Proceedings of symposia on dynamic experimental methods have also appeared [54, 57]. The texts by Kolsky [32] and Goldsmith [19] also contain portions on experimental methods. Technical articles covering experimental work appear, of course, in many technical journals, but the journal *Experimental Mechanics* is exclusively devoted to this topic.

C.1. Methods for producing stress waves

The simplest and most common method of producing stress waves is by the mechanical impact of one solid against another. The early investigations of

Hopkinson [25] used a falling weight to dynamically load a wire. A swinging pendulum has been used in many instances (see, for example, Frocht [17]) to dynamically load a structure. A difficulty associated with these schemes is that the stress waveform imparted to the structure is not easily characterized by simple mathematical functions, and thus leads to more complex problems in comparing theoretical and experimental results.

In the case of studies involving longitudinal stress waves in rods, the longitudinal, collinear impact of a striker bar, produces 'predictable' stress waves. As shown in the analyses of Chapter 2, if the striker bar and impact bar are of the same material and cross-section, simple rod theory predicts a rectangular pulse, having a pulse length of twice the length of the striker bar. In practice, the waveform produced departs somewhat from this shape. If the impact is indeed collinear and the impact bars are perfectly flat, the stress wave has oscillatory characteristics, as shown in some of the results of Chapter 8. The oscillating results from excitation of the higher modes of the bar. In order to avoid pulse oscillations, the contact surfaces of the bars are slightly rounded. This yields a pulse having no overshoot, but with a more gradual rise time, as shown in some of the results of Chapter 2.

A spherical ball is used as the impacting object in many stress-wave studies. In the case of longitudinal stress-wave studies in rods, the pulse shape is quite simple, as shown in some of the results of Chapter 2, and is analysable on the basis of a combined Hertzian contact and simple rod theory, also as shown in Chapter 2. A number of studies on the transverse impact of beams and plates and impact on a half-space have used the spherical ball as the impact geometry. The impact situation is not so easily analysed for these configurations, however.

A wide variety of techniques is used to propel the impacting objects into collision with a load. Gravity is, of course, the simplest means and has been used in experiments involving falling weights, swinging pendulums, and dropping balls. In the case of longitudinal rod impact experiments, the striker rod must be suspended by two sets of thin strings or wires to maintain alignment at impact. The experimental apparatus used by Becker [1] and Kuo [34] are typical of arrangements of this type. Compressed air, hydraulic pistons, elastic springs, and slingshot-type arrangements have also been used to propel bars into impact. Explosives and the impact of a swinging pendulum have also been used to initiate the motion of a striker bar. The use of solenoids to accelerate ferromagnetic rods into impact has also been reported.

Electromechanical phenomena are used to produce stress pulses in solids, particularly in applications where high stress levels are not necessary. The piezoelectric effect, whereby certain materials exhibit mechanical strain when subjected to an electrical field and the inverse effect, is widely used. Piezoelectric materials include many natural materials, such as quartz and Rochelle salt, man-made crystals, such as ammonium dihydrogen phosphate (ADP) and lithium sulphate, and a number of ceramics, such as barium titanate and lead zirconate titanate. The characteristics of the stress pulse produced in a material by a piezoelectric transducer depends on several things, including the geometry of the transducer, the means of coupling the transducer to the solid, and the waveform of the applied electrical signal. It is dependent also on the nature of the mechanical response to the electrical signal. Thus particular cuts of crystals or polarized ceramics may vibrate in thickness expansion, shear, or torsional modes to the applied electrical signal. Useful references on the behaviour of piezoelectric materials and transducers are Mason [39, 40] and Berlincourt, Curran, and Jaffe [4]. Redwood [48, 49, 50] has presented several articles on the nature of the mechanical or electrical outputs of transducers to, respectively, electrical or mechanical inputs. A survey by Bradfield [5] on types of ultrasonic transducers is also most helpful. White [56] describes a class of thin-layer transducers that use the piezoelectric effect.

The magneto-strictive effect, whereby a mechanical pulse is produced in a ferromagnetic material subjected to a transient magnetic field, is also used to produce stress pulses in a material. May [41, Section VII] reviews some of the considerations related to these devices. This technique of pulse production was used in the study by Britton and Langley [6] reported on in Chapter 3.

Explosives of various types are frequently used to initiate stress pulses in materials. Lead azide is the solid explosive most commonly used in laboratory-scale studies of stress waves. Kolsky [27, 31] and Dally [8, 9] have used this technique for launching pulses in rods and slabs and in photo-elastic specimens. Many other applications are reported in the literature. Kolsky [28] has reported on certain difficulties associated with the electro-magnetic pulse generated by the chemical disassociation in the explosion. The use of sprayed silver acetylide–silver nitrate to initiate a laboratory-scale explosive pulse over a broad area has been reported by Neville and Hoese [43], and Hoese, Langner, and Baker [24]. The explosion of gaseous mixtures has been used for pulse production by Davies [11]. The sharp acoustic shock-front produced by a spark-gap discharge may also be used to initiate low-amplitude stress waves in solids, and was the technique used by Press and Oliver [47], reviewed in Chapter 8.

Exploding wires and foils are also used to produce stress pulses directly or to provide the energy for launching projectiles into impact with solids. The explosion actually results from the nearly instantaneous vaporization arising from the discharge of electrical energy from a capacitor bank through the wire or foil. The use of exploding wire to radially load a cylinder has been reported by Fyfe [18]. An extensive documentation of this phenomenon has been given by Moore and Chase [42]. The use of exploding foil in launching thin striker or flyer plates into planar impact for anelastic wave-propagation studies has been reported by several investigators, including Berkowitz and Cohen [3] and Dueweke [14].

The use of a shock tube for producing very uniform step pressure pulses has been reported by Fox and Curtis [16] and is reviewed in Chapter 8. The rapid unloading produced by fracture is often used, directly or indirectly, to initiate stress waves. Thus, in crack propagation studies such as reported by Wells and Post [55] and Beebe [2], the stress waves produced by the propagating crack are the phenomena of interest. In other cases, the rapid unloading due to fracture of an axially loaded bar is used to produce an impulsive moment in a beam, such as reported by Stephenson and Wilhoit [53] or an impulsive torque in a rod, such as reported by Nicholas and Campbell [44]. Stress pulses have also been produced by the rapid deposition of heat energy into a solid. Thus, Percival and Cheney [45] have used this principle in a scheme involving the deposition of luminous energy from a Q-switched laser into a coloured glass rod. The rapid heating of the rod then initiates a pulse into a metallic rod.

C.2. Methods for detecting stress waves

Prior to the development of sensitive transducing and recording devices, experimental investigations of stress-wave phenomena were hampered by the lack of techniques for measuring the propagated waveform. The pressure bar developed by B. Hopkinson in 1914 used 'momentum traps' in the form of small pellets loosely coupled to a rod to study the characteristics of longitudinal stress pulses. The general principle of this method has been described in Chapter 2. The same principle has found more recent application in measuring intense shock waves in solids and is described by Goldsmith [19].

The electrical-resistance strain gauge has probably been more widely used than any other device in the study of stress-wave propagation. The principle of the gauge is based on the fact that a slight change in length of a metallic wire is accompanied by a slight change in the electrical resistance of the wire. By forming extremely fine wire or foil into a compact grid, or gauge, and cementing it to a specimen, static or dynamic strains in the specimen result in changes in electrical resistance of the gauge. Connecting the gauge into a potentiometric or Wheatstone-bridge circuit enables resistance changes to be determined in terms of voltage changes. Proper calibration enables the mechanical strains to be established. Strain gauges are also made from piezoelectrically active materials. Such gauges are much more sensitive to strain, but also more fragile than wire and foil gauges. Thorough expositions on the various types of strain gauges, strain-gauge circuitry, and recording techniques are contained in the earlier cited texts by Dove and Adams [13] and Dally and Reilly [10]. Reference is also made to the work of Perry and Lissner [46].

The capacitance and inductance effects are also used to detect stress waves, with the former finding more numerous applications. The principle employed in capacitance transducers is that a slight change in distance between two charged surfaces results in a voltage change between the surfaces due to the change in capacitance. The capacitance device will effectively average the displacement over the area of the capacitor plate, so that contributions from stress-wave components having wavelengths of the order of the transducer size or smaller may be absent from the records. The inductance effect has probably found less use in the detection of stress waves. Malvern [38] employed this principle to measure elastic–plastic waves in rods. In this application, fine copper wires were wrapped around an aluminum rod. The stress-wave induced motion of the wire loops in a magnetic field yielded a voltage signal proportional to the particle velocity at that point in the rod.

The piezoelectric effect is used to detect stress waves as well as induce stress waves. Thus, a mechanical stress wave impinging on a piezoelectric crystal or ceramic element will produce an electrical signal across the poles of the element. A simple piezoelectric disc bonded to the surface of a structure can serve in this capacity. Usually the construction is more elaborate, with various types of backings used in order to prevent resonances of the element from obscuring the basic signal of the stress wave. In other applications, piezoelectric elements are sandwiched between cylindrical rods to monitor the passage of a stress pulse or are interposed between a striking element, such as a rod or spherical ball, and the target in order to measure the applied force.

Photo-elasticity has been widely used in the detection of stress waves. This technique is based on the principle that many transparent materials have the property of birefringence. This is the property whereby the electric-field vector of a beam of polarized light, upon entering a stressed, transparent specimen, is resolved into two components along the axes of principal stress. The two light components are retarded by differing amounts during passage through the specimen and, upon recombining outside the specimen, will form interference patterns, or fringes, owing to the difference in retardation of each of the light-vector components. When monochromatic light is used as an illumination source, the interference patterns are simply a series of light and dark fringes. When white light is employed, the patterns are quite picturesque, but less useful, multi-coloured bands of light. The basic law governing interpretation of photo-elastic data is the stress–optic law, given by $\sigma_1 - \sigma_2 = nf_\sigma/h$, where σ_1, σ_2 are the principal stresses, n is the fringe order, h is the specimen thickness, and f_σ is the fringe constant of the material and is dependent on the photo-elastic properties of the material and the wavelength of the light source.

One of the main advantages of photo-elasticity is that data on the entire stress field in a specimen may be obtained, whereas devices such as strain gauges and piezoelectric and capacitance transducers give the behaviour only in a very small region. The only technique for recording such entire-field data is by photography, so that various high-speed photographic recording techniques play an important role here. High-speed framing cameras are used, as are multiple-flash, multiple-lens devices. Illumination techniques vary from continuous illumination used for framing cameras to spark-gap or Q-spoiled laser sources for the intermittent illumination used in the multiple source-image systems. Other techniques for recording dynamic photo-elastic data include streak photography and the use of photocells to record light intensity. Both of these record only point information, however, so that the entire-field advantages of photo-elasticity are, to a large extent, lost.

Many other techniques are used for detecting stress waves, including interferometric techniques, Moiré fringes, and diffraction gratings. This last, for example, has found extensive use in measuring elastic–plastic waves in rods. Holography has found application in detecting the steady vibrations of structures. However, from the standpoint of the experimental studies reviewed in this book, the basic detection techniques reviewed in the foregoing describe the most common methods used.

References to Appendices A, B, C.

1. Becker, E. C.H. Transient loading technique for mechanical impedance measurement. In *Experimental techniques in shock and vibration* (Ed. W. J. Worley). ASME, New York (1962).
2. Beebe, W. M. *An experimental investigation of dynamic crack propagation in plastics and metals.* Tech. Rep. AFML-TR-66-249 Air Force Materials Laboratory (1966).
3. Berkowitz, H. M. and Cohen, L. J. *A study of plate-slap technology*, Part I. A critical evaluation of plate-slap technology. Tech. Rep. AFML-TR-69-106 Air Force Material Laboratories (1969).
4. Berlincourt, D. A., Curran, D. R., and Jaffe, H. Piezoelectric and Piezomagnetic materials and their function in transducers. In *Physical acoustics* (Ed. W. P. Mason), Vol. 1, Part A. Chap. 3. Academic Press, New York (1964).
5. Bradfield, G. Ultrasonic transducers. In *Ultrasonics,* Part A, pp. 112–23, Part B pp. 177–89 (1970).
6. Britton, W. G. B. and Langley, G. O. Stress pulse dispersion in curved mechanical waveguides. *J. Sound Vib.* **7,** 417–30 (1968).
7. Churchill, R. V. *Operational mathematics.* McGraw-Hill, New York (1958).
8. Dally, J. W. A dynamic photoelastic study of a doubly loaded half-plane. *Develop. Mech.* **4,** 645–64 (1968).
9. —— and Lewis, D. (III). A photoelastic analysis of propagation of rayleigh waves past a step change in elevation. *Bull. seism. Soc. Am.* **58,** 539–63 (1968).
10. —— and Riley, W. F. *Experimental stress analysis.* McGraw-Hill, New York (1965).
11. Davies, R. M. A critical study of the Hopkinson pressure bar. *Phil. Trans. R. Soc.* **A240,** 375–457 (1948).
12. Doebelin, E. O. *Measurement systems: application and design.* McGraw-Hill, New York (1966).
13. Dove, R. C. and Adams, P. H. *Experimental stress analysis and motion measurement.* Charles Merrill Books, Columbus, Ohio (1964).
14. Dueweke, P. W. A technique for launching intermediate velocity thin plastic sheets. *Rev. scient. Instrum.* **41,** 539–41 (1970).

15. ERDELYI, A. (Ed.) *Tables of Integral Transforms* (Bateman Manuscript Project). McGraw-Hill, New York (1954).
16. FOX, G. and CURTIS, C. W. Elastic strain produced by sudden application of pressure to one end of a cylindrical bar—II. Experimental observations. *J. acoust. Soc. Am.* **30,** 559–63 (1958).
17. FROCHT, M. M. Studies in dynamic photoelasticity with special emphasis on the stress-optic law, in *Stress Wave Propagation in Materials* (Ed. N. Davids). Interscience, New York (1960).
18. FYFE, I. M. Plane-strain plastic wave propagation in a dynamically loaded hollow cylinder. *Symposium on the mechanical behavior of materials under dynamic loads*, San Antonio, Texas. Sponsored by U.S. Army Research Office and Southwest Research Institute (1967).
19. GOLDSMITH, W. *Impact: the theory and physical behaviour of colliding solids* Edward Arnold, London (1960).
20. HARRIS, C. M. and CREDE, E. *Shock and vibration handbook* Vols. I, II, and III. McGraw-Hill, New York (1961).
21. HETENYI, M. (Ed.) *Handbook of experimental stress analysis*. John Wiley and Sons, New York (1950).
22. HILLIER, K. W. A review of the progress in the measurement of dynamic elastic properties, in *Stress wave propagation in materials* (Ed. N. Davids), pp. 183–98 Interscience, New York (1960).
23. ——. The measurement of dynamic elastic properties, in *Progress in Solid Mechanics*, (Eds. I. N. Sneddon and R. Hill), Vol. II. North-Holland Publishing Company, Amsterdam (1961).
24. HOESE, F. O., LANGNER, C. G., and BAKER, W. E. Simultaneous initiation over large areas of a spray-deposited explosive. *Exp. Mech.* **25,** 392–7 (1968).
25. HOPKINSON, N. *Collected scientific papers*, Vol. ii. (1872).
26. KEAST, D. N. *Measurements in mechanical dynamics*. McGraw-Hill, New York (1967).
27. KOLSKY, H. An investigation of the mechanical properties of materials at very high rates of loading. *Proc. phys. Soc., Lond.* **B62,** 676–700 (1949).
28. ——. Electromagnetic waves emitted on detonation of explosives. *Nature, Lond.* **173,** 77 (1954).
29. ——. The propagation of stress waves in viscoelastic solids. *Appl. Mech. Rev.* **2,** 465 (1958).
30. ——. Experimental wave-propagation in solids, in *Structural Mechanics* (Eds. J. N. Goodier and N. Hoff), pp. 233–62. Pergamon Press, Oxford (1960).
31. ——. Viscoelastic waves, in *Stress wave propagation in materials* (Ed. N. Davids), pp. 59–60. Interscience, New York (1960).
32. ——. *Stress waves in solids*. Dover Publications, New York (1963).
33. ——. Experimental studies in stress wave propagation. *Proc. Vth U.S. natn. Congr. appl. Mech.* pp. 21–36 (1965).
34. KUO, S. S. Beam subjected to eccentric longitudinal impact, *Exp. Mech.* **18,** 102–8 (1961).
35. LANGHAAR, H. L. *Energy methods in applied mechanics*. John Wiley and Sons, New York (1962).

36. LOVE, A. E. H. *A treatise on the mathematical theory of elasticity*. Dover Publications, New York (1944).
37. MAGRAB, E. B. and BLOMQUIST, D. S. *The measurement of time-varying phenomena*. Wiley–Interscience, New York (1971).
38. MALVERN, L. E. Experiment studies of strain-rate effects and plastic-wave propagation in annealed aluminum. In *Behavior of materials under dynamic loading*, (Ed. N. J. Huffington (Jr.)), pp. 81–92. ASME, New York (1965).
39. MASON, W. P. *Electromechanical transducers and wave filters* (2nd edn). Van Nostrand Company, New York (1948).
40. ——. *Piezoelectric crystals and their application to ultrasonics*. Van Nostrand Company, New York (1950).
41. MAY, J. E. (Jr.) Guided wave ultrasonic delay lines. In *Physical acoustics* (Ed. W. P. Mason), Vol. 1, Part A, Chap. 6. Academic Press, New York (1964).
42. MOORE, H. K. and CHACE, W. G. (Eds.) *Exploding wires* Vol. 3. Plenum Press, New York (1964).
43. NEVILLE, G. E. (Jr.) and HOESE, F. O. Impulsive loading using sprayed silver acetylide–silver nitrate, *Exp. Mech.* **5**, 294–8 (1965).
44. NICHOLAS, T. and CAMPBELL, J. D. *The development and use of a torsional split Hopkinson bar for experiments in dynamic plasticity*. Tech. Rep. AFML-TR-71-32, Air Force Materials Laboratory (May 1971).
45. PERCIVAL, C. M. and CHENEY, J. A. Thermally generated stress waves in a dispersive elastic rod. *Exp. Mech.* **26**, 49–57 (1969).
46. PERRY, C. C. and LISSNER, H. R. *The strain gage primer*. McGraw-Hill, New York (1955).
47. PRESS, F. and OLIVER, J. Model study of air-coupled surface waves. *J. acoust. Soc. Am.* **27**, 43–6 (1955).
48. REDWOOD, M. Transient performance of a piezoelectric transducer. *J. acoust. Soc. Am.* **33**, 527–36 (1961).
49. ——. Piezoelectric generation of an electrical impulse. *J. acoust. Soc. Am.* **33**, 1386–90 (1961).
50. ——. Experiments with the electrical analog of a piezoelectric transducer. *J. acoust. Soc. Am.* **36**, 1872–80 (1964).
51. SNEDDON, I. N. *Fourier transforms*. McGraw-Hill, New York (1951).
52. SOKOLNIKOFF, I. S. *Mathematical theory of elasticity*. McGraw-Hill, New York (1956).
53. STEPHENSON, J. G. and WILHOIT, J. C. (Jr.) An experimental study of bending impact waves in beams. *Exp. Mech.* **22**, 16–21 (1965).
54. *Symposium on advanced experimental techniques in the mechanics of materials*, San Antonio, Texas. Sponsored by Air Force Office of Scientific Research and Southwest Research Institute (1970).
55. WELLS, A. A. and POST, D. *The dynamic stress distribution surrounding a running crack—a photoelastic analysis*. Nav. Res. Laboratory, NRL-4935, Washington, D.C. (1957).
56. WHITE, D. L. The depletion layer and other high-frequency transducers using fundamental modes, in *Physical acoustics* (Ed. W. P. Mason), Vol. I, Part B, Chap. 13. Academic Press, New York (1964).
57. WORELY, W. J. (Ed.) *Experimental techniques in shock and vibration*. ASME. New York (1962).

Author index

Subject index

A CATALOG OF SELECTED
DOVER BOOKS
IN SCIENCE AND MATHEMATICS

A CATALOG OF SELECTED

DOVER BOOKS

IN SCIENCE AND MATHEMATICS

QUALITATIVE THEORY OF DIFFERENTIAL EQUATIONS, V.V. Nemytskii and V.V. Stepanov. Classic graduate-level text by two prominent Soviet mathematicians covers classical differential equations as well as topological dynamics and ergodic theory. Bibliographies. 523pp. 5⅜ × 8½. 65954-2 Pa. $10.95

MATRICES AND LINEAR ALGEBRA, Hans Schneider and George Phillip Barker. Basic textbook covers theory of matrices and its applications to systems of linear equations and related topics such as determinants, eigenvalues and differential equations. Numerous exercises. 432pp. 5⅜ × 8½. 66014-1 Pa. $8.95

QUANTUM THEORY, David Bohm. This advanced undergraduate-level text presents the quantum theory in terms of qualitative and imaginative concepts, followed by specific applications worked out in mathematical detail. Preface. Index. 655pp. 5⅜ × 8½. 65969-0 Pa. $12.95

ATOMIC PHYSICS (8th edition), Max Born. Nobel laureate's lucid treatment of kinetic theory of gases, elementary particles, nuclear atom, wave-corpuscles, atomic structure and spectral lines, much more. Over 40 appendices, bibliography. 495pp. 5⅜ × 8½. 65984-4 Pa. $11.95

ELECTRONIC STRUCTURE AND THE PROPERTIES OF SOLIDS: The Physics of the Chemical Bond, Walter A. Harrison. Innovative text offers basic understanding of the electronic structure of covalent and ionic solids, simple metals, transition metals and their compounds. Problems. 1980 edition. 582pp. 6⅛ × 9¼. 66021-4 Pa. $14.95

BOUNDARY VALUE PROBLEMS OF HEAT CONDUCTION, M. Necati Özisik. Systematic, comprehensive treatment of modern mathematical methods of solving problems in heat conduction and diffusion. Numerous examples and problems. Selected references. Appendices. 505pp. 5⅜ × 8½. 65990-9 Pa. $11.95

A SHORT HISTORY OF CHEMISTRY (3rd edition), J.R. Partington. Classic exposition explores origins of chemistry, alchemy, early medical chemistry, nature of atmosphere, theory of valency, laws and structure of atomic theory, much more. 428pp. 5⅜ × 8½. (Available in U.S. only) 65977-1 Pa. $10.95

A HISTORY OF ASTRONOMY, A. Pannekoek. Well-balanced, carefully reasoned study covers such topics as Ptolemaic theory, work of Copernicus, Kepler, Newton, Eddington's work on stars, much more. Illustrated. References. 521pp. 5⅜ × 8½. 65994-1 Pa. $11.95

PRINCIPLES OF METEOROLOGICAL ANALYSIS, Walter J. Saucier. Highly respected, abundantly illustrated classic reviews atmospheric variables, hydrostatics, static stability, various analyses (scalar, cross-section, isobaric, isentropic, more). For intermediate meteorology students. 454pp. 6⅛ × 9¼. 65979-8 Pa. $12.95

RELATIVITY, THERMODYNAMICS AND COSMOLOGY, Richard C. Tolman. Landmark study extends thermodynamics to special, general relativity; also applications of relativistic mechanics, thermodynamics to cosmological models. 501pp. 5⅜ × 8½. 65383-8 Pa. $12.95

APPLIED ANALYSIS, Cornelius Lanczos. Classic work on analysis and design of finite processes for approximating solution of analytical problems. Algebraic equations, matrices, harmonic analysis, quadrature methods, much more. 559pp. 5⅜ × 8½. 65656-X Pa. $12.95

SPECIAL RELATIVITY FOR PHYSICISTS, G. Stephenson and C.W. Kilmister. Concise elegant account for nonspecialists. Lorentz transformation, optical and dynamical applications, more. Bibliography. 108pp. 5⅜ × 8½. 65519-9 Pa. $4.95

INTRODUCTION TO ANALYSIS, Maxwell Rosenlicht. Unusually clear, accessible coverage of set theory, real number system, metric spaces, continuous functions, Riemann integration, multiple integrals, more. Wide range of problems. Undergraduate level. Bibliography. 254pp. 5⅜ × 8½. 65038-3 Pa. $7.95

INTRODUCTION TO QUANTUM MECHANICS With Applications to Chemistry, Linus Pauling & E. Bright Wilson, Jr. Classic undergraduate text by Nobel Prize winner applies quantum mechanics to chemical and physical problems. Numerous tables and figures enhance the text. Chapter bibliographies. Appendices. Index. 468pp. 5⅜ × 8½. 64871-0 Pa. $10.95

ASYMPTOTIC EXPANSIONS OF INTEGRALS, Norman Bleistein & Richard A. Handelsman. Best introduction to important field with applications in a variety of scientific disciplines. New preface. Problems. Diagrams. Tables. Bibliography. Index. 448pp. 5⅜ × 8½. 65082-0 Pa. $11.95

MATHEMATICS APPLIED TO CONTINUUM MECHANICS, Lee A. Segel. Analyzes models of fluid flow and solid deformation. For upper-level math, science and engineering students. 608pp. 5⅜ × 8½. 65369-2 Pa. $13.95

ELEMENTS OF REAL ANALYSIS, David A. Sprecher. Classic text covers fundamental concepts, real number system, point sets, functions of a real variable, Fourier series, much more. Over 500 exercises. 352pp. 5⅜ × 8½. 65385-4 Pa. $9.95

PHYSICAL PRINCIPLES OF THE QUANTUM THEORY, Werner Heisenberg. Nobel Laureate discusses quantum theory, uncertainty, wave mechanics, work of Dirac, Schroedinger, Compton, Wilson, Einstein, etc. 184pp. 5⅜ × 8½. 60113-7 Pa. $4.95

INTRODUCTORY REAL ANALYSIS, A.N. Kolmogorov, S.V. Fomin. Translated by Richard A. Silverman. Self-contained, evenly paced introduction to real and functional analysis. Some 350 problems. 403pp. 5⅜ × 8½. 61226-0 Pa. $8.95

PROBLEMS AND SOLUTIONS IN QUANTUM CHEMISTRY AND PHYSICS, Charles S. Johnson, Jr. and Lee G. Pedersen. Unusually varied problems, detailed solutions in coverage of quantum mechanics, wave mechanics, angular momentum, molecular spectroscopy, scattering theory, more. 280 problems plus 139 supplementary exercises. 430pp. 6½ × 9¼. 65236-X Pa. $11.95

ORDINARY DIFFERENTIAL EQUATIONS, Morris Tenenbaum and Harry Pollard. Exhaustive survey of ordinary differential equations for undergraduates in mathematics, engineering, science. Thorough analysis of theorems. Diagrams. Bibliography. Index. 818pp. 5⅜ × 8½. 64940-7 Pa. $15.95

STATISTICAL MECHANICS: Principles and Applications, Terrell L. Hill. Standard text covers fundamentals of statistical mechanics, applications to fluctuation theory, imperfect gases, distribution functions, more. 448pp. 5⅜ × 8½. 65390-0 Pa. $9.95

ORDINARY DIFFERENTIAL EQUATIONS AND STABILITY THEORY: An Introduction, David A. Sánchez. Brief, modern treatment. Linear equation, stability theory for autonomous and nonautonomous systems, etc. 164pp. 5⅜ × 8¼. 63828-6 Pa. $4.95

THIRTY YEARS THAT SHOOK PHYSICS: The Story of Quantum Theory, George Gamow. Lucid, accessible introduction to influential theory of energy and matter. Careful explanations of Dirac's anti-particles, Bohr's model of the atom, much more. 12 plates. Numerous drawings. 240pp. 5⅜ × 8½. 24895-X Pa. $5.95

ORDINARY DIFFERENTIAL EQUATIONS, I.G. Petrovski. Covers basic concepts, some differential equations and such aspects of the general theory as Euler lines, Arzel's theorem, Peano's existence theorem, Osgood's uniqueness theorem, more. 45 figures. Problems. Bibliography. Index. xi + 232pp. 5⅜ × 8½. 64683-1 Pa. $6.95

GREAT EXPERIMENTS IN PHYSICS: Firsthand Accounts from Galileo to Einstein, edited by Morris H. Shamos. 25 crucial discoveries: Newton's laws of motion, Chadwick's study of the neutron, Hertz on electromagnetic waves, more. Original accounts clearly annotated. 370pp. 5⅜ × 8½. 25346-5 Pa. $8.95

INTRODUCTION TO PARTIAL DIFFERENTIAL EQUATIONS WITH APPLICATIONS, E.C. Zachmanoglou and Dale W. Thoe. Essentials of partial differential equations applied to common problems in engineering and the physical sciences. Problems and answers. 416pp. 5⅜ × 8½. 65251-3 Pa. $9.95

BURNHAM'S CELESTIAL HANDBOOK, Robert Burnham, Jr. Thorough guide to the stars beyond our solar system. Exhaustive treatment. Alphabetical by constellation: Andromeda to Cetus in Vol. 1; Chamaeleon to Orion in Vol. 2; and Pavo to Vulpecula in Vol. 3. Hundreds of illustrations. Index in Vol. 3. 2,000pp. 6⅛ × 9¼. 23567-X, 23568-8, 23673-0 Pa., Three-vol. set $41.85

ASYMPTOTIC EXPANSIONS FOR ORDINARY DIFFERENTIAL EQUATIONS, Wolfgang Wasow. Outstanding text covers asymptotic power series, Jordan's canonical form, turning point problems, singular perturbations, much more. Problems. 384pp. 5⅜ × 8½. 65456-7 Pa. $9.95

AMATEUR ASTRONOMER'S HANDBOOK, J.B. Sidgwick. Timeless, comprehensive coverage of telescopes, mirrors, lenses, mountings, telescope drives, micrometers, spectroscopes, more. 189 illustrations. 576pp. 5⅜ × 8¼. 24034-7 Pa. $9.95

SPECIAL FUNCTIONS, N.N. Lebedev. Translated by Richard Silverman. Famous Russian work treating more important special functions, with applications to specific problems of physics and engineering. 38 figures. 308pp. 5⅜ × 8½.
60624-4 Pa. $7.95

OBSERVATIONAL ASTRONOMY FOR AMATEURS, J.B. Sidgwick. Mine of useful data for observation of sun, moon, planets, asteroids, aurorae, meteors, comets, variables, binaries, etc. 39 illustrations. 384pp. 5⅜ × 8¼. (Available in U.S. only)
24033-9 Pa. $5.95

INTEGRAL EQUATIONS, F.G. Tricomi. Authoritative, well-written treatment of extremely useful mathematical tool with wide applications. Volterra Equations, Fredholm Equations, much more. Advanced undergraduate to graduate level. Exercises. Bibliography. 238pp. 5⅜ × 8½.
64828-1 Pa. $6.95

CELESTIAL OBJECTS FOR COMMON TELESCOPES, T.W. Webb. Inestimable aid for locating and identifying nearly 4,000 celestial objects. 77 illustrations. 645pp. 5⅜ × 8½.
20917-2, 20918-0 Pa., Two-vol. set $12.00

MODERN NONLINEAR EQUATIONS, Thomas L. Saaty. Emphasizes practical solution of problems; covers seven types of equations. ". . . a welcome contribution to the existing literature. . . ."—*Math Reviews.* 490pp. 5⅜ × 8½. 64232-1 Pa. $9.95

FUNDAMENTALS OF ASTRODYNAMICS, Roger Bate et al. Modern approach developed by U.S. Air Force Academy. Designed as a first course. Problems, exercises. Numerous illustrations. 455pp. 5⅜ × 8½.
60061-0 Pa. $8.95

INTRODUCTION TO LINEAR ALGEBRA AND DIFFERENTIAL EQUATIONS, John W. Dettman. Excellent text covers complex numbers, determinants, orthonormal bases, Laplace transforms, much more. Exercises with solutions. Undergraduate level. 416pp. 5⅜ × 8½.
65191-6 Pa. $9.95

INCOMPRESSIBLE AERODYNAMICS, edited by Bryan Thwaites. Covers theoretical and experimental treatment of the uniform flow of air and viscous fluids past two-dimensional aerofoils and three-dimensional wings; many other topics. 654pp. 5⅜ × 8½.
65465-6 Pa. $15.95

INTRODUCTION TO DIFFERENCE EQUATIONS, Samuel Goldberg. Exceptionally clear exposition of important discipline with applications to sociology, psychology, economics. Many illustrative examples; over 250 problems. 260pp. 5⅜ × 8½.
65084-7 Pa. $6.95

LAMINAR BOUNDARY LAYERS, edited by L. Rosenhead. Engineering classic covers steady boundary layers in two- and three-dimensional flow, unsteady boundary layers, stability, observational techniques, much more. 708pp. 5⅜ × 8½.
65646-2 Pa. $15.95

LECTURES ON CLASSICAL DIFFERENTIAL GEOMETRY, Second Edition, Dirk J. Struik. Excellent brief introduction covers curves, theory of surfaces, fundamental equations, geometry on a surface, conformal mapping, other topics. Problems. 240pp. 5⅜ × 8½.
65609-8 Pa. $6.95

ROTARY-WING AERODYNAMICS, W.Z. Stepniewski. Clear, concise text covers aerodynamic phenomena of the rotor and offers guidelines for helicopter performance evaluation. Originally prepared for NASA. 537 figures. 640pp. 6⅛ × 9¼.
64647-5 Pa. $14.95

DIFFERENTIAL GEOMETRY, Heinrich W. Guggenheimer. Local differential geometry as an application of advanced calculus and linear algebra. Curvature, transformation groups, surfaces, more. Exercises. 62 figures. 378pp. 5⅜ × 8½.
63433-7 Pa. $7.95

INTRODUCTION TO SPACE DYNAMICS, William Tyrrell Thomson. Comprehensive, classic introduction to space-flight engineering for advanced undergraduate and graduate students. Includes vector algebra, kinematics, transformation of coordinates. Bibliography. Index. 352pp. 5⅜ × 8½. 65113-4 Pa. $8.95

A SURVEY OF MINIMAL SURFACES, Robert Osserman. Up-to-date, in-depth discussion of the field for advanced students. Corrected and enlarged edition covers new developments. Includes numerous problems. 192pp. 5⅜ × 8½.
64998-9 Pa. $8.95

ANALYTICAL MECHANICS OF GEARS, Earle Buckingham. Indispensable reference for modern gear manufacture covers conjugate gear-tooth action, gear-tooth profiles of various gears, many other topics. 263 figures. 102 tables. 546pp. 5⅜ × 8½.
65712-4 Pa. $11.95

SET THEORY AND LOGIC, Robert R. Stoll. Lucid introduction to unified theory of mathematical concepts. Set theory and logic seen as tools for conceptual understanding of real number system. 496pp. 5⅜ × 8¼. 63829-4 Pa. $8.95

A HISTORY OF MECHANICS, René Dugas. Monumental study of mechanical principles from antiquity to quantum mechanics. Contributions of ancient Greeks, Galileo, Leonardo, Kepler, Lagrange, many others. 671pp. 5⅜ × 8½.
65632-2 Pa. $14.95

FAMOUS PROBLEMS OF GEOMETRY AND HOW TO SOLVE THEM, Benjamin Bold. Squaring the circle, trisecting the angle, duplicating the cube: learn their history, why they are impossible to solve, then solve them yourself. 128pp. 5⅜ × 8½.
24297-8 Pa. $3.95

MECHANICAL VIBRATIONS, J.P. Den Hartog. Classic textbook offers lucid explanations and illustrative models, applying theories of vibrations to a variety of practical industrial engineering problems. Numerous figures. 233 problems, solutions. Appendix. Index. Preface. 436pp. 5⅜ × 8½. 64785-4 Pa. $8.95

CURVATURE AND HOMOLOGY, Samuel I. Goldberg. Thorough treatment of specialized branch of differential geometry. Covers Riemannian manifolds, topology of differentiable manifolds, compact Lie groups, other topics. Exercises. 315pp. 5⅜ × 8½.
64314-X Pa. $8.95

HISTORY OF STRENGTH OF MATERIALS, Stephen P. Timoshenko. Excellent historical survey of the strength of materials with many references to the theories of elasticity and structure. 245 figures. 452pp. 5⅜ × 8½. 61187-6 Pa. $10.95

GEOMETRY OF COMPLEX NUMBERS, Hans Schwerdtfeger. Illuminating, widely praised book on analytic geometry of circles, the Moebius transformation, and two-dimensional non-Euclidean geometries. 200pp. 5⅜ × 8¼. 63830-8 Pa. $6.95

MECHANICS, J.P. Den Hartog. A classic introductory text or refresher. Hundreds of applications and design problems illuminate fundamentals of trusses, loaded beams and cables, etc. 334 answered problems. 462pp. 5⅜ × 8½. 60754-2 Pa. $8.95

TOPOLOGY, John G. Hocking and Gail S. Young. Superb one-year course in classical topology. Topological spaces and functions, point-set topology, much more. Examples and problems. Bibliography. Index. 384pp. 5⅜ × 8¼. 65676-4 Pa. $7.95

STRENGTH OF MATERIALS, J.P. Den Hartog. Full, clear treatment of basic material (tension, torsion, bending, etc.) plus advanced material on engineering methods, applications. 350 answered problems. 323pp. 5⅜ × 8½. 60755-0 Pa. $7.50

ELEMENTARY CONCEPTS OF TOPOLOGY, Paul Alexandroff. Elegant, intuitive approach to topology from set-theoretic topology to Betti groups; how concepts of topology are useful in math and physics. 25 figures. 57pp. 5⅜ × 8½. 60747-X Pa. $2.95

ADVANCED STRENGTH OF MATERIALS, J.P. Den Hartog. Superbly written advanced text covers torsion, rotating disks, membrane stresses in shells, much more. Many problems and answers. 388pp. 5⅜ × 8½. 65407-9 Pa. $9.95

COMPUTABILITY AND UNSOLVABILITY, Martin Davis. Classic graduate-level introduction to theory of computability, usually referred to as theory of recurrent functions. New preface and appendix. 288pp. 5⅜ × 8½. 61471-9 Pa. $6.95

GENERAL CHEMISTRY, Linus Pauling. Revised 3rd edition of classic first-year text by Nobel laureate. Atomic and molecular structure, quantum mechanics, statistical mechanics, thermodynamics correlated with descriptive chemistry. Problems. 992pp. 5⅜ × 8½. 65622-5 Pa. $18.95

AN INTRODUCTION TO MATRICES, SETS AND GROUPS FOR SCIENCE STUDENTS, G. Stephenson. Concise, readable text introduces sets, groups, and most importantly, matrices to undergraduate students of physics, chemistry, and engineering. Problems. 164pp. 5⅜ × 8½. 65077-4 Pa. $5.95

THE HISTORICAL BACKGROUND OF CHEMISTRY, Henry M. Leicester. Evolution of ideas, not individual biography. Concentrates on formulation of a coherent set of chemical laws. 260pp. 5⅜ × 8½. 61053-5 Pa. $6.00

THE PHILOSOPHY OF MATHEMATICS: An Introductory Essay, Stephan Körner. Surveys the views of Plato, Aristotle, Leibniz & Kant concerning propositions and theories of applied and pure mathematics. Introduction. Two appendices. Index. 198pp. 5⅜ × 8½. 25048-2 Pa. $6.95

THE DEVELOPMENT OF MODERN CHEMISTRY, Aaron J. Ihde. Authoritative history of chemistry from ancient Greek theory to 20th-century innovation. Covers major chemists and their discoveries. 209 illustrations. 14 tables. Bibliographies. Indices. Appendices. 851pp. 5⅜ × 8½. 64235-6 Pa. $17.95

THE FOUR-COLOR PROBLEM: Assaults and Conquest, Thomas L. Saaty and Paul G. Kainen. Engrossing, comprehensive account of the century-old combinatorial topological problem, its history and solution. Bibliographies. Index. 110 figures. 228pp. 5⅜ × 8½. 65092-8 Pa. $6.00

CATALYSIS IN CHEMISTRY AND ENZYMOLOGY, William P. Jencks. Exceptionally clear coverage of mechanisms for catalysis, forces in aqueous solution, carbonyl- and acyl-group reactions, practical kinetics, more. 864pp. 5⅜ × 8½. 65460-5 Pa. $18.95

PROBABILITY: An Introduction, Samuel Goldberg. Excellent basic text covers set theory, probability theory for finite sample spaces, binomial theorem, much more. 360 problems. Bibliographies. 322pp. 5⅜ × 8½. 65252-1 Pa. $8.95

LIGHTNING, Martin A. Uman. Revised, updated edition of classic work on the physics of lightning. Phenomena, terminology, measurement, photography, spectroscopy, thunder, more. Reviews recent research. Bibliography. Indices. 320pp. 5⅜ × 8¼. 64575-4 Pa. $7.95

PROBABILITY THEORY: A Concise Course, Y.A. Rozanov. Highly readable, self-contained introduction covers combination of events, dependent events, Bernoulli trials, etc. Translation by Richard Silverman. 148pp. 5⅜ × 8¼. 63544-9 Pa. $5.95

THE CEASELESS WIND: An Introduction to the Theory of Atmospheric Motion, John A. Dutton. Acclaimed text integrates disciplines of mathematics and physics for full understanding of dynamics of atmospheric motion. Over 400 problems. Index. 97 illustrations. 640pp. 6 × 9. 65096-0 Pa. $17.95

STATISTICS MANUAL, Edwin L. Crow, et al. Comprehensive, practical collection of classical and modern methods prepared by U.S. Naval Ordnance Test Station. Stress on use. Basics of statistics assumed. 288pp. 5⅜ × 8½. 60599-X Pa. $6.00

WIND WAVES: Their Generation and Propagation on the Ocean Surface, Blair Kinsman. Classic of oceanography offers detailed discussion of stochastic processes and power spectral analysis that revolutionized ocean wave theory. Rigorous, lucid. 676pp. 5⅜ × 8½. 64652-1 Pa. $16.95

STATISTICAL METHOD FROM THE VIEWPOINT OF QUALITY CONTROL, Walter A. Shewhart. Important text explains regulation of variables, uses of statistical control to achieve quality control in industry, agriculture, other areas. 192pp. 5⅜ × 8½. 65232-7 Pa. $6.95

THE INTERPRETATION OF GEOLOGICAL PHASE DIAGRAMS, Ernest G. Ehlers. Clear, concise text emphasizes diagrams of systems under fluid or containing pressure; also coverage of complex binary systems, hydrothermal melting, more. 288pp. 6½ × 9¼. 65389-7 Pa. $10.95

STATISTICAL ADJUSTMENT OF DATA, W. Edwards Deming. Introduction to basic concepts of statistics, curve fitting, least squares solution, conditions without parameter, conditions containing parameters. 26 exercises worked out. 271pp. 5⅜ × 8½. 64685-8 Pa. $7.95

DE RE METALLICA, Georgius Agricola. The famous Hoover translation of greatest treatise on technological chemistry, engineering, geology, mining of early modern times (1556). All 289 original woodcuts. 638pp. 6¾ × 11.
60006-8 Pa. $17.95

SOME THEORY OF SAMPLING, William Edwards Deming. Analysis of the problems, theory and design of sampling techniques for social scientists, industrial managers and others who find statistics increasingly important in their work. 61 tables. 90 figures. xvii + 602pp. 5⅜ × 8½. 64684-X Pa. $15.95

THE VARIOUS AND INGENIOUS MACHINES OF AGOSTINO RAMELLI: A Classic Sixteenth-Century Illustrated Treatise on Technology, Agostino Ramelli. One of the most widely known and copied works on machinery in the 16th century. 194 detailed plates of water pumps, grain mills, cranes, more. 608pp. 9 × 12. (EBE)
25497-6 Clothbd. $34.95

LINEAR PROGRAMMING AND ECONOMIC ANALYSIS, Robert Dorfman, Paul A. Samuelson and Robert M. Solow. First comprehensive treatment of linear programming in standard economic analysis. Game theory, modern welfare economics, Leontief input-output, more. 525pp. 5⅜ × 8½. 65491-5 Pa. $13.95

ELEMENTARY DECISION THEORY, Herman Chernoff and Lincoln E. Moses. Clear introduction to statistics and statistical theory covers data processing, probability and random variables, testing hypotheses, much more. Exercises. 364pp. 5⅜ × 8½. 65218-1 Pa. $8.95

THE COMPLEAT STRATEGYST: Being a Primer on the Theory of Games of Strategy, J.D. Williams. Highly entertaining classic describes, with many illustrated examples, how to select best strategies in conflict situations. Prefaces. Appendices. 268pp. 5⅜ × 8½. 25101-2 Pa. $5.95

MATHEMATICAL METHODS OF OPERATIONS RESEARCH, Thomas L. Saaty. Classic graduate-level text covers historical background, classical methods of forming models, optimization, game theory, probability, queueing theory, much more. Exercises. Bibliography. 448pp. 5⅜ × 8¼. 65703-5 Pa. $12.95

CONSTRUCTIONS AND COMBINATORIAL PROBLEMS IN DESIGN OF EXPERIMENTS, Damaraju Raghavarao. In-depth reference work examines orthogonal Latin squares, incomplete block designs, tactical configuration, partial geometry, much more. Abundant explanations, examples. 416pp. 5⅜ × 8¼.
65685-3 Pa. $10.95

THE ABSOLUTE DIFFERENTIAL CALCULUS (CALCULUS OF TENSORS), Tullio Levi-Civita. Great 20th-century mathematician's classic work on material necessary for mathematical grasp of theory of relativity. 452pp. 5⅜ × 8½.
63401-9 Pa. $9.95

VECTOR AND TENSOR ANALYSIS WITH APPLICATIONS, A.I. Borisenko and I.E. Tarapov. Concise introduction. Worked-out problems, solutions, exercises. 257pp. 5⅜ × 8¼. 63833-2 Pa. $6.95

TENSOR CALCULUS, J.L. Synge and A. Schild. Widely used introductory text covers spaces and tensors, basic operations in Riemannian space, non-Riemannian spaces, etc. 324pp. 5⅜ × 8¼. 63612-7 Pa. $7.95

A CONCISE HISTORY OF MATHEMATICS, Dirk J. Struik. The best brief history of mathematics. Stresses origins and covers every major figure from ancient Near East to 19th century. 41 illustrations. 195pp. 5⅜ × 8½. 60255-9 Pa. $7.95

A SHORT ACCOUNT OF THE HISTORY OF MATHEMATICS, W.W. Rouse Ball. One of clearest, most authoritative surveys from the Egyptians and Phoenicians through 19th-century figures such as Grassman, Galois, Riemann. Fourth edition. 522pp. 5⅜ × 8½. 20630-0 Pa. $9.95

HISTORY OF MATHEMATICS, David E. Smith. Nontechnical survey from ancient Greece and Orient to late 19th century; evolution of arithmetic, geometry, trigonometry, calculating devices, algebra, the calculus. 362 illustrations. 1,355pp. 5⅜ × 8½. 20429-4, 20430-8 Pa., Two-vol. set $21.90

THE GEOMETRY OF RENÉ DESCARTES, René Descartes. The great work founded analytical geometry. Original French text, Descartes' own diagrams, together with definitive Smith-Latham translation. 244pp. 5⅜ × 8½. 60068-8 Pa. $6.95

THE ORIGINS OF THE INFINITESIMAL CALCULUS, Margaret E. Baron. Only fully detailed and documented account of crucial discipline: origins; development by Galileo, Kepler, Cavalieri; contributions of Newton, Leibniz, more. 304pp. 5⅜ × 8½. (Available in U.S. and Canada only) 65371-4 Pa. $8.95

THE HISTORY OF THE CALCULUS AND ITS CONCEPTUAL DEVELOPMENT, Carl B. Boyer. Origins in antiquity, medieval contributions, work of Newton, Leibniz, rigorous formulation. Treatment is verbal. 346pp. 5⅜ × 8½. 60509-4 Pa. $7.95

THE THIRTEEN BOOKS OF EUCLID'S ELEMENTS, translated with introduction and commentary by Sir Thomas L. Heath. Definitive edition. Textual and linguistic notes, mathematical analysis. 2,500 years of critical commentary. Not abridged. 1,414pp. 5⅜ × 8½. 60088-2, 60089-0, 60090-4 Pa., Three-vol. set $29.85

GAMES AND DECISIONS: Introduction and Critical Survey, R. Duncan Luce and Howard Raiffa. Superb nontechnical introduction to game theory, primarily applied to social sciences. Utility theory, zero-sum games, n-person games, decision-making, much more. Bibliography. 509pp. 5⅜ × 8½. 65943-7 Pa. $11.95

THE HISTORICAL ROOTS OF ELEMENTARY MATHEMATICS, Lucas N.H. Bunt, Phillip S. Jones, and Jack D. Bedient. Fundamental underpinnings of modern arithmetic, algebra, geometry and number systems derived from ancient civilizations. 320pp. 5⅜ × 8½. 25563-8 Pa. $7.95

CALCULUS REFRESHER FOR TECHNICAL PEOPLE, A. Albert Klaf. Covers important aspects of integral and differential calculus via 756 questions. 566 problems, most answered. 431pp. 5⅜ × 8½. 20370-0 Pa. $7.95